Unconventional Tight Reservoir Simulation: Theory, Technology and Practice

Qiquan Ran

Unconventional Tight Reservoir Simulation: Theory, Technology and Practice

Science Press
Beijing

Springer

Qiquan Ran
Research Institute of Petroleum
Exploration and Development
Beijing, China

ISBN 978-981-32-9847-7 ISBN 978-981-32-9848-4 (eBook)
https://doi.org/10.1007/978-981-32-9848-4

Jointly published with Science Press
The print edition is not for sale in China (Mainland). Customers from China (Mainland) please order the print book from: Science Press.
ISBN of the Co-Publisher's edition: 978-7-03-065332-1

This Springer imprint is published by the registered company Springer Nature Singapore Pte Ltd.
The registered company address is: 152 Beach Road, #21-01/04 Gateway East, Singapore 189721, Singapore

Foreword

Continental tight oil and gas reservoirs in China have enormous development potential because of their abundance and significant hydrocarbon in-place. They are important targets of oil and gas exploration and development, and will play an increasingly important role in oil and gas production in the future. Compared with conventional reservoirs and tight reservoirs of marine facies in North America, continental tight reservoirs in China are more complex because: i. they possess stronger micro- and macro-heterogeneities on multiple scales from pores to fractures; ii. fluid flow needs to be described by multi-continuum multi-scale models; iii. horizontal wells and multi-stage hydraulic fractures add to the complexity of the reservoir. Conventional reservoir simulation methodologies are hardly applicable to these tight reservoirs. Unconventional reservoir simulation for tight oil and gas reservoir still requires development of innovative and physics-based methods and strategies.

With the complex features of continental tight oil and gas reservoirs in mind, the authors in this book presented a complete set of simulation theories, techniques, and software development strategies for tight oil and gas reservoirs. It reflects the authors' years of efforts in research, investigation, and development. The concept of a discrete multiple-interacting medium was first introduced to model the characteristics of multi-scale heterogeneities from nano/micro-pores to complex networks of natural/hydraulic fractures. Starting from this concept, a limitation of traditional theories for flows in continuous dual- and multi-porosity media which properties must be continuously varied in space were discussed and lifted and a theory that allows media properties to be discontinuous or discrete was developed. Afterwards, mathematical and numerical models for fluid flow in this complex media were developed. On these models, a complete numerical simulator for tight oil and gas reservoirs was built. This simulator synthesized five technological components: i. modeling of discrete natural/hydraulic fractures, discontinuous/discrete multiple-interacting media to handle effective properties of multi-scale heterogeneous media; ii. numerical simulation techniques for flows in discontinuous/discrete multiple-interacting media; iii. coupled flow-geomechanics to simulate geomechanical effects in complex reservoirs; iv. adaptive identification of flow regimes in multi-scale multiple-interacting media in consideration of complex flow mechanisms; v. coupling of multiphase flows from reservoirs to wellbores. Integration of these components generated a very useful tool, as demonstrated by the successful applications to match and predict actual field production from unconventional tight oil and gas reservoirs.

To summarize, the authors' monograph *Unconventional Tight Reservoir Simulation: Theory, Technology and Practice*, is a systematic summary of their theoretical and technical achievements, field application experiences, and critical understandings from practice. This content-rich and practical monograph is a strong contribution to the theory and practice of solving fluid flow through complex multiple-interacting media, and it contains several technological breakthroughs in numerical simulation techniques. I believe that the publication of this monograph will promote the development of better and physics-faithful unconventional tight oil and gas reservoir simulators to help effective and efficient production of these important resources.

December 2017

Zhangxing (John) Chen
University of Calgary Professor
International Outstanding Reservoir Simulation Expert
Member of Canadian National Academy of Engineering
Calgary, Canada

Preface

Unconventional tight oil and gas resources are huge. With the continuous advance of unconventional oil and gas development technology, continuous innovation of production management mode, and continuous reduction of investment costs, unconventional tight oil and gas resources have great commercial development values. The realization of the efficient development of unconventional tight oil and gas has important strategic significance for the sustainable and stable development of China's oil and gas industry.

Compared with conventional reservoirs, the reservoir conditions and fluid flow characteristics of unconventional tight reservoirs are more complex because: i. they have stronger micro- and macro-heterogeneities on multiple scales from nano/micro-pores to natural/artificial fractures; ii. the physical and chemical properties of fluids in different-scale pores and fractures are special and different in their occurrence, fluid properties and phase characteristics; iii. horizontal wells and multi-stage hydraulic fractures add to the complexity of these reservoirs in terms of flow mechanisms, flow regimes and coupling production mechanisms; iv. the physical parameters, flow regimes and flow mechanisms of different scales pores and fractures undergo significant dynamic changes during fracturing, injection and production. The effective development of unconventional tight oil and gas is greatly restricted by these characteristics and factors. Meanwhile, compared with marine tight reservoirs in North America, continental tight reservoirs in China have more complex geological conditions, lower resource grades, worse mobility and more difficulty of development. Current reservoir modeling and simulation methods are impossible to meet the requirements of unconventional tight reservoirs in terms of complex geological conditions, special physicochemical properties and flow mechanisms. Therefore, it is urgent to develop simulation theory and technology suitable for unconventional tight oil and gas.

At present, the research on numerical simulation theory and technology for unconventional tight oil and gas development is still in early stage. The authors focus on the key problems on reservoir modeling and simulation of unconventional tight oil and gas reservoirs. Combining the overseas advanced concepts and methods of unconventional oil and gas development with the features of continental tight reservoirs in China, the authors carried out multi-disciplinary joint research on development geology, fracturing, petroleum engineering, mathematics and computer science. Through great effort, the latest understanding of the geology, physicochemical properties, flow mechanism and development characteristics of unconventional tight oil and gas reservoirs have been gained. Based on these understanding, the simulation theory for unconventional tight oil and gas is innovatively developed, the geology modeling and simulation method for unconventional tight oil and gas are innovatively built, and a software integrating geology modeling and simulation with strong practicability and outstanding features is independently developed.

This book is a refinement and summary of the latest research results on simulation for unconventional tight oil and gas development. It is the world's first monograph on theory and method of multiple media simulation for unconventional tight oil and gas reservoirs, filling the gaps in this field. This book systematically introduces the multiple media simulation theory for unconventional tight oil and gas reservoirs, mathematical model of fluid flow through discontinuous multiple media, geology modeling technology of different scale discrete multiple media, and flow-geomechanics coupling simulation method for multiple scales, multiple flow regimes and multiple media. Combined with the development of typical tight oil and gas reservoirs, the practical application of simulation theory, technology and software for unconventional tight oil and gas numerical is introduced. This book plays an important guiding role and reference for the efficient development of unconventional tight oil and gas.

Organized by Professor Qiquan Ran, the editor board of the book *Unconventional Tight Reservoir Simulation: Theory, Technology and Practice* is established. Professor Qiquan Ran is responsible for the structure and overall layout of the whole book. All the chapters in this book are completed by all the members of editor board and then the whole book is finalized by Professor Qiquan Ran. The preparation of this book has received strong support and assistance from the leaders of the Ministry of Science and Technology, the Science and Technology Management Department of China National Petroleum Corporation, the China Petroleum Group Science and Technology Research Institute, Peking University, Northeast Petroleum University and other relevant experts. The project expert group gave careful guidance and technical checks. Changqing Oilfield, Xinjiang Oilfield, Jilin Oilfield, Daqing Oilfield, Southwest Oil and Gas Field Branch and other companies provided support and assistance to the application of the project results. Experts such as Shangping Guo, Yinan Qiu, Ronggai Zhu, Weiyao Zhu, Hanqiao Jiang, Bin Gong, Yuan Di, Chensong Zhang gave valuable revision suggestions for the preparation of this book. The "National High Technology Research and Development Program of China" group members Yongjun Wang, Lin Yan, Min Tong, Yuanhui Sun, Qi Wei, Ran Li, Wang Lin, Linzhi Zhao, Min Zhong, Dawei Yuan, Fuli Chen, Xiaobo Li, Baohua Wang, Jianfang Wang and Xianing Li did a lot of work. Some of the charts in this book quote the research results of the relevant technical personnel, and we would like to express our sincere gratitude.

Due to the complexity of unconventional tight reservoirs, the rapid development of simulation techniques and the limited knowledge and ability of the authors, the mistakes in this book are inevitable, and readers are welcome to give us advice and criticism.

Beijing, China
December 2019

Editor Board

Contents

Contributors

Qiquan Ran Research Institute of Petroleum Exploration and Development, Beijing, China

Mengya Xu Research Institute of Petroleum Exploration and Development, Beijing, China

Zhiping Wang Research Institute of Petroleum Exploration and Development, Beijing, China

Hui Peng Research Institute of Petroleum Exploration and Development, Beijing, China

Ning Li Research Institute of Petroleum Exploration and Development, Beijing, China

Jiaxin Dong Research Institute of Petroleum Exploration and Development, Beijing, China

Lifeng Liu Research Institute of Petroleum Exploration and Development, Beijing, China

Jiangru Yuan Research Institute of Petroleum Exploration and Development, Beijing, China

Pingliang Fang Engineering Technology R & D Company Limited, CNPC, Beijing, China

Huan Wang Petroleum Exploration and Production Research Institute, SINOPEC, Beijing, China

Guihua Tan Beijing ZHITAN Technology Co., Ltd., Beijing, China

1 Development Characteristics of Tight Oil and Gas Reservoirs

Considering the widespread reserves all over the world, tight oil and gas reservoirs have a huge potential. Currently, the development mode of horizontal well and hydraulic fracturing has obtained significant advances. However, the effective development of tight oil and gas reservoirs still face a big challenge due to the low productivity, quick declining, low recovery factors, high costs and low economical profits.

The lithology and lithofacies types of tight reservoirs are complex and diverse. The distribution of sand bodies, physical space and internal properties also change quickly. The differences of lithology, physical properties and oil-bearing ability between different sections of a reservoir, inter wells and even within a single well are large. There differ a lot in natural fractures at different scales, and the distribution of fluid saturation and fluid viscosity. The macroscopic heterogeneity and multi-scale characteristic are significant. In a reservoir, pores and fractures at different scales coexist in a discontinuous distribution in space. The physical properties, fluid composition, occurrence state, flow mechanisms and flow state in a pore/fracture media at different scales are significantly different. It has a strong micro heterogeneity and a significant characteristic of multiple media. For horizontal wells in tight reservoirs, different well completion methods have different contact relations between horizontal well and reservoir. During the processes of hydraulic fracturing and refracturing, a complex fracture network and flow relation is created. Furthermore, the flow mechanisms in the pore/fracture media are also different for the different development methods of primary recovery, water injection, imbibition recovery, and gas injection. Tight reservoirs has the following complex characteristics: i. strong macroscopic heterogeneity and microscopic heterogeneity in tight reservoirs; ii. remarkable characteristics of multiple scales and multiple media; iii. complex fluid composition in different pore-fracture media and complex fracture network; iv. complex flow mechanisms and flow regimes; v. complex flow relationships. Based on these characteristics, how can one carry out a fine geological modeling and numerical simulation to optimize and predict the performance of tight oil and gas reservoirs? The conventional theory and technology of reservoir geo-modeling and numerical simulation are no longer effective. Thus, it is urgent to develop a new theory and new technology of geo-modeling and numerical simulation methods for unconventional tight reservoirs.

1.1 Reservoir Characteristics of Tight Oil and Gas Reservoirs

1.1.1 The Macroscopic Heterogeneity of Tight Reservoirs

Tight reservoirs are distributed widely. Due to the comprehensive effect of deposition, diagenesis and tectonics, the lithology, lithofacies, sand bodies, physical properties, oil-bearing properties and reservoir types are significantly heterogeneous. In tight reservoirs, the natural fractures are widely distributed. The fracture network of hydraulic fracturing and refracturing is very complex. In addition, natural and hydraulic fractures coexist. They have an obvious characteristic of heterogeneity and multiple scales. The geostress and its spatial distribution in different reservoir regions and parts are significantly various. The spatial distribution of crude oil viscosity and density is also strongly heterogeneous. How to couple this macroscopic heterogeneity and multi-scale characteristics into the process of geo-modeling and numerical simulation is a problem that needs to be solved at present.

1. Distribution of reservoir lithology and lithofacies

The lithology and lithofacies types are complex and diverse. They are affected by the factors of basin nature, tectonic characteristic, sedimentary environment and diagenetic evolution. The vertical and horizontal characteristics change quickly, the distribution stability is poor, and the

Q. Ran, *Unconventional Tight Reservoir Simulation: Theory, Technology and Practice*,
https://doi.org/10.1007/978-981-32-9848-4_1

heterogeneity is strong. For the horizontal distribution, the reservoir lithology and lithofacies change quickly and discontinuously. The difference is significant. For the distribution between wells, the same reservoir lithology and lithofacies change discontinuously. The difference is large. For the distribution in different segments along horizontal well, the differences of reservoir lithology and lithofacies are large. Furthermore, the macroscopic heterogeneity is high.

2. Distribution of sand bodies

A massive distribution of sand bodies is controlled by the sedimentary facies belt of a reservoir. For horizontal direction, it presents a sheet-like distribution, potato-shaped distribution, irregular distribution and banded distribution. For vertical direction, it shows a tabular or asymmetrical lenticular distribution and an overlapped sand and mud "hamburger" distribution. In tight oil reservoirs, a horizontal well can go through many different sand bodies with different lengths.

3. Distribution of reservoir physical properties

Under the effect of sedimentary and diagenesis, reservoir physical properties and their distribution differ greatly in the horizontal and vertical directions. Along the source direction, the sand body is thicker. The porosity and permeability are distributed in a sheet-like shape. But in other sections, the reservoir properties are much worse. They have a strong heterogeneity.

(1) Porosity distribution

In tight reservoirs, the differences of storage space and developed degree of different lithology and lithofacies are obvious, and they result in a large change in porosity. In tight oil reservoir, because of the heterogeneity of lithology, lithofacies and sand bodies, the porosity distribution differs significantly, and the macroscopic heterogeneity is much severe. On the other side, due to the difference of lithology and lithofacies between wells, the porosity distribution is also different. The porosity distribution along the horizontal wellbore is also non-negligible.

(2) Permeability distribution

Compared with conventional reservoirs, a small pore and throat size in unconventional ones can cause a rapid decrease in reservoir permeability. It is far below that of the conventional reservoirs. Because of the effect of lithology, lithofacies and reservoir type, the reservoir permeability in tight oil reservoirs can change quickly within a wide range. The macroscopic heterogeneity is strong. Moreover, the permeability distribution along horizontal wellbore also changes greatly.

4. Distribution of sand body thickness

A tight reservoir is controlled by its lithology and lithofacies. The sand body thickness changes fast in a reservoir and the difference is large. The macroscopic heterogeneity of the sand body thickness is mainly manifested in several aspects. First, there is a wide distribution of sand bodies, but it is controlled by the changes of lithology and lithofacies. Then, due to the lateral discontinuous distribution, the changes of sand body thickness are fast in tight oil reservoirs, the difference is large, and heterogeneity is significant.

5. Distribution of reservoir types

Based on the properties of reservoir lithology, lithofacies and pore structure, the classification criteria of tight reservoirs are proposed. From this classification, a tight reservoir is classified into four types (Table 1.1). The discontinuous distribution of different reservoir types is controlled by the lithology, lithofacies and sand body distribution. Type I is mainly distributed in the main section of sand bodies. Then, around the boundary of a sand body, it gradually converts to type II and type III. Type IV is mainly distributed in the middle section between sand bodies. In a reservoir, the different reservoir types have a significant heterogeneity.

1.1.1.1 Distribution of Natural and Hydraulic Fractures in Tight Reservoirs

Natural fractures, hydraulic fractures and re-fractured fractures can form a complex fracture network in tight reservoirs. In a reservoir, natural fractures and hydraulic fractures

Table 1.1 Classification of tight reservoirs

Physical properties	Classification			
	I	II	III	IV
Lithology	Fine sandstone	Fine sandstone and siltstone		Siltstone
Porosity/%	>12	8–12	5–8	<5
Permeability/mD	>0.12	0.08–0.12	0.05–0.08	0.01–0.05
Pore-throat radius/μm	>0.15	0.1–0.15	0.05–0.1	<0.05

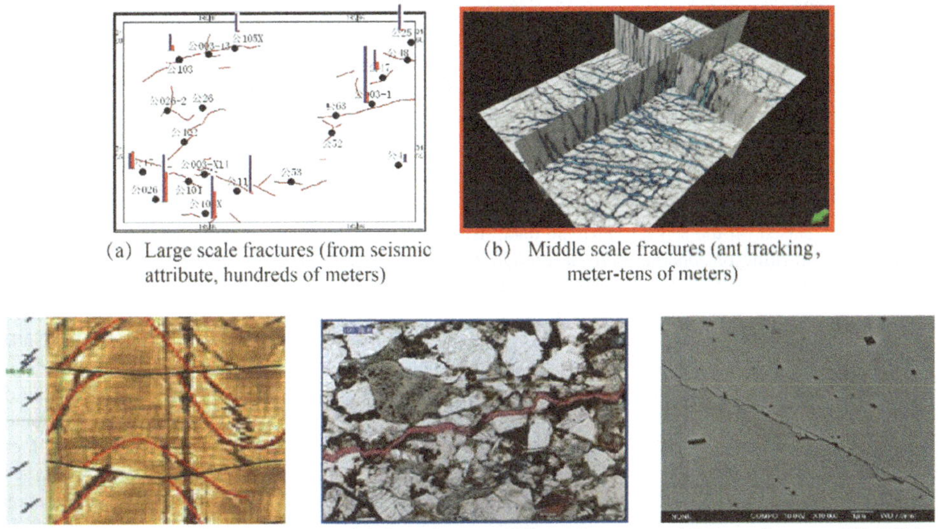

(a) Large scale fractures (from seismic attribute, hundreds of meters)

(b) Middle scale fractures (ant tracking, meter-tens of meters)

(c) Small fractures (imaging logging, centimeter-meter)

(d) Micro-fracture (thin section analysis, micrometer)

(e) Nano-fractures (FE-SEM, nanometer)

Fig. 1.1 Distribution of natural fractures at different scales in tight reservoir

present the characteristics of a discontinuous distribution, and the geometrical scales and physical properties also differ a lot. They show a significant feature of heterogeneity and multiple scales.

1. Natural fractures

Natural fractures in a tight reservoir have different types and different scales. The spatial distribution of these natural fractures at different scales shows a discontinuous characteristic. The geometrical parameters among them differ significantly. A surface distribution is significantly heterogeneous and has the characteristics of multiple scales. Because of the differences of lithology, lithofacies, reservoir thickness and rock properties, the development degree, fracture density, fracture number, cluster number and fracture direction are different, and the differences of spatial distributions are also non-negligible. Based on the geometrical parameters of extension distance and fracture width, it can be classified into five types, large scale fractures, middle scale fractures, small scale fracture, micro-fractures and nano-fractures (Fig. 1.1 and Table 1.2).

① Large scale fractures are based on 3D seismic data. Through a hydraulic interpretation method, highly reliable fractures with a large scale and wide distribution can be obtained. Generally, their extension length is higher than 100 m and width is higher than 10 mm. This type of fractures has a large scale and a high flow ability in a reservoir. But the proportion is relatively small.

② Middle scale fractures are based on seismic attribute data. Through a tracking or coherent body technique, the deterministic fractures between wells with a middle scale can be obtained. Their extension length is 10–100 m and width is 1–10 mm. For this type of fractures, the facture density is high; the extension distance is far; the flow ability in a reservoir is also relatively high. Compared with large scale fractures, the proportion is increased.

③ Small scale fractures are based on the data of coring, conventional logging and imaging logging. This type of fractures is the quantified fracture network around a well and the hard-described fracture network between wells. The extension length is 0.1–10.0 m and width is 0.1–1.0 mm. The extension distance and flow ability in a reservoir is not too high. But the proportion is relatively high.

④ Micro-fractures are based on a core observation and thin section analysis. Their extension length is 0.005–0.1 m and width is 0.001–0.1 mm. It connects pore systems at different scales. The extension distance is small. The proportion is high.

⑤ Nano-fractures are based on thin section and SEM data. The extension length is smaller than 0.005 m and width is smaller than 0.001 mm. It can connect pore systems at different scales. The extension distance is small. The proportion is higher.

Table 1.2 Classification and characteristics of natural fractures at different scales

Fracture	Interpretation method	Geometrical characteristics	Spatial distribution and geometrical parameters
Large scale	Based on the 3D seismic data, using a hydraulic interpretation method, the highly reliable fractures with a large scale and wide distribution can be obtained	The extension length is higher than 100 m and width is higher than 10 mm	① The spatial distribution and geometrical parameters are deterministic ② Spatial distribution parameters: trends, direction, trajectory data; geometrical parameters: length, width, height
Middle scale	Based on the seismic attribute data, using ant tracking or coherent body technique, the deterministic fractures between wells with a middle scale can be obtained	The extension length is 10–100 m and width is 1–10 mm	① Using the high-precision analysis technique of seismic attribute data, the spatial distribution between wells and geometrical parameters of middle scale fractures are deterministic ② Under the situation of insufficient resolution for the data between well, the regulation of spatial distribution and the value ranges of geometrical parameters are deterministic. But the fracture number is non-deterministic; the specific spatial position (location, trends, direction) and geometrical parameters (length, width, height) are non-deterministic
Small scale	It is based on the data of coring, conventional logging and imaging logging. This type of fracture is the quantified fracture network around the well and the hard-described fracture network between wells	The extension length is 0.1–10.0 m and width is 0.1–1.0 mm	① The spatial distribution and geometrical parameters are deterministic around wellbore ② The regulation of spatial distribution between wells and the value ranges of geometrical parameters are deterministic. The spatial location and specific geometrical parameters of fractures are non-deterministic
Micro scale	Micro-fracture is based on the core observation and thin section analysis	The extension length is 0.005–0.1 m and width is 0.001–0.1 mm	For nano/micro-fracture, the regulation of spatial distribution and the value ranges of geometrical parameters are deterministic. The spatial position and specific geometrical parameters are non-deterministic
Nano scale	Nano-fracture is based on the thin section and SEM data	The extension length is smaller than 0.005 m and width is smaller than 0.001 mm	

2. Hydraulic fractures

Hydraulic fractures refer to a complex fracture network system of different-scale fractures under the effect of hydraulic stress or induction stress and natural fractures (Fig. 1.2). It can be described, identified and evaluated by the methods of seismic data, fracture monitoring and production performance data. Natural fractures and hydraulic fractures at different scales are discretely distributed in a reservoir. The heterogeneity is very significant, and the differences of geometrical scales are also high. From the geometrical scale and mechanical mechanism of hydraulic fractures, it can be classified into main fractures, branch fractures and shear micro fractures (Table 1.3).

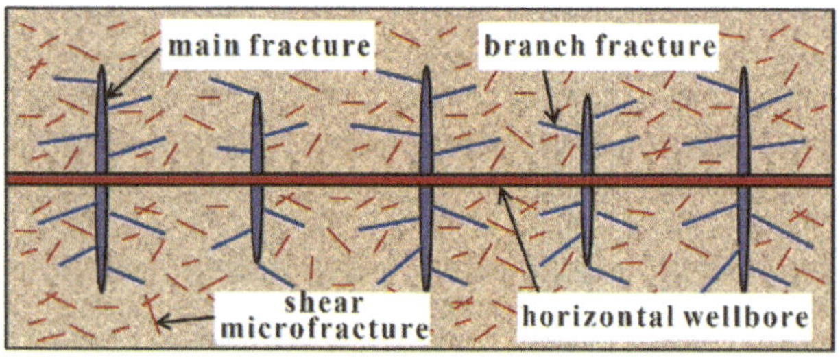

Fig. 1.2 Distribution of hydraulic fracture in tight reservoir

① Main fractures refer to the fractures which are perpendicular or heterotopic to a wellbore direction. The fracture length ranges from 300 to 600 m. The fracture aperture is higher than 1.8 mm. It is the main flow path, and the fracture conductivity is high.

② Branch fractures are the secondary small scale fractures which connect to the main fractures. The fracture length ranges from 1 to 100 m. The fracture aperture is 0.2–1.8 mm. It is the secondary flow path, and caused by small size proppants. The conductivity of branch fractures is high, and the connectivity between branch fracture and main fracture is also high. But the conductivity is highly affected by a sand amount.

③ Shear micro fractures are the micro scale fractures which are closed to the main fractures and controlled by the stress state in a fracturing region. They are also far from the main fractures. The fracture aperture is less than 0.2 mm. The conductivity is low. They are mainly

Table 1.3 Classification and characteristics of hydraulic fractures

Type	Mechanical mechanism	Geometrical and spatial distribution	Propped characteristics	Physical properties and conductivity
Main fracture	Tension	Length ranges from 300 to 600 m. Aperture is higher than 1.8 mm	Propped by the proppant of 30–50 mesh. Conductivity can sustain for a long time	Main flow path, and the fracture conductivity is high
Branch fracture	Tension-Shear	Length ranges from 1 to 100 m. Fracture aperture is 0.2–1.8 mm	Effectively propped by the proppant of 100 mesh	Secondary small scale fractures which connect to the main fractures. The conductivity is high, and the connectivity between branch and main fractures is also high. But the conductivity is highly affected by the sand amount
Shear micro-fracture	Shear	Discrete distribution, aperture is less than 0.2 mm	Self-propped, shear slip	The conductivity is low. They are mainly used to enhance the permeability of matrix. The connectivity is low. And the flow back of fracturing fluid is difficult

used to enhance the permeability of matrix. The connectivity is low and the flow back of a fracturing fluid is difficult.

1.1.1.2 Distribution of Reservoir Stress

Under the effects of a stratigraphic structure, lithology, reservoir depth and development activity, the geostress and its distribution in different areas and regions are much different in tight reservoirs (Ge and Lin 1998; Zhao 2015; Zeng and Wang 2005). Because of the large depth of a tight reservoir and the high hardness of rocks, the geostress is always high. On the other hand, under the effects of tectonic movement and large heterogeneity, a geostress value, geostress direction, and geostress difference in different locations change fast. Simultaneously, tight oil and gas reservoirs generally experience the processes of hydraulic fracturing, water injection and gas injection. The activities are results of the changes in reservoir pressure and temperature. Thus, the distribution of a geostress field is changed. The non-uniformity of a geostress field is also enhanced. In some cases, the direction and shape of hydraulic fractures are also affected.

1.1.1.3 Distribution of Reservoir Fluids

A tight reservoir has an extensive oil bearing area. But under the effects of micro facies, sand bodies and physical properties, the difference of an oil bearing condition in some regions is large. i. In a reservoir, the oil bearing condition changes quickly in a plane, and the difference is large. The distribution range of oil saturation is wide. The oil saturation in the main bodies of river channels is high, and it reduces as it is close to the boundary region. Simultaneously, the difference of oil saturation between wells is large. ii. The oil bearing condition for different wellbore segments in a same horizontal well differs significantly. The oil saturation from the well-logging interpretation changes greatly. The macroscopic heterogeneity is strong. iii. The movable fluid saturation is controlled by the multiple factors of a geological condition, initial oil saturation, a pore structure and fluid properties. Thus, the movable fluid saturation in surface, inter well and inter well-segment differs significantly.

In tight oil reservoirs, the oil viscosity and density in different basins or in different regions of a same basin differ significantly. The surface distribution of oil viscosity and density changes quickly and differs greatly. Among the wells in a same block, oil viscosity and density are also different. Surface heterogeneity is strong. Under the effects of a basin type, tectonic activity, source rock and its corresponding preservation condition, a tight reservoir has the characteristics of abnormal pressure. The pressure coefficient ranges widely and changes greatly. Simultaneously, because of the oil properties, reservoir temperature and pressure, in tight oil reservoirs, the initial GOR (gas-oil ratio) in different blocks and inter well regions also differs greatly.

1.1.2 The Microscopic Heterogeneity and Characteristics of Pore-Fracture Media at Different Scales

Micro-pores and nano-pores and natural fractures at different scales can be observed in a tight reservoir. From microscopic aspect, they have the significant feature of multiple media. The pore-fracture systems at different scales present a discontinuous discrete distribution, and the microscopic heterogeneity is non-negligible. The geometrical and physical properties for the pore-fracture systems with different scales have significant differences. The fluid composition, properties, an occurrence state, phase behavior, flow mechanisms and a flow regime are also different. How does one describe and quantitatively represent the pore-fracture systems at different scales? How does one geologically model

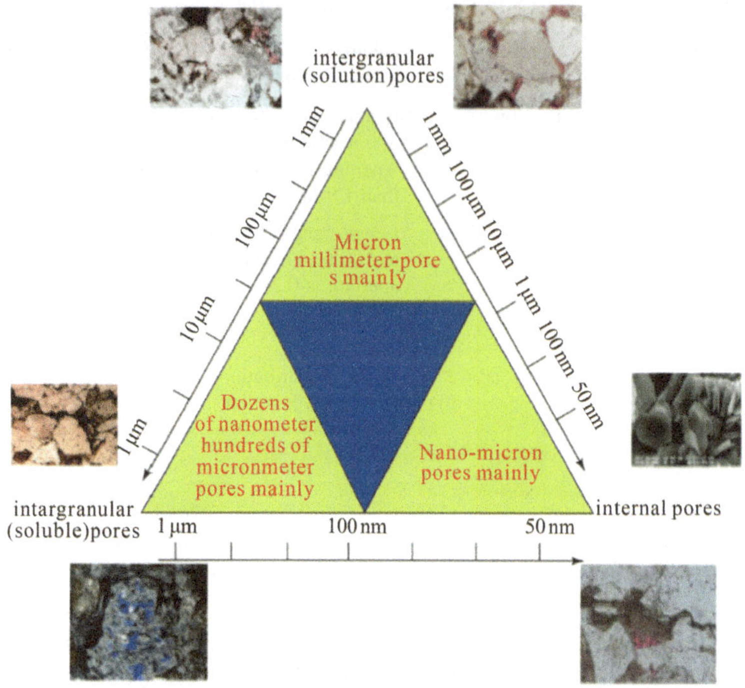

Fig. 1.3 Types of storage space in tight reservoir

the pore-fracture systems at different scales? How does one simulate the fluid flow and recovery performance in the pore-fracture systems at different scales? These questions are the urgent concerns that we encountered.

1.1.2.1 Geometry and Property Characteristics of Pores at Different Scales

Pores in unconventional reservoirs are tight, and nano-, nano/micro- and micro-scale porous media are developed. Porous media can be divided into pores and throats by different structures. Pore is the storage space of fluids, and throat is the main path way for fluid flow in rocks. The size and distribution of pores and throats are the main factors to affect the characteristics of fluid storage and flow. The composition and distribution patterns of pores at different scales are quite different; the pores at different scales are discrete and discontinuous in space, and the spatial distribution patterns are quite different; the geometric and physical parameters of porous media at different scales are greatly different.

1. Pore types of tight reservoir

For a terrestrial tight reservoir, under the effects of sedimentation, diagenesis, tectonics, and densification at different stages, the types of storage space show an obvious diversity and complexity. A tight reservoir mainly develops intergranular (solution) pores, intragranular (soluble) pores and internal pores (Fig. 1.3).

2. Pore geometry of tight reservoir

Unconventional reservoirs develop nano/micro- and micro-scale pores and throats, and their composition and distribution patterns are quite different. The pores and throats at different scales are discrete and discontinuous in space, and the spatial distribution patterns are quite different.

(1) Pore size

Compared with conventional reservoirs, the pore size of tight reservoirs is significantly lower than that of conventional reservoirs due to the combined internal and external factors such as sedimentation, diagenesis, and late densification (Fig. 1.4). The main pore sizes range from 0.01 to 10 μm.

Tight reservoirs develop porous media with nano-, nano/micro- and micro-scales. The composition and quantity distribution of pores at different scales can be obtained by an adsorption method, constant-rate mercury injection, high-pressure mercury injection, and a quantitative thin analysis (Fig. 1.5). There are large differences in the composition and quantity distribution of pores at different scales in different lithologies, different lithofacies, and different types of reservoirs (Fig. 1.6). At the same time, the pores at

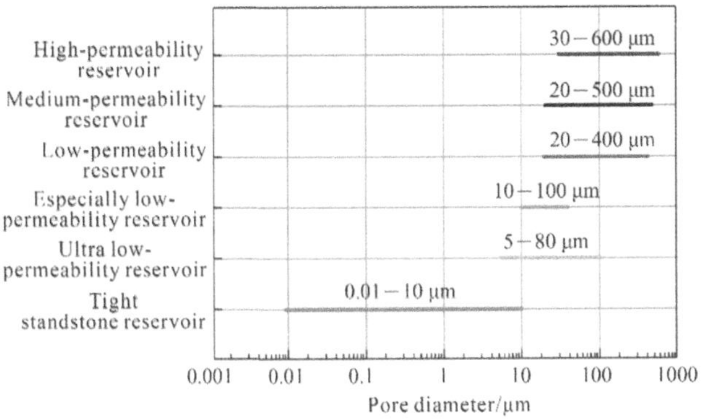

Fig. 1.4 Pore sizes of tight and conventional reservoirs

different scales are discretely and discontinuously distributed in space, and the spatial distribution patterns are quite different, showing the obvious characteristics of microscopic heterogeneity and multiple media. From the point of view of development geology, according to the pore size of tight reservoir, the pores can be divided into five categories: macro-pores (>100 μm), mid-pores (20–100 μm), small pores (10–20 μm), micro-pores (1–10 μm) and nano-pores (<1 μm) (Table 1.4).

(2) Throat size

Compared with conventional reservoirs, the throat size of a tight reservoir decreases dramatically (Fig. 1.7), and the throat diameter ranges from 0.005 μm to 1 μm.

Through constant-rate mercury injection, high pressure mercury injection and other experimental methods, capillary pressure curves of different throats and the histograms of different throat radius distributions can be obtained (Fig. 1.8). It indicates that tight reservoirs have throats with nano-, nano/micro- and micro-scales. The composition and quantity distribution of throats at different scales in different lithologies, different lithofacies and different types of reservoirs are quite different. The microscopic heterogeneity is extremely strong. Combining the characteristics of fluid flow in throats at different scales in tight reservoirs, the throats can be divided into major throats (>10 μm), middle throats (3–10 μm), narrow throats (1–3 μm), micro-throats (500 nm–1 μm) and nano-throats (<500 nm) (Table 1.5).

(3) Physical properties of pores at different scales

Compared with conventional reservoirs, the throat size in tight reservoir decreases dramatically, resulting in a sharp decrease in reservoir permeability. The permeability of tight reservoirs is mainly distributed from 0.001 mD to 0.2 mD, which is much lower than that of conventional reservoirs. At the same time, due to the difference of a pore radius and a throat radius for pores at different scales in tight reservoirs, the porosity and permeability of porous media at different scales are also greatly different (Table 1.6). Different lithologies, different lithofacies, and different types of reservoirs are composed of different numbers of pores with different scales, and, therefore, the porosity, permeability, and heterogeneity are significantly different.

1.1.2.2 Geometry and Attribute Characteristics of Fractures at Different Scales

Tight reservoirs have not only macro-scale fractures, but also many microscopic fractures at nano/micro-scales. In a reservoir, these fractures at different scales show the characteristics of discrete and discontinuous distribution. Spatial distribution patterns and geometrical scales are also different. There are large differences in the physical properties of fractures at different scales, and the heterogeneity is extremely strong.

Fig. 1.5 Pore scale distribution of tight reservoirs

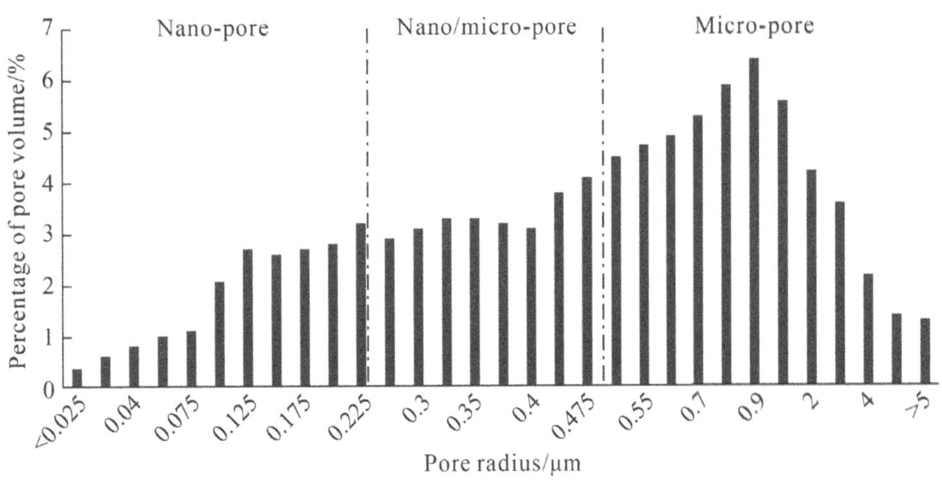

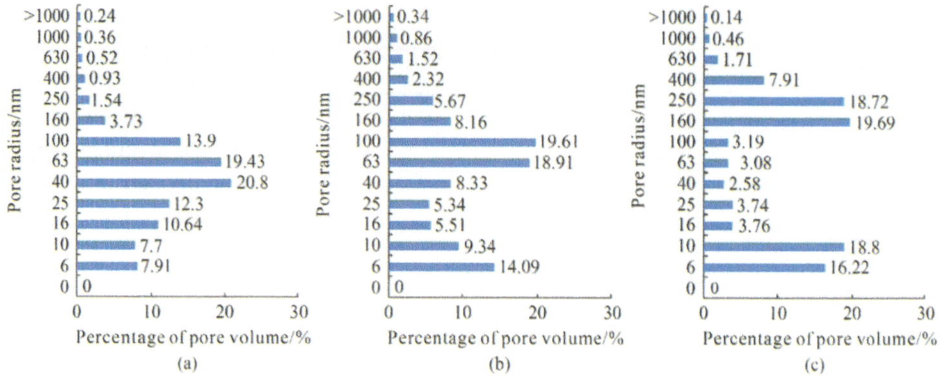

Fig. 1.6 Pore composition and quantity distribution at different scales for different lithology/reservoir

Table 1.4 Standard table of different scale porosity classification

Porosity classification	Macro-pore	Mid-pore	Small pore	Micro-pore	Nano-pore
Pore diameter/μm	>100	100–20	20–10	10–1	<1
Pore image	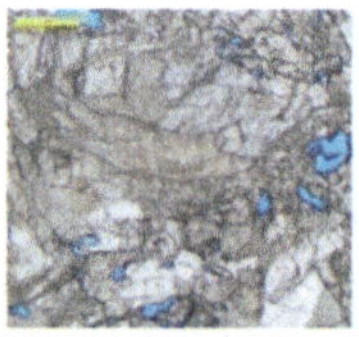	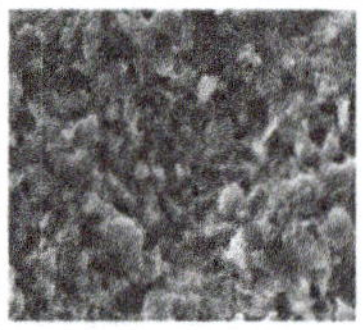		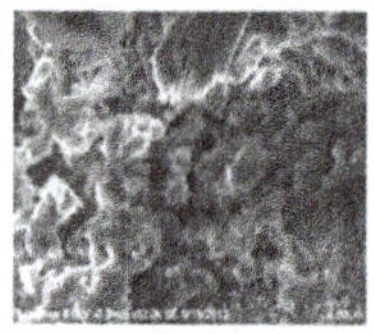	

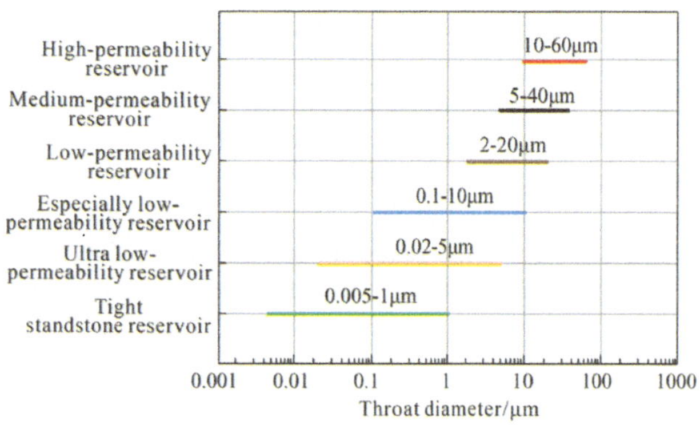

Fig. 1.7 Comparison of throat size of tight reservoirs and conventional reservoirs

1. Natural and hydraulic fracture types in tight reservoirs

Natural fractures in tight reservoirs are mainly affected by tectonic stress, sedimentation, diagenesis, crystallisation, and dissolution. Thus there are three types of fractures (Fig. 1.9): ectonic fractures, diagenetic fractures (interlayer fractures, suture fractures, turtle fractures, cleavage fractures, and intercrystallite fractures) and corrosion fractures. At the same time, due to hydrodraulic fracturing/re-fracturing, hydraulic fractures at different scales are formed. Thus, a complex fracture network of natural fractures and hydraulic fractures is created. It results in a complex flow behavior of oil and gas in reservoirs.

2. Geometrical characteristics of fractures in tight reservoirs

Tight reservoirs have a large number of fractures with different scales and obvious characteristics of multiple scale media, and the difference of geometry sizes between different media in tight reservoirs is great. According to the above features, based on the fluid flow behavior, fractures can be classified into five categories by the fracture aperture: large fractures (>100 μm), middle fractures (10 to 100 μm), small fractures (1 to 10 μm), micro-fractures (500 to 1 μm), and nano-fractures (<500 nm) (Table 1.7).

3. Characteristics of physical properties of fractures at different scales

In tight reservoirs, the geometrical parameters of media with different scales fracture are different. It leads to the difference of physical properties-permeability and conductivity of fractures. The larger the fracture aperture is, the higher the permeability and conductivity are (Table 1.8).

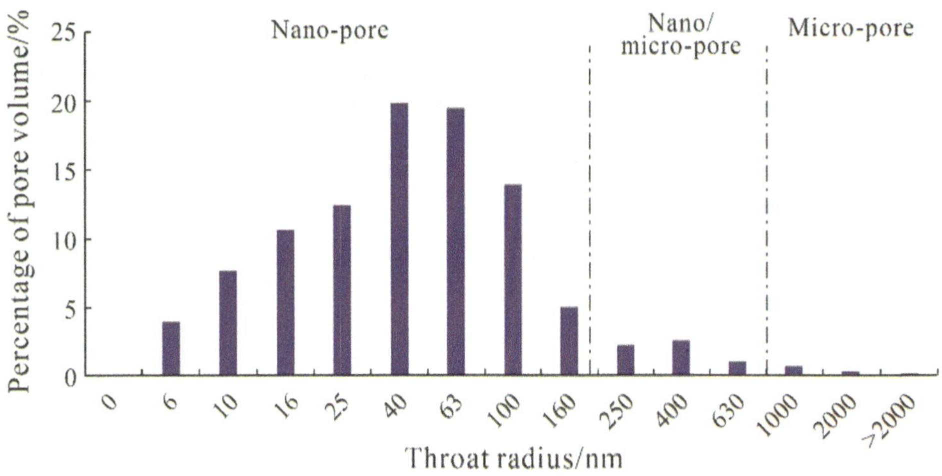

Fig. 1.8 Distribution of pore throat in tight reservoir

Table 1.5 Criterion for throat classification

Throat classification	Major throat	Middle throat	Narrow throat	Micro-throat	Nano-throat
Throat diameter/μm	>10	10–3	3–1	1–0.5	<0.5
Image		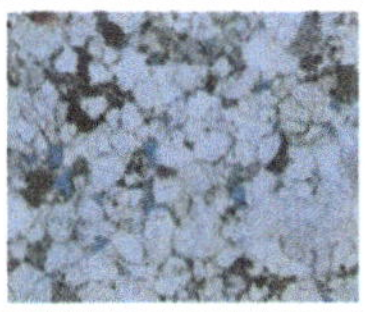			

Table 1.6 Physical properties of pore at different scales

Throat classification	Throat diameter/μm	Permeability/mD
Major throat	>10	>2.3
Middle throat	3–10	0.5–2.3
Narrow throat	1–3	0.18–0.5
Micro-throat	0.5–1	0.05–0.18
Nano-throat	<0.5	<0.05

1.1.2.3 Fluid Flow Through Different Scale Pores and Fractures Media

The pore size in conventional reservoirs is relatively homogeneous. Thus, the differences of fluid properties and flow mechanisms are small. For unconventional tight reservoirs, because of the large differences in a geometrical scale and physical properties, the composition, properties, an occurrence state, and phase behavior of fluids in those pores at different scales are much different. The flow mechanisms and flow regimes are also different.

1. Fluid composition in pores at different scales

The pore size in conventional reservoirs is relatively homogeneous. Thus, the differences of fluid composition are small. However, in tight reservoirs, the oil composition in porous media at different scales is different, and the fractions of light components and heavy components are also different. The fraction of heavy components in macrospores is high, and the fraction of light components in small pores is high. For small pores, the fraction of light components is higher (Fig. 1.10) (Nojabaei and Johns 2013).

2. Occurrence state of oil, gas and water in porous media at different scales

The occurrence state of fluids is affected by pore size, pore structure, specific surface area, fluid composition and properties, wettability and capillarity. The occurrence state of fluids in the porous media at different scales is different (Table 1.9).

According to the occurrence state of fluids in tight reservoirs, the crude oil can be divided into four types:

(a) Structural fractures (b) Diagenetic fractures (c) Dissolving fractures

Fig. 1.9 Fracture types in tight reservoir

Table 1.7 Criteria for the classification of fractures at different scales

Fracture classification	Large fractures	Middle fractures	Small fractures	Micro fractures	Nano fractures
Fracture aperture/μm	>100	100–10	10–1	1–0.5	<0.5

Table 1.8 Physical properties of fractures at different scales

Fracture classification	Fracture aperture/μm	Permeability/mD	Conductivity/(mD·m)
Large fractures	>100	>10000	>1
Middle fractures	100–10	100–10000	0.001–1
Small fractures	10–1	1–100	0.000001–0.001
Micro fractures	1–0.5	0.25–1	0.000000125–0.000001
Nano fractures	<0.5	<0.25	<0.000000125

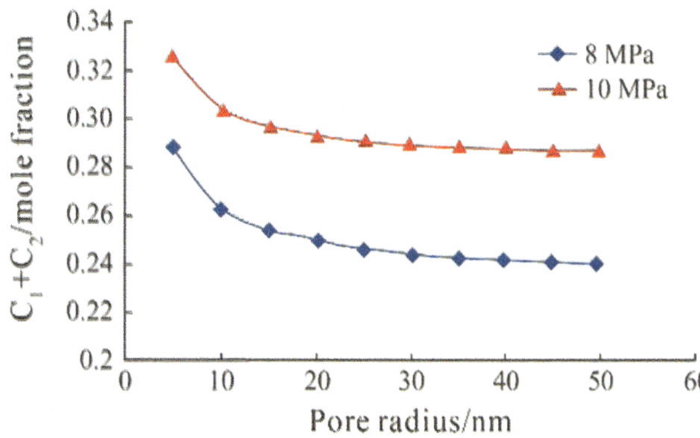

Fig. 1.10 The fraction of light components in nano-pores

movable oil, capillary oil, thin film oil and adsorbed oil. In large and middle scale pores, the pore physical properties are better; the pore size is higher; the capillary pressure is smaller. Therefore, the crude oil is mainly movable oil. In small and micropores, a specific surface area is larger, and the capillary pressure is also higher. Thus, the crude oil is under the state of capillary oil. In nano-pores, pore throat is smaller; the surface area is larger; the crude oil has strong adsorption capacity. Thus, the crude oil is mainly under the states of adsorbed oil and thin film oil.

Natural gas is mainly under the states of free gas and adsorbed gas. In large and middle pores, the fluid-wall inter action force is small; the adsorption capacity is weak; the surface area is small. Therefore, the content of adsorption gas is less than free gas, and the occurrence state of natural gas is mainly free gas. In small pores, gas molecules are strongly affected by pore wall. The adsorption energy is higher, and the surface area is also higher. So the content of adsorbed gas increases, and natural gas is under the states of adsorbed gas and free gas. In nano pores, pore throat is extremely small. The fluid-pore wall interaction is very high; the adsorption capacity is stronger; the surface area is also higher. Furthermore, the content of adsorbed gas is higher than free gas, and natural gas is under the state of adsorbed gas.

3. Oil saturation in porous media at different scales

In tight reservoirs, the size, porosity, permeability, surface effect and occurrence state of pores and throats at different scales have an important influence on the original oil saturation. The geometrical parameters and physical parameters of pores at different scales are different, and the oil-bearing properties are also different.

In tight reservoirs, the pore-throat scale of large and middle pores are relatively higher. The physical properties are better; the storage ability is strong; the surface area is small; the capillary pressure is also small. The crude oil is

Table 1.9 Occurrence state of crude oil and natural gas in tight reservoirs

Fluid type	Pore type	Influencing factors of occurrence state	Occurrence state
Crude oil/natural gas	Large and middle pores	Good physical properties, large pore size, small capillary pressure/wall adsorption force and small specific surface area	Movable oil/free gas
	Small and micro pores	Good physical properties, smaller pore size, smaller capillary pressure/wall adsorption force and larger specific surface area	Capillary oil/adsorbed gas and free gas
	Nano pore	Poor physical properties, small pore size, high capillary pressure/wall adsorption force and large specific surface area	Adsorbed oil, film oil/adsorbed gas

Table 1.10 Oil saturation in porous media at different scales

Pore type	Influencing factors of oil saturation	Irreducible water saturation	Original oil saturation
Large and middle pores	Large aperture, small specific surface area, small capillary pressure, and movable oil	<30%	>70%
Small and micro-pores	The pore size is small, the specific surface area is large, the capillary pressure is large, and the capillary oil is the main one	30%–50%	50%–70%
Nano-pores	The pore size is very small, the specific surface area is very large, the capillary pressure is large, and the adsorption oil and film oil are the main ones	>50%	<50%

mainly under the state of movable oil. For this type of pores, the irreducible water saturation is smaller than 30%, and the original oil saturation is higher than 70%. In small and micro pores, both the surface area and capillary pressure are larger. The oil is capillary oil. The irreducible water saturations 30%–50%, and the original oil saturation is 50%–70%. The pore size of nanopores is extremely small. The physical properties are poor, and the storage capacity is small. Simultaneously, both the specific surface area and capillary pressure are higher. In this type of pores, crude oil is mainly under the states of adsorbed oil and thin film oil. The irreducible water saturations greater than 50%, and the original oil saturation is less than 50% (Table 1.10).

4. Oil viscosity and density in porous media at different scales

In tight reservoirs, the fluid compositions in pores with different scales are different. The fluid properties are also different. In large pores, a pore radius is large, and the content of heavy components is high. Both the oil viscosity and oil density are large. In small pores, a pore radius is small, and the content of light components is high. Simultaneously, the oil viscosity and oil density are small. As the pore size decreases, both the oil viscosity and oil density decrease (Fig. 1.11 and Fig. 1.12) (Nojabaei and Johns 2013).

5. Phase behavior of fluids in porous media at different scales

The phase behavior in tight oil reservoirs and conventional oil reservoir has a large difference. In conventional oil reservoirs, the difference of phase behavior of fluids is small. But in tight oil reservoirs, the bubble point pressure, dew point pressure and phase behavior are quite different. In large pores, the bubble point is high and the dew point pressure is low. In small pores, the bubble point is low and the dew point pressure is high (Figs. 1.13 and 1.14) (Dong et al. 2016a). As a pore size decreases, the bubble point pressure is reduced, and the dew point pressure is increased. During the development process, the changes and transition of phase behavior in these pores at different scales are also different. From phase envelop, in large pores, the two-phases region is wider. But in small pores, the two-phases region is small (Fig. 1.15) (Dong et al. 2016b). During the primary development, as the pressure decreases, the cured oil in large pores is degasses firstly. Thus, the fluid is converted from two-phases flow to three-phases flow. It is produced firstly. For a CO_2 injection process, as the pressure increases, the CO_2 in large pores is changed from the gas phase to the liquid phase firstly. It is easy to reach the miscible state, and it is better for the production process.

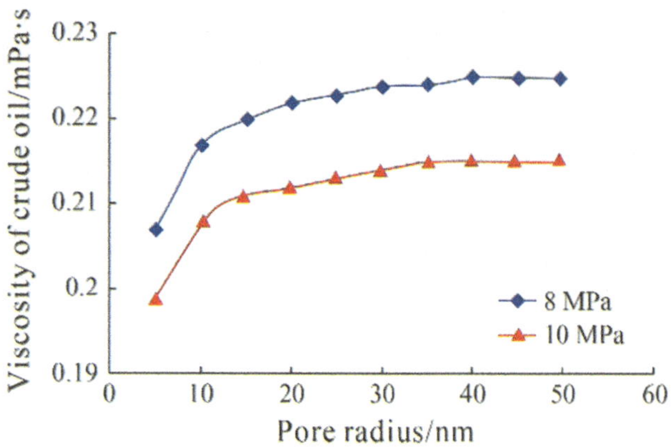

Fig. 1.11 Oil viscosity in nano-pores

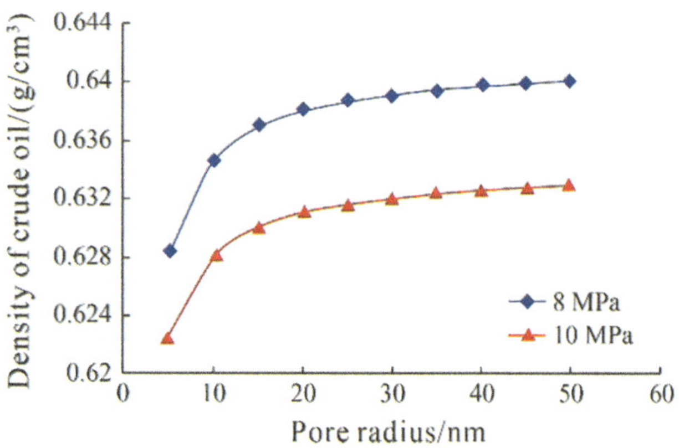

Fig. 1.12 Oil density in nano-pores

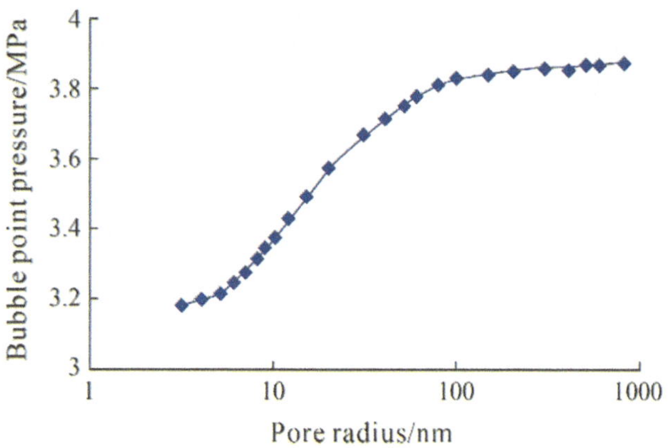

Fig. 1.13 Bubble point pressures in nano-pores

6. Characteristics of fluid flow and producing degree in media at different scales

The flow behavior of fluids in media with different scales pores and fractures is mainly reflected on capillary pressure and relative permeability curves. Media with different scale pores and fractures have different geometrical scales, and the physical properties are also different. They cause the difference of flow behavior of fluids in different scale media. The flow behavior of fluids is affected both by capillary pressure and relative permeability curves. The producing degree of fluids in different scale media is different.

(1) Capillary pressure and available producing fluids

The characteristics of capillary pressure in media with different scale pores and fractures are different. i. In porous media, as the pore radius increases, the capillary pressure is reduced. The smaller the pore radius is, the larger the capillary pressure is (Fig. 1.16) (Dong et al. 2016c). Firstly, in large and middle scale pores, the capillary pressure is small. For a production process, the pressure drop which should balance the capillary pressure is small. Thus, the fluid flow ability is good. In small and micro pores, the capillary pressure is reduced. The pressure drop which should balance the capillary pressure is smaller. The fluid flowability is reduced. In nano-pores, the capillary pressure is the largest. Once converting into a production process, pressure drop which should balance the capillary pressure is high. The fluid flowability is low (Fig. 1.17a). ii. Compared with porous media, capillary pressure in media with fractures is generally lower (Liu et al. 2000). There is no capillary pressure in large scale fractures. Small scale fractures have a small capillary pressure. In micro fractures and nano-scale fractures, the capillary pressure is close to the capillary pressure in the large scale pores (Fig. 1.17b). The fluid flowability in media with fractures is better than that in porous media.

(2) Relative permeability curves and available producing fluids

Because of the effects of geometrical scale, physical properties and fluid properties, the permeability curves of porous media at different scales also show the different characteristics. i. For porous media at different scales, as the geometric scale increases, the physical properties are improved; the irreducible water saturation is reduced; the residual oil saturation is reduced; the region of two-phase flow is increased; the fluid flowability is increased (Fig. 1.18a). ii. For media with different scales fractures, the irreducible water saturation and residual oil saturation are lower than those in porous media. The region of two-phases flow is wider than that in porous media, and the fluid flowability is higher than that in porous media. In large scale fractures, the relative permeability is close to a linear relationship. The endpoint of relative permeability reaches 1.0. The relative

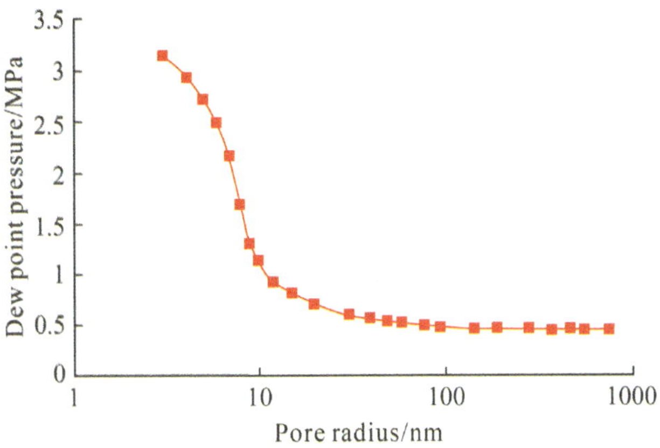

Fig. 1.14 Dew point pressures in nano-pores

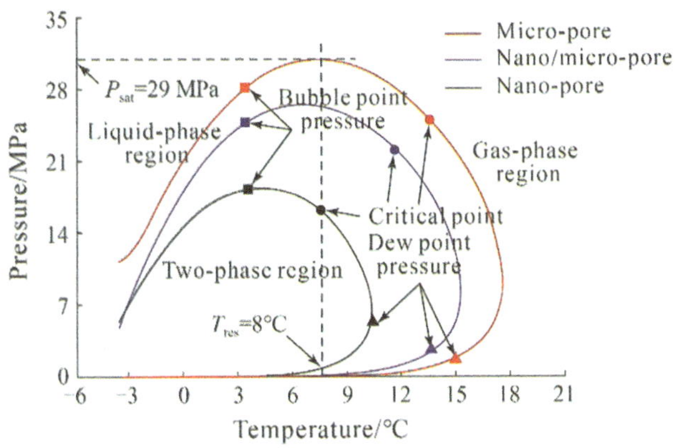

Fig. 1.15 Phase envelope in porous media at different scales

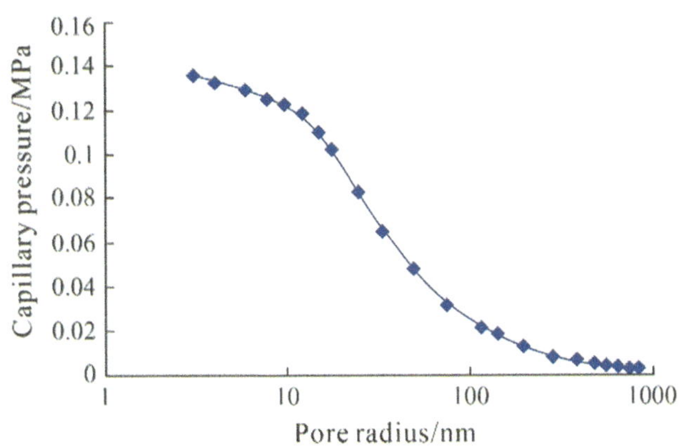

Fig. 1.16 Capillary pressure of fluids in porous media at different scales (gas)

permeability curves of micro- and nano-fractures are similar to those in large scale porous media, but they are better than large scale porous media (Fig. 1.18b) (Wang 2014a; Pan et al. 2016) (Table 1.11).

7. Flow regimes and flow mechanisms in different scale media

In tight reservoirs, the flow regimes in media with different scales pores and fractures can be classified into four types, high-speed nonlinear flow, linear flow, quasi-linear flow and low-speed nonlinear flow. Due to the difference of a geometrical size, property parameters and fluid properties, the flow regime of fluids in media with different scales pores and fractures is different (Fig. 1.19a). In the same medium, the pressure gradient is changed in the different development stages. It also leads to a great difference in a flow regime (Fig. 1.19b).

During the development process, the mechanisms of viscous flow, compaction, fluid expansion and gravity drainage are important. Simultaneously, in tight oil reservoirs, the mechanisms of dissolved gas and imbibition are also non-negligible. For tight gas reservoirs, the mechanisms of slippage, diffusion and adsorption are also important. The flow mechanisms of media with different scale spores and fractures in different production stages are also different.

8. Performance of pore-fracture media at different scales

In tight reservoirs, because of the difference of geometrical scale and physical properties, the different pore-fracture media show the performance characteristics of multiple media in the pressure buildup curves (Fig. 1.20). During the pressure buildup process, it shows a characteristic of joint bulidup for the media with different scales pores and fractures. The pressure in media with fractures is recovered firstly. Also, the pressure in large fractures is recovered firstly, and then it is the small scale fractures and micro-fractures. After that, the pressure is transferred to the large pores, small pores and nano/micro-pores. The pressure in different scale porous media is recovered gradually (Jia et al. 2016). Due to the large differences of geometrical properties and physical properties in different scale porous media, as the pressure is transferred from media with high permeability to media with low permeability, the pressure buildup curve presents an obvious step phenomenon. During the processes of pressure transfer and joint bulidup in media with different scale pores and fractures media, it shows a phenomenon of multiple steps (Wang and Yao 2009; Yin 1983; Zhang et al. 2010).

1.2 Development Mode and Characteristics of Tight Oil and Gas Reservoirs

Tight oil and gas reservoirs have a tight formation, low permeability, poor physical properties, low well productivity and even no natural productivity. Reservoir lithology, lithofacies,

Fig. 1.17 Capillary pressure of fluids in pores and fractures at different scales

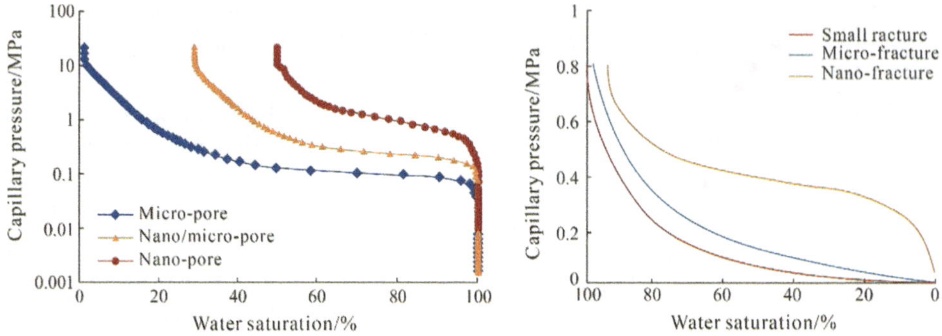

(a) Capillary pressure of fluids in porous media at different scales (crude oil)

(b) Capillary pressure of fluids in media with different scales fractures (crude oil)

Fig. 1.18 Relative permeability curves of pores and fractures at different scales

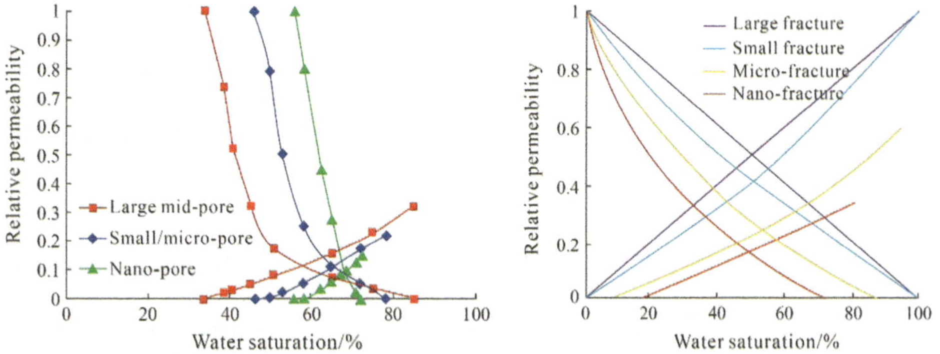

(a) Relative permeability curves in porous media

(b) Relative permeability curves in media with fractures

Table 1.11 The fluid flowability in porous media at different scales

Pore type	Occurrence state	Irreducible water saturation	Residual oil saturation	Range of oil and gas flow zone	Flow availability
Large and middle pore	mobile oil	<30%	20%–30%	50%–60%	Good
Small/micro-pore	Capillary oil	30%–50%	25%–35%	25%–35%	Middle
Nano-pore	Film oil, adsorbed oil	>50%	30%–40%	5%–10%	Bad

Fig. 1.19 Fluid flow curves of different media

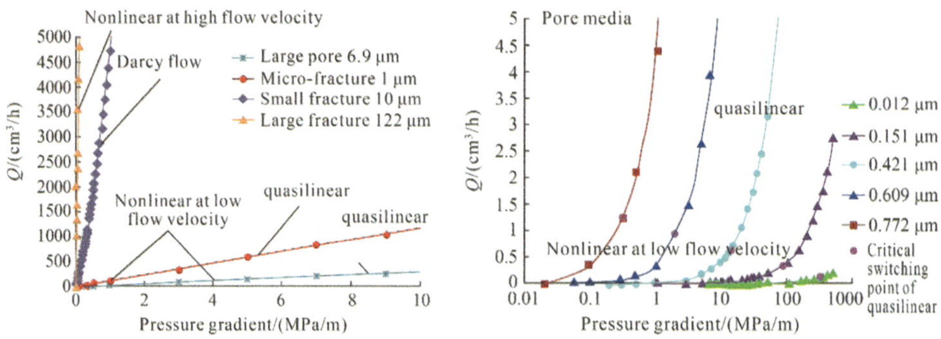

(a) Comparison of fluid flow curves between media with fractures and porous media

(b) Comparison of fluid flow curves between different scale porous media

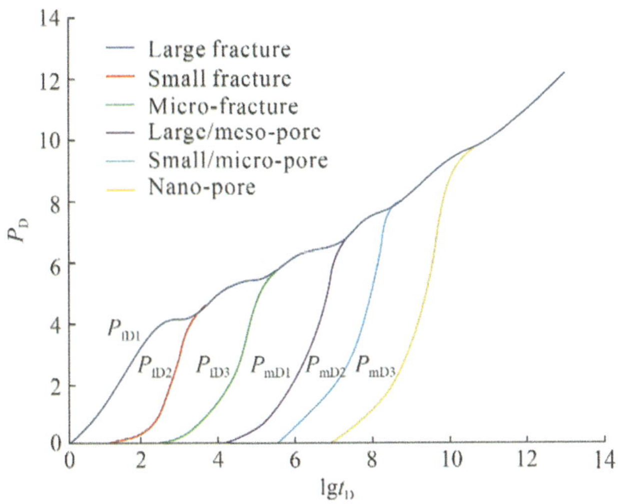

Fig. 1.20 Pressure build-up curves of porous medium at different scales

sand bodies and physical properties change fast. Different scale natural fractures can be found and their spatial distribution is greatly different. The macro heterogeneity is also high. Media with different scale pores and fractures have greatly different physical properties. The fluid composition and occurrence state are also different. The characteristics of micro-heterogeneity and multiple media are significant which lead to a great difference in fluid flow mechanisms and recovery mechanisms. According to the development concept of unconventional oil and gas reservoirs, the development mode of shale gas reservoirs is used as a reference to develop a new method of improving the well productivity and recovery performance for tight oil and gas reservoirs (Chen et al. 2012; Jin et al. 2013). The new technique of horizontal well drilling plus hydraulic fracturing is adopted to achieve the effective development of tight oil and gas reservoirs.

1.2.1 Development Mode with Horizontal Wells

For tight reservoirs, the development performance of horizontal wells is affected by wellbore direction, trajectory and horizontal wellbore length under the conditions of complex geological characteristics and a given development technique. The key problem to be solved for horizontal wells in tight reservoirs is how to develop the geological models based on the reservoir lithology, lithofacies, sand body, reservoir type, physical property, distribution of natural fractures, geostress and its spatial distribution, rock brittleness, oil-bearing condition and fluid distribution. Meanwhile, how to optimize the horizontal well location, direction, trajectory and wellbore length by the numerical simulation technique top redict the development performance is another key problem.

1.2.1.1 Development Purpose with Horizontal Well

Tight reservoirs have tight matrix and poor physical properties. The production and development of vertical wells is low and inefficient. Horizontal well development can effectively increase well control reserves, increase single well production and cumulative production. But the reservoir lithology, lithofacies, sand body, physical property, oil-bearing condition, and reservoir type change quickly and vary greatly. The spatial distribution of natural fractures is directional, macro heterogeneity is strong, and the development performance of horizontal wells is affected by factors such as a sweet spot and well position, horizontal well direction, trajectory and horizontal wellbore length. The purpose is mainly reflected in the following aspects:

① Drill more sand bodies, improve a drilling ratio of high-quality reservoirs, especially for the reservoirs of type I and type II.
② Drill more natural fractures, increase the contact area between a horizontal well and natural fractures, and improve the oil drainage area.
③ Increase the horizontal wellbore length, improve the drilling ratio of high-quality reservoirs and natural fractures, and expand the contact area between a horizontal well and an effective reservoir.

1.2.1.2 Main Controlling Factors for the Development Effect Using Horizontal Wells

1. Sweet spot and well location optimization

A tight reservoir can distribute in a large area. However, the lithology, lithofacies, sand body, reservoir type and physical properties change quickly, and the macroscopic heterogeneity is strong. The reservoirs have wide-spread natural fractures, but the spatial distribution is different. The geostress and the spatial distribution of different regions and parts

Table 1.12 "Sweet spot" grading standard for tight reservoirs

Evaluation index system		Grading standard		
		Type I	Type II	Type III
Physical property	Porosity/%	>12	8–12	5–8
	Permeability/mD	>0.12	0.08–0.12	0.05–0.08
Fractures	Fracture density/(m^{-1})	>10	3–10	<3
Brittleness	Brittleness index	>50	30–50	<30
Fluid property	Oil saturation/%	>70	50–70	<50
	Mobility/(mD/mPa·s)	>0.1	0.05–0.1	<0.05
	Gas oil ratio/(m^3/m^3)	>60	30–60	<30
	Reservoir pressure coefficient	>1.2	1–1.2	<1

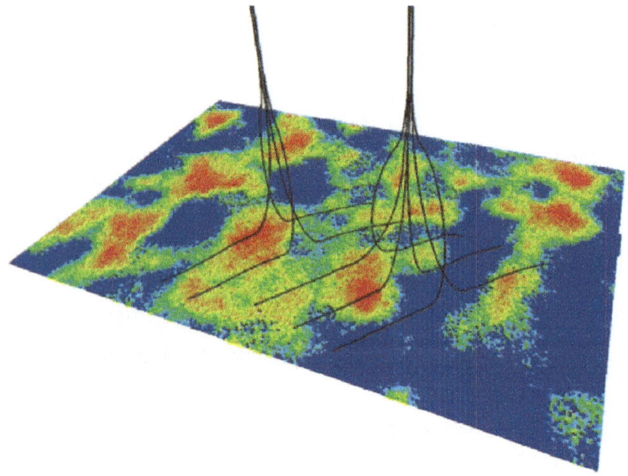

Fig. 1.21 Well location optimization and deployment

are distinct. The lithology and rock brittleness differ greatly, which can lead to a great difference in the rock compressibility. The oil saturation and fluid properties (viscosity, gas and oil ratio) are different, which lead to the difference of fluid flowability. Meanwhile, the pressure coefficient varies greatly in different regions and parts. All the factors can affect the productivity of a well. Therefore, the "sweet spot" grading standard for tight reservoirs is developed (Table 1.12). Through the optimization of a sweet spot, it can guide the well position deployment, improve the drilling success ratio and improve the high-quality reservoir drilling ratio (Fig. 1.21), thus improve the single well productivity and cumulative production.

2. Horizontal well direction optimization

The direction of a horizontal well is mainly affected by the sand body distribution, natural fractures, geostress direction and other related factors. Firstly, horizontal wells are deployed along the sand body distribution direction, which can improve the drilling ratio of high-quality reservoirs, especially for the type I and type II. Secondly, horizontal wells are vertical or oblique to natural fractures, which can increase the drilling ratio of natural fractures and contact area with fractures. Thirdly, horizontal wells are vertical or oblique to the direction of the maximum main stress, which is beneficial to the reservoir of natural fractures. Therefore, the well productivity and cumulative oil production can be increased by the optimization of a horizontal well direction.

3. Horizontal well trajectory optimization

The horizontal well trajectory in tight reservoirs is generally optimized in high-quality reservoirs with a high thickness, good continuity, a natural fracture wide-spread distribution, good physical properties, a good oil-bearing condition and strong compressible property. It can improve the drilling ratio of high-quality reservoirs, type I reservoir and type II reservoir. Finally, the fracturing effectiveness can be improved. The maximum controllable oil reserves can be obtained and the well productivity is improved (Fig. 1.22).

4. Horizontal wellbore length optimization

The horizontal wellbore length is optimized under the conditions of a given distribution of natural fractures and high-quality reservoirs. The scales of sand bodies and the distribution of effective reservoirs are also considered. By

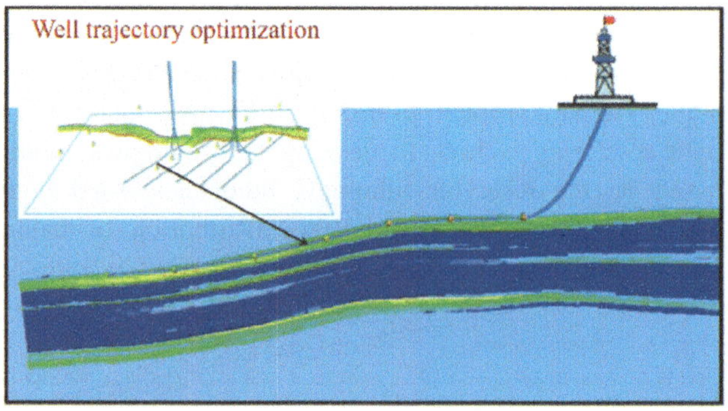

Fig. 1.22 Schematic diagram of well trajectory optimization

optimizing the horizontal wellbore length, the drilling ratio of natural fractures and high-quality reservoirs can be improved. The contact area of a horizontal well and a reservoir can be expanded, and the controllable oil reserves and the well productivity can be improved.

1.2.2 Stimulation Mode of Hydraulic Fracturing

Under the given conditions of the geological characteristics of tight reservoirs and a fracturing technique, the effect of a stimulated reservoir is affected by the fracture network complexity, the size of matrix rock separated by fracture network and SRV (stimulated reservoir volume). According to the reservoir lithology, brittleness, natural fracture and spatial distribution, and characteristics of geostress, how does one rationally characterize the fracture number, position, shape, distribution and geometric parameters of hydraulic fractures in a geological model? Simultaneously, how does one optimize a complex fracture network, the size of matrix rock and the SRV through the numerical simulation technique and predict the performance? These are the important issues that we face.

1.2.2.1 Purpose of Hydraulic Fracturing

A tight reservoir has poor physical properties. There is no natural productivity without fracturing. It only can be developed effectively with hydraulic fracturing. The SRV effect is mainly affected by the fracture network complexity, matrix rock size and SRV. Therefore, the purpose of SRV is mainly reflected in the following aspects:

① To form an effective complex fracture network, increase the number of fractures, expand the contact area between reservoir matrix and fractures, and improve the drainage capacity of matrix rock.
② To form smaller matrix rock separated by a fracture network, shorten the matrix flow distance, and improve the producing degree of a matrix system.
③ To expand the SRV formed by fracturing, increase the controllable oil reserves of horizontal wells for SRV, and increase the output and cumulative production of horizontal wells.

1.2.2.2 Process of Hydraulic Fracturing

Hydraulic fracturing refers to the reservoir stimulation technique that produces a complex fracture network through fracturing. Through the processes of multiple cluster perforation of horizontal wells, a high production rate, large liquid quantity, large sand quantity, low viscosity liquid, and steering material and technology, we can effectively connect natural fractures and rock bedding. Thus, the secondary fractures and even multistage secondary fractures are formed in the lateral direction of the main fractures which can form a fracture network system and maximize the contact area between fracture surface and oil-gas reservoir matrix. Simultaneously, the oil-gas flow distance from matrix to fractures in each direction is reduced. Finally, it can significantly increase the SRV, improve the total permeability of a reservoir, and thus expand the SRV of the reservoir in three different directions (Saldungaray and Palisch 2012; Wu et al. 2011, 2012), as shown in Fig. 1.23.

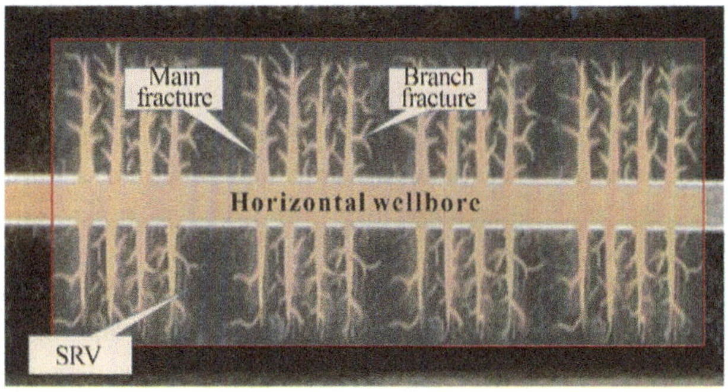

Fig. 1.23 Schematic diagram of hydraulic fracturing of horizontal well

Hydraulic fracturing can be divided into three stages: fracturing, soaking and flow back (Fig. 1.24).

① During the process of fracturing, the method of a large liquid volume, a large sand volume and a large discharge capacity is usually adopted to the separate fracturing step by step from toe end to heel end. With the injection of fracturing fluid, the pore pressure increases sharply, and the matrix pore expands elastically and even plastic fails. When the fluid pressure is higher than the fracturing pressure of reservoir rock, hydraulic fractures are formed. With the injection of fracturing fluid, the fractures expand. When the pressure reaches the extension pressure, the fracture ends break and the fractures extend forward. When the fluid

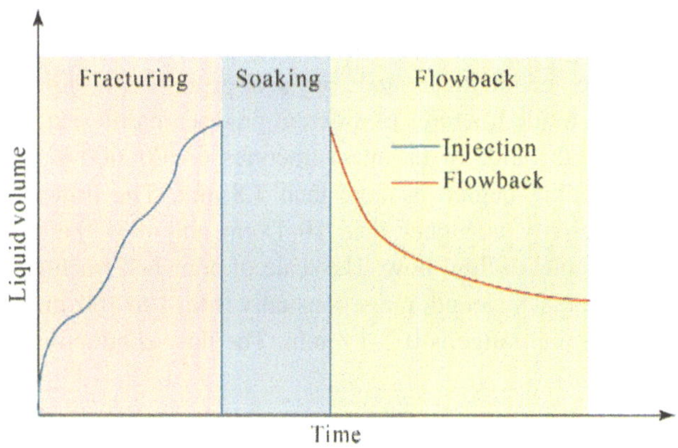

Fig. 1.24 SRV process for tight reservoir

pressure is larger than the opening pressure in natural fractures, the natural fractures open, and it is expanded with the fracturing fluid injection. When the pressure reaches the extension pressure, natural fractures extend forward. The fracturing fluid and propping agent in reservoir effectively play the role of replenishing reservoir energy.

② During the process of soaking, the fracturing fluid in hydraulic fractures flows into matrix pores and natural fractures. The pressure in hydraulic fractures decreased slowly, and the pressure of matrix pores and natural fractures increased.

③ During the flowback process, with the discharge of fracturing fluid, the reservoir pressure drops sharply, the matrix pores shrink and deform, the width of natural fracture is narrowed and even closed, and the proppants in hydraulic fractures are effectively supported and maintain a high conductivity in fractures.

1.2.2.3 Morphology Distribution and Geometrical Parameters of Hydraulic Fractures

① A complex fracture network is developed by the hydraulic/natural fractures at different scales formed after hydraulic fracturing in tight reservoirs, and the morphology and its distribution of fractures are complex. The more the number of fractures is, the greater the contact area of the fracture is, the better the connectivity of fractures is, the better the effect of complex fracture network is.

② The more complex the fracture network, the smaller the matrix rock. Thus, the fluid flow distance is reduced and the unlocking degree of matrix rock blocks is increased. The recovery performance is improved.

③ The complex fracture network system formed by different scale fractures form SRV. The larger the SRV volume, the greater the controllable oil reserves of a horizontal well and the higher the productivity of oil wells (Fig. 1.25).

④ The geometrical and physical parameters for the hydraulic fractures of different grades are different. The length range of the main fractures is 300–600 m. The opening degree is more than 1.8 mm. The flow conductivity is higher than 10 D·cm and it is the main channel of fluid flow. The scale of branched fractures is small. The length range is usually from 1 to 100 m. The opening range is 0.2–1.8 mm. The flow conductivity is 1–10 D·cm and is the secondary flow channel. The opening degree of microfractures and shearing fractures is less than 0.2 mm, and the flow conductivity is less than 1 D·cm.

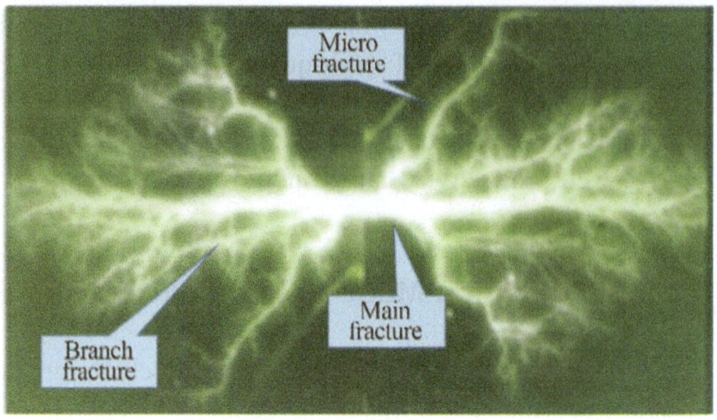

Fig. 1.25 Schematic diagram of fracture network of SRV for tight reservoir

The hydraulic fracturing in tight oil and gas reservoirs further improves the flow capacity of reservoirs, and increases the well control range, and effectively increases the single well productivity.

1.2.3 Flow Behavior Between Horizontal Wells and Reservoirs

The contact mode and flow behavior between reservoir media and wellbore are complex under the development mode of horizontal well drilling and hydraulic fracturing in tight oil and gas reservoirs. The geomechanics effect of reservoir media is significant, and the flow regime and flow mechanisms between different media and horizontal wells change because of a large pressure difference around the wellbore. The internal flow modes of horizontal wellbore and the flow variation are also complex. How to describe the complex flow relationships through mathematical models, indicate them in numerical simulation works, and improve the numerical simulation accuracy are the key issues that need to be solved for the simulation of horizontal well development (Darishchev et al. 2013).

1.2.3.1 Flow Behavior Between Different Reservoir Media and Wellbore Under Different Well Completion Methods

In tight reservoir, the contact and flow relationship between perforation well completion and open hole well completion are different under the development mode of horizontal well

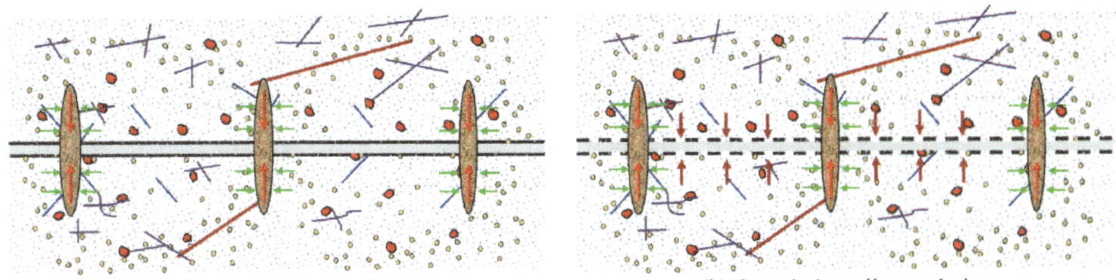

Fig. 1.26 The flow relationship between horizontal wells and reservoirs under different completions

drilling and hydraulic fracturing. For perforation well completion, the reservoir matrix pores and natural fractures cannot directly contact with horizontal wellbore. The fluids in the pores with different sizes must flow into hydraulic fractures firstly and then flow into the wellbore under the pressure in fractures. For open hole well completions, in addition to hydraulic fractures, the pores and natural fractures at different scales in reservoir are in direct contact with horizontal wellbore. Fluids in the pores with different sizes can flow through hydraulic fractures to the horizontal wellbore and can directly flow into the wellbore under the production pressure drop (Fig. 1.26).

1.2.3.2 Coupling Flow Behavior Between Different Reservoir Media and Horizontal Wells

① During the processes of injection and production in tight oil and gas reservoirs, the pressure of the near-wellbore zone changes drastically, and the geomechanics effect is prominent. The media deformation at different scales leads to a large change in physical properties and well indexes, which have a great impact on the well productivity.

During the fracturing and injection process, the pore pressure rapidly increases, the pores in matrix expand and deform, natural fractures open, and hydraulic fractures start to fracture and extend, resulting in an increase in permeability, a rapid increase in well index. Thus, the fluid exchanging capacity between pore-fracture media and wellbore is greatly enhanced. For the process of production, with the pore pressure decreases, pore throats at different scales are prone to shrink, natural fractures are easily deformed or closed, and hydraulic fractures are easily broken or embedded due to elastic deformation of proppants. The well index is significantly reduced, making the fluid exchanging capacity between pore-fracture media and wellbore rapidly weakened. Dynamic changing in hydraulic fractures, natural fractures, and porous media also have a serious impact on well productivity.

② During the process of injection and production of tight oil and gas reservoirs, the near-wellbore zone differential pressure changes greatly, and the flow regimes and flow mechanisms through different media at different stages vary greatly, affecting the flow characteristics of media at different scales to the bottom of well, and have a great impact on well productivity.

In tight reservoirs, when the fluids flow from large-scale fractures to wellbore, because the fractures are larger, the flow resistance is smaller and the flow state is dominated by high-speed nonlinear flow. When the fluids flow from small-scale fractures to wellbore or large-scale fractures, the resistance of fluid flow is relatively large due to the small fractures open, and the flow regime is mainly based on the pseudo-linear flow. When the fluids flow from the pores of reservoir matrix to the wellbore or fractures, the flow regime is dominated by low-velocity nonlinear flow due to poor reservoir properties, small pore throats, and high flow resistance. The changes of flow regime in porous media at different scales seriously affect the well productivity.

1.2.3.3 Internal Flow Pattern Within Horizontal Wells

Within horizontal wellbore, fluid flow is affected by friction, gravity and other factors, resulting in a pressure loss during the fluid exchanging process between different wellbore sections. At the same time, the single-phase flow and multiple phases flow exist in the horizontal wellbore, and there are different flow states (bubble flow, mist flow, and slug flow) when the oil-gas-water multiphase fluid flows occur in the wellbore of horizontal well. Friction resistance, pressure loss and different flow regimes have a great influence on the calculation of well productivity.

1.2.4 Production Mode

Tight reservoirs have the strong macroscopic heterogeneity and microscopic heterogeneity, and they present the characteristics of multiple scales and multiple media. The fluid composition, properties and occurrence state in pore-fracture

Table 1.13 The recovery mechanism of different development modes

Development mode	Recovering pore-fracture media type	Occurrence state of recovering crude oil	Recovery mechanism
Depletion Development (re-fracturing)	Micro-pore, mid-pore, large fractures, small fractures	Movable oil	Viscous flow under pressure differential, elastic displacement
Water huff-n-puff	Micro-pores, micro-pores, Micro-fractures	Movable oil and capillary oil	Energy supplement, viscous flow, imbibition
Gas injection development	Nano/micro-pores, nano-pores, Nano-fractures	Capillary oil, adsorbed oil	Energy supplement, interfacial tension reduction and viscosity reduction, viscous flow
Recovery method of changing the occurrence state and rock-fluid properties	Nan-/micro-pores, nano-pores, Nano-fractures	Adsorbed oil and film oil	Wettability changing, interfacial tension changing, occurrence state changing, viscosity reduction, viscous flow

media at different scales are greatly differs. Therefore, different development methods need to be applied to extract the hydrocarbons under different occurrence states in the pore-fracture media at different scales.

For tight gas reservoirs, the development mode of natural depletion is mainly used. For tight oil reservoirs, the development modes can be classified into the natural depletion mode in the early stage and the development mode for the improvement of unlocking degree and recovery factor in the middle and later stages (Table 1.13).

Currently, the primary natural depletion recovery process is still the main method for tight oil reservoirs, using the large pressure-drop/pressure-control method to unlock the movable oil in large/small fractures and large/middle pores. The main recovery mechanisms include viscous flow under pressure drop and elastic displacement. After the rapid depletion of energy and a rapid decline in liquid production, the reservoir energy can be supplemented by the multiple cycle re-fracturing method, and then the remaining movable oil in fractured regions and in previous un-fractured regions can be effectively unlocked.

The development mode to increase producing degree and recovery rate mainly includes water huff-n-puff, gas injection and in-situ oil upgrading. i. Water huff-n-puff has the dual effects of recovering reservoir energy and effective displacement. Under the effects of pressure drop and capillary imbibition, part of the movable oil and the capillary oil in some small pores, micro-pores and micro-fractures can be unlocked. ii. Gas injection works by recovering energy and overcoming the capillary pressure to unlock the capillary oil in nano/micro-pores and nano-fractures. Simultaneously, it also reduces interfacial tension and oil viscosity to unlock the adsorbed oil in micro-pores, nano-pores and nano-fractures. iii. The in-situ oil upgrading process is to unlock the thin film oil and adsorbed oil in micro-pores, nano-pore and nano-fractures by changing the wettability of rocks, reducing the interfacial tension and the capillary pressure (by injecting the modifier), changing the physical properties and occurrence state of oil and reducing the oil viscosity.

Different development modes and recovery methods can be used to unlock the crude oil in different types of media with pores and fractures and different occurrence states. For them, the recovery mechanisms are also different. How to apply a mathematical model to describe the complicated recovery mechanisms, how to apply the numerical simulation technique to optimize the technical strategies of different development modes and recovery methods, and to predict the recovery performance and recovery indicators are the main problems that we face currently.

1.2.4.1 Development Mode of Natural Depletion

Tight oil reservoirs mainly apply the development mode of horizontal well drilling, hydraulic fracturing and depletion process currently. This development mode has a high initial productivity and a rapid decline rate (Fig. 1.27). The

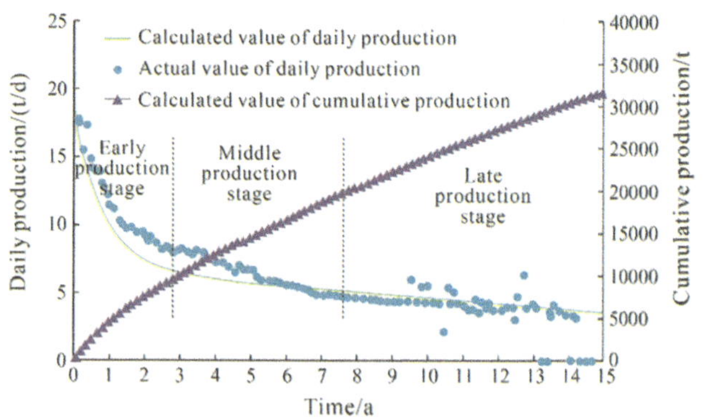

Fig. 1.27 Actual production curve in tight reservoir

productivity of each well is basically unstable, and the main development characteristics are as follows:

① The recovery performance is characterized by "three stages". The initial well productivity is very high, but it declines rapidly. In the middle stage, well productivity is reduced to 50%–70% of the initial productivity and declining process is slowed down. In later stage, the well productivity is very low, and the decline rate is small. The development time is long.
② In tight reservoirs, the heterogeneities of sand bodies, physical properties and oil-bearing conditions are strong. The development degree of natural fractures is different, and the difference of fracturing effect between different wells is great, which leads to the big difference in the initial productivity and the cumulative production of each well.
③ In the early stage of depletion process, the duration time of high productivity is short, and development time in the later stage is longer. The unlocking range and oil reserves are limited, and the cumulative oil production and recovery factor are low.

For the natural depletion recovery process, the oil displacement mechanisms in different production stages are also different (Fig. 1.28).

In the early stage, the reservoir pressure decreases rapidly, and the recovery of crude oil mainly depends on the elastic drive of fracturing fluid and the elastic expansion of rock and fluid. In the middle stage, it is mainly controlled by the elastic expansion of rock and fluid, followed by the elastic displacement of fracturing fluid and dissolved gas drive. In the later stage, it mainly depends on the dissolved gas drive, followed by the elastic drive of rock and fluid and imbibition.

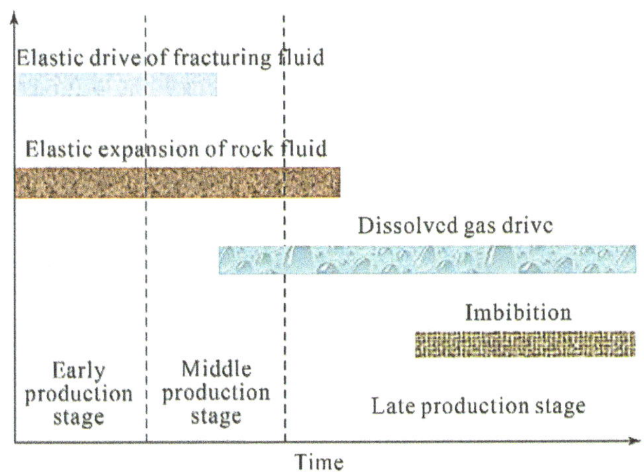

Fig. 1.28 Displacement mechanisms in different production stages of tight oil reservoirs

For the natural depletion development of tight oil reservoirs, according to the effects on the well productivity and cumulative oil production of displacement mechanisms, stress sensitivity of pores and fractures media at different scales, fluid properties and GOR, water saturation and water cut, and reservoir matrix supplement capacity, we can apply two production modes at different working strategies of increasing pressure-drop and controlling pressure-drop.

1. Increasing pressure-drop

For the development of tight oil reservoirs with three kinds of conditions: i. a weak stress sensitivity of a reservoir matrix, natural fractures and hydraulic fractures with a high proppant concentration and high strength in hydraulic fractures; ii. a low GOR, a low movable water saturation and a low initial water cut; iii. a relatively good matrix physical property, a high degree of micro-fractures development, a strong supplement capacity of reservoir matrix during the development process, we can use the production strategy of high pressure-drop to gain a high well productivity. It comes to achieve the production goal of getting a high initial oil production rate, obtaining a high recovery rate in the early stage and a low recovery rate in the late stage, and finally reaching a short investment payoff period and a good production profit in the late stage (Fig. 1.29).

The performance curves of tight oil reservoirs can be classified into three types: high rate and rapid declining, medium rate and slow declining, and low rate and slow declining (Table 1.14).

2. Controlling pressure-drop

For the development of tight oil reservoirs with three kinds of conditions: i. a strong stress sensitivity of a reservoir matrix, natural fractures and hydraulic fractures with a low proppant concentration and low strength in hydraulic fractures; ii. a high GOR, a high movable water saturation and a high initial water cut; iii. a relatively poor matrix physical

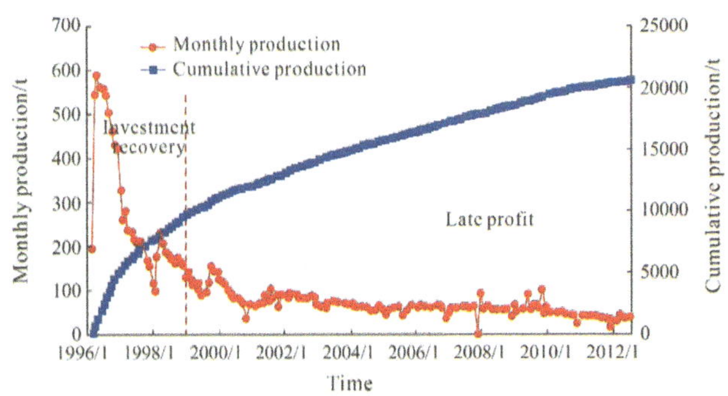

Fig. 1.29 Performance curve by the method of increasing the pressure-drop

Table 1.14 Characteristics of typical performance curves of tight oil reservoirs by the method of increasing the pressure-drop

Type	Dynamic characteristics	Analysis of dynamic law	Typical curve
Rapid decline of high productivity	The initial production rate is high, and the production rate decreases rapidly. The production rate becomes low and stable in the late stage	The natural fracture is developed, the effect of hydraulic fracturing is good, the production rate of oil well is high in the early stage, but due to the fast drop of pressure, resulting in the deformation and closing of pores and fractures, the insufficient capacity of matrix supplement, the production decline is fast, in the early stage	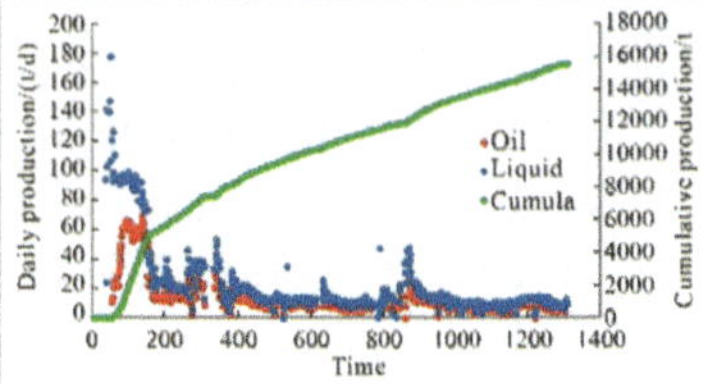
		The later production is low, but the material property of the reservoir is good because of the large range of fracture communication and the reservoir's physical property, so the matrix recharge can basically achieve dynamic balance	
Slow decline of middle productivity	The initial production rate is not high, and the production rate decreases slowly. The production rate becomes low and stable in the late stage	The natural fracture is relatively undeveloped, the effect of hydraulic fracturing is not very good, the early production of the oil well is not high, the fracture communication range is relatively small, the reservoir physical property is relatively poor, the substrate supply is relatively insufficient and the later production is low	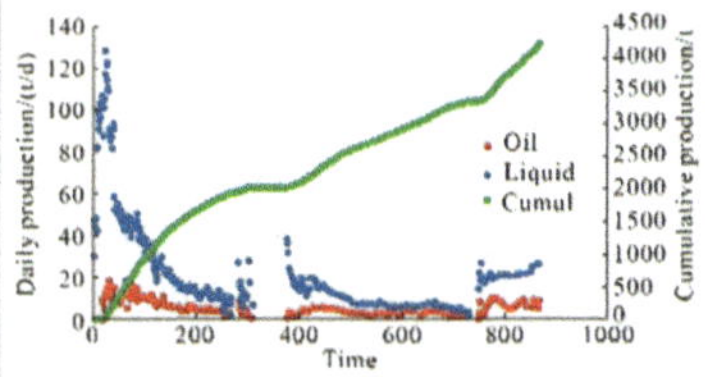
Slow decline of low productivity	The initial production is relatively low, and the production rate decreases slowly. The period of low and stable oil production is longer	The natural fracture is not developed, the effect of hydraulic fracturing is poor, the early production of the oil well is low, the communication range of the fracture is small, the physical property of the reservoir is poor, the supply capacity of the matrix is weak, and the later production is low	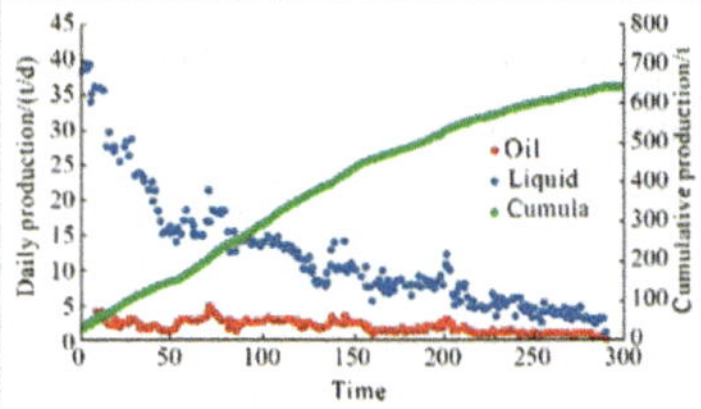

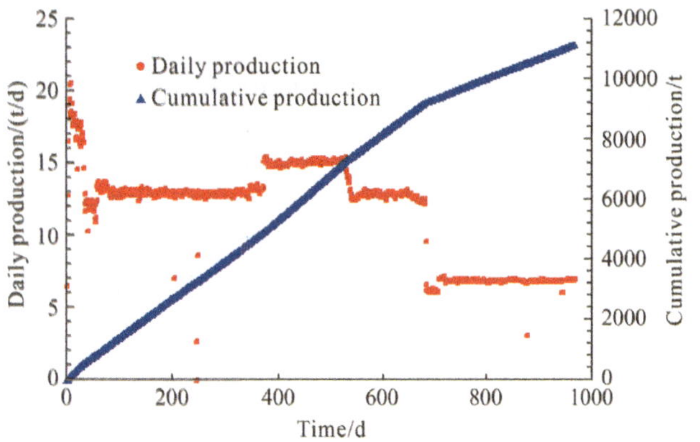

Fig. 1.30 Performance curve by the method of controlling pressure-drop

property, a low degree of micro-fractures development, a weak supplement capacity of reservoir matrix during the development process, we can use the production strategy of controlling pressure-drop. It reduces the damage on physical properties caused by the deformation of media with pores and fractures, reduce the oil viscosity and reduce the water cut. Finally, a dynamic balance between the matrix supplement and fracture flow can be obtained. A reasonable productivity and recovery rate can be gained to slow down the declining process and increase the cumulative oil production and recovery factor (Fig. 1.30). The development performance curves of tight oil reservoirs by the method of controlling pressure-drop can be classified into three types: high rate and slow declining, medium rate and slow declining, and low rate and slow declining (Table 1.15).

3. Multiple cycle re-fracturing method

Through the initial depletion recovery process, the movable oil in big pores and large fractures has been unlocked in the controllable reservoir range. But due to the decrease of pressure and energy, fracture closure and rapid decline of oil production, re-fracturing can be considered in order to improve the recovery process. The purposes of re-fracturing are as follows:

Table 1.15 Characteristics of typical performance curve of tight oil reservoirs by the method of controlling pressure-drop

Type	Dynamic characteristics	Analysis of dynamic law	Typical curve
Slow decline of high productivity	The initial production rate is high, and the production rate decreases slowly. The accumulative oil production is relatively high	In the case of controlling the production pressure difference, the output is relatively high, but because that reduces the damage to the material property and the effect of degassing to the flow, the production decline is slow, the reservoir physical property is better, the communication range is relatively large, and the dynamic balance of the supply and fracture exploitation of the matrix can be realized, and the later production can be achieved. The decline was small, and maintained relatively low and stable productivity	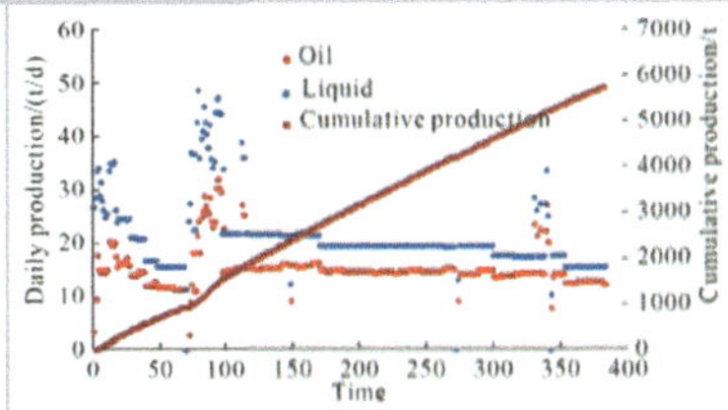
Slow decline of middle productivity	The initial production rate is not high, and the production rate decreases slowly. The accumulative oil production is relatively high	Due to the relatively undeveloped natural fractures and moderate hydraulic fracturing, the productivity is moderate in this case, the reservoir physical property is relatively poor, the fracture communication range is relatively limited, the output is relatively low, and the decline is slow	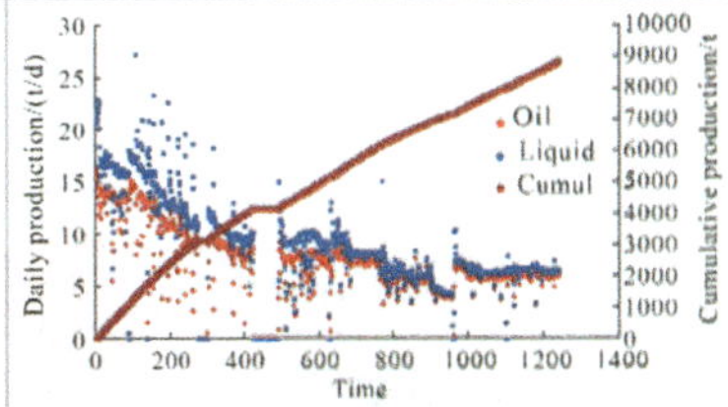
Slow decline of low productivity	The initial production rate is relatively low, and the production rate decreases slowly. The period of low and stable oil production is much longer, and the accumulative oil production is low	Because natural fractures and hydraulic fracturing are not developed and under the condition of controlling production pressure difference, the early production of oil well is low, the communication range of fracture is small, the physical property of the reservoir is poor, the supply capacity of the matrix is weak, and the later production is low	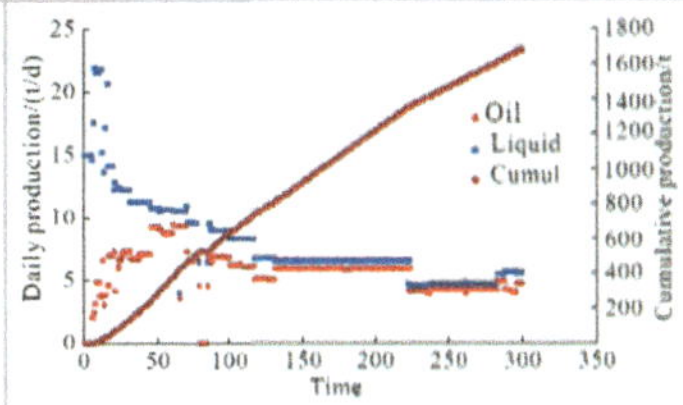

① For fractures with a good fracturing effect but closed due to the pressure drop decreases, hydraulic fracture can be opened again, and the fracture conductivity can be recovered by re-fracturing.

② For fractures with a poor fracturing effect, the range is limited and there is a high remaining oil saturation, the fracturing effect can be improved, and the unlocking range and recovery factor can be improved through re-fracturing.

③ For the regions that have not been fractured, oil reserves have not been effectively unlocked, and hydraulic fractures can be formed through re-fracturing to improve the recoverable range and reserve unlocking degree (Fig. 1.31).

By re-fracturing, the oil production rate can be recovered to 60%–80% of the initial rate and the well productivity can be increased. By repeated re-fracturing, the cumulative oil production, recoverable range and unlocking degree can be improved (Fig. 1.32).

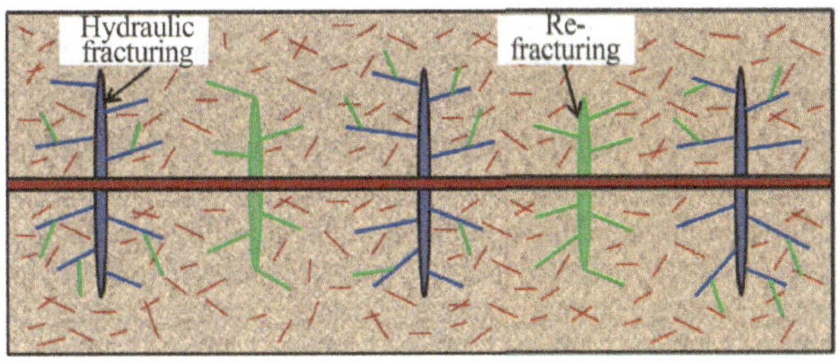

Fig. 1.31 Schematic diagram of re-fracturing

1.2.4.2 The Development Mode of Improving the Producing Degree and Recovery Factor

The macro heterogeneity and micro heterogeneity of tight reservoirs are strong. Due to the difference of rock composition and surface properties, the fluid composition, physical properties and occurrence state of porous media with different scales are different. Movable oil, capillary oil and film oil have different proportions in the pore-fracture media with different scales (Fig. 1.33).

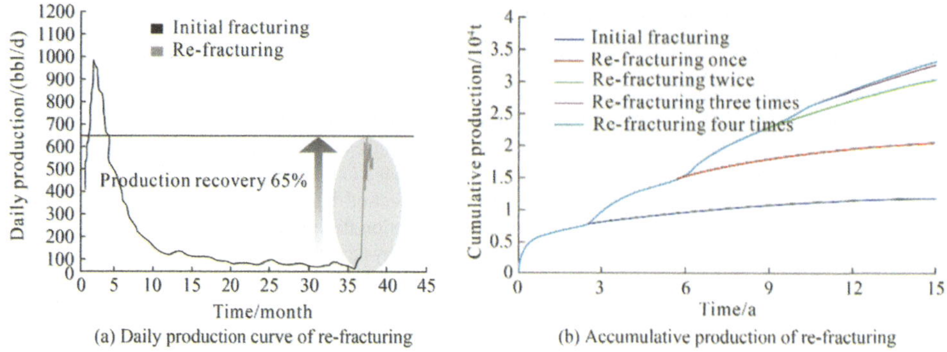

Fig. 1.32 Results of repeated fracturing production

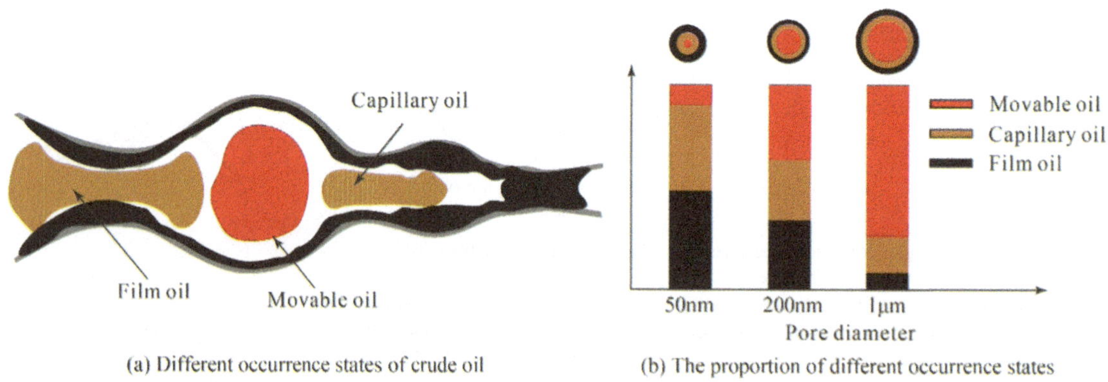

Fig. 1.33 Occurrence states and different proportions of crude oil in porous media at different scales

There are great differences in the unlocking conditions and unlocking degree for the crude oil in the pore-fracture media at different scales. Through the depletion method of hydraulic fracturing/re-fracturing, the unlocking degree under the macro heterogeneous condition is improved, and the movable oil in large scale pores and fractures can be effectively unlocked. However, a large amount of residual oil exists in small pores and micro pores, mainly including capillary oil and film oil. Therefore, how to effectively unlock the crude oil in small pores and micro- and nano-pores to improve the producing degree and recovery factor is very important. The ways to increase the producing degree and recovery factor in tight oil reservoirs are as follows.

1. Water huff-n-puff and imbibition

Water huff-n-puff has the dual effects of recovering reservoir energy for the effective displacement and imbibition, and it is an effective development method of recovering reservoir energy. The recovery mechanism is to inject water, soak and produce in a same well. Under the effect of pressure drop, water is firstly injected into fractures with different scales and large or mid-pores. After soaking, imbibition occurred under the effect of capillary pressure. That is, the injected water enters small pores and nano/micro-pores, the oil is replaced by water, and the crude oil comes into large pores and the different scale fractures from small pores and nano/micro-pores. Then, the oil and the water injected previously are re-distributed in a reservoir and they are replaced out after a well is opened. This method is to unlock the movable oil in small pores, nano/micro-pores and the movable oil and capillary oil in micro-fractures under the effects of pressure drop and capillary pressure.

2. Gas injection

Gas injection mainly refers to the method of injecting CO_2, N_2, natural gas or other gases into a tight reservoir to enhance oil recovery. In this method, the recovery factor can be improved through the method of cyclic gas injection. i. It can recover reservoir energy, increase pressure gradient and overcome capillary pressure. ii. When CO_2 is dissolved in crude oil, the oil viscosity is reduced and the flow rate of oil is increased. At the same time, the elastic drive caused by the volume expansion of oil and the dissolved gas drive caused by the pressure drop and CO_2 degassing are also important.

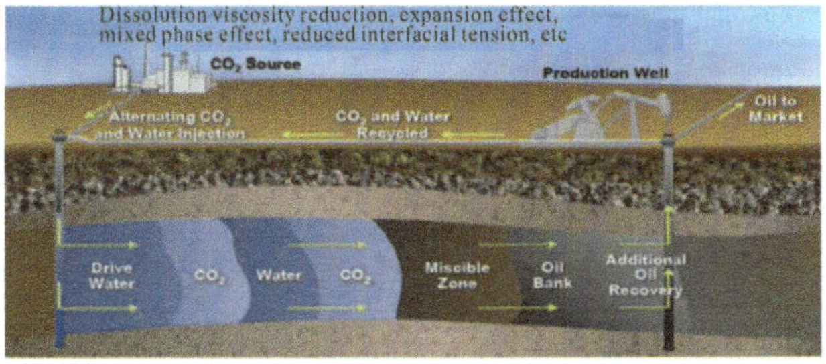

Fig. 1.34 The mechanism of CO_2 oil displacement

iii. It can reduce the interfacial tension and the oil flow resistance, and overcome the capillary pressure by means of pressure-drop. iv. The injected gas can enter the nano/micro-pores and thus unlock the capillary oil and adsorption oil in nano/micro-pores. Hence it can improve the displacement efficiency effectively, thereby improving the unlocking degree and recovery factor (Fig. 1.34).

3. Recovery mode of changing the occurrence state and rock-fluid properties

For nano/micro-pores in tight oil reservoirs, adsorbed oil and thin film oil are the main occurrence states. Conventional recovery methods are difficult to apply, and the method of changing the occurrence state and rock-fluid properties can be applied. We can change the wettability of reservoir rock, the contact angle, the occurrence state, the nature of crude oil and the interfacial tension and also reduce the oil viscosity and capillary pressure (Fig. 1.35) by injecting a modifying agent into reservoir. Thus, the thin film oil and adsorbed oil in porous media can be effectively recovered to gain a high displacement efficiency and recovery factor (Liu et al. 2016).

For tight oil reservoirs with poor oil properties, high viscosity and poor flowability (0.014–0.35 mD/cp), it is required to apply the methods of in-situ heating and in-situ oil upgrading for the effective development. In-situ heating can improve the oil quality, change the oil properties and occurrence state, and produce the light oil or condensate oil in reservoirs. Finally, the oil flowability can be improved to unlock the capillary oil and adsorbed oil in nano/micro-pores. Simultaneously, fractures and the associated gas formed during the heating process are also beneficial to the flow process of oil and gas. It can effectively improve the displacement efficiency and recovery factor (Zou et al. 2015; Zhang et al. 2015).

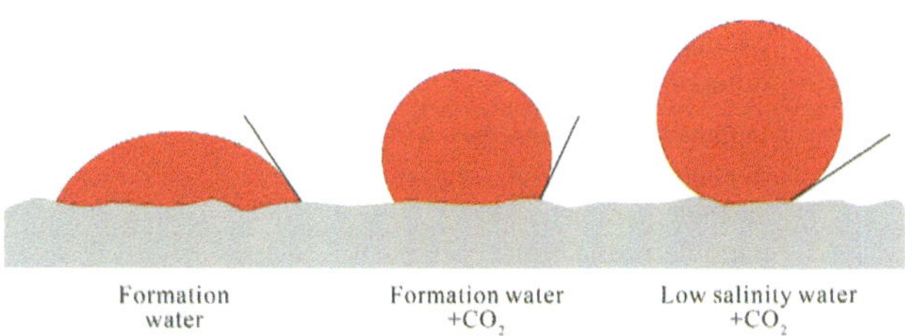

Fig. 1.35 Schematic diagram of changing wettability and occurrence state of reservoir

1.3 Problems and Requirements for the Numerical Simulation Process in Unconventional Tight Reservoirs

With the object shift of oil and gas field development from conventional reservoirs to unconventional tight oil and gas reservoirs, the development of tight oil and gas reservoirs faces the problems of macro heterogeneity and micro heterogeneity of reservoirs, multiple scale and multiple media, complex flow relationships in the mode of horizontal wells drilling and hydraulic fracturing, complex flow mechanisms in the pores and fractures media at different scales under the unconventional development methods of tight reservoirs, and many other challenges. How to characterize the complex characteristics of tight reservoirs and establish the corresponding geological models, and how to describe, simulate and predict the recovery performance by mathematical models are the main problems and challenges for the current numerical simulation. The traditional numerical simulation theory and technology are difficult to apply. It is urgent to adopt the unconventional ideas to innovate the numerical simulation theory and technology for unconventional tight reservoirs (Table 1.16).

1.3.1 Problems and Requirements for the Geo-Modeling Technologies of Unconventional Tight Oil and Gas Reservoirs

For unconventional tight reservoirs, there are several preliminary issues and requirements in geological modeling.

1.3.1.1 Requirements for the Geo-Modeling Technologies of Discontinuous Discrete Multiple Media with Multiple Scales

The distributions of lithology, lithicofacies, and formation types in unconventional tight reservoirs are very different. There are extreme macro- and micro-heterogeneity, the characteristics of multiple scale and multiple pore-fracture media at different scales. The porous media at different scales are discretely distributed in space. Simultaneously, the geometrical parameters and physical parameters of different porous media have great differences, and they have the property of discontinuous changing. A conventional technology of continuous geo-modeling is only suitable for

Table 1.16 Problems and requirements in numerical simulation of unconventional tight reservoirs

Geological and development characteristics		Conventional numerical simulation theory and technology			Numerical simulation theory and technology of unconventional tight reservoirs
			Status quo	Inadequacies	Demands
Macro heterogeneity	Complexity and diversity of lithomorphic and lithology in tight reservoirs	Geological modeling	Single/dual medium continuous geo-modeling technology	Unable to describe reservoir macroscopic heterogeneity, multi-scale characteristics	Discontinuous discrete multiple media modeling technology for pore and fracture media with different scales
	Strong heterogeneity of space distribution of sand body and physical properties, and internal properties				
	Multi-scale features of natural fractures relatively developed				
	Large differences in in-situ stress distribution		Conventional rock mechanics and in-situ stress modeling techniques	Difficult to characterize micro-heterogeneity, multiple media characteristics	
	Large difference in plane distribution of fluid saturation and viscosity		Fluid continuous distribution modeling technology	Failure to characterize the distribution characteristics of reservoir rock mechanics and in-situ stress field in reservoir and fluid parameter field during fracturing, injection and production processes	
	Strong heterogeneity between blocks, wells, and single wells		Static three-dimensional modeling technology	Inability to describe and characterize dynamic changes in geometrical and physical properties of multi-scale media, fluid parameters, reservoir temperature, in-situ stress, and rock mechanics parameters field during the fracturing, injection and production processes	Rock mechanics and in-situ stress modeling technology during fracturing-injection-production processes
					Discontinuous discrete fluid modeling technology for pore and fracture media with different scales
					Real-time dynamic four-dimensional geological modeling technology
Micro heterogeneity	Coexist and discontinuously spatial distribution of nano/micro-pores, natural/hydraulic fractures at different scales	mathematical model	Continuous single/double media multiphase flow theory	Unable to describe and characterize the discontinuous discrete distribution of pore and fracture media with different scales, multiple media characteristics, and coupling flow between different media	Multiphase flow theory through Discontinuous, multi-scale, multiple media
	There are significant differences in the physical properties of the media at different scales pores and fractures, in the fluid compositions, in the occurrence states, in the flow mechanisms, and in the flow regimes			Difficult to describe the dynamic changes of attribute parameters of multiple media with different sizes of pores and fractures under the coupling effect of fracturing, injection and production	
				Impossible to describe the complex flow mechanism and the changes of flow	Coupled mathematical model of multi-phase flow and geomechanics in pores and

(continued)

Table 1.16 (continued)

Geological and development characteristics		Conventional numerical simulation theory and technology			Numerical simulation theory and technology of unconventional tight reservoirs
			Status quo	Inadequacies	Demands
				regime in pores and fractures media at different scales difficult to describe the coupling flow between the reservoir medium and wellbore in a complex fracture network during hydraulic fracturing process	fractures media at different sizes under the coupling effects of fracturing, injection and production
					Multiphase flow mathematical model theory through multi-scale, multi-flow regime, multiple media
					Coupling multiphase flow mathematical model between horizontal well and reservoir medium under the hydraulic fracturing mode
Development mode	Complexity of contact relationships and flow characteristics between horizontal wells and reservoir	Simulation technology	Continuous single/dual medium numerical simulation technology	Difficult to simulate the effects of multiple pore/fractures media at different scales, flow behavior between media, fluid properties on the production performance	Numerical simulation technology through discontinuous, multi-scale, multiple media
	Hydraulic fracturing and re-fracturing caused from complex fractures network				
	Complexity off low mechanisms in natural depletion method, water huff-n-puff, imbibition recovery, gas injection, recovery method of changing the occurrence state and rock-fluid properties			Only simulate the fracturing, injection and production processes	
				Unable characterize and simulate the dynamic variation of parameters and the coupling flow mechanism of pore/fracture media at different scales during hydraulic fracturing, injection and production of horizontal wells	
				Impossible to recognize and simulate the fluid flow regime and the dynamic variation of complex flow mechanism in pore/fracture media at different scales	
				Difficult to simulate the flow mechanism, and the dynamics changes of phase state, composition,	Coupled fluid flow and geomechanicalnumerical simulation technology under the

(continued)

Table 1.16 (continued)

Geological and development characteristics		Conventional numerical simulation theory and technology			Numerical simulation theory and technology of unconventional tight reservoirs
			Status quo	Inadequacies	Demands
				flow pattern through pore/fracture media at different scales using the current black oil model/component model under different development methods such as gas injection, recovery method of changing the occurrence state and rock-fluid properties	fracturing-injection-production processes
					Adaptive flow regime recognition and numerical simulation technology of complex flow mechanisms
					Complex phase state numerical simulation technology of multiple media at different scales and multicomponents under development methods
					Efficient solution technology of unconventional tight reservoirs

geological modeling process of a continuous distribution of single media or dual media and a continuous change of physical parameters (Jia et al. 2016). It is imperative to breakthrough the geo-modeling technology of discontinuous discrete multiple media with multiple scales to achieve the shifts from the continuous geo-modeling to the discrete geo-modeling, from the dual-mediageo-modeling to the geo-modeling of discrete fractures and discrete multiple media, from the reservoir-scale geo-modeling to the equivalent geo-modeling of micro-scale, small-scale to reservoir-scale.

1.3.1.2 Requirements for the Geo-Modeling Technologies of Geostress Field and Rock Mechanics Parameters

In the geological modeling process of conventional reservoir, generally, the models of the geostress field and rock mechanics parameters are not developed. However, for unconventional tight oil and gas reservoirs, the spatial distribution of geostress field and rock mechanics parameters changes rapidly and widely (Ge and Lin 1998; Zeng and Wang 2005). At the same time, the geostress field and rock mechanics parameters dynamically change during the development process. It greatly influences the effectiveness and effects of hydraulic fracturing and re-fracturing. Therefore, during the development of unconventional tight oil and gas reservoirs, it is imperative to innovate and promote the geo-modelling technology of geostress field and rock mechanics parameters in unconventional tight reservoirs to guide the optimization and design of horizontal well hydraulic fracturing and refracturing and develop the coupled numerical simulation of the pressure field and fluidflow field and to correctly predict the well productivity and development indicators.

1.3.1.3 Requirements for the Geo-Modelling Technologies of Discontinuous Fluid Distribution in Multiple Media at Different Scales

Generally conventional oil and gas reservoirs have the obvious oil-water and gas-water interfaces. And the fluids are continuously distributed and fluid parameters are generally constant or continuously changed (Wang and Yao 2009). But for unconventional tight oil and gas reservoirs, they generally do not have the obvious oil-water and gas-water interfaces. And the spatial distribution of fluids in these reservoirs varies widely and discontinuously. Meanwhile, the discontinuous discrete distribution of pore-fracture media at different scales, and the occurrence states, compositions, and fluid parameters of fluids in the media are quite different and show a discontinuous change. Conventional technology based on the continuous fluid geo-modeling cannot meet the discontinuously changing characteristics. It is imperative to develop the geo-modeling technologies of discontinuous fluid distribution in multiple media at different scales which are suitable for unconventional tight oil and gas reservoirs.

1.3.1.4 Requirements for the Real-Time Dynamic 4D Geological Modeling Technology

During the development of unconventional tight oil and gas reservoirs, due to the effects of fracturing, injection and production processes, it will cause the re-distribution of fluid flow field, geostress field and temperature, resulting in the dynamic changes of geometrical parameters and physical properties of pore-fracturemedia at different scales, fluid properties and parameters, and rock mechanics properties and parameters. The conventional geological model is a static 3D geological modelling technology (Zhao et al. 2007), which is not suitable for modeling the dynamic changes of geological model, fluid distribution model, geostress model, and temperature model in unconventional tight oil and gas reservoirs. Therefore, it is necessary to innovate and develop the real-time dynamic 4D geological modeling technology, which is suitable for unconventional tight oil and gas reservoirs, to achieve the transition from static 3D geological modeling to real-time dynamic 4D geological modeling.

1.3.2 Problems and Requirements in Numerical Simulation Theory of Unconventional Tight Reservoirs

For unconventional tight oil and gas reservoirs, there are several preliminaryissues and requirements in the numerical simulation theory.

1.3.2.1 Requirements for the Development of Fluid Flow Theory in Discontinuous Multiple Media at Different Scales

Unconventional tight oil and gas reservoirs develop the natural and hydraulic fractures with multiple scales and the nano/micro-porous media. Fractures at different scales are discontinuously distributed and show a prominent characteristics of multiple scales. Meanwhile, the differences of geometrical parameters, physical parameters and fluid parameters at multi-scale pores are very large and they are characterized by discontinuous changes.

In addition, the flow mechanisms in pore-fracture media at different scales are complex and diverse, and shows the discontinuous flow characteristics. The characteristics of multiple media are significant.

However, the conventional continuous media flow theory and method can only describe the reservoirs with the continuous distribution of single media or dual media, with the continuous changes of physical parameters and with the continuous fluid flow (Song et al. 2013). It is difficult to describe the unconventional tight oil and gas reservoirs with the continuous multiple scales and multiple media characteristics of pore-fracture media with different scales.

And the discontinuous changes of physical parameters and the discontinuous flow characteristics of fluids are also included. Therefore, it is necessary to breakthrough the conventional continuous dual media flow theory, to innovate and develop the fluid flow theory and method in discontinuous multiple media at multiple scales.

1.3.2.2 Requirements for the Mathematical Models of Discontinuous Multiple Media at Multiple Scales and Multiple Flow States

For the development methods of natural depletion, cyclic water stimulation and imbibition oil recovery, considering the characteristics of multiple media at different scales, discontinuous discrete distributions, discontinuous changes of physical parameters and discontinuous fluid flow, it is necessary to develop the mathematical models of discontinuous multiple media with multiple scales and multiple flow states, including the establishment of mathematical models of oil and gas complex flow mechanisms in pore-fracture media with nano- and micro-scale, the mathematical models of nonlinear fluid flow in dynamically changing discrete fracture media with large-scale, the mathematical models of fluid exchanging behavior of different flow mechanisms between discontinuous pore media at different scales and the mathematical models of fluid exchanging behavior between reservoir media withmultiple scales and wellbore.

For the development methods of gas injection, in-situ oil upgrading and others, conventional numerical simulation models are difficult to embody the behavior of multiple components and complex phase changes in pore-fracture media at different scales, the occurrence state of oil and gas in multiple media with nano/micro-scales. And the changes of reservoirs and fluid interface properties and fluid properties and the dynamic changes of their parameters are also included (Pan et al. 2016). Therefore, it is necessary to innovate and develop a mathematical model that can describe the multiple components, the complex phase state, the hydrocarbon occurrence state, the interface properties between reservoir and fluids, the changes of fluid properties and the dynamic changes of their parameters based on the multiple media under different development methods.

1.3.2.3 Requirements for the Mathematical Models of Geomechanics Coupling Effect in Pore-Fracture Media at Different Scales

During the development process of unconventional tight oil and gas reservoirs, under the multiple field coupling effects of fluid flow field, geostress field and temperature field, the dynamic changes of pore-fracture media at different scales are extremely strong, and they have a great influence on the behavior of fluid flow and the production performance. The geomechanics coupling models are divided into two

Table 1.17 Comparison of multiphase flow mathematical models between conventional continuous media and unconventional discontinuous multiple media

Classification of feature		Conventional fluid flow theory and mathematical model	Discontinuous multiple media flow theory and mathematical model
Characteristics of multiple media	Scale features	Single scale	Multiple scales
	Distribution characteristics	Continuous distribution, continuous change of parameters	Discontinuous discrete distributions, discontinuous changes of parameters
	Multimedia number	Single media, dual media (matrix and fracture)	Multiple media (pores and nano/micro-fractures with different scales, natural/hydraulic discrete fractures with different scales)
Characteristics of coupled geomechanics flow	Change of multimedium	Stress sensitivity leads to dynamic changes in physical parameters	The coupling effects of fracture injection and production leads to the dynamics changes of geometrics and physical parameters
Characteristics of recovery mechanism	Recovery mechanism	Single mechanism	The mechanisms of different media are different, and the mechanisms of the same media at different stages are different
	Flow regime	Single flow regime	The same media has different multiple flow regimes, different media has different multiple flow regimes, the same media has different flow regimes in different stages
	Development method	Conventional development methods (depletion method, water injection)	Unconventional development methods (depletion method, water huff-n-puff, imbibition recovery, gas injection, recovery method of changing the occurrence state and rock-fluid properties)
	Composition/phase state	The component, parameters and changing law of phase behavior in the same media are the same	Large difference in component, parameters and changing law of phase behavior in the pore/fracture media with different scales

categories in general. The first one is to establish the pseudo geomechanics coupling mathematical model through the consideration of the porosity and permeability as the function of pore pressure changes (Yao et al. 2016). Firstly, it aims at the pseudo geomechanics coupling mathematical model for the transition from single/dual media to pore-fracture media at different scales. Secondly, it aims at the pseudo geomechanics coupling mathematical model for the transition from the conventional development process to the integration of fracturing, injection, and production. The second one is to establish the fully coupled numerical simulation model considering the fluid flow field, geostress field and temperature field. Firstly, it aims at the fully coupled numerical simulation model establishing fully coupled numerical simulation model of the pore-fracture media with different scales, the fluid flow field and geostress field during the fracturing-injection-production process. Secondly, it is to establish the coupled numerical simulation model for the integrated fracturing-injection-production processes and the multiple fields of fluid flow field, geostress field and temperature field.

1.3.2.4 Requirements for the Mathematical Model of Coupling Flow Behavior between Hydraulic-Fracturedhorizontal Well and Reservoir Media

For tight oil and gas reservoirs, under the development mode of horizontal well and hydraulic fracturing, natural/hydraulic fractures at different scales and the contacting methods and flowing relationship between porous media and wellbore and the flow relationship are complex. The pressure difference around wellbore changes greatly, which leads to the prominent geomechanics coupling effect of reservoir media, and the diverse change of flow mechanisms between different media and horizontal wellbores. And the changes of flow methods and flow regimes inside a horizontal wellbore are also extremely complex. The conventional mathematic model for fluid flow is difficult to describe these complex flow relationships (Wu et al. 2011). Therefore, it is necessary to develop the coupling fluid flow mathematical model for horizontal wells and reservoir media under the mode of hydraulic fracturing (Table 1.17).

1.3.3 Problems and Requirements in Numerical Simulation Technologies of Unconventional Tight Oil and Gas Reservoirs

For the development method of horizontal wells and hydraulic fracturing in unconventional tight reservoirs, there are several preliminary issues and requirements in numerical simulation theory.

1.3.3.1 Requirements for the Numerical Simulation Technologies of Discontinuous Multiple Media with Multiple Scales

Unconventional tight reservoirs develop the natural/hydraulic fractures with multiple scales and nano/micro-scale porous media. The large-scale natural/hydraulic fractures and micro/small scale pore-fracture media are discontinuously distributed in space. Meanwhile, the differences of geometrical parameters, physical parameters, and fluid parameters at pore-fracture media with multiple scales are very large. They show the characteristics of discontinuous changing. In addition, the flow mechanisms are very complex and diverse, showing the discontinuous flow characteristics. For the problems of strong macroscopic and microscopic heterogeneity, prominent features of multiple scales and multiple media in unconventional tight oil and gas reservoirs, it is difficult to adopt the conventional numerical simulation technologies and the mathematical methods of continuous function based on the continuous media (Wang and Yao 2009). Therefore, it is necessary to break through the conventional continuous numerical simulation technologies for dual media, to innovate and develop the numerical simulation technologies of discontinuous multiple media with multiple scales.

1.3.3.2 Requirements for the Integrated Numerical Simulation Technologies of Geomechanics Coupling Effect of Fracturing-Injection-Production Processes in Unconventional Tight Oil and Gas Reservoirs

During the processes of fracturing, injection, and production of unconventional tight oil and gas reservoirs, hydraulic and natural fractures and porous media at different scales will generate a significant deformation under the geomechanics coupling effects between fluidflow field and geostress field. It causes the shifts of geometrical scales and attributes parameters of pore-fracture media, and leads to the changes of well conductivity and well index between reservoir media. Finally, the recovery performance and productivity characteristics are significantly affected. At the same time, due to the temperature difference between fracturing fluid/injected liquid and reservoir, or the application of in-situ heating method to reduce fluid viscosity, the reservoir temperature distribution is changed. It results in the changes of fluid properties and reservoir thermal stress (Zhang et al. 2015). Therefore, the numerical simulation technologies of unconventional tight oil and gas reservoirs needs to be focused on: firstly, to realize the pseudo-coupled numerical simulation technology for the transition from single/dual media to multiple pore-fracture media with multiple scales; secondly, to achieve the transition of the pseudo coupling numerical simulation technology from the normal injection-production process to the integrated fracturing-injection-production process; thirdly, to finish the transition of the numerical simulation technology of the pseudo coupling simulation where the porosity and permeability is the function of pore pressure to the fully coupled numerical simulation of multiple pore-fracture media with multiple scales, fluid flow field and geostress field.

fourthly, to develop the numerical simulation technology of integrated fracturing-injection-production processes and the multi-field coupling of fluidflow field-geostress field-temperature field.

1.3.3.3 Requirements for the Numerical Simulation Technologies of Adaptive Fluid-Regime Recognition and Complex Flow Mechanisms in Multiple Media of Tight Oil and Gas Reservoirs

In unconventional tight oil and gas reservoirs, the differences of geometrical characteristics, attribute parameters and fluid properties in the pore-fracture media at different scales cause the characteristics of different flow regimes. Meanwhile, the flow regimes of different pore-fracture media in different production stages and different production workflows (pressure gradients) are also significantly different. Thus the reservoir performance and production characteristics are affected. Simultaneously, because of the large differences in fluid composition, occurrence state, and fluid properties of oil and gas in pore-fracture media with different scales, the oil displacement mechanisms and fluid flow mechanisms of tight oil and gas in the media at different scales are complex and varied.

The conventional numerical simulation technology can simulate the single flow regime and conventional flow mechanisms. But, it cannot simulate the multiple flow regimes and complex flow mechanisms in different media and different production stages (Pan et al. 2016). Therefore, according to the identification parameters and criteria of the critical flow regimes for oil and gas in pore-fracture media with different scales, the flow regime of oil and gas can be automatically identified by the geometric scales of media, fluid properties and pressure gradients. And the kinetics equation suited with the flow regimes of oil and gas and fluid flow mechanisms is selected. Eventually, the numerical simulation technologies of adaptive fluid-regime recognition and complex flow mechanisms in multiple media of tight oil and gas reservoirs are developed.

1.3.3.4 Requirements for the Numerical Simulation Technologies of Multiple Components, Complex Phase Behavioe and Dynamic Changes of Fluid Parameters in Multiple Media

For unconventional tight oil and gas reservoirs, currently the methods of natural depletion, water injection and imbibition oil recovery are mainly adopted to extract the movable oil in different size fractures and large/middle pores. The discontinuous multiple media numerical simulation technology based on the black oil model can be used to simulate there covery performance and predict the indicators.

However, for the development methods such as gas injection and in-situ oil upgrading, they aim to improve the producing degree and recovery factor of adsorbed oil and thin film oil in the nano/micro-pore-fracture media in unconventional tight reservoirs by changing the occurrence states of hydrocarbons, the interface properties between fluids and reservoir rocks and the fluid properties.

The current mathematical models of numerical simulation cannot describe the changes of multiple components, complex phase behavior in pore-fracture media with different scales, dynamic changes of occurrence states of hydrocarbons, interface properties between fluids and reservoir rocks, fluid properties and parameters (Wang 2014b). Therefore, it is urgent to develop the numerical simulation technologies of multiple components, complex phase behavioe and dynamic changes of fluid parameters in multiple media under different development methods of unconventional tight oil and gas reservoirs.

1.3.3.5 Requirements for the Efficient Solution Technologies of Numerical Simulation in Unconventional Tight Oil and Gas Reservoirs

For the numerical simulation process of unconventional tight oil and gas reservoirs, there are many technical difficulties, including huge storage space, matrix solving difficulty and long time consuming etc (Jia et al. 2016). Three main reasons for this problem are as follows. Firstly, due to the multiple media in unconventional oil and gas reservoirs, there are multiple components, multiple flow regimes, geomechanics effect, complex structure wells and many variables needed to be solved, which leads to the complexity of mathematical models and Jacobi equation. Secondly, due to the complex fracture networks, horizontal wells and boundary conditions which are handled by an unstructured well pattern, this further increases the invalid grid blocks of numerical simulation process and the complexity of matrix form. Thirdly, due to the integration of fracturing-injection-production processes, the geomechanics effect and the utilization of fine grids for multiple media and internal/external boundary conditions, this triggers a series of issues, like a large calculation scale and a long calculation time.

In view of the above problems, three main technical requirements are as follows. Firstly, it is necessary to develop a complex matrix compression and storage technology which becomes more and more heterogeneous and fragmented in parallel environments, in response to those requirements of a large storage space, an optimal storage structure, and a large-scale computing platform. Secondly, considering the characteristics of multiphase flow mathematical model, complex coefficient matrix, and unequal distribution of eigenvalues, it is required to develop the special efficient pre-conditional sub-processing techniques to reduce the difficulty of solving complex multiphase flow mathematical models and to increase the efficiency and robustness of numerical simulation. Thirdly, for the issues of large calculation scale and long calculation time, which come from the complexity of the multiphase flow mathematical model and complex coefficient matrix, the integrated process of fracturing-injection-production, the geomechanics effect and the utilization of fine grids for multiple media and internal/external boundary conditions, it is imperative to develop the parallel architecture technologies that are increasingly adapted to the algorithms, data structures, and implementation methods. These will increase the computational scale of unconventional numerical simulation and enhance the speed and accuracy of solution.

References

Chen M, Qian B, Ou Z et al (2012) Exploration and practice of volume fracturing in shale gas reservoir of Sichuan Basin, China. SPE155598:1–11

Darishchev A, Rouvroy P, Lemouzy P (2013) On simulation of flow in tight and shale gas reservoirs. SPE163990:1–17

Dong X, Chen Z et al (2016c) Phase equilibria of confined fluids in the nano-pores of tight and shale rocks considering the effect of capillary pressure and adsorption film. Ind Eng Chem Res 55 (3):798–811

Dong X, Liu H et al (2016a) Phase behavior of multicomponent hydrocarbons in organic nano-pores under the effects of capillary pressure and adsorption film. SPE-180237-MS, SPE Low Permeability Symposium, Co, USA

Dong X, Liu H, Chen Z (2016c) The effect of capillary condensation on the phase behavior of hydrocarbon mixtures in the organic nano-pores. Sci Technol, Petrol. https://doi.org/10.1080/10916466.1209683

Ge HK, Lin YS (1998) Distribution of In-situ stresses in oilfield. Fault-Block Oil & Gas Field 5(5):1–5

Jia YL, Sun GF, Nie RS et al (2016) Flow model and well test curves for quadruple-media reservoirs. Lithologic Reservoirs 28(1):123–127

Jin L, Zhu C, Ou Y Y et al (2013) Successful fracture stimulation in the first joint appraisal shale gas project in China. IPTC16762:26–28

Liu JJ, Liu XG, Hu YR et al (2000) The equivalent continuum media model of fracture sand stone reservoir. J Chongqing University (Natural Science Edition), 23(Extra): 158–160

Liu X, An F, Chen QH et al (2016) Analysis of the EOR techniques for tight oil reservoirs: taking bakken-formation as an example. Petrol Geol Oilfield Dev Daqing 35(6):164–169

Nojabaei B, Johns RT (2013) Effect of capillary pressure on phase behavior in tight rocks and shales. SPE J 16(3):281–289

Pan Y, Wang PR, Song DW et al (2016) Experimental researches on relative permeability curve of complex fracture system. J South Petrol Univ (Sci Technol Edition), 38(4):110–116

Saldungaray P, Palisch T (2012) Hydraulic fracture optimization in unconventional reservoirs. SPE151128:1–15

Song Y, Jiang L, Ma XZ (2013) Formation and distribution characteristics of unconventional oil and gas reservoirs. J Palaeogeogr 15 (5):605–614

Wang G (2014a) Study on the characteristics of Chang 7 Tight reservoir in Heshui Area. Southwest Petroleum University, Chengdu

Wang PR (2014b) Experimental study on oil-water mixing flow characteristics of fractured complex medium reservoirs. South Petrol Univ, Chengdu

Wang ZS, Yao J (2009) Analysis of characteristics for transient pressure of multi-permeability reservoirs. Well Testing 18(1):10–15

Wu Q, Xu Y, Wang TF et al (2011) The revolution of reservoir stimulation: an introduction of volume fracturing. Nat Gas Ind 31 (4):7–11

Wu Q, Xu Y, Wang X et al (2012) Volume fracturing technology of unconventional reservoirs: connotation, design optimization and implementation. Petroleum Expl Dev 39(3):377–384

Yao J, Sun ZX, Zhang K et al (2016) Scientific engineering problems and development trends in unconventional oil and gas reservoirs. Petrol Sci Bull 01(1):128–142

Yin D (1983) On the multiple porosity medium model and its application on pressure build-up curve analysis. Petrol Expl Dev 3:59–64

Zeng LB, Wang GW (2005) Distribution of Earth stress in Kuche Thrust Belt Tarim Basin. Petrol Expl Dev 32(3):59–60

Zhang DL, Li JL, Wu YS (2010) Influencing factors of the numerical well test model of the triple-continuum in fractured vuggy reservoir. J South Petrol Univ (Sci Technol Edn), 32(6):113–120

Zhang YH, Lu BP, Chen Z et al (2015) Technical Strategy thinking for developing continental tight oil in China. Petrol Drill Techn 43(1):1–6

Zhao B (2015) Study on the characteristics of Chang 7 reservoir in longdong area erdos extension basins. North Petrol Univ

Zhao D, Chen ZM, Cai XL et al (2007) Analysis of distribution rule of geostress in China. Chin J Rock Mechan Eng 26(6):1265–1271

Zou CN, Zhu RK, Bai B et al (2015) Significance, geologic characteristics, resource potential and future challenges of tight oil and shale oil. Bull Mineral Petrol Geochem 34(1):3–17

2 Flow and Recovery Mechanisms in Tight Oil and Gas Reservoirs

Because of the presence of nano/micro-scale pores and natural/hydraulic fractures, tight oil and gas reservoirs are characterized with multiple scale and multiple media. There is a large difference for flow regimes, multiphase flow mechanisms and oil displacement mechanisms in the pore-fracture media at different scales. In this chapter, the definition and classification of multiple media are clearly performed, which forms a method to divide the multiple media in tight oil and gas reservoirs. Multiphase flow behavior, oil displacement mechanisms and fluid characteristics of tight oil and gas reservoirs are also revealed. The changing law of oil unlocking capacity and drainage radius of reservoir matrix at different scales are analyzed. The coupled flow behavior and production mechanisms of porous media at different scales are explained. It lays a foundation for the establishment of the numerical simulation theory and mathematical models in the pore-fracture media at different scales.

2.1 Classification and Characteristics of Multiple Media at Different Scales in Tight Reservoirs

The system of pores and natural/hydraulic fractures in tight reservoirs is characterized with different scales. There is a big difference of the spatial distribution and geometric scale of the pores and fractures at different scales. Its storage space, flow capacity and related physical parameters vary greatly. In the pore-fracture media at different scales, the fluid composition and occurrence state are also different. During the production process, the flow regime and fluid flow mechanism also have a large difference, which causes the difference of oil and gas utilization degree and contributions to the total production in those pores and fractures. Therefore, it is necessary to classify reservoir media from the geometrical and property characteristics of pores and fractures, the occurrence state of fluids, fluid flow mechanisms and flow characteristics in porous media. Dividing pores and fractures with similar characteristics or similar changing behavior as one type of media, the porous space in tight oil and gas reservoirs can be finally characterized as a multiple scale media.

As the development of oil and gas fields continues, the types of oil and gas reservoirs are becoming more and more diversified. The conventional continuous porous media theory (Ge 2003) is also developed from the initial continuous single media to the continuous dual media and continuous multiple media. Because of the presence of pores and natural/hydraulic fractures, the unconventional reservoir has shown a discontinuous and leaping-change feature in the spatial distribution and characteristic parameters of pore-fracture media at different scales. An application of the conventional continuous media theory no longer satisfys the requirement of an actual field work. Thus, the theories and methods for the media with discrete fractures, discrete multiple media and other discontinuous multiple media are developed (Table 2.1).

2.1.1 Definition and Classification of Multiple Media at Different Scales

2.1.1.1 Continuous Multiple Media

1. Continuous single media

(1) Continuous porous media

Hydrocarbon reservoir is a system which is composed of several types of porous media. In this porous media system, the geometrical/attribute features and changing regulation are similar in the different spatial regions. They can be characterized by a continuous function or equation. It is called as the continuous porous media.

Q. Ran, *Unconventional Tight Reservoir Simulation: Theory, Technology and Practice*,
https://doi.org/10.1007/978-981-32-9848-4_2

Table 2.1 Classification of media in tight reservoirs

Media type			Reservior type	Property discription of different media
Continuous media	Single media		Conventional porous reservoir; unconventional tight reservoirs	The reservoirs only have the similar porous media. The geometrical features and attribute features are similar, and the changing behavior can be characterized by the continuous function
	Dual media	Dual-porosity, single-permeability	Fractured reservoir; low permeablity/tight reservoir	Reservoir developes the porous media and natural/hydraulic fractures meida. The geometrical and physical properties change a lot for porous media and fracture. Fracture permeability is much higher than the porous media (negligible). There is no flow inside the reservoir matrix system, only fluid flow happens in the fracture system. Properties of porous media and fracture can be characterized by continuous functions
		Dual-porosity, dual-permeability	Fractured reservoir; low permeablity/tight reservoir	Both the pores and natural/hydraulic fractures exist in reservoir. It has a certain storage capacity and percolation capacity. Fluid flow happens in both the reservoir matrix and fractures. The fluid can exchange between matix and fractures. The properties of pore and fractures can be characterized by a continuous function
	Multiple scale porous media		Fractured and vugged carbonate reservoir; unconventional reservior	The pore/cavity system and fracture system at different scales form a continuous multi-scale porous media. Fluid flow happens in each system, and fluid exchanges between pore system and fracture system. Properties of porous media and fractures can be characterized by a continuous function
Discontinuous media	Media with discrete fracture		Fractured reservior; unconventional reservior	The natural fractures and hydraulic fractures at different scales distribute randomly, and it is discontinuous and discrete. It is also a multi-scale system. The discontinuous and discrete fractures with different spatial positions and different sizes compose the media with decrete fractures.Its partial distribution, geometric characteristics, physical parameters and fluid flow can be characterized by a discrete continuous function
	Discrete multiple scale porous media		Fractured carbonate reservoir; unconventional reservior	For the pores and microfractures at different scales, the pores and fractures with similar features and changing behavior can be classifie as multiple media. Due to the random and interactive characteristics of poresystem, they exhibit a discontinuous discrete distribution and can be characterized by a discontinuous discrete function
	Hybrid discrete multiple scale porous media		Fractured carbonate reservoir; unconventional reservior	When macroscopic large-scale natural/hydraulic fractures, small-scale natural/hydraulic fractures and pores co-exist, the porous media at different scales have extremely strong randomness in spatial distribution as well as discontinuous discrete features, which have both media with discrete fractures and discrete multiple media. It is a hybrid discrete multi-scale porous media

A conventional porous oil reservoir usually has only one single type of pores. Pores are similar in shape and size, and their distribution in space is also continuous; therefore, they behave as a continuous single media.

For an unconventional tight reservoir, if only the pores with a similar shape and similar size are developed in a region where natural/hydraulic fractures do not exist, it also belongs to the continuous single media.

(2) Continuous fluid

A fluid is continuously distributed in the porous media system. Its properties can be the same ones (Fig. 2.1①) or vary continuously in different reservoir regions. For the latter case, the changing behavior is similar (Fig. 2.1②) and can be characterized by a continuous function or equation; it can be called as the continuous fluid. In unconventional reservoirs, the fluid in pore-fracture media at different scales is distributed discontinuously, and in different reservoirs regions, the fluid properties change with a discontinuous and leapfrog type (Fig. 2.1③).

(3) Continuous flow

Fluids are continuously distributed in the porous media system of a reservoir, and the flow characteristics and flow mechanisms are similar. They can be characterized by the

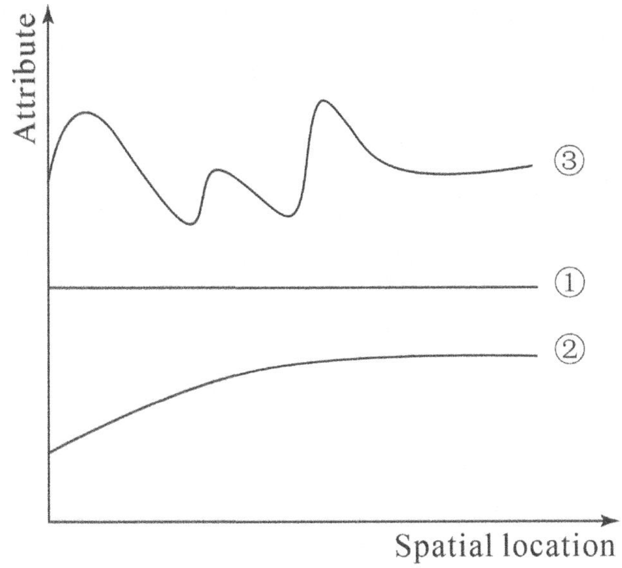

Fig. 2.1 Schematic on the changes of fluid properties with spatial location

same or a similar continuous function or equation and are called as the continuous flow.

(4) Flowing field of continuous media

The porous media system of oil and gas reservoir can be characterized by the system with a continuous porous media and a continuous fluid distribution. A fluid has the continuous and similar flow characteristics in continuous porous media. In continuous system, the porous media, fluid properties and flowing behavior of fluids in porous media can be characterized by a continuous function or equation. The flowing field in continuous system is called as the flowing field of continuous media.

For the conventional porous oil reservoir, porous media generally belongs to the single porous media with the similar geometrical properties. The fluid properties are similar and present a continuous distribution. The flow mechanism of them is single and belongs to the continuous Darcy flow. This single porous media is called as the continuous single media, where the flowing field is called as the flowing filed in a continuous single media.

2. Continuous dual media

As the oil and gas exploration technology develops, the objective has gradually changed from the reservoir with pores to the reservoir with fractures and the tight and low-permeability reservoirs, from the development model of conventional vertical well to horizontal well plus hydraulic fracturing. The reservoir media becomes more and more complicated, and changes from a single porous media to a dual porous media with pores and fractures. The porous system and the natural/hydraulic fracture system are the main types of pore space in tight reservoirs. They present the orders of magnitude differences on the geometrical scales. From the reservoir storage capacity, the porosity of porous media system is much higher than that of the fracture system. From transportation capacity, the permeability of the porous system is much smaller than that of the fracture system. The fracture system and porous system are regarded as the two continuous media fields in space and overlap with each other to form a dual media (Barenblatt et al. 1960; Warren and Root 1963).

Reservoirs are generally composed of two consecutive pore and fracture systems. Both of them have the continuous fluid distribution, and the reservoir fluid has the continuous and similar flow characteristics in each system. The fluid can exchange between them. Porous media, fracture media, any properties of fluid, and fluid flow properties can be characterized by a continuous function or equation. The fluid flow field in these two continuous systems is called as a continuous dual-media flowing field.

(1) Dual-porosity single-permeability

For the tight reservoirs with the development mode of hydraulic fracturing, there are two types of systems of porous media and natural/hydraulic fractures media. Both the reservoir matrix system and the fracture system have the same porosity. However, the porosity of the former is greater than that of the latter. The fracture system permeability is higher than that in the pore system. The permeability in reservoir matrix can be negligible. There is no flow inside the reservoir matrix, and fluid flow only happens within the fracture system.

(2) Dual-porosity dual-permeability

A tight reservoir has a pore system and a fracture system. Both of them have a certain capacity of fluid storage and fluid transportation. The reservoir matrix system has a larger storage capacity and lower fluid flow-ability, and it is the fluid source of the fracture system. The fracture system is the main channel of fluid flow, but has a lower storage capacity. A fluid can flow in the reservoir matrix system and fracture system, and the fluid exchanging behavior between them can happen.

3. Continuous multiple media

As the developed oil reservoir has gradually shifted from the conventional fractured reservoirs to the fractured carbonate reservoirs and unconventional reservoirs, the reservoir storage spaces have evolved from relatively simple pores and fractures to more and more complicated pores, vugs and fractures. As a result, the development mechanisms and flow characteristics of these reservoirs are more complicated, and it is difficult to describe them with the conventional dual media. Therefore, the triple, quadruple and multiple media theories have been developed.

There are two classification methods for the triple-porosity media in carbonate reservoir with fractures and cavities. The first method is to divide the porous system into two types according to the differences in porosity and permeability. One has a better connectivity with fracture system, and the other has a poor connectivity with fracture system. These two types of porous system and the fracture system make up a triple media system (Warren and Root 1963). The other method is to divide the small-scaled pores, large-scaled pores and fractures into three independent and interconnected triple media systems based on the differences in geometry, porosity and permeability (Wu and Ge 1983; Abdassah and Ershaghi 1986; Liu et al. 2003; Yao et al. 2004; Yao and Wang 2007; Wu et al. 2007). Similarly, according to the difference of size, porosity and permeability in the porous geometrical scale, a fracture system can be classified into large fractures and micro-fractures, and porous media can be classified into large pores and small pores to form a quadruple media system (Kang 2010).

For the carbonate reservoirs with different scale pores and different scale fractures and the unconventional tight reservoirs with nano/micro-pores and natural/hydraulic fractures, according to the porous geometrical scale, attribute parameters and flow characteristics, the pore and cavity system with different scales and the nano/micro-pore system with different scales can be classified into the pore-cavity media with multiple different scales (Yin 1983). At the same time, the natural/hydraulic fractures at different scales can be called as the fracture media at different scales. All of them form the multiple media system (Liu and Guo 1982; Yao et al. 2014).

A continuous multiple media system is composed of porous media at different scales and fracture media at different scales. Each media has a continuous fluid distribution. The fluid also has the continuous and similar flow characteristics. Reservoir fluids can exchange between them. Any properties of reservoir media and fluids and the fluid flow behavior can be characterized by a continuous function or equation. The fluid flow field in these multiple continuous systems is called as a continuous multiple media flow field.

2.1.1.2 Discontinuous Multiple Media

1. Conventional fluid flow theory of continuous multiple media is not suitable for the discontinuous multiple media

Whether it is a fracture-cavity carbonate reservoir or an unconventional tight reservoir, the distribution of pores and fractures in space is strongly heterogeneous. The developing degree, density, scale and size of pore, vug and fracture in different reservoirs regions are significant different. In space, it shows a non-uniform, discontinuous discrete distribution. In the same area, the pores, vugs, and fractures with different sizes and scales coexist, and a multiple scale feature is much obvious. The characteristics and parameters of reservoirs and fluid flow behavior in pores, vugs and fractures at different scales are very different. There is no continuous distribution and cannot be characterized by a continuous function.

The continuous media system can characterize the pore system at different scales as a N-space superposition of continuous media field. It is considered that the pores, vugs and fractures have the same scale, size and shape, but cannot reflect the characteristics of a multiple scale. Pores, vugs, and fractures can be connected through a quasi-steady-state interporosity flow function. At the same time, the exchanging coefficient between them is also difficult to determine. Within the same system and between the different systems, there is no flow difference, and it cannot reflect the difference between different flow regimes and flow mechanisms.

Therefore, for such reservoir media with pores, vugs and fractures at different scales, the property parameters and flow characteristics cannot be characterized by the continuous functions. Thus, they cannot be handled by a continuous media. The flow theory for the traditional continuous multiple media is no longer applicable, and it is necessary to break through the theory of continuous media and innovatively develop the fluid flow theory through porous media of discontinuous multiple media (Table 2.2).

2. Discontinuous multiple media

The media of pores, vugs and fractures at different scales have the characteristics of jumping mutations in reservoirs. It shows a discrete distribution. However, the different types of media show an interactive distribution in space. The geometrical characteristics with different scales and the different types of media, property parameters and flow characteristics are difficult to apply a continuous function or equation to characterize. It can only be described using a discontinuous, segmented or distributed discrete function and equation. This type of reservoir media with different scales is called as the discontinuous multiple media.

According to the spatial distribution characteristics of porous media at different scales and the size of different media, the large-scale natural/hydraulic fractures are treated as the media with discrete fractures. The micro and small scale natural/hydraulic fractures and nano/micro-pores at different scales are treated as the discrete multiple media.

Table 2.2 Differences of discontinuous multiple media and continuous multiple media

	Continuous multiple media	Discontinuous multiple media
Definition	A true porous media system is replaced by a number of consecutive characteristic unit cells (both multiple media and inner-fluid) in which any property (whether porous or fluid-specific) can be described by a continuous equation	The discontinuous feature units are used to describe the oil and gas reservoirs with discontinuous feature units. The discontinuous feature units present a jump mutation distribution in spatial distribution. The property parameters and flow characteristics between different unit bodies vary greatly. It can only be characterized by the discontinuous, piecewise or distributed functions and equations
Spatial distribution characteristics	Pores, vugs and fractures are well developed with high density, well connectivity, and distributed continuously and uniformly. Due to the continuous distribution of pores, vugs and fractures, it can be characterized by the continuous functions and can be treated as the continuous media	There are great differences in the development, density and connectivity of pores, vugs and fractures, which are non-uniform and discontinuous in space. Due to the discrete distribution of pores, vugs and fractures, it cannot be characterized by the continuous function and can only be treated as the discontinuous media
Multiscale characteristics	Continuous media will be different sizes of pores, vugs and fractures simplified to have the same size and shape of the media, cannot fully reflect the discontinuity and multi-scale characteristics of media with pores, caves and fractures	According to the actual location of the pore, vug and fracture and the different scale features, the discontinuous media can be processed to reflect the anisotropy, discontinuity and multiple scale of different reservoir media
Property characteristics	It is considered that the property parameters of pores, vugs and fractures at different scales are continuously distributed and can be characterized by continuous functions. The fluids are continuously distributed in the porous media system of the reservoir. The fluid properties are the same or continuous in different parts of the reservoir space, and their regulars are similar	Due to the discontinuous discrete distribution of pores, vugs and fractures at different scales, the property parameters also have discrete features and are difficult to be characterized by continuous functions. The fluid is discontinuous in the porous media at different scales and its fluid properties change dramatically in different parts of reservoir space
Fluid flow characteristics	Different reservoir media have the same and single flow mechanism and characteristics, which can be characterized by continuous function, but such treatment can not reflect the difference of flow characteristics and the coupling characteristics between different media. At the same time, how to calculate the fluid flow between different media is lack of effective theory and method, especially for multi-phase interporosity flow problem. Therefore, the calculated results differ greatly from the actual ones	Different reservoir media have different flow regimes and percolation mechanisms, which can only be characterized by discontinuous piecewise functions, which can fully reflect the differences in flow characteristics and the coupling characteristics between different media. At the same time, there's no need to calculate the interporosity flow behavior between different media, but through the size of the conductivity to characterize, the calculated results are in good agreement with the actual data

When the large-scale natural/hydraulic fractures, micro-scale natural/hydraulic fractures, and nano/micro-pores coexist, they can be considered as the mixed discrete multiple media.

(1) Discrete fracture media

Fractures refer to the discontinuous surfaces formed in rock due to the geological performance or hydraulic operation. A fracture distribution is affected by the geological conditions and reservoir heterogeneity, and has a strong randomness. A fracture system at different scales is discontinuous and discrete in space, and their sizes have the significant multiple scale characteristics.

The spatial location, distribution characteristics, geometrical morphology and characteristics of large-scale natural/hydraulic fractures in reservoirs can be explicitly described. At the same time, the distribution of fractures at different scales and their geometrical characteristics, physical parameters and fluid flow behavior can also be separately characterized by a discontinuous discrete function or equation. The different spatial locations and different sizes of thus discontinuous non-discrete fractures can form the discrete fracture media (Karimi-Fard and Firoozabadi 2003; Arnaud et al. 2004), as shown in Fig. 2.2.

(2) Discrete multiple media

The pores and micro/small-fractures at different scales in tight reservoirs have a strong randomness due to the influence of geological conditions and reservoir heterogeneity. Pore-fracture systems at different scales have a strong interactive distribution. However, they are significantly different because of the different geometrical dimensions, scales, physical parameters and flow characteristics between the pore and fracture systems. In a reservoir, they have the characteristics of a discontinuous and discrete distribution. It is difficult to describe them using a conventional continuous function, and only a discrete function is applicable.

According to the geometrical shape, scale size, physical parameters and flow characteristics of pore-fracture systems

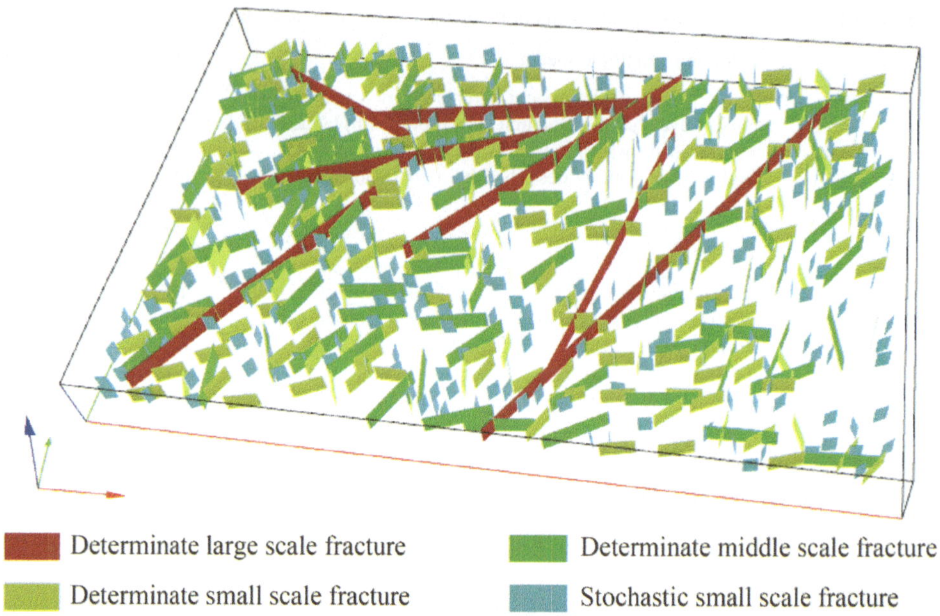

Fig. 2.2 Schematic diagram of discrete fractures

at different scales, based on the principle of similar variations or similar characteristics, the pore-fracture systems at different scales can be called as the multiple media. At the same time, because of the randomness and interaction of spatial distribution, it has a discontinuous discrete distribution, and only a discontinuous distribution function can be used to describe the geometrical characteristics, physical parameters and flow characteristics. The media with a discontinuous discrete distribution, unique geometrical characteristics, attribute parameters and flow characteristics can be called as the discrete multiple media (Snow 1968; Noorishad and Mehran 1982; Kim and Deo 2000; Yao et al. 2010).

From the spatial relationship and flow behavior of different porous media, the multiple media can be classified into nested distribution type multiple media, interactive discrete multiple media, and randomly distributed discrete multiple media, as shown in Figs. 2.3, 2.4 and 2.5.

Fig. 2.3 Nested distribution type multiple media

(3) Mixed discrete multiple media

In actual reservoirs, the macroscopic large-scale natural/hydraulic fractures, micro/small-scale natural/hydraulic fractures and pores coexist, and there is strong randomness for the spatial distribution and discontinuous distribution. Therefore, there are both media with discrete fractures and discrete multiple media, which are generally presented as the mixed discrete multiple media, as shown in Fig. 2.6.

Fig. 2.4 Interactive discrete multiple media

Fig. 2.5 Random discrete multiple media

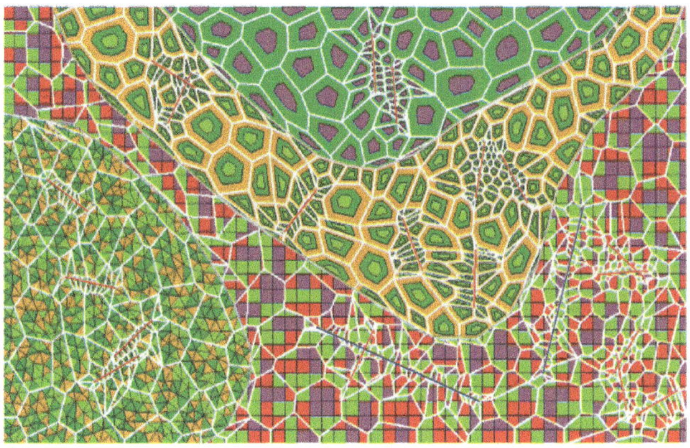

Fig. 2.6 Diagram of the mixed discrete multiple media

2.1.2 Classification Method of Multiple Media

For porous reservoirs and fracture oil and gas reservoirs, the fluid storage space can be classified into porous media and fracture media. For complex fractured-vug reservoirs, from the types and geometrical dimension of storage space, it can be classified into porous media, vug media and fracture media. For unconventional tight oil and gas reservoirs, there are many types of fluid storage space, with a wide scale distribution and a larger difference. Furthermore, the fluid occurrence state is also changed because of the different geometrical scales. It is required to break the limitation of the classification method by the geometrical scale. Combined with the physical and flow characteristics of media at different scales, the indicators and classification method for multiple media are developed.

2.1.2.1 Classification of Multiple Media from the Characteristics of Pore-Fractures Media

1. Based on the types of storage space

The types of reservoir space developed in unconventional tight reservoirs are diverse. According to the principle of same or similar types of reservoir space (Xie et al. 2002), the reservoir space can be classified into porous media and fracture media. Among them, the porous media include primary intergranular pores and corrosion vugs, and the fracture media include natural fractures and hydraulic fractures. However, this method of classification does not consider the differences in geometrical dimensions between the same media.

2. Based on the geometry

In the case of the same reservoir space, there is a great difference in the geometrical scale between the same porous media and fracture media, which needs further subdivision. According to the principle of the same or similar geometric scale (Yao et al. 2014), a carbonate pore system can be classified into large-scale vug, dissolved pores and reservoir matrix pores. The porous system in tight reservoir can be further subdivided into the porous media at different scales such as millimeter pores, micropores, and nanopores. The natural/hydraulic fracture system can be classified into large scale, middle scale, small scale and micro scale fracture media. However, this method does not take into account the differences between the physical characteristics of the same media.

3. Based on the reservoir attribute properties

Due to the differences of pore structure, fracture roughness and supporting conditions, there are differences in porosity and permeability in the same type of reservoir space and scale range. According to the principle of the same or similar attribute characteristics (Li 1999), porous media can be classified into high porosity media, middle porosity media and low porosity media. According to the permeability, porous media can be classified into high permeability media, middle permeability media, and low permeability media. However, this classification method does not consider the differences in fluid properties and flow characteristics between the same media.

2.1.2.2 Classification of Multiple Media from the Fluid Flow Characteristics

1. Based on the properties and occurrence state of fluids

In the case of similar pore-fracture media, due to the difference in fluid properties (viscosity and density), and occurrence state (movable oil in a free state, capillary oil controlled by capillary force, and film oil in an adsorption state) in different media, fluids in different media have the different development methods and operating conditions. Therefore, it is necessary to classify the different media according to the characteristics of those media, fluid properties and occurrence states of fluids (Wang et al. 1993; Sun et al. 2014).

2. Based on the producible state

Due to the differences in fluid properties and occurrence state in different media, there are major differences in the fluid producibility between different media. In general, capillary pressure curves can be used to reflect the fluid producibility, and relative permeability curves can reflect the range and size of movable fluid. Different media have different characteristics of capillary pressure curves and relative permeability curves. Therefore, multiple media can be classified according to the characteristics of capillary pressure curves and relative permeability curves.

3. Based on the flow regimes and flow mechanisms

Due to the different development methods and formation energy, there will be great differences in fluid pressure gradients between different reservoir parts and different media. At the same time, different media are affected by the geometrical characteristics and fluid properties, and the flow mechanisms and flow regimes are complicated and diverse. Therefore, the multiple media can be classified according to the flow regimes and flow mechanisms (Yang et al. 2008). Different mathematical models are used for different media to characterize the flow characteristics and simulate the recovery performance of reservoirs.

2.1.2.3 Classification of Multiple Media from the Characteristics of Pore-Fracture Media and Fluid Flow

Based on the characteristics of reservoir space, geometrical scale, attribute characteristics, fluid properties, occurrence state, producibility, flow regimes, fluid mechanisms, and other characteristics and indexes of porous media at different scales, the method to classify the multiple media at different scales has some shortcomings. Therefore, according to the characteristics of pore-fracture media and fluid flow, multiple media can be classified comprehensively to reflect the differences of reservoir types, geometrical scales and properties among different media. The differences in fluid properties, occurrence states, producibility, flow regimes and flow mechanisms between different media preferably reflect the actual performance characteristics of pore-fracture multiple media at different scales. Figure 2.7 is a flow chart for the classification of multiple media.

2.2 Flow Regimes and Flow Mechanisms

In tight reservoirs, there are great differences for the fluid occurrence state and fluid properties in pore-fracture media at different scales. Under the different pressure gradients, the fluid flow in different scale media presents the different types of fluid regimes as well as complicated and diverse flow mechanisms, which are characterized by the flow characteristics of multiple scales, multiple media, multiple fluid regimes and complicated flow mechanisms.

2.2.1 Definition of Flow Regime

A flow regime in tight oil and gas reservoirs refers to the flow morphology of hydrocarbons in the multiple media of tight reservoirs (Ge et al. 2001). Tight reservoirs have the characteristics of nano/micro-scale pores, complex natural/hydraulic fracture networks at different scales and multiple media. The flow regimes of hydrocarbons can be classified into three types: high-speed nonlinear flow, pseudolinear flow and low-speed nonlinear flow (Fig. 2.8).

2.2.2 Classification of Flow Regimes and Flow Mechanisms

Flow regimes of fluids in tight reservoirs can be classified based on the threshold pressure gradient, pseudolinear critical pressure gradient and high-speed nonlinear critical pressure gradient (Fig. 2.9, Table 2.3). A threshold pressure gradient refers to the pressure gradient at which a fluid must balance the resistance caused by the adsorption film or hydration film on rock surface (Zheng et al. 2016). Only when the displacement pressure gradient is greater than the threshold pressure gradient, the fluid can flow in reservoir. When the displacement pressure gradient increases to a certain value, a flow regime transition from low-speed nonlinear flow to pseudolinear flow can be observed. The

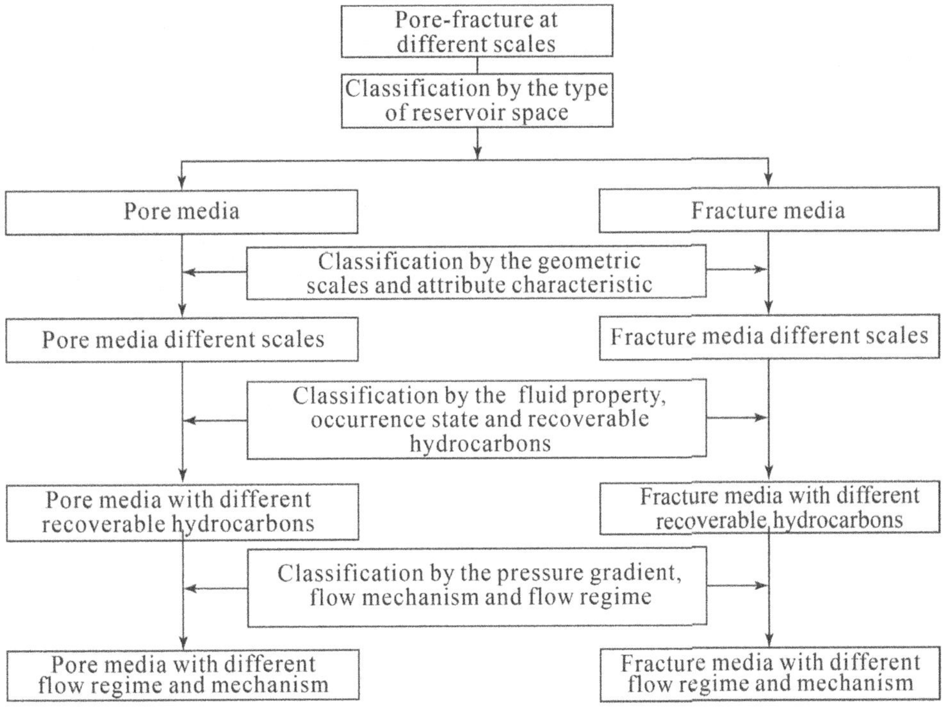

Fig. 2.7 A flow chart for the classification of multiple media

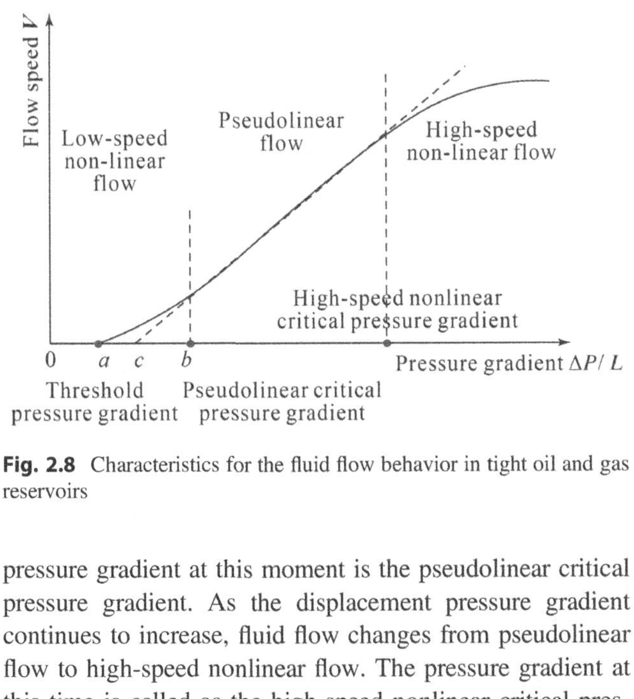

Fig. 2.8 Characteristics for the fluid flow behavior in tight oil and gas reservoirs

pressure gradient at this moment is the pseudolinear critical pressure gradient. As the displacement pressure gradient continues to increase, fluid flow changes from pseudolinear flow to high-speed nonlinear flow. The pressure gradient at this time is called as the high-speed nonlinear critical pressure gradient.

Due to the presence of nano/micro/milli-scale pore-fracture media in tight reservoirs, the flow regimes of fluids in different media are different under the same pressure gradient. The flow regimes of fluids at different development stages are different in the same media. The flow regimes of fluids with different properties in a same media are also different (Fig. 2.9).

2.2.2.1 High-Speed Nonlinear Flow and Flow Mechanisms

When the flow velocity increases to a certain value and the displacement pressure gradient is greater than the high-speed nonlinear critical pressure gradient, the flow velocity and pressure gradient are no longer linear, and the pseudolinear flow regulation is damaged. It is called as the high-speed nonlinear flow. There are different scales of reservoir matrix pores and natural fractures in tight sandstone reservoirs, which form a complicated fracture network with hydraulic fractures in different fracturing patterns. The conditions of high-speed nonlinear flow in fractures and pores at different scales are different (Fig. 2.10). For large fractures, based on the characteristics of large width, strong conductivity and high flow velocity, the pressure gradient can easily reach the high-speed nonlinear critical pressure gradient, which makes it easy to form the high-speed nonlinear flow. For micro-fractures and reservoir matrix throats, both the flow conductivity and fluid velocity are low. But, under the condition of a high pressure drop around the wellbore, as the pressure gradient exceeds the high-speed nonlinear critical pressure gradient, a high-speed nonlinear flow can be formed in reservoir.

Because of the high oil viscosity and slow flow behavior of crude oil in tight reservoirs, the pressure gradient required to reach the high-speed non-linear flow process is high in fracture media. The equation of the high-speed nonlinear critical pressure gradient is

$$G_d = 6.0537 \cdot w_f^{-0.571} \quad (2.1)$$

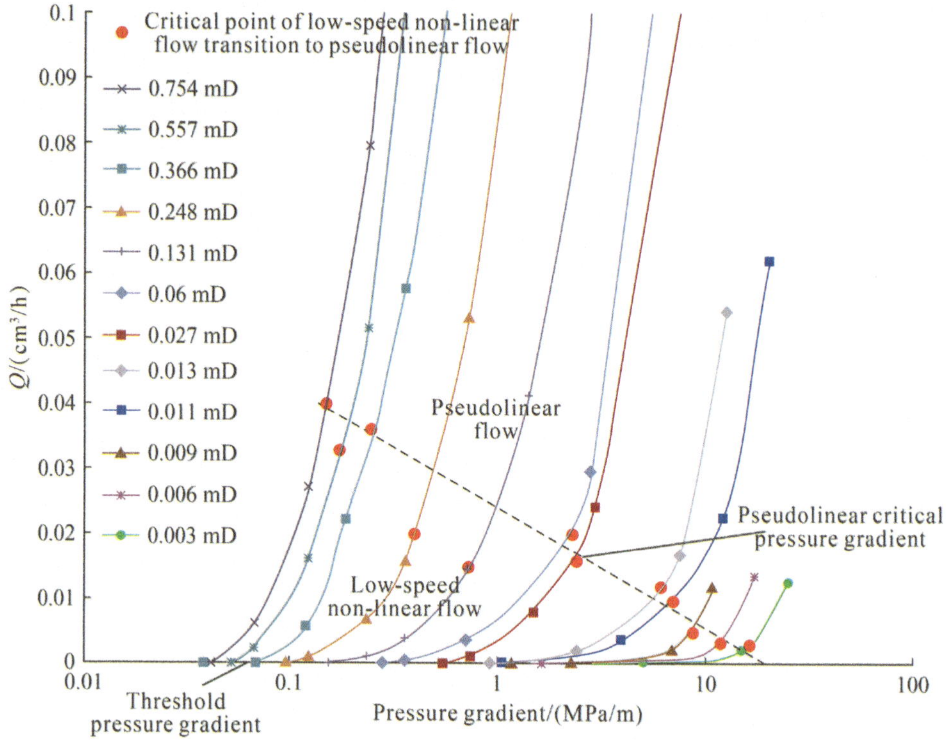

Fig. 2.9 Changes of flow regime in different media and different pressure gradient in tight oil reservoirs

Table 2.3 Stage classification of fluid flow in tight oil and gas reservoirs

Flow stage	Flow regime	Flow conditions
First stage	Oil and gas do not flow	Displacement pressure gradient $\leq$ Threshold pressure gradient
Second stage	Low-speed non-linear flow	Threshold pressure gradient < Displacement pressure gradient $\leq$ Pseudolinear critical pressure gradient
Third stage	Pseudolinear flow	Pseudolinear critical pressure gradient < Displacement pressure gradient < High-speed nonlinear critical pressure
Fourth stage	High-speed nonlinear flow	Displacement pressure gradient $\geq$ High speed nonlinear critical pressure gradient

For tight gas reservoirs, the viscosity is low and velocity is high. Thus, the pressure gradient required to reach the high-speed nonlinear flow is low in fracture media. The equation of the high-speed nonlinear critical pressure gradient is

$$G_{\mathrm{d}} = 8.65 \times 10^{-2} \cdot w_{\mathrm{f}}^{-0.386} \quad (2.2)$$

where, G_{d} is a high-speed nonlinear critical pressure gradient, MPa/m, and w_{f} is the fracture aperture, μm.

The Forchheimer equation (Forchheimer 1901) is usually used to describe the kinetics characteristic of high-speed nonlinear flow.

$$-\nabla P_{\mathrm{p}} = \frac{\mu_{\mathrm{p}} v_{\mathrm{p}}}{K_{\mathrm{F,m}}} + 10^{-3} \cdot \beta_{\mathrm{F,m}} \rho_{\mathrm{p}} v_{\mathrm{p}} |v_{\mathrm{p}}| \qquad |\nabla P_{\mathrm{p}}| \geq G_{\mathrm{pd}} \quad (2.3)$$

where, ∇P is the pressure gradient, MPa/m; subscript $p = o, g, w$, where o represents the oil phase, g represents the gas phase, w represents the water phase; v is the fluid flow velocity, m/s; $K_{\mathrm{F,m}}$ indicates fracture or reservoir matrix permeability, mD; μ_{p} is the viscosity of the p-phase fluid, mPa·s; $\beta_{\mathrm{F,\ m}}$ is the high-speed nonlinear coefficient in fractures or reservoir matrix, m^{-1}; subscript F represents the fractures and m represents the reservoir matrix; ρ_{p} is the p-phase fluid density, g/cm^3.

The high-speed nonlinear coefficient β_{F} is a key parameter to describe the effect of high-speed non-linear flow on the fluid flow behavior. There are some differences between the determination method of high-speed nonlinear coefficients in reservoir matrix and fractures.

1. High-speed nonlinear coefficients in reservoir matrix pores

High-speed nonlinear coefficients are highly related to the pore structure. As the pore throat radius increases, the tortuosity of pores decreases, the permeability increases, and

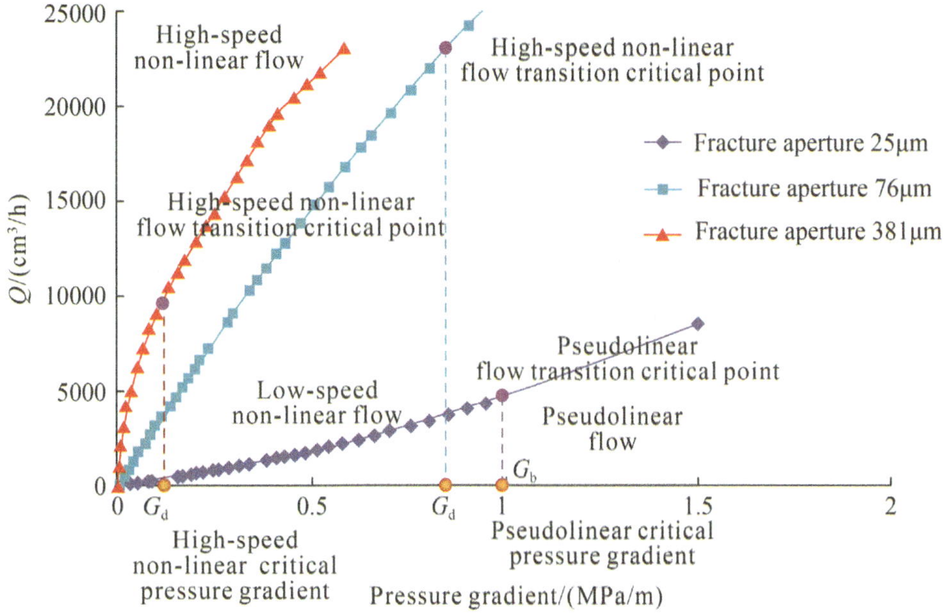

Fig. 2.10 Characteristics for the fluid flow behavior in fracture media in tight oil reservoirs

the high-speed nonlinear coefficient decreases. Thus, the high-speed nonlinear flow effect of fluid in the reservoir matrix pores increases. The high-speed nonlinear coefficient in the reservoir matrix pores has been (Fig. 2.11) experimentally determined (permeability 0.03 mD to 100 mD) (Noman and Archer 1987).

$$\beta_m = \frac{4.19 \times 10^{11}}{K_m^{1.57}} \tag{2.4}$$

where, K_m is the reservoir matrix permeability, mD; β_m is the reservoir matrix high-speed nonlinear coefficient, m^{-1}.

2. High-speed nonlinear coefficient in the fracture system

The calculation method of the high-speed nonlinear coefficient in fractures can be obtained through the experimental measurement (fracture permeability 0.1 mD to 10 mD) (Pascal et al. 1980).

$$\beta_F = \frac{4.8 \times 10^{12}}{K_F^{1.176}} \tag{2.5}$$

where, K_F is the fracture permeability, mD, β_F is the fracture high-speed nonlinear coefficient, m^{-1}.

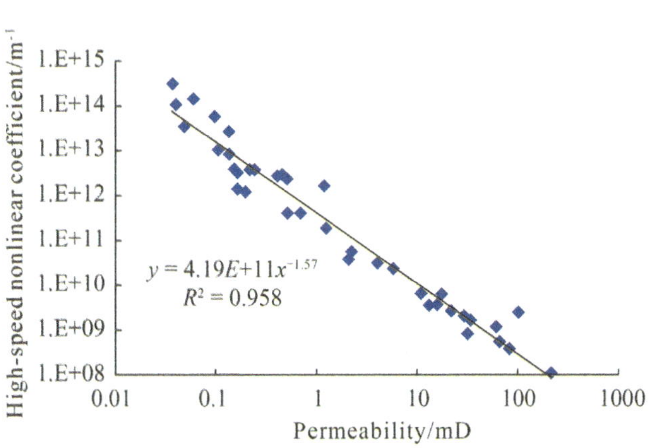

Fig. 2.11 Relationship between high-speed nonlinear coefficients and permeability

2.2.2.2 Pseudolinear Flow and Mechanisms

When the pressure gradient is higher than the pseudolinear critical pressure gradient and less than the high-speed nonlinear critical pressure gradient, the flow velocity and the pressure gradient are in a pseudolinear relationship. The flow behavior curve will not pass the coordinate origin and a threshold pressure gradient can be found (Fig. 2.12). The kinetics equation can be expressed as follows:

$$\begin{cases} v_p = -\frac{K_{f,m}}{\mu_p}\nabla P_p & \text{Linear flow} \\ v_p = -\frac{K_{f,m}}{\mu_p}\left(\nabla P_p - G_{pc}\right) \; G_{pb} \le \left|\nabla P_p\right| \le G_{pd} & \text{Pseudolinear flow} \end{cases} \tag{2.6}$$

where, $K_{f,m}$ is the micro-fracture or reservoir matrix permeability, mD; f is a micro-fracture and m is the reservoir matrix; G_{pc} is a threshold pressure gradient, MPa/m; G_{pb} is a pseudolinear critical pressure gradient, MPa/m.

2.2.2.3 Low-Speed Nonlinear Flow

When the pressure gradient is less than the threshold pressure gradient, the fluid cannot balance the additional resistance created by the interface adsorption layer or hydration film and it cannot flow. When the pressure gradient is higher

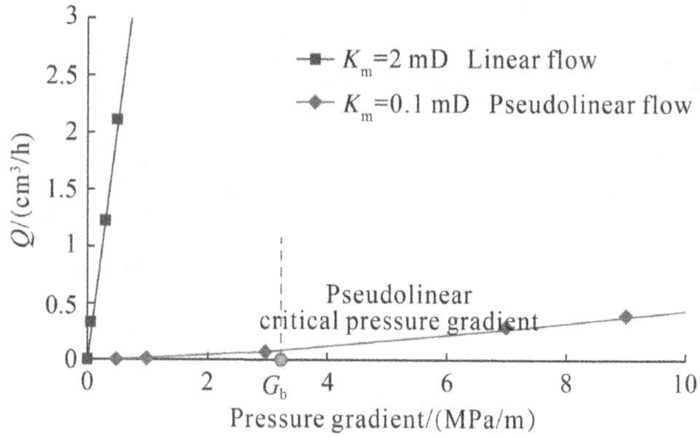

Fig. 2.12 Characteristics for the pseudolinear fluid flow behavior in tight oil reservoirs

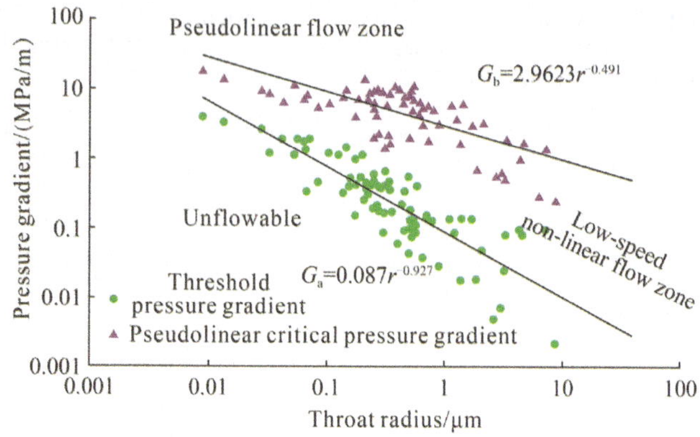

Fig. 2.14 Plate for the limitation of the low-speed non-linear flow behavior in tight oil reservoirs

than the threshold pressure gradient and smaller than the pseudolinear critical pressure gradient, the fluid velocity is low. The velocity and pressure gradient deviate from the linear relationship, and it shows the characteristics of low-speed nonlinear flow (Fig. 2.13).

The experimental results show that the fluid flow is significantly affected by the pore structure. Under the low displacement pressure gradient, the fluid cannot flow when the pore throat radius is small. When the pressure gradient is higher than the threshold pressure gradient, the fluids in larger throats will start flow first and then the fluids in smaller throats. As the pressure gradient increases, the fluid regimes in larger throats will convert first (Fig. 2.14). The threshold pressure gradient, the pseudolinear critical pressure gradient and the throat radius show a power function relationship. It can be expressed by

$$G_a = 0.087 \cdot r^{-0.927} \tag{2.7}$$

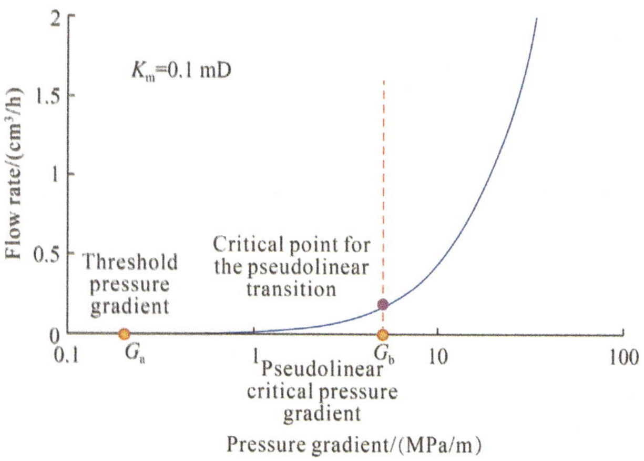

Fig. 2.13 Curves for the low-speed non-linear flow behavior in tight oil reservoirs

$$G_b = 2.9623 \cdot r^{-0.491} \tag{2.8}$$

where, G_a is the threshold pressure gradient, MPa/m; G_b is the pseudolinear critical pressure gradient, MPa/m; r is the pore throat radius, μm.

1. Characteristics and mechanisms of low-speed nonlinear flow

In tight sandstone reservoirs, there are threshold pressure gradients in both nano/micro-fractures and reservoir matrix pores regardless of oil-water two-phase flow or single oil phase flow. Only when the pressure gradient is higher than the threshold pressure gradient, the fluid can balance the resistance and start to flow. Fluid flow shows the low-speed nonlinear flow. The kinetics equation is

$$\begin{cases} \mathbf{v}_o = 0 & |\nabla P_o| < G_{oa} \\ \mathbf{v}_o = -\frac{K_{m,f}}{\mu_o}(\nabla P_o - G_{oa})^n & G_{ob} \geq |\nabla P_o| \geq G_{oa} \end{cases} \tag{2.9}$$

where, $K_{m,f}$ is the permeability of reservoir matrix or micro-fractures and nano-fractures, mD; m represents the reservoir matrix, f represents micro-fractures or nano-fractures. G_a is the threshold pressure gradient, MPa/m. o represents the oil phase. The linear flow coefficient is related to the pore throat structure and the fluid properties of reservoir.

The threshold pressure gradient is related to the reservoir physical properties, pore structure and water saturation. With the permeability decreases, the threshold pressure gradient increases. And the smaller the permeability is, the higher the threshold pressure gradient is. The tighter the reservoir is, the smaller the pore throat is, the higher the water saturation and the threshold pressure gradient are.

Experimental results show that the pore system of tight rocks is composed of small/nano-pores, and the fluid cannot flow until it balances the threshold pressure gradient. As the pressure gradient increases, the fluid in more pores will start

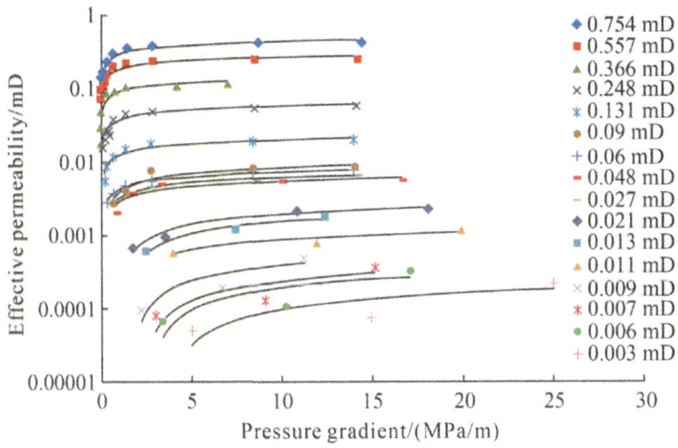

Fig. 2.15 Plate for the effective permeability versus displacement pressure gradient in tight oil reservoirs

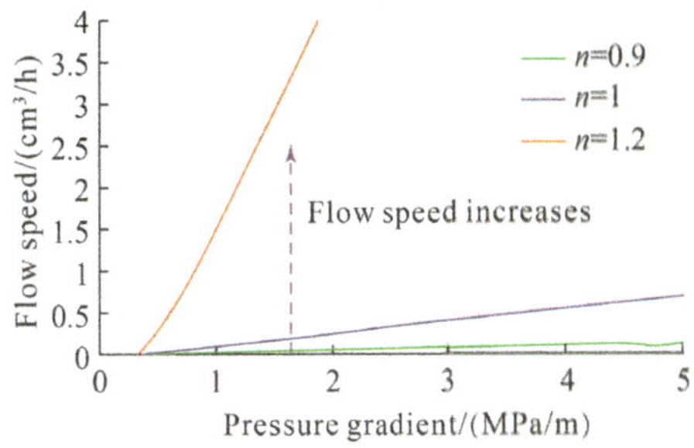

Fig. 2.16 Effect of non-linear flow index on the low-speed nonlinear flow behavior

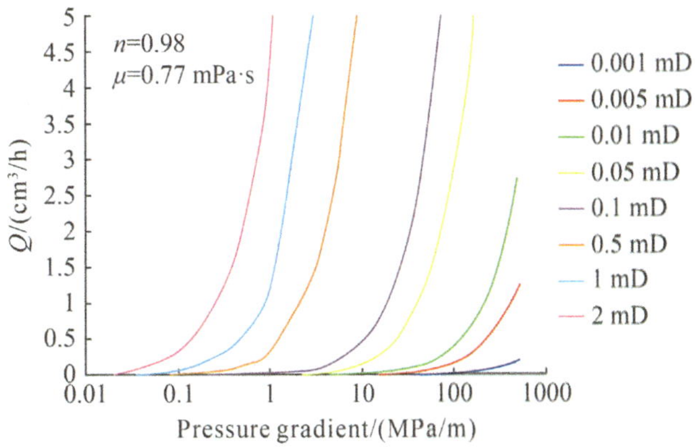

Fig. 2.17 Plate for the non-linear flow behavior in the media with different permeabilities

to flow and the core permeability becomes higher. Therefore, the core effective permeability is increasing (Fig. 2.15). The effective permeability of tight rock is

$$K_m = A_m \cdot \ln|\nabla P| + B_m \quad (2.10)$$

where, A_m and B_m are the coefficients, $A_m = 0.0326 \cdot (K_{m\infty}/\mu_o)^{1.0942}$, $B_m = 0.3436 \cdot (K_{m\infty}/\mu_o)^{1.8427}$, K_m is the effective permeability of reservoir matrix, mD; $K_{m\infty}$ is the gas permeability of reservoir matrix, mD. ∇P is the pressure gradient, MPa/m; μ_o is the viscosity of the crude oil, mPa·s.

The threshold pressure gradients for the porous media with different proportions at different scales are different. The surface effect caused by the fluid flow is also different. It results in the different values of the nonlinear flow index *n*. The greater the value of *n is*, the weaker the surface effect is. Therefore, the fluid will flow easily. As the value of *n* reduces, the surface effect is enhanced. Thus, the fluid will flow difficultly. Experimental results show that the value of *n* is between 0.9 and 1.2. Under the same pressure gradient, the fluid velocity increases with the increase of *n* in the low-speed nonlinear flow stage (Fig. 2.16).

For the tight rock with different permeabilities, as the nonlinear flow index *n* is the same, with the permeability decreases, the threshold pressure gradient increases, the low-speed nonlinear flow extends and the curve moves towards to right (Fig. 2.17). For the rock with a same permeability, the nonlinear flow index *n* is the same. As the fluid viscosity increases, the surface effect of the fluid is increased. With the initial pressure gradient increases, the fluid velocity decreases and the low-speed nonlinear flow lengthens (Fig. 2.18).

2. Characteristics and mechanisms of low-speed non-linear flow in tight gas reservoirs

For tight gas reservoirs, the gas-water two-phase flow exists in the nano/micro-scale reservoir matrix pores of water-contained tight sandstone gas reservoirs. The flow behavior of gas is significantly affected by water. It shows a threshold pressure gradient. Therefore, the low-speed nonlinear flow mechanisms of the water-contained tight sandstone gas reservoirs include threshold pressure gradient, slippage effect and diffusion effect. For tight sandstone gas reservoirs, the gas flow in the nano/micro-scale reservoir matrix pores is a single-phase gas flow, and there is no threshold pressure gradient. In the case of low permeability and low pressure, only the slippage and diffusion effects work (Fig. 2.19).

The experimental results (Zhou et al. 2003; Liu et al. 2008) show that the gas flow in the water-tight sandstone core shows the characteristics of transition composite flow behavior. Extending the initial section of the flow curve to the horizontal axis, it shows the threshold pressure gradient

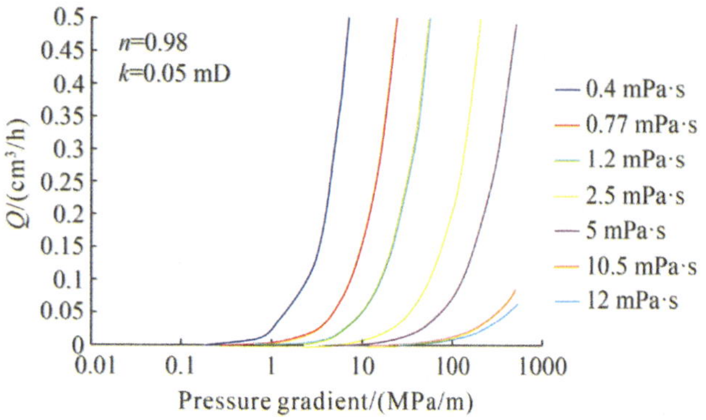

Fig. 2.18 Plate for non-linear flow behavior in a same media with different oil viscosities

of gas flow. The flow is mainly affected by the threshold pressure gradient. With the increase of pressure gradient, the gas flow velocity increases, the flow curve extends to the longitudinal axis, and there is an "initial flow velocity". Gas flow is mainly affected by the slippage effect (when the pressure is low, it is affected by the diffusion effect). The flow curve shows a turning point, and the turning point of the curve is the inflection point of gas flow (Fig. 2.19a). As the pressure gradient further increases, the gas flow behaves as a pseudolinear flow. The gas flow curve of sandstone core is convex type, and the extension of curve has no intersection with the horizontal axis. It represents the irrespective of the influence of threshold pressure gradient. In the low-speed nonlinear flow, when the pressure is low (the pressure gradient is very low), the gas mainly appears as the diffusion movement of gas molecules, which means the diffusion effect dominates. As the pressure gradient increases, the gas transforms from molecular diffusion to macroscopic molecular flow, which is the slippage effect dominates (Fig. 2.19b).

The performance characteristic equation of low-speed nonlinear flow considers the threshold pressure gradient, diffusion effect and slippage effect in water-contained sandstone tight gas reservoirs is as follows (Cui et al. 2009; Darabi et al. 2012):

$$\begin{cases} v_g = 0 & |\nabla P_g| < G_{ga} \\ v_g = -\dfrac{K_{m,f}}{\mu_g}(\nabla P_g - G_{ga})^n + \dfrac{K_{m,f}}{\mu_g}\left(\dfrac{b}{\bar{P}_g} + \dfrac{32\sqrt{2RT}\mu_g}{3r\sqrt{\pi M}\bar{P}_g}\right)\nabla P_g & G_{gb} \ge |\nabla P_g| \ge G_{ga} \end{cases} \tag{2.11}$$

The performance characteristic equation of low-speed nonlinear flow considers the diffusion effect and slippage effect in no-water-contained sandstone tight gas reservoirs is as follows:

$$v_g = -\frac{K_{m,f}}{\mu_g}\left(1 + \frac{b}{\bar{P}_g} + \frac{32\sqrt{2RT}\mu_g}{3r\sqrt{\pi M}\bar{P}_g}\right)\nabla P_g \qquad |\nabla p_g| \le G_{gb} \tag{2.12}$$

where, v_g represents the gas flow rate, m/s; $K_{m,f}$ represents the permeability of the reservoir matrix or nano-fractures, mD; $\bar{P}_g$ represents the average formation pressure, MPa; b is the slip factor, MPa. T is the absolute temperature, K; r is the throat radius, μm; M is the gas molar mass, g/mol; R is the gas molar constant, J/mol·K; μ_g is the gas viscosity, mPa·s.

The results of the threshold pressure gradient test show that there is a clear negative correlation between the threshold pressure gradient and the core permeability, $G_{ga} = a \cdot K^{-b_w}$ (Fig. 2.20). Coefficient a and water saturation S_w is an exponential function; coefficient b_w and water saturation is a linear relationship (Fig. 2.21). Based on the analysis of the threshold pressure gradients of 18 tight sandstone cores, the relationship between the threshold pressure gradient and

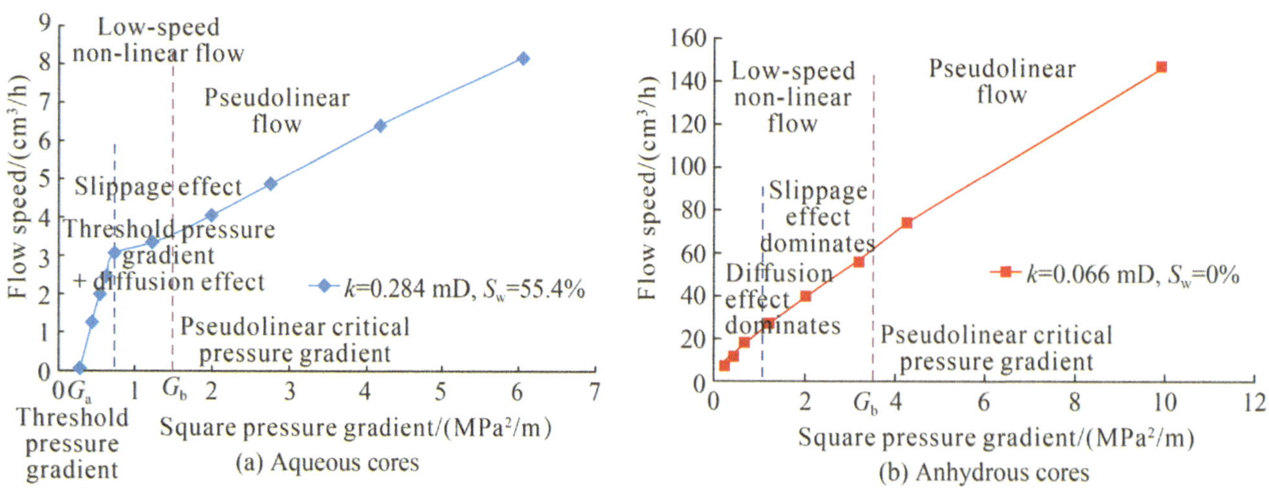

Fig. 2.19 Characteristics of low-speed non-linear flow in tight gas cores

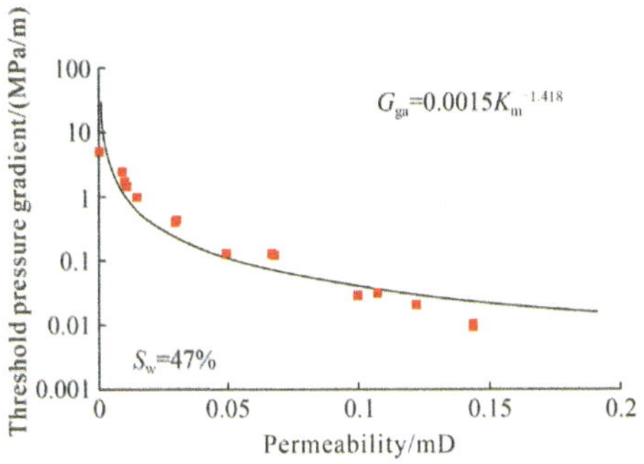

Fig. 2.20 Relationship between gas threshold pressure gradient and permeability

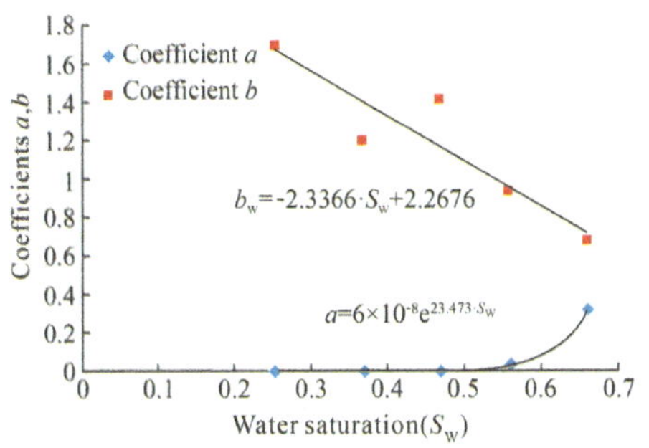

Fig. 2.21 Relationship between the coefficients a, b_w and water saturation

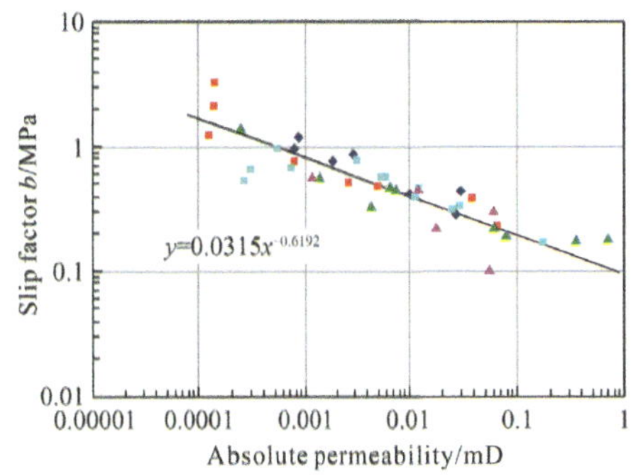

Fig. 2.22 Relationship between slip factor b and permeability $K_{m\infty}$

the permeability and water saturation of tight sandstone gas reservoir is obtained as follows:

$$G_{ga} = a \cdot K^{-b_w} = 6 \times 10^{-8} e^{23.473 S_w} \cdot K_m^{(2.3366 S_w - 2.2676)} \tag{2.13}$$

The experimental data of the slippage tests of 25 tight sandstone cores are analyzed. The results show that there is a clear negative correlation between the permeability and the slippage factor. The lower the permeability is, the greater the slip factor is and the stronger the slippage effect is (Fig. 2.22). At the same time, the smaller the pore throat radius is, the larger slip factor is and the greater the slippage effect is. Under the same permeability, as the proportion of small throats increases, the slippage effect is enhanced (Civan 2010).

$$b = 0.0315\, K_{m\infty}^{-0.6192} \tag{2.14}$$

2.3 Mechanisms of Displacement Processes

The pore-fracture media at different scales can be observed in tight oil and gas reservoirs. The displacement mechanisms of different media are different. In general, for tight oil and gas reservoirs, they include the conventional mechanisms and their own special mechanisms (Tables 2.4 and 2.5). Conventional mechanisms include viscous flow, compaction, fluid expansion and gravity. The special mechanisms for tight oil reservoirs include imbibition and dissolved gas drive. The special displacement mechanisms for tight gas reservoirs include slippage, diffusion and desorption.

During the recovery process of tight oil and gas reservoirs, the main displacement mechanisms of different tight oil and gas reservoirs are different in different stages. In the same stage, the displacement mechanism is also different at the different distances from the wellbore (Figs. 2.23 and 2.24). Viscous flow and gravity displacement work in the whole recovery process of tight oil and gas reservoirs. The viscous flow and compaction are strong in the early stage. As the recovery process continues, the viscous flow and compaction are gradually weakened. The pressure drops quickly in the early stage, and the effects of compaction and fluid elastic expansion are strong. In the latter stage, the pressure drop is small, and the effects of compaction and fluid elastic expansion are reduced. For tight oil reservoirs, when the pressure drops below the bubble point pressure, the effect of dissolved gas flooding is gradually increased. At the same time, as the formation pressure decreases, the imbibition effect increases. For tight gas reservoirs, the effect of gas desorption works as the formation pressure decreases to the desorption pressure. As the formation pressure further reduces, the slippage effect and gas diffusion are enhanced.

2.3.1 Conventional Displacement Mechanisms of Tight Oil and Gas Reservoirs

2.3.1.1 Viscous Flow

The viscous flow in tight oil and gas reservoirs refers to the macroscopic continuous flow in porous media under the effect of pressure difference and viscous force (Pan 2012). The viscous flow in tight oil and gas reservoirs can be

Table 2.4 Comparison on the displacement mechanisms for tight oil and gas reservoirs

Conventional mechanisms (early stage of production)	Special mechanisms of tight oil (later stage of production)	Special mechanisms of tight gas (later stage of production)
Viscous flow	Dissolved gas drive	Desorption
Compaction effect	Imbibition	Slipping
Fluid expansion		Diffusion
Gravity		

Table 2.5 The mechanisms of tight oil and gas reservoirs

Oil displacement mechanism	Mechanism	Basic equation	Schematic diagram
Viscous flow	Under the effect of pressure difference, the fluid is affected by the viscous force and flows in a macroscopic and continuous manner in the porous media	$v_{\text{viscous,p}} = -\frac{K\cdot K_{rp}}{\mu_p}\nabla\left(P_i - P_j\right)_p$ $p = o, g, w$	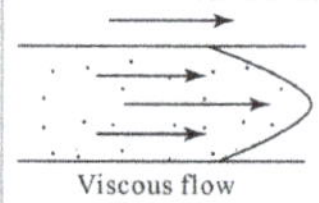
Gravity drive	Gravity forces the fluid in the pores to flow outward	$v_{\text{gravity,p}} = -\frac{K\cdot K_{rp}}{\mu_p}\cdot\nabla\left[\rho_p\cdot g\cdot\left(D_i - D_j\right)\right]$ $p = o, g, w$	
Compaction effect	As the production proceeds, the fluid is recovered, the pore pressure of the reservoir decreases, the effective stress increases, and the pores and fractures compact, deform and shrink, resulting in the reduction of the pore volume and the outflow of fluids in the rock pores	$C_R = -\frac{1}{V}\left(\frac{\partial V}{\partial P}\right)_{\text{compaction}} = \frac{1}{\phi}\left(\frac{\partial\varphi}{\partial P}\right)$ $\phi = \phi_0\cdot\left[1 + C_R(P - P_0)\right]$	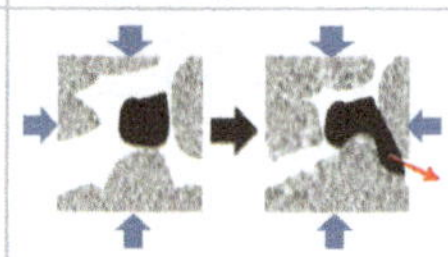
Fluid elastic expansion	With the recovery of oil and gas, the pore pressure continuously decreases. Due to the influence of the fluid elasticity, the crude oil (natural gas) in the pores expands in volume to drive out the fluid in the pores	$C_f = -\frac{1}{V}\left(\frac{\partial V}{\partial P}\right)_{\text{fluid expansion}} = \frac{1}{\rho}\left(\frac{\partial\rho}{\partial P}\right)$ $\rho_p = \frac{\rho_0}{B_p} = \rho_0\cdot\left[1 + C_f(P - P_0)\right]$ $p = o, g, w$	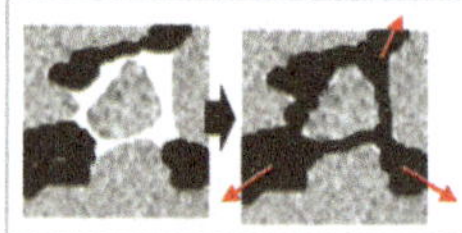
Dissolved gas flooding (tight oil)	When the formation pressure drops below the saturation pressure of the crude oil, the dissolved gas in the crude oil gradually separates out as the pressure decreases, and the elastic expansion of the dissolved gas drives the crude oil in the pores to be exhausted	$\rho_{gd} = \frac{\rho_{gsc}\cdot R_{g,o}(P)}{B_o}$	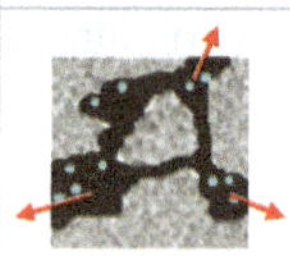
Imbibition (Tight oil)	Under tight oil volume fracturing, due to the capillary force, water imbibes into the pores of the reservoir matrix, and the water imbibed into the reservoir matrix replaces the oil to seep into the fracture, displacing the oil in the fracture by the pressure difference	$v_{\text{imbibition}} = -\frac{K\cdot K_{rp}}{\mu_p}\left(P_{\text{cow},i} - P_{\text{cow},j}\right)_p$ $p = o, w$	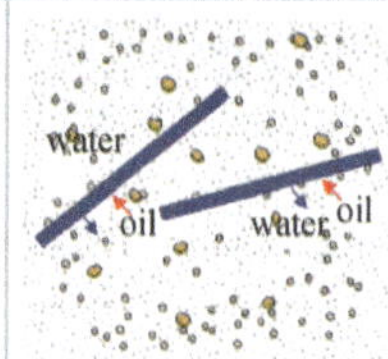
Slippage effect (Tight gas)	Under the conditions of low porosity, low permeability and low pressure, the flow velocity of natural gas on the pore wall surface is non-zero, which increases the fluid flow in the pore channel and is discharged from the pore under the pressure gradient	$v_{\text{slippage,g}} = -\frac{K}{\mu_g}\frac{b}{P_g}\nabla P_g$	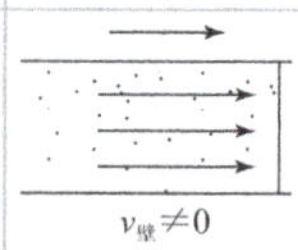
Diffusion (Tight Gas)	Under the conditions of low porosity, low permeability and low pressure, the chances of collisions between gas molecules and pore-wall molecules are greatly increased and non-directional movement occurs. The gas is discharged from the nano-scale pores under the pressure gradient	$v_{\text{diffusion,g}} = -\frac{K}{\mu_g}\frac{32\sqrt{2RT}\mu_g}{3r\sqrt{\pi M P_g}}\nabla P_g$	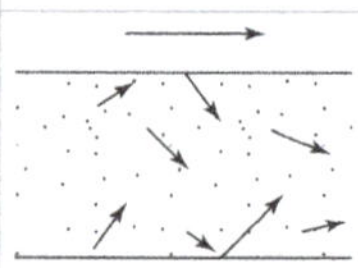
Desorption (Tight Gas)	With the decrease of formation pressure, when the formation pressure drops below the critical desorption pressure of the adsorbed gas, the adsorbed gas is converted from the adsorbed state to the free state and discharged from the pores	$P_{cd} = \frac{V_a}{b_L(V_L - V_a)}$	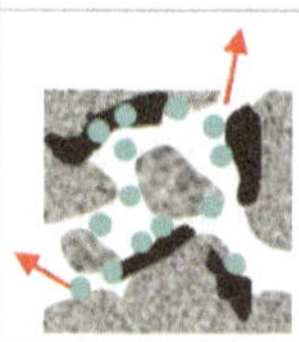

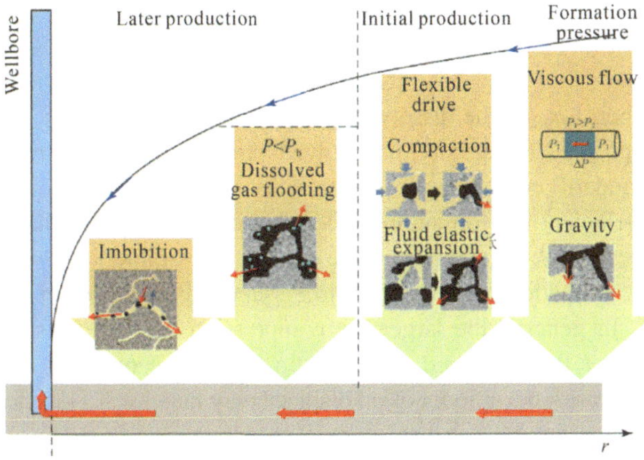

Fig. 2.23 Diagram for the mechanisms of tight oil reservoirs

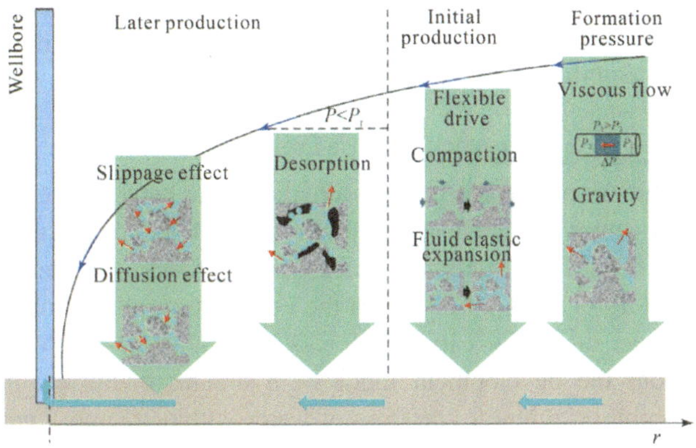

Fig. 2.24 Diagram for the mechanisms of tight gas reservoirs

classified into four flow states: high-speed nonlinear flow, linear flow, pseudolinear flow and low-speed nonlinear flow. Due to the different behavior of fluid flow in each flow state, the kinetic equation can be characterized by

$$\begin{cases} -\nabla P_p = \frac{\mu_p v_p}{K_{f,m}} + \beta_{f,m}\rho_p v_p |v_p| & |\nabla P_p| \geq G_{pd} & \text{High speed nonlinear flow} \\ v_p = -\frac{K_{f,m}}{\mu_p}\nabla P_p & & \text{Linear flow} \\ v_p = -\frac{K_{f,m}}{\mu_{o,g}}(\nabla P_p - G_{pc}) & G_{pb} \leq |\nabla P_p| \leq G_{pd} & \text{Pseudolinear flow} \\ v_p = -\frac{K_{m,f}}{\mu_{o,g}}(\nabla P_p - G_{pa})^n & G_{pb} \geq |\nabla P_p| \geq G_{pa} & \text{Low speed nonlinear flow} \\ v_p = 0 & |\nabla P_p| < G_{pa} & \text{Non flow} \end{cases} \quad (2.15)$$

2.3.1.2 Compaction

Compaction effect is that as the recovery process continues, the reservoir fluids are produced, and the reservoir pressure is decreased. Thus, the effective stress is increased, and the pores and fractures compact, deform and shrink. It results in the reduction of pore volume and drives the fluid out of pores (He 1994). Due to rock compaction, a large fraction of fluids in the pores is displaced out. The basic equations for rock compaction and displacement are as follows:

$$C_R = -\frac{1}{V}\left(\frac{\partial V}{\partial P}\right)_{\text{compassion}} = \frac{1}{\phi}\left(\frac{\partial \phi}{\partial P}\right) \quad (2.16)$$

$$\phi = \phi_0 \cdot [1 + C_R(P - P_0)] \quad (2.17)$$

where, C_R is the compressibility of rock pore volume, 1/MPa; V is the volume of rock pores, m^3; P_0 indicates the original formation pressure of reservoir, MPa; P is the current formation pressure, MPa; ϕ_0 is the porosity of reservoir under the original formation pressure, dimensionless; φ is the porosity under the current formation pressure, dimensionless.

2.3.1.3 Fluid Elastic Expansion

Fluid elastic expansion is that with the extraction of fluid in tight oil and gas reservoirs, the pore pressure decreases continuously. Due to the elastic effect of fluid, the crude oil (natural gas) in pores expands and drives the fluid in pores to be exhausted outward (He 1994; Xiao 2015). The basic equation of elastic fluid expansion is as follows:

$$C_f = -\frac{1}{V}\left(\frac{\partial V}{\partial P}\right)_{\text{fluidexpansion}} = \frac{1}{\rho}\left(\frac{\partial \rho}{\partial P}\right) \quad (2.18)$$

$$\rho_p = \frac{\rho_0}{B_p} = \rho_0 \cdot [1 + C_f(P - P_0)] \quad (2.19)$$

where, C_f is the fluid volumetric compressibility, MP_a^{-1}; ρ_0 is the fluid density under the original formation pressure, g/cm^3; ρ_p is the fluid density under the current formation pressure, g/cm^3. B_p is the fluid volume coefficient at the current formation pressure P, m^3/m^3; p = o, g, w, respectively, represent the oil, gas and water.

The reservoir elastic energy can be obtained by multiplying the reservoir total compressibility and the difference between formation pressure and saturation pressure. When the reservoir total compressibility remains the same, the greater the difference between formation pressure and saturation pressure is, the greater the elastic energy of reservoir is. The reservoir total compressibility is the sums of rock compressibility, effective oil compressibility and effective irreducible water compressibility. The elastic productivity can be obtained by multiplying the elastic energy and the geological oil reserves. It indicates the total oil production volume under the effect of formation elasticity.

2.3.1.4 Gravity Drainage

The fluids in tight oil and gas reservoirs can flow within different media under the effect of gravity, and this kind of mechanism is called as gravity drainage. Gravity drainage includes two forms. The first one is the fluid flow behavior in different media under the effect of oil gravity. The other is the changes of fluid flow behavior and spatial distribution caused by the density differences of oil and gas (or oil and water). It forces the fluid flow behavior in different media.

Considering the effect of gravity drainage, the fluid kinetic equation can be characterized by

$$\boldsymbol{v}_{\mathrm{g,p}} = -\frac{K \cdot K_{\mathrm{rp}}}{\mu_{\mathrm{p}}} \cdot \nabla\left[\rho_{\mathrm{p}} \cdot g \cdot \left(D_i - D_j\right)\right] \quad (2.20)$$

where, $\boldsymbol{v}_{\mathrm{g,p}}$ is the gravity flow velocity of an oil-gas-water system, m/s; p = o, g, w, respectively, represents the oil, gas, water three-phase; K is the reservoir permeability, mD; K_{rp} is the relative permeability of oil, gas and water, dimensionless; g is the gravity acceleration, 9.8 m/s^2; D is the depth from a reference plane, that is, the direction of gravity coordinates, m; ρ_{p} is the density of oil, gas and water, g/cm^3; μ_{p} is the viscosity of oil, gas and water, mPa·s.

2.3.2 Special Mechanisms of Tight Oil Reservoirs

2.3.2.1 Dissolved Gas Flooding

Dissolved gas flooding is when the local reservoir pressure drops below the saturation pressure, the dissolved gas in crude oil gradually separates out with the pressure drops, and the elastic expansion of dissolved gas can displace the crude oil in pores. The dissolved gas flooding is based on the natural energy. The amount of oil production volume depends on the factors such as formation rock compressibility, crude oil compressibility, irreducible water compressibility, dissolved gas-oil ratio and pressure drop. The basic equation of dissolved gas flooding is

$$\rho_{\mathrm{gd}} = \frac{\rho_{\mathrm{gsc}} \cdot R_{\mathrm{g,o}}(P)}{B_{\mathrm{o}}} \quad (2.21)$$

where, ρ_{gd} is the dissolved gas drive gas density, g/cm^3.ρ_{gsc} represents the density of dissolved gas drive gas, g/cm^3 under standard conditions; B_{o} represents the crude oil volume coefficient of formation under the condition of formation pressure P; $R_{\mathrm{g,o}}$ represents the dissolved gas-oil ratio at which the formation pressure drops to P (standard condition) m^3/m^3.

In general, the larger the compression coefficient is, the higher the dissolved gas-oil ratio is. As the pressure drop increases, the amount of oil displacement increases. The effect of pressure drop on displacement process is manifested on two aspects. On one hand, more formation energy can be released to produce more crude oil. On the other hand, it will lead to the serious degassing performance around the well bottom. During production, gas firstly flows into the well bottom, resulting in gas interporosity flow. Thus, the oil production rate declines, and the formation energy is consumed out rapidly.

2.3.2.2 Imbibition

Capillary pressure is the pressure difference between wetting and non-wetting phases on both sides of the meniscus in capillary tube. Imbibition refers to the process by which a wetting phase fluid displaces the non-wetting phase fluid in porous media by capillary forces. For tight oil reservoirs, under the development mode of hydraulic fracturing, the imbibition refers to the flow of water within fracture under the pressure gradient. At the same time, due to the capillary force, the water imbibes into reservoir matrix, and the water imbibed into the reservoir matrix will replace the crude oil to flow into fractures and then displace the crude oil in fractures (Fig. 2.25a) (Wang et al. 2007; Zhang 2013).

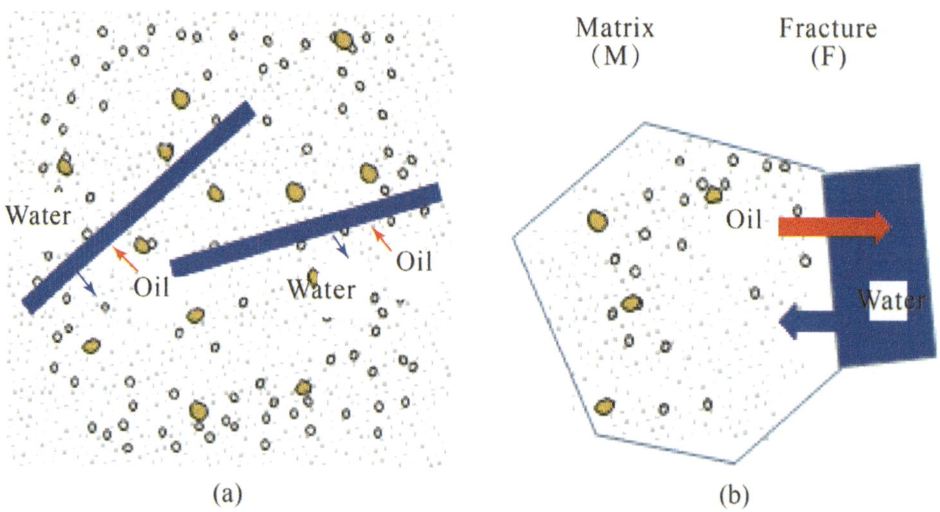

Fig. 2.25 Schematic diagram of oil and water imbibition behavior between different media in tight oil reservoirs

The basic equation of tight oil imbibition is

$$v_{\text{im}} = -\frac{K \cdot K_{\text{rp}}}{\mu_{\text{p}}} \left(P_{\text{cow},i} - P_{\text{cow},j}\right)_{\text{p}} \quad (2.22)$$

where, v_{im} represents the imbibition velocity, m/s; P_{cow} represents the capillary force of an oil-water system, MPa; p = o, w is the oil phase or water phase; μ_{p} refers to the viscosity of oil phase or aqueous phase, mPa·s.

In the same porous media (water wet media), due to the capillary pressure difference between oil and water phases, it will lead to the wetting phase (water phase) imbibes into the pores, so that the pressure inside the media increases. Under the pressure difference, the non-wetting phase (oil phase) is driven out. Due to the difference between imbibition and displacement processes and mechanisms, the corresponding capillary pressure curves differ (Fig. 2.26).

In different media (fracture media and porous media (Fig. 2.25b, large pore media and small pore media)), due to the capillary pressure difference ($P_{\text{cow}} = 0$ for fracture media, $P_{\text{cow}} \neq 0$ and $\Delta P_{\text{cow}} \neq 0$ for porous media. $P_{\text{cow,large}} < P_{\text{cow,small}}$ and $P_{\text{cow}} \neq 0$ for large pore media and small pore media). The water phase is flow from the media with low capillary force (fracture media or large pore media) to the media with high capillary force (porous media or small pore media). It results in the increase of pressure within media. Under the pressure difference, the oil phase is displaced from porous media or small pore media. Due to the difference of media and the different mechanisms of imbibition and displacement, there is a difference in the capillary force curve (Fig. 2.27) and the oil-water relative permeability curve between different media (Figs. 2.28 and 2.29).

Due to the combined effect of pressure-difference drive and capillary force, the effect of imbibition displacement can perform differently in different recovery methods, depletion development, water flooding, periodic water injection and continuous waterflooding etc. Therefore, under these different displacement methods, the impact of imbibition displacement on production performance should be considered.

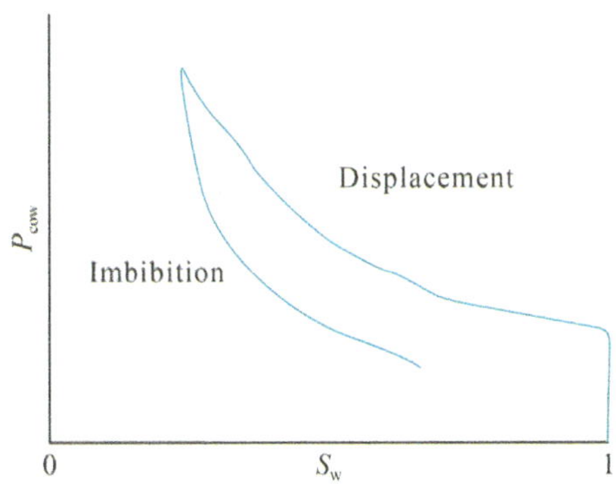

Fig. 2.26 Oil-water displacement curve and imbibition curve of the same media

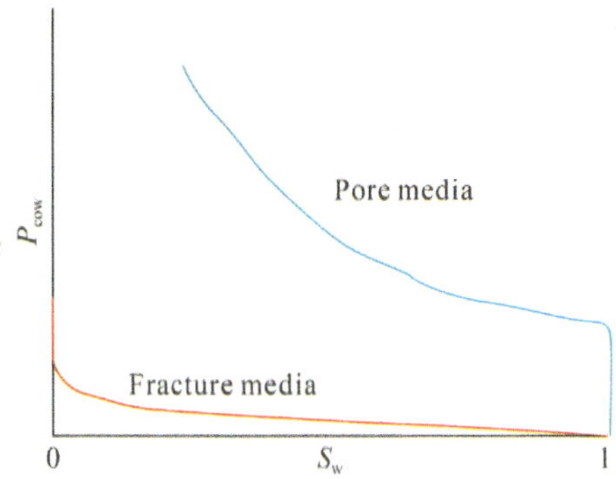

Fig. 2.27 Comparison on the capillary pressure curves between different media

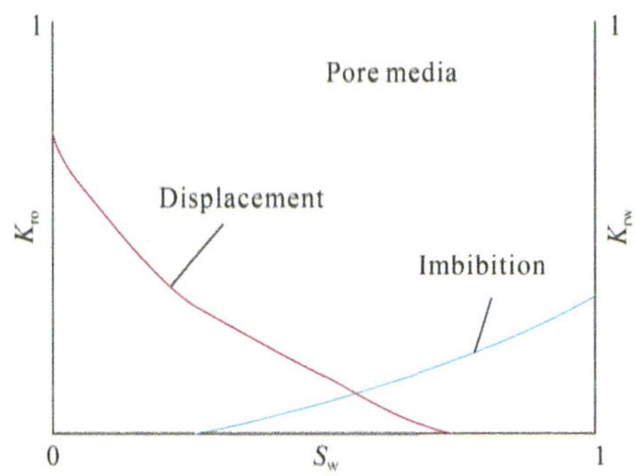

Fig. 2.28 The oil-water relative permeability cure of imbibition behavior and displacement process in porous media

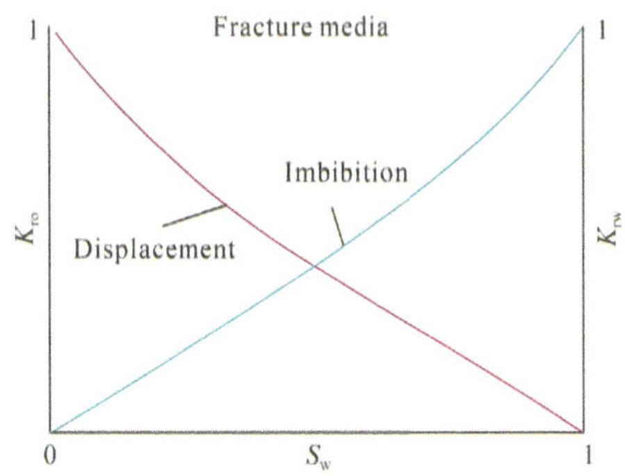

Fig. 2.29 The oil-water relative permeability cure of imbibition behavior and displacement process in fracture media

2.3.3 Special Displacement Mechanisms of Tight Gas Reservoirs

2.3.3.1 Slippage Effect

The pores and fractures at different scales can be observed in tight sandstone gas reservoirs, these pores are connected through pore throats with various shapes, and the pore-throat configuration relationship is very complicated. Under the conditions of low porosity, low permeability and low pressure, the flow velocity of natural gas in the pore wall surface is not zero, which increases the fluid flow capacity in the pore throat. The gas can be discharged from the pores under the effect of pressure gradient. This phenomenon is called as slippage effect (Liu et al. 2006; Yao et al. 2009).

According to Klinkenberg's theory for slippage, the equation of performance characteristic considering the slippage effect is obtained.

$$v_{\mathrm{sl,g}} = -\frac{K_{\mathrm{m}\infty}}{\mu_{\mathrm{g}}}\frac{b}{\bar{P}_{\mathrm{g}}}\nabla P_{\mathrm{g}} \tag{2.23}$$

where, $v_{\mathrm{sl,g}}$ is the gas flow velocity under the effect of slippage, m/s; $K_{\mathrm{m}\infty}$ represents the gas permeability of the reservoir matrix, mD; $\bar{P}_{\mathrm{g}}$ represents the average pore pressure, MPa; b is the slippage factor, MPa.

The equation considering the combined effect of Darcy flow and the slippage effect is

$$v_{\mathrm{g}} = -\frac{K}{\mu_{\mathrm{g}}}\left(1+\frac{b}{\bar{P}_{\mathrm{g}}}\right)\nabla P_{\mathrm{g}} \tag{2.24}$$

The slippage factor b reflects the strength of slippage effect for gas flow, which is related to the rock pore structure, gas properties and the average pore pressure (Wang 2016; Firouzi et al. 2014a).

$$b = \frac{4c\bar{\lambda}}{r}\bar{P}_{\mathrm{g}} = \frac{2\sqrt{2}c\kappa T}{\pi D_{\mathrm{g}}^2}\frac{1}{r} \tag{2.25}$$

where, c is a proportional constant which is approximately 1; r is the throat radius, μm; κ is the Boltzmann gas constant, 1.38×10^{-23} J/K; T is the absolute temperature, K; D_{g} is the molecular diameter of gas, μm.

2.3.3.2 Diffusion

Tight sandstone gas reservoir has an ultra-low water saturation, and gases in reservoir can exist in adsorbed state and free state. Under the original condition, the reservoir permeability is very low, the pore throat is small, and the corresponding capillary pressure is very high. Therefore, the pressure gradient needs to be increased so that the gases in reservoir can normally flow in pore. When the pore throat is small and the gas flows from pores to micro-fractures, a concentration difference will be formed. Due to the diffusion effect, gas flowability in reservoir matrix increases. The flowability under diffusion effect is even comparable compared with the effect of pressure drops. Therefore, the diffusion effect in the pores cannot be ignored.

Tight sandstone reservoirs develop nanoscale reservoir matrix pores with small throats. Under the conditions of low porosity, low permeability and low pressure, the collision probability between gas molecular and pore wall increases greatly, and thus the orientation movement occurs. The gas flowing behavior does not completely follow the Darcy's law, and it can be expelled out from the nano-scale pores under pressure gradient. This phenomenon is diffusion effect, that is, the Knudsen diffusion. The kinetic equation for Knudsen diffusion is (Yao et al. 2013; Wu et al. 2015, Firouzi et al. 2014b, Sheng et al. 2014):

$$v_{\mathrm{di,g}} = -\frac{K_{\mathrm{m}\infty}}{\mu_{\mathrm{g}}}\frac{32\sqrt{2RT}\mu_{\mathrm{g}}}{3r\sqrt{\pi M}\bar{P}_{\mathrm{g}}}\nabla P_{\mathrm{g}} \tag{2.26}$$

The kinetic equation considering the combined effects of Darcy flow and Knudsen diffusion is

$$v_{\mathrm{g}} = -\frac{K_{\mathrm{m}\infty}}{\mu_{\mathrm{g}}}\left(1+\frac{32\sqrt{2RT}\mu_{\mathrm{g}}}{3r\sqrt{\pi M}\bar{P}_{\mathrm{g}}}\right)\nabla P_{\mathrm{g}} \tag{2.27}$$

where, $v_{\mathrm{di,g}}$ represents the diffusion rate, m/s; ∇P_{g} is the gas pressure gradient, MPa/m; $\bar{P}_{\mathrm{g}}$ is the average formation pressure, MPa; T represents the absolute temperature, K; $r = d/2$, d is the throat diameter, μm; M represents the molar mass of gas, g/mol; R is the gas molar constant, J/mol·K; $K_{\mathrm{m}\infty}$ represents the permeability of the reservoir matrix gas measurement, mD; μ_{g} represents the gas viscosity, mPa·s.

2.3.3.3 Desorption

Tight sandstone gas reservoirs develop nano-scale reservoir matrix pores with small throats and large specific surface areas. The occurrence state of gas mainly includes free state and adsorbed state. During the development of tight gas reservoirs, once the formation pressure decreases below the critical desorption pressure of adsorbed gas, the adsorbed gas changes from adsorbed state to free state. This phenomenon is called gas desorption. According to the Langmuir isotherm equation, the kinetic equation of gas desorption is (Wang 2013)

$$q_{\mathrm{d}} = \frac{V_{\mathrm{L}}}{1+b_{\mathrm{L}}P} \tag{2.28}$$

where, q_{d} represents the desorption amount, $\mathrm{cm^3/g}$; V_{L} represents the maximum adsorption capacity in the desorption process, $\mathrm{cm^3/g}$; b_{L} is the desorption constant, $\mathrm{MPa^{-1}}$; P represents the formation pressure, MPa.

V_{L} indicates the maximum adsorption capacity of reservoir, and the physical meaning is the adsorption capacity of reservoir rocks when methane is saturated at a given temperature.

The desorption constant b_{L} is the function of adsorption rate, desorption rate and adsorption heat, the expression is

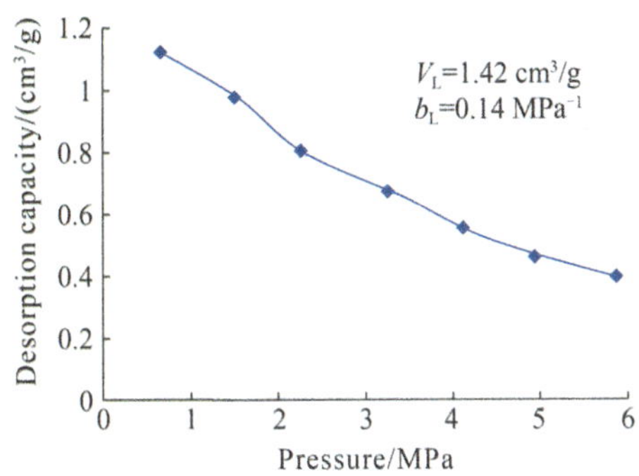

Fig. 2.30 The isothermal desorption curve of tight gas reservoirs

$$b_{\mathrm{L}} = b_0 \exp\left(\frac{Q}{RT}\right) \quad (2.29)$$

where, b_0 represents the ratio of the adsorption rate and desorption rate; Q represents the heat of adsorption, J; R is the gas constant, 8.314 J/(mol·K); T represents the formation temperature, K.

V_{L}, b_{L} can be accurately determined by isothermal adsorption/desorption experiments. Transform the desorption equation as the straight line type:

$$q_{\mathrm{d}} = V_{\mathrm{L}} - b_{\mathrm{L}} q_{\mathrm{d}} P \quad (2.30)$$

By testing the equilibrium amount of methane desorption of tight rocks under the same temperature and different pressure conditions, a straight line between q_{d} and P can be plotted (Fig. 2.30). Thus, from the intercept and slope of the straight line, the maximum adsorption capacity V_{L} and the desorption constant b_{L} can be obtained.

The critical desorption pressure is the pressure at which the gases on the nanopore surface starts to desorb. It is also the pressure at which the measured gas content corresponds to the adsorption isotherm (Xing et al. 2013), as shown in Fig. 2.31. The formula is

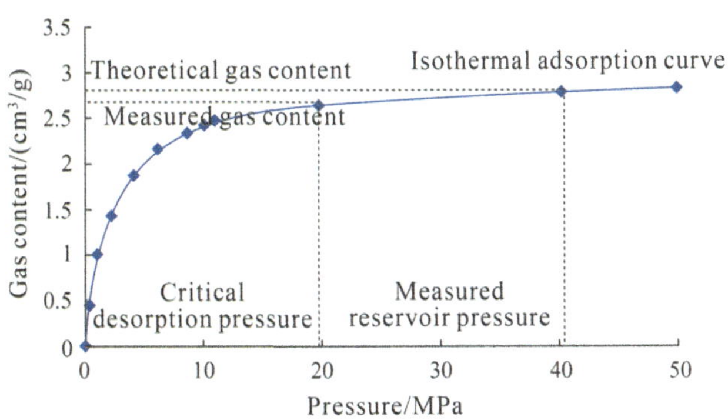

Fig. 2.31 Diagram for the desorption curve and adsorption curve of tight gas reservoirs

$$P_{\mathrm{cd}} = \frac{V_{\mathrm{a}}}{b_{\mathrm{L}}(V_{\mathrm{L}} - V_{\mathrm{a}})} \quad (2.31)$$

where, P_{cd} represents the critical desorption pressure, MPa; V_{a} represents the measured tight gas content, cm^3/g.

The desorption of gas on the rock surface of tight reservoir matrix particles is mainly affected by formation pressure, pore size of reservoir rock, gas saturation, gas composition and temperature.

2.4 Oil-Producing Capacity for the Porous Media at Different Scales

The oil-producing capacity of porous media at different scales depends on the geometry of porous media (pore radius and throat radius size) at different scales, physical parameters (porosity and permeability), fluid properties (viscosity, density, dissolved gas-oil ratio), flow regimes and mechanisms (Fig. 2.32). Under a given oil-producing capacity of the reservoir matrix, the size of matrix rock and the pressure drops affect the production capacity. The pressure in tight reservoirs is non-instantaneous, and a matrix mobilization radius increases with time. The variation law determines the producing range of tight oil matrix rocks.

2.4.1 Oil-Producing Capacity of Reservoir Matrix

The oil-producing capacity of reservoir matrix refers to the ability of fluid in reservoir matrix flow into fractures per unit area and unit pressure drop, reflecting the size of reservoir matrix supplement. According to different methods such as laboratory tests, actual production data and numerical simulation, the oil-producing capability of tight reservoir matrix

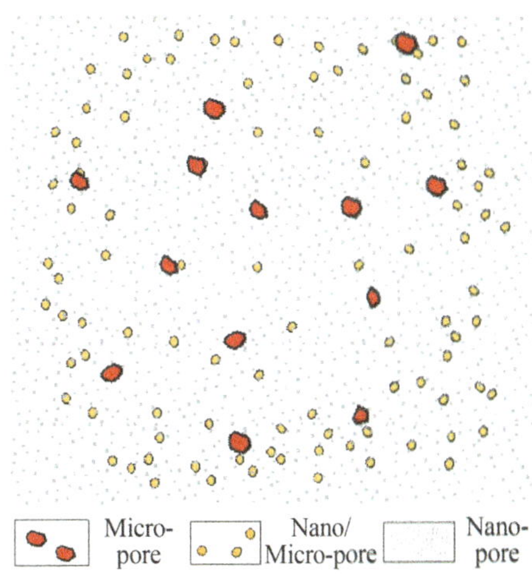

Fig. 2.32 Porous media at different scales in reservoir matrix

is analyzed synthetically and the sensitive factors and changing laws are studied.

2.4.1.1 Measured by Fluid Flow Experiments

Using the core samples from tight sandstone reservoirs in Lianggaoshan tight sandstone formation in Sichuan and Chang 7 tight sandstone formation in Changqing, the oil-producing capacity of different cores was calculated by the method of nonlinear fluid flow experiment. It is based on the experimentally measured flow velocities, the pressure drops across the core and the core cross-sectional area and length (Figs. 2.33 and 2.34).

From the experiment results, it can be seen that the oil-producing capacity of Lianggaoshan tight reservoir matrix and Chang 7 tight reservoir matrix are mainly distributed between 12×10^{-5} and 140×10^{-5} (m^3/d)/(MPa·m^2). At the same time, the oil-producing capacity increases with the increase of permeability.

2.4.1.2 Calculation of Oil-Producing Capacity of Reservoir Matrix by Using the Recovery Performance Data

According to the actual recovery performance data of several horizontal wells in Chang 7 tight oil reservoir in Changqing, the oil-producing capacity of reservoir matrix can be calculated according to the production area, pressure drops and the hydraulic fracture area from statistical analysis (Table 2.6).

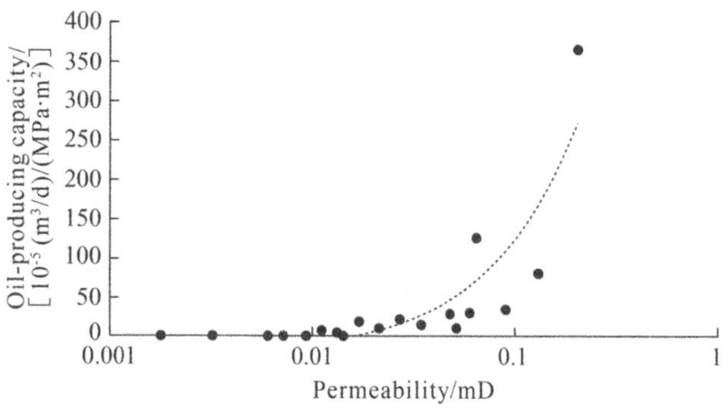

Fig. 2.33 The oil-producing capacity curve of sandstone reservoir matrix in Lianggaoshan, Sichuan

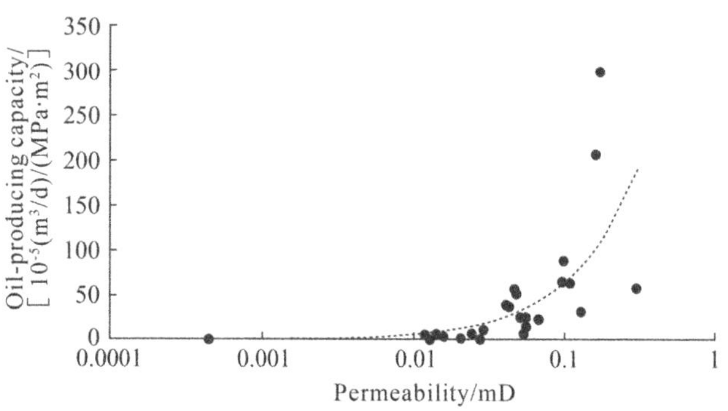

Fig. 2.34 The oil-producing capacity curve of sandstone reservoir matrix in Chang 7, Changqing

Table 2.6 Oil-producing capacity results of reservoir matrix in single well

Well	Production rate/(m^3/d)	Pressure drops/MPa	Fracture area/m^2	Oil-producing capacity of reservoir matrix/ [(m^3/d)/ (MPa·m^2)]
YPHW1	13	15	188064	5.2×10^{-5}
YPHW 6	13.4	4	265920	20×10^{-5}
YPHW 7	11.6	7.7	432000	4×10^{-5}
YPHW 8	14.7	9.4	487488	3.7×10^{-5}
YPHW 9	17.2	8.1	659520	3×10^{-5}

It can be seen from the results that the oil-producing capacity of Chang 7 tight oil reservoir matrix is between 3.0×10^{-5} and 20×10^{-5} (m^3/d)/(MPa·m^2). When the properties of tight reservoirs are relatively good, the oil-producing capacity of reservoir matrix can reach 20×10^{-5} (m^3/d)/(MPa·m^2). When the reservoir properties are relatively poor, the oil-producing capacity of reservoir matrix is only 3.0×10^{-5}(m^3/d)/(MPa·m^2).

2.4.1.3 Oil-Producing Capacity of Reservoir Matrix by Numerical Simulation Method

Using numerical simulation method, the physical properties (porosity, permeability and oil saturation) of reservoir matrix, fluid properties parameters (viscosity, oil-gas ratio) and the distribution of porous media at different scales are simulated, the influence on the oil-producing capacity of reservoir matrix is analyzed, and the changing law is also analyzed.

1. Effect of physical properties of reservoir matrix on the oil-producing capacity

(1) Permeability

Permeability is a parameter that characterizes the flowability of reservoir itself. It is related to the pore geometry, pore throat radius, pore quantity distribution and spatial distribution of pores. Permeability directly affects the oil-producing capacity of reservoir matrix.

By simulating the oil-producing capacity of reservoir matrix with different permeabilities (Fig. 2.35), permeability has a significant effect on the oil-producing capacity of reservoir matrix. With the permeability increases, the oil-producing capacity of reservoir matrix increases. When the permeability increases from 0.02 mD to 0.15 mD, the

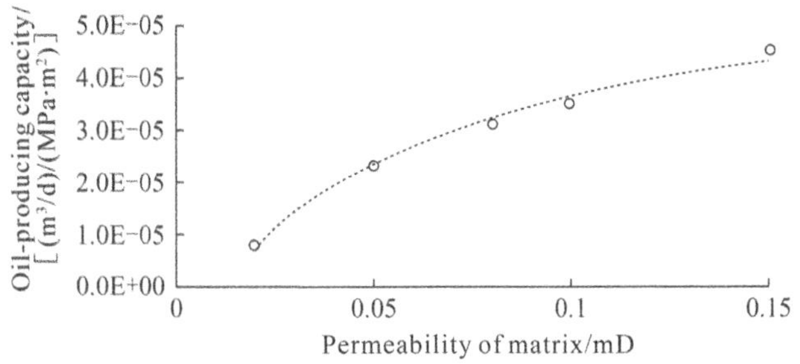

Fig. 2.35 Reservoir matrix permeability versus oil-producing capacity

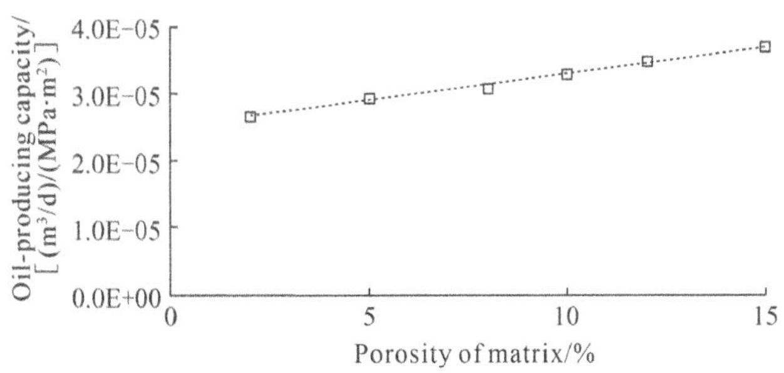

Fig. 2.36 Reservoir matrix porosity versus oil-producing capacity curve

oil-producing capacity of reservoir matrix increases from 0.84×10^{-5}(m^3/d)/(MPa·m^2) to 4.74×10^{-5}(m^3/d)/(MPa·m^2).

(2) Porosity

Porosity indicates the size of reservoir matrix capacity. it reflects the replenishment ability of reservoir matrix.

By simulating the oil-producing capacity of reservoir matrix at different porosities (Fig. 2.36), the porosity has an important influence on the oil-producing capacity of reservoir matrix. With the porosity increases, the oil-producing capacity of reservoir matrix also increases correspondingly and linearly. When the porosity increases from 2% to 10%, the oil-producing capacity of reservoir matrix increases from 2.63×10^{-5} (m^3/d)/(MPa·m^2) to 3.45×10^{-5} (m^3/d)/(MPa·m^2).

(3) Oil saturation

Oil saturation indicates the difference of oil-containing capacity of reservoir matrix pores, but the oil saturation in the pores at different scales in reservoir matrix varies greatly.

Based on simulation results (Fig. 2.37), as the oil saturation increases, the oil-producing capacity of reservoir matrix increases, but when it reaches to a certain value, the increasing tendency slows down. This is because when the oil saturation increases, the oil flowability and available producing range also increase, resulting in the increase of oil-producing capacity of reservoir matrix.

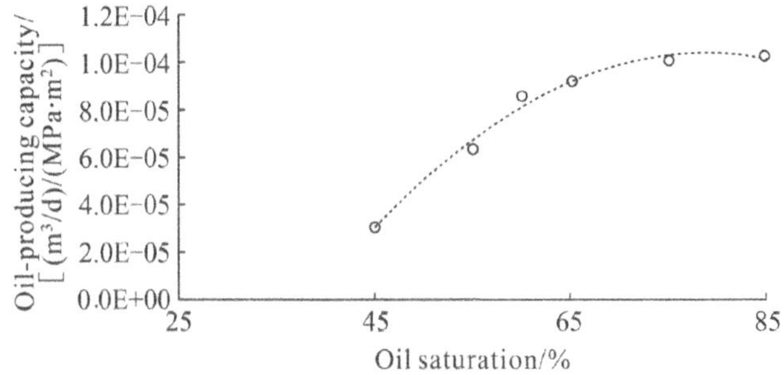

Fig. 2.37 Reservoir matrix oil saturation versus oil-producing capacity

2. Effect of fluid properties on the oil-producing capacity

(1) Viscosity

Viscosity indicates the fluid flowability in pores, therefore fluid viscosity directly affects the oil-producing capacity of reservoir matrix.

Based on simulation results (Fig. 2.38), as the oil viscosity increases, the oil-producing capacity of reservoir matrix decreases. When the oil viscosity in tight oil reservoirs reduces from 11 mPa·s to 0.5 mPa·s, the oil-producing capacity of reservoir matrix can be increased more than 5 times. Obviously, oil viscosity has a great influence on the oil-producing capacity of reservoir matrix. For the low-mobility tight oil reservoirs, the reduction of oil viscosity can effectively enhance the oil-producing capacity of reservoir matrix and improve the development performance.

(2) Dissolved gas-oil ratio

The dissolved gas-oil ratio indicates the degassing amount when the reservoir pressure reduces to the bubble point pressure. The degassing amount directly affects the changes of formation energy and the changes of the oil viscosity after

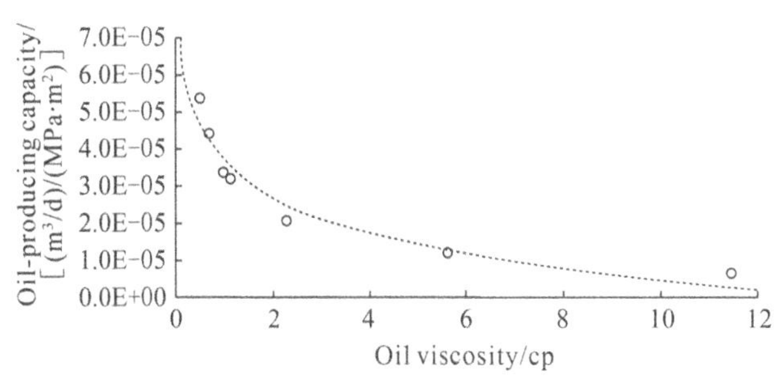

Fig. 2.38 Effect of oil viscosity on the oil-producing capacity

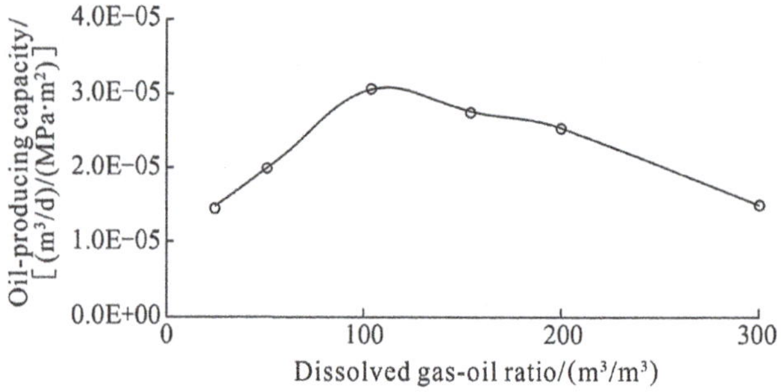

Fig. 2.39 Effect of dissolved gas-oil ratio on the oil-producing capacity

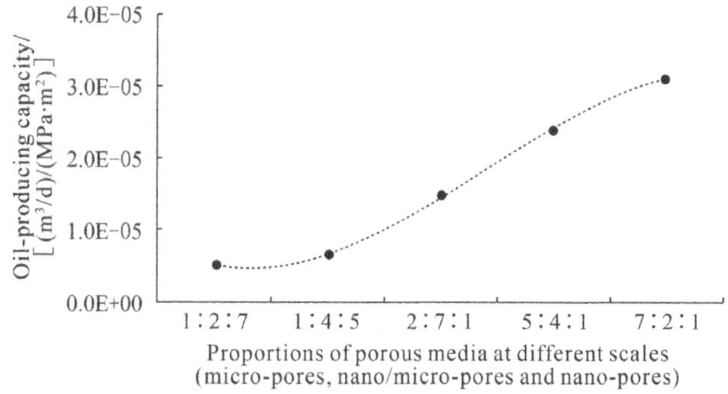

Fig. 2.40 Effect of proportions of porous media at different scales on oil-producing capacity

degassing, thus affects the oil-producing capacity of reservoir matrix.

Based on simulation results (Fig. 2.39), with the increases of dissolved gas-oil ratio, the oil-producing capacity of reservoir matrix per unit area increases, then decreases. When the dissolved gas-oil ratio increases within a certain range, the dissolved gas flooding energy increases, and the oil-producing capacity of reservoir matrix increases. When the dissolved gas-oil ratio exceeds a certain range, the degassing amount increases, the oil viscosity is greatly increased, thus the fluid flowability is reduced. Therefore, the oil-producing capacity of reservoir matrix decreases.

3. Effect of porous media at different scales on the oil-producing capacity of reservoir matrix

Tight reservoirs usually develop pores with different sizes, such as micropores, nano/micro-pores and nanopores. The physical properties, oil-bearing condition, and accessibility of porous media at different scales are different (Du et al. 2016; Sun et al. 2016; Castellarini et al. 2015). The composition and quantity distribution of pores at different scales directly affect the oil-producing capacity of reservoir matrix.

Based on simulation results (Fig. 2.40), with the proportion of large-scale micropores in reservoirs increases and the proportion of small-scale nanopores decreases, the oil-producing capacity of reservoir matrix increases. When the proportion of micropores increases from 10% to 70%, the oil-producing capacity of reservoir matrix increases nearly 5 times.

It can be seen that the composition and quantity distribution of porous media at different scales affect the physical properties, oil-bearing condition and availability of reservoir matrix, resulting in the differences of the oil-producing capacity of reservoir matrix.

4. Effect of reservoir matrix size on the well productivity

Based on the effects of natural/hydraulic fractures at different scales, tight reservoirs are usually separated into the matrix blocks with different size (Moradi and Jamiolahmady 2015). Under a given oil-producing capacity of reservoir matrix, the complex network of natural/hydraulic fractures at different scales is more developed, the reservoir matrix is smaller, the numbers of fracture is larger, the fracture area is larger, and the productivity is much higher.

Based on the simulation results of production performance under different reservoir matrix sizes (Figs. 2.41, 2.42, 2.43, 2.44 and 2.45): i. The oil-producing capacity of reservoir matrix varies with different matrix size, but it is relatively stable. ii. The smaller reservoir matrix is, the larger

Fig. 2.41 "Sugar Lump" Rock-Size 100 × 100 m

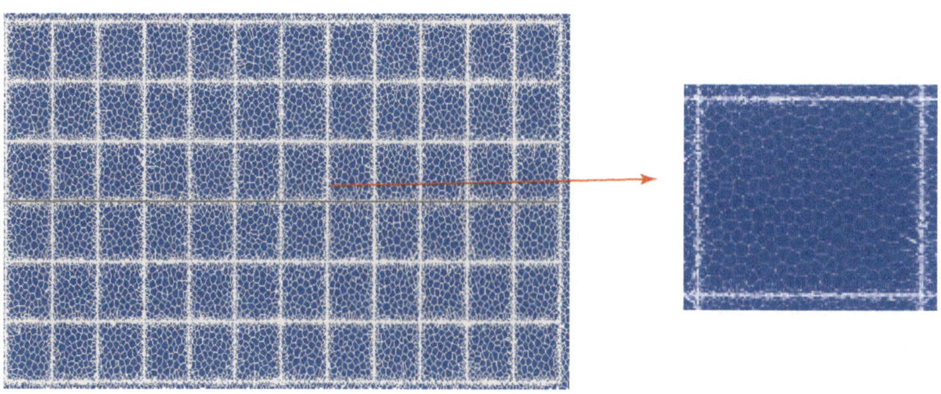

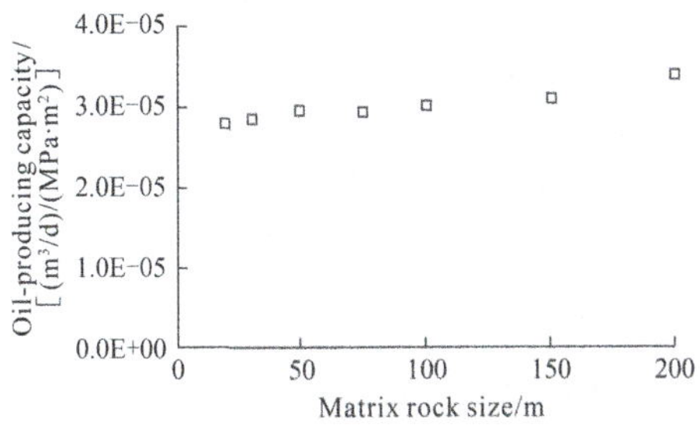

Fig. 2.42 Effect of reservoir matrix rock size on the oil-producing capacity

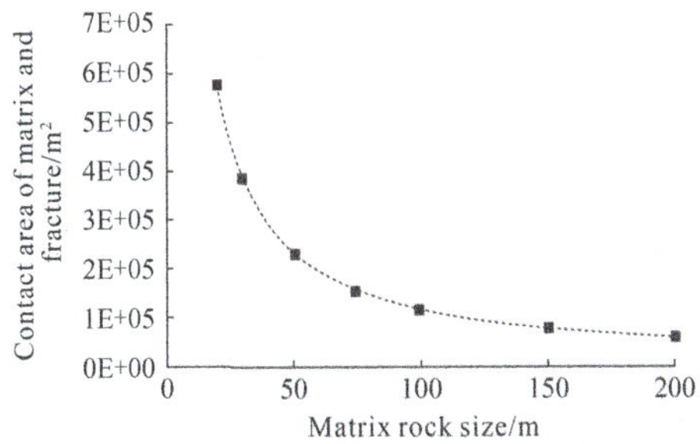

Fig. 2.43 Effect of reservoir matrix rock size on the contact area of reservoir matrix and fracture

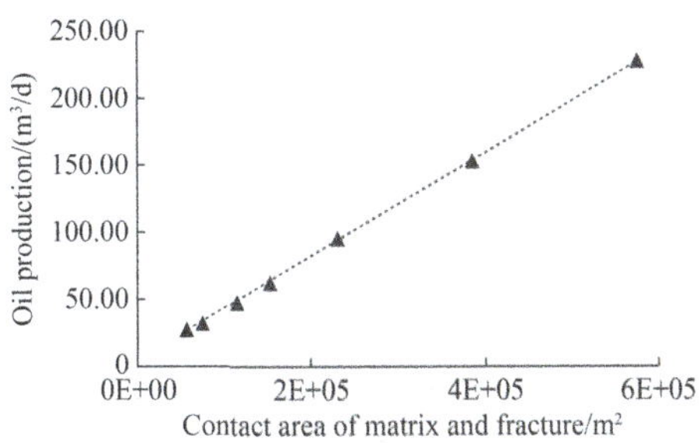

Fig. 2.44 Effect of contact area on the oil production (pressure drops is 15 MPa)

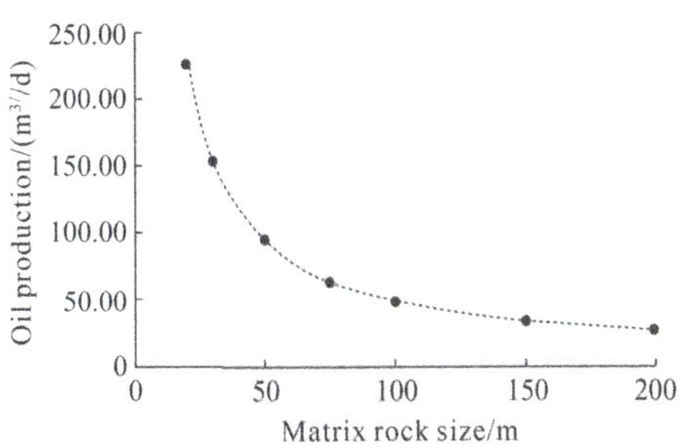

Fig. 2.45 Effect of reservoir matrix rock size on the oil production (pressure drops is 15 MPa)

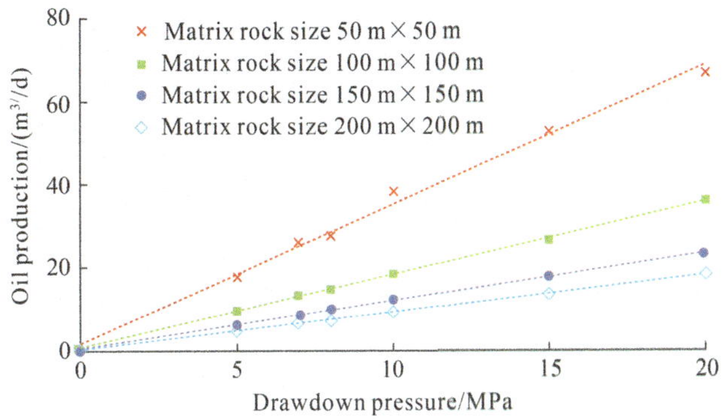

Fig. 2.46 Effect of pressure drops on well productivity

the contact area between matrix rock and fractures is. Simultaneously, as the flow distance from reservoir matrix to fracture is reduced, the flow resistance is reduced. Therefore, the well productivity is increased and the recovery performance is improved.

5. Effect of pressure drops on the well productivity

Under a given oil-producing capacity of reservoir matrix and a same size of reservoir matrix block, the productivity under different pressure drops is simulated (Fig. 2.46). Results show that with the pressure drops increase, the well productivity is increased.

In summary, the oil-producing capacity of reservoir matrix depends on the physical properties of reservoir, oil-bearing condition, fluid properties, and the composition and quantity distribution of reservoir matrix pores at different scales. Under a given oil-producing capacity of reservoir matrix, the complexity of fracture network (the size of reservoir matrix block) and the reasonable pressure drops are the key factors to improve the single-well productivity.

2.4.2 Oil-Drainage Area of Reservoir Matrix

Nano/micro-scale pores can be observed in the matrix of tight reservoirs, which leads to the small pore throats and poor reservoir properties (Zhao et al. 2012). The pressure propagation in tight reservoirs is different from that of the conventional reservoirs (Mirzayev and Jensen 2016). In conventional reservoirs, the pressure propagates instantaneously to the boundary and the oil-drainage radius of reservoir matrix is constant (Fig. 2.47a). In tight reservoirs, pressure propagation has a non-transient effect. With the extension of propagation time, the oil-drainage radius of reservoir matrix gradually increases (Zhu et al. 2010) (Fig. 2.47b).

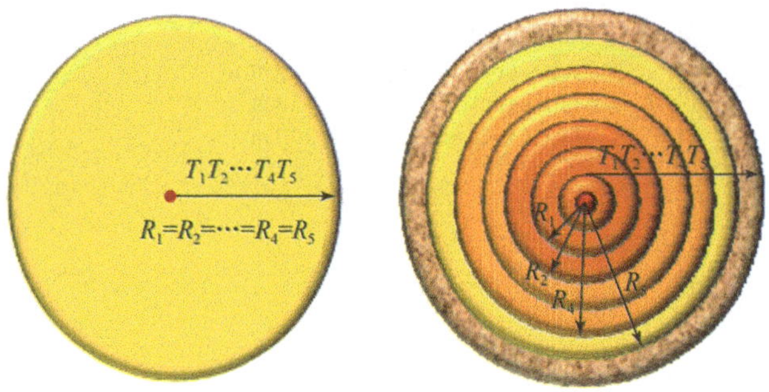

Fig. 2.47 Comparisons of the oil-drainage radius of different types of reservoir matrix

2.4.2.1 Analytical Method to Calculate the Oil-Drainage Radius of Reservoir Matrix

Aiming at the characteristics of tight reservoirs and the non-transient propagation behavior of the oil-drainage radius of reservoir matrix, the calculation model for the oil-drainage radius of tight reservoir matrix is developed by considering the complex nonlinear flow mechanisms such as threshold pressure gradient and stress sensitivity of reservoir matrix.

$$R(t) = \sqrt{r_{\mathrm{w}}^2 + \frac{4K_{\mathrm{m0}}t}{\mu C_{\mathrm{t}}} \Bigg/ \left[\alpha_{\mathrm{m}} + \frac{2\alpha_{\mathrm{m}} e^{-\alpha_{\mathrm{m}}(p_{\mathrm{e}}-p)} \cdot G \cdot R(t) \cdot \ln\frac{R(t)}{r_{\mathrm{w}}}}{1 - e^{-\alpha_{\mathrm{m}}(p_{\mathrm{e}}-p_{\mathrm{wf}})} - \alpha_{\mathrm{m}} e^{-\alpha_{\mathrm{m}}(p_{\mathrm{e}}-\bar{p})} \cdot G \cdot (R(t) - r_{\mathrm{w}})}\right]} \tag{2.32}$$

where, $R\ (t)$ represents the reservoir matrix oil-drainage radius at time t, m; K_{m0} represents the initial permeability of the reservoir matrix, mD; α_{m} represents the reservoir matrix stress sensitivity coefficient, dimensionless.

Using this model, the changing law for the oil-drainage radius of reservoir matrix during the recovery process of tight oil reservoirs can be calculated (Fig. 2.48). During the recovery process of tight oil reservoirs, the effective radius of reservoir matrix changes with time. The longer the production time is, the larger the oil-drainage radius is. The oil-drainage radius of reservoir matrix is 56 m after 3 years production. Until 10 years production, the oil-drainage radius of reservoir matrix has reach 84 m.

It can be seen that during the recovery process of unconventional tight oil reservoirs, the oil-drainage radius changes non-transiently and it indicates the effective producing range of reservoir matrix (Stimpson and Barrufet 2016).

2.4.2.2 Numerical Simulation Method to Calculate the Oil-Drainage Radius of Reservoir Matrix

As the tight reservoirs are separated into the matrix blocks with different sizes by regular fractures, the regular reservoir matrix blocks and fractures are usually alternately distributed in reservoirs, and the rock matrix is surrounded by fractures. During the development process, hydrocarbons in matrix blocks can flow into fractures and the crude oil in fractures can flow into wellbore. Based on simulation results (Fig. 2.49), the oil-drainage radius of reservoir matrix increases with the development time. The oil-drainage radius of reservoir matrix is 62 m after 3 years production. The oil-drainage radius of reservoir matrix reaches 125 m after 10 years production. During the development process, the pressure drops propagates gradually and non-instantaneously from fracture surface to inter space of matrix blocks. As the production time increases, the distance between the front of pressure propagation and fracture surface is increased, and the oil-drainage radius of reservoir matrix is increased.

As the tight reservoirs are separated into the matrix blocks with different sizes and shapes by irregular natural/hydraulic fractures, the irregular reservoir matrix blocks and the fractures at different scales are alternately distributed in reservoirs. Based on simulation results (Fig. 2.50), as the matrix block reduces, the fluid flow distance is reduced and the oil unlocking degree is increased. As the matrix block increases, the flow distance is increased and the oil unlocking degree is decreased. Thus, the smaller the matrix block size, the better the recovery performance.

2.5 Coupled Recovery Mechanisms of Pore-Fracture Media at Different Scales

Pore-fracture media at different scales has been observed in tight oil and gas reservoirs. The pore-fracture media at different scales are discontinuously and discretely distributed in reservoirs. Their geometrical and physical properties, fluid

Fig. 2.48 Schematic diagram for the effective oil-drainage range of reservoir matrix

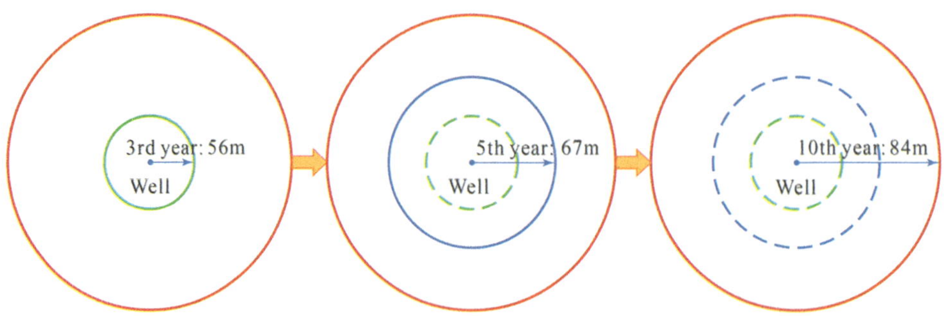

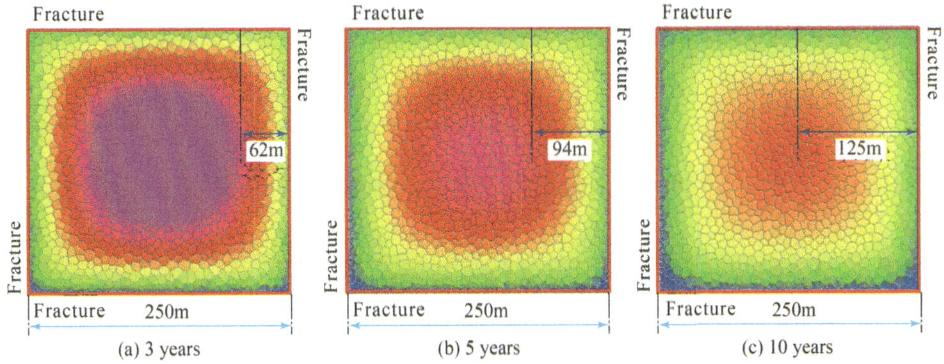

Fig. 2.49 Schematic for the effective oil-drainage range of regular matrix blocks

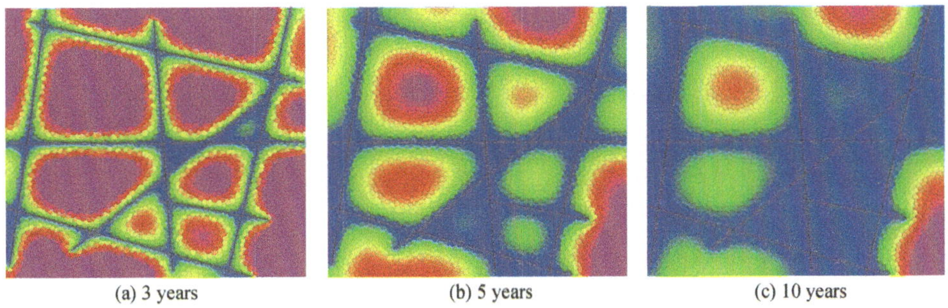

Fig. 2.50 Schematic for the effective oil-drainage range of irregular matrix blocks

properties and flow characteristics are quite different, and there is a complicated coupled flow behavior. Therefore, during the development process, there is a strong coupling mechanism of pore-fracture media at different scales. Simultaneously, the contributions to production and the effects at different production stages of pore-fracture media at different scales are also different.

2.5.1 Coupled Flow Behavior Between Multiple Media

2.5.1.1 Discontinuous and Discrete Spatial Distribution of Pore-Fracture Media at Different Scales

In tight oil and gas reservoirs, the existences of nano/micro-pores at different scales and complex natural/hydraulic fracture network have been observed. They have the characteristics of heterogeneity, multiple scale and multiple media. There are many types of pore-fracture media with different scale, which are discontinuous and discrete in space and are interactively distributed with each other. The pore-fracture media at different scales have their own geometrical and physical parameters, oil-bearing conditions and fluid parameters. It has a large difference and presents the characteristics of discontinuous change in reservoir (Fig. 2.51).

2.5.1.2 Coupled Flow Behaviors of Pore-Fracture Media with Different Scale

Under the conditions of discrete, interactive distributions and discontinuous changes of attribute parameters of pore-fracture media at different scales, there is a strong coupled flow behavior due to the large difference in flow regimes and mechanisms (Fig. 2.52).

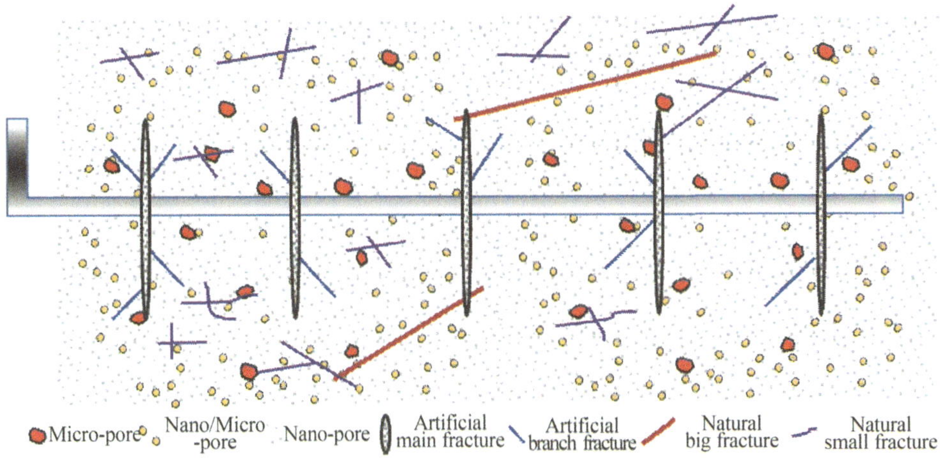

Fig. 2.51 Spatial distribution of multiple media in tight reservoirs

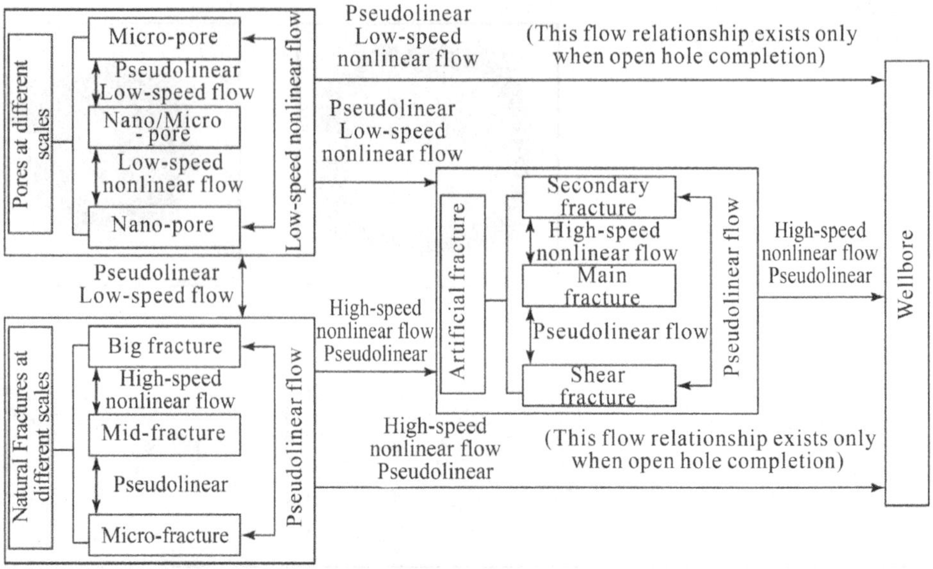

Fig. 2.52 Schematic diagram of the coupled flow behavior between multiple media

1. Flow behavior in different scale pores

For porous media at different scales, due to the effects of narrow pore throats, small geometrical scales and poor physical properties, the flow regimes in porous media at different scales are dominated by the low-speed nonlinear flow. For a large pressure gradients, the pseudolinear flow can happen between the relatively large scale micro-pores and nano/micro-pores which are at a relatively large scale.

2. Flow behavior in different scale natural fractures

For natural fractures at different scales, due to the large differences of fracture scale and flow regime, the large/middle-scale fractures are dominated by high-speed nonlinear flow. But for small fractures, pseudolinear flow dominates.

3. Flow behavior in different scale hydraulic fractures

Due to the difference of geometrical scale, there are differences in flow regimes between hydraulic fractures at different scales. Main fractures have a large aperture and a high flow rate. It is usually dominated by the high-speed nonlinear flow. For secondary fractures and shear small fractures, the aperture is small and the flow rate is also low. It is usually dominated by the pseudolinear flow.

4. Flow behavior in different scale pores, natural/hydraulic fractures and wellbores

In the different scale pore, the fluids can flow into natural fractures at different scales through the low-speed nonlinear and pseudolinear flow. In natural fractures, the fluids can flow to hydraulic fractures through the high-speed nonlinear and pseudolinear flow. In hydraulic fractures, the fluids can flow to wellbore through the high-speed nonlinear and pseudolinear flow. For an open hole well completion, the fluids in pores and natural fractures at different scales can directly flow to wellbore.

The pores and natural/hydraulic fractures at different scales are discontinuous and interactively distributed in reservoir. During the recovery process of unconventional tight reservoirs, there is a coupled flow behavior for the many flow regimes between pore-fracture media at different scales and wellbores.

2.5.1.3 The Effect of Pore-Fracture Media at Different Scales in Recovery Process

During recovery process, the effect of pore-fracture media at different scales and their contribution to the production process are different. Only a suitable match between different scale fractures and reservoir matrix pores can effectively unlock the tight oil and gas reservoirs (Table 2.7).

The extension length of large/middle-scale fractures is high. The controllable oil reserves of single well are large. Simultaneously, they have a strong conductivity, which directly affects the well productivity in the early production stage and the scale of oil production.

Small scale fractures have a short extension length and a limited controllable oil range. They can effectively connect the reservoir matrix pores and the large/middle-scale fractures. Thus, the oil reserve in reservoir matrix blocks can be effectively unlocked.

The nano/micro-pores at different scales are the basis of oil reserves, which mainly play the role of fluid supplement in later stage. The unlocking ratio of different pores and contribution to well productivity are different, which mainly affects the production in the middle and later stages.

Table 2.7 The effect of pore-fracture media at different scales

Media type	The role of different media
Large and middle scale fractures	Extended far distance, single-fracture control and control of the large reserves; diversion ability, affecting the initial production and production scale
Small scale fractures	Short extension, limited control, connectivity to the reservoir matrix and large fractures, the effective use of reservoir matrix
Nano/micro-pores at different scales	The basis of reserves, play a replenishment role; different porosity effect on the contribution of different productivity, affecting the late production level

2.5.2 Coupled Recovery Mechanisms in Different Production Stages

In unconventional tight oil and gas reservoirs, there is a large difference for the oil-producing capacity of pore-fracture media at different scales. The coupled flow behavior and coupled production mechanism are very different. And during different production stages, the function and contribution of pore-fracture media at different scales are also different. Based on mechanism analysis and numerical simulation, it shows that only a suitable match between porous media at different scales and natural/hydraulic fractures at different scales can lead to the effective development of tight oil and gas reservoirs.

2.5.2.1 Coupled Recovery Mechanisms of Pore-Fracture Media at Different Scales

According to the characteristics of recovery performance of tight oil reservoirs, the recovery process can be classified into three stages, the rapid decline of high well productivity in the early stage, the middle transition stage and the slow decline of low well productivity in the later stage (Fig. 2.53) (Zou et al. 2013; Du et al. 2014). At different stages, the pore-fracture media at different scales has different effects and the coupled recovery mechanisms are also different. The recovery performance and development rules are different (Clarkson and Pedersen 2011) (Table 2.8, Fig. 2.54).

1. Rapid decline of high productivity in the early stage

At this stage, the effective pore-fracture media include hydraulic fractures, large and middle scale natural fractures and nano/micro-pores around the fractures.

The crude oil in nano/micro-pores firstly flows into the fractures and then flows into the wellbore through fractures. The coupled recovery mechanisms in this stage mainly include pressure drops (viscous effect, gravity effect), fracture deformation and fracture closure (compaction) (Xiong et al. 2016; Salam 2016). The typical characteristics of recovery performance in this stage are high productivity and rapid decline, and the contribution of this stage to the cumulative production is about 42%.

2. Middle transition stage

At this stage, the small and middle scale natural fractures and nano/micro-pores are effective. The main mechanisms in this stage include pressure drops, fracture deformation,

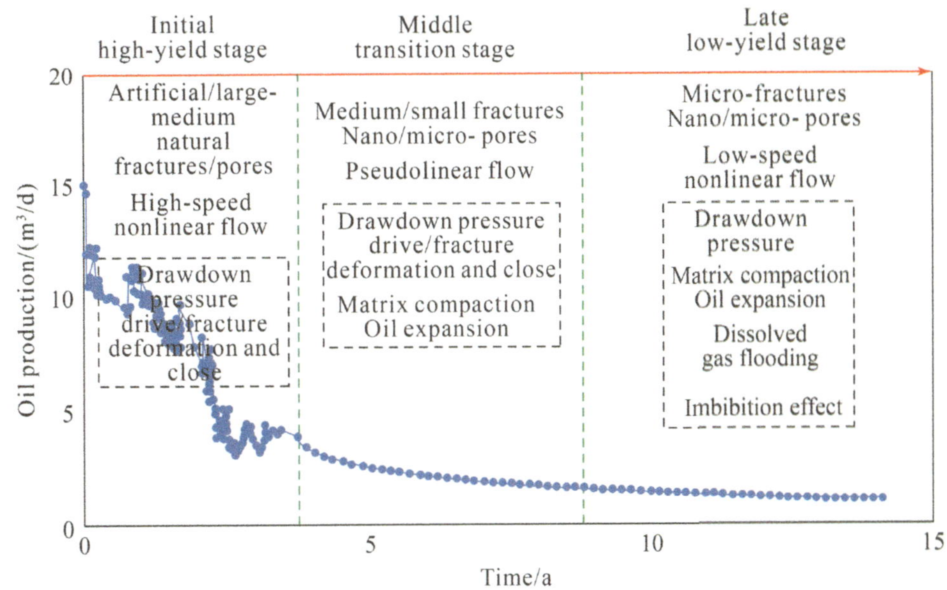

Fig. 2.53 Production performance curve of different scale media

Table 2.8 Coupled recovery mechanisms of multiple media in different stages

Stage	Pore-fracture media and its function	Coupled recovery mechanisms	Characteristic of recovery performance
Rapid decline of high productivity in the early stage	Hydraulic fractures: connect wellbores and nearby pore-fracture media	Pressure drops (viscous effect, gravity effect)	High initial productivity, rapid decline, the cumulative production contribution: 42%
	Large and middle fractures: connect wellbore, hydraulic fractures and nearby pores. Nano/micro-pores around the fractures	Fracture deformation and close (compaction)	
Middle transition stage	Small and middle scale fractures: connect hydraulic/natural fractures and nano/micro-pores	Pressure drops (viscous effect, gravity effect)	Lower middle productivity, slower decline, the cumulative production contribution: 33%
	Nano/micro-pores: supplement	Fracture deformation and close (compaction) Compaction of reservoir matrix pores	
		Crude oil expansion	
Slow decline of low productivity in the later stage	Micro-fractures: connect large/middle fractures and nano/micro-pores to achieve a dynamic balance	Effect of pressure drops weakened	Lower later productivity, slowly decline, the cumulative production contribution: 25%
	Nano/micro-pores: supplement	Compaction of reservoir matrix pore weakened	
		Crude oil expansion Dissolved gas flooding Imbibition effect	

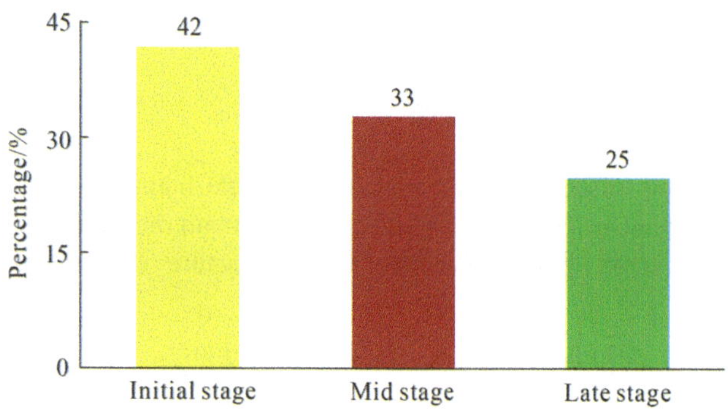

Fig. 2.54 Percentage of cumulative oil production at different stages

fracture closure, reservoir matrix compaction and expansion of crude oil. Compared with the early stage, the productivity decreases slowly in this stage, and the contribution to the cumulative production is about 33%.

3. Slow decline of low productivity in the later stage

At this stage, the micro scale natural fractures and nano/micro-pores are effective. The supplement of nano/micro-pores and the flow behavior in micro scale fractures is a dynamic equilibrium. At this stage, the effect of pressure drops and reservoir matrix compaction are weakened. The main driving mechanisms include crude oil expansion, dissolved gas drive and imbibition. The well productivity in this stage is lower, and it decreases slowly. The contribution to the cumulative production is about 25%.

2.5.2.2 The Effect of Pore-Fracture Media at Different Scales

In unconventional tight oil and gas reservoirs, the pore-fracture media at different scales occupy different proportions in oil reserves. During recovery process, the effect is significantly different. The recovery process of

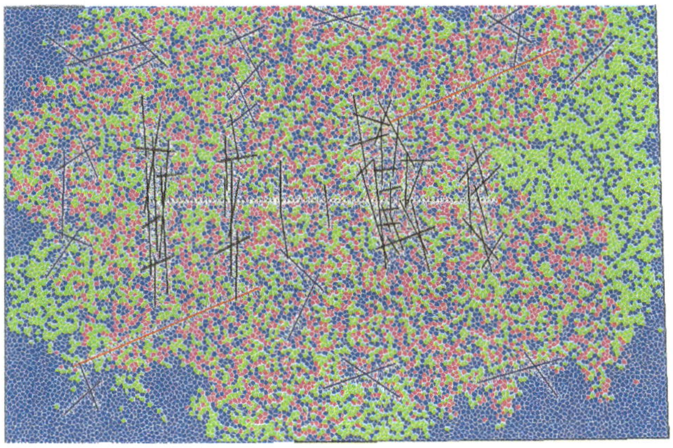

Fig. 2.55 Spatial distribution of pore-fracture media at different scales (blue: nano-pore; green: nano/micro-pore; red: micro-pore)

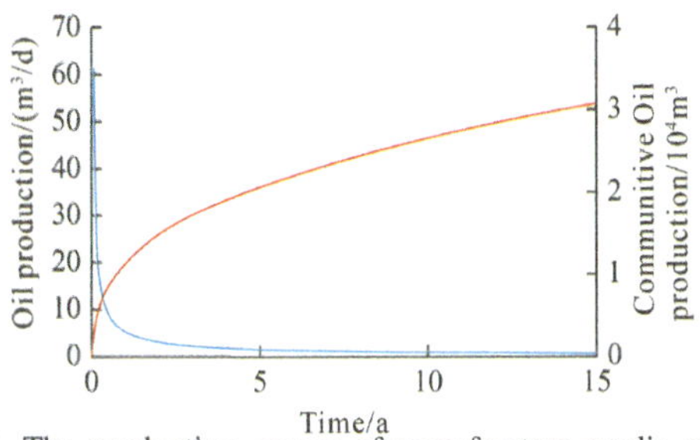

Fig. 2.56 The production curves of pore-fracture media at different scales

pore-fracture media at different scales is simulated. Thus, the effect of pore-fracture media at different scales during recovery process and the recovery factory are also analyzed (Figs. 2.55, and 2.56).

In tight oil and gas reservoirs, the oil reserves are mainly distributed in reservoir matrix pores, which accounts for about 96.57% in the whole reservoir matrix reserves (micropores account for 21.44%, nano/micro-pores account for 32.13% and nanopores account for 46.27%). The total oil reserve in fracture media is only 3.43% (including 5.72% for large scale fractures, 8.57% for middle scale fractures and 85.71% for micro scale fractures) (Table 2.9).

After development for 15 years, the ratio of cumulative oil production from reservoir matrix pores is 72.9% (micro-pores: 47.56%; nano/micro-pores: 36.44%; nano-pores: 16%). The fraction of cumulative oil production from fractures is 27.1% (large fractures: 9.31%; middle fractures: 11.93%; micro-fractures: 78.76%) (Table 2.10).

Because the effect of pore-fracture media at different scales in the recovery process is different, the recovery factor is quite different (the recovery factor of pore-fracture media at different scales is the ratio of cumulative oil production in different media and the original geological reserves in different media). Fracture mainly aims to communicate and they have a high conductivity. The hydrocarbons in fractures are easy to unlock and their geological reserves are small. Therefore, the recovery factor is as high as 59.86%. Fore porous media, the physical properties are poor, and fractures are generally required for the communication of porous media. The producing degree of porous media is low, and geological reserves are high. Therefore, the recovery factor is only 5.70%. For fracture media, large and middle scale fractures have a strong communication and a high recovery factor. The recovery factor in large scale factures is 97.50%, and that in middle scale factures, it is 83.33%. The conductivity of micro scale fractures is lower, and thus, they have a lower recovery factor, 55%. For porous media in reservoir matrix, the physical properties of large scale micropores are better, and fluid flowability is high. Thus, the recovery factor can reach 12.65%. in comparison, the physical properties of nano/micro-pores are poor, and the recovery factor is 6.47%. The physical properties of nanopores are poor, and the recovery factor is only 1.97% (Table 2.10).

Table 2.9 The fraction of oil reserve in the media at different scales

Media type	Large fracture	Middle fracture	Micro-fracture	Total in fracture media	Micro-pore	Nano/micro-pore	Nano-pore	Total in reservoir matrix media
Reserves/10^4 m^3	0.08	0.12	1.2	1.4	8.46	12.68	18.26	39.46
Fraction/%	5.71	8.57	85.71	3.43	21.44	32.13	46.27	96.57

Table 2.10 Oil production, recovery factor and fraction of media at different scales

Media type	Large fracture	Middle fracture	Micro-fracture	Total in fracture media	Micro-pore	Nano/micro–pore	Nano-pore	Total in reservoir matrix media
Production/10^4 m^3	0.078	0.1	0.66	0.838	1.07	0.82	0.36	2.25
Fraction/%	9.31	11.93	78.76	27.10	47.56	36.44	16.00	72.90
Recovery factor/%	97.50	83.33	55.00	59.86	12.65	6.47	1.97	5.7

Table 2.11 Contribution of pore-fracture media at different scales to oil production at different stages

Stage	Production of different stages/10^4 m^3						Stage oil production	Contribution of different media/%					
	Large fracture	Middle fracture	Micro-fracture	Micro-pore	Nano/micro-pore	Nano-pore		Large fracture	Middle fracture	Micro-fracture	Micro-pore	Nano/micro-pore	Nano-pore
Initial	0.074	0.024	0.25	0.52	0.36	0.08	1.308	5.66	1.83	19.11	39.76	27.52	6.12
Middle	0.002	0.041	0.22	0.35	0.28	0.09	0.983	0.20	4.17	22.38	35.61	28.48	9.16
Later	0.002	0.035	0.19	0.2	0.18	0.19	0.797	0.25	4.39	23.84	25.09	22.58	23.84

2.5.2.3 The Effect and Contribution of Pore-Fracture Media at Different Scales During Different Production Stages

During the different recovery stages of tight oil and gas reservoirs, the effect of different pore-fracture media is different from their contribution to the oil production (contribution refers to the ratio of cumulative oil production of pore-fracture media at different scales in this stage and the cumulative oil production of all the media in this stage). Based on the results of recovery performance simulation, the contribution of pore-fracture media at different scales during different stages are statistically analyzed (Table 2.11).

1. Contribution to oil production of pore-fracture media at different scales during different stages

Large scale fractures have a high initial productivity and a large contribution to oil production. In the middle and later stages, as the oil reserves decrease and fractures closes, the contribution of large-scale fractures is getting smaller and smaller. The middle scale fracture comes to affect the productivity in the initial stage. In the middle and later stages, as the expansion of pressure effective range, the contribution is increased gradually. The microscale fracture mainly aims to communicate the reservoir matrix pores. Then, through the large and middle scale fractures, they can contribute to the oil production. From initial stage to later stage, the contribution to oil production is gradually increased.

In initial stage, the micropores in reservoir matrix pores are the main oil-producing media, and then the contribution of micropores is reduced gradually. The contribution of nano/micro-scale porous media to oil production is relatively low initially, and then increases in the middle stage and declines in the later period. In initial stage, the contribution of nanopores is very small. In the middle and later stages, it is increased and even plays the major role for oil production (Fig. 2.57).

2. Difference of recovery factor for the pore-fracture media at different scales during different stages

Hydrocarbons in the large fractures are rapidly produced under a large pressure drops in the initial stage, and the recovery factor can reach 90% above. In the middle and later stages, the large fractures mainly perform as the flow path, and the recovery factor is low. For middle scale fractures, the recovery factor is low in the early stage. In the middle stage, as the pressure spreads, the recovery factor is increased. In the later stage, because of the reduction of oil reserves, the recovery factor is reduced. For micro scale fractures, they have a relatively high recovery factor in the initial stage. That is because that the strength of micro scale fractures is high around the large and middle scale fractures. Thus, the oil reserves in this kind of fractures can be unlocked firstly. In the middle and later stages, the micro scale fractures far from the large and middle scale fractures are gradually

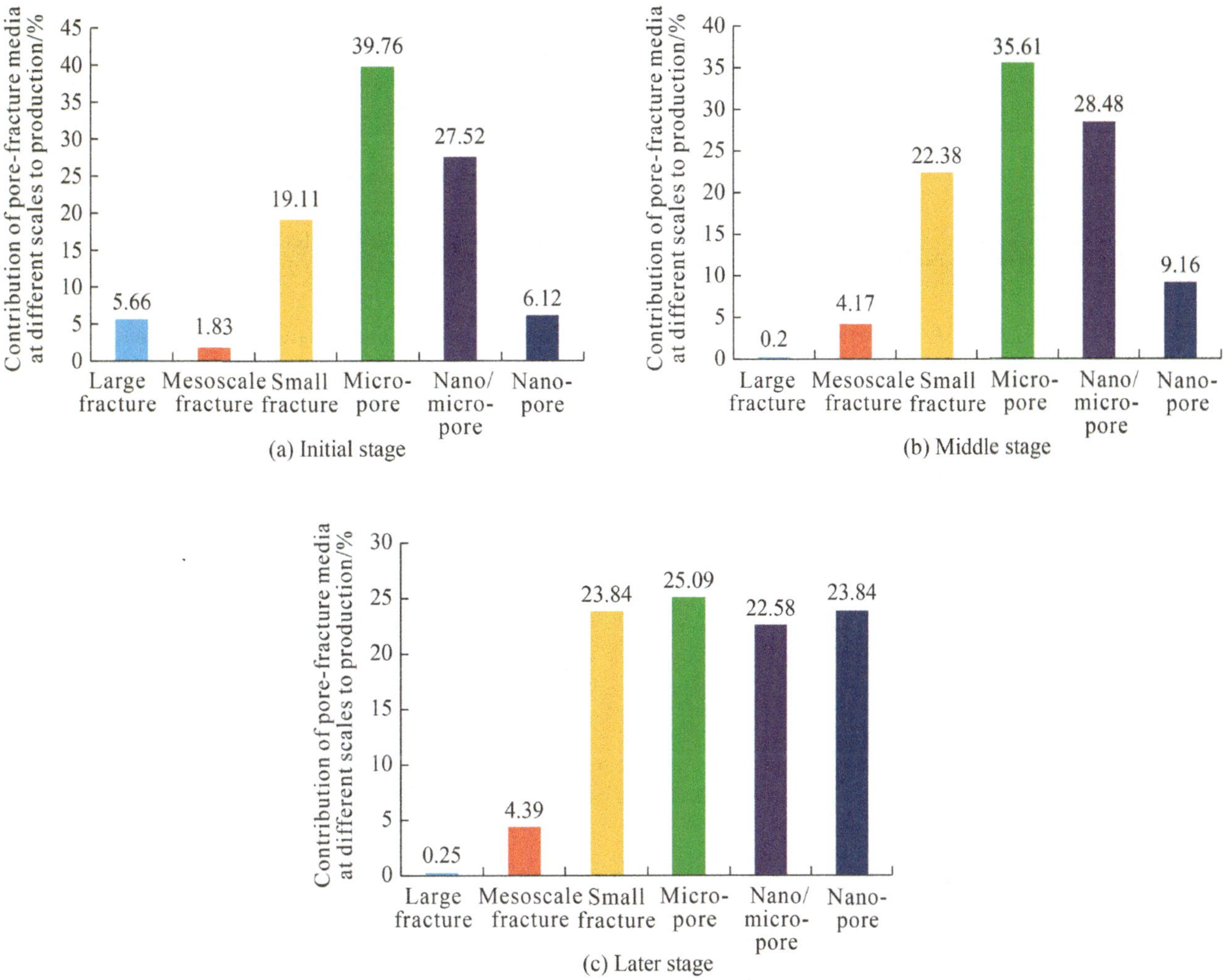

Fig. 2.57 Contribution of pore-fracture media at different scales to oil production at different stages

Table 2.12 The recovery performance of fracture media at different scales in different stages

Stage	Initial oil reserves of fracture media at different scales/10^4 m^3			Oil production of fracture media at different scales/10^4 m^3			Recovery factors of fracture media at different scales/%		
	Large	Middle	Micro	Large	Middle	Micro	Large	Middle	Micro
Initial	0.08	0.12	1.2	0.074	0.024	0.25	92.50	20.00	20.83
Mid-term	0.006	0.096	0.95	0.002	0.041	0.22	2.5	34.17	18.33
Later	0.004	0.055	0.73	0.002	0.035	0.19	2.5	29.17	15.83
Total	0.08	0.12	1.2	0.078	0.1	0.66	97.50	83.33	55.00

Table 2.13 The recovery performance of porous media at different scales in different stages

Stage	Initial oil reserves of porous media at different scales/10^4 m^3			Oil production of porous media at different scales/10^4 m^3			Recovery factors of porous media at different scales/%		
	Micro	Nano/micro	Nano	Micro	Nano/micro	Nano	Micro	Nano/micro	Nano
Initial	8.46	12.68	18.26	0.52	0.36	0.08	6.15	2.84	0.44
Mid-term	7.94	12.32	18.18	0.35	0.28	0.09	4.14	2.21	0.49
Later	7.65	12.04	18.09	0.2	0.18	0.19	2.36	1.42	1.04
Total	8.46	12.68	18.26	1.07	0.82	0.36	12.65	6.47	1.97

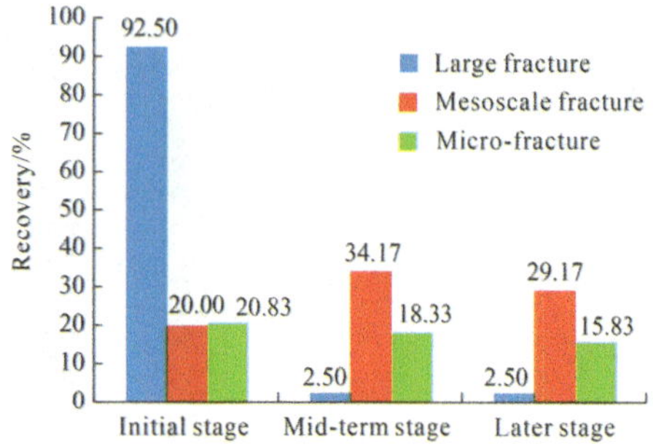

Fig. 2.58 The recovery factor of fracture media at different scales in different stages

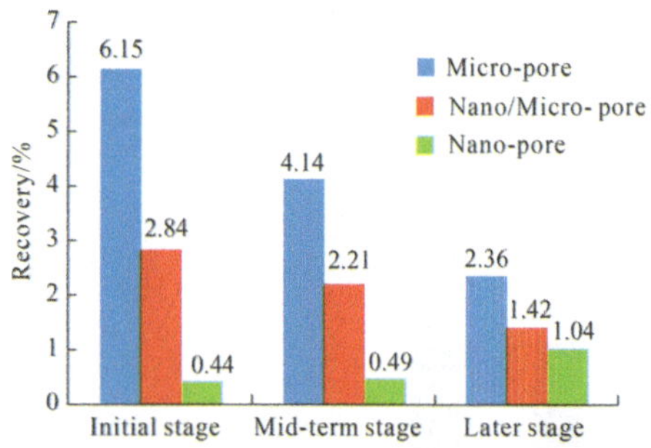

Fig. 2.59 The recovery factor of porous media at different scales in different stages

unlocked. And as the pressure reduces, the recovery factor is reduced gradually.

The porous media at different scales is the basis of oil reserves, and they play the important supplement role during production process. The scales of micro-pores and nano/micro-pores are relatively large, and the physical properties and oil-bearing conditions are also better. It is easy to develop and will preferably unlock the hydrocarbons in the early stage. Moreover, the recovery factor is also high. In the later stage, as the pressure reduces, the recovery factor is reduced gradually. For small scale nanopores, the physical properties and oil-bearing conditions are poor, and it is hard to develop. In the later stage, as the pressure spreads, the hydrocarbons are unlocked and the recovery factor is low (Tables 2.12 and 2.13, Figs. 2.58 and 2.59).

For oil drainage range, under the condition of a large pressure drop in the early stage, the oil reserves in reservoir matrix pores around the hydraulic fractures and large-scale natural fractures are unlocked firstly (Fig. 2.60a). In the middle stage (Fig. 2.60b), with the expansion of pressure affected range, small/middle scale and microscale fractures and the nearby reservoir matrix pores are gradually unlocked. In the later stage (Fig. 2.60c), the producible range is further expanded and more connected porous matrix pores can be unlocked.

Fig. 2.60 Pressure distribution of pore-fracture media at different scales in different stages

(a) Initial production stage

(b) Mid-term production stage

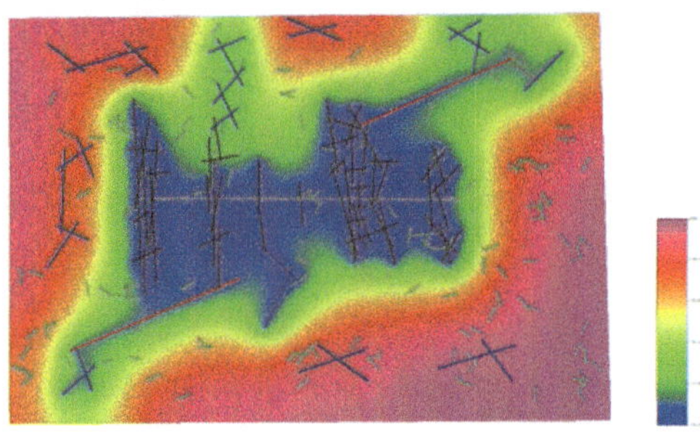

(c) Later production stage

References

Abdassah D, Ershaghi I (1986) Triple-porosity systems for representing naturally fractured reservoirs. SPE Form Eval 1(2):113–127

Arnaud L, Remy B, Bernard B (2004) Hydraulic characterization of faults and fractures using a dual media discrete fracture network simulator. SPE 88675, The 11th Abu Dhabi international petroleum exhibition and conference, Abu Dhabi, U.A.E

Barenblatt GI, ZheltovIu P, Kochina IN (1960) Basic concept in the theory of homogeneous liquids in fissured rocks. J Appl Math Mech (USSR) 24:1286–1303

Castellarini PA, Garbarino F, Garcia MN, Sorenson F (2015) How rock properties understanding from micro to macro scale affect productivity profile of tight reservoirs: Neuquén, Argentina. 13CONGRESS-2015-100

Civan F (2010) Effective correlation of apparent gas permeability in low-permeability porous media. Tran Porous Med 82(2):375–384

Clarkson CR, Pedersen PK (2011) Production analysis of western canadian unconventional light oil plays. SPE149005

Cui X, Bustin AM, Bustin R (2009) Measurements of gas permeability and diffusivity of tight reservoir rocks, different approaches and their applications. J Geofluids 9(3):208–223

Darabi H, Ettehad A, Javadpour F et al (2012) Gas flow in ultra-tight shale strata. J Fluid Mech 710(November):641–658

Du JH et al (2016) China terrestrial tightening oil. Petroleum Industry Press, Beijing

Du JH, He HQ, Yang T, Li JZ, Huang FX, GuoB C, Yan WP (2014) Progress in China's tight oil exploration and challenges. China Petrol Explor 19(1):1–8

Firouzi M, Alnoaimi K, Kovscek A et al (2014a). Klinkenberg effect on predicting and measuring helium permeability in gas shales. Int J Coal Geol 123(1): 62-68

Firouzi M, Ruppa EC, Liu CW et al (2014b) Molecular simulation and experimental characterization of the nanoporous structures of coal and gas shale. Int J Coal Geol. 121(1): 123-128

Forchheimer P (1901) Wasserbewegungdurch bode. ZVDI 27(45):26–30

Ge JL (2003) The modern mechanics of fluids flow in oil reservoir, Vol 1. Petroleum Industry Press,Beijing, pp 14-20

Ge JL, Ning ZF, Liu YT et al (2001) The modern mechanics of fluids flow in oil reservoir. Petroleum Industry Press, Beijing

He GS (1994) Reservoir physics. Petroleum Press, Beijing

Kang ZJ (2010) Mathematic Model for Flow Coupling of Crevice-Cave type Carbonate Reservoir. Petrol Geol Oilfield Develop Daqing 29 (1):29–32

Karimi-Fard Firoozabadi A (2003) Numerical simulation of water injection in fractured media using the discrete-fracture model and the Galerkin method. SPE Res Eval Eng 6(2):117–126

Kim JG, Deo MD (2000) Finite element, discrete-fracture model for multiphase flow in porous media. AIChE J 46(6):1120–1130

Li DP (1999) Low permeability reservoir development technology. Petroleum Industry Press, Beijing

Liu J, Bodvarsson G, Wu YS (2003) Analysis of flow behavior in fractured lithophysal reservoirs. J Contam Hydrol 62–63:189–211

Liu DH, Liu ZS (2004). Reservoir engineering foundation. Petroleum Industry Press, Beijing.

Liu XX, Hu Y, Zhu B et al (2006) Study on low-velocity non-Darcy gas percolation mechanism and characteristics. Spec Oil Gas Res 13 (6):43–46

Liu XX, Zhong B, Hu Y et al (2008) Experiment on Gas Seepage Mechanism in Low-Permeability Gas Reservoirs. Nat Gas Ind 28 (4):130–132

Liu CQ, Guo SP (1982) Research progress of multi-media flow. Adv Mech 12(4):360–364

Mirzayev M, Jensen JL (2016) Measuring inter-well communication using the capacitance model in tight reservoirs. SPE180429-MS

Moradi D, Jamiolahmady M (2015) Novel approach for predicting multiple fractured horizontal wells performance in tight reservoirs. SPE 175446-MS

Noorishad J, Mehran M (1982) An upstream finite element method for solution of transient transport equation in fractured porous media. Water Resour Res 18(3):58–96

Noman R, Archer JS (1987) The effect of pore structure on non-Darcy gas flow in some low-permeability reservoir rocks. SPE/DOE 16400

Pan JS (2012) Fundamentals of gas dynamics. National Defense Industry Press, Beijing

Pascal H, Quillian, RG, Kingston J (1980) Analysis of vertical fracture length and non-Darcy flow coefficient using variable rate tests. Paper SPE 9438 presented at the 1980 SPE annual technical conference and exhibition, Dallas. Sept 21–24

Sun YY, Song XM, Ma DS (2014) Characteristics of Remaining Oil Micro-Distribution in Laojunmiao Oilfield after Waterflooding. Xinjiang Petrol Geol 35(3):311–314

Sun ZD, Jia CZ, Li XF (2016) Unconventional oil & gas exploration and development. Petroleum Industry Press, Beijing

Salam AR (2016) Performance-based comparison for fractal configured tight and shale gas reservoirs with and without non-darcy flow impact. SPE-AFRC-2554119-MS

Sheng M, Li GS, Huang ZW et al (2014) Shale gas transient flow model with effects of surface diffusion. Acta Petrolei Sinica 35 (2):347–352

Snow D (1968) Rock-fracture spacing, openings and porosities. J Soil Mech Founda Div ASCE 94:73-91

Stimpson BC, Barrufet MA (2016) Effects of confined space on production from tight reservoirs. SPE 181686-MS

Wang HL, Xu WY, Chao ZM et al (2016) Experimental study on slippage effects of gas flow in compact rock. Chinese J Geotech Eng 38(5):777–785

Wang R, Yue XA, You Y et al (2007) Cyclic waterflooding and imbibition experiments for fractured low-permeability reservoirs. J Xi'an Shiyou Univ (Nat Sci Ed) 22(6):56-59

Wang WF (2013) Research on porous flow and numerical simulation of shale gas reservoir. Southwest Petroleum University, Chengdu

Wang WF, Liu ZR, Jin Q (1993) Study on fluid properties in reservoir description. J Univ Petrol China (Nat Sci Ed),(6):12-17

Warren JE, Root PJ (1963) The behavior of naturally fracture reservoirs. SPE 426

Wu YS, Ge JL (1983) The transient flow in naturally fractured reservoirs with three-porosity systems. Acta MechanicaSinica 1:81–85

Wu YS, Ehlig-Economides C, Qin G et al (2007) A triple-continuum pressure-transient model for a naturally fractured vuggy reservoir. SPE 110044, presented at the 2007 SPE annual technical conference and exhibition held in Anaheim, California, USA 11–14 November 2007

Wu KL, Li XF, Chen ZX et al (2015) Gas transport behavior through micro fractures of shale and tight gas reservoirs. Chinese J Theo Appl Mech 47(6):955–964

Xiao QH (2015) The reservoir evaluation and porous flow mechanism for typical tight oilfields. University of Chinese Academy of Sciences

Xie QB, Han DX, Zhu XM (2002) Reservoir space feature and evolution of the volcanic rocks in the Santanghu basin. Petrol Expl Dev 29(1):84–86

Xing X, Hu WS, Ji L et al (2013) Based on the isothermal adsorption experiments of shale gas content calculation method. Sci Technol Eng 13(16):4659–4662

Xiong, Y, Yu JB, Sun HX, Yuan JG, Huang ZQ, Wu YS (2016) Colorado school of mines a new non-darcy flow model for low velocity multiphase flow in tight reservoirs. SPE180072-MS

Yang J, Kang YL, Li QG, Zhang H (2008) Characters of micro-structure and percolation in tight sandstone gas reservoirs. Adv Mech 38(2):229–236

Yao GJ, Peng HL, Xiong Y et al (2009) Research on characteristics of gas flowing in low permeability sandstone gas reservoirs. Petrol Geol Recovery Effi 16(4):104–105,108

Yao J, Wang ZS (2007) Theory and method for well test interpretation in fractured-vuggy carbonate reservoirs. China Petroleum University Press, Dongying

Yao J, Sun H, Fan D Y et al (2013) Transport mechanisms and numerical simulation of shale gas reservoirs. J China Univ Petrol (Ed Nat Sci) 37(1):91–98

Yao J, Dai WH, Wang ZS (2004) Well test interpretation method for triple Media Reservoir with Variable Wellbore Storage.J Univ Petrol China (Ed Nat Sci) 28(1):46–51

Yao J, Huang CQ, Wang ZS et al (2010) Mathematical model of fluid flow in fractured vuggy reservoirs based on discrete fracture-vug network. Acta PetroleiSinica 31(5):15–20

Yao J, Huang CQ et al (2014) Fractured vuggy carbonate reservoir simulation. China Petroleum University Press,Beijing, pp 48–51

Yin D (1983) On the multiple porosity media model and its application on pressure build-up curve analysis. Petrol Explor Dev 16(3):59–64

Zheng M, Li JZ, Wu XZ et al (2016) Physical modeling of oil charging in tight reservoirs. Petroleum Explor Dev 43(2):219–227

Zhou KM, Li N, Yuan XL (2003) Gas percolation mechanism of low permeability reservoirs with low speed under residual water conditions. Nat Gas Ind 23(6):103–106

Zhu WY, Liu JZ, Song H Q, Sun YK, Wang M (2010) Calculation of effective startup degree of non-darcy flow in low or ultra-low permeability reservoirs. ActaPetroleiSinica 31(3):453–457

Zhang X (2013) Study on the imbibition law of low-permeability sandstone reservoirs. Sinopec Press, Beijing

Zhao ZZ, Du JH, Zou CN et al (2012) Tight oil and gas. Petroleum Industry Press, Beijing

Zou CN, Zhang GS, Yang Z, Tao SZ, Hou LH, Zhu RK, Yuan XJ, Ran QQ, Li DH, Wang ZP (2103) Geological concepts, characteristics, resource potential and key techniques of unconventional hydrocarbon: on unconventional petroleum geology. Petrol Explor Dev 40(4):385–399

3 Mathematical Model of Multiphase Flow in Multiple Media at Different Scales

Tight reservoirs have the natural/hydraulic fractures and nano/micro-scale porous media at different scales. The pore-fracture media at different scales have their own geometrical and attribute characteristics with continuous/discontinuous distributions. Fluids in such media system have large differences in composition, properties and occurrence states, and they have their own flow regimes and flow mechanisms. Parameters for multiple media at different scales and fluids have the continuous/discontinuous variation characteristics. Simultaneously, the processes of fracturing, injection and exploitation with a strong flow-geomechanics coupling process can result in the dynamic changes of geometrical parameters, attribute parameters, conductivity between media and well productivity indexes. It is difficult to describe the characteristics of multiple scales and multiple media, also including the multiple flow regimes, complicated flow mechanisms and flow-geomechanics coupling effects in pore-fracture media at different scales by the conventional dual media mathematical modeling method. Therefore, the newly proposed method not only break the conventional multiphase flow theory for dual media but also innovatively extended the discontinuous numerical simulation theory for multiple media and develop the corresponding mathematical model of multiphase flow. It makes the transitions from single media and dual media to multiple media, from single regime flow to multiple regimes and multiple mechanisms, and also from the fluid flow theory in continuous porous media to fluid flow theory in discontinuous porous media (Table 3.1).

3.1 Mathematical Model of Multiphase Flow in Multiple Media for Tight Oil Reservoirs

Nano/micro-pores and natural/hydraulic fractures at different scales can be found in tight reservoirs. They have the characteristics of multiple scales and multiple media. In reservoirs, different media show a continuous/discontinuous distribution in space, and their properties are also continuous/discontinuous. Simultaneously, in different media, the fluids also have different compositions, physical properties and occurrence states. And the flow regimes and flow mechanisms are also different. There are the complicated characteristics of interporosity flow between different media. From the characteristics of tight oil reservoirs above, a mathematical model for fluid flow in the multiple media of tight oil reservoirs (Han et al. 1993; Chen et al. 2006; Wu 2016) can be established. The model consists of flow items, source/sink items, and cumulative items.

$$\underbrace{-\nabla\cdot(\rho_{\mathrm{p}}\boldsymbol{v}_{\mathrm{p}})}_{\text{flow}} + \underbrace{q_{\mathrm{p}}^{\mathrm{W}}}_{\text{source/sink}} = \underbrace{\frac{\partial(\phi S_{\mathrm{p}}\rho_{\mathrm{p}})}{\partial t}}_{\text{cumulation}} \quad (3.1)$$

where, the subscript p = o, g, w, the subscripts o, g, and w refer to the oil, gas, and water phases; ρ_{p} is the density of the p phase, g/cm^3; ϕ is the porosity, dimensionless; S_{p} is the saturation of the p phase, dimensionless; t is time, s; $q_{\mathrm{p}}^{\mathrm{W}}$ is the production rate of the p phase,g; v is the flow velocity, m/s.

$$\boldsymbol{v}_{\mathrm{p}} = -\frac{KK_{\mathrm{rp}}}{\mu_{\mathrm{p}}}\nabla\phi_{\mathrm{p}} = -\frac{KK_{\mathrm{rp}}}{\mu_{\mathrm{p}}}\nabla(P-\rho gD)_{\mathrm{p}} \quad (3.2)$$

where, K is the absolute permeability, mD; μ_{p} is the viscosity of the p phase, mPa • s; K_{rp} is the relative permeability of the p phase, dimensionless; ϕ is the potential function, Mpa; ∇P is the pressure gradient, MPa/m; g is the acceleration of gravity, m/s^2; D is the vertical depth, m.

3.1.1 Mathematical Model of Multiphase Flow in Continuous Single Media

For the porous tight oil reservoirs, only porous media can be observed. A porous media system usually contains a single porous media with similar geometrical and attribute

Q. Ran, *Unconventional Tight Reservoir Simulation: Theory, Technology and Practice*,
https://doi.org/10.1007/978-981-32-9848-4_3

Table 3.1 Comparison of models for multiplemedia at different scales

Multiple media model			Diagram	Model description	Application Condition
Multiple media model	Single media model		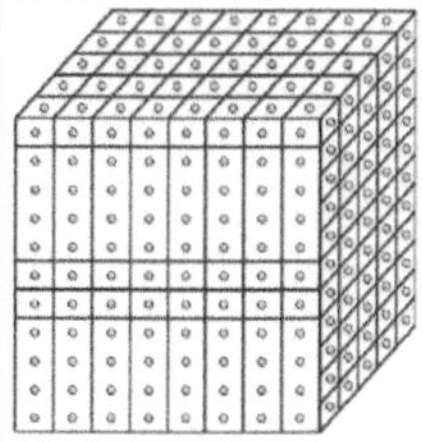	① Only one system of single-porosity media ② The physical properties, geometrical and flow characteristics of porous media and fluids are continuously changing, which can be described by continuity equations	Tight reservoirs of porous media, and have single porous media with the similar geometry and properties. The properties of fluids in these reservoirs are similar and the fluids distribute continuously in reservoirs. The flow mechanism is simple and mainly obeys the continuous Darcy flow
	Dual media model	Dual porosity single permeability model	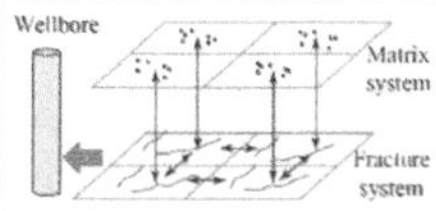	① Two systems of the matrix and the fracture are in this model. The matrix system has properties of porous media and the fracture system has properties of media with fractures ② The physical properties, geometrical and flow characteristics of the same system continuously change, and they can be described by continuity equations. These two systems are overlapped in space to form the dual media ③ The mass exchange of fluids occurs between the matrix system and the fractures system. The fluid flow does not occur between different matrix systems, and between matrix systems and wellbores. However, the fluid flow occurs between different fracture systems, and between fracture systems and wellbores	Tight reservoirs have fractures and pores, while the pore permeability is quite small so the fluid is hard to flow through matrix Geometrical morphology, properties and flow characteristics of the matrixes and the fractures continuously change. The flow mechanism is simple and mainly obeys the continuous Darcy flow
		Dual porosity dual permeability model	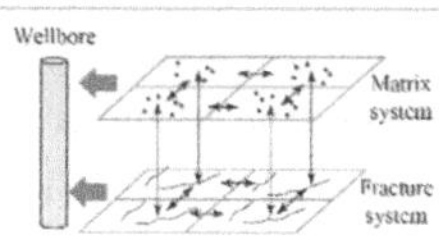	① Two systems of the matrix and the fracture are in this model. The matrix system has properties of porous media and the fracture system has properties of media with fractures ② The physical properties, geometrical and flow characteristics of the same system continuously change, and they can be described by continuity equations. These two systems are overlapped in space to form the dual media ③Flow occurs not only inside the two systems respectively but also between the matrix system and the fractures system. Fluid exchange occurs between the matrix system and the wellbore, the fractures	Fractures and pores are developed in tight reservoirs, and the pore throat has certain conductivity ability, multiphase flows slowly in the matrix system Geometrical morphology, properties and flow characteristics of the matrix system and the fractures system continuously change, respectively. The flow mechanism is simple and mainly the continuous Darcy flow

(continued)

Table 3.1 (continued)

Multiple media model		Diagram	Model description	Application Condition
			system and wellbore, respectively	
	Continuous multiple media model	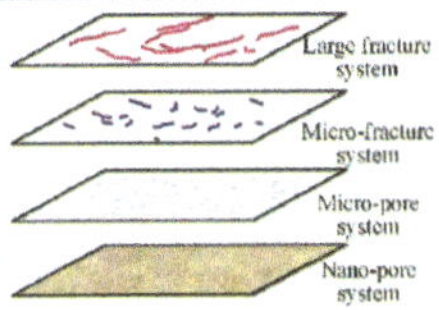	① In this model, the reservoir is classified into many systems of pores and fractures at different scales. The same system has its special properties which continuously distribute in space and can be described by a continuity equation. These systems are overlapped in space to form the multiple media ② Continuous fluid distribution within each system. Fluid has continuous and similar flow characteristics in each system ③ The fluid exchange occurs between different systems and it can be described by mass exchange term in mathematical model	Tight reservoirs have pores and natural/hydraulic fractures at different scales. All systems have conductivity ability Geometrical morphology, properties and flow characteristics of the systems are continuously changing, respectively. The flow mechanism is simple and mainly obeys the continuous Darcy flow
Discontinuous multiple media model	Discontinuous multiple media model	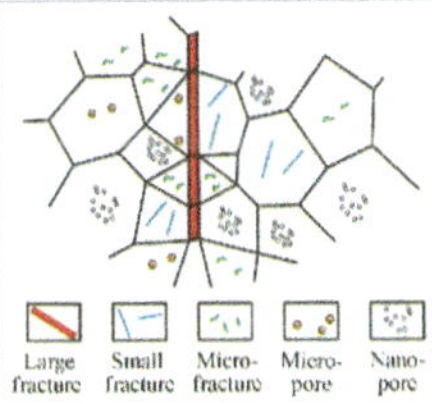	① Only one system developed in this model. This system is consisted of media at different scales pores and fractures. These media are interacted with each other, and discontinuously and discretely distributed in reservoirs ② The pores and fractures at different scales are treated as discrete multiple media at different scales. The properties and flow characteristics of each media discontinuously change ③ The fluid flow through different media have various flow regimes and mechanisms, which can be described as different equations, respectively ④ The fluid flow between different media is described by flow term. The term of interporosity flow does not exist	Tight reservoirs have media at different scales pores and fractures. These media are interacted with each other Geometrical morphology and properties characteristics of different media discontinuously change There are large differences of flow mechanisms and flow regimes between different media. The flow exchanging behavior can occur between all these media

characteristics. The fluid properties are similar and show a continuous distribution. The flow mechanism is relatively simple, and the main flow process is a continuous Darcy flow. This type of tight reservoir with a single pore media can be simplified to a continuous single medium model (Table 3.1). It has a continuous distribution of porous media, and the oil-gas-water three-phase fluids coexist. Porous media and fluid properties are continuously changed in reservoir. It can be characterized by the continuity equations.

A mathematical model for the three-phases flow of oil-gas-water in a single continuous medium is developed for the porous tight oil reservoirs.

(1) Flow equation of the oil phase

All of the oil components are composed of oil phase fluids, and the flow equation is

$$\nabla \cdot \left(\rho_o \frac{KK_{ro}}{\mu_o} \nabla P_o - \rho_o \frac{KK_{ro}}{\mu_o} \rho_{og} g \nabla D \right) + q_o^W = \frac{\partial}{\partial t} (\phi S_o \rho_o) \tag{3.3}$$

(2) Flow equation of the gas phase

The gas components are composed of free gas components in the gas phase and dissolved gas components in the oil phase, and the flow equation is

$$\begin{aligned} &\nabla \cdot \left[\begin{matrix} (\rho_g \frac{KK_{rg}}{\mu_g} \nabla P_g - \rho_g \frac{KK_{rg}}{\mu_g} \rho_g g \nabla D) \\ + (\rho_{gd} \frac{KK_{ro}}{\mu_o} \nabla P_o - \rho_{gd} \frac{KK_{ro}}{\mu_o} \rho_{og} g \nabla D) \end{matrix} \right] + q_g^W \\ &= \frac{\partial}{\partial t} [\phi (S_o \rho_{gd} + S_g \rho_g)] \end{aligned} \tag{3.4}$$

(3) Flow equation of the water phase

All of the water components are composed of water phase fluids, and the flow equation is

$$\begin{aligned} &\nabla \cdot \left(\rho_w \frac{KK_{rw}}{\mu_w} \nabla P_w - \rho_w \frac{KK_{rw}}{\mu_w} \rho_w g \nabla D \right) + q_w^W \\ &= \frac{\partial}{\partial t} (\phi S_w \rho_w) \end{aligned} \tag{3.5}$$

In this model, the flow term describes the multiphase flow behavior under the effects of viscous flow, gravity drainage, dissolved gas drive, elastic expansion. The source/sink term describes the injection rate and production rate. The cumulative term describes the changes of fluid accumulation under the effects of various mechanisms.

3.1.2 Mathematical Model of Multiphase Flow in Continuous Dual Media

For the tight reservoirs with porous media and fractures, both of them have their own geometrical shapes, attribute characteristics, and flow characteristics. Each of them is continuously distributed in reservoirs, and the attribute properties are also continuously changed. The flow mechanism is relatively simple, and the main flow process is a continuous Darcy flow. This system can be simplified into a dual-media model (Barrenblatt et al. 1960; Warren and Root 1963). It is composed of a matrix system and a fracture system. The matrix system has a unique porous media system, and the fracture system has a unique system of media with fractures. The same system can be described by a continuous equation in space, and both of them can form a dual-media system in a reservoir.

According to the internal multiphase flow process in porous media and its relationship with wellbore, it can be classified into a dual-porosity single-permeability model and a dual-porosity dual-permeability model.

3.1.2.1 Mathematical Model of Multiphase Flow in the Dual Media of Dual-Porosity and Single-Permeability

For tight reservoirs with a nano-scale porous matrix, a small pore throat radius and poor physical properties, during the recovery process, the fluids in matrix system are hard to flow, and cannot directly flow from matrix to wellbore. Multiphase flow occurs firstly from matrix system to fracture system, and then directly goes to the wellbore from the fracture system.

This kind of tight reservoir can be simplified into a dual-media system of dual-porosity and single-permeability. In this model, multiphase flow does not occur in the matrix system, and no multiphase flow occurs between matrix and wellbore. Fluids can flow from the matrix system to the fracture system and then to the wellbore (Table 3.1). In view of the above characteristics, the multiphase mathematical model in the dual-porosity single-permeability system can be established:

(1) The pore system

Equation for the oil phase flow:

$$-\tau_{omf} = \frac{\partial}{\partial t} (\phi \rho_o S_o)_m \tag{3.6}$$

Equation for the gas phase flow:

$$-\tau_{gmf} - \tau_{gdmf} = \frac{\partial}{\partial t} [\phi (\rho_g S_g + \rho_{gd} S_o)]_m \tag{3.7}$$

Equation for the water phase flow:

$$-\tau_{cow,mf} = \frac{\partial}{\partial t} (\phi \rho_w S_w)_m \tag{3.8}$$

(2) The fracture system

Equation for the oil phase flow:

$$-\nabla \cdot (\rho_o v_o)_f + \tau_{omf} + q_{of}^W = \frac{\partial}{\partial t} (\phi \rho_o S_o)_f \tag{3.9}$$

Equation for the gas phase flow:

$$-\nabla \cdot (\rho_g \boldsymbol{v}_g + \rho_{gd} \boldsymbol{v}_o)_f + \tau_{gmf} + \tau_{gdmf} + q_{gf}^W = \frac{\partial}{\partial t}[\phi(\rho_g S_g + \rho_{gd} S_o)]_f \quad (3.10)$$

Equation for the water phase flow:

$$-\nabla \cdot (\rho_w \boldsymbol{v}_w)_f + \tau_{cow,mf} + q_{wf}^W = \frac{\partial(\phi \rho_w S_w)_f}{\partial t} \quad (3.11)$$

where, τ_{cow} is the fluid exchanging term of the water phase flow equation with the consideration of the imbibition between the media systems; the subscripts f, m refer to the fracture system and matrix system, respectively; ρ_{gd} is the density of solution gas, g/cm^3.

In the model, the flow term shows that the fracture system can flow internally and there is no flow in the matrix system. The fluid exchanging term reflects the behavior of inter-porosity flow between matrix and fracture. The source/sink not only shows the injection rate and the production rate, but also manifests the flow behavior between fracture and wellbore. But there is no flow between matrix and wellbore (Table 3.2).

3.1.2.2 Mathematical Model of Multiphase Flow in the Dual Media with Dual-Porosity and Dual-Permeability

For the tight reservoirs where small pores and micro-pores are developed in matrix, and the pore-throat system also has a certain conductivity, during the recovery process, fluids can flow in the matrix system. Simultaneously, multiphase flow occurs not only from matrix to fracture, but also from matrix to wellbore.

This kind of tight reservoir can be simplified into a dual-porosity dual-permeability model. In this model, reservoir fluids can flow in the matrix and fracture systems, and between the matrix and fracture systems. Simultaneously, the reservoir fluids can also exchange between the matrix/fracture system and wellbore (Table 3.1). In view of the characteristics above, the mathematical model of multiphase flow in the dual-porosity dual-permeability system can be established:

(1) The pore system

Equation for the oil phase flow:

$$-\nabla \cdot (\rho_o \boldsymbol{v}_o)_m - \tau_{omf} + q_{om}^W = \frac{\partial}{\partial t}(\phi \rho_o S_o)_m \quad (3.12)$$

Equation for the gas phase flow:

$$-\nabla \cdot (\rho_g \boldsymbol{v}_g + \rho_{gd} \boldsymbol{v}_o)_m - \tau_{gmf} - \tau_{gdmf} + q_{gm}^W = \frac{\partial}{\partial t}[\phi(\rho_g S_g + \rho_{gd} S_o)]_m \quad (3.13)$$

Equation for the water phase flow:

$$-\nabla \cdot (\rho_w \boldsymbol{v}_w)_m - \tau_{cow,mf} + q_{wm}^W = \frac{\partial}{\partial t}(\phi \rho_w S_w)_m \quad (3.14)$$

(2) The fracture system

Equation for the oil phase flow:

$$-\nabla \cdot (\rho_o \boldsymbol{v}_o)_f + \tau_{omf} + q_{of}^W = \frac{\partial}{\partial t}(\phi \rho_o S_o)_f \quad (3.15)$$

Equation for the gas phase flow:

$$-\nabla \cdot (\rho_g \boldsymbol{v}_g + \rho_{gd} \boldsymbol{v}_o)_f + \tau_{gmf} + \tau_{gdmf} + q_{gf}^W = \frac{\partial}{\partial t}[\phi(\rho_g S_g + \rho_{gd} S_o)]_f \quad (3.16)$$

Equation for the water phase flow:

$$-\nabla \cdot (\rho_w \boldsymbol{v}_w)_f + \tau_{cow,mf} + q_{wf}^W = \frac{\partial(\phi \rho_w S_w)_f}{\partial t} \quad (3.17)$$

In the model, the flow term reflects the multiphase flow behavior in the matrix and fracture systems. The fluid exchanging term shows the fluid exchanging behavior between the matrix and fracture systems. The source/sink term shows the injection rate and production rate and also indicates the multiphase flow behavior among matrix, fractures and wellbore (Table 3.3).

3.1.3 Mathematical Model of Multiphase Flow in Continuous Multiple Media

3.1.3.1 Mathematical Model of Multiphase Flow in Multiple Media Based on the Dual Porosity Model

For the tight oil reservoirs where the different scale pores and natural/hydraulic fractures (small pores, nano-pores, large scale fractures and micro-scale fractures) are developed, the media can interact with each other in the same system, and the geometrical and attribute parameters are continuously distributed. Also, both the two systems can superpose with each other to form the multiple media. During a recovery process, the fluids can flow within the matrix system, from matrix to fractures, and also from matrix to wellbore directly. This kind of tight reservoir can be simplified into a multiple media model of dual porosity system. In this model, multiphase flow is allowed within matrix and fracture systems, and between the matrix and fracture systems. Simultaneously, fluid exchanging behaviors are also allowed between the matrix-fracture systems and wellbore. For the characteristics above, a mathematical

Table 3.2 The mass exchange of fluids in dual-porosity single permeability model

Fluids		Flow term within the system		Fluid mass exchanging term between systems	Source/sink	
		Matrix system	Fracture system	Matrix-fracture	Matrix-wellbore	Fracture-wellbore
Oil		0	$-\nabla\cdot(\rho_o\boldsymbol{v}_o)_f$	$\tau_{omf}=\frac{\alpha_{fm}K_m\rho_o K_{rom}}{\mu_o}(\phi_{om}-\phi_{of})$	0	q_{of}^W
Gas	Free gas	0	$-\nabla\cdot(\rho_g\boldsymbol{v}_g)_f$	$\tau_{gmf}=\frac{\alpha_{fm}K_m\rho_g K_{rgm}}{\mu_g}(\phi_{gm}-\phi_{gf})$	0	q_{gf}^W
	Solution gas	0	$-\nabla\cdot(\rho_{gd}\boldsymbol{v}_o)_f$	$\tau_{gdmf}=\frac{\alpha_{fm}K_m\rho_{gd}K_{rom}}{\mu_o}(\phi_{om}-\phi_{of})$	0	
Water		0	$-\nabla\cdot(\rho_w\boldsymbol{v}_w)_f$	$\tau_{cow,mf}=\frac{\alpha_{fm}K_m\rho_w K_{rwm}}{\mu_w}\cdot\left[\Delta P_{o,mf}-\rho_w g\Delta D_{mf}-(P_{cow,m}-P_{cow,f})\right]$	0	q_{wf}^W

Table 3.3 The mass exchange of fluids in dual porosity dual permeability model

Fluids		Flow term within the system		Fluid mass exchanging term between systems	Source/sink	
		Matrix system	Fracture system	Matrix-fracture	Matrix system	Fracture system
Oil		$-\nabla\cdot(\rho_o\boldsymbol{v}_o)_m$	$-\nabla\cdot(\rho_o\boldsymbol{v}_o)_f$	$\tau_{omf}=\frac{\alpha_{fm}K_m\rho_o K_{rom}}{\mu_o}(\phi_{om}-\phi_{of})$	q_{om}^W	q_{of}^W
Gas	Free gas	$-\nabla\cdot(\rho_g\boldsymbol{v}_g)_m$	$-\nabla\cdot(\rho_g\boldsymbol{v}_g)_f$	$\tau_{gmf}=\frac{\alpha_{fm}K_m\rho_g K_{rgm}}{\mu_g}(\phi_{gm}-\phi_{gf})$	q_{gm}^W	q_{gf}^W
	Solution gas	$-\nabla\cdot(\rho_{gd}\boldsymbol{v}_o)_m$	$-\nabla\cdot(\rho_{gd}\boldsymbol{v}_o)_f$	$\tau_{gdmf}=\frac{\alpha_{fm}K_m\rho_{gd}K_{rom}}{\mu_o}(\phi_{om}-\phi_{of})$		
Water		$-\nabla\cdot(\rho_w\boldsymbol{v}_w)_m$	$-\nabla\cdot(\rho_w\boldsymbol{v}_w)_f$	$\tau_{cow,mf}=\frac{\alpha_{fm}K_m\rho_w K_{rwm}}{\mu_w}\cdot\left[\Delta P_{o,mf}-\rho_w g\Delta D_{mf}-(P_{cow,m}-P_{cow,f})\right]$	q_{wm}^W	q_{wf}^W

model of multiphase flow in the multiple media of dual-porosity system can be developed (Fig. 3.1).

1. Pore systems
   (1) Small pores (m1)

Equation for the oil phase flow:

$$-\nabla\cdot(\rho_o\boldsymbol{v}_o)_{m2(m1),m1}-\tau_{om1,f}+q_{om1}^W=\frac{\partial}{\partial t}(\phi\rho_o S_o)_{m1} \tag{3.18}$$

Equation for the gas phase flow:

$$\begin{aligned}&-\nabla\cdot(\rho_g\boldsymbol{v}_g+\rho_{gd}\boldsymbol{v}_o)_{m2(m1),m1}-\tau_{gm1,f}-\tau_{gdm1,f}+q_{gm1}^W\\&=\frac{\partial}{\partial t}\left[\phi(\rho_g S_g+\rho_{gd}S_o)\right]_{m1}\end{aligned} \tag{3.19}$$

Equation for the water phase flow:

$$-\nabla\cdot(\rho_W\boldsymbol{v}_W)_{m2(m1),m1}-\tau_{cow,m1,f}+q_{wm1}^W=\frac{\partial}{\partial t}(\phi\rho_w S_w)_{m1} \tag{3.20}$$

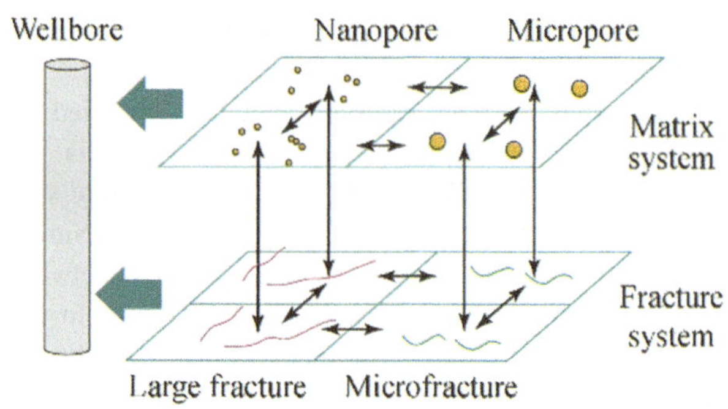

Fig. 3.1 A schematic diagram for the multiple media based on dual-permeability model

(2) Nano-pores (m2)

Equation for the oil phase flow:

$$-\nabla\cdot(\rho_o\boldsymbol{v}_o)_{m1(m2),m2}-\tau_{om2,F}+q_{om2}^W=\frac{\partial}{\partial t}(\phi\rho_o S_o)_{m2} \tag{3.21}$$

Equation for the gas phase flow:

$$-\nabla\cdot(\rho_g\mathbf{v}_g+\rho_{gd}\mathbf{v}_o)_{m1(m2),m2}-\tau_{gm2,F}-\tau_{gdm2,F}+q_{gm2}^{W}=\frac{\partial}{\partial t}\left[\phi(\rho_gS_g+\rho_{gd}S_o)\right]_{m2} \tag{3.22}$$

Equation for the water phase flow:

$$-\nabla\cdot(\rho_w\mathbf{v}_w)_{m1(m2),m2}-\tau_{cow,m2,F}+q_{wm2}^{W}=\frac{\partial}{\partial t}(\phi\rho_wS_w)_{m2} \tag{3.23}$$

2. Fracture systems

(1) Large fractures (F)

Equation for the oil phase flow:

$$-\nabla\cdot(\rho_o\mathbf{v}_o)_{f(F),F}+\tau_{om2,F}+q_{of}^{W}=\frac{\partial}{\partial t}(\phi\rho_oS_o)_F \tag{3.24}$$

Equation for the gas phase flow:

$$-\nabla\cdot(\rho_g\mathbf{v}_g+\rho_{gd}\mathbf{v}_o)_{f(F),F}+\tau_{gm2,F}+\tau_{gdm2,F}+q_{gF}^{W}=\frac{\partial}{\partial t}\left[\phi(\rho_gS_g+\rho_{gd}S_o)\right]_F \tag{3.25}$$

Equation for the water phase flow:

$$-\nabla\cdot(\rho_w\mathbf{v}_w)_{f(F),F}+\tau_{cow,m2,F}+q_{wF}^{W}=\frac{\partial(\phi\rho_wS_w)_F}{\partial t} \tag{3.26}$$

(2) Micro-fractures (f)

Equation for the oil phase flow:

$$-\nabla\cdot(\rho_o\mathbf{v}_o)_{F(f),f}+\tau_{om1,f}+q_{of}^{W}=\frac{\partial}{\partial t}(\phi\rho_oS_o)_f \tag{3.27}$$

Equation for the gas phase flow:

$$-\nabla\cdot(\rho_g\mathbf{v}_g+\rho_{gd}\mathbf{v}_o)_{F(f),f}+\tau_{gm1,f}+\tau_{gdm1,f}+q_{gf}^{W}=\frac{\partial}{\partial t}\left[\phi(\rho_gS_g+\rho_{gd}S_o)\right]_f \tag{3.28}$$

Equation for the water phase flow:

$$-\nabla\cdot(\rho_w\mathbf{v}_w)_{F(f),f}+\tau_{cow,m1,f}+q_{wf}^{W}=\frac{\partial(\phi\rho_wS_w)_f}{\partial t} \tag{3.29}$$

where, the subscript F refers to the large fracture system; f refers to the micro-fracture system; m refers to the matrix system.

In this model, the flow term indicates that the multiphase flow in different scale matrix system and different scale fracture system. The fluid exchanging term shows the exchanging behavior between matrix and fracture systems. The source/sink term indicates the injection rate and production rate, and also shows the multiphase flow behavior between the matrix-fracture system and wellbore (Table 3.4).

3.1.3.2 Mathematical Model of Multiphase Flow in Continuous Multiple Media

For the tight oil reservoirs where multiple pores and natural/hydraulic fractures are developed and their geometrical and attribute characteristics have a great difference, the same media and the internal fluids are continuously distributed in the reservoirs, and the geometrical and attribute characteristics also continuously change in the reservoirs. Different types of media can overlap with each other in space. The mass exchanging behavior also happens between different media. During the recovery process, the fluids can flow between multiple media, and also flow from media to wellbore directly. This kind of tight reservoir can be simplified to a continuous multiple media model, and each of them can be treated as an independent continuous system. Fluids are continuously distributed and communicate with each other. The spatial distribution of parameters for the same system can be described by a continuous equation. The different scale pore-fracture systems overlap with each other in space, and they can be considered as a continuous media field which is superimposed with each other. The fluid exchanging capacity between different systems (i.e., between different media) can be reflected by the exchanging rate (Table 3.1). Therefore, a numerical simulation model for the multiphase flow in an N media is composed of N mathematical models (Peng 2006); that is, each media system has a set of independent equations and each equation must have the fluid exchanging term. The multiphase flow equation has the following expression:

$$-\nabla\cdot(\rho_i\mathbf{v}_i)+\sum_{j=1,i\neq j}^{N}\tau_{ji}=\frac{\partial(\rho_i\phi_iS_i)}{\partial t} \tag{3.30}$$

where, i and j refer to the different media $i, j = 1,\ldots, N$; τ_{ji} is the fluid exchanging rate from medium j to medium i.

Generally, the fluid exchanging rate of fluid p in any two continuous media can be expressed in the form of Darcy's law:

$$\tau_{pij}=\frac{\alpha_{ij}\rho_pX_{cp}K_jK_{rpj}}{\mu_p}\left(\Phi_j-\Phi_i\right)_p \tag{3.31}$$

where, α_{ij} is the shape factor of media, dimensionless; X_{cp} is the mole fraction of component c in phase p, dimensionless.

The multiphase flow process described in this formula includes both the multiphase flow in a same media and the

Table 3.4 The fluids' mass exchange of flow through multiple media based on dual-permeability model

Flow objects			Fluids			
			Oil	Gas		Water
				Free gas	Solution gas	
Multiphase flow term within the system	Matrix system	Micro-pore	$-\nabla\cdot(\rho_o\boldsymbol{v}_o)_{m2,m1}=\nabla\cdot\left(\rho_o\frac{K_{m1}K_{rom1}}{\mu_o}\nabla\Phi_{o,m2m1}\right)$	$-\nabla\cdot(\rho_g\boldsymbol{v}_g)_{m2,m1}=\nabla\cdot\left(\rho_g\frac{K_{m1}K_{rgm1}}{\mu_g}\nabla\Phi_{g,m2m1}\right)$	$-\nabla\cdot(\rho_{gd}\boldsymbol{v}_o)_{m2,m1}=\nabla\cdot\left(\rho_{gd}\frac{K_{m1}K_{rom1}}{\mu_o}\nabla\Phi_{o,m2m1}\right)$	$-\nabla\cdot(\rho_w\boldsymbol{v}_w)_{m2,m1}=\nabla\cdot\left(\rho_w\frac{K_{m1}K_{rwm1}}{\mu_w}\nabla\Phi_{w,m2m1}\right)$
		Nano-pore	$-\nabla\cdot(\rho_o\boldsymbol{v}_o)_{m1,m2}=\nabla\cdot\left(\rho_o\frac{K_{m2}K_{rom2}}{\mu_o}\nabla\Phi_{o,m1m2}\right)$	$-\nabla\cdot(\rho_g\boldsymbol{v}_g)_{m1,m2}=\nabla\cdot\left(\rho_g\frac{K_{m2}K_{rgm2}}{\mu_g}\nabla\Phi_{g,m1m2}\right)$	$-\nabla\cdot(\rho_{gd}\boldsymbol{v}_o)_{m1,m2}=\nabla\cdot\left(\rho_{gd}\frac{K_{m2}K_{rom2}}{\mu_o}\nabla\Phi_{o,m1m2}\right)$	$-\nabla\cdot(\rho_w\boldsymbol{v}_w)_{m1,m2}=\nabla\cdot\left(\rho_w\frac{K_{m2}K_{rwm2}}{\mu_w}\nabla\Phi_{w,m1m2}\right)$
	Fracture system	Large fracture	$-\nabla\cdot(\rho_o\boldsymbol{v}_o)_{f,F}=\nabla\cdot\left(\rho_o\frac{K_F K_{roF}}{\mu_o}\nabla\Phi_{o,fF}\right)$	$-\nabla\cdot(\rho_g\boldsymbol{v}_g)_{f,F}=\nabla\cdot\left(\rho_g\frac{K_F K_{rgF}}{\mu_g}\nabla\Phi_{g,fF}\right)$	$-\nabla\cdot(\rho_{gd}\boldsymbol{v}_o)_{f,F}=\nabla\cdot\left(\rho_{gd}\frac{K_F K_{roF}}{\mu_o}\nabla\Phi_{o,fF}\right)$	$-\nabla\cdot(\rho_w\boldsymbol{v}_w)_{f,F}=\nabla\cdot\left(\rho_w\frac{K_F K_{rwF}}{\mu_w}\nabla\Phi_{w,fF}\right)$
		Micro-facture	$-\nabla\cdot(\rho_o\boldsymbol{v}_o)_{F,f}=\nabla\cdot\left(\rho_o\frac{K_f K_{rof}}{\mu_o}\nabla\Phi_{o,Ff}\right)$	$-\nabla\cdot(\rho_g\boldsymbol{v}_g)_{F,f}=\nabla\cdot\left(\rho_g\frac{K_f K_{rgf}}{\mu_g}\nabla\Phi_{g,Ff}\right)$	$-\nabla\cdot(\rho_{gd}\boldsymbol{v}_o)_{F,f}=\nabla\cdot\left(\rho_{gd}\frac{K_f K_{rof}}{\mu_o}\nabla\Phi_{o,Ff}\right)$	$-\nabla\cdot(\rho_w\boldsymbol{v}_w)_{F,f}=\nabla\cdot\left(\rho_w\frac{K_f K_{rwf}}{\mu_w}\nabla\Phi_{w,Ff}\right)$
Fluid mass exchanging term between systems	Micro-pores-micro-facture		$\tau_{om1f}=\frac{\alpha_{m1f}K_{m1}\rho_o K_{rom1}}{\mu_o}(\Phi_{om1}-\Phi_{of})$	$\tau_{gm1f}=\frac{\alpha_{m1f}K_{m1}\rho_g K_{rgm1}}{\mu_g}(\Phi_{gm1}-\Phi_{gf})$	$\tau_{gdm1f}=\frac{\alpha_{m1f}K_{m1}\rho_{gd}K_{rom1}}{\mu_o}(\Phi_{om1}-\Phi_{of})$	$\tau_{cow,m1,f}=\frac{\alpha_{fm1}K_{m1}\rho_w K_{rwm1}}{\mu_w}\cdot\left[\Delta P_{o,m1f}-\rho_w g\Delta D_{m1f}-(P_{cow,m1}-P_{cow,f})\right]$
	Nano-pores-large fracture		$\tau_{om2F}=\frac{\alpha_{m2F}K_{m2}\rho_o K_{rom2}}{\mu_o}(\Phi_{om2}-\Phi_{oF})$	$\tau_{gm2F}=\frac{\alpha_{m2F}K_{m2}\rho_g K_{rgm2}}{\mu_g}(\Phi_{om2}-\Phi_{oF})$	$\tau_{gdm2F}=\frac{\alpha_{m2F}K_{m2}\rho_{gd}K_{rom2}}{\mu_o}(\Phi_{om2}-\Phi_{oF})$	$\tau_{cow,m2F}=\frac{\alpha_{Fm2}K_{m2}\rho_w K_{rwm2}}{\mu_w}\cdot\left[\Delta P_{o,m2F}-\rho_w g\Delta D_{m2F}-(P_{cow,m2}-P_{cow,F})\right]$
Source/sink	Micro-pore-wellbore		q_{om1}^{W}	q_{gm1}^{W}		q_{wm1}^{W}
	Nano-pore-wellbore		q_{om2}^{W}	q_{gm2}^{W}		q_{wm2}^{W}
	Large fracture-wellbore		q_{oF}^{W}	q_{gF}^{W}		q_{wF}^{W}
	Micro-facture-wellbore		q_{of}^{W}	q_{gf}^{W}		q_{wf}^{W}

fluid exchanging process between the different media. The multiphase flow term is an important symbol for the multiple media model (Liu and Guo 1982).

(1) Small pores (m1)

Equation for the oil phase flow:

$$-\nabla\cdot(\rho_o\boldsymbol{v}_o)_{m1}-\tau_{om1F}-\tau_{om1f}+\tau_{om2m1}+q_{om1}^{W}=\frac{\partial}{\partial t}(\phi\rho_o S_o)_{m1} \tag{3.32}$$

Equation for the gas phase flow:

$$-\nabla\cdot(\rho_g\boldsymbol{v}_g+\rho_{gd}\boldsymbol{v}_o)_{m1}-\tau_{gm1F}-\tau_{gm1f}+\tau_{gm2m1}-\tau_{gdm1F}-\tau_{gdm1f}+\tau_{gdm2m1}+q_{gm1}^{W}=\frac{\partial}{\partial t}\left[\phi(\rho_g S_g+\rho_{gd}S_o)\right]_{m1} \tag{3.33}$$

Equation for the water phase flow:

$$-\nabla\cdot(\rho_w\boldsymbol{v}_w)_{m1}-\tau_{cow,m1F}-\tau_{cow,m1f}+\tau_{cow,m2m1}+q_{wm1}^{W}=\frac{\partial}{\partial t}(\phi\rho_w S_w)_{m1} \tag{3.34}$$

(2) Nano-pores (m2)

Equation for the oil phase flow:

$$-\nabla\cdot(\rho_o\boldsymbol{v}_o)_{m2}-\tau_{om2F}-\tau_{om2f}-\tau_{om2m1}+q_{om2}^{W}=\frac{\partial}{\partial t}(\phi\rho_o S_o)_{m2} \tag{3.35}$$

Equation for the gas phase flow:

$$-\nabla\cdot(\rho_g\boldsymbol{v}_g+\rho_{gd}\boldsymbol{v}_o)_{m2}-\tau_{gm2F}-\tau_{gm2f}-\tau_{gm2m1}-\tau_{gdm2F}-\tau_{gdm2f}-\tau_{gdm2m1}+q_{gm2}^{W}=\frac{\partial}{\partial t}\left[\phi(\rho_g S_g+\rho_{gd}S_o)\right]_{m2} \tag{3.36}$$

Equation for the water phase flow:

$$-\nabla\cdot(\rho_w\boldsymbol{v}_w)_{m2}-\tau_{cow,m2F}-\tau_{cow,m2f}-\tau_{cow,m2m1}+q_{wm2}^{W}=\frac{\partial}{\partial t}(\phi\rho_w S_w)_{m2} \tag{3.37}$$

(3) Large fractures (F)

Equation for the oil phase flow:

$$-\nabla\cdot(\rho_o\boldsymbol{v}_o)_F+\tau_{om1F}+\tau_{om2F}+\tau_{ofF}+q_{of}^{W}=\frac{\partial}{\partial t}(\phi\rho_o S_o)_F \tag{3.38}$$

Equation for the gas phase flow:

$$-\nabla\cdot(\rho_g \boldsymbol{v}_g + \rho_{gd}\boldsymbol{v}_o)_F + \tau_{gm1F} + \tau_{gm2F} + \tau_{gfF} + \tau_{gdm1F} + \tau_{gdm2F} + \tau_{gdfF} + q_{gF}^W = \frac{\partial}{\partial t}[\phi(\rho_g S_g + \rho_{gd} S_o)]_F \tag{3.39}$$

Equation for the water phase flow:

$$-\nabla\cdot(\rho_w \boldsymbol{v}_w)_F + \tau_{cow,m1F} + \tau_{cow,m2F} + \tau_{cow,fF} + q_{wF}^W = \frac{\partial(\phi\rho_w S_w)_F}{\partial t} \tag{3.40}$$

(4) Micro-fractures (f)

Equation for the oil phase flow:

$$-\nabla\cdot(\rho_o \boldsymbol{v}_o)_f + \tau_{om1f} + \tau_{om2f} - \tau_{ofF} + q_{of}^W = \frac{\partial}{\partial t}(\phi\rho_o S_o)_f \tag{3.41}$$

Equation for the gas phase flow:

$$-\nabla\cdot(\rho_g \boldsymbol{v}_g + \rho_{gd}\boldsymbol{v}_o)_f + \tau_{gm1f} + \tau_{gm2f} - \tau_{gfF} + \tau_{gdm1f} + \tau_{gdm2f} - \tau_{gdfF} + q_{gf}^W = \frac{\partial}{\partial t}[\phi(\rho_g S_g + \rho_{gd} S_o)]_f \tag{3.42}$$

Equation for the water phase flow:

$$-\nabla\cdot(\rho_w \boldsymbol{v}_w)_f + \tau_{cow,m1f} + \tau_{cow,m2f} - \tau_{cow,fF} + q_{wf}^W = \frac{\partial(\phi\rho_w S_w)_f}{\partial t} \tag{3.43}$$

In this model, the flow term indicates the multiphase flow in each medium. The fluid exchanging term indicates the exchanging behavior in the continuous multiple media system. The source/sink term indicates the injection rate and production rate, and also shows the multiphase flow behavior between multiple media and wellbore (Table 3.5).

3.1.4 Mathematical Model of Multiphase Flow in Discontinuous Multiple Media

In tight reservoirs, the media of pores, nano/micro-fractures and natural/hydraulic discrete fractures are discontinuously distributed in the reservoirs. The geometrical and attribute parameters of each media are discontinuously changed. The flow mechanisms and the flow regimes of different media are different. Fluid exchanging behavior can happen between any different media. The different distribution features of multiple media affect the coupling flow between media and the production performance.

This kind of tight reservoir can be simplified to a discontinuous multiple media model, and this model has only one system. This system can be classified into a number of independent, not superposed units. The different units represent different media, and the unit distribution is the same with the actual space distribution of multiple media. The flow behavior between different units represents the complicated multiphase flow between multiple media (Table 3.1).

A discontinuous multiple media model is a system which is adjacent with each other but has the discontinuous variation of attribute parameters. Therefore, the discontinuous multiple media model just only has one set of equations. Every unit in this model can be characterized by the corresponding equation, and multiphase flow mechanisms in different units are different. Both the flow regimes and equations are also different. According to the discontinuous multiple media theory, the mathematical model of discontinuous multiple media based on a feature element is as follows:

$$\sum_{j=1}^{N} \varepsilon_{i,j}(\rho_p \boldsymbol{v}_p)_{j,i} + q_{pi}^W = \frac{\partial}{\partial t}(V\phi \sum_{p} S_p \rho_p X_{cp})_i \tag{3.44}$$

where, i, j refer to different units; ε_{ij} is the geometric operator between unit i and unit j; V is the volume of a grid block, cm^3; ϕ is the porosity, dimensionless.

3.1.4.1 Integrated Model of Multiphase Flow in Discontinuous Multiple Media

For the tight oil reservoirs where the multiple media at different scales are developed, including large fractures (F), small fractures (f1), micro-fractures (f2), small pores (M), micro-pores (m1), and nano-pores (m2) (Wu and Ge 1983; Wang and Pan 2016), the multiphase flow model for a discontinuous multiple medium which can describe the different scale pores and different scale fractures is established, as follows:

(1) Large fractures (F)

Equation for the oil phase flow:

$$\sum_{J=1}^{n_1} \varepsilon_{J,F}(\rho_o \boldsymbol{v}_o)_{J,F} + q_{oF}^W = \frac{\partial}{\partial t}(V\phi S_o \rho_o)_F \tag{3.45}$$

Equation for the gas phase flow:

$$\sum_{J=1}^{n_1} \varepsilon_{J,F}(\rho_g \boldsymbol{v}_g + \rho_{gd}\boldsymbol{v}_o)_{J,F} + q_{oF}^W = \frac{\partial}{\partial t}[V\phi(\rho_g S_g + \rho_{gd} S_o)]_F \tag{3.46}$$

Equation for the water phase flow:

$$\sum_{J=1}^{n_1} \varepsilon_{J,F}(\rho_w \boldsymbol{v}_w)_{J,F} + q_{wF}^W = \frac{\partial}{\partial t}(V\phi S_w \rho_w)_F \tag{3.47}$$

Table 3.5 The fluids' exchange term in model of flow through multiple media

Flow objects		Fluids			
		Oil	Gas		Water
			Free gas	Solution gas	
Multiphase flow term within the system	Micro-pore	$-\nabla \cdot (\rho_o \boldsymbol{v}_o)_{m1}$	$-\nabla \cdot (\rho_g \boldsymbol{v}_g)_{m1}$	$-\nabla \cdot (\rho_{gd} \boldsymbol{v}_o)_{m1}$	$-\nabla \cdot (\rho_W \boldsymbol{v}_W)_{m1}$
	Nano-pore	$-\nabla \cdot (\rho_o \boldsymbol{v}_o)_{m2}$	$-\nabla \cdot (\rho_g \boldsymbol{v}_g)_{m2}$	$-\nabla \cdot (\rho_{gd} \boldsymbol{v}_o)_{m2}$	$-\nabla \cdot (\rho_W \boldsymbol{v}_W)_{m2}$
	Large fracture	$-\nabla \cdot (\rho_o \boldsymbol{v}_o)_F$	$-\nabla \cdot (\rho_g \boldsymbol{v}_g)_F$	$-\nabla \cdot (\rho_{gd} \boldsymbol{v}_o)_F$	$-\nabla \cdot (\rho_W \boldsymbol{v}_W)_F$
	Micro-facture	$-\nabla \cdot (\rho_o \boldsymbol{v}_o)_f$	$-\nabla \cdot (\rho_g \boldsymbol{v}_g)_f$	$-\nabla \cdot (\rho_{gd} \boldsymbol{v}_o)_f$	$-\nabla \cdot (\rho_W \boldsymbol{v}_W)_f$
Fluid mass exchanging term between systems	Micro-pore-nano-pore	$\tau_{om2m1} = \frac{\alpha_{m2m1} K_{m2} \rho_o K_{rom2}}{\mu_o} (\Phi_{om2} - \Phi_{om1})$	$\tau_{gm2m1} = \frac{\alpha_{m2m1} K_{m2} \rho_g K_{rgm2}}{\mu_g} (\Phi_{gm2} - \Phi_{gm1})$	$\tau_{gdm2m1} = \frac{\alpha_{m2m1} K_{m2} \rho_{gd} K_{rom2}}{\mu_o} (\Phi_{om2} - \Phi_{om1})$	$\tau_{cow,m2,m1} = \frac{\alpha_{fm2} K_{m2} \rho_W K_{rwm2}}{\mu_W} \cdot \left[\Delta P_{o,m2m1} - \rho_W g \Delta D_{m2m1} - (P_{cow,m2} - P_{cow,m1})\right]$
	Micro-pore-micro-facture	$\tau_{om1f} = \frac{\alpha_{m1f} K_{m1} \rho_o K_{rom1}}{\mu_o} (\Phi_{om1} - \Phi_{of})$	$\tau_{gm1f} = \frac{\alpha_{m1f} K_{m1} \rho_g K_{rgm1}}{\mu_g} (\Phi_{gm1} - \Phi_{gf})$	$\tau_{gdm1f} = \frac{\alpha_{m1f} K_{m1} \rho_{gd} K_{rom1}}{\mu_o} (\Phi_{om1} - \Phi_{of})$	$\tau_{cow,m1,f} = \frac{\alpha_{fm1} K_{m1} \rho_W K_{rwm1}}{\mu_W} \cdot \left[\Delta P_{o,m1f} - \rho_W g \Delta D_{m1f} - (P_{cow,m1} - P_{cow,f})\right]$
	Micro-pore-large fracture	$\tau_{om1F} = \frac{\alpha_{m1f} K_{m1} \rho_o K_{rom1}}{\mu_o} (\Phi_{om1} - \Phi_{oF})$	$\tau_{gm1F} = \frac{\alpha_{m1f} K_{m1} \rho_g K_{rgm1}}{\mu_g} (\Phi_{gm1} - \Phi_{gF})$	$\tau_{gdm1F} = \frac{\alpha_{m1f} K_{m1} \rho_{gd} K_{rom1}}{\mu_o} (\Phi_{om1} - \Phi_{oF})$	$\tau_{cow,m1,F} = \frac{\alpha_{Fm1} K_{m1} \rho_W K_{rwm1}}{\mu_W} \cdot \left[\Delta P_{o,m1F} - \rho_W g \Delta D_{m1F} - (P_{cow,m1} - P_{cow,F})\right]$
	Nano-pore-large fracture	$\tau_{om2F} = \frac{\alpha_{m2F} K_{m2} \rho_o K_{rom2}}{\mu_o} (\Phi_{om2} - \Phi_{oF})$	$\tau_{gm2F} = \frac{\alpha_{m2F} K_{m2} \rho_g K_{rgm2}}{\mu_g} (\Phi_{om2} - \Phi_{oF})$	$\tau_{gdm2F} = \frac{\alpha_{m2F} K_{m2} \rho_{gd} K_{rom2}}{\mu_o} (\Phi_{om2} - \Phi_{oF})$	$\tau_{cow,m2F} = \frac{\alpha_{Fm2} K_{m2} \rho_W K_{rwm2}}{\mu_W} \cdot \left[\Delta P_{o,m2F} - \rho_W g \Delta D_{m2F} - (P_{cow,m2} - P_{cow,F})\right]$
	Nano-pore-Micro-facture	$\tau_{om2f} = \frac{\alpha_{m2f} K_{m2} \rho_o K_{rom2}}{\mu_o} (\Phi_{om2} - \Phi_{of})$	$\tau_{gm2f} = \frac{\alpha_{m2f} K_{m2} \rho_g K_{rgm2}}{\mu_g} (\Phi_{gm2} - \Phi_{gf})$	$\tau_{gdm2f} = \frac{\alpha_{m2f} K_{m2} \rho_{gd} K_{rom2}}{\mu_o} (\Phi_{om2} - \Phi_{of})$	$\tau_{cow,m2,f} = \frac{\alpha_{fm2} K_{m2} \rho_W K_{rwm2}}{\mu_W} \cdot \left[\Delta P_{o,m2f} - \rho_W g \Delta D_{m2f} - (P_{cow,m2} - P_{cow,f})\right]$
Source/sink	Micro-pore-wellbore	q_{om1}^W	q_{gm1}^W		q_{wm1}^W
	Nano-pore-wellbore	q_{om2}^W	q_{gm2}^W		q_{wm2}^W
	Large fracture-wellbore	q_{of}^W	q_{gF}^W		q_{wF}^W
	Micro-facture-wellbore	q_{of}^W	q_{gf}^W		q_{wf}^W

(2) Small fractures (f1)

Equation for the oil phase flow:

$$\sum_{J=1}^{n_2} \varepsilon_{J,\mathrm{f1}}(\rho_\mathrm{o}\boldsymbol{v}_\mathrm{o})_{J,\mathrm{f1}} + q_{\mathrm{of1}}^{\mathrm{W}} = \frac{\partial}{\partial t}(V\phi S_\mathrm{o}\rho_\mathrm{o})_{\mathrm{f1}} \tag{3.48}$$

Equation for the gas phase flow:

$$\sum_{J=1}^{n_2} \varepsilon_{J,\mathrm{f1}}(\rho_\mathrm{g}\boldsymbol{v}_\mathrm{g} + \rho_\mathrm{gd}\boldsymbol{v}_\mathrm{o})_{J,\mathrm{f1}} + q_{\mathrm{of1}}^{\mathrm{W}} = \frac{\partial}{\partial t}\left[V\phi\left(\rho_\mathrm{g}S_\mathrm{g} + \rho_\mathrm{gd}S_\mathrm{o}\right)\right]_{\mathrm{f1}} \tag{3.49}$$

Equation for the water phase flow:

$$\sum_{J=1}^{n_2} \varepsilon_{J,\mathrm{f1}}(\rho_\mathrm{w}\boldsymbol{v}_\mathrm{w})_{J,\mathrm{f1}} + q_{\mathrm{of1}}^{\mathrm{W}} = \frac{\partial}{\partial t}(V\phi S_\mathrm{w}\rho_\mathrm{w})_{\mathrm{f1}} \tag{3.50}$$

(3) Micro-fractures (f2)

Equation for the oil phase flow:

$$\sum_{J=1}^{n_3} \varepsilon_{J,\mathrm{f2}}(\rho_\mathrm{o}\boldsymbol{v}_\mathrm{o})_{J,\mathrm{f2}} + q_{\mathrm{of2}}^{\mathrm{W}} = \frac{\partial}{\partial t}(V\phi S_\mathrm{o}\rho_\mathrm{o})_{\mathrm{f2}} \tag{3.51}$$

Equation for the gas phase flow:

$$\sum_{J=1}^{n_3} \varepsilon_{J,\mathrm{f2}}(\rho_\mathrm{g}\boldsymbol{v}_\mathrm{g} + \rho_\mathrm{gd}\boldsymbol{v}_\mathrm{o})_{J,\mathrm{f2}} + q_{\mathrm{gf2}}^{\mathrm{W}} = \frac{\partial}{\partial t}\left[V\phi\left(\rho_\mathrm{g}S_\mathrm{g} + \rho_\mathrm{gd}S_\mathrm{o}\right)\right]_{\mathrm{f2}} \tag{3.52}$$

Equation for the water phase flow:

$$\sum_{J=1}^{n_3} \varepsilon_{J,\mathrm{f2}}(\rho_\mathrm{w}\boldsymbol{v}_\mathrm{w})_{J,\mathrm{f2}} + q_{\mathrm{wf2}}^{\mathrm{W}} = \frac{\partial}{\partial t}(V\phi S_\mathrm{w}\rho_\mathrm{w})_{\mathrm{f2}} \tag{3.53}$$

(4) Mesopores (M)

Equation for the oil phase flow:

$$\sum_{J=1}^{n_4} \varepsilon_{J,\mathrm{M}}(\rho_\mathrm{o}\boldsymbol{v}_\mathrm{o})_{\mathrm{J,M}} + q_{\mathrm{oM}}^{\mathrm{W}} = \frac{\partial}{\partial t}(V\phi S_\mathrm{o}\rho_\mathrm{o})_{\mathrm{M}} \tag{3.54}$$

Equation for the gas phase flow:

$$\sum_{J=1}^{n_4} \varepsilon_{J,\mathrm{M}}(\rho_\mathrm{g}\boldsymbol{v}_\mathrm{g} + \rho_\mathrm{gd}\boldsymbol{v}_\mathrm{o})_{J,\mathrm{M}} + q_{\mathrm{gM}}^{\mathrm{W}} = \frac{\partial}{\partial t}\left[V\phi\left(\rho_\mathrm{g}S_\mathrm{g} + \rho_\mathrm{gd}S_\mathrm{o}\right)\right]_{\mathrm{M}} \tag{3.55}$$

Equation for the water phase flow:

$$\sum_{J=1}^{n_4} \varepsilon_{J,\mathrm{M}}(\rho_\mathrm{w}\boldsymbol{v}_\mathrm{w})_{J,\mathrm{M}} + q_{\mathrm{wM}}^{\mathrm{W}} = \frac{\partial}{\partial t}(V\phi S_\mathrm{w}\rho_\mathrm{w})_{\mathrm{M}} \tag{3.56}$$

(5) Micro-pores (m1)

Equation for the oil phase flow:

$$\sum_{J=1}^{n_5} \varepsilon_{J,\mathrm{m1}}(\rho_\mathrm{o}\boldsymbol{v}_\mathrm{o})_{J,\mathrm{m1}} + q_{\mathrm{om1}}^{\mathrm{W}} = \frac{\partial}{\partial t}(V\phi S_\mathrm{o}\rho_\mathrm{o})_{\mathrm{m1}} \tag{3.57}$$

Equation for the gas phase flow:

$$\sum_{J=1}^{n_5} \varepsilon_{J,\mathrm{m1}}(\rho_\mathrm{g}\boldsymbol{v}_\mathrm{g} + \rho_\mathrm{gd}\boldsymbol{v}_\mathrm{o})_{J,\mathrm{m1}} + q_{\mathrm{gm1}}^{\mathrm{W}} = \frac{\partial}{\partial t}\left[V\phi\left(\rho_\mathrm{g}S_\mathrm{g} + \rho_\mathrm{gd}S_\mathrm{o}\right)\right]_{\mathrm{m1}} \tag{3.58}$$

Equation for the water phase flow:

$$\sum_{J=1}^{n_5} \varepsilon_{J,\mathrm{m1}}(\rho_\mathrm{w}\boldsymbol{v}_\mathrm{w})_{\mathrm{J,m1}} + q_{\mathrm{wm1}}^{\mathrm{W}} = \frac{\partial}{\partial t}(V\phi S_\mathrm{w}\rho_\mathrm{w})_{\mathrm{m1}} \tag{3.59}$$

(6) Nano-pores (m2)

Equation for the oil phase flow:

$$\sum_{J=1}^{n_6} \varepsilon_{J,\mathrm{m2}}(\rho_\mathrm{W}\boldsymbol{v}_\mathrm{W})_{J,\mathrm{m2}} + q_{\mathrm{om2}}^{\mathrm{W}} = \frac{\partial}{\partial t}(V\phi S_\mathrm{o}\rho_\mathrm{o})_{\mathrm{m2}} \tag{3.60}$$

Equation for the gas phase flow:

$$\sum_{J=1}^{n_6} \varepsilon_{J,\mathrm{m2}}(\rho_\mathrm{g}\boldsymbol{v}_\mathrm{g} + \rho_\mathrm{gd}\boldsymbol{v}_\mathrm{o})_{J,\mathrm{m2}} + q_{\mathrm{gm2}}^{\mathrm{W}} = \frac{\partial}{\partial t}\left[V\phi\left(\rho_\mathrm{g}S_\mathrm{g} + \rho_\mathrm{gd}S_\mathrm{o}\right)\right]_{\mathrm{m2}} \tag{3.61}$$

Equation for the water phase flow:

$$\sum_{J=1}^{n_6} \varepsilon_{J,\mathrm{m2}}(\rho_\mathrm{w}\boldsymbol{v}_\mathrm{w})_{J,\mathrm{m2}} + q_{\mathrm{wm2}}^{\mathrm{W}} = \frac{\partial}{\partial t}(V\phi S_\mathrm{w}\rho_\mathrm{w})_{\mathrm{m2}} \tag{3.62}$$

where, F, f1, f2 refer to the large fracture system, the small fracture system and the micro-fracture system; M, m1, m2 refer to the meso-pore system, the micro-pore system and the nano-pore system, respectively.$(\rho_\mathrm{w} v_{\mathrm{cow}})_{i,j}$ is the formula of the water phase flow which considers the imbibition between different media, and the detailed information can be found in the model of multiple media.

3.1.4.2 Mathematical Model of Multiphase Flow Mechanisms in Different Media

Nano/micro-porous media can be observed in tight reservoirs. Their physical properties are poor. The interface effect and micro-scale effect are strong. Simultaneously, there is a complicated natural-hydraulic fracture network. During different stages, for different types of fluids, the mechanisms of flow indifferent scales (pore and fracture) media in tight oil reservoirs are different. In addition, a flow regime also changes. The possible flow regimes include low speed nonlinear flow, quasi-linear flow, and high speed nonlinear flow. Due to the different media and fluid properties, the critical pressure gradient of the corresponding flow regime is different, and the changes of pressure gradient and media spatial location also affect the flow regime.

1. Porous media at different scales

The porous media at different scales can be observed in tight reservoirs (Fig. 3.2), and the flow mechanisms of different media are also different. The flow behavior of oil and gas fluids in matrix pores is a low speed and nonlinear flow affected by pseudo-linear flow and threshold pressure gradient.

(1) Quasi-linear flow

For the flow behavior of oil and gas in porous media of tight oil reservoirs, the pressure gradient is between the quasi linear critical pressure gradient and the high speed nonlinear critical pressure gradient, and the flow regime is quasi linear flow. The flow equation is:

$$\sum_{j=1}^{N}\left\{\frac{A_{\mathrm{m}i,\mathrm{m}j}\boldsymbol{n}_{\mathrm{m}i,\mathrm{m}j}}{L_{\mathrm{m}i,\mathrm{m}j}}K_{\mathrm{m}i,\mathrm{m}j}\sum_{\mathrm{p}}\frac{\rho_{\mathrm{p}}X_{\mathrm{cp}}K_{\mathrm{rp}}}{\mu_{\mathrm{p}}}\left[(P_{\mathrm{m}i}-\rho g D_{\mathrm{m}i})-(P_{\mathrm{m}j}-\rho g D_{\mathrm{m}j})-G_{\mathrm{c,m}i,\mathrm{m}j}\right]_{\mathrm{p}}\right\} + q_{\mathrm{pm}i}^{\mathrm{W}} = \frac{\partial}{\partial t}\left[V\phi\sum_{\mathrm{p}}(S_{\mathrm{p}}\rho_{\mathrm{p}}X_{\mathrm{cp}})\right]_{\mathrm{m}i} \tag{3.63}$$

$$G_{\mathrm{c,m}i,\mathrm{m}j} = \mathrm{c}\left(L_{\mathrm{m}i} + L_{\mathrm{m}j}\right) \tag{3.64}$$

where, $A_{\mathrm{m}i,\mathrm{m}j}$ is the contact area of adjacent grids i, j, m^2; $L_{\mathrm{m}i,\mathrm{m}j}$ is the actual distance from the center of a grid centroid to an adjacent grid contact surface,m; $\boldsymbol{n}_{\mathrm{m}i,\mathrm{m}j}$ is the orthogonal vector between grid blocks, dimensionless.

(2) Low speed nonlinear flow

For the flow behavior of oil and gas in porous media of tight oil reservoirs, the pressure gradient is greater than the threshold pressure gradient and less than the quasi linear critical pressure gradient, and the flow regime is the low speed nonlinear seepage which is affected by the threshold pressure gradient. The flow equation is

$$\sum_{j=1}^{N}\left\{\frac{A_{\mathrm{m}i,\mathrm{m}j}\boldsymbol{n}_{\mathrm{m}i,\mathrm{m}j}}{L_{\mathrm{m}i,\mathrm{m}j}}K_{\mathrm{m}i,\mathrm{m}j}\sum_{\mathrm{p}}\frac{\rho_{\mathrm{p}}X_{\mathrm{cp}}K_{\mathrm{rp}}}{\mu_{\mathrm{p}}}\left[(P_{\mathrm{m}i}-\rho g D_{\mathrm{m}i})-(P_{\mathrm{m}j}-\rho g D_{\mathrm{m}j})-G_{\mathrm{a,m}i,\mathrm{m}j}\right]_{\mathrm{p}}^{n*}\right\} + q_{\mathrm{pm}i}^{\mathrm{W}} = \frac{\partial}{\partial t}\left[V\phi\sum_{\mathrm{p}}(S_{\mathrm{p}}\rho_{\mathrm{p}}X_{\mathrm{cp}})\right]_{\mathrm{m}i} \tag{3.65}$$

$$G_{\mathrm{a,m}i,\mathrm{m}j} = a\left(L_{\mathrm{m}i} + L_{\mathrm{m}j}\right) \tag{3.66}$$

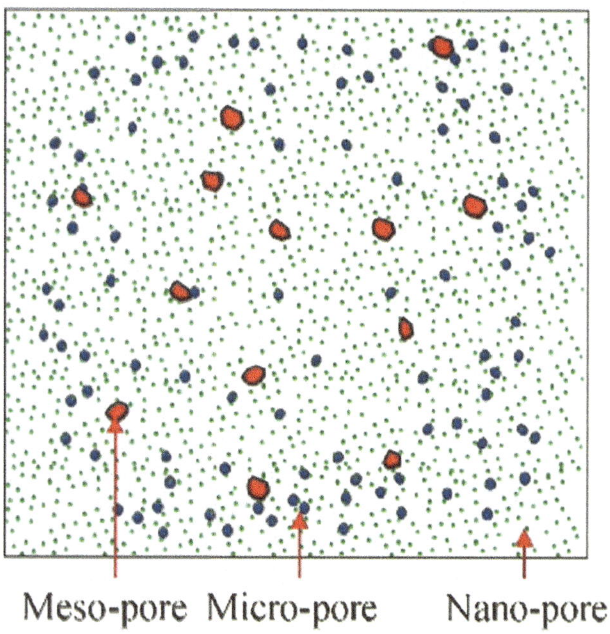

Fig. 3.2 Pore systems at different scales in tight reservoirs

2. Fracture media at different scales

In tight reservoirs, the natural fractures and hydraulic fractures created under different fracturing methods comprise a complicated fracture network. The flow mechanisms and flow regimes of fluids in these different scale fractures are different. There are the flow regimes of high speed nonlinear flow, quasi linear flow and low speed nonlinear flow (Fig. 3.3).

(1) High speed nonlinear flow

For the flow behavior of oil and gas in media with fractures of tight oil reservoirs, the pressure gradient is greater than the high speed nonlinear critical pressure gradient, and the flow regime is high speed nonlinear flow. According to the Forchheimer equation (Forchheimer 1901), the flow equation is

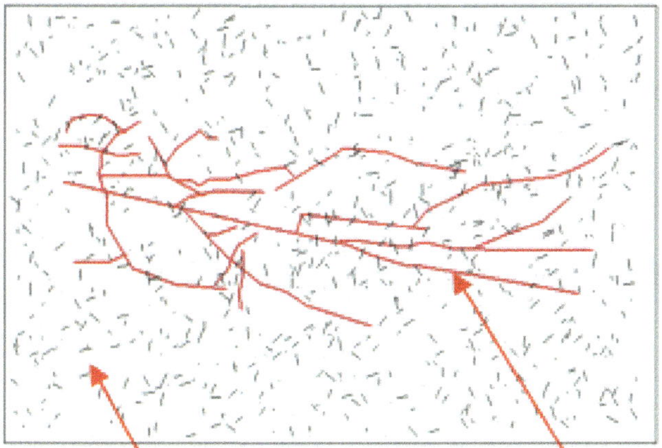

Fig. 3.3 Fracture systems at different scales in tight reservoirs

$$\sum_{j=1}^{N}\left\{\frac{A_{fi,fj}\boldsymbol{n}_{fi,fj}}{L_{fi,fj}}K_{fi,fj}F_{ND}\sum_{p}\frac{\rho_p X_{cp}K_{rp}}{\mu_p}\left[(P_{fi}-\rho g D_{fi})-(P_{fj}-\rho g D_{fj})\right]_p\right\}+q_{pfi}^{W}=\frac{\partial}{\partial t}\left[V\phi\sum_{p}(S_p\rho_p X_{cp})\right]_{fi} \tag{3.67}$$

$$F_{ND}=\frac{1}{1+\frac{\beta\rho_p q_p K}{A\mu_p}} \tag{3.68}$$

where, F_{ND} is the high-speed nonlinear flow equation.

(2) Quasi-linear flow

For the flow behavior of oil and gas in media with fractures of tight oil reservoirs, the pressure gradient is between the quasi-linear critical pressure gradient and the high speed nonlinear critical pressure gradient, and the flow regime is quasi linear flow. The flow equation is

$$\sum_{j=1}^{N}\left\{\frac{A_{fi,fj}\boldsymbol{n}_{fi,fj}}{L_{fi,fj}}K_{fi,fj}\sum_{p}\frac{\rho_p X_{cp}K_{rp}}{\mu_p}\left[(P_{fi}-\rho g D_{fi})-(P_{fj}-\rho g D_{fj})-G_{c,fi,fj}\right]_p\right\}+q_{pfi}^{W}=\frac{\partial}{\partial t}\left[V\phi\sum_{p}(S_p\rho_p X_{cp})\right]_{fi} \tag{3.69}$$

$$G_{c,fi,fj}=c\left(L_{fi}+L_{fj}\right) \tag{3.70}$$

(3) Low speed nonlinear flow

For the flow behavior of oil and gas in media with nano/micro-scale fractures of tight oil reservoirs, the pressure gradient is greater than the threshold pressure gradient and less than the quasi linear critical pressure gradient, the flow regime is the low speed nonlinear flow which is affected by the threshold pressure gradient. The flow equation is

$$\sum_{j=1}^{N}\left\{\frac{A_{fi,fj}\boldsymbol{n}_{fi,fj}}{L_{fi,fj}}K_{fi,fj}\sum_{p}\frac{\rho_p X_{cp}K_{rp}}{\mu_p}\left[(P_{fi}-\rho_p g D_{fi})-(P_{fj}-\rho_p g D_{fj})-G_{a,fi,fj}\right]_p^{n*}\right\}+q_{pfi}^{W}=\frac{\partial}{\partial t}\left[V\phi\sum_{p}(S_p\rho_p X_{cp})\right]_{fi} \tag{3.71}$$

$$G_{a,fi,fj}=a\left(L_{fi}+L_{fj}\right) \tag{3.72}$$

3.1.4.3 Mathematical Model of Multiphase Flow Between Multiple Media

1. Calculation model of fluid exchanging behavior between multiple media

The model of fluid exchanging behavior between media at different scales is composed of the flow capacity of different media and the effect of different flow mechanisms on the flow capacity. In this equation, the flow behavior between multiple media, and the flow behavior between media and wellbore are shown in Table 3.6.

$$\varepsilon_{i,j}(\rho_p\boldsymbol{v}_p)_{i,j}=\left(\frac{A_{i,j}\boldsymbol{n}_{i,j}}{L_{i,j}}K_{i,j}\right)\sum_{p}\left[\frac{\rho_p X_{cp}K_{rp}}{\mu_p}(\Phi_i-\Phi_j)_p\right]=C_{i,j}F_{i,j} \tag{3.73}$$

2. Calculation model of transmissibility between different media

The flowability of different media can be characterized by transmissibility. In discontinuous multiple media systems, multiphase flow is discontinuous, and different media also have different geometrical characteristics and attribute parameters. Therefore, in order to reflect the geometrical and attribute characteristics of media, the transmissibility calculation model Gong et al. (2006) between different elements is established:

$$C_{i,j}=\frac{A_{i,j}\boldsymbol{n}_{i,j}}{L_{i,j}}K_{i,j} \tag{3.74}$$

where, $A_{i,j}$ is the contact area between units i and j, m^2; $\boldsymbol{n}_{i,j}$ is the orthogonal vector between units i and j, dimensionless.

Table 3.6 The fluid mass exchanging term in the models of discontinuous multiple media

Adjacent units main units	The Fluid mass exchanging term between different media						The fluids' mass exchange between media and the wellbore
	Large fracture	Small fracture	Micro-fracture	Meso-pore	Micro-pore	Nano-pore	
Large fracture	–	$\varepsilon_{\mathrm{f1,F}}(\rho_{\mathrm{p}}\mathbf{v}_{\mathrm{p}})_{\mathrm{f1,F}}$	$\varepsilon_{\mathrm{f2,F}}(\rho_{\mathrm{p}}\mathbf{v}_{\mathrm{p}})_{\mathrm{f2,F}}$	$\varepsilon_{\mathrm{M,F}}(\rho_{\mathrm{p}}\mathbf{v}_{\mathrm{p}})_{\mathrm{M,F}}$	$\varepsilon_{\mathrm{m1,F}}(\rho_{\mathrm{p}}\mathbf{v}_{\mathrm{p}})_{\mathrm{m1,F}}$	$\varepsilon_{\mathrm{m2,F}}(\rho_{\mathrm{p}}\mathbf{v}_{\mathrm{p}})_{\mathrm{m2,F}}$	$q_{\mathrm{pF}}^{\mathrm{W}}$
Small fracture	$\varepsilon_{\mathrm{f1,F}}(\rho_{\mathrm{p}}\mathbf{v}_{\mathrm{p}})_{\mathrm{F,f1}}$	–	$\varepsilon_{\mathrm{f2,f1}}(\rho_{\mathrm{p}}\mathbf{v}_{\mathrm{p}})_{\mathrm{f2,f1}}$	$\varepsilon_{\mathrm{M,f1}}(\rho_{\mathrm{p}}\mathbf{v}_{\mathrm{p}})_{\mathrm{M,f1}}$	$\varepsilon_{\mathrm{m1,f1}}(\rho_{\mathrm{p}}\mathbf{v}_{\mathrm{p}})_{\mathrm{m1,f1}}$	$\varepsilon_{\mathrm{m2,f1}}(\rho_{\mathrm{p}}\mathbf{v}_{\mathrm{p}})_{\mathrm{m2,f1}}$	$q_{\mathrm{pf1}}^{\mathrm{W}}$
Micro-fracture	$\varepsilon_{\mathrm{f2,F}}(\rho_{\mathrm{p}}\mathbf{v}_{\mathrm{p}})_{\mathrm{F,f2}}$	$\varepsilon_{\mathrm{f2,f1}}(\rho_{\mathrm{p}}\mathbf{v}_{\mathrm{p}})_{\mathrm{f1,f2}}$	–	$\varepsilon_{\mathrm{M,f2}}(\rho_{\mathrm{p}}\mathbf{v}_{\mathrm{p}})_{\mathrm{M,f2}}$	$\varepsilon_{\mathrm{m1,f2}}(\rho_{\mathrm{p}}\mathbf{v}_{\mathrm{p}})_{\mathrm{m1,f2}}$	$\varepsilon_{\mathrm{m2,f2}}(\rho_{\mathrm{p}}\mathbf{v}_{\mathrm{p}})_{\mathrm{m2,f2}}$	$q_{\mathrm{pf2}}^{\mathrm{W}}$
Small pore	$\varepsilon_{\mathrm{M,F}}(\rho_{\mathrm{p}}\mathbf{v}_{\mathrm{p}})_{\mathrm{F,M}}$	$\varepsilon_{\mathrm{M,f1}}(\rho_{\mathrm{p}}\mathbf{v}_{\mathrm{p}})_{\mathrm{f1,M}}$	$\varepsilon_{\mathrm{M,f2}}(\rho_{\mathrm{p}}\mathbf{v}_{\mathrm{p}})_{\mathrm{f2,M}}$	–	$\varepsilon_{\mathrm{m1,M}}(\rho_{\mathrm{p}}\mathbf{v}_{\mathrm{p}})_{\mathrm{m1,M}}$	$\varepsilon_{\mathrm{m2,M}}(\rho_{\mathrm{p}}\mathbf{v}_{\mathrm{p}})_{\mathrm{m2,M}}$	$q_{\mathrm{pM}}^{\mathrm{W}}$
Micro-pore	$\varepsilon_{\mathrm{m1,F}}(\rho_{\mathrm{p}}\mathbf{v}_{\mathrm{p}})_{\mathrm{F,m1}}$	$\varepsilon_{\mathrm{m1,f1}}(\rho_{\mathrm{p}}\mathbf{v}_{\mathrm{p}})_{\mathrm{f1,m1}}$	$\varepsilon_{\mathrm{m1,f2}}(\rho_{\mathrm{p}}\mathbf{v}_{\mathrm{p}})_{\mathrm{f2,m1}}$	$\varepsilon_{\mathrm{m1,M}}(\rho_{\mathrm{p}}\mathbf{v}_{\mathrm{p}})_{\mathrm{M,m1}}$	–	$\varepsilon_{\mathrm{m1,m2}}(\rho_{\mathrm{p}}\mathbf{v}_{\mathrm{p}})_{\mathrm{m2,m1}}$	$q_{\mathrm{pm1}}^{\mathrm{W}}$
Nano-pore	$\varepsilon_{\mathrm{m2,F}}(\rho_{\mathrm{p}}\mathbf{v}_{\mathrm{p}})_{\mathrm{F,m2}}$	$\varepsilon_{\mathrm{m2,f1}}(\rho_{\mathrm{p}}\mathbf{v}_{\mathrm{p}})_{\mathrm{f1,m2}}$	$\varepsilon_{\mathrm{m2,f2}}(\rho_{\mathrm{p}}\mathbf{v}_{\mathrm{p}})_{\mathrm{f2,m2}}$	$\varepsilon_{\mathrm{m2,M}}(\rho_{\mathrm{p}}\mathbf{v}_{\mathrm{p}})_{\mathrm{M,m2}}$	$\varepsilon_{\mathrm{m1,m2}}(\rho_{\mathrm{p}}\mathbf{v}_{\mathrm{p}})_{\mathrm{m1,m2}}$	–	$q_{\mathrm{pm2}}^{\mathrm{W}}$

$L_{i,j}$ is the distance between the centroids of units i and j, m; $K_{i,j}$ is the absolute permeability between units i and j, mD.

3. Calculation model of fluid flow capacity under different flow mechanisms

The fluid exchanging behavior among matrix pores, large-scale fractures and small-scale fractures is controlled by the special mechanisms of viscous effect affected by threshold pressure gradient and imbibition effect. It shows the flow regimes of high speed nonlinear flow, quasi linear flow and low speed nonlinear flow. Therefore, the calculation models for the multiphase flowability under different mechanisms are developed. The models well describe the effect of flow regimes and flow mechanisms on the fluid exchanging behavior between different media.

$$F_{i,j} = \sum_{\mathrm{p}} \left[\frac{\rho_{\mathrm{p}} X_{\mathrm{cp}} K_{\mathrm{rp}}}{\mu_{\mathrm{p}}} (\Phi_i - \Phi_j)_{\mathrm{p}} \right] \tag{3.75}$$

The specific form is as follows:

(1) Mathematical model of multiphase flow in porous media at different scales

1) Quasi-linear flow

When the pressure gradient of multiphase flow between different porous media is between the quasi linear critical pressure gradient and the high speed nonlinear critical pressure gradient, the multiphase flow is quasi linear flow and the multiphase flow term can be expressed by

$$(F_{\mathrm{quasi-linear,c}})_{\mathrm{m}i,\mathrm{M}j} = \sum_{\mathrm{p}} \left\{ \frac{\rho_{\mathrm{p}} X_{\mathrm{cp}} K_{\mathrm{rp}}}{\mu_{\mathrm{p}}} \left[(P_{\mathrm{m}i} - \rho g D_{\mathrm{m}i}) - (P_{\mathrm{M}j} - \rho g D_{\mathrm{M}j}) - G_{\mathrm{c,m}i,\mathrm{M}j} \right]_{\mathrm{p}} \right\} \tag{3.76}$$

$$G_{\mathrm{c,m}i,\mathrm{M}j} = c\left(L_{\mathrm{m}i} + L_{\mathrm{M}j}\right) \tag{3.77}$$

2) Low speed nonlinear flow

When the pressure gradient of multiphase flow between different porous media is greater than the threshold pressure gradient and is less than the quasi linear critical pressure gradient, the multiphase flow is low speed nonlinear flow which is affected by the threshold pressure gradient, and the multiphase flow term can be expressed by

$$(F_{\mathrm{starting,c}})_{\mathrm{m}i,\mathrm{M}j} = \sum_{\mathrm{p}} \left\{ \frac{\rho_{\mathrm{p}} X_{\mathrm{cp}} K_{\mathrm{rp}}}{\mu_{\mathrm{p}}} \left[(P_{\mathrm{m}i} - \rho g D_{\mathrm{m}i}) - (P_{\mathrm{M}j} - \rho g D_{\mathrm{M}j}) - G_{\mathrm{a,m}i,\mathrm{M}j} \right]_{\mathrm{p}}^{n*} \right\} \tag{3.78}$$

$$G_{\mathrm{a,m}i,\mathrm{M}j} = a\left(L_{\mathrm{m}i} + L_{\mathrm{M}j}\right) \tag{3.79}$$

3) Imbibition effect

In tight reservoirs, the multiphase flow behavior between media at different scales is affected by the imbibition effect, which can be reflected by the water flow equation (Yin et al. 2004):

$$(F_{\mathrm{imbition,w}})_{\mathrm{m}i,\mathrm{M}j} = \frac{\rho_{\mathrm{w}} K_{\mathrm{rw}}}{\mu_{\mathrm{w}}} \left[(P_{\mathrm{o,m}i} - P_{\mathrm{o,M}j}) + \rho_{\mathrm{w}} g (D_{\mathrm{m}i} - D_{\mathrm{M}j}) - (P_{\mathrm{cow,m}i} - P_{\mathrm{cow,M}j}) \right] \tag{3.80}$$

(2) Fluid exchanging behavior between pores and different scales fractures

1) Quasi-linear flow

When the pressure gradient of multiphase flow in porous media is between the quasi linear critical pressure gradient and the high speed nonlinear critical pressure gradient, the multiphase flow is quasi linear flow, and the multiphase flow term can be expressed by

$$(F_{\text{quasi-linear,c}})_{mi,f(F)j} = \sum_{p}\left\{\frac{\rho_p X_{cp} K_{rp}}{\mu_p}\left[(P_{mi}-\rho g D_{mi})-(P_{f(F)j}-\rho g D_{f(F)j})-G_{c,mi,f(F)j}\right]_p\right\} \tag{3.81}$$

$$G_{c,mi,f(F)j} = c\left(L_{mi}+L_{f(F)j}\right) \tag{3.82}$$

2) Low speed nonlinear flow

When the pressure gradient of multiphase flow in porous media is greater than the threshold pressure gradient and is less than the quasi linear critical pressure gradient, the multiphase flow is the low speed non-linear flow which is affected by the threshold pressure gradient, and the multiphase flow term can be expressed by

$$(F_{\text{starting,c}})_{mi,f(F)j} = \sum_{p}\left\{\frac{\rho_p X_{cp} K_{rp}}{\mu_p}\left[(P_{mi}-\rho g D_{mi})-(P_{f(F)j}-\rho g D_{f(F)j})-G_{a,mi,f(F)j}\right]_p^{n*}\right\} \tag{3.83}$$

$$G_{a,mi,f(F)j} = a\left(L_{mi}+L_{f(F)j}\right) \tag{3.84}$$

3) Imbibition effect

In tight reservoirs, the multiphase flow behavior between matrix and fractures is affected by the imbibition effect. Thus, the water phase flow can be expressed by

$$(F_{\text{imbition,w}})_{mi,f(F)j} = \frac{\rho_w K_{rw}}{\mu_w}\left[(P_{o,mi}-P_{o,f(F)j}) -\rho_w g(D_{mi}-D_{f(F)j})-(P_{cow,mi}-P_{cow,f(F)j})\right] \tag{3.85}$$

(3) Fluid exchanging behavior between fractures at different scales

1) High speed nonlinear flow

When the pressure gradient of the oil-gas two-phase flow in large scale fractures is greater than that of high speed nonlinear critical pressure gradient, the multiphase flow is the high speed nonlinear flow, and the multiphase flow term can be expressed by

$$(F_{\text{highspeed,c}})_{fi,Fj} = F_{ND}\sum_{p}\left\{\frac{\rho_p X_{cp} K_{rp}}{\mu_p}\left[(P_{fi}-\rho g D_{fi})-(P_{Fj}-\rho g D_{Fj})\right]_p\right\} \tag{3.86}$$

2) Quasi-linear flow

When the pressure gradient of the oil-gas two-phase flow from small scale fractures to large scale fractures is between the quasi linear critical pressure gradient and the high speed nonlinear critical pressure gradient, the multiphase flow is on the quasi linear flow and the multiphase flow term can be expressed by

$$(F_{\text{quasi-linear,c}})_{fi,Fj} = \sum_{p}\left\{\frac{\rho_p X_{cp} K_{rp}}{\mu_p}\left[(P_{fi}-\rho g D_{fi})-(P_{Fj}-\rho g D_{Fj})-G_{c,fi,Fj}\right]_p\right\} \tag{3.87}$$

$$G_{c,fi,Fj} = c\left(L_{fi}+L_{Fj}\right) \tag{3.88}$$

3) Low speed nonlinear flow

When the pressure gradient of the oil-gas two-phase flow in fractures is greater than the threshold pressure gradient and is less than the quasi linear critical pressure gradient, the multiphase flow is the low speed nonlinear seepage which is affected by the threshold pressure gradient, and the multiphase flow term can be expressed by

$$(F_{\text{starting,c}})_{fi,Fj} = \sum_{p}\left\{\frac{\rho_p X_{cp} K_{rp}}{\mu_p}\left[(P_{fi}-\rho_p g D_{fi})-(P_{Fj}-\rho_p g D_{Fj})-G_{a,fi,Fj}\right]_p^{n*}\right\} \tag{3.89}$$

$$G_{a,fi,Fj} = a\left(L_{fi}+L_{Fj}\right) \tag{3.90}$$

4) Imbibition effect

In tight reservoirs, the multiphase flow behavior between different scale fractures is affected by the imbibition effect. Thus, the water phase flow can be expressed by

$$(F_{\text{imbibition, w}})_{fi,Fj} = \frac{\rho_w K_{rw}}{\mu_w}\left[(P_{o,fi}-P_{o,f(F)j}) -\rho_w g(D_{fi}-D_{Fj})-(P_{cow,fi}-P_{cow,Fj})\right] \tag{3.91}$$

3.2 Mathematical Model of Multiphase Flow in Multiple Media for Tight Gas Reservoirs

Similar to tight oil reservoirs, tight gas reservoirs have the characteristics of multiple scales and multiple media. The characteristics of discontinuous discrete distributions and discontinuous attribute parameters changing in different media can be observed. And, the flow regime and multiphase flow mechanisms are also complicated. The multiphase flow mechanism in tight gas reservoirs is significantly different from that in tight oil reservoirs, which is mainly affected by the mechanisms of threshold pressure gradient, slippage effect and diffusion effect. It shows many different flow regimes, including high speed nonlinear flow, quasi-linear flow and low speed nonlinear flow etc.

In tight gas reservoirs, considering the nonlinear flow behavior in porous media and natural/hydraulic fracture system at different scales, and the fluid exchanging behavior of the mechanisms of multiphase flow between multiple media at different scales, the mathematical models of multiphase flow in multiple media in tight gas reservoirs are developed. They include the models for continuous single media, continuous dual media, continuous multiple media and discontinuous multiple media.

3.2.1 Mathematical Model of Multiphase Flow in Continuous Single Media

For the tight gas reservoirs where only a single porous media is developed, it can be simplified to a continuous single media model. This model has a continuous distribution of single porous media. In this model, gas-water two-phase fluids coexist. The parameters of porous media and fluid properties change continuously. It can be characterized by the continuous equation.

According to the characteristics of tight gas reservoirs above, the mathematical model for gas-water two-phase flow in continuous single media is established:

(1) Equation for the gas phase flow:

$$\nabla \cdot \left(\rho_{\mathrm{g}}\frac{KK_{\mathrm{rg}}}{\mu_{\mathrm{g}}}\nabla P_{\mathrm{g}} - \rho_{\mathrm{g}}\frac{KK_{\mathrm{rg}}}{\mu_{\mathrm{g}}}\rho_{\mathrm{g}}g\nabla D\right) + q_{\mathrm{g}}^{\mathrm{W}} = \frac{\partial}{\partial t}(\phi S_{\mathrm{g}}\rho_{\mathrm{g}}) \tag{3.92}$$

(2) Equation for the water phase flow:

$$\nabla \cdot \left(\rho_{\mathrm{w}}\frac{KK_{\mathrm{rw}}}{\mu_{\mathrm{w}}}\nabla P_{\mathrm{w}} - \rho_{\mathrm{w}}\frac{KK_{\mathrm{rw}}}{\mu_{\mathrm{w}}}\rho_{\mathrm{w}}g\nabla D\right) + q_{\mathrm{w}}^{\mathrm{W}} = \frac{\partial}{\partial t}(\phi S_{\mathrm{w}}\rho_{\mathrm{w}}) \tag{3.93}$$

In the model, the flow term describes the flow behavior of fluids under the effects of viscous flow, gravity drive, and elastic expansion; the source/sink term represents the injection and production rates of wells; the cumulative term shows the cumulative rate under different recovery mechanisms.

3.2.2 Mathematical Model of Multiphase Flow in Continuous Dual Media

When a continuously distributed pore system and a fracture system are developed in tight reservoirs, they can be simplified to a continuous dual media model. The model is composed of a pore system and a fracture system. The pore system has the unique property of porous media, and the fracture system has the unique property of fracture media. In the same system, the continuous equation can be used to describe the spatial distribution of parameters. Two systems overlap with each other in space and can form the dual media.

According to the internal flow behavior in porous media and the flow behavior between porous media and wellbore, the model of flow in such media can be classified into a dual-porosity single-permeability model and a dual-porosity dual-permeability model.

3.2.2.1 Mathematical Model of Multiphase Flow in the Dual Media of Dual-Porosity and Single-Permeability

(1) The pore system

Equation for the gas phase flow:

$$-\tau_{\mathrm{gmf}} = \frac{\partial}{\partial t}\left(\phi\rho_{\mathrm{g}}S_{\mathrm{g}}\right)_{\mathrm{m}} \tag{3.94}$$

Equation for the water phase flow:

$$-\tau_{\mathrm{wmf}} = \frac{\partial}{\partial t}(\phi\rho_{\mathrm{w}}S_{\mathrm{w}})_{\mathrm{m}} \tag{3.95}$$

(2) The fracture system

Equation for the gas phase flow:

$$-\nabla \cdot (\rho_{\mathrm{g}}\boldsymbol{v}_{\mathrm{g}})_{\mathrm{f}} + \tau_{\mathrm{gmf}} + q_{\mathrm{gf}}^{\mathrm{W}} = \frac{\partial}{\partial t}\left(\phi\rho_{\mathrm{g}}S_{\mathrm{g}}\right)_{\mathrm{f}} \tag{3.96}$$

Table 3.7 The fluids' exchange term in mode of flow through dual porosity single permeability media

Fluids	Multiphase flow term within the system		Fluids' mass exchange term between systems	Source/sink	
	Matrix system	Fracture system	Matrix-fracture	Matrix-wellbore	Fracture-wellbore
Gas	0	$-\nabla \cdot (\rho_g \boldsymbol{v}_g)_f$	$\tau_{gmf} = \frac{\alpha_{fm} K_m \rho_g K_{rgm}}{\mu_g}(\Phi_{gm} - \Phi_{gf})$	0	q_{gf}^W
Water	0	$-\nabla \cdot (\rho_w \boldsymbol{v}_w)_f$	$\tau_{wmf} = \frac{\alpha_{fm} K_m \rho_w K_{rwm}}{\mu_w}(\Phi_{wm} - \Phi_{wf})$	0	q_{wf}^W

Table 3.8 The fluids' exchange term in model of flow through dual porosity dual permeability media

Fluids	Multiphase flow term within the system		Fluid mass exchanging term between systems	Source/sink	
	Matrix system	Fracture system	Matrix-fracture	Matrix-wellbore	Fracture-wellbore
Gas	$-\nabla \cdot (\rho_g \boldsymbol{v}_g)_m$	$-\nabla \cdot (\rho_g \boldsymbol{v}_g)_f$	$\tau_{gmf} = \frac{\alpha_{fm} K_m \rho_g K_{rgm}}{\mu_g}(\Phi_{gm} - \Phi_{gf})$	q_{gm}^W	q_{gf}^W
Water	$-\nabla \cdot (\rho_w \boldsymbol{v}_w)_m$	$-\nabla \cdot (\rho_w \boldsymbol{v}_w)_f$	$\tau_{wmf} = \frac{\alpha_{fm} K_m \rho_w K_{rwm}}{\mu_w}(\Phi_{wm} - \Phi_{wf})$	q_{wm}^W	q_{wf}^W

Equation for the water phase flow:

$$-\nabla \cdot (\rho_w \boldsymbol{v}_w)_f + \tau_{wmf} + q_{wf}^W = \frac{\partial (\phi \rho_w S_w)_f}{\partial t} \quad (3.97)$$

The multiphase flow between the pore system and the fracture system in the model is shown in Table 3.7.

3.2.2.2 Mathematical Model of Multiphase Flow in the Dual Media with Dual-Porosity and Dual-Permeability

In tight gas reservoirs, the mathematical model of multiphase flow in a dual-porosity dual-permeability system is:

(1) The pore system

Equation for the gas phase flow:

$$-\nabla \cdot (\rho_g \boldsymbol{v}_g)_m - \tau_{gmf} + q_{gm}^W = \frac{\partial}{\partial t}(\phi \rho_g S_g)_m \quad (3.98)$$

Equation for the water phase flow:

$$-\nabla \cdot (\rho_w \boldsymbol{v}_w)_m - \tau_{wmf} + q_{wm}^W = \frac{\partial}{\partial t}(\phi \rho_w S_w)_m \quad (3.99)$$

(2) The fracture system

Equation for the gas phase flow:

$$-\nabla \cdot (\rho_g \boldsymbol{v}_g)_f + \tau_{gmf} + q_{gf}^W = \frac{\partial}{\partial t}(\phi \rho_g S_g)_f \quad (3.100)$$

Equation for the water phase flow:

$$-\nabla \cdot (\rho_w \boldsymbol{v}_w)_f + \tau_{wmf} + q_{wf}^W = \frac{\partial (\phi \rho_w S_w)_f}{\partial t} \quad (3.101)$$

The flow behavior between pore system and fracture system in this model is shown in Table 3.8.

3.2.3 Mathematical Model of Multiphase Flow in Continuous Multiple Media

3.2.3.1 Mathematical Model of Multiphase Flow in Multiple Media Based on the Dual Porosity Model

For the tight gas reservoirs where pore system and natural/hydraulic fracture system at different scales are developed, within a same media system, the media system will interact with each other. The geometrical and attribute parameters are continuously changed, and the two media systems overlap with each other to form the multiple media. This kind of tight gas reservoir can be simplified to a multiple media model based on dual-porosity model. In this model, multiphase flow can be observed within the pore system and the fracture system, and between pore and fracture systems. The fluid exchanging behavior can be also observed between pore/fracture systems and wellbore. According to the characteristics above, a mathematical model of multiphase flow in multiple media based on a dual-porosity model can be developed:

1. Pore systems
   (1) Small pores (m1)

Equation for the gas phase flow:

$$-\nabla \cdot (\rho_g \boldsymbol{v}_g)_{m2,m1} - \tau_{gm1,f} + q_{gm1}^W = \frac{\partial}{\partial t}(\phi \rho_g S_g)_{m1} \quad (3.102)$$

Table 3.9 The fluids' exchange in model of flow through multiple media based on dual-permeability model

Flow objects			Fluids	
			Gas	Water
Multiphase flowterm within the system	Pore system	Micro-pore	$-\nabla\cdot(\rho_g \boldsymbol{v}_g)_{m2,m1} = \nabla\cdot\left(\rho_g \frac{K_{m1}K_{rgm1}}{\mu_g}\nabla\phi_{g,m2m1}\right)$	$-\nabla\cdot(\rho_w \boldsymbol{v}_w)_{m2,m1} = \nabla\cdot\left(\rho_w \frac{K_{m1}K_{rwm1}}{\mu_w}\nabla\phi_{w,m2m1}\right)$
		Nano-pore	$-\nabla\cdot(\rho_g \boldsymbol{v}_g)_{m1,m2} = \nabla\cdot\left(\rho_g \frac{K_{m2}K_{rgm2}}{\mu_g}\nabla\phi_{g,m1m2}\right)$	$-\nabla\cdot(\rho_w \boldsymbol{v}_w)_{m1,m2} = \nabla\cdot\left(\rho_w \frac{K_{m2}K_{rwm2}}{\mu_w}\nabla\phi_{w,m1m2}\right)$
	Fracture system	Large fracture	$-\nabla\cdot(\rho_g \boldsymbol{v}_g)_{f,F} = \nabla\cdot\left(\rho_g \frac{K_F K_{rgF}}{\mu_g}\nabla\phi_{o,fF}\right)$	$-\nabla\cdot(\rho_w \boldsymbol{v}_w)_{f,F} = \nabla\cdot\left(\rho_w \frac{K_F K_{rwF}}{\mu_w}\nabla\phi_{w,fF}\right)$
		Micro-fracture	$-\nabla\cdot(\rho_g \boldsymbol{v}_g)_{F,f} = \nabla\cdot\left(\rho_g \frac{K_f K_{rgf}}{\mu_g}\nabla\phi_{o,Ff}\right)$	$-\nabla\cdot(\rho_w \boldsymbol{v}_w)_{F,f} = \nabla\cdot\left(\rho_w \frac{K_f K_{rwf}}{\mu_w}\nabla\phi_{w,Ff}\right)$
Fluid mass exchanging term between systems			Micro-pore-micro-fracture	
			$\tau_{gm1f} = \frac{\alpha_{m1f}K_{m1}\rho_g K_{rgm1}}{\mu_g}(\phi_{gm1} - \phi_{gf})$	$\tau_{wm1f} = \frac{\alpha_{m1f}K_{m1}\rho_w K_{rwm1}}{\mu_w}(\phi_{wm1} - \phi_{wf})$
Nano-pore-large fracture			$\tau_{gm2F} = \frac{\alpha_{m2F}K_{m2}\rho_g K_{rgm2}}{\mu_g}(\phi_{om2} - \phi_{oF})$	$\tau_{wm2F} = \frac{\alpha_{m2F}K_{m2}\rho_w K_{rwm2}}{\mu_w}(\phi_{wm2} - \phi_{wF})$
Source/sink	Micro-pore-wellbore		q_{gm1}^{W}	q_{wm1}^{W}
	Nano-pore-wellbore		q_{gm2}^{W}	q_{wm2}^{W}
	Large fracture-wellbore		q_{gF}^{W}	q_{wF}^{W}
	Micro-fracture-wellbore		q_{gf}^{W}	q_{wf}^{W}

Equation for the water phase flow:

$$-\nabla\cdot(\rho_w \boldsymbol{v}_w)_{m2,m1} - \tau_{wm1,f} + q_{wm1}^{W} = \frac{\partial}{\partial t}(\phi\rho_w S_w)_{m1} \tag{3.103}$$

(2) Nano-pores (m2)

Equation for the gas phase flow:

$$-\nabla\cdot(\rho_g \boldsymbol{v}_g)_{m1,m2} - \tau_{gm2,F} + q_{gm2}^{W} = \frac{\partial}{\partial t}(\phi\rho_g S_g)_{m2} \tag{3.104}$$

Equation for the water phase flow:

$$-\nabla\cdot(\rho_w \boldsymbol{v}_w)_{m1,m2} - \tau_{wm2,F} + q_{wm2}^{W} = \frac{\partial}{\partial t}(\phi\rho_w S_w)_{m2} \tag{3.105}$$

2. Fracture systems

(1) Large fractures (F)

Equation for the gas phase flow:

$$-\nabla\cdot(\rho_g \boldsymbol{v}_g)_{f,F} + \tau_{gm2,F} + q_{gF}^{W} = \frac{\partial}{\partial t}(\phi\rho_g S_g)_F \tag{3.106}$$

Equation for the water phase flow:

$$-\nabla\cdot(\rho_w \boldsymbol{v}_w)_{f,F} + \tau_{wm2,F} + q_{wF}^{W} = \frac{\partial(\phi\rho_w S_w)_F}{\partial t} \tag{3.107}$$

(2) Micro-fractures (f)

Equation for the gas phase flow:

$$-\nabla\cdot(\rho_g \boldsymbol{v}_g)_{F,f} + \tau_{gm1,f} + q_{gf}^{W} = \frac{\partial}{\partial t}(\phi\rho_g S_g)_f \tag{3.108}$$

Equation for the water phase flow:

$$-\nabla\cdot(\rho_w \boldsymbol{v}_w)_{F,f} + \tau_{wm1,f} + q_{gf}^{W} = \frac{\partial(\phi\rho_w S_w)_f}{\partial t} \tag{3.109}$$

The flow behavior between the pore system and the fracture system in the model is shown in Table 3.9.

3.2.3.2 Mathematical Model of Multiphase Flow in Continuous Multiple Media

In the mathematical model of multiphase flow in continuous multiple media in tight gas reservoirs, each medium can be considered as an independent continuous system. The fluids are continuously distributed in the reservoir and can flow between different media. In the same system, a continuous equation can be used to describe the spatial distribution of parameters. In a different media system (i.e. between different media), the fluid exchanging behavior can be described by an exchange flux (Table 3.10).

(1) Small pores (m1)

Equation for the gas phase flow:

$$-\nabla \cdot (\rho_g \boldsymbol{v}_g)_{m1} - \tau_{gm1F} - \tau_{gm1f} + \tau_{gm2m1} + q_{gm1}^{W} = \frac{\partial}{\partial t}(\phi \rho_g S_g)_{m1} \quad (3.110)$$

Equation for the water phase flow:

$$-\nabla \cdot (\rho_w \boldsymbol{v}_w)_{m1} - \tau_{cow,m1F} - \tau_{cow,m1f} + \tau_{cow,m2m1} + q_{wm1}^{W} = \frac{\partial}{\partial t}(\phi \rho_w S_w)_{m1} \quad (3.111)$$

(2) Nano-pores (m2)

Equation for the gas phase flow:

$$-\nabla \cdot (\rho_g \boldsymbol{v}_g)_{m2} - \tau_{gm2F} - \tau_{gm2f} - \tau_{gm2m1} + q_{gm2}^{W} = \frac{\partial}{\partial t}(\phi \rho_g S_g)_{m2} \quad (3.112)$$

Equation for the water phase flow:

$$-\nabla \cdot (\rho_w \boldsymbol{v}_w)_{m2} - \tau_{cow,m2F} - \tau_{cow,m2f} - \tau_{cow,m2m1} + q_{wm2}^{W} = \frac{\partial}{\partial t}(\phi \rho_w S_w)_{m2} \quad (3.113)$$

(3) Large fractures (F)

Equation for the gas phase flow:

$$-\nabla \cdot (\rho_g \boldsymbol{v}_g)_F + \tau_{gm1F} + \tau_{gm2F} + \tau_{gfF} + q_{gF}^{W} = \frac{\partial}{\partial t}(\phi \rho_g S_g)_F \quad (3.114)$$

Equation for the water phase flow:

$$-\nabla \cdot (\rho_w \boldsymbol{v}_w)_F + \tau_{cow,m1F} + \tau_{cow,m2F} + \tau_{cow,fF} + q_{wF}^{W} = \frac{\partial(\phi \rho_w S_w)_F}{\partial t} \quad (3.115)$$

(4) Micro-fractures (f)

Equation for the gas phase flow:

$$-\nabla \cdot (\rho_g \boldsymbol{v}_g)_f + \tau_{gm1f} + \tau_{gm2f} - \tau_{gfF} + q_{gf}^{W} = \frac{\partial}{\partial t}(\phi \rho_g S_g)_f \quad (3.116)$$

Equation for the water phase flow:

$$-\nabla \cdot (\rho_w \boldsymbol{v}_w)_f + \tau_{cow,m1f} + \tau_{cow,m2f} - \tau_{cow,fF} + q_{wf}^{W} = \frac{\partial(\phi \rho_w S_w)_f}{\partial t} \quad (3.117)$$

3.2.4 Mathematical Model of Multiphase Flow in Discontinuous Multiple Media

In tight gas reservoirs, the media at different scales pores, nano/micro-fractures, and natural/hydraulic fractures are discontinuously discretely distributed in the reservoirs. The geometrical and attribute parameters are also discontinuously changed. This kind of tight gas reservoir can be simplified by a model for discontinuous multiple media. Only one system is introduced in this model and this system is classified into a number of independent units. Different units represent different media. The distribution of units is the same with the actual spatial distribution of multiple media. The multiphase flow between different units reflects the complex flow behavior between multiple media.

3.2.4.1 Integrated Model of Multiphase Flow in Discontinuous Multiple Media

For the multiple media at different scales (large fractures (F) small fractures (f1), micro-fractures (f2), small pores (M), micro-pores (m1), and nano-pores (m2)) in tight gas reservoirs, the mathematical model of multiphase flow in discontinuous multiple media at different scales pores and fractures is developed.

(1) Large fractures (F)

Equation for the gas phase flow:

$$\sum_{J=1}^{n_1} \varepsilon_{J,F}(\rho_g \boldsymbol{v}_g)_{J,F} + q_{oF}^{W} = \frac{\partial}{\partial t}(V \phi S_g \rho_g)_F \quad (3.118)$$

Equation for the water phase flow:

$$\sum_{J=1}^{n_1} \varepsilon_{J,F}(\rho_w \boldsymbol{v}_w)_{J,F} + q_{wF}^{W} = \frac{\partial}{\partial t}(V \phi S_w \rho_w)_F \quad (3.119)$$

Table 3.10 The fluids' exchange in model of flow through continuous multiple media

Flow objects		Fluids		
		Oil	Gas	Water
Multiphase flow term within the system	Micro-pore	$-\nabla \cdot (\rho_o \boldsymbol{v}_o)_{m1}$	$-\nabla \cdot (\rho_g \boldsymbol{v}_g)_{m1}$	$-\nabla \cdot (\rho_w \boldsymbol{v}_w)_{m1}$
	Nano-pore	$-\nabla \cdot (\rho_o \boldsymbol{v}_o)_{m2}$	$-\nabla \cdot (\rho_g \boldsymbol{v}_g)_{m2}$	$-\nabla \cdot (\rho_w \boldsymbol{v}_w)_{m2}$
	Large fracture	$-\nabla \cdot (\rho_o \boldsymbol{v}_o)_F$	$-\nabla \cdot (\rho_g \boldsymbol{v}_g)_F$	$-\nabla \cdot (\rho_w \boldsymbol{v}_w)_F$
	Micro-fracture	$-\nabla \cdot (\rho_o \boldsymbol{v}_o)_f$	$-\nabla \cdot (\rho_g \boldsymbol{v}_g)_f$	$-\nabla \cdot (\rho_w \boldsymbol{v}_w)_f$
Fluid mass exchanging term between systems	Micro-pore-Nano-pore	$\tau_{om2m1} = \frac{\alpha_{m2m1} K_{m2} \rho_o K_{rom2}}{\mu_o} (\Phi_{om2} - \Phi_{om1})$	$\tau_{gm2m1} = \frac{\alpha_{m2m1} K_{m2} \rho_g K_{rgm2}}{\mu_g} (\Phi_{gm2} - \Phi_{gm1})$	$\tau_{cow,m2,m1} = \frac{\alpha_{fm2} K_{m2} \rho_w K_{rwm2}}{\mu_w} \cdot \left[\Delta P_{o,m2m1} - \rho_w g \Delta D_{m2m1} - (P_{cow,m2} - P_{cow,m1})\right]$
	Micro-pore-micro-fracture	$\tau_{om1f} = \frac{\alpha_{m1f} K_{m1} \rho_o K_{rom1}}{\mu_o} (\Phi_{om1} - \Phi_{of})$	$\tau_{gm1f} = \frac{\alpha_{m1f} K_{m1} \rho_g K_{rgm1}}{\mu_g} (\Phi_{gm1} - \Phi_{gf})$	$\tau_{cow,m1,f} = \frac{\alpha_{fm1} K_{m1} \rho_w K_{rwm1}}{\mu_w} \cdot \left[\Delta P_{o,m1f} - \rho_w g \Delta D_{m1f} - (P_{cow,m1} - P_{cow,f})\right]$
	Micro-pore-large fracture	$\tau_{om1F} = \frac{\alpha_{m1f} K_{m1} \rho_o K_{rom1}}{\mu_o} (\Phi_{om1} - \Phi_{oF})$	$\tau_{gm1F} = \frac{\alpha_{m1f} K_{m1} \rho_g K_{rgm1}}{\mu_g} (\Phi_{gm1} - \Phi_{gF})$	$\tau_{cow,m1,F} = \frac{\alpha_{Fm1} K_{m1} \rho_w K_{rwm1}}{\mu_w} \cdot \left[\Delta P_{o,m1F} - \rho_w g \Delta D_{m1F} - (P_{cow,m1} - P_{cow,F})\right]$
	Nano-pore-large fracture	$\tau_{om2F} = \frac{\alpha_{m2F} K_{m2} \rho_o K_{rom2}}{\mu_o} (\Phi_{om2} - \Phi_{oF})$	$\tau_{gm2F} = \frac{\alpha_{m2F} K_{m2} \rho_g K_{rgm2}}{\mu_g} (\Phi_{om2} - \Phi_{oF})$	$\tau_{cow,m2F} = \frac{\alpha_{Fm2} K_{m2} \rho_w K_{rwm2}}{\mu_w} \cdot \left[\Delta P_{o,m2F} - \rho_w g \Delta D_{m2F} - (P_{cow,m2} - P_{cow,F})\right]$
	Nano-pore-micro-fracture	$\tau_{om2f} = \frac{\alpha_{m2f} K_{m2} \rho_o K_{rom2}}{\mu_o} (\Phi_{om2} - \Phi_{of})$	$\tau_{gm2f} = \frac{\alpha_{m2f} K_{m2} \rho_g K_{rgm2}}{\mu_g} (\Phi_{gm2} - \Phi_{gf})$	$\tau_{cow,m2,f} = \frac{\alpha_{fm2} K_{m2} \rho_w K_{rwm2}}{\mu_w} \cdot \left[\Delta P_{o,m2f} - \rho_w g \Delta D_{m2f} - (P_{cow,m2} - P_{cow,f})\right]$
Source/sink	Micro-pore-wellbore	q_{om1}^W	q_{gm1}^W	q_{wm1}^W
	Nano-pore-wellbore	q_{om2}^W	q_{gm2}^W	q_{wm2}^W
	Large fracture-wellbore	q_{oF}^W	q_{gF}^W	q_{wF}^W
	Micro-fracture-wellbore	q_{of}^W	q_{gf}^W	q_{wf}^W

(2) Small fractures (f1)

Equation for the gas phase flow:

$$\sum_{J=1}^{n_2} \varepsilon_{J,\mathrm{f1}} (\rho_{\mathrm{g}} \boldsymbol{v}_{\mathrm{g}})_{J,\mathrm{f1}} + q_{\mathrm{gf1}}^{\mathrm{W}} = \frac{\partial}{\partial t} (V\phi S_{\mathrm{g}} \rho_{\mathrm{g}})_{\mathrm{f1}} \tag{3.120}$$

Equation for the water phase flow:

$$\sum_{J=1}^{n_2} \varepsilon_{J,\mathrm{f1}} (\rho_{\mathrm{w}} \boldsymbol{v}_{\mathrm{w}})_{J,\mathrm{f1}} + q_{\mathrm{wf1}}^{\mathrm{W}} = \frac{\partial}{\partial t} (V\phi S_{\mathrm{w}} \rho_{\mathrm{w}})_{\mathrm{f1}} \tag{3.121}$$

(3) Micro-fractures (f2)

Equation for the gas phase flow:

$$\sum_{J=1}^{n_3} \varepsilon_{J,\mathrm{f2}} (\rho_{\mathrm{g}} \boldsymbol{v}_{\mathrm{g}})_{J,\mathrm{f2}} + q_{\mathrm{gf2}}^{\mathrm{W}} = \frac{\partial}{\partial t} (V\phi S_{\mathrm{g}} \rho_{\mathrm{g}})_{\mathrm{f2}} \tag{3.122}$$

Equation for the water phase flow:

$$\sum_{J=1}^{n_3} \varepsilon_{J,\mathrm{f2}} (\rho_{\mathrm{w}} \boldsymbol{v}_{\mathrm{w}})_{J,\mathrm{f2}} + q_{\mathrm{wf2}}^{\mathrm{W}} = \frac{\partial}{\partial t} (V\phi S_{\mathrm{w}} \rho_{\mathrm{w}})_{\mathrm{f2}} \tag{3.123}$$

(4) Small pores (M)

Equation for the gas phase flow:

$$\sum_{J=1}^{n_4} \varepsilon_{J,\mathrm{M}} (\rho_{\mathrm{g}} \boldsymbol{v}_{\mathrm{g}})_{J,\mathrm{M}} + q_{\mathrm{gM}}^{\mathrm{W}} = \frac{\partial}{\partial t} (V\phi S_{\mathrm{g}} \rho_{\mathrm{g}})_{\mathrm{M}} \tag{3.124}$$

Equation for the water phase flow:

$$\sum_{J=1}^{n_4} \varepsilon_{J,\mathrm{M}} (\rho_{\mathrm{w}} \boldsymbol{v}_{\mathrm{w}})_{J,\mathrm{M}} + q_{\mathrm{wM}}^{\mathrm{W}} = \frac{\partial}{\partial t} (V\phi S_{\mathrm{w}} \rho_{\mathrm{w}})_{\mathrm{M}} \tag{3.125}$$

(5) Micro-pores (m1)

Equation for the gas phase flow:

$$\sum_{J=1}^{n_5} \varepsilon_{J,\mathrm{m1}} (\rho_{\mathrm{g}} \boldsymbol{v}_{\mathrm{g}})_{J,\mathrm{m1}} + q_{\mathrm{gm1}}^{\mathrm{W}} = \frac{\partial}{\partial t} (V\phi S_{\mathrm{g}} \rho_{\mathrm{g}})_{\mathrm{m1}} \tag{3.126}$$

Equation for the water phase flow:

$$\sum_{J=1}^{n_5} \varepsilon_{J,\mathrm{m1}} (\rho_{\mathrm{w}} \boldsymbol{v}_{\mathrm{w}})_{J,\mathrm{m1}} + q_{\mathrm{wm1}}^{\mathrm{W}} = \frac{\partial}{\partial t} (V\phi S_{\mathrm{w}} \rho_{\mathrm{w}})_{\mathrm{m1}} \tag{3.127}$$

(6) Nano-pores (m2)

Equation for the gas phase flow:

$$\sum_{J=1}^{n_6} \varepsilon_{J,\mathrm{m2}} (\rho_{\mathrm{g}} \boldsymbol{v}_{\mathrm{g}})_{J,\mathrm{m2}} + q_{\mathrm{gm2}}^{\mathrm{W}} = \frac{\partial}{\partial t} (V\phi S_{\mathrm{g}} \rho_{\mathrm{g}})_{\mathrm{m2}} \tag{3.128}$$

Equation for the water phase flow:

$$\sum_{J=1}^{n_6} \varepsilon_{J,\mathrm{m2}} (\rho_{\mathrm{w}} \boldsymbol{v}_{\mathrm{w}})_{J,\mathrm{m2}} + q_{\mathrm{wm2}}^{\mathrm{W}} = \frac{\partial}{\partial t} (V\phi S_{\mathrm{w}} \rho_{\mathrm{w}})_{\mathrm{m2}} \tag{3.129}$$

3.2.4.2 Mathematical Model of Multiphase Flow Mechanisms in Different Media

In different development stages of tight gas reservoirs, the mechanisms of multiphase flow of different fluids in the media at different scales pores and fractures are different. The flow regime is also changed. The possible flow regimes include the low speed nonlinear flow, quasi linear flow and high speed nonlinear flow.

1. Porous media at different scales

Tight reservoirs develop the porous media at different scales. The flow process of gas in matrix pores is a low-speed nonlinear flow process which is controlled by a threshold pressure gradient. Under the conditions of low permeability and low pressure at the late stage, the flow process is significantly affected by slippage effect and diffusion effect. As the formation pressure drops below the critical gas desorption pressure, under the effect of desorption process, the adsorbed gas is transformed from adsorption state to free state (Li et al. 2013; Wu et al. 2009).

(1) Quasi-linear flow

For the gas flow process in tight gas reservoirs, the pressure gradient is between the quasi linear critical pressure gradient and the high speed nonlinear critical pressure gradient, and the flow regime is quasi linear flow in porous media. The flow equation is

$$\begin{aligned} & \sum_{j=1}^{N} \left\{ \frac{A_{\mathrm{m}i,\mathrm{m}j} \boldsymbol{n}_{\mathrm{m}i,\mathrm{m}j}}{L_{\mathrm{m}i,\mathrm{m}j}} K_{\mathrm{m}i,\mathrm{m}j} \frac{\rho_{\mathrm{g}} K_{\mathrm{rg}}}{\mu_{\mathrm{g}}} \left[P_{\mathrm{m}i} - \rho_{\mathrm{p}} g D_{\mathrm{m}i}) - (P_{\mathrm{m}j} - \rho_{\mathrm{p}} g D_{\mathrm{m}j}) - G_{\mathrm{c,m}i,\mathrm{m}j} \right]_{\mathrm{g}} \right\} \\ & \quad + q_{\mathrm{gm}i}^{\mathrm{W}} = \frac{\partial}{\partial t} (V\phi S_{\mathrm{g}} \rho_{\mathrm{g}})_{\mathrm{m}i} \end{aligned} \tag{3.130}$$

(2) Low speed nonlinear flow

For the gas flow process in tight gas reservoirs, the pressure gradient is greater than the threshold pressure gradient and less than the quasi linear critical pressure gradient, and the flow regime is the low velocity nonlinear flow. It is affected by the threshold pressure gradient, and the nonlinear index is 0.9–1.2. The flow equation is

$$\sum_{j=1}^{N}\left\{\frac{A_{\mathrm{m}i,\mathrm{m}j}\boldsymbol{n}_{\mathrm{m}i,\mathrm{m}j}}{L_{\mathrm{m}i,\mathrm{m}j}}K_{\mathrm{m}i,\mathrm{m}j}\frac{\rho_{\mathrm{g}}K_{\mathrm{rg}}}{\mu_{\mathrm{g}}}\left[P_{\mathrm{m}i}-\rho_{\mathrm{p}}gD_{\mathrm{m}i})-(P_{\mathrm{m}j}-\rho_{\mathrm{p}}gD_{\mathrm{m}j})-G_{\mathrm{a,m}i,\mathrm{m}j}\right]_{\mathrm{g}}^{n*}\right\}$$
$$+q_{\mathrm{gm}i}^{\mathrm{W}}=\frac{\partial}{\partial t}(V\phi S_{\mathrm{g}}\rho_{\mathrm{g}})_{\mathrm{m}i} \tag{3.131}$$

(3) Slippage effect

Under a low pressure condition, the gas flow process in nano/micro-porous media is affected by slippage effect (Yao et al. 2009). The low equation is (for the adjacent two grids, they have the same slippage factor B and critical pressure P_{mi})

$$\sum_{j=1}^{N}\left\{\frac{A_{\mathrm{m}i,\mathrm{m}j}\boldsymbol{n}_{\mathrm{m}i,\mathrm{m}j}}{L_{\mathrm{m}i,\mathrm{m}j}}K_{\mathrm{m}i,\mathrm{m}j}\frac{\rho_{\mathrm{g}}K_{\mathrm{rg}}}{\mu_{\mathrm{g}}}\left[\left(1+\frac{b}{P_{\mathrm{m}i}}\right)(P_{\mathrm{m}i}-\rho_{\mathrm{p}}gD_{\mathrm{m}i})-\left(1+\frac{b}{P_{\mathrm{m}j}}\right)(P_{\mathrm{m}j}-\rho_{\mathrm{p}}gD_{\mathrm{m}j})\right]_{\mathrm{g}}\right\}$$
$$+q_{\mathrm{gm}i}^{\mathrm{W}}=\frac{\partial}{\partial t}(V\phi S_{\mathrm{g}}\rho_{\mathrm{g}})_{\mathrm{m}i} \tag{3.132}$$

(4) Diffusion effect

Under a low pressure condition, the gas flow in nano/micro-porous media is affected by diffusion effect (Wu 2015). The flow equation is (for the adjacent two grids, they have the same diffusion coefficients)

$$\sum_{j=1}^{N}\left\{\frac{A_{\mathrm{m}i,\mathrm{m}j}\boldsymbol{n}_{\mathrm{m}i,\mathrm{m}j}}{L_{\mathrm{m}i,\mathrm{m}j}}K_{\mathrm{m}i,\mathrm{m}j}\frac{\rho_{\mathrm{g}}K_{\mathrm{rg}}}{\mu_{\mathrm{g}}}\left[\begin{array}{l}\left(1+\frac{32\sqrt{2}\sqrt{RT}\mu g}{3r\sqrt{\pi \mathrm{M}}P_{\mathrm{m}i}})(P_{\mathrm{m}i}-\rho_{\mathrm{p}}gD_{\mathrm{m}i}\right)\\-\left(1+\frac{32\sqrt{2}\sqrt{RT}\mu g}{3r\sqrt{\pi \mathrm{M}}P_{\mathrm{m}j}})(P_{\mathrm{m}j}-\rho_{\mathrm{p}}gD_{\mathrm{m}j}\right)\end{array}\right]_{\mathrm{g}}\right\}$$
$$+q_{\mathrm{gm}i}^{\mathrm{W}}=\frac{\partial}{\partial t}(V\phi S_{\mathrm{g}}\rho_{\mathrm{g}})_{\mathrm{m}i} \tag{3.133}$$

(5) Desorption effect

When the formation pressure drops below the desorption pressure, the gas is desorbed (Wang 2013), and the flow equation is

$$\sum_{j=1}^{N}\left\{\frac{A_{\mathrm{m}i,\mathrm{m}j}\boldsymbol{n}_{\mathrm{m}i,\mathrm{m}j}}{L_{\mathrm{m}i,\mathrm{m}j}}K_{\mathrm{m}i,\mathrm{m}j}\frac{\rho_{\mathrm{g}}K_{\mathrm{rg}}}{\mu_{\mathrm{g}}}\left[P_{\mathrm{m}i}-\rho_{\mathrm{p}}gD_{\mathrm{m}i})-(P_{\mathrm{m}j}-\rho_{\mathrm{p}}gD_{\mathrm{m}j})\right]_{\mathrm{g}}\right\}$$
$$+q_{\mathrm{gm}i}^{\mathrm{W}}=\frac{\partial}{\partial t}\left(V\phi S_{\mathrm{g}}\rho_{\mathrm{g}}+\rho_{\mathrm{g}}\rho_{\mathrm{R}}\frac{v_{\mathrm{L}}}{1+b_{\mathrm{L}}P_{\mathrm{g}}}\right)_{\mathrm{m}i} \tag{3.134}$$

2. Fracture media at different scales

In tight gas reservoirs, natural fractures and hydraulic fractures form a complex fracture network. The flow mechanisms and flow regimesof fluids in different scales fractures are significantly different. It shows different flow regimes, include high speed nonlinear flow, quasi linear flow and low speed nonlinear flow.

(1) High speed nonlinear flow

For the gas flow process in fractures media of tight gas reservoirs, as the pressure gradient is greater than the high speed nonlinear critical pressure gradient, the flow regime is the high speed nonlinear seepage. Based on the Forchheimer equation, the flow equation is

$$\sum_{j=1}^{N}\left\{\frac{A_{\mathrm{f}i,\mathrm{f}j}\boldsymbol{n}_{\mathrm{f}i,\mathrm{f}j}}{L_{\mathrm{f}i,\mathrm{f}j}}K_{\mathrm{f}i,\mathrm{f}j}F_{\mathrm{ND}}\frac{\rho_{\mathrm{g}}K_{\mathrm{rg}}}{\mu_{\mathrm{g}}}\left[(P_{\mathrm{f}i}-\rho_{\mathrm{p}}gD_{\mathrm{f}i})-(P_{\mathrm{f}j}-\rho_{\mathrm{p}}gD_{\mathrm{f}j})\right]_{\mathrm{g}}\right\}$$
$$+q_{\mathrm{gf}i}^{\mathrm{W}}=\frac{\partial}{\partial t}(V\phi S_{\mathrm{g}}\rho_{\mathrm{g}})_{\mathrm{f}i} \tag{3.135}$$

$$F_{\mathrm{ND}}=\frac{1}{1+\frac{\beta\rho_{\mathrm{g}}q_{\mathrm{g}}K}{A\mu_{\mathrm{g}}}} \tag{3.136}$$

(2) Quasi-linear flow

For the gas flow process in fracture media of tight gas reservoirs, as the pressure gradient is between the quasi linear critical pressure gradient and the high speed nonlinear critical pressure gradient, the flow regime is the quasi linear flow. The flow equation is

$$\sum_{j=1}^{N}\left\{\frac{A_{\mathrm{f}i,\mathrm{f}j}\boldsymbol{n}_{\mathrm{f}i,\mathrm{f}j}}{L_{\mathrm{f}i,\mathrm{f}j}}K_{\mathrm{f}i,\mathrm{f}j}\frac{\rho_{\mathrm{g}}K_{\mathrm{rg}}}{\mu_{\mathrm{g}}}\left[(P_{\mathrm{f}i}-\rho_{\mathrm{p}}gD_{\mathrm{f}i})-(P_{\mathrm{f}j}-\rho_{\mathrm{p}}gD_{\mathrm{f}j})-G_{\mathrm{c,f}i,\mathrm{f}j}\right]_{\mathrm{g}}\right\}$$
$$+q_{\mathrm{gf}i}^{\mathrm{W}}=\frac{\partial}{\partial t}(V\phi S_{\mathrm{g}}\rho_{\mathrm{g}})_{\mathrm{f}i} \tag{3.137}$$

(3) Low speed nonlinear flow

For the gas flow process in nano/micro-fractures, as the pressure gradient is greater than the threshold pressure gradient and less than the quasi linear critical pressure gradient, the flow regime is the low speed non-linear flow. It is affected by the threshold pressure gradient. The flow equation is

$$\sum_{j=1}^{N}\left\{\frac{A_{\mathrm{f}i,\mathrm{f}j}\boldsymbol{n}_{\mathrm{f}i,\mathrm{f}j}}{L_{\mathrm{f}i,\mathrm{f}j}}K_{\mathrm{f}i,\mathrm{f}j}\frac{\rho_{\mathrm{g}}K_{\mathrm{rg}}}{\mu_{\mathrm{g}}}\left[(P_{\mathrm{f}i}-\rho_{\mathrm{p}}gD_{\mathrm{f}i})-(P_{\mathrm{f}j}-\rho_{\mathrm{p}}gD_{\mathrm{f}j})-G_{\mathrm{a,f}i,\mathrm{f}j}\right]_{\mathrm{g}}^{n*}\right\} + q_{\mathrm{gf}i}^{\mathrm{W}} = \frac{\partial}{\partial t}(V\phi S_{\mathrm{g}}\rho_{\mathrm{g}})_{\mathrm{f}i} \quad (3.138)$$

3.2.4.3 Mathematical Model of Multiphase Flow Between Multiple Media

The calculation model for the fluid exchanging behavior between multiple media and the model for the conductivity between different media in tight gas reservoirs are similar to that of tight oil reservoirs. Therefore, they will not be presented in this section. The fluid exchanging behavior between matrix pores and small/large scale fractures is mainly influenced by the multiple mechanisms of threshold pressure gradient, slippage effect and diffusion effect. It shows the flow regimes of high speed nonlinear flow, quasi linear flow and low speed nonlinear flow. Therefore, the calculation model of the multiphase flowability under different mechanisms can be used to describe the effect of flow regime and flow mechanism on the fluid exchanging behavior between different media.

1. Multiphase flow model between porous media at different scales
   (1) Quasi-linear flow

When the pressure gradient of gas flow between different pore media is between the quasi linear critical pressure gradient and the high speed nonlinear critical pressure gradient, the flow regime is quasi linear flow, and the multiphase flow term can be expressed as follows:

$$(F_{\text{quasi-linear,g}})_{\mathrm{m}i,\mathrm{M}j} = \frac{\rho_{\mathrm{g}}K_{\mathrm{rg}}}{\mu_{\mathrm{g}}}\left[(P_{\mathrm{m}i}-\rho_{\mathrm{p}}gD_{\mathrm{m}i}) -(P_{\mathrm{M}j}-\rho_{\mathrm{p}}gD_{\mathrm{M}j})-G_{\mathrm{c,m}i,\mathrm{M}j}\right]_{\mathrm{g}} \quad (3.139)$$

(2) Low speed nonlinear flow

When the pressure gradient of gas flow between different pore media is greater than the threshold pressure gradient and less than the quasi linear critical pressure gradient, the flow regime is low-speed nonlinear flow. It is affected by the threshold pressure gradient, and the multiphase flow term can be expressed as follows:

$$(F_{\text{starting,g}})_{\mathrm{m}i,\mathrm{M}j} = \frac{\rho_{\mathrm{g}}K_{\mathrm{rg}}}{\mu_{\mathrm{g}}}\left[(P_{\mathrm{m}i}-\rho_{\mathrm{p}}gD_{\mathrm{m}i}) -(P_{\mathrm{M}j}-\rho_{\mathrm{p}}gD_{\mathrm{M}j})-G_{\mathrm{a,m}i,\mathrm{M}j}\right]_{\mathrm{g}}^{n^*} \quad (3.140)$$

(3) Slippage effect

Under the conditions of low permeability and low pressure, the multiphase flow term is affected by the slippage effect of gas flow. The specific expression is (for the adjacent two grids, they have the same slippage factor B and critical pressure P_{mi})

$$(F_{\text{slippage,g}})_{\mathrm{m}i,\mathrm{M}j} = \frac{\rho_{\mathrm{g}}K_{\mathrm{rg}}}{\mu_{\mathrm{g}}}\left[\left(1+\frac{b}{P_{\mathrm{m}i}}\right)(P_{\mathrm{m}i}-\rho_{\mathrm{p}}gD_{\mathrm{m}i})-\left(1+\frac{b}{P_{\mathrm{M}j}}\right)(P_{\mathrm{M}j}-\rho_{\mathrm{p}}gD_{\mathrm{M}j})\right]_{\mathrm{g}} \quad (3.141)$$

(4) Diffusion effect

Under the conditions of low permeability and low pressure, the multiphase flow term is affected by the gas diffusion. The specific expression is (for the adjacent two grids, they have the same diffusion coefficient)

$$(F_{\text{diffusion,g}})_{\mathrm{m}i,\mathrm{M}j} = \frac{\rho_{\mathrm{g}}K_{\mathrm{rg}}}{\mu_{\mathrm{g}}}\left[\left(1+\frac{32\sqrt{2}\sqrt{RT}\mu_{\mathrm{g}}}{3r\sqrt{\pi M}P_{\mathrm{m}i}}\right)(P_{\mathrm{m}i}-\rho_{\mathrm{p}}gD_{\mathrm{m}i})-\left(1+\frac{32\sqrt{2}\sqrt{RT}\mu_{\mathrm{g}}}{3r\sqrt{\pi M}P_{\mathrm{M}j}}\right)(P_{\mathrm{M}j}-\rho_{\mathrm{p}}gD_{\mathrm{M}j})\right]_{\mathrm{g}} \quad (3.142)$$

2. The fluid exchanging behavior from matrix to different scales fractures

(1) Quasi-linear flow

When the pressure gradient of gas flow from matrix to fractures is between the quasi linear critical pressure gradient and the high speed nonlinear critical pressure gradient, the flow regime is quasi linear flow. The multiphase flow term can be expressed as follows:

$$(F_{\text{quasi-linear,g}})_{\text{m}i,\text{f(F)}j} = \frac{\rho_{\text{g}} K_{\text{rg}}}{\mu_{\text{g}}} \left[(P_{\text{m}i} - \rho_{\text{p}} g D_{\text{m}i}) - (P_{\text{f(F)}j} - \rho_{\text{p}} g D_{\text{f(F)}j}) - G_{\text{c,m}i,\text{f(F)}j} \right]_{\text{g}} \quad (3.143)$$

(2) Low speed nonlinear flow

When the pressure gradient of gas flow from matrix to fractures is greater than the threshold pressure gradient and less than the quasi linear critical pressure gradient, the flow regime is low-speed nonlinear flow. It is affected by the threshold pressure gradient. The multiphase flow term can be expressed as follows:

$$(F_{\text{starting,g}})_{\text{m}i,\text{f(F)}j} = \frac{\rho_{\text{g}} K_{\text{rg}}}{\mu_{\text{g}}} \begin{bmatrix} (P_{\text{m}i} - \rho_{\text{p}} g D_{\text{m}i}) \\ -(P_{\text{f(F)}j} - \rho_{\text{p}} g D_{\text{f(F)}j}) - G_{\text{a,m}i,\text{f(F)}j} \end{bmatrix}_{\text{g}}^{n*} \quad (3.144)$$

(3) Slippage effect

Under the conditions of low permeability and low pressure, the multiphase flow process is affected by the slippage effect. The specific expression is (for the adjacent two grids, they have the same slippage factor B and critical pressure $P_{\text{m}i}$)

$$(F_{\text{slippage,g}})_{\text{m}i,\text{f(F)}j} = \left[\left(1 + \frac{b}{P_{\text{m}i}}\right)(P_{\text{m}i} - \rho_{\text{p}} g D_{\text{m}i}) - \left(1 + \frac{b}{P_{\text{f(F)}j}}\right)(P_{\text{f(F)}j} - \rho_{\text{p}} g D_{\text{f(F)}j}) \right]_{\text{g}} \quad (3.145)$$

(4) Diffusion effect

Under the conditions of low permeability and low pressure, the multiphase flow process is also affected by the diffusion effect of gas. The specific expression is (for the adjacent two grids, they have the same diffusion coefficient)

$$(F_{\text{diffusion,g}})_{\text{m}i,\text{f(F)}j} = \frac{\rho_{\text{g}} K_{\text{rg}}}{\mu_{\text{g}}} \begin{bmatrix} \left(1 + \frac{32\sqrt{2}\sqrt{RT}\mu_{\text{g}}}{3r\sqrt{\pi \text{M}} P_{\text{m}i}}\right)(P_{\text{m}i} - \rho_{\text{p}} g D_{\text{m}i}) \\ -\left(1 + \frac{32\sqrt{2}\sqrt{RT}\mu_{\text{g}}}{3r\sqrt{\pi \text{M}} P_{\text{f(F)}j}}\right)(P_{\text{f(F)}j} - \rho_{\text{p}} g D_{\text{f(F)}j}) \end{bmatrix}_{\text{g}} \quad (3.146)$$

3. The fluid exchanging behavior between fractures at different scales

(1) High speed nonlinear flow

When the pressure gradient of gas flow between large scale fractures is greater than the high speed nonlinear critical pressure gradient, the flow regime is high speed nonlinear flow. The specific expression is

$$(F_{\text{high speed,g}})_{\text{f}i,\text{F}j} = F_{\text{ND}} \frac{\rho_{\text{g}} K_{\text{rg}}}{\mu_{\text{g}}} \left[(P_{\text{f}i} - \rho_{\text{p}} g D_{\text{f}i}) - (P_{\text{F}j} - \rho_{\text{p}} g D_{\text{F}j}) \right]_{\text{g}} \quad (3.147)$$

(2) Quasi-linear flow

When the pressure gradient of gas flow from small scale fractures to large scale fractures is between the quasi linear critical pressure gradient and the high speed nonlinear critical pressure gradient, the flow regime is quasi linear flow. The multiphase flow term can be expressed as follows:

$$(F_{\text{quasi-linear,g}})_{\text{f}i,\text{F}j} = \frac{\rho_{\text{g}} K_{\text{rg}}}{\mu_{\text{g}}} \left[(P_{\text{f}i} - \rho_{\text{p}} g D_{\text{f}i}) - (P_{\text{F}j} - \rho_{\text{p}} g D_{\text{F}j}) - G_{\text{c,f}i,\text{F}j} \right]_{\text{g}} \quad (3.148)$$

(3) Low speed nonlinear flow

When the pressure gradient of gas flow between different scale fractures is greater than the threshold pressure gradient and less than the quasi linear critical pressure gradient, the flow regimeis low speed nonlinear flow. It is affected by the threshold pressure gradient. The multiphase flow term can be expressed as follows:

$$(F_{\text{starting,g}})_{\text{f}i,\text{F}j} = \frac{\rho_{\text{g}} K_{\text{rg}}}{\mu_{\text{g}}} \left[(P_{\text{f}i} - \rho_{\text{p}} g D_{\text{f}i}) - (P_{\text{F}j} - \rho_{\text{p}} g D_{\text{F}j}) - G_{\text{a,f}i,\text{F}j} \right]_{\text{g}}^{n^*} \quad (3.149)$$

References

Barrenblatt GI, Zehltov YP, Kochina IN (1960) Basic concepts in the theory of seepage of homogeneous liquids in fissured rocks. J Appl Math Mech 24(5):1286–1303

Chen ZX, Huan GR, Ma YL (2006) Computational methods for multiphase flows in porous media. Society for Industrial and Applied Mathematics, Dallas, TX

Forchheimer P (1901) Wasserbewegung durch bode. ZVDI 27(45):26–30

Gong B, Karimi-Fard M, Durlofsky LJ (2006) An upscaling procedure for constructing generalized dual-porosity/dual-permeability models from discrete fracture characterizations. In: SPE annual technical conference and exhibition, TX, USA

Han DK, Chen QL, Yan CZ (1993) Fundamental numerical simulation of reservoirs. Beijing: Petroleum Industry Press

Li N, Ran QR, Li JF et al (2013) A multiple-media model for simulation of gas production from shale gas reservoirs. In: SPE reservoir characterization and simulation conference and exhibition, Abu Dhabi, UAE

Liu CQ, Guo SP (1982) Research progress in multi-medium flow. Adv Mech 4:360–363

Peng XL, Du ZM, Qi ZL (2006) Analysis on the adaptability of multi-medium percolation model. J Oil Gas Technol 28(4):99–101

Wang WF (2013) Research on seepage and numerical simulation of shale gas reservoir. Southwest Petroleum University, Chengdu

Wang F, Pan ZQ (2016) Numerical simulation of chemical potential dominated fracturing multiphase flowback in hydraulically fractured shale gas reservoirs. Petrol Expl Devel 43(6):1–7

Warren JE, Root PJ (1963) The behavior of naturally fractured reservoirs. SPE J 245–255

Wu YS (2016) Multiphase flow in Porous and fractured reservoirs. Elsevier Inc.

Wu YS, Ge JL (1983) The transient flow in naturally fractured reservoirs with three-porosity systems. Acta Mech Sin 19(1):81–85

Wu Y, Moridis SG, Bai B et al (2009) A multi-media model for gas production in tight fractured reservoirs. In: SPE hydraulic fracturing technology conference, TX, USA

Wu KL, Li XF, Chen ZX et al (2015) Gas transport behavior through micro fractures of shale and tight gas reservoirs. Chin J Theor Appl Mech 47(6):955–964

Yao GJ, Peng HL, Xiong Y et al (2009) Research on characteristics of gas flowing in low permeability sandstone gas reservoirs. Petrol Geol Recov Effic 16(4):104–105, 108

Yin DY, Pu H, Wu YX (2004) Numerical simulation of imbibition oil recovery for low permeability fractured reservoir. J Hydrodyn 19 (4):440–445

4 Discretization Methods on Unstructured Grids and Mathematical Models of Multiphase Flow in Multiple Media at Different Scales

According to the complex geological conditions, macro-scale heterogeneity, and micro-scale multiple media characteristics of unconventional oil and gas reservoirs, different types of unstructured grids are used to divide the reservoir into different regions according to lithofacies, lithogoloy or reservoir type, to further divide each region into different reservoir elements, and also to partition each element into pore and fracture media at different scales, in order to characterize reservoir heterogeneities, complex geometry of natural/hydraulic fractures and horizontal wells, the multi-scale and multiple media characteristics of pore and fracture media at different scales. With all of these, grid partitioning and grid generation technology for numerical simulation is formed. Through the optimization of unstructured grid order, the establishment of grid connectivity list, and the calculation of transmissibility between grid blocks, the connectivity characterization technology for numerical simulation is generated. At the same time, the finite volume method which is suitable for unstructured grid and the mathematical model of multiphase flow through discontinuous multiple media is used to discretize the mathematical model for multiphase flow through discontinuous multiple pore and fracture media at different scales. Through the works above, the discretization technology is formed for the mathematical model for multiphase flow through multiple media.

4.1 Grid Partitioning and Grid Generation Technology for Numerical Simulation

The lithology, lithofacies, reservoir types, and physical properties for tight reservoirs are highly heterogeneous in spatial distribution. The geometric characteristics of natural/hydraulic fractures and horizontal wells vary greatly, and their morphology and distribution are also complicated. There are great differences in the quantitative composition and spatial distribution pattern for microscopic pore and fracture media. The oil and gas flow pattern and behavior through multiple media in different parts of the reservoir vary greatly.

In order to characterize the characteristics of heterogeneity, multi-scale, multiple media and flow behavior through pore and fracture media of tight reservoirs, different types of structured grids, unstructured grids, variable scale grids and hybrid grids are used to divide the whole simulated reservoir area into different regions, further into simulation elements, and finally into different pore and fracture media to characterize the reservoir heterogeneity, the irregular geometry and complex morphological characteristics of natural/hydraulic fractures at different scales and horizontal wells, the multi-scale features of pore and fracture media at different scales, and the multiple media characteristics with different quantitative and spatial distribution patterns. The numerical simulation gridding technology for multiple media in tight oil and gas reservoirs is established accordingly.

Numerical simulation gridding technology can be divided into a structural gridding technology and an unstructured gridding technology (Wang 2004) (Table 4.1).

For tight reservoirs with weak heterogeneity, a regular distribution of sand bodies, low density of fractures with a single type of geometry, geological boundaries and other internal or external boundary conditions are relatively simple and regular. This yields simple flow behavior with no obvious multi-scale and multiple media characteristics existing. Thus, if we can simulate a reservoir as a continuous single or dual media, structural grid can be used to process a simulation model. Structured grid generation is simple and fast, with its data structure being simple and regular. Simulation with structured grids has rapid convergence, good stability, and fast calculation speed (Zhang and Tan 2003). Structured grid technology has great limitations for the treatment of geological conditions in complex reservoirs, which cannot well reflect these complex boundary conditions. A regular grid has serious grid orientation effects and

Q. Ran, *Unconventional Tight Reservoir Simulation: Theory, Technology and Practice*,
https://doi.org/10.1007/978-981-32-9848-4_4

Table 4.1 Comparison between structured grids and unstructured grids

Category	Structured grids	Unstructured grids
Grid types	Rectangular grid, Corner point grid, Radial grid	Triangular grid, Quadrilateral grid, PEBI grid, CVFE grid
Grid characteristics	① Geometric characteristics: Structured grid is a type of regular grid, which is also quadrilateral grid with regular geometric morphology ② Neighboring relationship: All connections between grid points in a grid block have clear and regular topological relationships. The grid block has a fixed number of nodes. All internal points also have the fixed number of adjacent grid blocks (Wang 2004) ③ Grid characteristics: The distribution of nodes and cells in the structured grid is fixed and in a good order. The grid block's size, shape, and positions of the grid points are predetermined	① Geometric characteristics: Unstructured grid is a type of irregular grid with irregular geometric morphology ② Neighboring relationship: The connection of all grid points has an uncertain irregular topology. Different grid blocks have different number of nodes and different adjacent grids (Wang 2004) ③ Grid characteristics: The distribution of nodes and grid blocks within unstructured grid is arbitrary. The size, shape of grid blocks and position of grid points are changeable depending on our needs. Therefore, it has better flexibility than structured grid (Lu 2008)
Suitable conditions	(1) Reservoir and flow characteristics: ① The reservoirs, heterogeneity is weak and distribution of sand bodies is regular ② The reservoirs have low density of fractures with single type of geometry ③ The geological boundaries and other internal or external boundary conditions are relatively simple and regular (2) It yields simple behavior and single direction of flow (3) Well types: Vertical well (4) Media type: Continuous single/dual-porosity media	(1) Reservoir and flow characteristics: ① Large heterogeneity in each area and each element of the reservoir ② Various complex inner and outer boundary conditions (Xiang et al. 2006) ③ Complicated multi-scale natural and hydraulic fracture networks ④ Characteristics of multiple media for pores and fractures with different scalessc ⑤ Complicated flow characteristics in various media and around the wellbore (Xiang et al.2006) (2) Well types: vertical well, horizontal well, complex-structure well (3) Media type: Discontinuous and discrete multiple media
Advantages and disadvantages	Advantages: ① Structured grid generation is simple and fast ② The data structure is simple with strong regularity ③ Discretization of higher order nonlinear partial differential equations is very simple and convenient. It is also very easy to implement in the process of programming. Simulation with the structured grids has rapid convergence, good stability, and fast calculation speed (Xiang et al. 2006; Chen et al. 2010) Disadvantages: ① Structured grid technology has great limitations for the treatment of geological conditions in complex reservoirs, which cannot well reflect these complex boundary conditions. Particularly the treatment of the flow characteristics near the well is not ideal (Tang et al. 2007; Xiang et al. 2006; Chen et al. 2010) ② The regular grid has serious grid orientation effects and poor adaptability ③ Under the scenario of complex reservoirs and boundary conditions, there are many invalid grids, which generates large amount of ineffective calculations	Advantages: ① Fine characterization with high degree of details The unstructured grid can better characterize the complex geologic conditions, heterogeneity and various complicated boundaries of the reservoir and better approximate the fluid flow patterns, which more truly reflect the real conditions of the reservoir (Lu 2008) ② Strong grid flexibility The distribution of nodes and grid blocks of unstructured grid is arbitrary. It is possible to adapt irregular, complex-shaped geologic conditions by adjusting sizes and shapes of the grid blocks, and positions of the grid points. With local and global decomposition, plus refinement and amalgamation, such grid can adapt to the complex boundary conditions. Therefore, it has better flexibility than the structured grid (Lu 2008) ③ High simulation accuracies The unstructured grid generation method uses some criteria for optimization judgments during the grid generation process. So high-quality grid system with

(continued)

Table 4.1 (continued)

Category	Structured grids	Unstructured grids
		less invalid grids can be generated. The random data structure makes it convenient for grid self-adaption, which can better improve the computational efficiency of the grid system. At the same time, due to the flexibility of the unstructured grid, it is possible to approximate any complex geological conditions. Therefore, more accurate numerical solution can be obtained (Zhang et al. 1999) Disadvantages: ① Grid partitioning method is complicated and difficult, which has low computational speed ② Complex data structures and slow simulation speed Using fine grids to approximate complex geological conditions and arbitrary-shaped boundary conditions, we can improve the accuracy of the characterization. On the other hand, this will make the number of grids extremely large. Furthermore, the simulated calculations are also slow with heavy workload
Grid system schematics	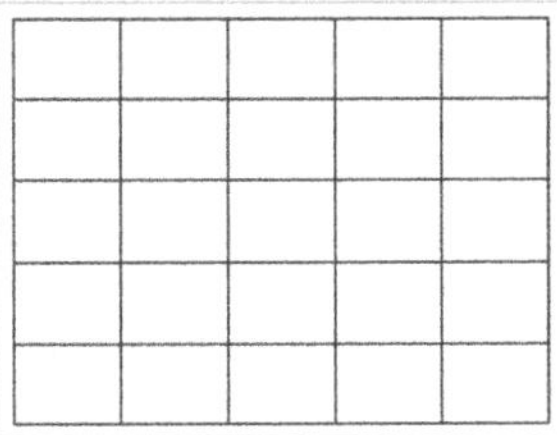	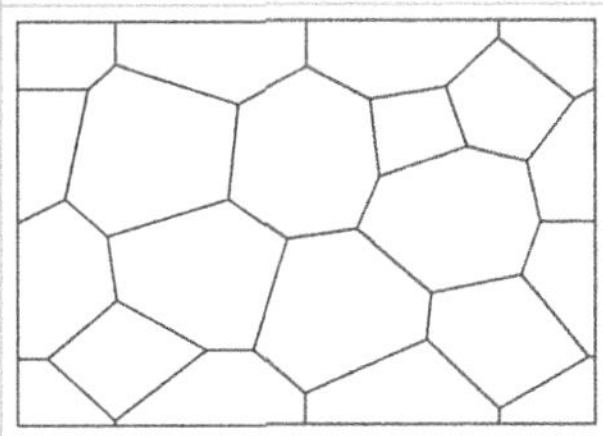

poor adaptability (Chen et al. 2010). At the same time, for complex reservoirs and boundary conditions, there are many invalid grids, which generate large amount of ineffective calculations.

For a tight reservoir with strong heterogeneity, complex geometries and complicated distribution patterns for natural/hydraulic fractures at different scales and horizontal wells, discontinuous and discrete distributions of multiple media with different scale pores and fractures, and significant multi-scale and multiple media features, unstructured grid technology can be used for processing a simulation model.

Using a single set of unstructured variable-scale grids and unstructured variable-scale hybrid grids, the differences of reservoir heterogeneity in different regions and different elements can be characterized. With this method, the complex geometry and morphological characteristics of natural/hydraulic fractures, horizontal wells, faults and geological boundary conditions can be described flexibly with high precision. Large-scale natural/hydraulic fractures can be treated as discontinuous discrete media. Micro to nano-scale pores and fractures with different quantitative composition and spatial distribution patterns can be treated as multiple media of discontinuous discrete distribution. Furthermore, this method can also flexibly describe the flow patterns and behavior for oil and gas at different locations, especially around wellbores. Unstructured grid technology's specialty is flexible treatment of internal and external boundary conditions of various complex morphologies, strong heterogeneous areas and elements, multi-scale and multiple media features, and complex flow patterns, with high degree of details and high simulation accuracies (Lu 2008). The unstructured grid technology has the following characteristics: treating the internal and external boundary conditions of various complex morphologies flexibly, dividing the reservoir into different regions and further into elements according to strong heterogeneity, characterizing the characterizations of multi-scale and multi-media and the complex flow patterns. This technology has high degree of details and high simulation accuracies (Lu 2008).

However, the grid generation method is complicated and difficult, with complex data structures. Furthermore, the simulation speed is slow.

4.1.1 Structured Grid Technology

For tight reservoirs with weak heterogeneity, simple geometries and a single medium, we can treat them as continuous single or double-porosity media and use structured grid technology to generate numerical simulation grids. A structured grid is a type of regular grid, which is also a quadrilateral grid with regular geometric morphology. In all the grid blocks, the connections between grid points have clear and regular topological relationships. A grid block has a fixed number of nodes in the gridded area. In addition, all internal points have the same adjacent grid blocks. The distribution of nodes and cells in a structured grid is fixed and in a good order. The size, shape of grid block and the position of grid points are predetermined.

4.1.1.1 Types of Structured Grids

Structured grids mainly include a rectangular grid, a corner point grid and a radial grid (Table 4.2).

1. Rectangular grids

(1) The definition and characteristics of rectangular grids

A rectangular grid is a type of global orthogonal grid, which is built on a Cartesian coordinate system (Lu 2008).

The characteristics of a rectangular grid are mainly reflected as follows. Each grid cell is a regular hexahedron with unequal side lengths. Each grid cell has 6 adjacent grid cells and 8 nodes. Each of them connected with the surrounding 7 adjacent grid nodes. The connection line between the centers of the two adjacent grid cells is perpendicular to the interface between the two grid cells. The same column of grid cells has the same length. Also, the same row of the grid cells also have the same width. All grid blocks can have different vertical heights. The grid block geometry can be completely determined by its length, width and height (Chang 1998).

Table 4.2 Types of structured grids and their characteristic descriptions

Category	Rectangular grid	Corner point grid	Radial grid
Definitions	Rectangular grid is a type of global orthogonal grid, which is built on a Cartesian coordinate system	Corner point grid is a type of non-orthogonal grid, which is built on a Cartesian coordinate system	The radial grid is fan-shaped grid, which is generated by considering radial distance, azimuth, and height in the cylindrical coordinate system
Grid characteristics	(1) Geometric characteristics: ① Each grid cell is a regular hexahedron with unequal side lengths ② The same column of grid cells has the same length. And the same row of the grid cells also have the same width. The vertical heights of all grid blocks can be different ③ The grid block geometry can be completely determined by its length, width and height (2) Adjacency status: ① Each grid cell has 6 adjacent grid cells ② Each grid cell has 8 nodes, each of them connected with the surrounding 6 adjacent grid nodes (3) Orthogonal Features: The connection line between the centers of the two adjacent grid cells is perpendicular to the interface between the two grid cells	(1) Geometric characteristics: ① Each grid cell is an irregular hexahedron with unequal side lengths ② All grid blocks can have different lengths, widths and heights ③ When the length, width and height of a grid cell are determined, the grid block's geometric configuration needs to be determined by the spatial coordinates of the 8 nodes of the grid cell, because the edges of the cell are not perpendicular to each other (2) Adjacency status: ① Each grid cell has 6 adjacent grid cells ② Each grid cell has 8 nodes, with each of them connected with the surrounding 7 adjacent grid nodes (Mao et al. 2012) (3) Orthogonal features: The connection line between the centers of the two adjacent grid cells is not perpendicular to the interface between the two grid cells	(1) Geometric characteristics: ① Each radial grid cell is an irregular fan-shaped hexahedron with unequal side lengths ② All grid blocks can have different radial distances, azimuth, and heights ③ The grid block can be completely determined by its radial length, position, arc length and height (2) Adjacency status: ① Each grid cell has 6 adjacent grid cells ② Each grid cell has 8 nodes with each of them connected to the surrounding 7 adjacent grid nodes (3) Orthogonal features: The connection line between the centers of the two grid cells is perpendicular to the interface between the two grid cells
Applicable conditions	① Simple reservoir geological conditions ② Weak reservoir heterogeneity ③ Regular distribution of sand bodies ④ Regular geological boundaries and other internal or external boundaries conditions (Wang et al. 2012) ⑤ Single directional flow or simple flow behavior	① Relatively complex reservoir geological conditions ② Strong reservoir heterogeneity ③ Irregular distribution of sand bodies ④ Irregular and complex reservoir boundaries ⑤ Relative complicated reservoir flow direction and pattern	① The characteristics of radial flow near the wellbore ② The characteristics of high flow velocity near the wellbore

(continued)

Table 4.2 (continued)

Category	Rectangular grid	Corner point grid	Radial grid
Advantages and disadvantages	Advantages: ① Rectangular grid method is simple and efficient in computational time ② Its discretized equations are easiest to be implemented (Lin 2010) ③ Due to the orthogonality of the grid system, its numerical simulation has fast convergence and good stability, which improves the simulation accuracy of global orthogonal grid Disadvantages: ① There is large numerical error and poor flexibility when the rectangular grid is used to describe the complex geological conditions, reservoir heterogeneity, complicated geometries and boundaries ② When there are many wells in the simulating area, it is difficult to place all of them in the center of the different grids, which will increase the error ③ In situations where there are faults, boundaries, sand body distributions, horizontal wells, and significant directional fluid flows, there is serious grid orientation effect (Xie et al. 2001) ④ When there are complex geological conditions or boundary conditions, there is no guarantee that every grid block is an effective cell in simulation, which increases the computational workload (Xie et al. 2001)	Advantages: ①The corner point grid can describe the complex geological conditions, the reservoir heterogeneity, the complex geometry and complicated boundaries to some extent. This method has high accuracy in characterizing the reservoir (Lu 2008) ② The corner point grid approximates irregular and complex shape geological features together with complex boundary conditions by changing the size, shape, and node positions of the grid block. The flexibility of such grid is relatively strong Disadvantages: ① The corner grid is extremely irregular in shape if there is no constraint for its shape. At the same time, using the parameters at the centers of the grids to represent the property parameters of the entire grid block will bring large error, which reduces the simulation accuracy ② The corner point grids are non-orthogonal grids. So calculating the transmissibility of the corner point grid in the simulation is very complicated, which causes the simulation accuracy to be poor (Wang et al. 2012) ③ Corner point grid cannot accurately describe the radial flow characteristics near the wellbore ④ The grid partitioning method is relatively complex and difficult	Advantages: ① The radial grid is adopted in the vertical well area, which is the most consistent with the radial flow characteristics near the oil and water wells (Lin 2010) ② The use of radial grid allows the grid volume to change greatly and rapidly, which starts from fine grid near the well to coarse grid far from the well. Thereby, we can simulate the changing characteristics of flow from the low velocity flow away from the wellbore to the fast velocity flow near the wellbore. This method can make smooth transitions, thus achieving smooth computations and rapid convergence of the simulations (An et al. 2007) ③ By forming hybrid grid with other types of grid, the radial grids are used to simulate radial flow characteristics near the wellbore. At the same time, using other grids to describe the complex geology and boundary conditions at the periphery of the wells can precisely characterize those problems (Lin 2010) Disadvantages: ① The radial grids cannot be used to describe the complex geological features and boundary conditions. They are not suitable to describe other flow characteristics except the radial flow. Radial grids are only suitable for single wells and local near well simulations. So there is great limitation in the scope of its applications
Grid system schematics	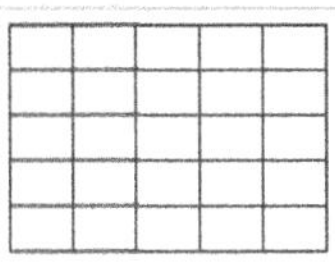	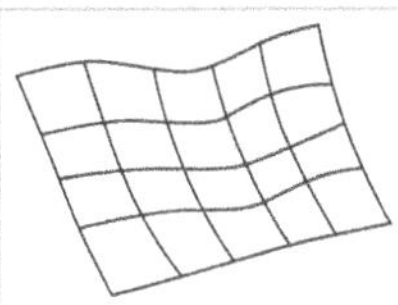	

(2) Applicable conditions for rectangular grids

Rectangular grids are suitable for such situations where reservoir geological conditions are simple, the heterogeneity is weak, the distribution of sand bodies is regular, the geological boundaries and other internal or external boundary conditions are regular, and the flow is single directional or the flow behavior is simple (Wang et al. 2012).

The advantages of a rectangular grid are as follows. The rectangular grid method is simple and also efficient in computational time. Its discretized equations are easiest to be implemented. Due to the orthogonality of a grid system, its numerical simulation has fast convergence and good stability, which improves the simulation accuracy of global orthogonal grid (Lin 2010).

The disadvantages of a rectangular grid are as follows. There is a large numerical error and poor flexibility when the rectangular grid is used to describe complex geological conditions, reservoir heterogeneity, complicated geometries and boundaries. When there are many wells in a simulating area, it is difficult to place all of them in the centers of different grid blocks, which will increase the error. In situations where there are faults, boundaries, sand body distributions, horizontal wells, and significant directional fluid flows, there is a serious grid orientation effect (Lu 2008). When there are complex geological conditions or boundary conditions, there is no guarantee that every grid block is an effective cell in simulation, which increases the computational workload (Lin 2010).

2. Corner point grid

(1) The definition and characteristics of corner point grid

A corner point grid is a type of non-orthogonal grid, which is built on a Cartesian coordinate system.

The characteristics of a corner point grid are mainly reflected as follows. Each grid cell is an irregular hexahedron with unequal side lengths. Each grid cell has 6 adjacent grid cells and 8 nodes. And each grid node connects with the surrounding 6 adjacent grid nodes. The connection line between the centers of the two adjacent grid cells is not perpendicular to the interface between the two grid cells. All grid blocks can have different lengths, widths and heights. When the length, width and height of a grid cell are determined, the grid block geometric configuration needs to be determined by the spatial coordinates of the 8 nodes of the grid cell, because the edges of the cell are not perpendicular to each other (Mao et al. 2012).

(2) Applicable conditions for corner point grids

The usage of corner point grids is suitable for the following complex situations: relatively complex reservoir geological conditions, strong reservoir heterogeneity, irregular distribution of sand bodies, irregular and complex reservoir boundaries, and relatively complicated reservoir flow direction and pattern.

The advantages of a corner point grid are as follows. The corner point grid can describe the complex geological conditions, the reservoir heterogeneity, the complex geometry and complicated boundaries to some extent. This method has high accuracy in characterizing the reservoir. The corner point grid approximates irregular and complex shape geological features together with complex boundary conditions by changing the size, shape, and node positions of the grid block. The flexibility of such a grid is relatively strong (Lu 2008).

The disadvantages of a corner point grid are as follows. The corner grid is extremely irregular in shape if there is no constraint for its shape. At the same time, using the parameters at the centers of grid blocks to represent the property parameters of the entire grid blocks will bring a large error, which reduces the simulation accuracy (Lu 2008). The corner point grids are non-orthogonal grids. Thus, calculating the transmissibility in the corner point grid in the simulation is very complicated, which causes the simulation accuracy to be poor (Wang et al. 2012). A corner point grid cannot accurately describe the radial flow characteristics near a wellbore. Finally, the grid partitioning method is relatively complex and difficult.

3. Radial grid

(1) The definition and characteristics of radial grid

A radial grid is a fan-shaped grid, which is generated by considering radial distance, azimuth, and height in a cylindrical coordinate system.

The characteristics of a radial grid are mainly reflected as follows. Each grid cell is an irregular fan-shaped hexahedron with unequal side lengths. Each grid cell has 6 adjacent grid cells and 8 nodes. And each grid node connects with the surrounding 6 adjacent grid nodes. The connection line between the centers of two grid cells is perpendicular to the interface between the two grid cells. All grid blocks can have different radial distances, azimuth, and heights. A grid block can be completely determined by its radial length, position, arc length and height.

(2) Applicable conditions for radial grids

Radial grids are suitable for the characteristics of radial flow and high velocity flow near a wellbore.

The advantages of radial point grids are as follows. The radial grid is adopted in a vertical well area, which is the most consistent with the radial flow characteristics near oil and water wells. The use of a radial grid allows a grid volume to change quickly and rapidly, which starts from a fine grid near a well to a coarse grid far from the well. Thereby, we can simulate the changing characteristics of flow from the low velocity flow away from the wellbore to the fast velocity flow near the wellbore. This method can make smooth transitions, thus achieving smooth computations and rapid convergence of simulations. By forming a hybrid grid with other types of grids, the radial grids are used to simulating radial flow characteristics near a wellbore. At the same time, using other grids to describe the complex geology and boundary conditions at the periphery of the wells can precisely characterize those problems (Lin 2010).

The disadvantage of radial grids is that they cannot be used to describe the complex geological features and boundary conditions. They are not suitable to describe other flow characteristics except the radial flow. Radial grids are only suitable for single wells and local near well simulations. So, there is a great limitation in the scope of its applications.

4.1.1.2 Structured Grid Partitioning and Generation Technology

Structured grid partitioning and generation technology is an important part for the numerical simulation technology. The quality of grid partitioning is an important factor which affects the level of fine description of reservoir geology, accuracy of numerical simulation results, simulation stability and computational speed.

1. Rectangular grid partitioning and generation technology

According to the grid partitioning principle, the process of rectangular grid partitioning and generation technology can be formed. Here are the main steps.

① We determine the scope of the simulation area.
② We determine the direction of the grid. First, we can define the origin of a coordinate system, and then define the direction of the grid system, according to the boundary conditions, direction of sand body distribution and direction of flow.
③ We determine the size of the grid blocks. In the horizontal plane, we determine the grid sizes according to certain number of grid blocks between the wells. At the same time, we can also determine the grid sizes in different directions confirmed based on the anisotropy of the planar reservoir. And in the vertical direction, We determine the size of the vertical grid blocks according to the reservoir thickness and the stratigraphic requirements.
④ We determine the number of grids. We optimize and decide the number of grids, according to the requirements for the size of the grid blocks, the storage capacity and computational speed of the computer.
⑤ We use 3D geometry algorithms to generate the rectangular grid cells (Fig. 4.1).
First of all, along the x direction, we determine the positions of the grid lines in x direction according to the different length of each grid block. At the same time, along the y direction, we determine the positions of the grid lines in the y direction according to the different width of each grid block. Then, we generate a planar grid system based on the positions of the grid lines along the x and y directions. Finally, along the z direction, we gradually generate each grid cell according to the different height of each grid block and calculate the spatial coordinates of different nodes of each grid cell.
⑥ We calculate the geometric parameters for each grid cell in space.

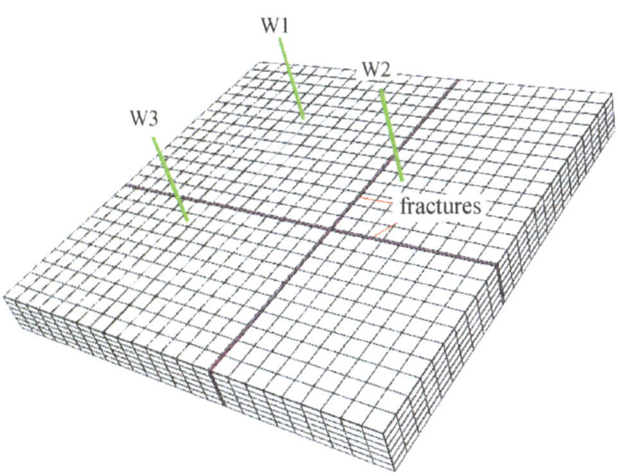

Fig. 4.1 3D rectangular grid cell spatial distributions

2. Corner point grid partitioning and generation echnology

A corner point grid is a type of "deformed" grid obtained by appropriately stretching, compressing and distorting the node positions of a regular hexahedral grid. By specifying the coordinates for the corner points of each grid block, the complex shape and boundary conditions of a reservoir can be accurately described. According to the principles of corner point partitioning, we highlight the influence of complex boundaries and sand body distribution patterns on grid partitioning and grid morphology. The technical process of corner point grid partitioning and generation is formed (Fig. 4.2). Here are the main steps.

① We determine the scope of the simulation area.
② We determine the direction of the grid. We define the origin of a coordinate system and define the direction of the grid system according to the boundary conditions, direction of sand body distribution and direction of flow.
③ We determine the size of the grid blocks. We determine the grid size according to the number of grids between the wells in the horizontal plane, with the grid sizes in different directions confirmed based on the anisotropy of the planar reservoir. Close to such spatial constraints as faults, fractures and boundaries, the grid is appropriately refined based on the requirements of accurate description of the geological conditions. We determine the vertical size of the grid according to the reservoir sickness and the stratigraphic requirements in the vertical direction.
④ We determine the number of grids. According to such a principle that we deploy a larger number of grids in the area where there are great heterogeneity in reservoir horizontal anisotropy and also severe constraints such as faults, fractures, and boundaries, and fewer grid blocks in the area where there are less heterogeneity in reservoir horizontal anisotropy and also relatively simple geological conditions, the number of grids is optimized and determined based on the grid sizes.

Fig. 4.2 Workflow for corner point grid partitioning and generation

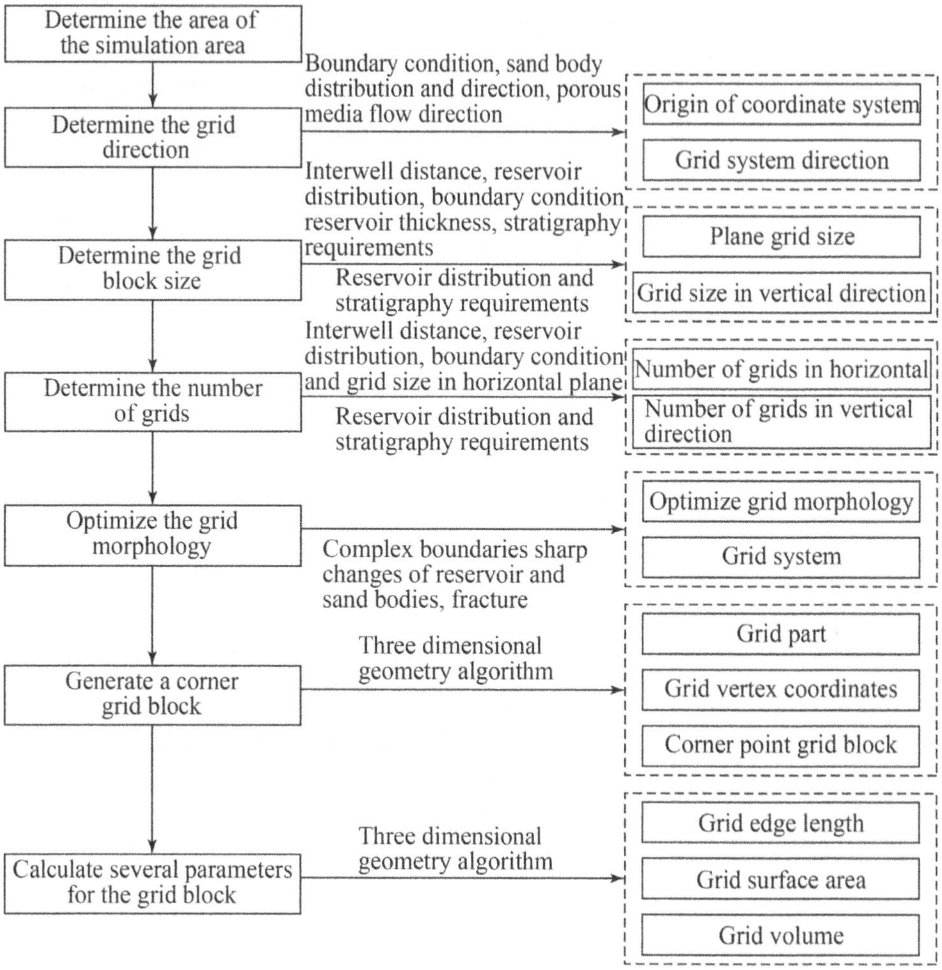

⑤ We determine the shape of the grid. By adjusting the node positions of regular hexahedral cells, we change the size and shape of the grid block, so that we can approach the irregular and complex shape for geological conditions and complicated boundary conditions. In this way, the grid shape not only meets the grid number and size requirements, but also meets the requirements for describing complex boundaries and geological conditions.

⑥ We generate corner grid blocks. According to the grid size, gird number and grid geometry, by using the 3D geometry algorithms, we can partition the simulation area into 3D corner point grids. We generate corner point grid blocks by calculating the coordinates of different nodes for each grid cell (Fig. 4.3) (Lu 2008).

⑦ We calculate the geometric parameters of each grid cell.

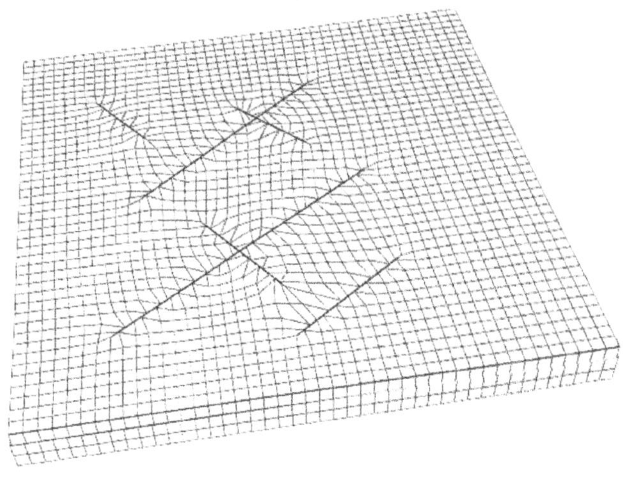

Fig. 4.3 3D corner point grid spatial distributions

3. Radial grids partitioning and generation technology

A radial grid is a fan-shaped grid system, which is partitioned by radial distance, azimuth, and height in a cylindrical coordinate system. According to the radial grid partitioning principle, the process of radial grid partitioning and generation technology can be formed. Here are the main steps.

① Determining the Scope of the Simulation Area.
② Determining the Origin of Coordinate System. We determine the origin of a coordinate system based on the well locations. The starting and ending azimuths are determined based on the scope of the simulated area.
③ Determining the Size of Grid Blocks. In the horizontal plane, the lengths of the grid at different radial positions are determined based on the distribution of sand bodies and the heterogeneity of the reservoir. Also, the arc lengths of different grids are determined based on the reservoir anisotropy. In the vertical direction, the vertical height of the grid is determined based on the reservoir thickness and stratigraphic requirements.
④ Determing the Number of Grids.
According to the requirements of a grid size and computational speed, we can optimize and determine the number of grids.
⑤ Generating the Radial Grid Cells by 3D Geometry Algorithms (Fig. 4.4).
We determine the positions of radial grid lines based on the radial length of each grid block. We determine the positions of the peripheral grid lines based on the arc length of each grid block in different directions and positions. Then, we can partition the planar grid based on the positions of the radial and peripheral grid lines. In the vertical direction, we gradually generate each grid cell according to the different height of each grid block, and calculate the spatial coordinates of different nodes for each grid cell.
⑥ Calculating the Geometric Parameters of Each Grid Cell in Space.

4.1.2 Unstructured Grid Technology

An unstructured grid is a type of irregular grid, in which the connection of all grid points has an uncertain irregular topology. The inner points within the area of a grid system do not share the same adjacent grids. In addition, the number of nodes of different grid blocks are different. The distribution of nodes and grid blocks of an unstructured grid is arbitrary. The size and the shape of grid blocks and position of grid points are changeable depending on needs. Therefore, it has better flexibility than the structured grid.

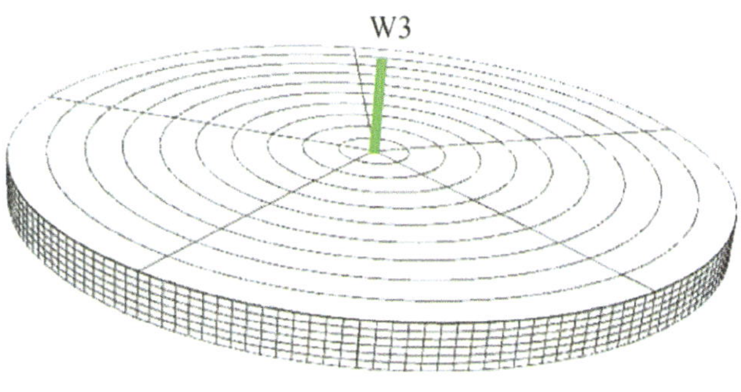

Fig. 4.4 3D radial grid spatial distributions

An unstructured grid applies to complex reservoirs and flow characteristics through the multiple media, including great heterogeneity in different regions and further in simulation elements, various complex inner and outer boundary conditions, multi-scale natural and hydraulic complex fracture networks, the characteristics of multi-medium in different porous, complicated flow characteristics in various media and around wellbores. Suitable well types include vertical wells, horizontal wells, and complex-structure wells. Suitable medium type is mainly discontinuous and discrete multiple medium (Chen et al. 2010).

4.1.2.1 Types of Unstructured Grid

Unstructured grids include mainly triangular grids, quadrilateral grids, PEBI grid and CVFE grid (Table 4.3).

1. Triangular grid

(1) Definition and characteristics of triangular grid

A triangular grid is a type of unstructured grid in a Cartesian coordinate system. In the plane, the grid consists of irregular triangles. And in space, every grid element is irregular tetrahedron with unequal side lengths.

The characteristics of a triangular grid are shown mainly in the following aspects: each grid block is irregular tetrahedron; the side lengths of the grid can be different to each other; the spatial coordinates of the four vertices for a grid block determine the geometric characteristics and spatial distribution of the grid. Each grid block has four adjacent grid blocks. Each grid block has four nodes, connecting with different number of adjacent grid nodes. If the connection line between the centers of two adjacent grid blocks is oblique to the interface between the two grid blocks that means it is non-orthogonal grid system (Yang 2005).

(2) Applicable conditions of triangular grid

As the basis for other complex grid systems, a triangular grid has the best flexibility. It can be flexible in representing complex geological boundaries, fractures, faults and complicated reservoir characteristics. It is hard to describe the multi-directional feature and anisotropy of a reservoir, due to a limited number of edges and faces in a triangular grid. The non-orthogonality of a triangular grid leads to large errors while computing transmissibility and flow terms. Also, the accuracy and efficiency of simulation are poor. Therefore, this grid system is generally not used in reservoir simulations (Lu 2008).

Table 4.3 Unstructured grid types and their characteristics

Classification	Triangular grid	Quadrilateral grid	PEBI grid	CVFE grid
Definition	Triangular grid is a type of unstructured grid in Cartesian coordinate system. The grid consists of irregular triangles in 2D space, and irregular tetrahedrons of different shapes in 3D space	Quadrilateral grid is a type of unstructured grid in the Cartesian coordinate system. The grid is comprised of irregular quadrangles in 2D space and of irregular hexahedron in 3D space	PEBI grid is a type of unstructured grid that is polygon grid constructed by successively connecting the circumcenters of triangles, based on triangular grid partitioning. Interface of two adjacent PEBI grid blocks must be perpendicular to and equally dividing the connection line of the two adjacent grid centers	CFVE grid is a type of unstructured grid based on triangular grid partitioning. It is such polygon grid constructed by connecting the gravity centers with the midpoint of each edge for different triangles
Grid characteristics	(1) Geometric characteristics: ① Each grid block is irregular quadrilateral ② The side lengths of the grid can be different to each other ③ The spatial coordinates of the four nodes for the grid block determine the geometric characteristics and spatial distribution of the grid (2) Connectivity characteristics: ① Each grid block has four adjacent grid blocks ② Each grid block has four nodes, connecting with different number of adjacent grid nodes (3) Orthogonality characteristics: It is non-orthogonal grid system, because the connection line between the centers of two adjacent grid blocks is oblique to the interface between the two grid blocks	(1) Geometric characteristics: ①Each grid block is irregular tetrahedron ② The side lengths of the grid can be different to others ③ The spatial coordinates of eight nodes of a grid block determine the geometric characteristics of this grid block (2) Connectivity characteristics: ① Each grid block has six adjacent grid blocks ② Each grid block has eight nodes connected with six adjacent grid nodes (3) Orthogonality characteristics: It is non-orthogonal grid system, because the connection line between the centers of two adjacent grid blocks is oblique to the interface between the two grid blocks	(1) Geometric characteristics: ① Each grid block is an irregular polyhedron ② The side lengths of a PEBI grid can be different to each other ③ The spatial coordinates of corresponding nodes of a PEBI grid block determine the geometric characteristics and spatial distribution of that grid block (2) Connectivity characteristics: ① Each grid block has different number of corresponding adjacent grid blocks ② Each grid block has different number of nodes which connect with different number of other nodes of adjacent grids (3) Orthogonality characteristics: Interface of two adjacent PEBI grid blocks must be perpendicular to and equally dividing the connection line of the two adjacent grid centers. PEBI grid is a type of partially orthogonal grid	(1) Geometric characteristics: ① Each grid block is an irregular polyhedron. There are more faces for each block than the PEBI grid ② The side lengths of each grid can be different to each other ③ The spatial coordinates of corresponding nodes of a grid block determine the geometric characteristics and the spatial distribution of this grid block. Each CVFE grid has more vertices than PEBI grid (2) Connectivity characteristics: ① Each grid block has different number of corresponding adjacent grid blocks. The neighbor number is higher than that of the PEBI grid ② Each grid block has different number of nodes which connect with different number of other corresponding adjacent grid blocks nodes (3) Orthogonality characteristics: Interface of two adjacent CVFE grid blocks is not perpendicular to the connection line of the two grid centers. Thus, CVFE grid is a non-orthogonal grid system
Applicable conditions	① As the basis for other complex grid systems, triangular grid has the best flexibility. It can be flexible in representing complex geological boundaries, fractures, faults and	① By adjusting spatial locations of the irregular quadrangle's four nodes and irregular hexahedron's eight nodes, quadrilateral grid can represent faults, fractures, geological boundaries and	① PEBI grid is more flexible than quadrilateral grid in describing geometric shapes of reservoirs. It can more precisely describe and approximate the actual reservoir conditions	① The number of edges and faces in CVFE grid is several times greater than that in PEBI grid. Compared with PEBI grid, CVFE grid has better flexibility that brings more

(continued)

Table 4.3 (continued)

Classification	Triangular grid	Quadrilateral grid	PEBI grid	CVFE grid
	complicated reservoir characteristics ② It is hard to describe the multi-directional feature and anisotropy of reservoir, due to limited number of edges and faces in a triangular grid ③ The non-orthogonality of triangular grid leads to large errors when computing transmissibility and flow terms. And the accuracy and efficiency of simulation are poor. Therefore, this grid system is generally not used in reservoir simulations	formation distributions, which more precisely describes the irregular boundary conditions and complicated reservoir characteristics ② Quadrilateral grid has disadvantages in dealing with wellbore characteristics. Well location is not guaranteed to be right in the center of the grid block, because of the irregular shape of quadrilateral grid. And the grid is unable to accurately describe the characteristics of radial flow near wellbores ③ Being a distorted grid, quadrilateral grid has an irregular shape and distortions without limit. So it is hard to use parameters at the center of the grid to represent the parameters of the entire grid block. Therefore, there is considerable computational error ④ The non-orthogonality of quadrilateral grid leads to severe difficulty in transmissibility calculations between grid blocks, which is also the same for errors in flow term calculations. Furthermore, the non-orthogonality results in unpleasant simulation convergence behavior and low simulation accuracies, and prolongs the simulation time	(fractures, faults, complex boundary conditions and complex reservoir characteristics) ② PEBI grid is flexible in constraining the reservoir model with relatively regular shape. PEBI grid has high quality grid morphology, which is also convenient for local grid refinement ③ PEBI grid has many edges and interfaces that can describe the anisotropic characteristics of the reservoir and reduce grid orientation effects by using full tensor form of permeability ④ Meanwhile, wellbore is guaranteed to be in the center of the grid. The properties in the center of the grid can represent properties of the whole grid ⑤ PEBI grid satisfies the requirements of orthogonality of the grid in the finite difference method due to its great orthogonality. This makes it relatively easy to conduct finite difference discretization of flow equations ⑥ The orthogonality of PEBI grid leads to high computing accuracies for transmissibility and flow term calculations, thus enhancing the efficiency and accuracy of numerical simulations	precise description of the complex reservoir morphology and boundaries ② CVFE grid can describe anisotropy better, since it has the most edges and faces, which reduces grid orientation effect more effectively than other types of grid systems, by using full tensor form of permeability ③ Meanwhile, wellbore is always guaranteed to be in the center of grid. And the properties in the center of grids can represent those of the whole grids ④ However, there is serious difficulty in computing transmissibility and flow terms with CVFE grid because CVFE grid is a type of non-orthogonal grid. Furthermore, the efficiency and accuracy of computation is unpleasant ⑤ CVFE grid is less used in reservoir numerical simulation for its relatively severe limitations
Grid block schematics	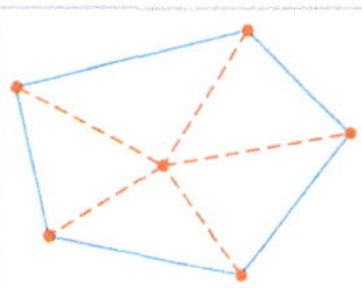	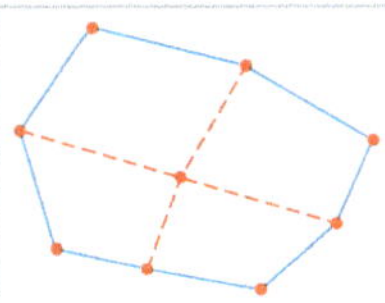	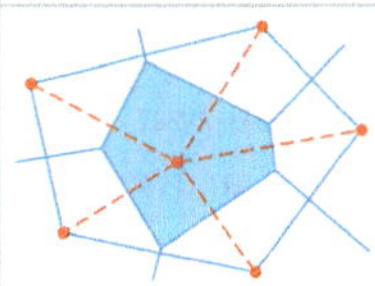	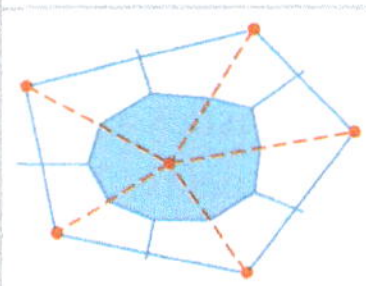

2. Quadrilateral grid

(1) Definition and characteristics of quadrilateral grid

A quadrilateral grid is a type of unstructured grid in a Cartesian coordinate system. The grid is comprised of irregular quadrangles in 2D space and of irregular hexahedron in 3D space.

The characteristics of a quadrilateral grid are shown mainly in the following aspects. Each grid block is an irregular tetrahedron. The side lengths of the grid can be different to others. In addition, the spatial coordinates of eight nodes of a grid block determine the geometric characteristics of this grid block. Each grid block has six adjacent grid blocks with eight nodes connected with six adjacent grid nodes. The connection line between the centers of two grid blocks is oblique to the interface of these two grid blocks, which means it is a non-orthogonal grid system (Lu 2008).

(2) Applicable conditions of quadrilateral grid

By adjusting spatial locations of an irregular quadrangle's four nodes and irregular hexahedron's eight nodes, a quadrilateral grid can represent faults, fractures, geological boundaries and formation distribution, which more precisely describes irregular boundary conditions and complicated reservoir characteristics. A quadrilateral grid has disadvantages in dealing with wellbore characteristics. A well location is not guaranteed to be right in the center of a grid block, because of the irregular shape of a quadrilateral grid. Also, the grid is unable to accurately describe the characteristics of radial flow near wellbores. Being a distorted grid, a quadrilateral grid has an irregular shape and distortions without limit. Thus, it is hard to use parameters at the center of a grid block to represent the parameters of the entire grid block. Therefore, there is a considerable computational error. The non-orthogonality of a quadrilateral grid leads to severe difficulty in transmissibility calculations between grid blocks, which is also the same for errors in flow term calculations. Furthermore, the non-orthogonality results in unpleasant simulation convergence behavior and low simulation accuracies, and prolongs the simulation time (Lu 2008).

3. PEBI grid

(1) Definition and characteristics of PEBI grid

A PEBI grid is a type of unstructured grid that is made of a polygon grid constructed by successively connecting the circumcenters of triangles, based on a triangular grid partitioning. An interface of two adjacent PEBI grid blocks must be perpendicular to and equally dividing the connection line of the two adjacent grid centers.

Characteristics of a PEBI grid are shown mainly in the following aspects. Each grid block is an irregular polyhedron. The side lengths of a PEBI grid can be different to each other. Moreover, the spatial coordinates of the different number of nodes of a PEBI grid block determine the geometric characteristics and spatial distribution of that grid block. Each grid block has different numbers of adjacent grid blocks and nodes, which are connected with other nodes of adjacent grids. An interface of two adjacent PEBI grid blocks must be perpendicular to and equally dividing the connection line of the two adjacent grid centers. A PEBI grid is a type of partially orthogonal grid (Lin 2010).

(2) Applicable conditions of PEBI grid

A PEBI grid is more flexible than a quadrilateral grid in describing geometric shapes of reservoirs. It can more precisely describe and approximate the actual reservoir conditions (fractures, faults, complex boundary conditions and complex characteristics). A PEBI grid is flexible in constraining a reservoir model with a relatively regular shape. It has high quality grid morphology, which is also convenient for local grid refinement. A PEBI grid has many edges and interfaces that can describe the anisotropic characteristics of the reservoir and reduce grid orientation effects by using full tensor form of permeability. Meanwhile, a wellbore is guaranteed to be in the center of a grid. The properties in the center of the grid can represent those of the whole grid. A PEBI grid satisfies the requirements of orthogonality of the grid in the finite difference method due to its great orthogonality. This makes it relatively easy to conduct finite difference discretization of flow equations. The orthogonality of a PEBI grid leads to high computing accuracy for transmissibility and flow term calculations, thus enhancing the efficiency and accuracy of numerical simulations (Tang et al. 2007; Wang et al. 2012).

4. CVFE grid

(1) Definition and characteristics of CVFE grid

A CFVE grid is a type of unstructured grid based on triangular grid partitioning. It is such a polygonal grid constructed by connecting the gravity centers with the midpoint of different triangles (Xie et al. 2001).

The characteristics of a CVFE grid are shown mainly in following aspects. Each grid block is an irregular polyhedron. There are more faces of each CVFE grid than a PEBI grid. The side lengths of each grid can be different to each other. The spatial coordinates of the corresponding nodes of a grid block determine the geometric characteristics of this grid block. Each CVFE grid has more vertices than a PEBI grid. Each grid block has a different number of corresponding adjacent grid blocks. The neighboring number is greater than that of the PEBI grid. Each grid block has such vertices which connect with vertices of other corresponding adjacent grid blocks. An interface of two adjacent CVFE grid blocks is not perpendicular to the connection line of the two grid centers. Thus, a CVFE grid is a non-orthogonal grid system.

(2) Applicable conditions of CVFE grid

The number of edges and faces in a CVFE grid is several times greater than that in a PEBI grid. Compared with a

PEBI grid, a CVFE grid has better flexibility that brings a more precise description of the complex reservoir morphology and boundaries. A CVFE grid can describe anisotropy better, since it has the most edges and faces, which reduces a grid orientation effect more effectively than other types of grid systems, by using full tensor form of permeability. Meanwhile, a wellbore is always guaranteed to be in the center of a grid. Also, the properties in the center of grids can represent those of the whole grids. However, there is a serious difficulty in computing transmissibility and flow terms with a CVFE grid because it is a type of non-orthogonal grid. Furthermore, the efficiency and accuracy of computation is unpleasant. Thus, a CVFE grid is less used in reservoir numerical simulation for its relatively severe limitations (Lu 2008).

4.1.2.2 Unstructured Grid Partitioning and Generation Technology

Complex reservoir numerical simulation needs to precisely describe complex geological boundaries, fractures, faults and complicated characteristics of reservoir morphology and to ensure the computational efficiency and accuracy of simulation calculations. Therefore, it is necessary to use a more flexible unstructured grid partitioning technology to spatially discretize the simulation region and establish an optimized reservoir numerical simulation grid (Chen et al. 2010; Lu 2008).

1. Triangular grid partitioning and generation technology

In the already known node set V, connecting adjacent nodes can generate a triangular grid system. However, there are different combinations to form multiple types of triangular grid systems. Among them, a triangular grid system that satisfies the Delaunay conditions is the optimal triangular grid system. In addition, such optimal grid system is unique. A Delaunay triangular grid has two fundamental properties. First, the empty circumcircle property: in the Delaunay triangular grid formed by the node set V, the circumcircle of each triangle does not contain any other nodes in the node set V. Second, the maximum value of minimum angle: in all the triangular grid systems that formed by the node set V, the smallest angle of the triangle in the Delaunay triangular grid system is the largest. Therefore, the Delaunay triangular grid can ensure that the triangles in the grid system best satisfy the approximate equilaterality (equiangularity), with the shapes of these grids being the fullest. At the same time, high-quality Delaunay triangular grid is a guarantee for the generation of high-quality quadrilateral, PEBI grid and CVFE grid systems (Lu 2008).

The specific implementation of the Bowyer-Watson algorithm for generating a Delaunay triangular grid is as follows (Yang et al. 2015) (Fig. 4.5).

(1) Determining constraint conditions

A simulation area boundaries, reservoir geometry, faults, fractures, reservoir boundaries, and well locations can be used as constraints. These constraints can be divided into three types: point constraints (vertical wells), line constraints (horizontal wells, faults, fractures), and zonal constraints (sand body distributions).

(2) Deploying nodes

The deployment of the nodes directly affects the quality, scale of computations, simulation time, and solution accuracy of grid cells. The principles of deploying nodes include the following aspects:

i. We determine the appropriate node density based on reservoir conditions and requirements of grid sizes from simulation calculations. ii. According to the boundary conditions, geometry and direction of fluid flow, we determine the grid direction and then keep consistency with such direction. iii. According to the shape of the triangular grid, and at the same time satisfying the principles of equilaterality (equiangularity) and the triangular shapes being the fullest, we determine the positions of the nodes.

Under different constraints, according to different ways of node deployment, we deploy nodes. i. Rectangular nodes deployment is discussed as follows. For regional constraints with weak reservoir heterogeneity, rectangular distribution of nodes can be used to generate more uniform nodes. ii. Circular nodes deployment is discussed as follows. For point alike constraints such as vertical wells and according to the radial flow characteristics, the nodes can be generated in a circular layout manner. iii. Advancing nodes deployment is discussed as follows. Aiming at line alike boundary constraints such as boundaries, fractures, and faults, the deployment of nodes is gradually advancing towards the interior of a reservoir.

(3) Delaunay triangular grid partitioning

After completing the deployment of nodes according to different constraints, Delaunay triangular grid partitioning can be performed. The point-by-point interpolation method is commonly used for Delaunay triangular grid partitioning. The main steps are as follows.

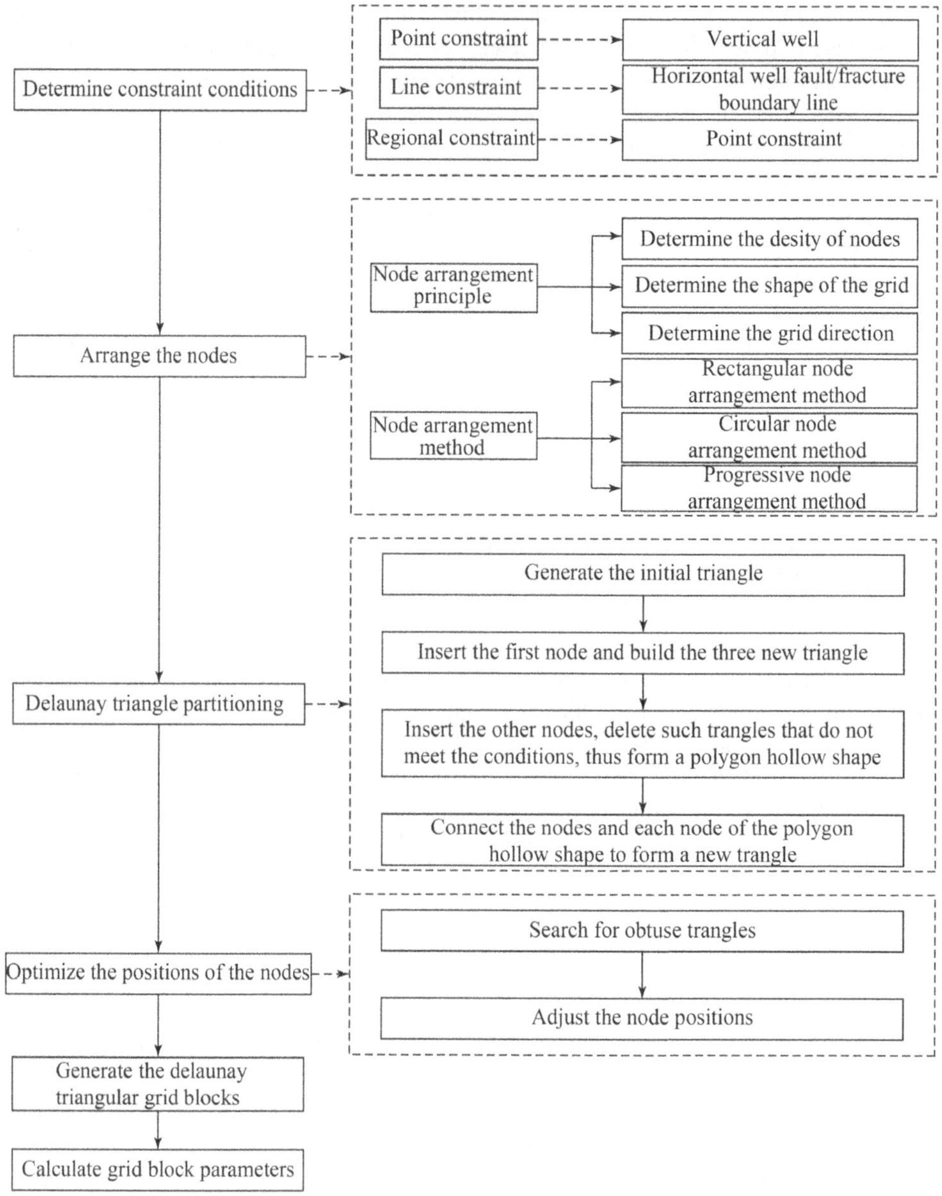

Fig. 4.5 Triangular grid generation workflow

① We determine an initial triangle containing all the nodes.

② We select a node from all the nodes. If this node falls within the circumcircle of the initial triangle, we delete the initial triangle and form a blank region containing the node. We connect this node with the three vertices of the blank region and construct three new triangles.

③ We select a node from all the nodes again and find out all the triangles whose circumcircles encompass this node. Then, we delete the triangles to form a polygon blank region.

④ This node is connected with each node of the polygon blank region to form a set of new triangles.

⑤ We repeat the process of ③ and ④ above until all nodes participate in forming a triangle.

(4) Finding obtuse triangles to optimize and adjust node positions

The presence of an obtuse triangle in grid partitioning will directly affect the quality of a Delaunay triangular grid. According to the cosine theorem, we can get the cosine value of the angles of a triangle and determine whether it is an obtuse triangle. If this happens, we adjust the position of the node so that this obtuse angle disappears to ensure the quality of the Delaunay triangular grid.

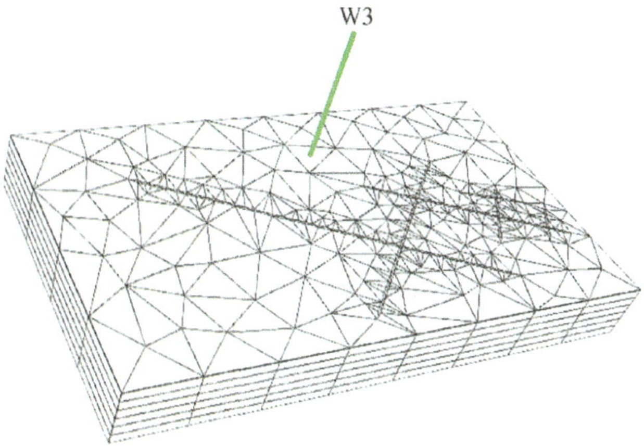

Fig. 4.6 Spatial distribution of three-dimensional triangular grids

(5) Generating delaunay triangular grid blocks (Fig. 4.6).
(6) Calculating the geometric parameters of each grid block in space.

2. Quadrilateral grid partitioning and generation technology

A quadrilateral grid can be generated through Delaunay triangular grids with relatively full shape and high quality. In this way, the quadrilateral grid generated can have superior quality. The main method is to remove the common edge of two adjacent triangles that satisfy certain conditions and directly synthesize the quadrilateral. The specific generation process is as follows (Lu 2008).

(1) Determination of constraint conditions

According to the simulation area boundaries, reservoir geometry, faults, fractures, and reservoir boundaries, we determine the line constraints and zonal constraints.

(2) Deploying nodes

We determine the appropriate node density, grid direction, and grid shape based on the principles of deploying nodes. According to different constraints, we adopt different deployment methods to deploy nodes.

Since a quadrilateral grid is very irregular, we try to adopt an evenly deployment strategy when deploying the nodes. Generally, only boundaries, faults, fractures, and other line constraints and zonal constraints are considered. However, a radial deployment strategy cannot be adopted. In addition, well locations are not considered as point constraints.

(3) Generating delaunay triangular grid

After completing the deployment of nodes according to different constraints, we perform Delaunay triangular grid partitioning. Also, obtuse triangles are found out to optimize the positions of nodes. Then, Delaunay triangular grid cells are generated.

(4) Generating quadrilateral grids

① We go through the edges of all triangles. If an edge is on a control line (border, fault, or fracture), it is marked as immovable. Otherwise, it is marked as movable.
② Triangles are merged one by one according to the order of the numbering of their movable edges from small to large. The specific process is as follows.

First, we find the neighboring triangles of each movable edge of the triangles, which make up the triangle pairs.

Second, we evaluate the quadrilateral formed by each triangle pair. Then, we choose the quadrilaterals whose shapes are closest to rectangles. This forms the merger scheme to generate quadrilateral grid cells. Afterwards, we delete the corresponding two triangular grid cells (Fig. 4.7).

③ Special processing of grids: after the merger process, there may be triangles that cannot be merged. At this moment, some special processing is required. For triangles that cannot be merged, we find out their longest immovable edges, and then add nodes to the midpoints of the edges to make them quadrilateral grids.

(5) Calculating the geometric parameters of each grid block in space.

3. PEBI grid partitioning and generation technology

The common methods for generating a PEBI grid include an incremental method, indirect method, and dividing and conquering method. The indirect method is mainly based on the dual property of the PEBI grid and the Delaunay triangular grid. First, a Delaunay triangular grid network is

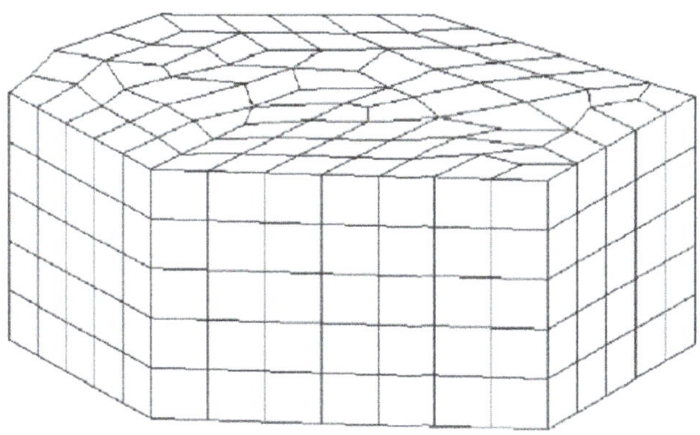

Fig. 4.7 Schematic diagram of spatial distribution of three-dimensional quadrilateral grid

generated. Then the perpendicular bisectors of the different edges of triangles are created to obtain the circumcenter of these triangles. Finally, PEBI grids are generated by connecting the circumcenters of adjacent triangles.

The specific implementation of the indirect method to generate a PEBI grid is as follows (Yang et al. 2015) (Fig. 4.8).

(1) Determining constraint conditions

According to the simulation area boundaries, reservoir geometry, faults, fractures, reservoir boundaries, and well locations, we determine the conditions of point constraints, line constraints and zonal constraints.

(2) Deploying nodes

We determine the appropriate node density, grid direction, and grid shape based on the principles of deploying nodes. According to different constraints, we adopt different deployment methods to deploy nodes.

(3) Generating delaunay triangular grid

After completing the deployment of nodes according to different constraints, we perform Delaunay triangular grid partitioning. Moreover, obtuse triangles are found out to optimize the positions of nodes. Then, Delaunay triangular grid cells are generated.

(4) Generating the PEBI grids

① Calculate the circumcenter of each triangle.

We find out the circumcenter of each triangle in Delaunay triangle grids.

② Determine the center of the PEBI girds.
③ Around the centers of the PEBI grids, PEBI grids are formed by sequentially connecting the circumcenters of adjacent triangles (Fig. 4.9).

The vertices of the original triangles are the centers of the PEBI grids. Furthermore, all the circumcenters of the triangles where the vertices are located constitute the vertices of the PEBI grid blocks.

④ Special processing of borders.

After performing the second step, disclosed PEBI grid cells are formed around the nodes around the borders. In addition, there are still some circumcenters of these triangles outside the borders. The formed grid cell borders distort the shape of the regional borders. We need to do special processing of these points around the borders.

(5) Calculating geometric parameters of each grid block in space

4. CVFE grid partitioning and generation technology

A CVFE grid can be generated based on the Delaunay triangle grid. First, the gravity center and the midpoints of the edges of the triangle are calculated. Then, the gravity centers and the midpoints of the edges are connected to form a CVFE grid.

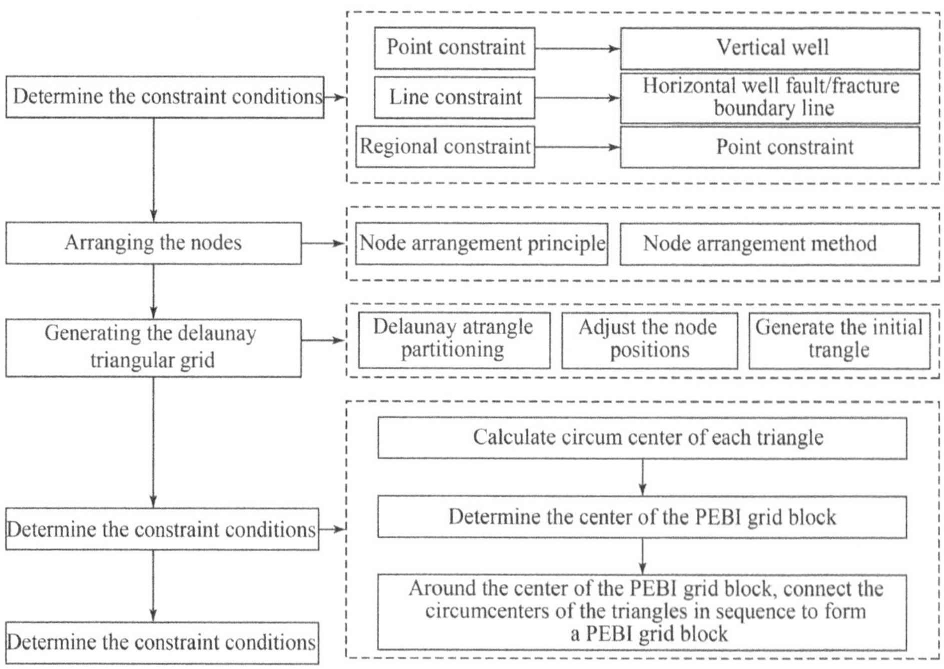

Fig. 4.8 PEBI grid generation flowchart

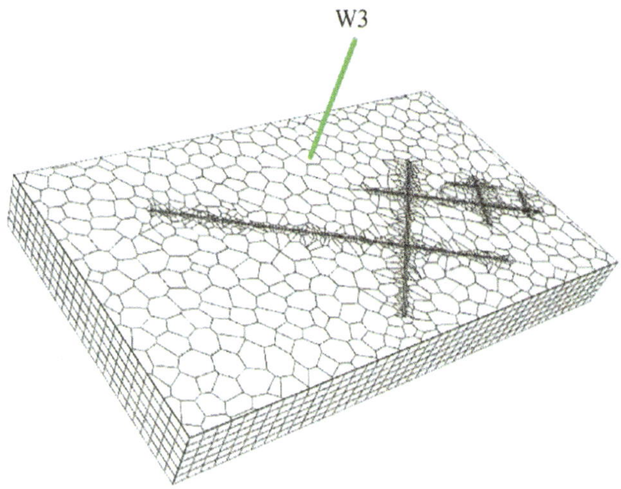

Fig. 4.9 Spatial distribution of three-dimensional PEBI grids

The specific implementation of CVFE grid generation is as follows (Lu 2008).

(1) Determination of constraint conditions

According to the simulation area boundaries, reservoir geometry, faults, fractures, reservoir boundaries, and well locations, we determine the conditions of point constraints, line constraints and zonal constraints.

(2) Deploying nodes

We determine the appropriate node density, grid direction, and grid shape based on the principles of deploying nodes. According to different constraints, we adopt different deployment methods to deploy nodes.

(3) Generating delaunay triangular grids

After completing the deployment of nodes according to different constraints conditions, we perform Delaunay triangular grid partitioning. Also, obtuse triangles are found out to optimize the positions of nodes. Then, Delaunay triangular grid cells are generated.

(4) Generating the CVFE grids

① Calculate the circucenter of each triangle and the midpoint of each triangle edges.

Find the circumcenter of each triangle in the Delaunay triangle grid and find out the midpoint positions of the triangle edges.

② Determine the centers of the CVFE grids.
③ Around the centers of the CVFE grids, we connect the circumcenters of the triangles and the midpoints of the edges in sequence to form the CVFE grids (Fig. 4.10). These CVFE grids are formed by connecting the circumcenters of adjacent triangles and the midpoints of the common edges of adjacent triangles. The vertices of the original triangles form the centers of the CVFE grids. All the circumcenters and the midpoints of the edges of the triangles which encompass this vertex constitute the vertices of the corresponding CVFE grid block.

(5) Calculating the geometric parameters of each grid block in space

4.1.2.3 Unstructured Multi-scale Grid Technology

For reservoir areas of vertical wells/horizontal wells, faults/boundaries, natural/hydraulic fractures, and strong heterogeneity, the same type of a grid (triangular, quadrilateral, PEBI, and CVFE) can be used. We change the size and shape of a single grid, as well as the deployment and shape of the overall grid system, to perform the grid partitioning. In this way, we accurately describe the heterogeneity characteristics, complex shapes, and boundaries of the reservoir.

1. Vertical well multi-scale grid technology

A PEBI grid is used to partition the simulation area of vertical wells. First, a center grid is determined according to the vertical well positions. Then, the well location is set to be in the center of the grid. The PEBI grid distribution is deployed around the center grid in a ring shape. Refined grids are used near a wellbore. At the same time, coarse grids are used in the reservoir area away from the wellbore. In addition, the radial flow characteristics near a vertical well are simulated by adjusting the grid scale and morphology (Fig. 4.11).

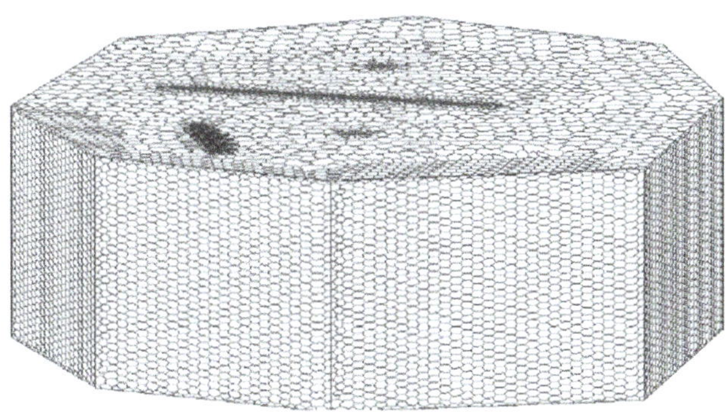

Fig. 4.10 Schematic diagram of spatial distribution of three-dimensional CVFE grids

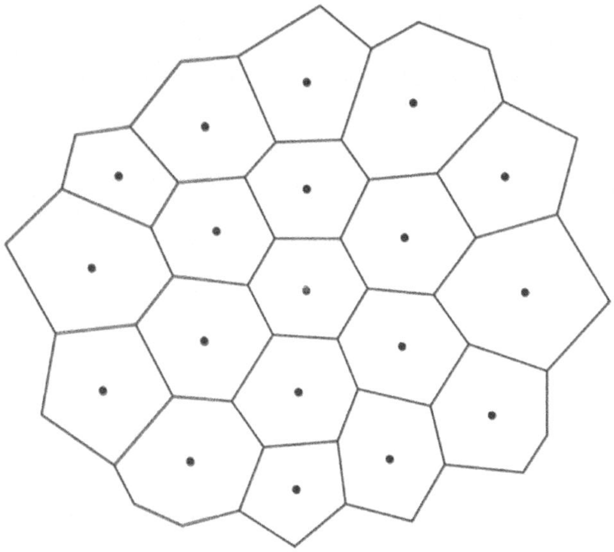

Fig. 4.11 Vertical well multi-scale grid partitioning diagram

2. Horizontal well multi-scale grid technology

A PEBI grid is used to partition the simulation area of horizontal wells (Fig. 4.12). First, according to the simulation requirements, the horizontal wells are discretized into several point sources. Then, the grids are partitioned along the horizontal well trajectory according to the point source locations. Furthermore, the point sources are placed at the centers of the grids. The PEBI grids are deployed around the horizontal well grid blocks in an elliptical ring strip shape. The horizontal well PEBI grids generated can be divided into three groups. First, refined grids are used near the end of the horizontal wells. Second, coarse grids are used farther away. Third, the grids around the end of the horizontal wells are distributed radially as a whole to simulate radial flow characteristics at the two ends of the horizontal wells. Similarly, refined grids are used near the horizontal wells in the horizontal section. Moreover, coarse grids are used farther away to simulate the linear flow characteristics near the horizontal section.

3. Fractures/faults multi-scale grid technology

We partition the internal and surrounding areas of the fractures/faults with PEBI grids of different sizes and shapes (Fig. 4.13). Along the direction of natural/hydraulic fractures and faults/boundaries, a refined grid is used to process the linear geometric characteristics inside the fractures/faults. Moreover, the PEBI grids are deployed around the fractures/faults in an elliptical ring-shaped zone. Then, the generated fracture/fault PEBI grids can be divided into three parts. First, refined grids are used near the two ends of the horizontal wells. Second, coarse grids are used farther away. Third, the grids around the two ends of the horizontal wells are distributed radially as a whole to simulate radial flow characteristics around the two ends. Similarly, refined grids are used near the fractures/faults and in the fracture/fault bodies. Furthermore, coarse grids are used at locations farther away to simulate the linear flow characteristics near the fractures/faults.

4. Multi-scale grid technology for complex reservoir areas

For different geological objects in a reservoir area, according to their local geometric characteristics and flow behavior through pore and fracture media of tight reservoirs, refined PEBI grids are used for the internal partitioning. For the reservoir region, we adjust the sizes of the grids, morphology and distribution patterns, based on the extent of its heterogeneity, in order to conduct grid partitioning. Through this, we describe the heterogeneity and flow characteristics of the internal part of the reservoir (Fig. 4.14).

4.1.3 Hybrid Grid Technology

Unconventional oil and gas reservoirs generally have multiple complex geological conditions (macroscopic strong heterogeneity in reservoir areas), boundary conditions (faults, boundaries), and large-scale natural/hydraulic fracture. Therefore, there are vertical wells, highly deviated wells, horizontal wells, and many other well types. In such complex reservoir areas, hybrid grids of different sizes and types are required for treating different corresponding objects. A hybrid grid is a type of unstructured grids comprised of many different types and sizes of grids. The optimized partitioning of grids by using the hybrid grid technology can not only retain the advantages of high-accuracy discretization and high computational speed of the structural grids, but also greatly reduce the number of grids and improve the speed of solving.

Using the hybrid grid technology, through the combination of different types of grids, and also through the adjustments of the sizes, shapes, and deployment of the grids, we can solve the grid partitioning problem of different objects and their complex flow behavior through pore and fracture media of tight reservoirs. First, for vertical wells, radial grids and PEBI/quadrilateral grids can be used to fully describe the radial flow characteristics around the wellbores and to achieve rapid changes of a grid volume from small to large. Second, track-shaped grids (the combination of strip grids and radial grids), PEBI grids and other hybrid grids can be used for horizontal wells to accurately describe the linear flow of the horizontal section and the radial flow at the two ends. Third, for large-scale natural/hydraulic fractures and faults/boundaries, we can use strip grids, triangular grids,

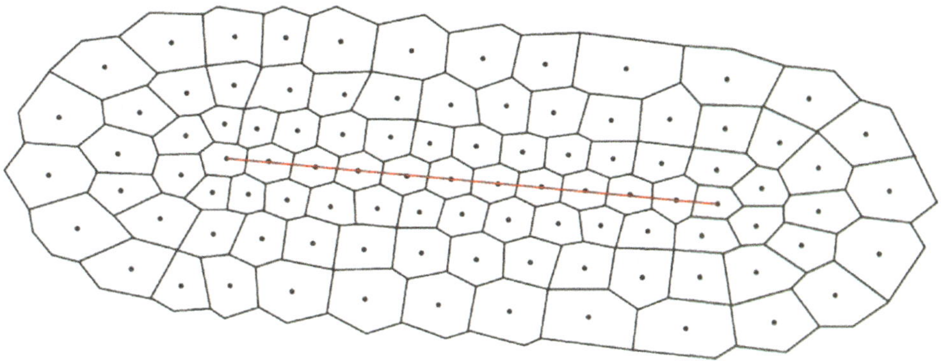

Fig. 4.12 Horizontal well multi-scale grid partitioning

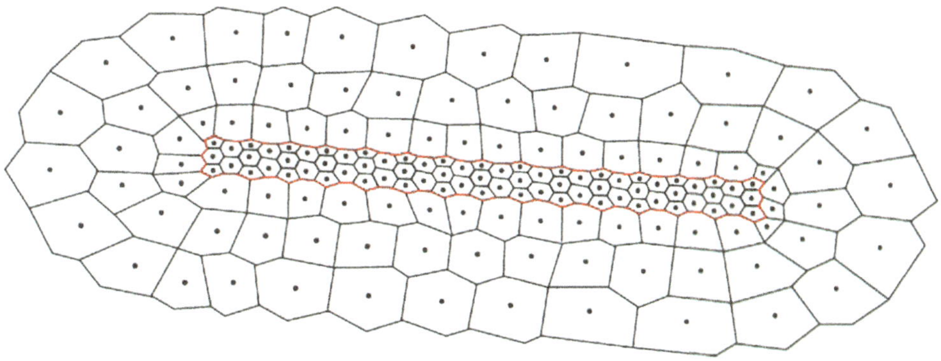

Fig. 4.13 Fracture/fault multi-scale grid partitioning diagram

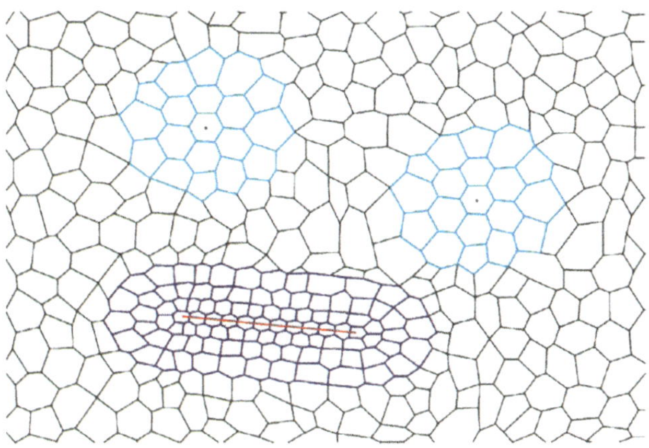

Fig. 4.14 Multi-scale grid partitioning diagram of complex reservoir areas

PEBI grids, and other hybrid grids, to accurately describe their linear characteristics and complex morphological characteristics. Fourth, for reservoir areas with strong macroscopic heterogeneity, rectangular grids, quadrilateral grids, PEBI grids, CVFE grids, and other hybrid grids can be used to describe the characteristics of complex reservoirs and flow behavior through pore and fracture media of tight reservoirs. Especially, we can divide the strong heterogeneity region into different regions, further into simulation elements. Fifth, for multiple media at different scales, nested hybrid grids and interactive hybrid grids can be used to deal with discontinuous discrete multiple media characteristics.

4.1.3.1 Different Types of Hybrid Grid Technology

1. Hybrid grid processing of vertical wells

Radial grids are used near wellbores to deploy the well locations in the centers of the radial grids. Refined grids are used near each well, while coarse grids are used further away. In this way, we achieve such transition of a grid volume from small to large. In addition, we use PEBI grids to describe the reservoir heterogeneity in the reservoir area away from the wellbores. By using this type of hybrid grid, the flow conditions near the wellbores can be accurately captured. In addition, the radial flow characteristics of the fluid flow around the wellbores can be simulated (Fig. 4.15).

2. Hybrid grid processing of horizontal wells

For a horizontal section, strip grids are used. Refined grids are used near each well, while coarse grids are used further away. In this way, we simulate the linear flow characteristics near the horizontal section. Around the two ends of a horizontal well, PEBI grids are used. Refined grids are used near the two ends, with coarse grids deployed further away. In

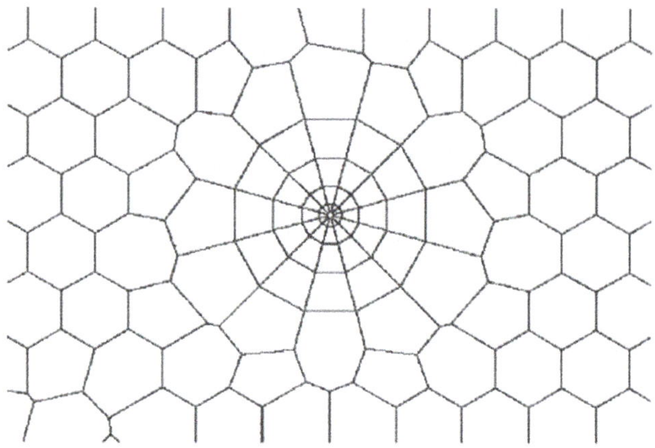

Fig. 4.15 Treatment by mixing of radial and PEBI grids for vertical wells

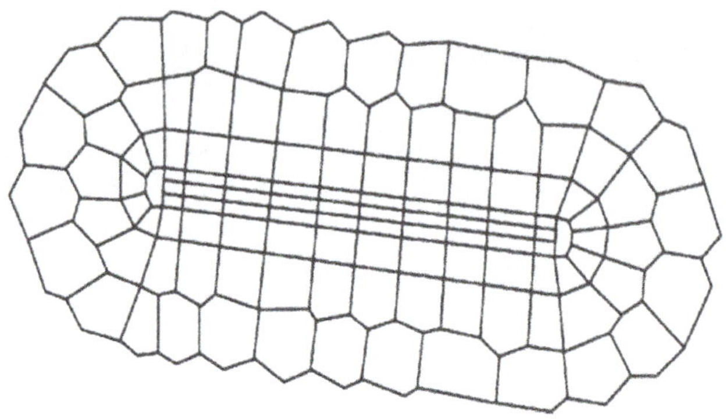

Fig. 4.16 Treatment by mixing of strip grids and PEBI grids for horizontal wells

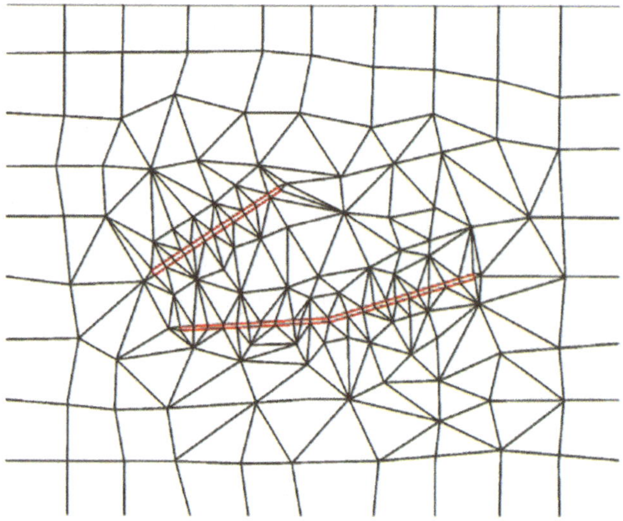

Fig. 4.17 Mixed processing of fracture/fault using strip grids, triangular grids, and quadrilateral grids

this way, we simulate the radial flow characteristics of the two ends. For reservoir areas far away from the horizontal wells, other grids such as PEBI grids and quadrilateral grids can be used to describe reservoir heterogeneities and complex morphological characteristics (Fig. 4.16).

3. Hybrid grid processing of fracture/fault

For natural/hydraulic fractures and faults/boundaries with linear geometrical characteristics, regular rectangular strip grids/irregular strip grids can be used to handle their linear geometrical characteristics. In addition, near the fractures/faults, more refined triangular grids/PEBI grids were used for partitioning. In reservoir areas far away from the fractures/faults, coarser quadrilateral grids/PEBI grids can be used to simulate reservoir heterogeneity and complex morphological characteristics (Fig. 4.17).

4. Hybrid grid processing of complex reservoir areas

In complex reservoir areas, there are complex geological and boundary conditions such as vertical wells, horizontal wells, natural/hydraulic fractures, faults/boundaries, and macroscopic heterogeneity, which can be processed with more types of more complex hybrid grids (Fig. 4.18).

4.1.3.2 Multiple Media Hybrid Grid Technology

Tight reservoirs include porous media and nano/micro-scale natural/hydraulic multiple media at different scales. Media with different scales pores and fractures have different geometric and property characteristics, fluid properties and flow characteristics. At the same time, this type of media is discontinuously and discretely distributed in space. In addition, they have different numbers and spatial distribution patterns. Besides, multi-scale features and multiple media characteristics are significant. Therefore, different types of hybrid grid technology can be used to represent the spatial distribution characteristics, number of medium types, media geometry and property characteristics, and quantitative distribution characteristics of the multiple media at different scales.

According to the spatial distribution and flow behavior of multiple media at different scales, multiple media hybrid grids can be divided into nested hybrid grids and interactive hybrid grids.

1. Multiple media nested hybrid grid

For the media with different scale pores and fractures in the reservoir area with a nested-distribution-type flow relationship, multiple media nested hybrid grids can be processed in space. First, according to the differences in the spatial and quantitative distributions of nano/micro-fractures and pores

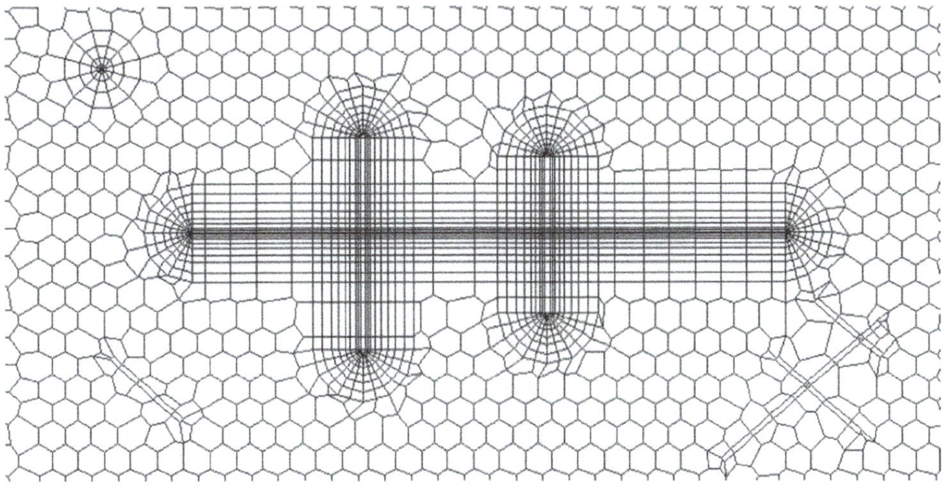

Fig. 4.18 Mixed processing with strip rectangular grids, radial grids, and PEBI grids

at different scales in the reservoir area, the hybrid grids of PEBI grids, quadrilateral grids, and triangular grids are used to divide the reservoir area into several different cells with heterogeneity characteristics. Then, based on the number of media types and volume percentage of the multiple media at different scales in each cell, the nested grids at different media are partitioned by using the same grid type as the cell. Due to the differneces between the number of media types and the volume percentage of each cell, the number and size of grids in each cell are different (Fig. 4.19).

2. Multiple media interactive hybrid grid

For the media with different scale pores and fractures in the reservoir area with an interactive flow relationship, we can spatially process by using multiple media interactive hybrid grids. According to the difference of spatial and quantitative distributions of multiple media at different scales in the reservoir area, different multiple media cells are divided by using hybrid grids such as PEBI grids, quadrilateral grids, and triangular grids. In each cell, according to the distribution patterns (strip distribution, ring distribution, and random distribution), the number of media types and the volume percentage of the porous medium with fractures at different scales, the cell is divided into different media using triangular grids and quadrilateral grids. The distribution patterns, number of medium types and volume percentage of media for each cell are different. Therefore, the distribution, number, and size of grids in each cell are different (Fig. 4.20).

4.2 Grid Connectivity Characterization Technology for Numerical Simulation

In response to the different objects, multiple media characteristics, complex morphology and large number of different types of grids within complex reservoir area, we developed grid ordering optimization technology for an unstructured grid. It can reduce the matrix memory usage in numerical simulation and improve the computational speed. According to the grid type and grid topological relationship, together with the geometric and physical characteristics of grids for different media, we can determine the neighboring relationship between the grids and calculate the connectivity between grids in adjacent media. Thus, the grid neighboring characterization and connectivity characterization of numerical simulation are formed.

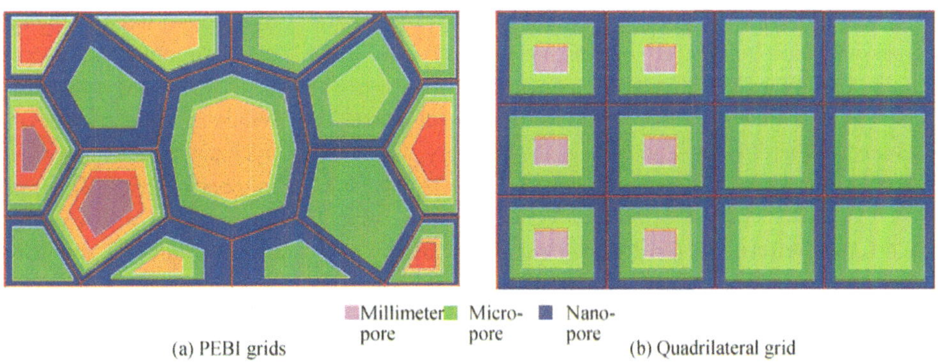

Fig. 4.19 Multiple media nested hybrid grid illustration

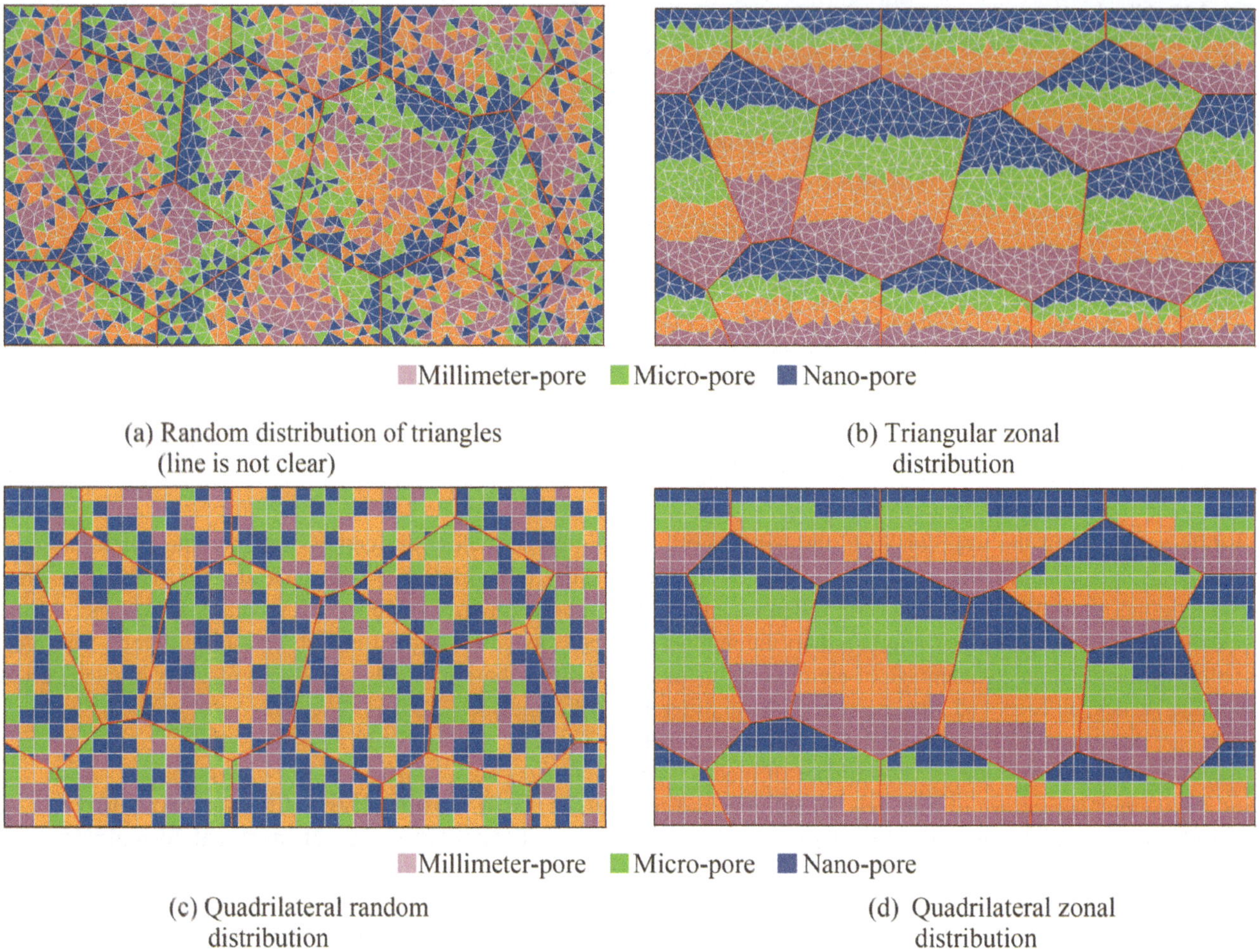

Fig. 4.20 Multiple media interactive hybrid grid diagram

4.2.1 Grid Ordering Technology for Numerical Simulation

For different objects in complex reservoir areas, different types of grids are used to partition a domain. At the same time, we can perform optimizations on the grid ordering. The grid ordering optimization technology sorts the grids in different regions, which is based on the impact of different objects on field development evolutions. In turn, grid ordering optimization and numbering are performed according to the local characteristics of different objects, grid types and morphology. By optimizing the grid ordering, a more reasonable distribution of non-zero elements in the coefficient matrix solved by numerical simulation can be achieved. Thereby, we can reduce the memory usage and increase the speed of solving.

4.2.1.1 Structured Grid Ordering Technology

The principle of structured grid ordering is regarding the coordinates of the origin as the starting point and sorting in increasing order in the *I*, *J* and *K* directions. Since there are a fixed number of grids in the *X*, *Y* and *Z* directions, this is a regular ordering method. For the rectangular grid partitioning, the main body direction *I* of the grid is made to be parallel to the directions of fractures, horizontal wells and flow direction of a reservoir. Therefore, first, the grids are successively ordered in increasing order along the main direction *I* of the grid distribution. Second, they are ordered along the *J* direction. Finally, they are ordered along the *K* direction. For a corner point grid, the overall principle and method are same as the rectangular grid ordering technology by considering the reservoir fractures, horizontal wells, and flow directions together. With such situations where the numbers of grids in the *X*, *Y* and *Z* directions keep unchanged, every row in the *X* direction has the same number of grids. The *Y* and *Z* directions are the same. The corner point grid can have different edge lengths. For the radial grids, we use the coordinates of the origin as the starting point and order the grid in increasing order along the radial, angular and *Z* directions. First, the grids are ordered

Table 4.4 Structured grid ordering table

<table>
<tr><th>Grid type</th><th>Ordering methods</th><th>Schematic diagram</th></tr>
<tr><td>Rectangular grid</td><td>① The grid ordering is used the coordinates of the origin as the starting point and sorted in increasing order in I, J and K directions
②For the rectangular grid partitioning, the main body direction I of the grid is made to be parallel to the directions of fractures, horizontal wells and flow directions of the reservoir. Therefore, first, the grids are successively ordered in increasing order along the main direction I of the grid distribution
③Second, they are ordered along the J direction. Finally, they are ordered along the K direction</td><td>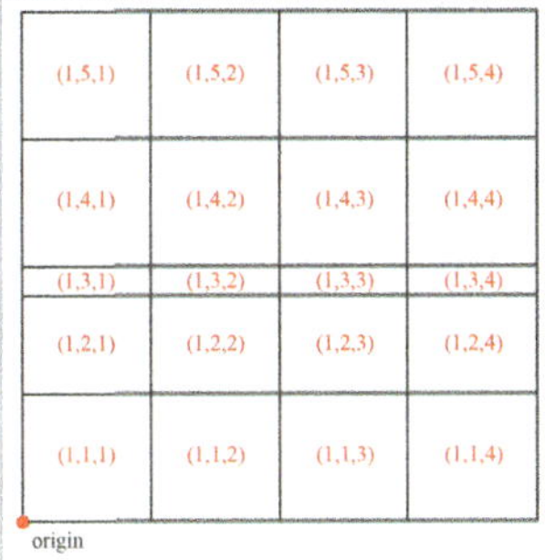
</td></tr>
<tr><td>Corner point grid</td><td>① The overall principle and method are same as the rectangular grid ordering technology by considering the reservoir fractures, horizontal wells, and flow directions, together
② With such situations that the numbers of grids in the X, Y and Z directions keep unchanged, every row in the X direction has the same number of grids. Y and Z directions are the same. The corner point grid can have different edge lengths</td><td>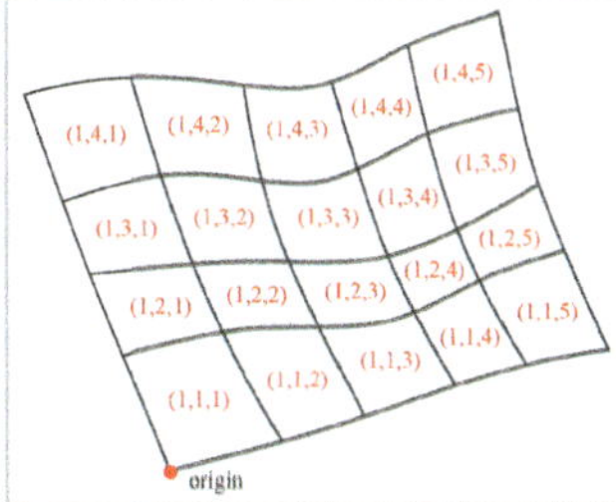
</td></tr>
<tr><td>Radial grid</td><td>① we use the coordinates of the origin as the starting point and order the grid in increasing order along the radial, angular and Z directions
②First, the grids are ordered in radial direction, following the rule that grid numbers increase from the center to the outside
③Second, they are ordered along the angular direction. Finally, the grids are ordered along the Z direction</td><td>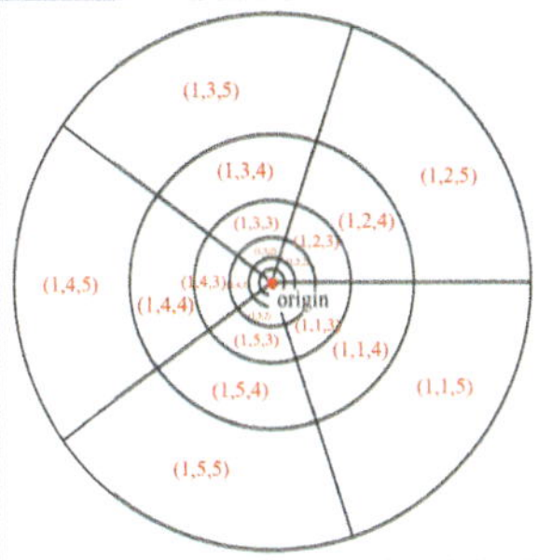
</td></tr>
</table>

in the radial direction, following the rule that grid numbers increase from the center to the outside. Second, they are ordered along the angular direction. Finally, the grids are ordered along the Z direction (Table 4.4).

4.2.1.2 Unstructured Grids Ordering Technology

There is a big difference between the principle of unstructured grid ordering and structural grid ordering. Generally speaking, the structured grid ordering belongs to the regular ordering scope, while the unstructured grid ordering belongs to the irregular ordering scope. Here is the principle of unstructured grid ordering. i. We optimize the ordering of different geologic objects according to the importance and local characteristics which influence the effectiveness of reservoir development. ii. We prioritize and order the geologic objects which have the greatest impact on effectiveness of reservoir development. We determine the starting positions and conduct the numbering job according to their geometry and flow behavior through pore and fracture media of tight reservoirs. iii. We conduct grid ordering for each object in sequence according to the importance of different geologic objects. The ordering is not limited by the *X*, *Y*, *Z* directions and the number of grids in each direction.

When a single type of unstructured grid (triangular, quadrilateral, PEBI, or CVFE grid) is used for grid partitioning, the ordering principle and ordering method are the same. For the same type of variable-scale grid, taking a PEBI grid as an example, we can do the grid partitioning and ordering for the different geologic objects such as horizontal wells, vertical wells, hydraulic fractures, natural fractures, faults and complex reservoir regions by adjusting the grid scale and morphology (Table 4.5).

1. Vertical wells variable scale grid ordering

We use a PEBI grid to partition the area around vertical wells. We simulate the radial flow characteristics near the vertical wells by adjusting the grid scale and grid morphology. We take the grid region where a wellbore is located

Table 4.5 Unstructured grid ordering table

Different objects	Ordering methods	Schematic diagram
Vertical wells	We take the grid where the wellbore is located as the starting point and first conduct the grid ordering in increasing order from the inside to the outside in the radial flow direction. Then we perform the grid ordering in increasing order counterclockwise in the angular direction	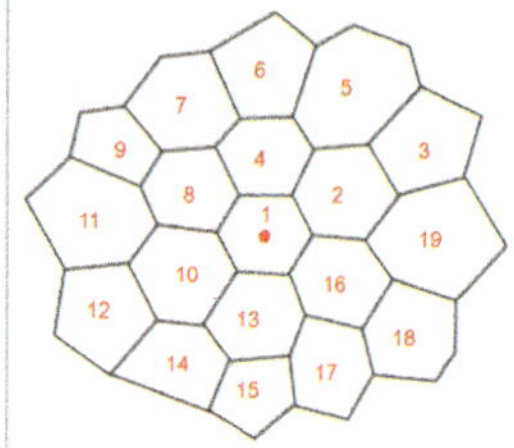
Horizontal wells	① First, in the radial flow region at the left end of the horizontal section, we conduct the grid ordering in increasing order from the inside to the outside in the radial flow direction. And then we perform grid ordering in increasing order counterclockwise in the angular direction ② Second, in the middle horizontal section, we conduct the grid ordering in increasing order along the linear flow direction from top to bottom, and then increase from left to right along the length direction of the well trajectory ③ At last, in the radial flow region at the right end of the horizontal section, we also perform grid ordering in increasing order from the inside to the outside in the radial flow direction. And then we conduct grid ordering in increasing order counterclockwise in the angular direction	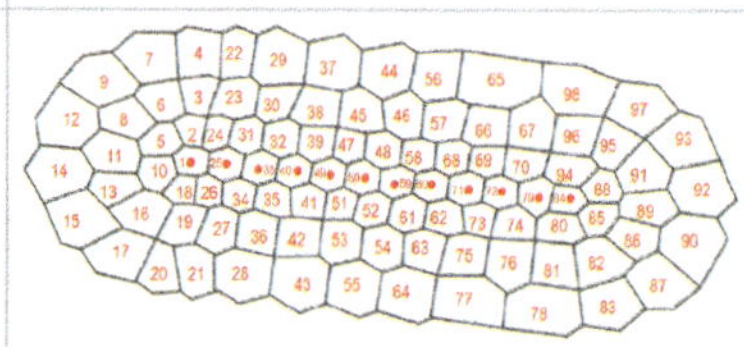
Fractures and faults	① We conduct the gird ordering in increasing order for the internal grids of the fractures and faults along the fracture direction ② In the radial flow region which is located at the left end of the fractures/faults, we perform grid ordering in increasing order from the inside to the outside in the radial flow direction. And then we conduct grid ordering in increasing order counterclockwise in the angular direction ③ In the upper part near the main body of the fractures, we perform the grid ordering in increasing order along the linear flow direction from top to bottom ④ We conduct grid ordering in increasing order in the lower part from top to down along the linear flow direction ⑤ In the radial flow region which is located at the right end of the fractures/faults, we also perform grid ordering in increasing order from the inside to the outside in the radial flow direction. Then we perform grid ordering in increasing order counterclockwise in the angular direction	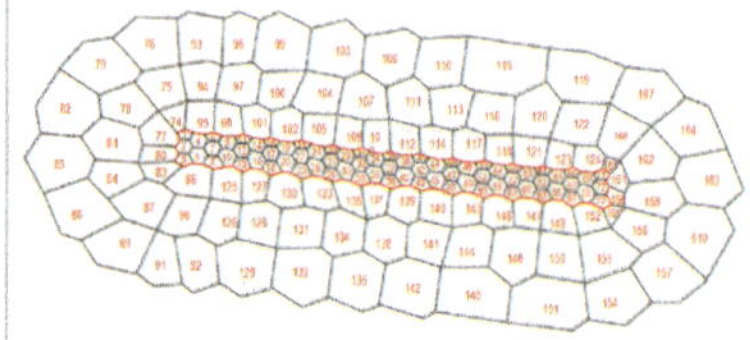
Complex reservoir regions	We following such sequence: horizontal wells, vertical wells, hydraulic fractures, natural fractures/faults and reservoir areas. For the grids located in the reservoir area, we conduct grid ordering in increasing order from bottom to top, and then from left to right	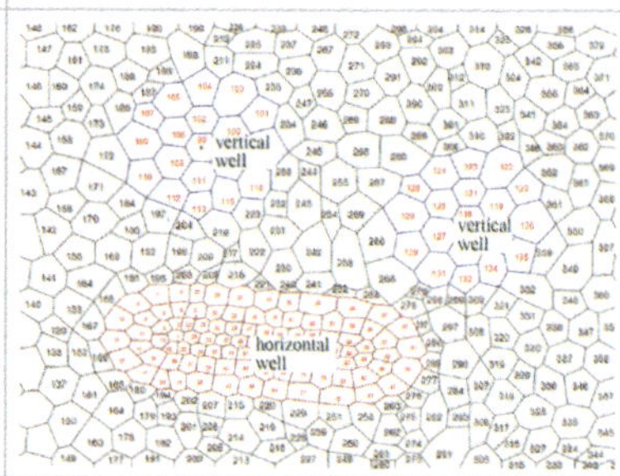

as the starting point and first conduct the grid ordering in increasing order from the inside to the outside in the radial flow direction. Then we perform the grid ordering in increasing order counterclockwise in the angular direction.

2. Horizontal wells variable scale grid ordering

We divide the grid of the horizontal well control area into three parts. Both ends of the horizontal section of the well simulate the radial flow characteristics of the horizontal well. In addition, the middle part simulates the linear flow characteristics of the horizontal well. First, in the radial flow region at the left end of the horizontal section, we conduct the grid ordering in increasing order from the inside to the outside in the radial flow direction. Then we perform grid ordering in increasing order counterclockwise in the angular direction. Second, in the middle horizontal section, we conduct the grid ordering in increasing order along the linear flow direction from top to bottom, and then from left to right along the length direction of the well trajectory. At last, in the radial flow region at the right end of the horizontal section, we also perform grid ordering in increasing order from the inside to the outside in the radial flow direction, and then we conduct grid ordering in increasing order counterclockwise in the angular direction.

3. Fractures and faults variable scale grid ordering

We partition the inside and vicinity areas of fractures and faults by using grids with different sizes and morphologies. First of all, we conduct the gird ordering in increasing order for the internal grids of the fractures and faults along the fracture direction. Then we perform grid ordering for the grids in the surrounding area. First, in the radial flow region which is located at the left of the fractures/faults, we perform grid ordering in increasing order from the inside to the outside in the radial flow direction. Then we conduct grid ordering in increasing order counterclockwise in the angular direction. Second, in the upper part near the main body of the fractures, we perform the grid ordering in increasing order along the linear flow direction from top to bottom. Then we conduct grid ordering in increasing order in the lower part from top to down along the linear flow direction. At last, in the radial flow region which is located at the right side of the fractures/faults, we also perform grid ordering in increasing order from the inside to the outside in the radial flow direction and perform grid ordering in increasing order counterclockwise in the angular direction.

4. Complex reservoir region variable scale grid ordering

We partition a reservoir area according to the distribution of different geologic objects in the reservoir area. Also, we optimize the ordering of different geologic objects according to the local characteristics and importance which affect the effects of reservoir development. We follow such sequence: horizontal wells, vertical wells, hydraulic fractures, natural fractures/faults and reservoir areas. For the grids located in the reservoir area, we conduct grid ordering in increasing order from bottom to top, and then from left to right. With all of these steps, we can finish the grid ordering in the complex reservoir region.

4.2.1.3 Hybrid Grid Ordering Technology

1. Different type-hybrid grid ordering technology

A different type-hybrid grid belongs to an unstructured grid. Its grid ordering is classified as a type of irregular ordering. The ordering principle of a different type-hybrid grid is identical with that of a single type-variable scale unstructured grid (Table 4.6).

(1) Vertical well hybrid grid ordering

We use radial grids near a wellbore and PEBI/quadrilateral grids around the wellbore to partition and perform ordering of the hybrid grid. First, for the radial grids near the wellbore, we conduct the grid ordering in increasing order from the inside to the outside in the radial flow direction. Then we perform grid ordering in increasing order counterclockwise in the angular direction. Finally, for the PEBI/quadrilateral grids around the wellbore, we conduct grid ordering in increasing order from bottom to top, and from left to right.

(2) Horizontal well hybrid grid ordering

We use PEBI grids at the two ends of a horizontal well, strip grids around the horizontal section and PEBI grids in the reservoir areas far away from the horizontal wells to partition the grid and perform grid ordering of such hybrid grid. First, for the PEBI grids located at the left end of the horizontal section, we conduct the grid ordering in increasing order from the inside to the outside in the radial flow direction. Then we perform grid ordering in increasing order counterclockwise in the angular direction. Second, for the strip grids around the horizontal section, we conduct the grid ordering in increasing order along the linear flow direction from top to bottom, and then from left to right along the length of the well trajectory. Third, for the PEBI grids located at the right end of the horizontal section, we conduct the grid ordering in increasing order from the inside to the outside in the radial flow direction. Then we perform grid ordering in increasing order counterclockwise in the angular direction. At last, for the PEBI/quadrilateral grids in the reservoir areas far away from the horizontal wells, we conduct grid ordering in increasing order from top to bottom, and from left to right.

(3) Fracture/fault hybrid grid ordering

We use regular strip rectangular grids/irregular strip grids inside fractures/faults, and use triangular grids around the fractures/faults and quadrilateral grids in reservoir areas far away from the fractures/faults to partition and order hybrid grid. First, for the strip grids in the fractures/faults, we conduct grid ordering in increasing order along the fracture direction. Then, for the triangular grids around the fractures/faults, we perform grid ordering in increasing order from bottom to top, and from left to right. At last, for the quadrilateral grids in a reservoir far away from the fractures/faults, we conduct grid ordering in increasing order from bottom to top, and from left to right.

Table 4.6 Hybrid grid ordering table

Different objects	Ordering principle	Schematic diagram
Vertical wells	① For the radial grids near the wellbore, we conduct the grid ordering in increasing order from the inside to the outside in the radial flow direction. And then we perform grid ordering in increasing order counterclockwise in the angular direction ② For the PEBI/quadrilateral grids around the wellbore, we conduct grid ordering in increasing order from bottom to top, and from left to right	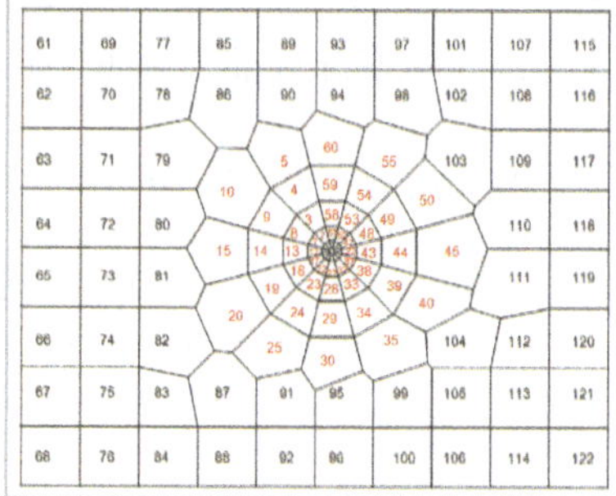
Horizontal wells	① For the PEBI grids located at the left end of the horizontal section, we conduct the grid ordering in increasing order from the inside to the outside in the radial flow direction. And then we perform grid ordering in increasing order counterclockwise in the angular direction ② For the strip grids around the horizontal section, we conduct the grid ordering in increasing order along the linear flow direction from top to bottom, and then from left to right along the length direction of the well trajectory ③ For the PEBI grids located at the right end of the horizontal section, we conduct the grid ordering in increasing order from the inside to the outside in the radial flow direction. And then we perform grid ordering in increasing order counterclockwise in the angular direction ④ For the PEBI grids in the reservoir areas far away from the horizontal wells, we conduct grid ordering in increasing order from bottom to top, and from left to right	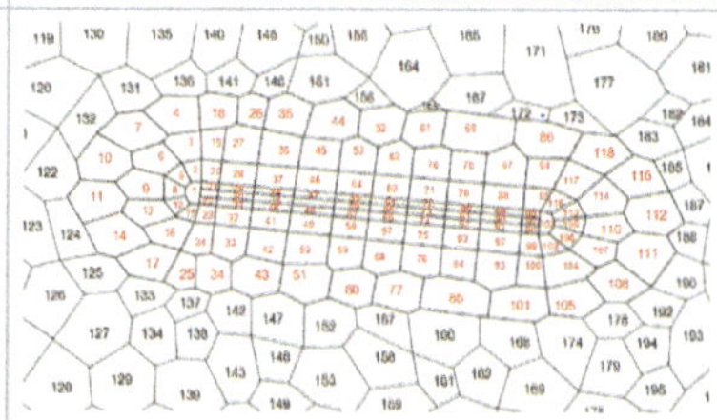
Fractures and faults	① For the strip grids in the fractures/faults, we conduct grid ordering in increasing order along the fracture direction ② For the triangular grids around the fractures/faults, we perform grid ordering in increasing order from top to bottom, and from left to right ③ For the quadrilateral grids in reservoir where far away from the fractures/faults, we conduct grid ordering in increasing order from top to bottom, and from left to right	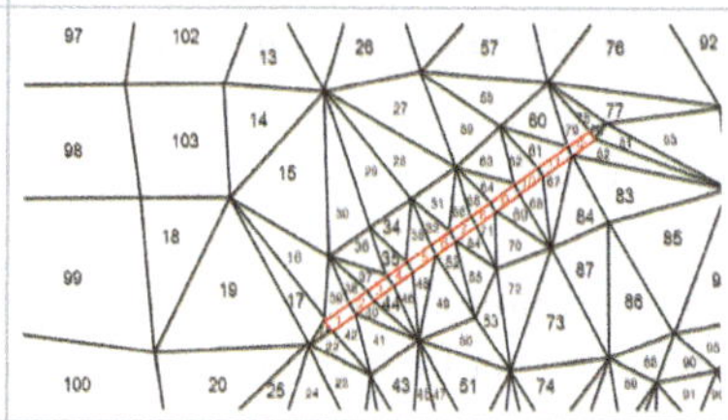
Complex reservoir regions	We follow such sequence: horizontal wells, vertical wells, hydraulic fractures, natural fractures/faults and reservoir areas. For the grids located in the reservoir areas, we perform grid ordering in increasing order from bottom to top, and from left to right	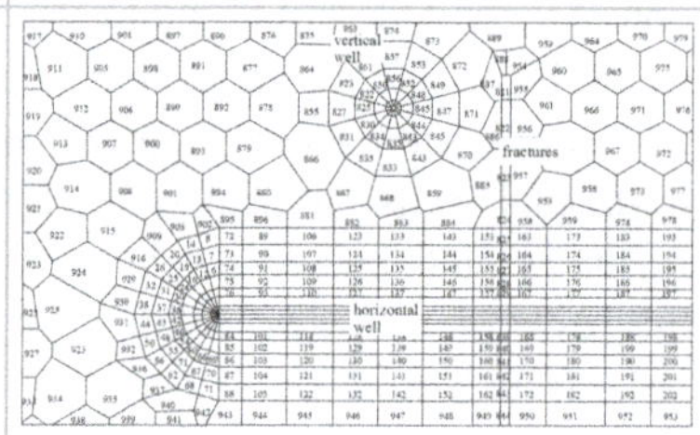

(4) Complex reservoir area hybrid grid ordering

We partition the grids according to the distribution of the different geologic objects in the reservoir area and optimize the grid ordering of different geologic objects according to the importance and local characteristics that influence the effects of reservoir development. We follow such sequence: horizontal wells, vertical wells, hydraulic fractures, natural fractures/faults and reservoir areas. For the grids located in the reservoir areas, we perform grid ordering in increasing order from bottom to top, and from left to right. With all of these steps, we can finish the grid ordering in the complex reservoir area.

2. Multiple media hybrid grid ordering technology

Multiple media hybrid grids can be composed of different types of grids such as PEBI grids, quadrilateral grids and triangular grids. Multiple media hybrid grids can also be divided into nested hybrid grids and interactive hybrid grids. We conduct the grid ordering considering the spatial distribution and flow behavior of the porous medium with

Table 4.7 Multiple media hybrid grid ordering table

Grid type	Ordering principle	Schematic diagram
Nested grid	The principle of ordering for the multiple media grids is consistent with the principle of reservoir region ordering with hybrid grids. We conduct grid ordering for the nested grid inside the element in increasing order from outside to inside, according to the nested-distribution-type mechanism	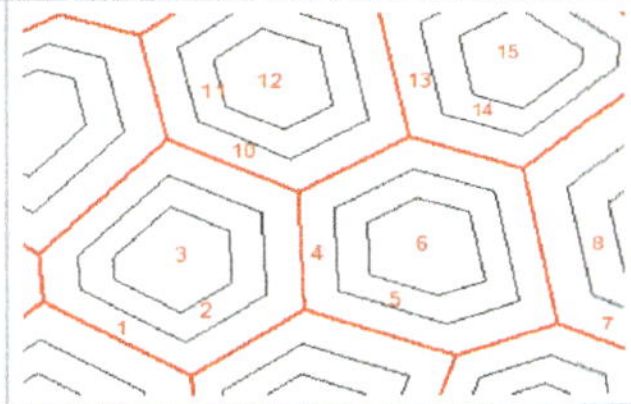
Interactive hybrid grid	The principle of ordering for the multiple media grids is consistent with the principle of reservoir region ordering with hybrid grids. We conduct grid ordering for the internal grids of the element in increasing order from left to right, and from top to bottom	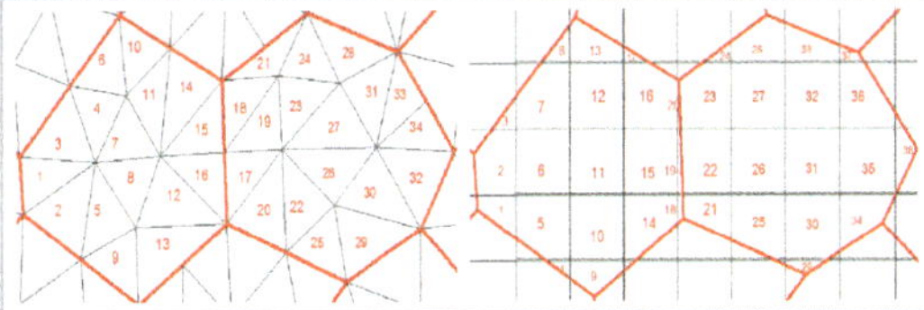
Nested and interactive hybrid grid	The principle of ordering for the multiple media grids is consistent with the principle of reservoir region ordering with hybrid grids. We conduct grid ordering for the nested grids and interactive hybrid grids inside the element according to the ordering rules of the nested hybrid grids and that of the interactive hybrid grids respectively	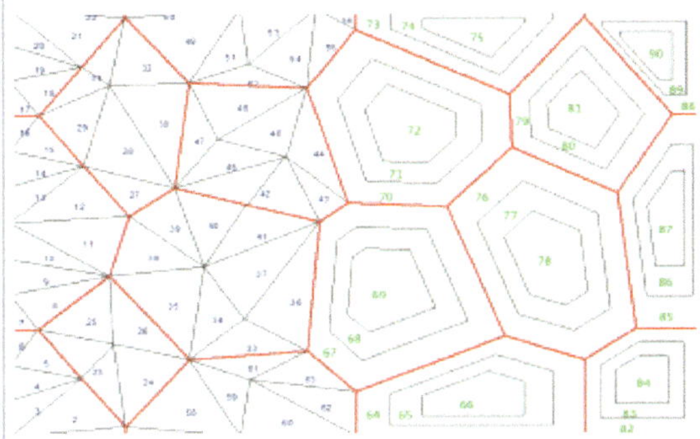

fractures. The multiple media hybrid grid ordering belongs to irregular ordering (Table 4.7).

(1) Multiple media nested hybrid grid ordering

For different reservoir areas, we can use different types of grids such as PEBI grids, quadrilateral grids and triangular grids for multiple media grid partitioning in each element. The same types of grids are used for nested grids partitioning and grid ordering within an element. The principle of ordering for the grids is consistent with the principle of reservoir region ordering with hybrid grids. We conduct grid ordering for the nested grid inside an element in increasing order from outside to inside, according to the nested-distribution-type mechanism.

(2) Multiple media interactive hybrid grid ordering

For different reservoir areas, we can use different types of grids such as PEBI grids, quadrilateral grids and triangular grids for multiple media grid partitioning in each element. Quadrilateral grids and triangular grids are used inside the element for interactive grid partitioning and grid ordering. The principle of ordering of grids is consistent with the principle of reservoir region ordering with hybrid grids. We conduct grid ordering for the internal grids of an element in increasing order from left to right, and from top to bottom.

(3) Nested and interactive hybrid grid ordering

For different reservoir areas, we can use different types of grids such as PEBI grids, quadrilateral grids and triangular grids for multiple media grid partitioning in each element. We conduct nested and interactive grid partitioning and grid ordering within the element. The principle of ordering of the grids is consistent with the principle of reservoir region ordering with hybrid grids. We conduct grid ordering for the nested grids and interactive grids inside the element according to the ordering rules of the nested hybrid grids and that of the interactive hybrid grids, respectively.

4.2.2 Grid Neighbor Characterization Technology for Numerical Simulation

Through grid partitioning, when there is a common contact surface between a grid and its surrounding grid, it is

Table 4.8 Structured grid connectivity list

Grid type	Neighboring relationship diagram	Single grid connectivity list	
		Neighboring relationships	Connectivity relationships
		Grid, neighboring grid	Grid, neighboring grid, transmissibility
Rectangular grid	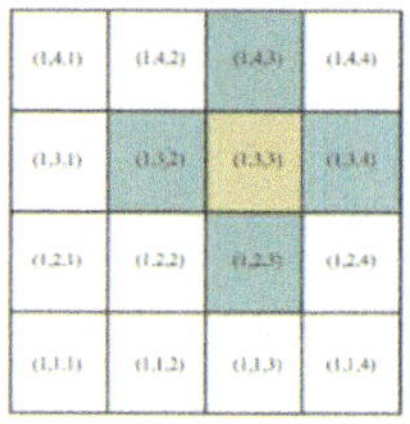	1.3.3, 1.3.2 1.3.3, 1.3.4 1.3.3, 1.2.3 1.3.3, 1.4.3	(1.3.3, 1.3.2, $T_{1.3.3,\ 1.3.2}$) (1.3.3, 1.3.4, $T_{1.3.3,\ 1.3.4}$) (1.3.3, 1.2.3, $T_{1.3.3,\ 1.2.3}$) (1.3.3, 1.4.3, $T_{1.3.3,\ 1.4.3}$)
Corner point grid	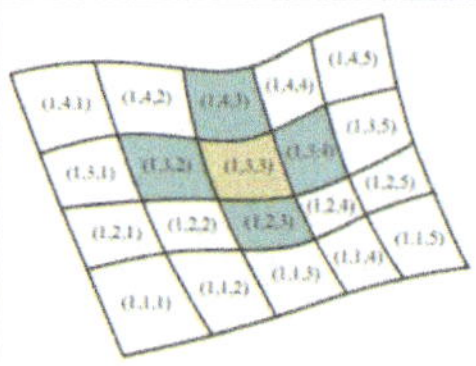	1.3.3, 1.3.2 1.3.3, 1.4.3 1.3.3, 1.3.4 1.3.3, 1.2.3	(1.3.3, 1.3.2, $T_{1.3.3,\ 1.3.2}$) (1.3.3, 1.4.3, $T_{1.3.3,\ 1.4.3}$) (1.3.3, 1.3.4, $T_{1.3.3,\ 1.3.4}$) (1.3.3, 1.2.3, $T_{1.3.3,\ 1.2.3}$)
Radial grid	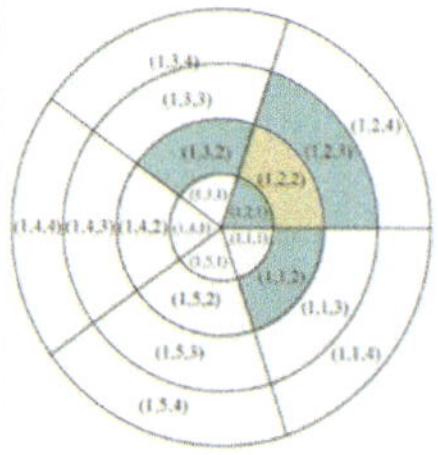	1.2.2, 1.2.1 1.2.2, 1.2.3 1.2.2, 1.3.2 1.2.2, 1.1.2	(1.2.2, 1.2.1, $T_{1.2.2,\ 1.2.1}$) (1.2.2, 1.2.3, $T_{1.2.2,\ 1.2.3}$) (1.2.2, 1.3.2, $T_{1.2.2,\ 1.3.2}$) (1.2.2, 1.1.2, $T_{1.2.2,\ 1.1.2}$)

considered to be connectivity between adjacent grids. On the basis of grid ordering, the connectivity between adjacent grids are usually characterized by a connectivity list. The grid connectivity list consists of the serial number of this grid, the serial number of all of its neighboring grids and the transmissibility between the grid and its neighboring grids. We build the connectivity list according to the following principles:

① Different types of grids have different neighboring relationships and corresponding connectivity lists.
② For neighboring grids that do not have flow in between, they are considered to be disconnected grids. Their connectivity relationship is not included in the connectivity list.
③ In the whole connectivity list, the same neighboring connection cannot repeat to show up.

4.2.2.1 Structured Grid Connectivity List

A structured grid is a type of regular grid, which also belongs to a quadrilateral grid. All connections between the grid blocks have clear and regular topological relationships. All grids have the same number of neighboring grids. In other words, every grid of the rectangular grid, corner point grid or radial grid has 4 neighboring grids in 2D. Moreover, its grid connectivity list for a single grid reflects 4 connections (Table 4.8).

4.2.2.2 Unstructured Grid Connectivity List

An unstructured grid is a type of irregular grid. Different types of grids have uncertain irregular neighboring relationships. Different types of grids have different numbers of neighboring grids. A triangular grid has 3 neighboring grids, and its grid connectivity list for a single grid reflects 3 connections. The quadrilateral grid has 4 neighboring grids, and its grid connectivity list for a single grid reflects 4 connections. The PEBI grid and CVFE grid have an arbitrary number of neighboring grids, and their grid connectivity list for a single grid reflects different numbers of connections (Table 4.9).

4.2.2.3 Hybrid Grid Connectivity List

1. Different types of hybrid grid connectivity list

Different types of hybrid grids have different neighboring relationships and different connectivity lists. A triangular grid, quadrilateral grid, PEBI grid and CVFE grid have different numbers of neighboring grids. The same grid type

Table 4.9 Unstructured grid connectivity list

Grid type	Neighboring relationships diagram	Single grid connectivity list	
		Neighboring relationships	Connectivity relationships
		Grid, neighboring grid	Grid, neighboring grid, transmissibility
Triangular grid	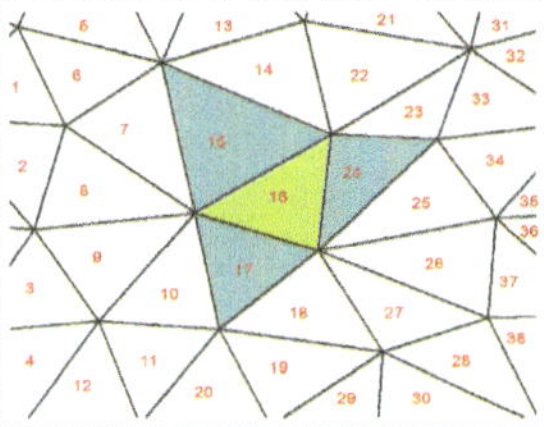	16, 15	(16, 15, $T_{16,\ 15}$)
		16, 17	(16, 17, $T_{16,\ 17}$)
		16, 24	(16, 24, $T_{16,\ 24}$)
Quadrilateral grid	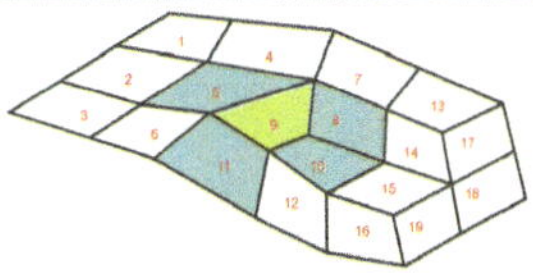	9, 5	(9, 5, $T_{9,\ 5}$)
		9, 8	(9, 8, $T_{9,\ 8}$)
		9, 10	(9, 10, $T_{9,\ 10}$)
		9, 11	(9, 11, $T_{9,\ 11}$)
PEBI grid	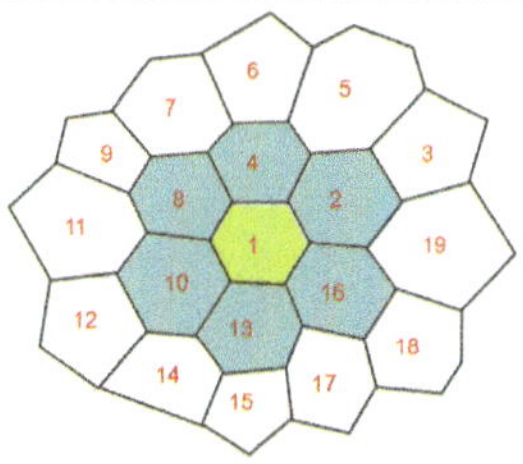	1, 2	(1, 2, $T_{1,\ 2}$)
		1, 4	(1, 4, $T_{1,\ 4}$)
		1, 8	(1, 8, $T_{1,\ 8}$)
		1, 10	(1, 10, $T_{1,\ 10}$)
		1, 13	(1, 13, $T_{1,\ 13}$)
		1, 16	(1, 16, $T_{1,\ 16}$)
CVFE grid	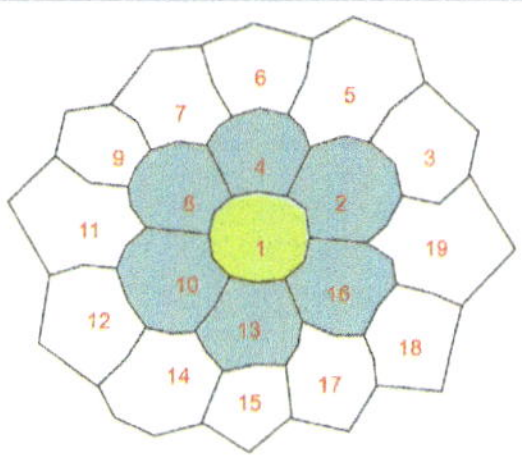	1, 2	(1, 2, $T_{1,\ 2}$)
		1, 4	(1, 4, $T_{1,\ 4}$)
		1, 8	(1, 8, $T_{1,\ 8}$)
		1, 10	(1, 10, $T_{1,\ 10}$)
		1, 13	(1, 13, $T_{1,\ 13}$)
		1, 16	(1, 16, $T_{1,\ 16}$)

(PEBI grid or CVFE grid) may have different numbers of edges and neighboring relationships in different situations. The corresponding changes occur for the connectivity lists. In addition, different objects have multiple types of connectivity lists because they are characterized in a hybrid way by having a variety of different types of grids (Table 4.10).

2. Multiple media hybrid grid connectivity list

(1) Nested grid connectivity list

Nested grids have three types of connectivity lists (Table 4.11). The innermost grid has only one outward neighboring grid, and its grid connectivity list reflects one connection. The middle layer grid has two neighboring grids, inward and outward, and the grid connectivity list reflects two connections. The connectivity list of the outermost grid has two neighboring grids, which are the inward neighboring grid and the outward neighboring grid for the outermost grid. The grid connectivity list reflects two connections.

(2) Interactive grid connectivity list

The interactive grid connectivity list (Table 4.11) depends on the grid type in the multiple media element. When the interior of an element is partitioned using triangular grids with 3 neighboring grids, the grid connectivity list reflects 3 connections. A quadrilateral grid has 4 neighboring grids, and the grid connectivity list reflects 4 connections.

4.2.3 Grid Connectivity Characterization Technology for Numerical Simulation

On the basis of grid ordering and connectivity lists, the strength of connectivity between neighboring grids is usually characterized by transmissibility. The magnitude of transmissibility mainly depends on the geometric morphology of different grids, the grid permeability and the orthogonality of the flow between grid blocks.

Table 4.10 Connectivity list for different types of hybrid grid

Different objects (hybrid grid types)	Neighboring relationships diagram	Single grid connectivity list	
		Neighboring relationships	Connectivity relationships
		Grid, neighboring grid	Grid, neighboring grid, transmissibility
Vertical wells	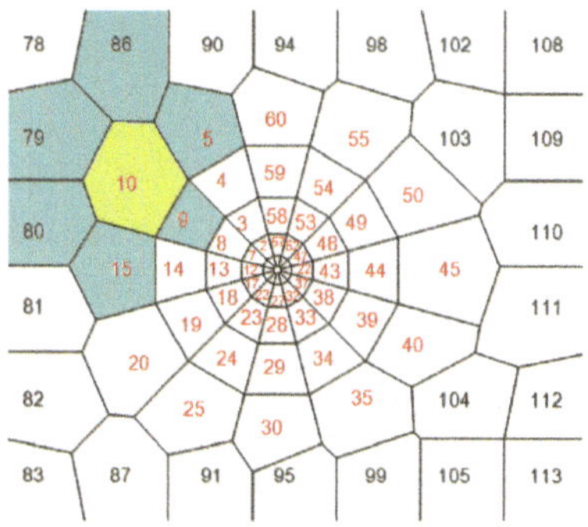	10, 5	(10, 5, $T_{10,\ 5}$)
		10, 9	(10, 9, $T_{10,\ 9}$)
		10, 15	(10, 15, $T_{10,\ 15}$)
		10, 79	(10, 79, $T_{10,\ 79}$)
		10, 80	(10, 80, $T_{10,\ 80}$)
		10, 86	(10, 86, $T_{10,\ 86}$)
Horizontal wells	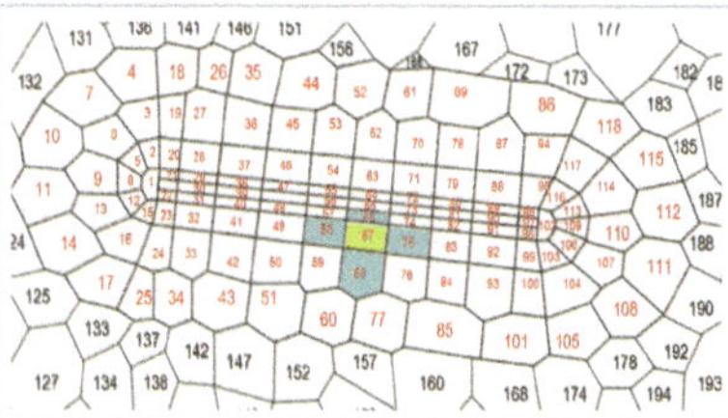	67, 66	(67, 66, $T_{67,\ 66}$)
		67, 68	(67, 68, $T_{67,\ 68}$)
		67, 58	(67, 58, $T_{67,\ 58}$)
		67, 75	(67, 75, $T_{67,\ 75}$)
Fractures/faults	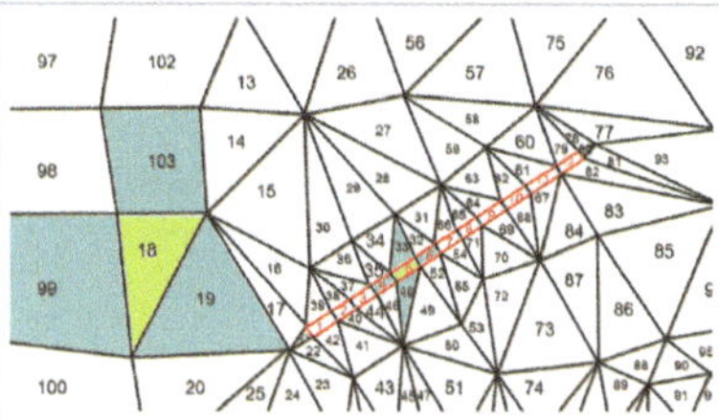	18, 19	(18, 19, $T_{18,\ 19}$)
		18, 99	(18, 99, $T_{18,\ 99}$)
		18, 103	(18, 103, $T_{18,\ 103}$)
		5, 4	(5, 4, $T_{5,\ 4}$)
		5, 6	(5, 6, $T_{5,\ 6}$)
		5, 33	(5, 33, $T_{5,\ 33}$)
		5, 48	(5, 48, $T_{5,\ 48}$)
Complex reservoir regions	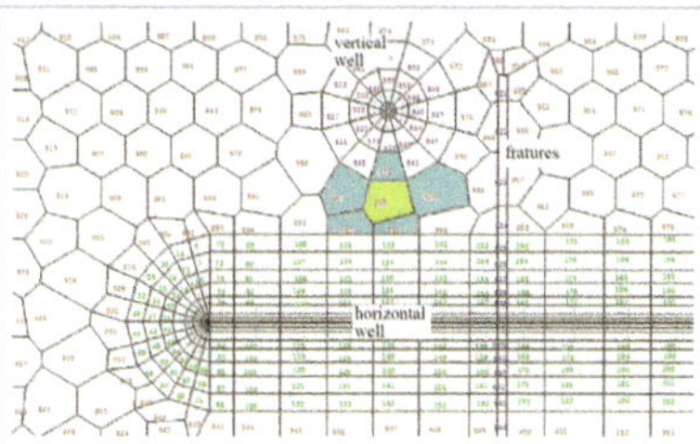	555, 539	(555, 539, $T_{555,\ 539}$)
		555, 567	(555, 567, $T_{555,\ 567}$)
		555, 569	(555, 569, $T_{555,\ 569}$)
		555, 582	(555, 582, $T_{555,\ 582}$)
		555, 583	(555, 583, $T_{555,\ 583}$)

4.2.3.1 Structured Grid Transmissibility Calculation

The calculation method for transmissibility in a structured grid can be divided into two types. i. A rectangular grid has regular geometric morphology. The flow between grid blocks is always orthogonal. We can use the regular grid transmissibility calculation formula for this instance (Table 4.12). ii. The radial grids and corner point grids have irregular geometric morphology. It is difficult to always maintain orthogonal flow between grid blocks. So, it is necessary to use irregular grid transmissibility calculation formulas for calculations (Table 4.12).

1. Regular grid transmissibility calculation formula

The morphology of every grid cell in a rectangular grid is regular. The grid center is both the centroid of the grid and also its circumcenter. The flow direction between grid blocks is orthogonal to the adjacent face between the blocks (Fig. 4.21).

Table 4.11 Multiple media hybrid grid connectivity list

Grid type	Neighboring relationships diagram	Single grid connectivity list	
		Neighboring relationships	Connectivity relationships
		Grid, neighboring grid	Grid, neighboring grid, transmissibility
Nested grid	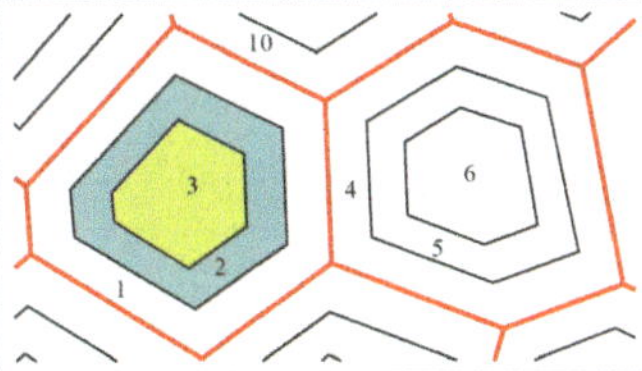	3, 2	(3, 2, $T_{3,\ 2}$)
	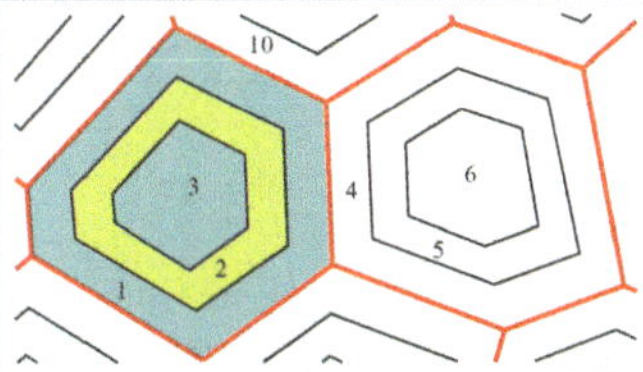	2, 1	(2, 1, $T_{2,\ 1}$)
		2, 3	(2, 3, $T_{2,\ 3}$)
	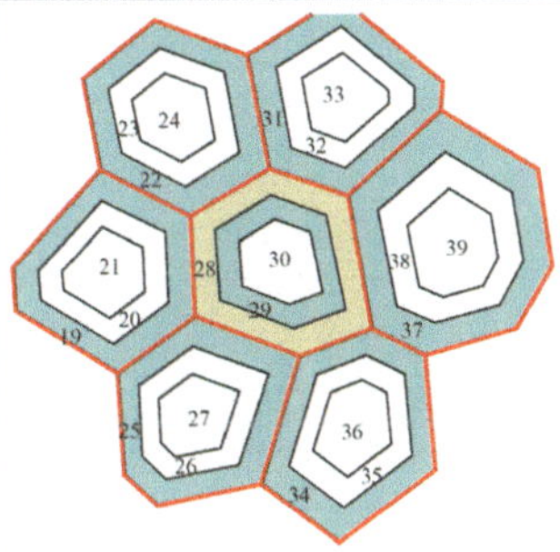	28, 19	(28, 19, $T_{28,\ 19}$)
		28, 29	(28, 29, $T_{28,\ 29}$)
Interactive grid	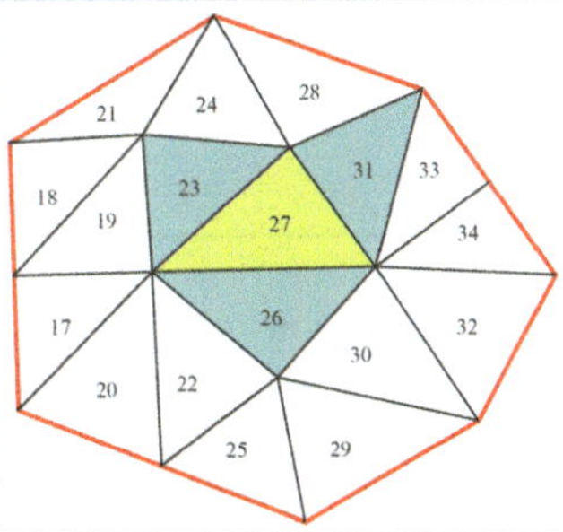	27, 23	(27, 23, $T_{27,\ 23}$)
		27, 26	(27, 26, $T_{27,\ 26}$)
		27, 31	(27, 31, $T_{27,\ 31}$)
	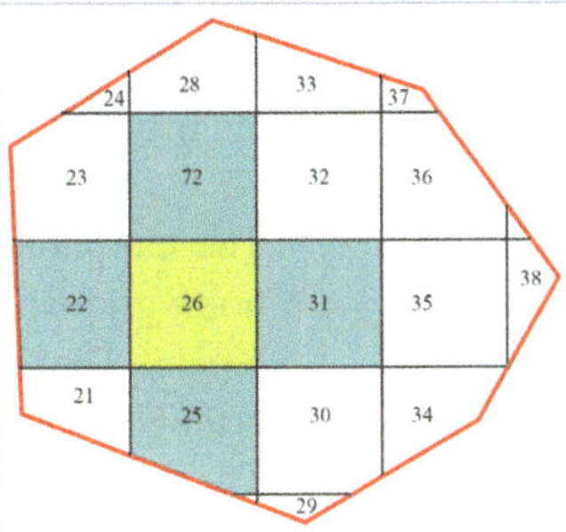	26, 22	(26, 22, $T_{26,\ 22}$)
		26, 25	(26, 25, $T_{26,\ 25}$)
		26, 27	(26, 27, $T_{26,\ 27}$)
		26, 31	(26, 31, $T_{26,\ 31}$)

Table 4.12 Structured grid transmissibility calculation table

Grid type	Transmissibility calculation diagram	Transmissibility calculation formula
Rectangular grid	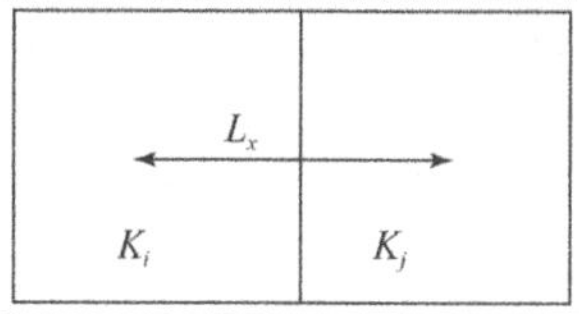	$T_{i,j} = \sigma \frac{2K_iK_j}{K_i + K_j}$ $\sigma = 4\left(\frac{1}{L_x^2} + \frac{1}{L_y^2} + \frac{1}{L_z^2}\right)$
Radial grid	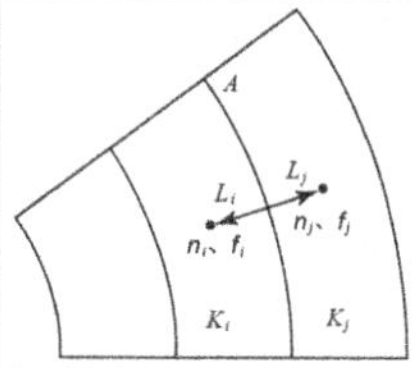	$T_{i,j} = \frac{\alpha_i\alpha_j}{\alpha_i + \alpha_j}$ $\alpha = A\frac{K}{L}\boldsymbol{n} \cdot \boldsymbol{f}$
Corner point grid	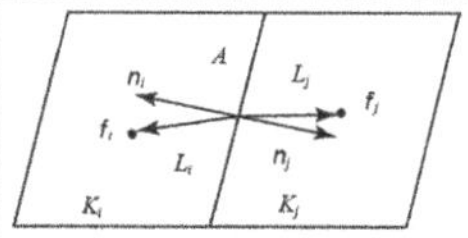	

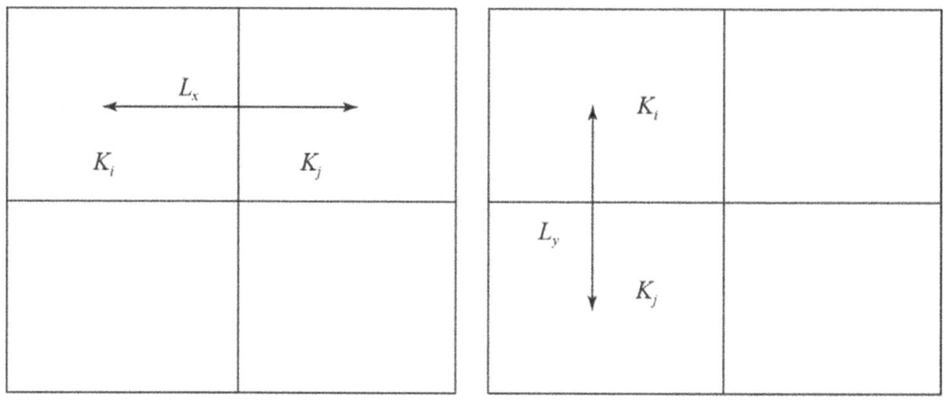

Fig. 4.21 Schematic diagram of regular grid transmissibility calculations

Transmissibility between regular grid blocks can be calculated based on the distance between the centers of the neighboring grids and the permeabilities of the different grids. Here is the calculation formula.

$$T_{i,j} = \sigma \frac{2K_iK_j}{K_i + K_j} \tag{4.1}$$

$$\sigma = 4\left(\frac{1}{L_x^2} + \frac{1}{L_y^2} + \frac{1}{L_z^2}\right) \tag{4.2}$$

The subscripts i and j are the neighboring grid index numbers. $T_{i,j}$ is the transmissibility between neighboring grid blocks i and j. K_i and K_j are the effective permeabilities of grid blocks i, j (mD). L_x is the distance between the center points of the neighboring grids i and j along the x direction (m). L_y is the distance between the center points of the neighboring grids i and j along the Y direction (m). L_z is the distance between the center points of the neighboring grids i and j along the Z direction (m).

2. Irregular grid transmissibility calculation formula

The morphology of every grid cell in a radial grid or corner point grid is irregular. A grid center is the centroid of a grid block. The flow direction between the grids is hardly orthogonal to the adjacent face between the grids (Fig. 4.12). The transmissibility of irregular grids can be calculated based on the distance from the center of the neighboring grid to the center point of the adjacent face, its normal angle and the permeabilities of the different grids. Here is the calculation formula.

$$T_{i,j} = \frac{\alpha_i\alpha_j}{\alpha_i + \alpha_j} \tag{4.3}$$

The shape factor of grid α is:

$$\alpha = A \cdot \frac{K}{L} \cdot \boldsymbol{n} \cdot \boldsymbol{f} \quad (4.4)$$

The subscripts i and j are the neighboring grid index numbers. $T_{i,j}$ is the transmissibility between neighboring grids i and j. K_i and K_j are the effective permeabilities of grids i, j (mD). A_{ij} is the actual contact area of the neighboring grids i and j (m^2). α_i and α_j are the shape factors of the grid i and j. L_i and L_j are the actual distances from the centers of gravity of grids i and j to the center of the neighboring grids' contact interface (m). $\boldsymbol{n} \cdot \boldsymbol{f}$ are the normal direction corrections for orthogonality of the unstructured grids.

4.2.3.2 Unstructured Grid Transmissibility Calculation

Different types of unstructured grids have different geometric morphology. Even if the type of a variable-scale unstructured grid is the same, there is still a big difference about the geometric morphology. The grid center is the centroid of a grid block. The flow direction between the grids is hardly orthogonal to the adjacent face between the grids (Fig. 4.22).

The transmissibility in unstructured grids can be calculated based on the distance from the centers of the neighboring grids to the center point of the adjacent face, its normal angle and the permeabilities of different grids. Here is the calculation formula (Table 4.13).

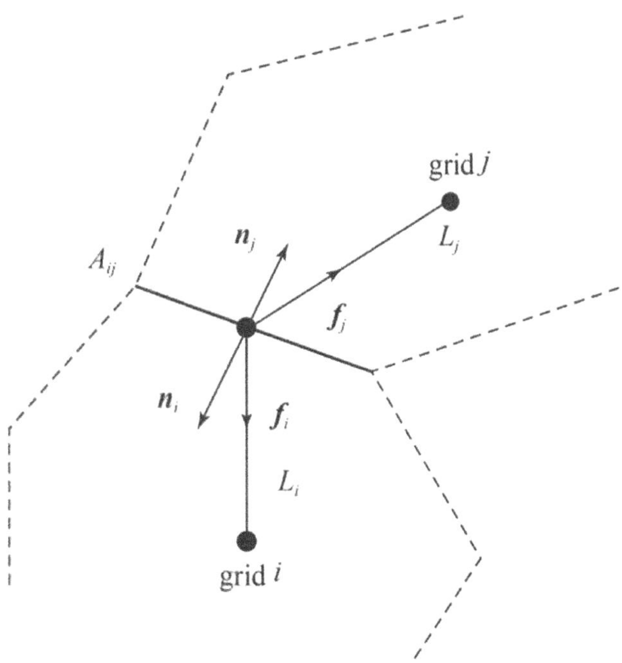

Fig. 4.22 Schematic diagram of unstructured grid transmissibility calculations

$$T_{i,j} = \frac{\alpha_i \alpha_j}{\alpha_i + \alpha_j} \quad (4.5)$$

The morphologies of the neighboring grids in triangular and quadrilateral grids are different. It is difficult for the flow direction between the grids to be orthogonal to the interface. The shape factor is calculated by formula (4.4). The PEBI grids and CVFE grids are orthogonal grids, with $\boldsymbol{n} \cdot \boldsymbol{f} = 1$. The shape factor is calculated by formula (4.6).

$$\alpha = A \cdot \frac{K}{L} \quad (4.6)$$

4.2.3.3 Hybrid Grid Transmissibility Calculation

1. Transmissibility calculation for different types of hybrid grid

In actual reservoir simulations, for complex geologic conditions and boundary conditions such as vertical wells, horizontal wells, natural/hydraulic fractures, faults/boundaries and macroscopic heterogeneities, we usually treat a reservoir with different types of hybrid grids. The morphologies of the different types of hybrid grids are very different. The flow direction between the grids is hardly orthogonal to the adjacent face between the grids. Different types of hybrid grid transmissibility calculations can use the calculations method of transmissibility between unstructured grids (Table 4.14).

2. Transmissibility calculation for multiple media hybrid grid

There are two ways to calculate the transmissibility for multiple media hybrid grid. One is the transmissibility calculation on nested hybrid grids with nested-distribution-type flow between neighboring grids. Another is the transmissibility calculation on interactive hybrid grids with arbitrary flow between neighboring grids.

(1) Transmissibility calculation for interactive hybrid grid

The transmissibility calculation in an interactive hybrid grid with arbitrary flow between neighboring grids can use the method of transmissibility calculation between unstructured grids (Table 4.15).

(2) Transmissibility calculation for nested hybrid grid

For the nested hybrid grids with nested-distribution-type flow between neighboring grids, we can use different transmissibility calculation formulas for grids in different

Table 4.13 Unstructured grid transmissibility calculation

Grid type	Transmissibility calculation diagram	Transmissibility calculation formula
Triangular grid	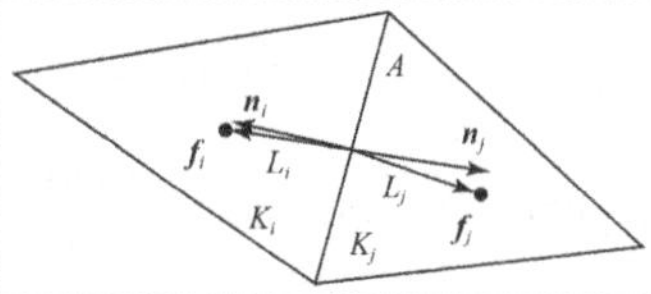	$T_{i,j} = \frac{\alpha_i \alpha_j}{\alpha_i + \alpha_j}$ $\alpha = A\frac{K}{L}\boldsymbol{n} \cdot \boldsymbol{f}$
Quadrilateral grid	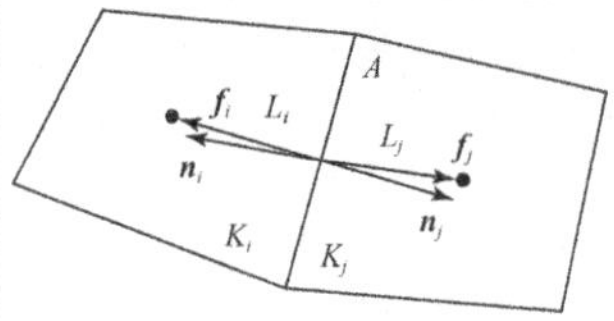	
PEBI grid	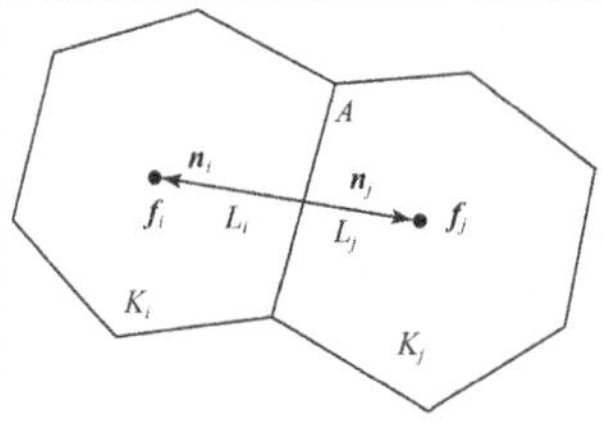	$T_{i,j} = \frac{\alpha_i \alpha_j}{\alpha_i + \alpha_j}$ $\alpha = A\frac{K}{L}$
CVFE grid	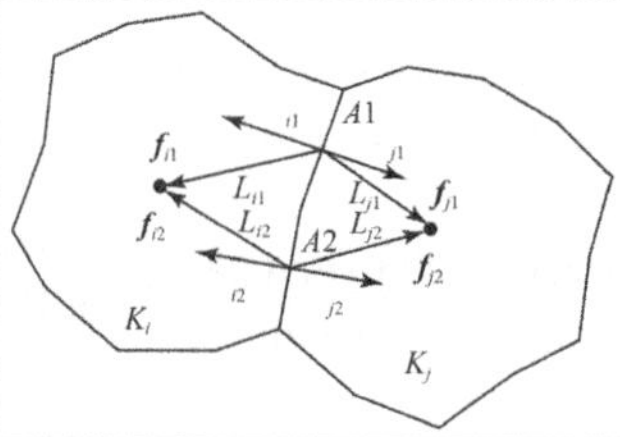	

Table 4.14 Transmissibility calculations for different types of hybrid grid

Grid type	Transmissibility calculation diagram	Transmissibility calculation formula
PEBI grid + radial grid	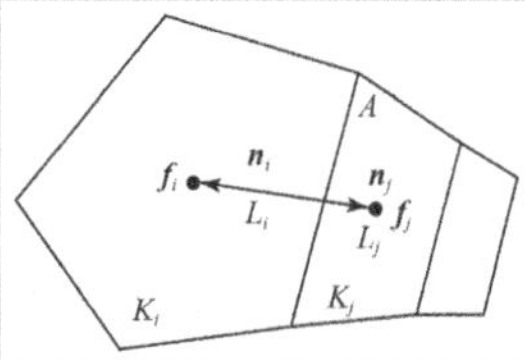	$T_{i,j} = \frac{\alpha_i \alpha_j}{\alpha_i + \alpha_j}$ $\alpha = A\frac{K}{L}\boldsymbol{n} \cdot \boldsymbol{f}$
PEBI grid + strip grid	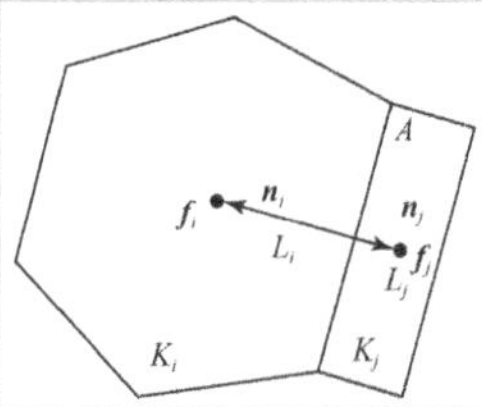	
Triangular grid + quadrilateral grid	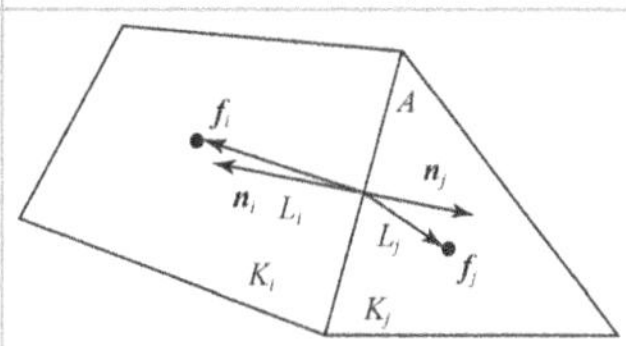	

Table 4.15 Multiple media hybrid grid transmissibility calculation

Grid type		Transmissibility calculation diagram	Transmissibility calculation formula
Interactive hybrid grid		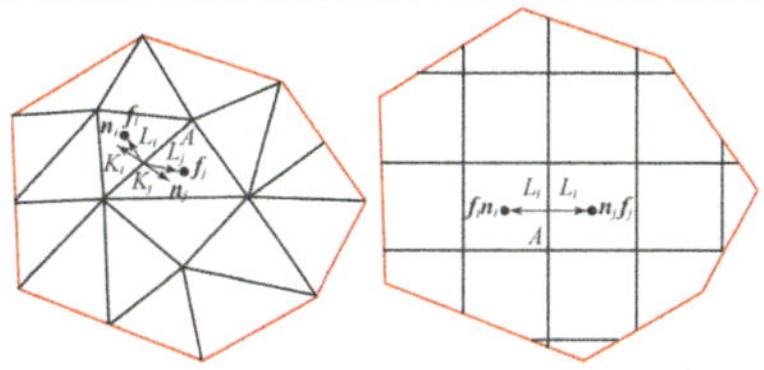	$T_{i,j} = \frac{\alpha_i \alpha_j}{\alpha_i + \alpha_j}$ $\alpha = A\frac{K}{L}\boldsymbol{n}\cdot\boldsymbol{f}$
Nested hybrid grid	Different media in the same element	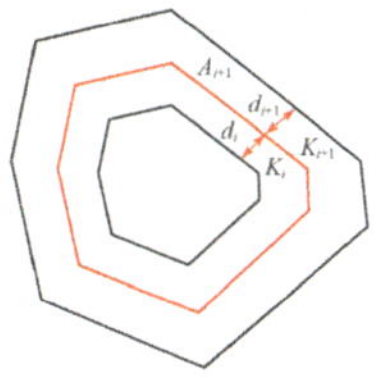	$T_{i,i+1} = \frac{4A_{i,i+1}}{d_i + d_{i+1}}\left(\frac{K_i K_{i+1}}{K_i + K_{i+1}}\right)$
	Between different elements	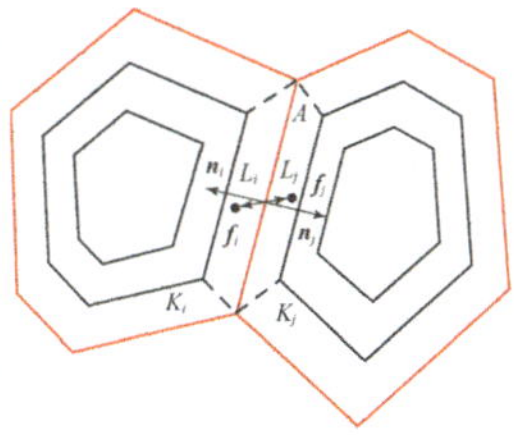	$T_{i,j} = \frac{\alpha_i \alpha_j}{\alpha_i + \alpha_j}$ $\alpha = A\frac{K}{L}\boldsymbol{n}\cdot\boldsymbol{f}$

media within the same element and for grids in different elements (Table 4.15).

① Transmissibility calculation between different media in the same element

Different media in the same element perform as nested-distribution-type flow. The media grids are nested within each other and ring-shaped distributed. Corresponding edges of the grids are parallel to each other. The width inside the same grid is equal everywhere (Fig. 4.23).

The transmissibility of different media in the same element can be calculated according to the adjacent area of the neighboring nested grids, the width of each grid and the permeabilities of different grids. Here is the calculation formula.

$$T_{i,i+1} = \frac{4A_{i,i+1}}{d_i + d_{i+1}}\left(\frac{K_i K_{i+1}}{K_i + K_{i+1}}\right) \tag{4.7}$$

$T_{i,i+1}$ is the transmissibility between grid i and grid i +1. K_i and K_{i+1} are the effective permeabilities of grid i and grid i +1, respectively (m^2). d_i and d_{i+1} are the widths of grid i and grid i +1, respectively. When the grid is the innermost grid, $d_1 = \frac{V_1}{A_{i\_J_1 J_2}}$ (m).

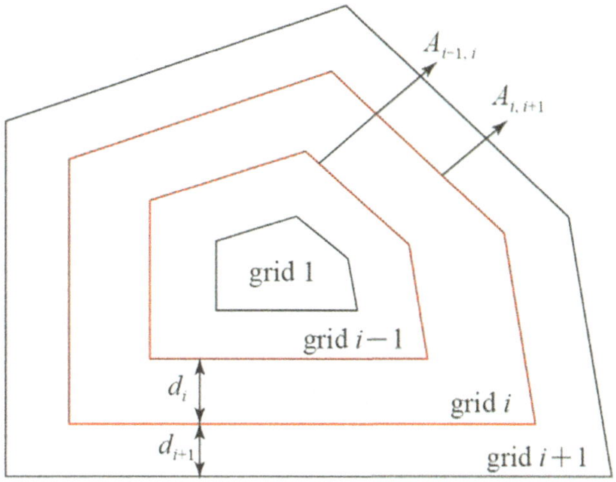

Fig. 4.23 Schematic diagram of transmissibility calculation among different media in the same element

② Inter-element transmissibility calculation

First, we build two adjacent trapezoid grids by using the two parallel edges of the two outermost adjacent grids of adjacent elements as the upper and lower bases (Fig. 4.24).

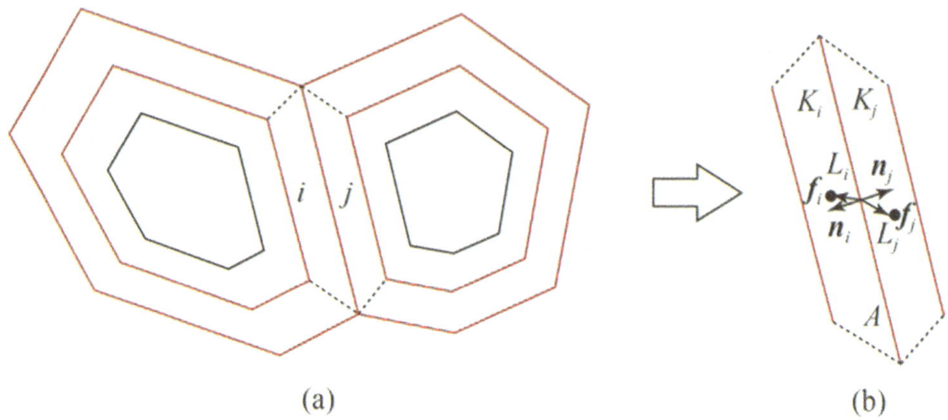

Fig. 4.24 Schematic diagram of transmissibility calculation between different elements

The transmissibility between two elements can be calculated according to the adjacent area of the two trapezoid grids, the distance from the centers of the grids to the center point of the adjacent interface and its normal angle, and the permeabilities of different grids. The transmissibility calculation formula is as follows:

$$T_{i,j} = \frac{\alpha_i \alpha_j}{\alpha_i + \alpha_j} \tag{4.8}$$

The trapezoid shape factor α is:

$$\alpha = A\frac{K}{L}\boldsymbol{n}\cdot\boldsymbol{f} \tag{4.9}$$

The subscripts i, j are the outermost grid index numbers of the neighboring nested grids. $T_{i,j}$ is the transmissibility between neighboring trapezoid grids, which is also the transmissibility between grids i and j. K_i and K_j are the effective permeabilities of grids i and j (mD). A_{ij} is the contact area of the neighboring trapezoid grids, (m^2). α_i and α_j are the trapezoid grid shape factors of the grids i and j. L_i and L_j are the actual distance from the center points of the trapezoid grid to the center of the neighboring grids' contact interface. $\boldsymbol{n}\cdot\boldsymbol{f}$ is the correction in the normal direction for the orthogonality of trapezoid grids.

4.3 The Discretization Technology of the Mathematical Model for Multiphase Flow in Multiple Media at Different Scales

On the basis of numerical simulation grid partitioning and generation techniques and grid connectivity characterization techniques, by using the finite volume method suitable for unstructured grids and discontinuous medium mathematical model of flow in porous media, we form the discretization technology for the mathematical models of flow through multiple media. We further conduct discretization processing of the flow through discontinuous multiple media mathematical models to establish the discrete numerical simulation models (Han et al. 1993).

Discrete processing of mathematical model of flow in porous media through the discretization technologies includes temporal discretization and spatial discretization. Among them, the temporal discretization means that the continuously changing flow process through pore and fracture media of tight reservoirs is obtained by flow states at a series of time points through the use of discretized equations. Spatial discretization refers to the process of flow in a series of space points through the discrete equations. A finite-difference method based on a structured grid is used for continuous media. Moreover, a finite volume method based on an unstructured grid is used for discontinuous media (Aziz and Settari 2004).

4.3.1 The Spatial Discretization Method of the Mathematical Model for Multiphase Flow in Multiple Media at Different Scales

4.3.1.1 Finite Difference Discretization Method Based on Structured Grid

For continuous media, a medium in the same system is continuously distributed. Its geometric characteristics, property characteristics, and flow characteristics are similar. The fluid is continuously distributed, with a single type of flow mechanism. The flow mainly consists of Darcy flow. For this type of continuous media, they can be divided into several regular grid cells by using a structural grid. On basis of this, the finite difference method is used to discretize a mathematical model of flow in porous media, which a discretized numerical model can be established.

The finite difference method divides the computational domain into several regular grids, which replaces the continuous computational domain with a finite number of grid cells. This is a method to achieve approximate numerical solutions that transforms the differential problem directly into an algebraic problem (Rao 2009). Therefore, the finite difference method based on a structured grid is suitable for discretization of a continuous porous medium flow model. It has stable computation and high accuracy.

4.3.1.2 Finite Volume Discretization Method Based on Unstructured Grids

For discontinuous media, multiple media at different scales are discontinuously discretely distributed in a system. The different media can be contacted arbitrarily. The geometric and property characteristics of different media are quite different. The parameters of the media have discontinuous changes. In addition, the mechanisms of flow in porous media and flow states of different media are complex and diverse. A fluid can arbitrarily flow between any media. For discontinuous multiple media, it is suitable to use an unstructured grid to divide them into several irregular grid cells for the purpose of discretization. Because of inflexibility of the traditional finite difference method for complex boundaries and discontinuous multiple media problems, it is difficult to apply the finite difference method. Although the finite element method is based on unstructured grids with the enhanced flexibility in spatial discretization, it only achieves mass conservation in the global area. It is impossible to achieve the requirement of mass conservation for fluid flow in a local grid, which makes it prone to numerical fluctuations (Zhang 2015). Therefore, for the problem of discontinuous multiple media processed by an unstructured grid, it is no longer suitable to use the traditional finite difference method and finite element method. The finite volume method based on an unstructured grid should be applied to perform discretization for the mathematical model of flow through discontinuous multiple media, which ultimately establishes a discrete numerical model.

The finite volume method is also called the control volume method. Its main idea is to divide a computation domain into several non-repetitive control volumes. Each control volume unit has a centroid. We integrate differential equations to be solved on each control volume. Then a set of discrete equations are obtained (Zha 2009). The finite volume method is applicable for both structured grids and unstructured grids. It not only has the flexibility in grid partitioning, but also can describe complex boundaries and geometric shapes. Moreover, the finite volume method under the condition of the unstructured grids can satisfy local mass conservation for fluid flow. Moreover, it has pleasant computational stability. So, it is a discretization method based on unstructured grids suitable for mathematic models of flow through discontinuous multiple media.

When the finite volume method and the finite difference method are used to discretize the mathematical models of flow in porous media, there are differences in the applicability of the two methods in terms of reservoir conditions, media types, and grid types. The mathematical principles and the discretization methods are different. Thus, there are large differences in the discrete forms for flow terms, source and sink terms, and accumulation terms in a numerical model (Table 4.16).

The differences between the two methods in the discrete forms mainly lie in the expression type differences for discretization of transmissibility for flow terms, a well index for source and sink terms, and a grid volume for accumulation terms, which is caused mainly by differences in grid types. However, the discrete expression forms that represent the flow mechanisms are the same approximately.

① A flow term is discussed here. Since the finite difference method is based on a structural grid, the discrete expression of its transmissibility reflects the clear directionality and the parameters of length, width and height of a regular grid cell. The finite volume method is based on an unstructured grid. Its form of conductivity discretization reflects the explicit orthogonality, the contact area, and the distance from the unit centroid to the center of the contact surface in irregular grids (Table 4.16).

② A source and sink term is discussed here. The discrete expression form that calculates the well index using the finite difference method clarifies the directionality and the length, width and height parameters of the regular grid block. The discrete expression form that calculates the well index using the finite volume method reflects the explicit orthogonality and the distance from the centroid of the irregular grid block to the center of the contact surface (Table 4.16).

③ An accumulation term is discussed here. The discrete expression form of the finite difference method for computing the accumulation term demonstrates the parameters of length, width, and height of a regular grid cell. The discrete expression form of the finite volume method for computing the accumulation term reflects the grid volume of each irregular grid cell (Table 4.16).

4.3.2 Finite Volume Discretization Method of the Mathematical Model for Multiphase Flow in Discontinuous Multiple Media at Different Scales

When the finite volume method is used to discretize a mathematical model for flow through discontinuous multiple media, the discrete expressions are mainly reflected in such two aspects of geometric characteristics of the discontinuous multiple media, and the mechanisms of

Table 4.16 Comparison between finite difference method and finite volume method

Media type	Continuous media	Discontinuous discrete media
Media characteristics	The media in the same system is continuously distributed. Their geometric characteristics, property characteristics, and flow characteristics are similar. The fluid is continuously distributed, with single flow mechanism. The flow mainly consists of Darcy flow	Multiple media at different scales are discontinuously and discretely distributed in the system. The differences in geometric and property characteristics between different media are large, with the parameters discontinuously varying. There are mechanisms of flow in porous media and flow states. The different media can be contacted arbitrarily, with fluid flow between them
Type of grid blocks	Structural grid: divide the computational domain into several regular grid cells	Unstructured grid: partition the computational domain into several irregular control volume cells
Schematics of the grid	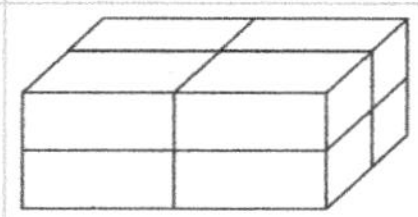	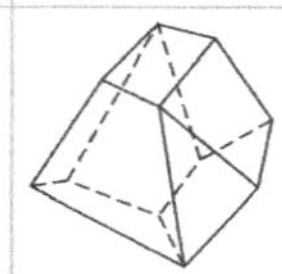
Discretization methods	Finite difference method	Finite volume method
Center of computation	The center of the grid cell	The centroid of control volume unit
Basic principles	Based on the derivation of differential equation, the finite difference approximation formed by using the value of the unknowns at the center of the grid cell is used in place of each derivative of the partial differential equation. Then, we can discretize the partial differential equation which represents the continuous variation of the variables into a finite number of algebraic equations. Then, we use iterative method to solve the set of finite difference equations	Based on the derivation of integral equation, the continuous solution domain is partitioned into a number of non-repeating finite control volume units. Within each control volume unit, the differential equations are volume-integrated to convert the nonlinear partial differential equations into set of linear equations to be solved (Zhang 2015)
Applicable conditions and features	① Applicable reservoir conditions: suitable for situations with weak heterogeneity, regular distribution of sand bodies, sparse fractures with single type of geometric morphology, relatively regular reservoir boundaries and other internal and external boundary conditions ② Applicable grid conditions: simple format and versatile regular structured grid ③ Discrete methods: With simple data structure, strong regularity, fast convergence, good stability, fast computational speed	① Applicable reservoir conditions: suitable for situations with strong heterogeneity in different regions futher in simulation elements, with such internal and external boundary conditions with various complex morphologies, with complex fracture network of natural/hydraulic fractures at different scales, and with multiple media characteristics of pores and fractures at different scales ② Applicable grid conditions: suitable for structural grid and unstructured grid at the same time, with strong flexibility, less ineffective grids, high degree of fine description, and high simulation accuracy ③ Discretization method: The complex and versatile data structure can satisfy the local mass conservation of fluid flow with great computational stability

(continued)

Table 4.16 (continued)

General formula of discrete equations		$\sum_m T\cdot\left(F_p^{\textcircled{1}}\cdot F_p^{\textcircled{2}}\right)+A_{p,i}^{C}=\mathrm{WI}_i\cdot F_{p,i}^{W}$	$\sum_j T_{ij}\cdot\left(F_{p,ij}^{\textcircled{1}}\cdot F_{p,ij}^{\textcircled{2}}\right)+A_{p,i}^{C}=\mathrm{WI}_i\cdot F_{p,i}^{W}$
Discrete equations	Flow term	$\sum_m T\left(F_p^{\textcircled{1}}\cdot F_p^{\textcircled{2}}\right)=\sum_{m=i+\frac{1}{2},i-\frac{1}{2}} T_X\left(F_{p,m}^{\textcircled{1}}\cdot F_{p,i+1,j}^{\textcircled{2}}\right)+\sum_{m=j+\frac{1}{2},j-\frac{1}{2}} T_Y\left(F_{p,m}^{\textcircled{1}}\cdot F_{p,j+1,j}^{\textcircled{2}}\right)+\sum_{m=k+\frac{1}{2},k-\frac{1}{2}} T_Z\left(F_{p,m}^{\textcircled{1}}\cdot F_{p,k+1,k}^{\textcircled{2}}\right)$ $T_X=\sum_{m=i+\frac{1}{2},i-\frac{1}{2}}\frac{\Delta y_j\Delta z_k}{\Delta x_m}$ $T_Y=\sum_{m=j+\frac{1}{2},j-\frac{1}{2}}\frac{\Delta x_i\Delta z_k}{\Delta y_m}$ $T_Z=\sum_{m=k+\frac{1}{2},k-\frac{1}{2}}\frac{\Delta x_i\Delta y_j}{\Delta z_m}$ $F_{p,m}^{\textcircled{1}}=\left(\rho_p\frac{KK_{rp}}{\mu_p}\right)_m^{n+1}$ $F_{p,i+1,i}^{\textcircled{2}}=\left(P_{i+1}^{n+1}-P_i^{n+1}\right)-\gamma_{pg,m}^{n+1}\left(D_{i+1}-D_i\right)$ $F_{p,j+1,j}^{\textcircled{2}}=\left(P_{j+1}^{n+1}-P_j^{n+1}\right)-\gamma_{pg,m}^{n+1}\left(D_{j+1}-D_j\right)$ $F_{p,k+1,k}^{\textcircled{2}}=\left(P_{k+1}^{n+1}-P_k^{n+1}\right)-\gamma_{pg,m}^{n+1}\left(D_{k+1}-D_k\right)$	$\sum_j T_{ij}\cdot F_{p,ij}=\sum_j T_{ij}\cdot\left(F_{p,ij}^{\textcircled{1}}\cdot F_{p,ij}^{\textcircled{2}}\right)$ $T_{i,j}=\frac{\alpha_i\alpha_j}{\alpha_i+\alpha_j}$ $\alpha=A\frac{K}{L}\boldsymbol{n}\cdot\boldsymbol{f}$ $F_{p,ij}^{\textcircled{1}}=\left(\frac{K_{rp}\rho_p}{\mu_p}\right)^{n+1}$ $F_{p,i,j}^{\textcircled{2}}=\left(P_j^{n+1}-P_i^{n+1}\right)-\gamma_{pg,m}^{n+1}\left(D_j-D_i\right)$
	Source and sink term	$\mathrm{WI}_i\cdot F_{p,i}^{W}=\mathrm{WI}_i\cdot\left(F_{p,i}^{W\textcircled{1}}\cdot F_{p,i}^{W\textcircled{2}}\right)$ $\mathrm{WI}=\frac{2\pi\sqrt{K_xK_y}\Delta z}{\ln\left(\frac{r_e}{r_w}\right)+s}$ $r_e=0.28\frac{\left(\left(\frac{K_y}{K_x}\right)^{\frac{1}{2}}\Delta x^2+\left(\frac{K_x}{K_y}\right)^{\frac{1}{2}}\Delta y^2\right)^{\frac{1}{2}}}{\left(\frac{K_y}{K_x}\right)^{\frac{1}{4}}+\left(\frac{K_x}{K_y}\right)^{\frac{1}{4}}}$ 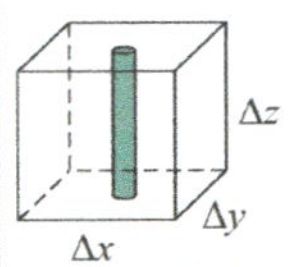 $F_{p,i}^{W\textcircled{1}}=\left(\frac{K_{rp}\rho_p}{\mu_p}\right)^{n+1}$ $F_{p,i}^{W\textcircled{2}}=\left[\left(P_{p,i}-\rho_p gD_i\right)-\left(P_{wfk}-\rho_p gD_k{}^{W}\right)\right]^{n+1}$	$\mathrm{WI}_i\cdot F_{p,i}^{W}=\mathrm{WI}_i\cdot\left(F_{p,i}^{W\textcircled{1}}\cdot F_{p,i}^{W\textcircled{2}}\right)$ $\mathrm{WI}=\frac{2\pi KV^{1/3}}{\ln\left(\frac{r_e}{r_w}\right)+s}$ $r_e=0.2V^{1/3}$ 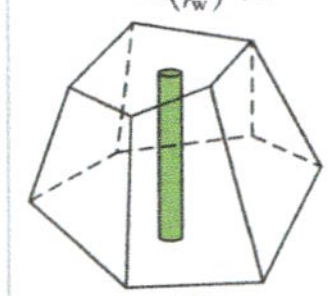 $F_{p,i}^{W\textcircled{1}}=\left(\frac{K_{rp}\rho_p}{\mu_p}\right)^{n+1}$ $F_{p,i}^{W\textcircled{2}}=\left[\left(P_{p,i}-\rho_p gD_i\right)-\left(P_{wfk}-\rho_p gD_k{}^{W}\right)\right]^{n+1}$
	Accumulation term	$A_{p,i}^{C}=\frac{\Delta x_i\Delta y_j\Delta z_k}{\Delta t}\left[\left(\phi\rho_p S_p\right)^{n+1}-\left(\phi\rho_p S_p\right)^{n}\right]$	$A_{p,i}^{C}=\frac{V_i}{\Delta t}\left[\left(\phi S_p\rho_p\right)^{n+1}-\left(\phi S_p\rho_p\right)^{n}\right]$

complex fluid flow in multiple states. The discrete expressions reflecting the geometric features are basically the same, but the discrete expressions for the mechanisms of complex fluid flow in multiple states are very different. In the tight oil and gas production process, the flow mechanisms can be divided into conventional flow mechanisms and unconventional flow mechanisms. The conventional flow mechanisms include viscous flow, gravity drive, rock compression, fluid elastic expansion, and solution gas drive. The flow states for conventional flow usually manifest as a single type of Darcy flow. The unconventional flow mechanisms include threshold pressure gradients, slippage effects, diffusion effects, desorption, and imbibition. The flow states can be considered as high-speed nonlinear flow, pseudo-linear flow, low-speed nonlinear flow and other complex flow states. When the finite volume discretization method is performed, a discrete numerical model for the unconventional flow mechanisms is quite different from a discrete model for the conventional flow mechanisms (Karimi-Fard et al. 2003).

4.3.2.1 Finite Volume Discretization Numerical Model Based on Conventional Flow Mechanisms

For the tight oil and gas reservoirs with coexistence of oil, gas and water, the conventional flow mechanisms mainly include viscous flow, gravity drive, rock compression, fluid elastic expansion, and solution gas drive. The flow characteristics are usually expressed as a single type of Darcy flow. The finite volume discretization method is used to form the finite volume discretization numerical model based on the conventional flow mechanisms.

According to the principle of mass conservations, the general formula of the finite volume discretization numerical model based on the conventional flow mechanisms is

$$\sum_j T_{ij}\cdot F_{\mathrm{c},ij} + \mathrm{WI}_i \cdot F^{\mathrm{W}}_{\mathrm{c},i} = A^{\mathrm{C}}_{\mathrm{c},i} \tag{4.10}$$

The discretized numerical model consists of three parts: the flow term, the source and sink term, and the accumulation term. The detailed expression is as follows.

1. Expressions flow term, accumulation term, and source and sink term for oil component

(1) Flow term

The flow term consists of two parts, which are the transmissibility that reflects the ability to flow for different media, and the influence of different flow mechanisms on the flowability.

$$\begin{aligned}
&\sum_j T_{ij}\cdot F_{\mathrm{o},ij} = \sum_j T_{ij}\cdot\left(F^1_{\mathrm{o},ij}\cdot F^2_{\mathrm{o},ij}\right)\\
&F^1_{\mathrm{o},ij} = \left(\frac{K_{\mathrm{ro}}}{\mu_{\mathrm{o}}}\frac{\rho_{\mathrm{osc}}}{B_{\mathrm{o}}}\right)^{n+1} = \rho_{\mathrm{Osc}}\begin{bmatrix} (\frac{K_{\mathrm{ro}}}{\mu_{\mathrm{o}}B_{\mathrm{o}}})^l - (\frac{K_{\mathrm{ro}}}{B_{\mathrm{o}}}\frac{\partial \mu_{\mathrm{o}}}{\partial P} + \frac{K_{\mathrm{ro}}}{\mu_{\mathrm{o}}}\frac{\partial B_{\mathrm{o}}}{\partial P})^l \\ \delta P_{\mathrm{o},i|j} + (\frac{1}{\mu_{\mathrm{o}}B_{\mathrm{o}}})^l(\frac{\partial K_{\mathrm{ro}}}{\partial S_{\mathrm{w}}}\delta S_{\mathrm{w},i|j} + \frac{\partial K_{\mathrm{ro}}}{\partial S_{\mathrm{g}}}\delta S_{\mathrm{g},i|j})^l \end{bmatrix}\\
&F^2_{\mathrm{o},ij} = \left[(P_{\mathrm{o},j} - \rho_{\mathrm{o}} g D_j) - (P_{\mathrm{o},i} - \rho_{\mathrm{o}} g D_i)\right]^{n+1}\\
&= (P^l_{\mathrm{o},j} - P^l_{\mathrm{o},i} + \delta P_{\mathrm{o},j} - \delta P_{\mathrm{o},i}) - (\frac{\rho_{\mathrm{osc}} + \rho_{\mathrm{gsc}} R_{\mathrm{g,o}}}{B_{\mathrm{o}}})^l g(D_j - D_i)\\
&- [\frac{\rho_{\mathrm{gsc}}}{B_{\mathrm{o}}}\frac{\partial R_{\mathrm{g,o}}}{\partial P} - (\rho_{\mathrm{osc}} + \rho_{\mathrm{gsc}} R_{\mathrm{g,o}})\frac{\partial B_{\mathrm{o}}}{\partial P}]^l g(D_j \delta P_{\mathrm{o},j} - D_i \delta P_{\mathrm{o},i})
\end{aligned} \tag{4.11}$$

where, ρ_{osc} represents the density of the oil component at the ground standard condition, g/cm^3; ρ_{gsc} is the density of the gas component at the standard condition on the ground, g/cm^3; $R_{\mathrm{g,o}}$ denotes the solution gas-oil ratio, cm^3/cm^3; B_{o} is volume factor for the oil component, dimensionless; T_{ij} indicates the transmissibility between i and j grids.

(2) Accumulation term

$$\begin{aligned}
A^{\mathrm{C}}_{\mathrm{o},i} &= \frac{V_i}{\Delta t}(\phi^{n+1} S^{n+1}_{\mathrm{o}} \frac{\rho_{\mathrm{osc}}}{B^{n+1}_{\mathrm{o}}} - \phi^n S^n_{\mathrm{o}} \frac{\rho_{\mathrm{osc}}}{B^n_{\mathrm{o}}})_i\\
&= \frac{V_i}{\Delta t}\rho_{\mathrm{osc}}\begin{bmatrix} (\frac{\phi S_{\mathrm{o}}}{B_{\mathrm{o}}})^l - (\frac{\phi S_{\mathrm{o}}}{B_{\mathrm{o}}})^n + (\frac{S_{\mathrm{o}}}{B_{\mathrm{o}}}\frac{\partial\phi}{\partial P} - \phi S_{\mathrm{o}}\frac{\partial B_{\mathrm{o}}}{\partial P})^l \delta P_{\mathrm{o},i} \\ -(\frac{\phi}{B_{\mathrm{o}}})^l \delta S_{\mathrm{g},i} - (\frac{\phi}{B_{\mathrm{o}}})^l \delta S_{\mathrm{w},i} \end{bmatrix}
\end{aligned} \tag{4.12}$$

where the superscript n represents the n-th time step, which is dimensionless; $n + 1$ represents the $(n + 1)$-th time step, which is also dimensionless; l represents the dimensionless l-th Newton iteration step.

(3) Source and sink term

$$\mathrm{WI}_i \cdot F^{\mathrm{W}}_{\mathrm{o},i} = \mathrm{WI}_i \cdot \left(F^{\mathrm{W}①}_{\mathrm{o},i} \cdot F^{\mathrm{W}②}_{\mathrm{o},i}\right) \tag{4.13}$$

$$\begin{aligned}
F^{\mathrm{W1}}_{\mathrm{o},i} &= \left(\frac{K_{\mathrm{ro}}}{\mu_{\mathrm{o}}}\frac{\rho_{\mathrm{osc}}}{B_{\mathrm{o}}}\right)^{n+1}\\
&= \rho_{\mathrm{osc}}\begin{bmatrix} (\frac{K_{\mathrm{ro}}}{\mu_{\mathrm{o}}B_{\mathrm{o}}})^l - (\frac{K_{\mathrm{ro}}}{B_{\mathrm{o}}}\frac{\partial \mu_{\mathrm{o}}}{\partial P} + \frac{K_{\mathrm{ro}}}{\mu_{\mathrm{o}}}\frac{\partial B_{\mathrm{o}}}{\partial P})^l \delta P_{\mathrm{o},i} \\ + (\frac{1}{\mu_{\mathrm{o}}B_{\mathrm{o}}})^l(\frac{\partial K_{\mathrm{ro}}}{\partial S_{\mathrm{w}}}\delta S_{\mathrm{w},i} + \frac{\partial K_{\mathrm{ro}}}{\partial S_{\mathrm{g}}}\delta S_{\mathrm{g},i})^l \end{bmatrix}
\end{aligned} \tag{4.14}$$

$$\begin{aligned}
F^{\mathrm{W2}}_{\mathrm{o},i} &= \left[(P_{\mathrm{o},i} - \rho_{\mathrm{o}} g D_i) - (P_{\mathrm{wfk}} - \rho_{\mathrm{o}} g D^{\mathrm{W}}_{\mathrm{k}})\right]^{n+1}\\
&= -\begin{bmatrix} (P^l_{\mathrm{wfk}} - P^l_{\mathrm{o},i} + \delta P_{\mathrm{wfk}} - \delta P_{\mathrm{o},i}) - (\frac{\rho_{\mathrm{osc}} + \rho_{\mathrm{gsc}} R_{\mathrm{g,o}}}{B_{\mathrm{o}}})^l g(D^{\mathrm{W}}_{\mathrm{k}} \\ -D_i) - [\frac{\rho_{\mathrm{gsc}}}{B_{\mathrm{o}}}\frac{\partial R_{\mathrm{g,o}}}{\partial P} - (\rho_{\mathrm{osc}} + \rho_{\mathrm{gsc}} R_{\mathrm{g,o}})\frac{\partial B_{\mathrm{o}}}{\partial P}]^l g(D^{\mathrm{W}}_{\mathrm{k}} \delta P_{\mathrm{wfk}} - D_i \delta P_{\mathrm{o},i}) \end{bmatrix}
\end{aligned} \tag{4.15}$$

where, P_{wf} represents the bottom hole pressure, MPa; the superscript W means such term is related to a wellbore.

2. Expressions of flow term, accumulation term, and source and sink term of the gas component

(1) Flow term

$$\sum_j T_{ij} \cdot F_{g,ij} = \sum_j T_{ij} \cdot \left(F_{g,ij}^{\textcircled{1}} \cdot F_{g,ij}^{\textcircled{2}} + F_{go,ij}^{\textcircled{1}} \cdot F_{go,ij}^{\textcircled{2}}\right) \tag{4.16}$$

$$\begin{aligned} F_{g,ij}^{\textcircled{1}} &= \left(\frac{K_{rg}}{\mu_g}\frac{\rho_{gsc}}{B_g}\right)^{n+1} \\ &= \rho_{gsc} \cdot \left[\left(\frac{K_{rg}}{\mu_g B_g}\right)^l - \left(\frac{K_{rg}}{B_g}\frac{\partial \mu_g}{\partial P} + \frac{K_{rg}}{\mu_g}\frac{\partial B_g}{\partial P}\right)^l \delta P_{o,i|j} + \left(\frac{1}{\mu_g B_g}\frac{\partial K_{rg}}{\partial S_g}\right)^l \delta S_{g,i|j} \right] \end{aligned} \tag{4.17}$$

$$\begin{aligned} F_{g,ij}^{\textcircled{2}} &= \left[\left(P_{g,j} - \frac{\rho_{gsc}}{B_g} g D_j\right) - \left(P_{g,i} - \frac{\rho_{gsc}}{B_g} g D_i\right) \right]^{n+1} \\ &= P_{o,j}^l + P_{cog,j}^l - P_{o,i}^l - P_{cog,i}^l - \frac{\rho_{gsc} g (D_j - D_i)}{B_g^l} + \delta P_{o,j} - \delta P_{o,i} \\ &\quad + \frac{\partial P_{cog,j}}{\partial S_g} \delta S_{g,j} - \frac{\partial P_{cog,i}}{\partial S_g} \delta S_{g,i} + \frac{\partial B_g}{\partial P} \rho_{gsc} g (D_j \delta P_{o,j} - D_i \delta P_{o,i}) \end{aligned} \tag{4.18}$$

$$\begin{aligned} F_{go,ij}^{\textcircled{1}} &= \left(\frac{K_{ro}}{\mu_o}\frac{\rho_{gsc} R_{g,o}}{B_o}\right)^{n+1} \\ &= \rho_{gsc} \left[\begin{array}{c} \left(\frac{K_{ro} R_{g,o}}{\mu_o B_o}\right)^l + \left(\frac{K_{ro}}{\mu_o B_o}\frac{\partial R_{g,o}}{\partial P} - \frac{K_{ro} R_{g,o}}{B_o}\frac{\partial \mu_o}{\partial P} - \frac{K_{ro} R_{g,o}}{\mu_o}\frac{\partial B_o}{\partial P}\right)^l \delta P_{o,i|j} \\ + \left(\frac{R_{g,o}}{\mu_o B_o}\right)^l \left(\frac{\partial K_{ro}}{\partial S_w} \delta S_{w,i|j} + \frac{\partial K_{ro}}{\partial S_g} \delta S_{g,i|j}\right)^l \end{array} \right] \end{aligned} \tag{4.19}$$

$$\begin{aligned} F_{go,ij}^{\textcircled{2}} &= \left[(P_{o,j} - P_{o,i}) - \frac{(\rho_{osc} + \rho_{gsc} R_{g,o})}{B_o} g (D_j - D_i) \right]^{n+1} \\ &= P_{o,j}^l - P_{o,i}^l - \left(\frac{\rho_{osc} + \rho_{gsc} R_{g,o}}{B_o}\right)^l g (D_j - D_i) + \delta P_{o,j} - \delta P_{o,i} \\ &\quad - \left[\frac{\rho_{gsc}}{B_o}\frac{\partial R_{g,o}}{\partial P} - (\rho_{osc} + \rho_{gsc} R_{g,o}) \frac{\partial B_o}{\partial P}\right]^l g (D_j \delta P_{o,j} - D_i \delta P_{o,i}) \end{aligned} \tag{4.20}$$

(2) Accumulation term

$$\begin{aligned} A_g^C &= \frac{V_i}{\Delta t} \left[\begin{array}{c} \left(\phi^{n+1} S_g^{n+1} \frac{\rho_{gsc}}{B_g^{n+1}} - \phi^n S_g^n \frac{\rho_{gsc}}{B_g^n}\right) \\ + \left(\phi^{n+1} S_o^{n+1} \frac{\rho_{gsc} R_{g,o}^{n+1}}{B_o^{n+1}} - \phi^n S_o^n \frac{\rho_{gsc} R_{g,o}^n}{B_o^n}\right) \end{array} \right] \\ &= \frac{V_i}{\Delta t} \rho_{gsc} \left[\begin{array}{c} \left(\frac{\phi S_g}{B_g}\right)^l - \left(\frac{\phi S_g}{B_g}\right)^n + \left(\frac{\phi S_o R_{g,o}}{B_o}\right)^l - \left(\frac{\phi S_o R_{g,o}}{B_o}\right)^n \\ + \left(\frac{S_g}{B_g}\frac{\partial \phi}{\partial P} - \phi S_g \frac{\partial B_g}{\partial P}\right)^l \delta P_{o,i} + \left(\frac{\phi}{B_g}\right)^l \delta S_{g,i} \end{array} \right] \\ &\quad + \frac{V_i}{\Delta t} \rho_{gsc} \left[\begin{array}{c} \left(\frac{S_o R_{g,o}}{B_o}\frac{\partial \phi}{\partial P} + \frac{\phi S_o}{B_o}\frac{\partial R_{g,o}}{\partial P} - \phi S_o R_{g,o} \frac{\partial B_o}{\partial P}\right)^l \delta P_{o,i} \\ - \left(\frac{\phi R_{g,o}}{B_o}\right)^l \delta S_{w,i} - \left(\frac{\phi R_{g,o}}{B_o}\right)^l \delta S_{g,i} \end{array} \right] \end{aligned} \tag{4.21}$$

(3) Source and sink term

$$WI_i F_{g,ij}^{W} = WI_i \left(F_{g,ij}^{W\textcircled{1}} \cdot F_{g,ij}^{W\textcircled{2}} + F_{go,ij}^{W\textcircled{1}} \cdot F_{go,ij}^{W\textcircled{2}}\right) \tag{4.22}$$

$$\begin{aligned} F_{g,ij}^{W\textcircled{1}} &= \left(\frac{K_{rg}}{\mu_g}\frac{\rho_{gsc}}{B_g}\right)^{n+1} \\ &= \rho_{gsc} \left[\begin{array}{c} \left(\frac{K_{rg}}{\mu_g B_g}\right)^l - \left(\frac{K_{rg}}{B_g}\frac{\partial \mu_g}{\partial P} + \frac{K_{rg}}{\mu_g}\frac{\partial B_g}{\partial P}\right)^l \delta P_{o,i} \\ + \left(\frac{1}{\mu_g B_g}\frac{\partial K_{rg}}{\partial S_g}\right)^l \delta S_{g,i} \end{array} \right] \end{aligned} \tag{4.23}$$

$$\begin{aligned} F_{g,ij}^{W\textcircled{2}} &= -\left[\left(P_{wfk} - \frac{\rho_{gsc}}{B_g} g D_k^W\right) - \left(P_{g,i} - \frac{\rho_{gsc}}{B_g} g D_i\right) \right]^{n+1} \\ &= -\left[\begin{array}{c} P_{wfk}^l - P_{o,i}^l - P_{cog,i}^l - \frac{\rho_{gsc} g (D_k^W - D_i)}{B_g^l} + \delta P_{wfk} \\ -\delta P_{o,i} - \frac{\partial P_{cog,i}}{\partial S_g} \delta S_{g,i} + \frac{\partial B_g}{\partial P} \rho_{gsc} g (D_k^W \delta P_{wfk} - D_i \delta P_{o,i}) \end{array} \right] \end{aligned} \tag{4.24}$$

$$\begin{aligned} F_{go,ij}^{W\textcircled{1}} &= \left(\frac{K_{ro}}{\mu_o}\frac{\rho_{gsc} R_{g,o}}{B_o}\right)^{n+1} \\ &= \rho_{gsc} \left[\begin{array}{c} \left(\frac{K_{ro} R_{g,o}}{\mu_o B_o}\right)^l + \left(\frac{K_{ro}}{\mu_o B_o}\frac{\partial R_{g,o}}{\partial P} - \frac{K_{ro} R_{G,o}}{B_o}\frac{\partial \mu_o}{\partial P}\right. \\ \left. - \frac{K_{ro} R_{g,o}}{\mu_o}\frac{\partial B_o}{\partial P}\right)^l \delta P_{o,i} + \left(\frac{R_{g,o}}{\mu_o B_o}\right)^l \left(\frac{\partial K_{ro}}{\partial S_w} \delta S_{w,i} + \frac{\partial K_{ro}}{\partial S_g} \delta S_{g,i}\right)^l \end{array} \right] \end{aligned} \tag{4.25}$$

$$\begin{aligned} F_{go,ij}^{W2} &= -\left[(P_{wfk} - P_{o,i}) - \frac{(\rho_{osc} + \rho_{gsc} R_{g,o})}{B_o} g (D_k^W - D_i) \right]^{n+1} \\ &= -\left[\begin{array}{c} P_{wfk}^l - P_{o,i}^l - \left(\frac{\rho_{osc} + \rho_{gsc} R_{g,o}}{B_o}\right)^l g (D_k^W - D_i) + \delta P_{wfk} - \delta P_{o,i} \\ -\left[\frac{\rho_{gsc}}{B_o}\frac{\partial R_{g,o}}{\partial P} - (\rho_{osc} + \rho_{gsc} R_{g,o}) \frac{\partial B_o}{\partial P}\right]^l g (D_k^W \delta P_{wfk} - D_i \delta P_{o,i}) \end{array} \right] \end{aligned} \tag{4.26}$$

3. Expressions of flow term, accumulation term, and source and sink term of the water component

(1) Flow term

$$\sum_j T_{ij} \cdot F_{w,ij} = \sum_j T_{ij} \cdot \left(F_{w,ij}^{\textcircled{1}} \cdot F_{w,ij}^{\textcircled{2}}\right) \tag{4.27}$$

$$\begin{aligned} F_{w,ij}^{\textcircled{1}} &= \left(\frac{K_{rw}}{\mu_w}\frac{\rho_{wsc}}{B_w}\right)^{n+1} \\ &= \rho_{wsc} \left[\begin{array}{c} \left(\frac{K_{rw}}{\mu_w B_w}\right)^l - \left(\frac{K_{rw}}{B_w}\frac{\partial \mu_w}{\partial P} + \frac{K_{rw}}{\mu_w}\frac{\partial B_w}{\partial P}\right)^l \delta P_{o,i|j} \\ + \left(\frac{1}{\mu_w B_w}\frac{\partial K_{rw}}{\partial S_w}\right)^l \delta S_{w,i|j} \end{array} \right] \end{aligned} \tag{4.28}$$

$$\begin{aligned} F_{w,ij}^{\textcircled{2}} &= \left[\left(P_{w,j} - \frac{\rho_{wsc}}{B_w} g D_j\right) - \left(P_{w,i} - \frac{\rho_{wsc}}{B_w} g D_i\right) \right]^{n+1} \\ &= P_{o,j}^l - P_{o,i}^l - P_{cow,j}^l + P_{cow,i}^l - \frac{\rho_{wsc}}{B_w^l} g (D_j - D_i) + \left(1 + \frac{\partial B_w}{\partial P} \rho_{wsc} g D_j\right) \\ &\quad \delta P_{o,j} - \left(1 + \frac{\partial B_w}{\partial P} \rho_{wsc} g D_i\right) \delta P_{o,i} - \frac{\partial P_{cow,j}}{\partial S_w} \delta S_{w,j} + \frac{\partial P_{cow,i}}{\partial S_w} \delta S_{w,i} \end{aligned} \tag{4.29}$$

(2) Accumulation term

$$
\begin{aligned}
A_{\mathrm{w}}^{\mathrm{C}} &= \frac{V_i}{\Delta t}\left(\phi^{n+1} S_{\mathrm{w}}^{n+1} \frac{\rho_{\mathrm{wsc}}}{B_{\mathrm{w}}^{n+1}} - \phi^{n} S_{\mathrm{w}}^{n} \frac{\rho_{\mathrm{wsc}}}{B_{\mathrm{w}}^{n}}\right) \\
&= \frac{V_i}{\Delta t}\rho_{\mathrm{osc}}\begin{bmatrix} \left(\frac{\phi S_{\mathrm{w}}}{B_{\mathrm{w}}}\right)^l - \left(\frac{\phi S_{\mathrm{w}}}{B_{\mathrm{w}}}\right)^n + \left(\frac{S_{\mathrm{w}}}{B_{\mathrm{w}}}\frac{\partial \phi}{\partial P} - \phi S_{\mathrm{w}}\frac{\partial B_{\mathrm{w}}}{\partial P}\right)^l \delta P_{\mathrm{o},i} \\ + \left(\frac{\phi}{B_{\mathrm{w}}}\right)^l \delta S_{\mathrm{w},i} \end{bmatrix}
\end{aligned}
\tag{4.30}
$$

(3) Source and sink term

$$
WI_i \cdot F_{\mathrm{w},i}^{\mathrm{W}} = WI_i \cdot \left(F_{\mathrm{w},i}^{\mathrm{W}①} \cdot F_{\mathrm{w},i}^{\mathrm{W}②}\right) \tag{4.31}
$$

$$
\begin{aligned}
F_{\mathrm{w},i}^{\mathrm{W}①} &= \left(\frac{K_{\mathrm{rw}}}{\mu_{\mathrm{w}}}\frac{\rho_{\mathrm{wsc}}}{B_{\mathrm{w}}}\right)^{n+1} \\
&= \rho_{\mathrm{wsc}}\begin{bmatrix} \left(\frac{K_{\mathrm{rw}}}{\mu_{\mathrm{w}} B_{\mathrm{w}}}\right)^l - \left(\frac{K_{\mathrm{rw}}}{B_{\mathrm{w}}}\frac{\partial \mu_{\mathrm{w}}}{\partial P} + \frac{K_{\mathrm{rw}}}{\mu_{\mathrm{w}}}\frac{\partial B_{\mathrm{w}}}{\partial P}\right)^l \delta P_{\mathrm{o},i} \\ + \left(\frac{1}{\mu_{\mathrm{w}} B_{\mathrm{w}}}\frac{\partial K_{\mathrm{rw}}}{\partial S_{\mathrm{w}}}\right)^l \delta S_{\mathrm{w},i} \end{bmatrix}
\end{aligned}
\tag{4.32}
$$

$$
\begin{aligned}
F_{\mathrm{w},i}^{\mathrm{W}②} &= -\left[\left(P_{\mathrm{wfk}} - \frac{\rho_{\mathrm{wsc}}}{B_{\mathrm{w}}} g D_{\mathrm{k}}^{\mathrm{W}}\right) - \left(P_{\mathrm{w},i} - \frac{\rho_{\mathrm{wsc}}}{B_{\mathrm{w}}} g D_i\right)\right]^{n+1} \\
&= -\left[P_{\mathrm{wfk}}^l - P_{\mathrm{o},i}^l + P_{\mathrm{cow},i}^l - \frac{\rho_{\mathrm{wsc}}}{B_{\mathrm{w}}^l} g\left(D_{\mathrm{k}}^{\mathrm{W}} - D_i\right)\right] \\
&\quad - \begin{bmatrix} \left(1 + \frac{\partial B_{\mathrm{w}}}{\partial P}\rho_{\mathrm{wsc}} g D_{\mathrm{k}}^{\mathrm{W}}\right)\delta P_{\mathrm{wfk}} - \left(1 + \frac{\partial B_{\mathrm{w}}}{\partial P}\rho_{\mathrm{wsc}} g D_i\right)\delta P_{\mathrm{o},i} \\ - \frac{\partial P_{\mathrm{cow},j}}{\partial S_{\mathrm{w}}}\delta S_{\mathrm{w},j} + \frac{\partial P_{\mathrm{cow},i}}{\partial S_{\mathrm{w}}}\delta S_{\mathrm{w},i} \end{bmatrix}
\end{aligned}
\tag{4.33}
$$

1. Expressions for residuals in finite volume discretization numerical models

When we have $\phi_{\mathrm{o},i} > \phi_{\mathrm{o},j}$, $\phi_{\mathrm{g},i} > \phi_{\mathrm{g},j}$, and $\phi_{\mathrm{w},i} > \phi_{\mathrm{w},j}$, the expressions in residual in the finite volume discretization numerical model of the oil, gas and water components shown previously are as follows.

(1) Expression for the residual of the oil component

$$
\begin{aligned}
R_{\mathrm{o}i} &= (b_1 + b_2\delta P_{\mathrm{o},i} + b_3\delta S_{\mathrm{g},i} + b_4\delta S_{\mathrm{w},i}) \\
&\quad - \sum_j T_{ij}(a_1 + a_2\delta P_{\mathrm{o},j} + a_3\delta P_{\mathrm{o},i} + a_4\delta S_{\mathrm{g},i} + a_5\delta S_{\mathrm{w},i}) \\
&\quad + (c_1 + c_2\delta P_{\mathrm{wfk}} + c_3\delta P_{\mathrm{o},i} + c_4\delta S_{\mathrm{g},i} + c_5\delta S_{\mathrm{w},i})
\end{aligned}
\tag{4.34}
$$

(2) Expression for the residual of the gas component

$$
\begin{aligned}
R_{\mathrm{g}i} &= (b_1' + b_2'\delta P_{\mathrm{o},i} + b_3'\delta S_{\mathrm{g},i} + b_4'\delta S_{\mathrm{w},i}) \\
&\quad - \sum_j T_{ij}\cdot(a_1' + a_2'\delta P_{\mathrm{o},j} + a_3'\delta P_{\mathrm{o},i} + a_4'\delta S_{\mathrm{g},j} + a_5'\delta S_{\mathrm{g},i} + a_6'\delta S_{\mathrm{w},i}) \\
&\quad + (c_1' + c_2'\delta P_{\mathrm{wfk}} + c_3'\delta P_{\mathrm{o},i} + c_4'\delta S_{\mathrm{g},i} + c_5'\delta S_{\mathrm{w},i})
\end{aligned}
\tag{4.35}
$$

(3) Expression for the residual of the water component

$$
\begin{aligned}
R_{\mathrm{w}i} &= (b_1'' + b_2''\delta P_{\mathrm{o},i} + b_3''\delta S_{\mathrm{w},i}) \\
&\quad - \sum_j T_{ij}(a_1'' + a_2''\delta P_{\mathrm{o},j} + a_3''\delta P_{\mathrm{o},i} + a_4''\delta S_{\mathrm{w},j} + a_5''\delta S_{\mathrm{w},i}) \\
&\quad + (c_1'' + c_2''\delta P_{\mathrm{wfk}} + c_3''\delta P_{\mathrm{o},i} + c_4''\delta S_{\mathrm{w},i})
\end{aligned}
\tag{4.36}
$$

where, R_{ci} represents the residual of fluid component c in the i-th grid block.

2. Coefficient expressions corresponding to all the variables in the coefficient matrix in the discretized model

In the finite volume discretization numerical model, variables such as pressure and saturation are combined. The coefficient expression for each variable is shown in Table 4.17.

4.3.2.2 Finite Volume Discretization Numerical Simulation Model Based on Unconventional Flow Mechanism

Tight reservoirs contain multiple media of pores and fractures at different scales. The geometric and property characteristics of different media vary greatly. The fluid properties and occurrence state in different media are different. At the same time, in different development methods and different development stages, the differences in production conditions lead to great differences in the flow states and flow mechanisms between different media. The unconventional flow mechanisms include the threshold pressure gradient (TPG), slippage effect, diffusion effect, desorption effect, and imbibition effect. The flow state mainly manifests itself in many complex flow states, such as high-speed nonlinear flow, pseudo-linear flow, and low-speed nonlinear flow (Table 4.18). Using the finite volume discretization method, a finite volume discretization numerical simulation model based on the unconventional flow mechanisms is formed.

According to the principle of conservation of mass, the general formula of the finite volume discretization numerical model based on the unconventional flow mechanisms is the same as in Eq. 4.10. This discretized numerical model is also composed of three parts: the flow term, the source and sink term, and the accumulation term. The discretization expressions that embody the geometric characteristics are consistent with the numerical models based on conventional mechanism. However, due to the influence of

Table 4.17 The expressions of the coefficients corresponding to all the variables of the coefficient matrix of the discretized numerical model based on the conventional mechanisms of flow in porous media

Variable	Coefficients corresponding to all the variables in the coefficient matrix		
	Oil components	Gas components	Water components
$\delta P_{o,i}$	$b_2 - \sum_j T_{ij} a_3 + c_3$	$b'_2 - \sum_j T_{ij} a'_3 + c'_3$	$b''_2 - \sum_j T_{ij} a''_3 + c''_3$
$\delta S_{g,i}$	$b_3 - \sum_j T_{ij} a_4 + c_4$	$b'_3 - \sum_j T_{ij} a'_5 + c'_4$	–
$\delta S_{w,i}$	$b_4 - \sum_j T_{ij} a_5 + c_5$	$b'_4 - \sum_j T_{ij} a'_6 + c'_5$	$b''_3 - \sum_j T_{ij} a''_5 + c''_4$
$\delta P_{o,j}$	$-\sum_j T_{ij} a_2$	$-\sum_j T_{ij} a_2{}'$	$-\sum_j T_{ij} a''_2$
$\delta S_{g,j}$	–	$-\sum_j T_{ij} a_4{}'$	–
$\delta S_{w,j}$	–	–	$-\sum_j T_{ij} a''_4$
δP_{wfk}	c_2	c'_2	c''_2

Note The coefficients of the solution variables, such as constants, pressures, and saturations, etc., in the flow term, accumulation term, and the source/sink term, in the oil composition equation are a1 to a_5, b_1 to b_4, c_1 to c_5, respectively; the coefficients of the solution variables, such as the constants, pressure, and saturation, etc., in the flow term, accumulation term, and source/sink term in the gas composition equation are a'_1-a'_6, b'_1-b'_4, c'_1-c'_5, respectively; the coefficients of the solution variables, such as constants, pressures, and saturations in the flow term, accumulation term, and source and sink term are a''_1 to a''_5, b''_1 to b''_3, and c''_1 to c''_4

Table 4.18 Mathematical representation of different flow mechanisms in discretized numerical models based on unconventional flow mechanisms

Flow states and flow mechanism		Involved components: oil, gas, or water	Terms changed of the discretized model
High-speed nonlinear flow		Oil, gas	Flow term, source and sink term
Pseudo-linear flow		Oil, gas	
Low-speed nonlinear flow	Threshold pressure gradient	Oil, gas	
	Slippage effect	Gas	
	Diffusion effect	Gas	
	Desorption effect	Gas	Accumulation term
	Imbibition effect	Oil, water	Flow term
Flow-geomechanics coupling (pressure sensitivity)		Oil, gas, water	Flow term, source and sink term

unconventional flow mechanisms, the specific expression of a discretized numerical model has changed to a large extent. According to the differences of the flow states and the flow mechanisms, the finite volume discretization numerical models based on different unconventional flow mechanisms are, respectively, elaborated.

1. Discretized numerical model for high-speed nonlinear flow

Under large pressure gradients, flow of oil and gas in large-scale media with different scale pores and fractures can exhibit high-speed nonlinear flow, which can be reflected in the flow term and source and sink term of the oil and gas components in the discretized numerical simulation models. It reflects the influence of high-speed nonlinear flow on the flow dynamics and productivity of different media.

(1) Flow term

In the flow term, the transmissibility reflects the geometric and physical characteristics of different media and is not affected by changes in the flow mechanisms. Therefore, the effect of high-speed nonlinear flow on the flow term is mainly reflected in the influence of different flow mechanisms on the flow capability, which is specifically represented by the addition of high-speed nonlinear turbulence terms. Its expression is as follows:

$$F_{(\mathrm{ND})\mathrm{p},ij} = F_{\mathrm{ND}}{}^{n+1} \cdot F^{①}_{\mathrm{p},ij} \cdot F^{②}_{\mathrm{p},ij} \tag{4.37}$$

$$F^{①}_{\mathrm{p},ij} = \left(\frac{K_{\mathrm{rp}}\rho_{\mathrm{p}}}{\mu_{\mathrm{p}}}\right)_{i|j}^{n+1} \tag{4.38}$$

$$F^{②}_{\mathrm{p},ij} = (P_{\mathrm{p},j}{}^{n+1} - \rho_{\mathrm{p}}{}^{n+1} gD_j) - (P_{\mathrm{p},i}{}^{n+1} - \rho_{\mathrm{p}}{}^{n+1} gD_i) \tag{4.39}$$

$$F_{\mathrm{ND}} = \frac{-1 + \sqrt{1 + 4(\frac{\beta Tk}{A})(\Delta\phi_{\mathrm{p}})(\frac{\rho_{\mathrm{p}} K_{\mathrm{rp}}}{\mu_{\mathrm{p}}^2})}}{2(\frac{\beta Tk}{A})(\Delta\phi_{\mathrm{p}})(\frac{\rho_{\mathrm{p}} K_{\mathrm{rp}}}{\mu_{\mathrm{p}}^2})} \tag{4.40}$$

(2) Source and sink term

In the source and sink term, the well index reflects the geometric and physical characteristics of different well types and reservoir media and is not affected by changes in the flow mechanisms. Therefore, the influence of high-speed nonlinear flow on the source and sink term is mainly reflected in the influence of different flow mechanisms on the flow capability, which is specifically represented by the addition of high-speed nonlinear turbulence terms. Its expression is as follows:

$$F^{\mathrm{W}}_{(\mathrm{ND})\mathrm{p},i} = F_{\mathrm{ND}}{}^{n+1} \cdot F^{\mathrm{W}①}_{\mathrm{p},i} \cdot F^{\mathrm{W}②}_{\mathrm{p},i} \tag{4.41}$$

$$F^{\mathrm{W}①}_{\mathrm{p},i} = \left(\frac{K_{\mathrm{rp}}\rho_{\mathrm{p}}}{\mu_{\mathrm{p}}}\right)_{i}^{n+1} \tag{4.42}$$

$$F_{\mathrm{p},i}{}^{\mathrm{W}②} = (P_{\mathrm{p},i}{}^{n+1} - \rho_{\mathrm{p}}{}^{n+1} gD_i) - (P_{\mathrm{wfk}}{}^{n+1} - \rho_{\mathrm{p}}{}^{n+1} gD_{\mathrm{k}}{}^{\mathrm{W}}) \tag{4.43}$$

Among them, the subscript p = o, g.

2. Discretized numerical model for pseudo-linear flow

Oil and gas in tight reservoirs generally flow in micropores, small pores, and microfractures as pseudo-linear flow, which is reflected in the flow term and source and sink term of the discretized numerical models of the oil and gas components. It reflects the effect of the pseudo-linear flow on the flow dynamics and productivity of different media.

The influence of the pseudo-linear flow on the flow term and the source and sink term is mainly demonstrated in the effect of different flow mechanisms on the flowability, which is reflected as the pseudo-linear pressure term added in a discretized numerical model. The detailed expression is as follows.

(1) Flow term

$$F_{(\mathrm{li})\mathrm{p},ij} = F^{①}_{\mathrm{p},ij} \cdot F^{②}_{(\mathrm{li})\mathrm{p},ij} \tag{4.44}$$

$$F^{①}_{\mathrm{p},ij} = \left(\frac{K_{\mathrm{rp}}\rho_{\mathrm{p}}}{\mu_{\mathrm{p}}}\right)_{i|j}^{n+1} \tag{4.45}$$

$$F^{②}_{(\mathrm{li})\mathrm{p},ij} = (P_{\mathrm{p},j}{}^{n+1} - \rho_{\mathrm{p}}{}^{n+1} gD_j) - (P_{\mathrm{p},i}{}^{n+1} - \rho_{\mathrm{p}}{}^{n+1} gD_i) - G_{\mathrm{c},ij} \tag{4.46}$$

where

$$G_{\mathrm{c},ij} = c(L_i + L_j) \tag{4.47}$$

(2) Source and sink term

$$F^{\mathrm{W}}_{(\mathrm{li})\mathrm{p},i} = F^{\mathrm{W}①}_{\mathrm{p},i} \cdot F^{\mathrm{W}②}_{(\mathrm{li})\ \mathrm{p},i} \tag{4.48}$$

$$F^{\mathrm{W}①}_{\mathrm{p},i} = \left(\frac{K_{\mathrm{rp}}\rho_{\mathrm{p}}}{\mu_{\mathrm{p}}}\right)_{i}^{n+1} \tag{4.49}$$

$$F^{\mathrm{W2}}_{(\mathrm{li})\mathrm{p},i} = (P^{n+1}_{\mathrm{p},i} - \rho^{n+1}_{\mathrm{p}} gD_i) - (P^{n+1}_{\mathrm{wfk}} - \rho^{n+1}_{\mathrm{p}} gD^{\mathrm{W}}_{\mathrm{k}}) - G^{\mathrm{W}}_{\mathrm{c},ij} \tag{4.50}$$

$$G^{\mathrm{W}}_{\mathrm{c},i} = c^{\mathrm{W}}(L_i + r_{\mathrm{w}}) \tag{4.51}$$

where, the subscript c = o, g and p = o, g.

3. Discretized numerical model for low-speed nonlinear flow

Influenced by different flow mechanisms such as a threshold pressure gradient, slippage effect, diffusion effect, desorption effect, and imbibitions effect, all of the fluid flow is characterized by low-speed nonlinear flow. Different flow mechanisms have different expressions in discretized numerical models.

(1) Discretized numerical model considering threshold pressure gradient

A threshold pressure gradient is reflected in the flow term and source and sink term of the oil and gas components in the discretized numerical model, which reflects its influence on the flow dynamics and productivity of different media.

The influence of a threshold pressure gradient on flow term and source and sink term is mainly demonstrated in the influence of different flow mechanisms on the flowability, which is reflected in the threshold pressure term and non-linear indices added in the discretized numerical model. The specific expression is as follows:

① Flow Term

$$F_{(g)p,ij}=F_{p,ij}^{\textcircled{1}}\cdot F_{(g)p,ij}^{\textcircled{2}} \tag{4.52}$$

$$F_{p,ij}^{\textcircled{1}}=\left(\frac{K_{rp}\rho_p}{\mu_p}\right)_{ij}^{n+1} \tag{4.53}$$

$$F_{(g)p,ij}^{\textcircled{2}}=\left[(P_{p,j}{}^{n+1}-\rho_p{}^{n+1}gD_j)-(P_{p,i}{}^{n+1}-\rho_p{}^{n+1}gD_i)-G_{a,ij}\right]^{n^*} \tag{4.54}$$

where

$$G_{a,ij}=a(L_i+L_j) \tag{4.55}$$

② Source and Sink Term

$$F_{(g)p,i}^{W}=F_{p,i}^{W\textcircled{1}}\cdot F_{(g)\,p,i}^{W\textcircled{2}} \tag{4.56}$$

$$F_{p,i}^{W\textcircled{1}}=\left(\frac{K_{rp}\rho_p}{\mu_p}\right)_i^{n+1} \tag{4.57}$$

$$F_{(g)p,i}{}^{W\textcircled{2}}=\left[(P_{p,i}{}^{n+1}-\rho_p{}^{n+1}gD_i)-(P_{wfk}{}^{n+1}-\rho_p{}^{n+1}gD_k{}^{W})-G_{a,i}^{W}\right]^{n^*} \tag{4.58}$$

$$G_{a,i}^{W}=a^{W}(L_i+r_W) \tag{4.59}$$

where, the subscript p = o, g.

(2) Discretized numerical model considering slippage effect

The slippage effect is reflected in the flow term and source and sink term of the gas component in the discretized numerical model, which reflects the effect of slippage on the flow dynamics and productivity of different media in gas reservoirs under low permeability and low pressure conditions.

The effect of a slippage effect on the flow term and source and sink term is mainly demonstrated in the influence of different flow mechanisms on flowability. In a discretized numerical model, the slippage effect term changing with gas phase pressure is added. The specific expression is as follows.

① Flow Term

$$F_{(si)g,ij}=F_{g,ij}^{\textcircled{1}}\cdot F_{(si)g,ij}^{\textcircled{2}}+F_{go,ij}^{\textcircled{1}}\cdot F_{go,ij}^{\textcircled{2}} \tag{4.60}$$

$$F_{g,ij}^{\textcircled{1}}=\left(\frac{K_{rg}}{\mu_g}\frac{\rho_{gsc}}{B_g}\right)^{n+1} \tag{4.61}$$

$$F_{(si)\ g,ij}^{\textcircled{2}}=\left[(1+\frac{b}{P_{g,j}})(P_{g,j}-\frac{\rho_{gsc}}{B_g}gD_j)-(1+\frac{b}{P_{g,i}})(P_{g,i}-\frac{\rho_{gsc}}{B_g}gD_i)\right]^{n+1} \tag{4.62}$$

$$F_{go,ij}^{\textcircled{1}}=\left(\frac{K_{ro}}{\mu_o}\frac{\rho_{gsc}R_{g,o}}{B_o}\right)^{n+1} \tag{4.63}$$

$$F_{go,ij}^{\textcircled{2}}=\left[(P_{o,j}-P_{o,i})-\frac{(\rho_{osc}+\rho_{gsc}R_{g,o})}{B_o}g(D_j-D_i)\right]^{n+1} \tag{4.64}$$

② Source and Sink Term

$$F_{(si)g,i}^{W}=F_{g,i}^{W\textcircled{1}}\cdot F_{(si)g,i}^{W\textcircled{2}}+F_{go,i}^{W\textcircled{1}}\cdot F_{go,i}^{W\textcircled{2}} \tag{4.65}$$

$$F_{g,ij}^{W\textcircled{1}}=\left(\frac{K_{rg}}{\mu_g}\frac{\rho_{gsc}}{B_g}\right)^{n+1} \tag{4.66}$$

$$F_{(si)g,i}^{W\textcircled{2}}=\left[(1+\frac{b}{P_{g,i}})(P_{g,i}-\frac{\rho_{gsc}}{B_g}gD_i)-(P_{wfk}-\frac{\rho_{gsc}}{B_g}gD_k^{W})\right]^{n+1} \tag{4.67}$$

$$F_{go,ij}^{W\textcircled{1}}=\left(\frac{K_{ro}}{\mu_o}\frac{\rho_{gsc}R_{g,o}}{B_o}\right)^{n+1} \tag{4.68}$$

$$F_{go,ij}^{W\textcircled{2}}=\left[(P_{o,i}-P_{wfk})-\frac{(\rho_{osc}+\rho_{gsc}R_{g,o})}{B_o}g(D_i-D_k{}^{W})\right]^{n+1} \tag{4.69}$$

(3) Discretized numerical model considering diffusion effect

Similar to the slippage effect, the diffusion effect is demonstrated in the flow term and source & sink term of the gas component in the discretized numerical model, which reflects the influence of diffusion effect on the flow dynamics and productivity of different media in gas reservoirs under low permeability and low pressure conditions.

The influence of the diffusion effect on the flow term and source and sink term is mainly reflected in the influence of different flow mechanisms on the flowability. In the discretized numerical model, the diffusion effect term changing with gas phase pressure is added. The specific expression is as follows:

① Flow Term

$$F_{(\mathrm{di})\mathrm{g},ij} = F^{①}_{\mathrm{g},ij} \cdot F^{②}_{(\mathrm{di})\mathrm{g},ij} + F^{①}_{\mathrm{go},ij} \cdot F^{②}_{\mathrm{go},ij} \tag{4.70}$$

$$F^{①}_{\mathrm{g},ij} = \left(\frac{K_{\mathrm{rg}}}{\mu_{\mathrm{g}}}\frac{\rho_{\mathrm{gsc}}}{B_{\mathrm{g}}}\right)^{n+1} \tag{4.71}$$

$$F^{2}_{(\mathrm{di})\mathrm{g},ij} = \begin{bmatrix} (1+\frac{32\sqrt{2}\sqrt{RT}\mu_{\mathrm{g}}}{3r\sqrt{\pi M}P_{\mathrm{g},j}})(P_{\mathrm{g},j}-\frac{\rho_{\mathrm{gsc}}}{B_{\mathrm{g}}}gD_j) \\ -(1+\frac{32\sqrt{2}\sqrt{RT}\mu_{\mathrm{g}}}{3r\sqrt{\pi M}P_{\mathrm{g},i}})(P_{\mathrm{g},i}-\frac{\rho_{\mathrm{gsc}}}{B_{\mathrm{g}}}gD_i) \end{bmatrix}^{n+1} \tag{4.72}$$

$$F^{①}_{\mathrm{go},ij} = \left(\frac{K_{\mathrm{ro}}}{\mu_{\mathrm{o}}}\frac{\rho_{\mathrm{gsc}}R_{\mathrm{g,o}}}{B_{\mathrm{o}}}\right)^{n+1} \tag{4.73}$$

$$F^{②}_{\mathrm{go},ij} = \left[(P_{\mathrm{o},j}-P_{\mathrm{o},i}) - \frac{(\rho_{\mathrm{osc}}+\rho_{\mathrm{gsc}}R_{\mathrm{g,o}})}{B_{\mathrm{o}}}g(D_j-D_i)\right]^{n+1} \tag{4.74}$$

② Source and Sink Term

$$F^{\mathrm{W}}_{(\mathrm{di})\mathrm{g},i} = F^{\mathrm{W}①}_{\mathrm{g},i} \cdot F^{\mathrm{W}②}_{(\mathrm{di})\ \mathrm{g},i} + F^{\mathrm{W}①}_{\mathrm{go},i} \cdot F^{\mathrm{W}②}_{\mathrm{go},i} \tag{4.75}$$

$$F^{\mathrm{W}①}_{\mathrm{g},ij} = \left(\frac{K_{\mathrm{rg}}}{\mu_{\mathrm{g}}}\frac{\rho_{\mathrm{gsc}}}{B_{\mathrm{g}}}\right)^{n+1} \tag{4.76}$$

$$F^{\mathrm{W2}}_{(\mathrm{di})\mathrm{g},i} = \left[(1+\frac{32\sqrt{2}\sqrt{RT}\mu_{\mathrm{g}}}{3r\sqrt{\pi M}P_{\mathrm{g},i}})(P_{\mathrm{g},i}-\frac{\rho_{\mathrm{gsc}}}{B_{\mathrm{g}}}gD_i)-(P_{\mathrm{wfk}}-\frac{\rho_{\mathrm{gsc}}}{B_{\mathrm{g}}}gD^{\mathrm{W}}_{\mathrm{k}})\right]^{n+1} \tag{4.77}$$

$$F^{\mathrm{W}①}_{\mathrm{Go},ij} = \left(\frac{K_{\mathrm{ro}}}{\mu_{\mathrm{o}}}\frac{\rho_{\mathrm{Gsc}}R_{\mathrm{G,o}}}{B_{\mathrm{o}}}\right)^{n+1} \tag{4.78}$$

$$F^{\mathrm{W}②}_{\mathrm{Go},ij} = \left[(P_{\mathrm{o},i}-P_{\mathrm{wfk}}) - \frac{(\rho_{\mathrm{Osc}}+\rho_{\mathrm{Gsc}}R_{\mathrm{G,o}})}{B_{\mathrm{o}}}g(D_i-D_{\mathrm{k}}{}^{\mathrm{W}})\right]^{n+1} \tag{4.79}$$

(4) Discretized numerical model considering desorption effect

The desorption effect is demonstrated in the accumulation term of the gas component in the discretized numerical model, reflecting the effect of desorption on the gas volume in multiple media at different scales under low permeability and low pressure conditions. Such a change in gas volume within unit time due to the desorption effect is added in the discretized numerical model. Its expression is as follows:

$$\begin{aligned} A^{\mathrm{C}}_{(\mathrm{de})\mathrm{g}} &= \frac{1}{\Delta t}\left[V_i(\phi^{n+1}S^{n+1}_{\mathrm{g}}\frac{\rho_{\mathrm{gsc}}}{B^{n+1}_{\mathrm{g}}} - \phi^n S^n_{\mathrm{g}}\frac{\rho_{\mathrm{gsc}}}{B^n_{\mathrm{g}}})\right] \\ &+ \frac{1}{\Delta t}\left[V_i(\phi^{n+1}S^{n+1}_{\mathrm{o}}\frac{\rho_{\mathrm{gsc}}R^{n+1}_{\mathrm{g,o}}}{B^{n+1}_{\mathrm{o}}} - \phi^n S^n_{\mathrm{o}}\frac{\rho_{\mathrm{gsc}}R^n_{\mathrm{g,o}}}{B^n_{\mathrm{o}}})\right] \\ &+ \frac{1}{\Delta t}\left[V_{\mathrm{L}}\rho_{\mathrm{R}}(\frac{\rho_{\mathrm{gsc}}}{B^{n+1}_{\mathrm{g}}(1+b_{\mathrm{L}}P^{n+1}_{\mathrm{g}})} - \frac{\rho_{\mathrm{gsc}}}{B^n_{\mathrm{g}}(1+b_{\mathrm{L}}P^n_{\mathrm{g}})})\right] \end{aligned} \tag{4.80}$$

(5) Imbibition effect

The influence of the imbibition on the flow term is mainly demonstrated in the influence of different flow mechanisms on the flowability, which is reflected in the addition of the imbibition term in the discretized numerical model of the water component. The specific expression is as follows:

$$F_{(\mathrm{im})\mathrm{w},ij} = F^{①}_{\mathrm{w},ij} \cdot F^{②}_{(\mathrm{im})\ \mathrm{w},ij} \tag{4.81}$$

$$F^{①}_{\mathrm{w},ij} = \left(\frac{K_{\mathrm{rw}}\rho_{\mathrm{w}}}{\mu_{\mathrm{w}}}\right)^{n+1}_{ij} \tag{4.82}$$

$$F^{②}_{\mathrm{w},ij} = (P_{\mathrm{w},j}{}^{n+1} - \rho_{\mathrm{w}}{}^{n+1}gD_j) - (P_{\mathrm{w},i}{}^{n+1} - \rho_{\mathrm{w}}{}^{n+1}gD_i) - (P_{\mathrm{cow},j} - P_{\mathrm{cow},i}) \tag{4.83}$$

The finite volume method is used to discretize the mathematic model of flow through discontinuous multiple media to form a discretized numerical model. The comparison of discretized numerical model expressions based on unconventional flow mechanisms and conventional flow mechanisms are shown in Table 4.19.

1. Residual expression for finite volume discretization numerical model based on unconventional flow mechanisms

For the finite volume discretization numerical model with unconventional flow mechanisms, when $\phi_{\mathrm{o},\ i} > \phi_{\mathrm{o},\ j}$, $\phi_{\mathrm{g},\ i} > \phi_{\mathrm{g},\ j}$, $\phi_{\mathrm{w},\ i} > \phi_{\mathrm{w},\ j}$, the residual expression is as follows:

Table 4.19 Comparison of discretized numerical simulation models with conventional and unconventional flow mechanisms

Components	Mechanisms	Flow term	Source and sink term	Accumulation term
Oil components	Conventional flow	$\sum_j T_{ij}\cdot F_{o,ij}=\sum_j T_{ij}\cdot\left(F^{①}_{o,ij}\cdot F^{②}_{o,ij}\right)$ $F^{①}_{o,ij}=\left(\frac{K_{ro}}{\mu_o}\frac{\rho_{osc}}{B_o}\right)^{n+1}$ $F^{②}_{o,ij}=\left[(P_{o,j}-\rho_o gD_j)-(P_{o,i}-\rho_o gD_i)\right]^{n+1}$	$\mathrm{WI}_i\cdot F^{W}_{o,i}=\mathrm{WI}_i\cdot\left(F^{W①}_{o,i}\cdot F^{W②}_{o,i}\right)$ $F^{W①}_{o,i}=\left(\frac{K_{ro}}{\mu_o}\frac{\rho_{osc}}{B_o}\right)^{n+1}$ $F^{W②}_{o,i}=\left[(P_{o,i}-\rho_o gD_i)-(P_{wfk}-\rho_o gD_k{}^{W})\right]^{n+1}$	$A^{C}_{o,i}=\frac{V_i}{\Delta t}\left(\phi^{n+1}S^{n+1}_{o}\frac{\rho_{osc}}{B^{n+1}_{o}}-\phi^{n}S^{n}_{o}\frac{\rho_{osc}}{B^{n}_{o}}\right)_i$
	High-speed nonlinear flow	$F_{(ND)o,ij}=F_{ND}{}^{n+1}\cdot F^{①}_{o,ij}\cdot F^{②}_{o,ij}$ $F_{ND}=\left[-1+\sqrt{1+4\left(\frac{\beta Tk}{A}\right)(\Delta\phi_o)\left(\frac{\rho_o K_{ro}}{\mu_o^2}\right)}\right]\Big/\left[2\left(\frac{\beta Tk}{A}\right)(\Delta\phi_o)\left(\frac{\rho_o K_{ro}}{\mu_o^2}\right)\right]$	$F^{W}_{(ND)o,i}=F_{ND}{}^{n+1}\cdot F^{W①}_{o,i}\cdot F^{W②}_{o,i}$ $F_{ND}=\left[-1+\sqrt{1+4\left(\frac{\beta Tk}{A}\right)(\Delta\phi_o)\left(\frac{\rho_o K_{ro}}{\mu_o^2}\right)}\right]\Big/\left[2\left(\frac{\beta Tk}{A}\right)(\Delta\phi_o)\left(\frac{\rho_o K_{ro}}{\mu_o^2}\right)\right]$	–
	Pseudo-linear flow	$F_{(li)p,ij}=F^{①}_{p,ij}\cdot F^{②}_{(li)p,ij}$ $F^{②}_{(li)p,ij}=(P_{p,j}{}^{n+1}-\rho_p{}^{n+1}gD_j)-(P_{p,i}{}^{n+1}-\rho_p{}^{n+1}gD_i)-G_{c,ij}$ $G_{c,ij}=c\left(L_i+L_j\right)$	$F^{W}_{(li)p,i}=F^{W①}_{p,i}\cdot F^{W②}_{(li)\ p,i}$ $F_{(li)p,i}{}^{W②}=(P_{p,i}{}^{n+1}-\rho_p{}^{n+1}gD_i)-(P_{wfk}{}^{n+1}-\rho_p{}^{n+1}gD_k{}^{W})-G^{W}_{c,ij}$ $G^{W}_{c,i}=c^{W}\left(L_i+r_w\right)$	–
	Low-speed nonlinear flow: Threshold pressure gradient	$F_{(g)p,ij}=F^{①}_{p,ij}\cdot F^{②}_{(g)p,ij}$ $F^{②}_{(g)p,ij}=\left[(P_{p,j}{}^{n+1}-\rho_p{}^{n+1}gD_j)-(P_{p,i}{}^{n+1}-\rho_p{}^{n+1}gD_i)-G_{a,ij}\right]^{n^*}$ $G_{a,ij}=a\left(L_i+L_j\right)$	$F^{W}_{(g)p,i}=F^{W①}_{p,i}\cdot F^{W②}_{(g)\ p,i}$ $F_{(g)p,i}{}^{W②}=\left[(P_{p,i}{}^{n+1}-\rho_p{}^{n+1}gD_i)-(P_{wfk}{}^{n+1}-\rho_p{}^{n+1}gD_k{}^{W})-G^{W}_{a,i}\right]^{n^*}$ $G^{W}_{a,i}=a^{W}\left(L_i+r_w\right)$	–
Gas components	Conventional flow mechanism	$\sum_j T_{ij}\cdot F_{g,ij}=\sum_j T_{ij}\cdot\left(F^{①}_{g,ij}\cdot F^{②}_{g,ij}+F^{①}_{go,ij}\cdot F^{②}_{go,ij}\right)$ $F^{①}_{g,ij}=\left(\frac{K_{rg}}{\mu_g}\frac{\rho_{gsc}}{B_g}\right)^{n+1}$ $F^{②}_{g,ij}=\left[(P_{g,j}-\frac{\rho_{gsc}}{B_g}gD_j)-(P_{g,i}-\frac{\rho_{gsc}}{B_g}gD_i)\right]^{n+1}$ $F^{①}_{go,ij}=\left(\frac{K_{ro}}{\mu_o}\frac{\rho_{gsc}R_{g,o}}{B_o}\right)^{n+1}$ $F^{②}_{go,ij}=\left[(P_{o,j}-P_{o,i})-\frac{(\rho_{osc}+\rho_{gsc}R_{g,o})}{B_o}g(D_j-D_i)\right]^{n+1}$	$\mathrm{WI}_i F^{W}_{g,ij}=\mathrm{WI}_i(F^{W①}_{g,ij}\cdot F^{W②}_{g,ij}+F^{W①}_{go,ij}\cdot F^{W②}_{go,ij})$ $F^{W①}_{g,ij}=\left(\frac{K_{rg}}{\mu_g}\frac{\rho_{gsc}}{B_g}\right)^{n+1}$ $F^{W②}_{g,ij}=-\left[(P_{wfk}-\frac{\rho_{gsc}}{B_g}gD_k{}^{W})-(P_{g,i}-\frac{\rho_{gsc}}{B_g}gD_i)\right]^{n+1}$ $F^{W①}_{go,ij}=\left(\frac{K_{ro}}{\mu_o}\frac{\rho_{gsc}R_{g,o}}{B_o}\right)^{n+1}$ $F^{W②}_{go,ij}=-\left[(P_{wfk}-P_{o,i})-\frac{(\rho_{osc}+\rho_{gsc}R_{g,o})}{B_o}g(D_k{}^{W}-D_i)\right]^{n+1}$	$A^{C}_{g}=\frac{V_i}{\Delta t}\left[(\phi^{n+1}S^{n+1}_{g}\frac{\rho_{gsc}}{B^{n+1}_{g}}-\phi^{n}S^{n}_{g}\frac{\rho_{gsc}}{B^{n}_{g}})\right]$ $+\frac{V_i}{\Delta t}\left[(\phi^{n+1}S^{n+1}_{o}\frac{\rho_{gsc}R^{n+1}_{G,o}}{B^{n+1}_{o}}-\phi^{n}S^{n}_{o}\frac{\rho_{gsc}R^{n}_{g,o}}{B^{n}_{o}})\right]$
	High-speed nonlinear flow	$F_{(ND)g,ij}=F_{ND}{}^{n+1}\cdot F^{①}_{g,ij}\cdot F^{②}_{g,ij}$ $F_{ND}=\left[-1+\sqrt{1+4\left(\frac{\beta Tk}{A}\right)(\Delta\phi_g)\left(\frac{\rho_g K_{rg}}{\mu_g^2}\right)}\right]\Big/\left[2\left(\frac{\beta Tk}{A}\right)(\Delta\phi_g)\left(\frac{\rho_g K_{rg}}{\mu_g^2}\right)\right]$	$F^{W}_{(ND)g,i}=F_{ND}{}^{n+1}\cdot F^{W①}_{g,i}\cdot F^{W②}_{g,i}$ $F_{ND}=\left[-1+\sqrt{1+4\left(\frac{\beta Tk}{A}\right)(\Delta\phi_g)\left(\frac{\rho_g K_{rg}}{\mu_g^2}\right)}\right]\Big/\left[2\left(\frac{\beta Tk}{A}\right)(\Delta\phi_g)\left(\frac{\rho_g K_{rg}}{\mu_g^2}\right)\right]$	
	Pseudo-linear flow	$F_{(li)g,ij}=F^{①}_{g,ij}\cdot F^{②}_{(li)g,ij}+F^{①}_{go,ij}\cdot F^{②}_{(li)go,ij}$ $F^{②}_{(li)g,ij}=\begin{bmatrix}(P_{g,j}-\frac{\rho_{gsc}}{B_g}gD_j)-\\(P_{g,i}-\frac{\rho_{gsc}}{B_g}gD_i)-G_{c,ij}\end{bmatrix}^{n+1}$ $F^{②}_{(li)go,ij}=\begin{bmatrix}(P_{o,j}-P_{o,i})-\\\frac{(\rho_{osc}+\rho_{gsc}R_{g,o})}{B_o}g(D_j-D_i)-G_{c,ij}\end{bmatrix}^{n+1}$	$F^{W}_{(li)g,i}=F^{W①}_{g,i}\cdot F^{W②}_{(li)\ g,i}+F^{W①}_{go,ij}\cdot F^{W②}_{(li)go,ij}$ $F_{(li)g,i}{}^{W②}=(P_{g,i}{}^{n+1}-\rho_g{}^{n+1}gD_i)-(P_{wfk}{}^{n+1}-\rho_g{}^{n+1}gD_k{}^{W})-G^{W}_{c,ij}$ $F^{W②}_{(li)go,ij}=\begin{bmatrix}(P_{o,j}-P_{o,i})-\\\frac{(\rho_{osc}+\rho_{gsc}R_{g,o})}{B_o}g(D_j-D_i)-G_{c,ij}\end{bmatrix}^{n+1}$ $G^{W}_{c,i}=c^{W}(L_i+r_w)$	–

(continued)

Table 4.19 (continued)

Components	Mechanisms		Flow term	Source and sink term	Accumulation term
			$G_{c,ij} = c(L_i + L_j)$		
	Low-speed nonlinear flow	Threshold pressure gradient	$F_{(g)g,ij} = F^{①}_{g,ij} \cdot F^{②}_{(g)g,ij} + F^{①}_{go,ij} \cdot F^{②}_{(g)go,ij}$ $F^{②}_{(g)g,ij} = \left[(P_{g,j}{}^{n+1} - \rho_g{}^{n+1} g D_j) - (P_{g,i}{}^{n+1} - \rho_g{}^{n+1} g D_i) - G_{a,ij}\right]^{n^*}$ $F^{②}_{(g)go,ij} = \left[(P_{o,j} - P_{o,i}) - \frac{(\rho_{osc} + \rho_{gsc} R_{g,o})}{B_o} g(D_j - D_i) - G_{a,ij}\right]^{n^*}$ $G_{a,ij} = a(L_i + L_j)$	$F^{W}_{(g)g,ij} = F^{W①}_{g,ij} \cdot F^{W②}_{(g)g,ij} + F^{W①}_{go,ij} \cdot F^{W②}_{(g)go,ij}$ $F_{(g)g,i}{}^{W②} = \left[(P_{g,i}{}^{n+1} - \rho_g{}^{n+1} g D_i) - (P_{wfk}{}^{n+1} - \rho_g{}^{n+1} g D_k{}^{W}) - G^{W}_{a,i}\right]^{n^*}$ $F^{W②}_{(g)go,ij} = \left[(P_{o,i} - P_{wfk}) - \frac{(\rho_{osc} + \rho_{gsc} R_{g,o})}{B_o} g(D_i - D_k{}^{W}) - G_{a,ij}\right]^{n^*}$ $G^{W}_{a,i} = a^{W}(L_i + r_w)$	–
		Slippage effect	$F_{(si)g,ij} = F^{①}_{g,ij} \cdot F^{②}_{(si)g,ij} + F^{①}_{go,ij} \cdot F^{②}_{go,ij}$ $F^{②}_{(si)\ g,ij} = \begin{bmatrix} (1 + \frac{b}{P_{g,j}})(P_{g,j} - \frac{\rho_{gsc}}{B_g} g D_j) \\ -(1 + \frac{b}{P_{g,i}})(P_{g,i} - \frac{\rho_{gsc}}{B_g} g D_i) \end{bmatrix}^{n+1}$ $F^{②}_{go,ij} = \left[(P_{o,j} - P_{o,i}) - \frac{(\rho_{osc} + \rho_{gsc} R_{g,o})}{B_o} g(D_j - D_i)\right]^{n+1}$	$F^{W}_{(si)g,i} = F^{W①}_{g,i} \cdot F^{W②}_{(si)g,i} + F^{W①}_{go,i} \cdot F^{W②}_{go,i}$ $F^{W②}_{(si)\ g,i} = \begin{bmatrix} (1 + \frac{b}{P_{g,i}})(P_{g,i} - \frac{\rho_{gsc}}{B_g} g D_i) \\ -(P_{wfk} - \frac{\rho_{gsc}}{B_g} g D_k{}^{W}) \end{bmatrix}^{n+1}$ $F^{W②}_{go,ij} = \left[(P_{o,i} - P_{wfk}) - \frac{(\rho_{osc} + \rho_{gsc} R_{g,o})}{B_o} g(D_i - D_k{}^{W})\right]^{n+1}$	–
		Diffusion	$F_{(di)g,ij} = F^{①}_{g,ij} \cdot F^{②}_{(di)g,ij} + F^{①}_{go,ij} \cdot F^{②}_{go,ij}$ $F^{②}_{(di)\ g,ij} = \begin{bmatrix} (1 + \frac{32\sqrt{2}\sqrt{RT}\mu g}{3r\sqrt{\pi M} P_{g,j}})(P_{g,j} - \frac{\rho_{gsc}}{B_g} g D_j) \\ -(1 + \frac{32\sqrt{2}\sqrt{RT}\mu g}{3r\sqrt{\pi M} P_{g,i}})(P_{g,i} - \frac{\rho_{gsc}}{B_g} g D_i) \end{bmatrix}^{n+1}$	$F^{W}_{(di)g,i} = F^{W①}_{g,i} \cdot F^{W②}_{(di)\ g,i} + F^{W①}_{go,i} \cdot F^{W②}_{go,i}$ $F^{W②}_{(di)\ g,i} = \begin{bmatrix} (1 + \frac{32\sqrt{2}\sqrt{RT}\mu g}{3r\sqrt{\pi M} P_{g,i}})(P_{g,i} - \frac{\rho_{gsc}}{B_g} g D_i) \\ -(P_{wfk} - \frac{\rho_{gsc}}{B_g} g D_k{}^{W}) \end{bmatrix}^{n+1}$	–
		Desorption	–	–	$A^{C}_{(de)g} = \frac{1}{\Delta t}\left[V_i(\phi^{n+1} S_g{}^{n+1} \frac{\rho_{gsc}}{B_g{}^{n+1}} - \phi^{n} S_g{}^{n} \frac{\rho_{gsc}}{B_g{}^{n}})\right]$ $+ \frac{1}{\Delta t}\left[V_i(\phi^{n+1} S_o{}^{n+1} \frac{\rho_{gsc} R_{g,o}{}^{n+1}}{B_o{}^{n+1}} - \phi^{n} S_o{}^{n} \frac{\rho_{gsc} R_{g,o}{}^{n}}{B_o{}^{n}})\right]$ $+ \frac{1}{\Delta t}\left[V_L \rho_R (\frac{\rho_{gsc}}{B_g{}^{n+1}(1 + b_L P_g{}^{n+1})} - \frac{\rho_{gsc}}{B_g{}^{n}(1 + b_L P_g{}^{n})})\right]$
Water component	Conventional flow mechanism		$\sum_j T_{ij} \cdot F_{w,ij} = \sum_j T_{ij} \cdot (F^{①}_{w,ij} \cdot F^{②}_{w,ij})$ $F^{①}_{w,ij} = (\frac{K_{rw}}{\mu_w} \frac{\rho_{wsc}}{B_w})^{n+1}$ $F^{②}_{w,ij} = \left[(P_{w,j} - \frac{\rho_{wsc}}{B_w} g D_j) - (P_{w,i} - \frac{\rho_{wsc}}{B_w} g D_i)\right]^{n+1}$	$\mathrm{WI}_i \cdot F^{W}_{w,i} = \mathrm{WI}_i \cdot (F^{W①}_{w,i} \cdot F^{W②}_{w,i})$ $F^{W①}_{w,i} = (\frac{K_{rw}}{\mu_w} \frac{\rho_{wsc}}{B_w})^{n+1}$ $F^{W②}_{w,i} = -\left[(P_{wfk} - \frac{\rho_{wsc}}{B_w} g D_k{}^{W}) - (P_{w,i} - \frac{\rho_{wsc}}{B_w} g D_i)\right]^{n+1}$	$A^{C}_{w} = \frac{V_i}{\Delta t}(\phi^{n+1} S^{n+1}_{w} \frac{\rho_{wsc}}{B^{n+1}_{w}} - \phi^{n} S^{n}_{w} \frac{\rho_{wsc}}{B^{n}_{w}})$
	Low-speed nonlinear flow	Imbibition	$F_{(im)w,ij} = F^{①}_{w,ij} \cdot F^{②}_{(im)\ w,ij}$ $F^{①}_{w,ij} = (\frac{K_{rw} \rho_w}{\mu_w})^{n+1}_{i/j}$ $F^{②}_{w,ij} = (P_{w,j}{}^{n+1} - \rho_w{}^{n+1} g D_j) - (P_{w,i}{}^{n+1} - \rho_w{}^{n+1} g D_i) - (P_{cow,j} - P_{cow,i})$	–	–

(1) The residual expression for the oil component

$$\begin{aligned} R_{oi} = & (b_1 + b_2\delta P_{o,i} + b_3\delta S_{g,i} + b_4\delta S_{w,i}) \\ & - \sum_j T_{ij}(a_{(me)1} + a_{(me)2}\delta P_{o,j} + a_{(me)3}\delta P_{o,i} + a_{(me)4}\delta S_{g,i} + a_{(me)5}\delta S_{w,i}) \\ & + (c_{(me)1} + c_{(me)2}\delta P_{wfk} + c_{(me)3}\delta P_{o,i} + c_{(me)4}\delta S_{g,i} + c_{(me)5}\delta S_{w,i}) \end{aligned} \tag{4.84}$$

(2) The residual expression for the gas component

$$\begin{aligned} R_{gi} = & (b'_{(me)1} + b'_{(me)2}\delta P_{o,i} + b'_3\delta S_{g,i} + b'_4\delta S_{w,i}) \\ & - \sum_j T_{ij} \cdot (a'_{(me)1} + a'_{(me)2}\delta P_{o,j} + a'_{(me)3}\delta P_{o,i} + a'_{(me)4}\delta S_{g,j} + a'_{(me)5}\delta S_{g,i} + a'_{(me)6}\delta S_{w,i}) \\ & + (c'_{(me)1} + c'_{(me)2}\delta P_{wfk} + c'_{(me)3}\delta P_{o,i} + c'_{(me)4}\delta S_{g,i} + c'_{(me)5}\delta S_{w,i}) \end{aligned} \tag{4.85}$$

(3) The residual expression for the water component

$$\begin{aligned} R_{wi} = & (b''_1 + b''_2\delta P_{o,i} + b''_3\delta S_{w,i}) \\ & - \sum_j T_{ij}(a''_1 + a''_2\delta P_{o,j} + a''_3\delta P_{o,i} + a''_4\delta S_{w,j} + a''_5\delta S_{w,i}) \\ & + (c''_1 + c''_2\delta P_{wfk} + c''_3\delta P_{o,i} + c''_4\delta S_{w,i}) \end{aligned} \tag{4.86}$$

2. The coefficient expression corresponding to each variable of the coefficient matrix in the discretization model based on unconventional flow mechanisms

For the coefficient expression of each variable in the discretized numerical model with unconventional flow mechanisms, see Table 4.20.

3. Finite volume discretization numerical model based on flow-geomechanics coupling mechanics

During the process of fracturing, injection and production, due to the dynamic changes in pore pressure and effective stress, the geometric characteristics and property parameters of multiple media at different scales will change, with the well index and transmissibility between the different porous media also changed, which affects the reservoir dynamics and production characteristics. Therefore, the flow-geomechanics coupling mechanism is demonstrated in the flow term and the source and sink term of the oil, gas, and water components in the discretized numerical model, which reflects the effects of flow-geomechanics coupling interactions on flow dynamics and productivity in different media.

(1) Finite volume discretization numerical model

① Flow term

The influence of flow-geomechanics coupling interactions on the flow term is reflected in the dynamic changes in transmissibility. After considering the flow-geomechanics coupling effect, the permeability becomes a function of pressure, and its expression is as follows:

$$K_{ps} = K_0 e^{\gamma(\overline{P} - P_i)} \tag{4.87}$$

The function of the average pressure dependent on the oil, gas, and water phase pressure is as follows:

$$\overline{P} = S_o P_o + S_g P_g + S_w P_w \tag{4.88}$$

The transmissibility expression affected by flow-geomechanics coupling is as follows:

$$T_{(ps)ij}(P_i, P_j) = \frac{K_{ps,i}(P_i)\beta_i K_{ps,j}(P_j)\beta_j}{\left[K_{ps,i}(P_i)\beta_i + K_{ps,j}(P_j)\beta_j\right]} \tag{4.89}$$

② Source and sink term

The influence of flow-geomechanics coupling interactions on the source and sink term is reflected in the well index. After considering the flow-geomechanics interactions, the well index becomes a function of pressure. Its expression is as follows:

$$WI_{ps}(P) = \frac{2\pi K_{(ps)} V^{1/3}}{\ln\left(\frac{r_e}{r_w}\right) + s} \tag{4.90}$$

Therefore, in the discretized numerical model which is affected by flow-geomechanics coupling interactions, changes in the flow term and the source and sink term are described in detail in Table 4.21.

(2) Finite-volume discretization numerical model residual expression based on flow-geomechanics coupling mechanisms

Table 4.20 The coefficient expression corresponding to each variable of the coefficient matrix in the discretized simulation model based on the unconventional flow mechanisms

Components	Mechanisms		Coefficient corresponding to each variable in the coefficient matrix						
			$\delta P_{o,i}$	$\delta S_{g,i}$	$\delta S_{w,i}$	$\delta P_{o,j}$	$\delta S_{g,j}$	$\delta S_{w,j}$	δP_{wfk}
Oil components	Conventional flow mechanism		$b_2 - \sum_j T_{ij}a_3 + c_3$	$b_3 - \sum_j T_{ij}a_4 + c_4$	$b_4 - \sum_j T_{ij}a_5 + c_5$	$-\sum_j T_{ij}a_2$	–	–	c_2
	High-speed nonlinear flow		$b_2 - \sum_j T_{ij}a_{(ND)3} + c_{(ND)3}$	$b_3 - \sum_j T_{ij}a_{(ND)4} + c_{(ND)4}$	$b_4 - \sum_j T_{ij}a_{(ND)5} + c_{(ND)5}$	$-\sum_j T_{ij}a_{(ND)2}$	–	–	c_2
	Pseudo-linear flow		$b_2 - \sum_j T_{ij}a_{(li)3} + c_{(li)3}$	$b_3 - \sum_j T_{ij}a_{(li)4} + c_{(li)4}$	$b_4 - \sum_j T_{ij}a_{(li)5} + c_{(li)5}$	$-\sum_j T_{ij}a_{(li)2}$	–	–	c_2
	Low-speed nonlinear flow	Threshold pressure gradient	$b_2 - \sum_j T_{ij}a_{(g)3} + c_{(g)3}$	$b_3 - \sum_j T_{ij}a_{(g)4} + c_{(g)4}$	$b_4 - \sum_j T_{ij}a_{(g)5} + c_{(g)5}$	$-\sum_j T_{ij}a_{(g)2}$	–	–	c_2
Gas components	Conventional flow mechanism		$b'_2 - \sum_j T_{ij}a'_3 + c'_3$	$b'_3 - \sum_j T_{ij}a'_5 + c'_4$	$b'_4 - \sum_j T_{ij}a'_6 + c'_5$	$-\sum_j T_{ij}a_2{}'$	$-\sum_j T_{ij}a_4{}'$	–	c'_2
	High-speed nonlinear flow		$b'_2 - \sum_j T_{ij}a'_{(ND)3} + c'_{(ND)3}$	$b'_3 - \sum_j T_{ij}a'_{(ND)5} + c'_{(ND)4}$	$b'_4 - \sum_j T_{ij}a'_{(ND)6} + c'_{(ND)5}$	$-\sum_j T_{ij}a_{(ND)2}{}'$	$-\sum_j T_{ij}a_{(ND)4}{}'$	–	c'_2
	Pseudo-linear flow		$b'_2 - \sum_j T_{ij} \cdot a'_{(li)3} + c'_{(li)3}$	$b'_3 - \sum_j T_{ij} \cdot a'_{(li)5} + c'_{(li)4}$	$b'_4 - \sum_j T_{ij} \cdot a'_{(li)6} + c'_{(li)5}$	$-\sum_j T_{ij} \cdot a_{(li)2}{}'$	$-\sum_j T_{ij} \cdot a_{(li)4}{}'$	–	c'_2
	Low-speed nonlinear flow	Threshold pressure gradient	$b'_2 - \sum_j T_{ij} \cdot a'_{(g)3} + c'_{(g)3}$	$b'_3 - \sum_j T_{ij} \cdot a'_{(g)5} + c'_{(g)4}$	$b'_4 - \sum_j T_{ij} \cdot a'_{(g)6} + c'_{(g)5}$	$-\sum_j T_{ij} \cdot a_{(g)2}{}'$	–	–	c'_2
		Slippage effect	$b'_2 - \sum_j T_{ij} \cdot a'_{(si)3} + c'_{(si)3}$	$b'_3 - \sum_j T_{ij} \cdot a'_{(si)5} + c'_{(si)4}$	$b'_4 - \sum_j T_{ij} \cdot a'_{(si)6} + c'_{(si)5}$	$-\sum_j T_{ij} \cdot a_{(si)2}{}'$	$-\sum_j T_{ij} \cdot a_{(si)4}{}'$	–	c'_2
		Diffusion effect	$b'_2 - \sum_j T_{ij} \cdot a'_{(di)3} + c'_{(di)3}$	$b'_3 - \sum_j T_{ij} \cdot a'_{(di)5} + c'_{(di)4}$	$b'_4 - \sum_j T_{ij} \cdot a'_{(di)6} + c'_{(di)5}$	$-\sum_j T_{ij} \cdot a_{(di)2}{}'$	$-\sum_j T_{ij} \cdot a_{(di)4}{}'$	–	c'_2
		Desorption effect	$b'_{(de)2} - \sum_j T_{ij} \cdot a'_3 + c'_3$	$b'_3 - \sum_j T_{ij} \cdot a'_5 + c'_4$	$b'_4 - \sum_j T_{ij} \cdot a'_6 + c'_5$	$-\sum_j T_{ij} \cdot a_2{}'$	$-\sum_j T_{ij} \cdot a_4{}'$	–	c'_2
Water components	Conventional flow mechanism		$b''_2 - \sum_j T_{ij} \cdot a''_3 + c''_3$	–	$b''_3 - \sum_j T_{ij} \cdot a''_5 + c''_4$	$-\sum_j T_{ij} \cdot a''_2$	–	$-\sum_j T_{ij} \cdot a''_4$	c''_2
	Low-speed nonlinear flow	Imbibition effect	$b''_2 - \sum_j T_{ij}a''_3 + c''_3$	–	$b''_3 - \sum_j T_{ij}a''_5 + c''_4$	$-\sum_j T_{ij}a''_2$	–	$-\sum_j T_{ij}a''_4$	c''_2

Note The coefficients for the solution variables such as constant, pressure and saturation in the flow term, accumulation term, and source & sink term in the equations of oil, gas and water components are consistent with Table 4.17. ND, linearity, g, slippage, diffusion, and desorption in the subscript are the related coefficients of the changing solution variables such as constant, pressure and saturation in the flow term, accumulation term, and source & sink term after considering the high-speed nonlinear flow, the pseudo-linear flow, the threshold pressure gradient, the slippage effect, the diffusion effect, the desorption effect, respective

Table 4.21 Comparison of the discretized numerical models considering flow-geomechanics coupling interactions

Variableterm	Components	Conventional mechanisms	Flow-geomechanics coupling
Flow term	Oil component	$\sum_j T_{ij} \cdot F_{o,ij} = \sum_j T_{ij} \cdot \left(F_{o,ij}^{①} \cdot F_{o,ij}^{②}\right)$	$\sum_j T_{(ps)ij}{}^{n+1} \cdot F_{o,ij} = \sum_j T_{ij} \cdot \left(F_{o,ij}^{①} \cdot F_{o,ij}^{②}\right)$
	Gas component	$\sum_j T_{ij} \cdot F_{g,ij} = \sum_j T_{ij} \cdot \left(F_{g,ij}^{①} \cdot F_{g,ij}^{②} + F_{go,ij}^{①} \cdot F_{go,ij}^{②}\right)$	$\sum_j T_{(ps)ij}{}^{n+1} \cdot F_{g,ij} = \sum_j T_{(ps)ij}{}^{n+1} \cdot \left(F_{g,ij}^{①} \cdot F_{g,ij}^{②} + F_{go,ij}^{①} \cdot F_{go,ij}^{②}\right)$
	Water component	$\sum_j T_{ij} \cdot F_{w,ij} = \sum_j T_{ij} \cdot \left(F_{w,ij}^{①} \cdot F_{w,ij}^{②}\right)$	$\sum_j T_{(ps)ij}{}^{n+1} \cdot F_{w,ij} = \sum_j T_{(ps)ij}{}^{n+1} \cdot \left(F_{w,ij}^{①} \cdot F_{w,ij}^{②}\right)$
Source and sink term	Oil component	$\mathrm{WI}_i \cdot F_{o,i}^{W} = \mathrm{WI}_i \cdot \left(F_{o,i}^{W①} \cdot F_{o,i}^{W②}\right)$	$\mathrm{WI}_{(ps)i}{}^{n+1} \cdot F_{o,i}^{W} = \mathrm{WI}_{(ps)i}{}^{n+1} \cdot \left(F_{o,i}^{W①} \cdot F_{o,i}^{W②}\right)$
	Gas component	$\mathrm{WI}_i F_{g,ij}^{W} = \mathrm{WI}_i (F_{g,ij}^{W①} \cdot F_{g,ij}^{W②} + F_{go,ij}^{W①} \cdot F_{go,ij}^{W②})$	$\mathrm{WI}_{(ps)i}{}^{n+1} F_{g,ij}^{W} = \mathrm{WI}_{(ps)i}{}^{n+1} (F_{g,ij}^{W①} \cdot F_{g,ij}^{W②} + F_{go,ij}^{W①} \cdot F_{go,ij}^{W②})$
	Water component	$\mathrm{WI}_i \cdot F_{w,i}^{W} = \mathrm{WI}_i \cdot \left(F_{w,i}^{W①} \cdot F_{w,i}^{W②}\right)$	$\mathrm{WI}_{(ps)i}{}^{n+1} \cdot F_{w,i}^{W} = \mathrm{WI}_{(ps)i}{}^{n+1} \cdot \left(F_{w,i}^{W①} \cdot F_{w,i}^{W②}\right)$

Table 4.22 Coefficient expression corresponding to each variable of the coefficient matrix in the discretized numerical model considering flow-geomechanics coupling interactions

Variable	Coefficient corresponding to each variable in the coefficient matrix		
	Oil components	Gas components	Water components
$\delta P_{o,i}$	$b_2 - \sum_j T_{(ps)ij} a_{(ps)3} + c_{(ps)3}$	$b'_2 - \sum_j T_{(ps)ij} a'_{(ps)3} + c'_{(ps)3}$	$b''_2 - \sum_j T_{(ps)ij} a''_{(ps)3} + c''_{(ps)3}$
$\delta S_{g,i}$	$b_3 - \sum_j T_{(ps)ij} a_{(ps)4} + c_{(ps)4}$	$b'_3 - \sum_j T_{(ps)ij} a'_{(ps)5} + c'_{(ps)4}$	–
$\delta S_{w,i}$	$b_4 - \sum_j T_{(ps)ij} a_{(ps)5} + c_{(ps)5}$	$b'_4 - \sum_j T_{(ps)ij} a'_{(ps)6} + c'_{(ps)5}$	$b''_3 - \sum_j T_{(ps)ij} a''_{(ps)5} + c''_{(ps)4}$
$\delta P_{o,j}$	$-\sum_j T_{(ps)ij} a_{(ps)2}$	$-\sum_j T_{(ps)ij} a_{(ps)2}{}'$	$-\sum_j T_{(ps)ij} a''_{(ps)2}$
$\delta S_{g,j}$	–	$-\sum_j T_{(ps)ij} a_{(ps)4}{}'$	–
$\delta S_{w,j}$	–	–	$-\sum_j T_{(ps)ij} a''_{(ps)4}$
δP_{wfk}	$c_{(ps)2}$	$c'_{(ps)2}$	$c''_{(ps)2}$

Note The pressure sensitivity in the subscript mean related coefficients of the solution variables such as constant, pressure and saturation in the flow term, accumulation term, and source and sink term after considering the flow-geomechanics coupling interactions

For the finite volume discretization numerical model considering flow-geomechanics coupling interactions, when $\phi_{o,i} > \phi_{o,j}$, $\phi_{g,i} > \phi_{g,j}$, $\phi_{w,i} > \phi_{w,j}$, the residual expression is given below.

① Oil component residual expression

$$\begin{aligned} R_{oi} = {} & (b_1 + b_2 \delta P_{o,i} + b_3 \delta S_{g,i} + b_4 \delta S_{w,i}) \\ & - \sum_j T_{(ps)ij} (a_{(ps)1} + a_{(ps)2} \delta P_{o,j} + a_{(ps)3} \delta P_{o,i} + a_{(ps)4} \delta S_{g,i} + a_{(ps)5} \delta S_{w,i}) \\ & + (c_{(ps)1} + c_{(ps)2} \delta P_{wfk} + c_{(ps)3} \delta P_{o,i} + c_{(ps)4} \delta S_{g,i} + c_{(ps)5} \delta S_{w,i}) \end{aligned} \tag{4.91}$$

② Gas component residual expression

$$\begin{aligned} R_{gi} = {} & (b'_1 + b'_2 \delta P_{o,i} + b'_3 \delta S_{g,i} + b'_4 \delta S_{w,i}) \\ & - \sum_j T_{(ps)ij} \cdot (a'_{(ps)1} + a'_{(ps)2} \delta P_{o,j} + a'_{(ps)3} \delta P_{o,i} + a'_{(ps)4} \delta S_{g,j} + a'_{(ps)5} \delta S_{g,i} + a'_{(ps)6} \delta S_{w,i}) \\ & + (c'_{(ps)1} + c'_{(ps)2} \delta P_{wfk} + c'_{(ps)3} \delta P_{o,i} + c'_{(ps)4} \delta S_{g,i} + c'_{(ps)5} \delta S_{w,i}) \end{aligned} \tag{4.92}$$

③ Water component residual expression

$$\begin{aligned} R_{wi} = {} & (b''_1 + b''_2 \delta P_{o,i} + b''_3 \delta S_{w,i}) \\ & - \sum_j T_{(ps)ij} (a''_{(ps)1} + a''_{(ps)2} \delta P_{o,j} + a''_{(ps)3} \delta P_{o,i} + a''_{(ps)4} \delta S_{w,j} + a''_{(ps)5} \delta S_{w,i}) \\ & + (c''_{(ps)1} + c''_{(ps)2} \delta P_{wfk} + c''_{(ps)3} \delta P_{o,i} + c''_{(ps)4} \delta S_{w,i}) \end{aligned} \tag{4.93}$$

4. The coefficient expression corresponding to each variable of the coefficient matrix in the discretized numerical model based on flow-geomechanics coupling mechanisms

In the finite volume discretization numerical model considering flow-geomechanics coupling interactions, such variables as pressure and saturation are combined. The coefficient expression for each variable is shown in Table 4.22.

References

An YS, Wu XD, Han GQ (2007) Application of numerical simulation of complex wells based on hybrid PEBI grid. J China Univ Pet (Natural Science Edition) 06:60–63

Aziz K, Settari A (2004) Numerical simulation of reservoirs. Shiyi Yuan, translated by Wang J L. Beijing: Petroleum Industry Press

Chang SQ (1998) Study on the method of grid partitioning in numerical simulation of 3D flow. J Wuhan Autom Univ 02:3–7

Chen JM, Wang XF, Wei WB, Weng WP (2010) Overview of grid technology application in reservoir numerical simulation. China Sci Technol Expo 22:2–3

Han DK, Chen QL, Yan CZ (1993) Fundamentals of numerical simulation of reservoirs. Petroleum Industry Press, Beijing

Karimi-Fard M, Durlofsky LJ, Aziz K (2003) An efficient discrete fracture model applicable for general purpose reservoir simulators. SPE Reservoir Simulation Symposium, Texas, USA

Lin CY (2010) Research, development and application based on PEBI grid reservoir numerical simulator. University of Science and Technology of China

Lu QJ (2008) Research on the generation technology of complex grid system in reservoir numerical simulation. China University of Petroleum

Mao XP, Zhang ZT, Qian Z (2012) Analysis of geological model expressed by corner grid model and its application in simulation of hydrocarbon accumulation process. Geol J 36(03):265–273

Rao SW (2009) Numerical simulation of two-phase porous media flow in low permeability reservoirs. Southwest Petroleum University

Tang Y, Chen W, Duan YG, Fang QT, Chen XJ (2007) Study on numerical simulation of reservoir based on Voronoi grid technology. J Southwest Pet Univ (S1):22–24+7

Wang FJ (2004) Computational fluid dynamics analysis: principles and applications of CFD software. Tsinghua University Press

Wang DG, Hou J, Xing XJ, Zhang XS, Zhong HJ (2012) Improved PEBI grid generation method based on frontier propulsion. Comput Phys 29(05):675–683

Xiang ZP, Zhang LH, Chen LH, Chen P, Su W, Ma L (2006) Generation algorithm of PEBI grid in arbitrary constrained planar domain of reservoir. J Southwest Pet Inst (02):32–35+7

Xie HB, Ma YL, Qi GR, Guo SP (2001) Study on numerical simulation method of unstructured grid reservoirs. J Pet (01):63–66+4–3

Yang Q (2005) Qualified delaunay triangulation grid technology. Electronics Industry Press

Yang YL, Jing J, Yang ZJ, Xu TF, Wang FG (2015) Mechanism of grid of complex geological blocks in numerical simulation of multiphase flow. J Jilin Univ Eng Sci 45(04):1281–1287

Zha WS (2009) Reservoir numerical calculation based on PEBI network and its implementation. Ph.D. thesis, University of Science and Technology of China

Zhang RH (2015) Numerical simulation of fractured reservoirs based on finite volume method. Southwest Petroleum University

Zhang J, Tan JJ (2003) Generation and optimization of 3D unstructured grid. Aeronaut Comput Technol 04:31–34

Zhang LP, Yang YJ, Wo C, Zhang HX, Gao SZ (1999) The automatic generation technology and application of unstructured grids with three dimensional complex shapes. Chin J Comput Phys 05:552–558

5 Geological Modeling Technology for Tight Reservoir with Multi-scale Discrete Multiple Media

5.1 Geological Modeling Strategy for Discontinuous Multi-scale Discrete Multiple Media

The tight reservoir usually has large coverage area. There is great areal heterogeneity in reservoir lithology, lithofacies, formation thickness, physical properties and oil saturations, in different reservoir blocks, in areas between different wells, and also in different segments of a single well. Different scales of natural fracture develop in tight reservoirs Hydraulic fracturing also forms the complex hydraulic fracture at different scales, which has great macroscopic heterogeneity with prominent multi-scale features. Unconventional tight reservoirs consist of different scales of pore media from nano-sized to nano-micron to micron-sized. There is a large difference for pore with different scales in compositions and quantity distributions. Different scales of pores present discrete discontinuous distribution in space, with great heterogeneity in spatial distribution. There is a large difference for different scale pore media in geometric parameters and physical properties different scales pore media. Different scales pores present large microscopic heterogeneity and multiple media features. All of these features make the conventional geological modeling theory and technology not applicable in this scenario. According to the great areal heterogeneity in tight reservoirs' lithology, lithofacies and reservoir categories, the porous medium with fractures at different scales is developed. Therefore, there are great macroscopic heterogeneity, multi-scale features and microscopic multiple media. We develop a new technology for building a discontinuous multi-scale multiple media reservoir model, which can realize multi-scale integrated geological modeling through region decomposition based on large-scale heterogeneity, elements division based on small-scale heterogeneity, and multiple pores and fractures media division based on micro-scale heterogeneity. In addition, we have three major transformations. i. It is the transition from continuous modeling to discrete modeling. ii. It is the transition from dual medium modeling to discrete fracture and discrete multiple media modeling. iii. It is the transition from reservoir scale modeling to equivalent modeling of micro-scale, small scale and reservoir upscaling.

5.1.1 The Division of Representative Elements and Multiple Media Based on Multi-scale Heterogeneity

Tight formations have large macroscopic heterogeneity for lithology, lithofacies, reservoir types, development and scale of natural fractures, magnitude and direction of earth stress and fluid properties in spatial distribution. According to the scale and intensity of heterogeneity, the tight formations can be divided into the first level regions based on large scale heterogeneity, and the second level representative elements based on small scale heterogeneity. Representative elements can be divided into the third level multiple pores and fractures media based on the quantity and distribution of microscopic fractures with different scales.

5.1.1.1 The Division of Regions and Representative Elements Based on Macroscopic Heterogeneity

1. The division of first level regions based on large scale heterogeneity

According to the great macroscopic heterogeneity for lithology, lithofacies, reservoir types, development and scale of natural fracture, magnitude and direction of groud stress and flow properties in spatial distribution, tight formations can be divided into several first level regions. There is a large difference for geological conditions and heterogeneity in different regions and little difference in the same regions (Fig. 5.1).

Q. Ran, *Unconventional Tight Reservoir Simulation: Theory, Technology and Practice*,
https://doi.org/10.1007/978-981-32-9848-4_5

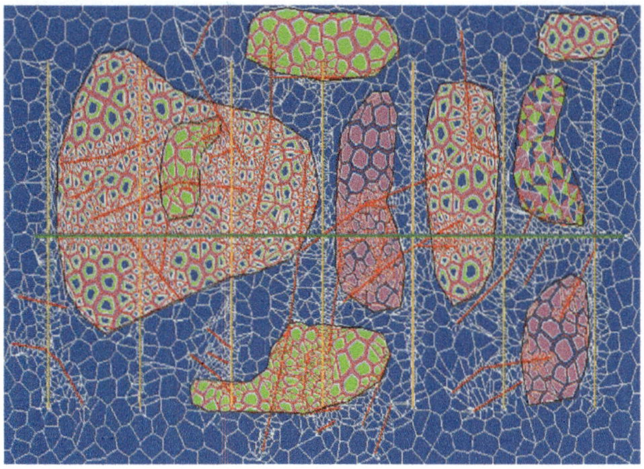

Fig. 5.1 First level domain decomposition based on macroscopic heterogeneity

2. The division of second level representative elements based on small scale heterogeneity

According to the differences of small-scale geological conditions and heterogeneity in the same regions, these regions can be divided into several second level representative elements. Large differences exist in different representative elements for small scale heterogeneity, and quantity and distribution of fractures with different scales, while these features are similar in the same representative elements (Fig. 5.2). Generally, the second level representative elements can be divided into discrete fracture representative elements and pores and fractures representative elements at micro and nano-scale based on the difference in the characteristics of large scale fractures and micro to nano-pores and fractures.

5.1.1.2 The Division of Third Levels of Multiple Pores and Fractures Media Based on Microscopic Heterogeneity

Different scales of pores and fractures exist in different representative elements of tight oil reservoirs. Therefore, there is a large difference in characteristics between different pores and fractures. A difference in characteristics also exists between different scales of pore space or between different scales of fractures, which is shown in the differences of geometric morphology and characteristics of properties for pores and fractures with different scales. These all lead to large differences in the occurrence state of flow, flow mechanisms, and flow characteristics. Therefore, we need to divide the pores and fractures with different scales within the same representative element into multiple pores and fractures media with different characteristics (Fig. 5.3).

Representative elements of discrete fractures are discretized into different scale discrete fracture media. Pore and fracture representative elements with micro and nano-scale are divided into nested distribution type (in the following, it is referred as NDT), multiple media and randomly interacting distribution type (in the following, it is referred as RIDT), multiple media based on the spatial distribution, quantity distribution and volume percentage of micro to nano-fractures and different scales of pores.

5.1.2 Geological Modeling Strategy for Discontinuous Multi-scale Discrete Multiple Media

Unconventional tight reservoirs usually have great heterogeneity. They can be divided into different regions based on large-scale heterogeneity, representative elements based on small-scale heterogeneity, and multiple pore and fracture media based on micro-scale heterogeneity. There are clear

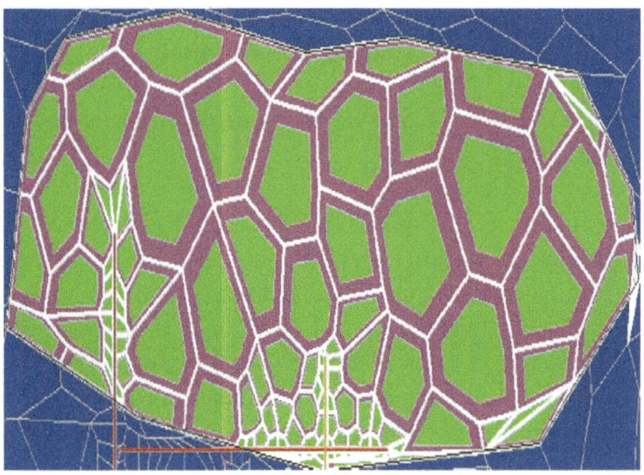

Fig. 5.2 Second level representative element determined based on small scale heterogeneity

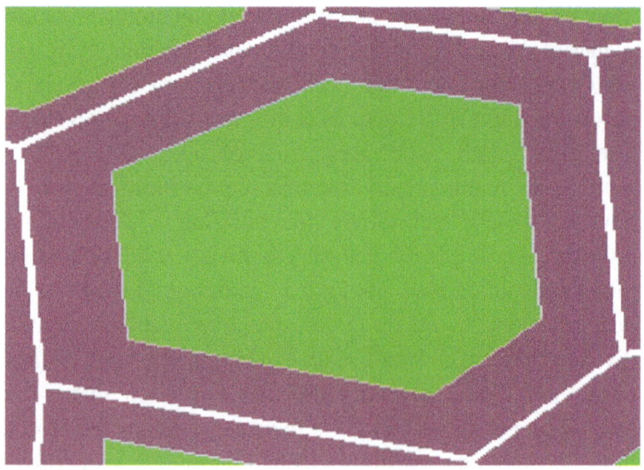

Fig. 5.3 Third level multiple media determined based on microscopic heterogeneity

Table 5.1 Discontinuous multi-scale discretized multiple media geological modeling technique

Modeling method	Media type	Method description	Geoligic model
Continuous modeling	Continuous media	The mathematical model of continuous function is used to describe the geological body. The spatial distribution of the continuous media is calculated by using the continuous function interpolation method such as the three-dimensional geological trend surface method and the Krigging method	Tectonic model Lattice model Lithofacies model
Discrete fracture modeling	discrete fracture media	According to the spatial distribution and geometric parameters of deterministic large and medium sized fractures, and the spatial distribution and range of geometric parameters of semi-deterministic small sized microscopic fractures, we use deterministic or constrained stochastic algorithms to generate the fracture body in space. Then, different discrete fractures are grouped into different discrete fracture elements. And discrete fractures are discretized into simulation grids, to form media with discrete fractures. Discrete fracture parameter models are established based on discrete fracture grids, including fracture width, porosity, permeability and conductivity, etc.	Geological model of discrete fractures
Discrete multiple media modeling	Discrete multiple media	The number of medium types is determined according to the quantity distribution and volume percentage of multiple media at different scales. Porous representative elements are divided into two multiple media distribution modes including NDT media and RIDT media, based on the spatial distribution of multiple media at different scales. Representative elements with micro and nano-scale effect are discretized through grid partitioning into nested and interactive second grids, which can generate discrete multiple media grids and discrete multiple media. We use discrete multiple media modeling techniques to build the models of geometric parameters (pore throat radius, etc.), property parameters (porosity, permeability, etc.) and pore media flow parameters, etc. for discrete multiple media	Geological model of discrete multiple media
Equivalent modeling of multiple media at different scales	Equivalent fractured pore media in different scales	According to the composition, quantity, and spatial distribution of porous medium with fractures at different scales within representative elements, the fractures and pores with different scales are discretized into media with discrete fractures and discrete multiple media, respectively, through unstructured grid method. The media with different scale fractures and pores in a representative element is homogenized to a single porous medium, based on the rules of volume equivalence and flow rate equivalence. We then build the geological model for the equivalent pore media with such equivalent pore media's geometric parameters, property parameters, and flow parameters through pore and fracture media of tight reservoirs, which the equivalent modeling is established. Equivalent modeling methods are divided into reservoir-scale equivalent modeling method and unconventional reservoir upscaling equivalent modeling method	Geological model of equivalent media

multi-scale features for macroscopic large, medium, and small scale plus microscopic micro and nano-scale natural and hydraulic fractures and the pores with different scales. Different scales of natural/hydraulic fractures and pores are discontinuously and discretely distributed in space. Their distribution characteristics and properties are difficult to be characterized by using continuous functions, which can only be described by discrete functions. Different scale pores and fractures have different geometry and property characteristics, fluid compositions and properties, occurrence state and flow mechanisms through pore and fracture media of tight reservoirs, with clear multiple media characteristics. Therefore, we need to use a discontinuous multi-scale discrete multiple media geological modeling technique for unconventional tight reservoirs to establish a discrete fracture geological model, a discrete multiple media geological model, and an equivalent media geological model (see Table 5.1 and Fig. 5.4). The overall modeling ideas are as follows.

5.1.2.1 Region Decomposition Based on Macroscopic Heterogeneity

According to the macroscopic heterogeneity for lithology, lithofacies and reservoir types of tight reservoirs, as well as quantity and spatial distribution differences existing in pores, natural fractures and hydraulic fractures at different scales, we develop dividing indicators and criteria of macro-scale heterogeneity to form such a dividing method for large scale

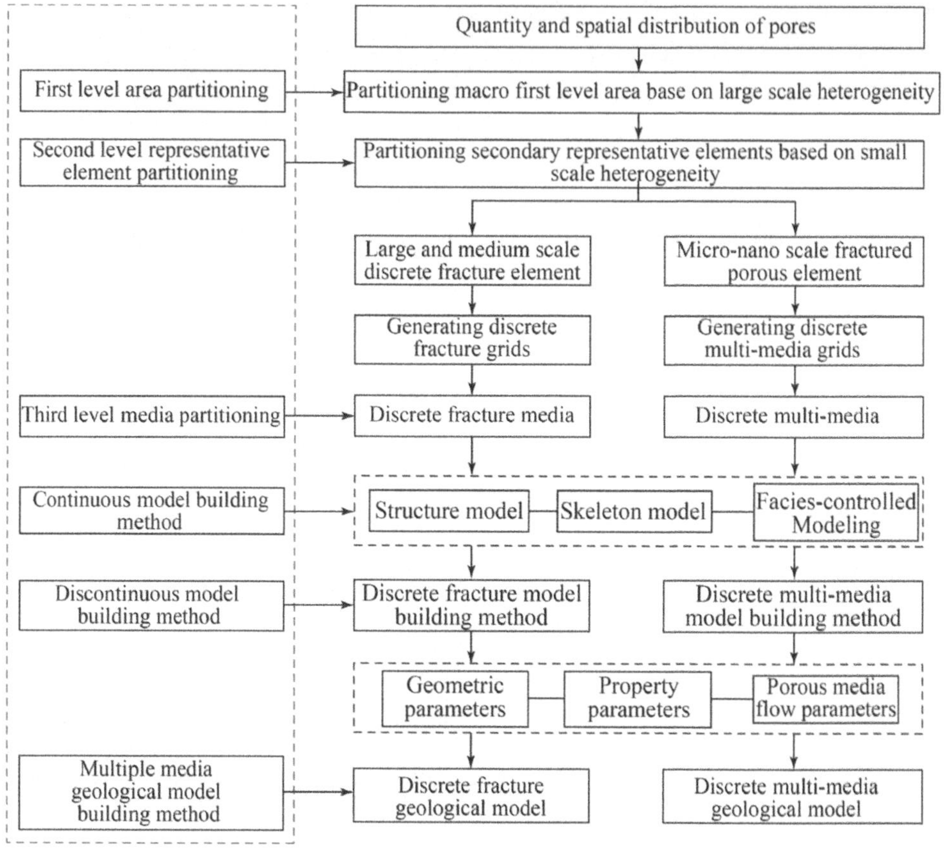

Fig. 5.4 Flow chart for discontinuous multi-scale discretized multiple media geological modeling

heterogeneity, which divides tight reservoirs into several first level regions with large differences in their characteristics.

5.1.2.2 Dividing of First Level Grid Based on Representative Elements of the Small Scale Heterogeneity

On the basis of the first level of division, we divide the second representative elements based on the small-scale heterogeneity, which includes discrete fracture representative elements and pore and fracture representative elements with micro and nano-scale.

① The discretely distributed large and medium scale natural/hydraulic fractures in the regions are treated as several discrete fracture representative elements.

② For the regions that only consist of discretely distributed micro and nano-scale fractures and pores, we develop subdivision indicators and criteria for small scale heterogeneity, which forms a division method for small scale heterogeneity representative elements based on differences in the quantity and spatial distribution of micro and nano-scale pores and fractures. These regions are divided into several representative elements with micro and nano-scale that have different characteristics of media with fractures and pores.

③ According to the distribution features of the discrete fracture representative elements and the micro and nano-pore and fracture representative elements, we select the appropriate unstructured grids. These regions are divided into the first level grids, which can represent both different discrete fracture representative elements and pore and fracture representative elements with micro and nano-scale effects.

5.1.2.3 Dividing of Second Grid Based on the Quantity and Spatial Distribution Features of Micro Scale Multiple Media

We divide multiple media based on the different characteristics in quantity and spatial distribution of fractures and pores at different scales in different discrete fracture representative elements and pore and fracture representative elements with micro and nano-scale. Gird partitioning forms the discrete second grid. In addition, we determine the media types for different grids. i. The large and medium scale discrete fracture elements are divided into several discrete fracture media with different geometric and property features according to the variation of geometric features. The second discrete fracture grid is generated through unstructured grid partitioning. Also, we determine the type of media with discrete fractures for each second discrete gird (hydraulic fractures/natural fractures). ii. Based on the differences in quantity and spatial distribution of micro and nano-fractures and pores with different scales, the pore and fracture representative elements with micro and nano-scale effects are divided into several discrete multiple media with different geometric features and property features. The second discrete multiple media grid is generated through unstructured grid partitioning. Furthermore, we determine a medium type for each second discrete grid (pores or fractures).

5.1.2.4 Continuous Modeling Method for Building First Level Grid Geological Model with Macroscopic Heterogeneity

According to the characteristics of the spatial continuous/discrete distribution of reservoir structure, depth, formation thickness, reservoir lithology, lithofacies and reservoir categories, we can use conventional continuous/discrete modeling methods to build a tectonic model, a lattice model and a face-controlled model of different representative elements.

1. Strategy of building a tectonic model

According to the changes of different layers' depths obtained by seismic interpretation and different layers' depth data of a single well obtained by well logging interpretation, we use three-dimensional geological trend surfaces, Kriging and other continuous function interpolation methods to generate the changes of different layers' structural depths in three-dimensional space. We also determine the relationship between depths at different locations within the same layer, and build a tectonic model of the first level grid, which generates the horizon grids' elevation depth.

2. Strategy of building a lattice model

Based on the first level grid structure model, we use stratigraphic sequence analysis data to build a high-precision sequence model between stratigraphic interfaces. Then, according to the need of accuracy for property modeling, we subdivide the stratigraphic sequence and build a geological lattice model to generate a 3D geological model grid system. After that, we calculate the parameters for the lattice model such as stratum thickness, sand layer thickness, oil reservoir thickness, and effective thickness.

3. Strategy of building a face-controlled model

Based on the three-dimensional reservoir lattice model, division on single wells, cross-well seismic prediction, and geological types are used as constraints. Then, we use geological statistics, Kriging, and other methods to build a face-controlled model of reservoir lithology, lithofacies or reservoir types in spatial distributions. In addition, we determine the reservoir lithology, lithofacies and reservoir types of different first level grid units to build the three-dimensional reservoir facies-controlled model.

5.1.2.5 Discontinuous Multi-Scale Discrete Multiple Media Geological Modeling Method for Building Multiple Media Geological Models of Second Grids

1. Strategy of building a discrete fracture geological model

We select appropriate unstructured grids for discretizing the large and medium scale discrete natural/hydraulic fractures. In addition, we discretize the large and medium scale fracture units (fracture scale > grid scale) into second grids to generate discrete fracture grids and discrete fracture media. Based on the deterministic discrete fracture modeling method, we obtain geometric parameters (fracture width, etc.), property parameters (porosity and permeability, etc.) and flow parameters in pore media (flow conductivity) for discrete fracture media in order to form a discrete fracture geological model.

2. Strategy of building a discrete multiple media geological model

For micro and nano-scale fractures and different scales of pores, there is a large characteristic difference in the quantity and spatial distribution for pores and fractures of different scales. When heterogeneity and multiple media characteristics are prominent, we decide the number of medium types of multiple media, according to the composition and quantity distribution of pores and fractures multiple media at different scales, and determine the volumetric percentage of different media. The poremedia are divided into NDT media and RIDT media, based on the spatial distribution of multiple media at different scales. By selecting appropriate unstructured girds, we discretize representative elements with micro and nano-scale into nested and interactive second grids based on the number of medium types, volumetric percentage and spatial distribution patterns. Moreover, we generate discrete multiple media grids and discrete multiple media (pores, micro and nano-scale fractures at different scales).We use the discrete multiple media modeling method to build the models and obtain corresponding geometric parameters (pore throat radius and fracture width, etc.), property parameters (porosity and permeability, etc.) and flow parameters in the pore media (relative permeability and capillary force, etc.) for the discrete multiple media in order to form the discrete multiple media geological model.

3. Strategy for building an equivalent media geological model

Unconventional tight reservoir macroscopic heterogeneity and multi-scale characteristics of natural/hydraulic fractures are obvious. While on the microscopic scale, within different scales of pore media and fracture media from nano to nano-micron to micron scales, microscopic heterogeneity and multiple media characteristics are prominent. In accordance with the unconventional modeling philosophy, regions decomposition based on large-scale heterogeneity, element division based on small-scale heterogeneity, and multiple pore and fracture media division based on micro-scale heterogeneity, we build geological models with different scales. By upscaling a refined geological model, refined grids of the discrete multiple media are coarsened into equivalent coarse grids within representative elements in order to increase the simulation speed. This reflects the effect of multiple media composition and distribution at different scales on the flow performance in pore and fracture media of representative elements. At the same time, by the upscaling equivalent modeling method, we build equivalent geological models with upscaling of different scales from microscopic scale to small scale to reservoir scale. Also, we upscale the microscopic scale and small scale heterogeneity to an equivalent reservoir scale geological model. This fully reflects the influence of different scale heterogeneities on the flow patterns and production performance.

5.2 Geological Modeling Technology for Discrete Natural/Hydraulic Fracture at Different Scales

Unconventional tight reservoirs contain natural/hydraulic fractures of different scales. According to the spatial distribution and ranges of geometric parameters of fractures at different scales, we use the deterministic and constrained stochastic generation technology of natural/hydraulic discrete fracture at different scales to generate the discrete fractures at different scales. We use unstructured grid technology to divide different discrete fractures into different discrete fracture elements. In addition, the grids are discretized to generate discrete fracture media. Considering the roughness and filling conditions of natural fractures, as well as the concentration, size, combination mode and stacking mode of hydraulic fracturing proppants, we innovatively develop discrete modeling techniques for natural/hydraulic fractures at different scales. Furthermore, we obtain geometric parameters (fracture width, etc.), property parameters (porosity and permeability, etc.) and flow parameters through fracture media of tight reservoirs (flow conductivity) for discrete fracture media, in order to form a discrete fracture geological model.

5.2.1 Geological Modeling Strategy for Discrete Natural/Hydraulic Fracture at Different Scales

5.2.1.1 Unconventional Tight Reservoirs Containing Natural/Hydraulic Fractures at Different Scales

Unconventional tight reservoirs contain the natural fractures at different scales. Among them, large-scale fractures are fractures with large-scale and strong regional reliability, which is obtained by manual interpretation based on three-dimensional seismic data volumes. Medium scale fractures are those fractures with relatively large-scale and high certainty among wells, which are identified by using ant tracking or a seismic coherence cube technique based on seismic attribute volumes (Lang and Guo 2013). Small scale fractures can be identified based on data such as coring, conventional well logging, and formation imaging logging. In addition, they are such small scale fractures that can be quantitatively characterized near wellbores, while their distribution difficult to be accurately identified between the wells. Microscopic fractures are fractures at the microscopic scale, which are obtained by core observations and thin section analysis. Nano scale fractures are the fracture at nano-scale, which are obtained by thin section analysis and scanning electron microscope.

By using hydraulic fracturing/refracturing technology, unconventional tight reservoirs generate hydraulic fractures at different scales. Hydraulic fractures can be identified and described by using micro seismic, fracturing monitoring, and production performance data. According to the mechanical mechanisms of forming hydraulic fractures and their corresponding scales, hydraulic fractures can be divided into three categories including main fractures, branch fractures and shear fractures.

5.2.1.2 Generation of Natural/Hydraulic Fractures at Different Scales

Based on the description, identification, and characterization results of different types of fractures (natural/hydraulic fractures) at different scales, we can obtain a spatial distribution of different types of fractures (including the groups of fractures, density/number of fractures, direction/inclination and trajectory) and geometric parameters of fractures at different scales (length, width, and height). For fractures of different types at different scales, due to the differences in their data sources and description as well as characterization methods, the spatial distribution and the reliability of geometric parameters are also different. Therefore, we have to use different generation methods (deterministic method and constrained stochastic method) to generate discrete fractures of different types and at different scales, based on the spatial

Table 5.2 Generation and modeling methods of natural/hydraulic fractures at different scales

	Conventional reservoirs	Unconventional reservoirs
Fracture types	Large scale discrete natural fractures	Natural fractures and hydraulic fractures at different scales
Fracture generation methods	Large scale fractures: deterministic method	① Large scale natural/hydraulic fractures: deterministic method ② Medium and small scale natural/hydraulic fractures: deterministic method and stochastic generation method ③ Micro and nano scale fractures: stochastic generation method
Fracture modeling	Natural fractures: using planar fracture model to build the model	① Natural fractures: discrete modeling considering fracture roughness and filling conditions ② Hydraulic fractures: discrete modeling considering proppants concentration, size, combination, and supporting mechanisms

distribution and the reliability of geometric parameters (Table 5.2).

5.2.1.3 Discrete Modeling of Natural/Hydraulic Fractures at Different Scales

We regard discrete fractures at different scales as discrete fracture elements at different scales. By unstructured grid partitioning technology, we discretize the discrete fracture elements (first level grids) to generate discrete fracture second grids, which forms discrete fracture media at different scales.

According to the discrete fracture second grid and the media types of the different grids, we use the calculation method of 3D geometry to build the geometric parameter models such as the area and volume of the fracture grids. Considering the difference between natural fractures and hydraulic fractures, we use the discrete media natural/hydraulic fracture modeling method to obtain the geometric parameters, property parameters, and the parameters model of flow in pore media with discrete fractures in different grids, in order to form a discrete fracture geological model (Fig. 5.5).

5.2.2 Generation Technique for Discrete Natural/Hydraulic Fracture at Different Scales

The spatial distribution patterns (including fracture system settings, density/number of fractures, directions/inclinations and trajectories) and geometric parameters (length, width and height) of macroscopic large-scale fractures in the regions are deterministic. According to the results from large-scale fracture descriptions, their spatial distribution and geometric parameters are given. By using spatial three-dimensional geometry deterministic algorithms, we generate the discrete fractures in order to form large scale discrete fracture elements (Liang et al. 2014).

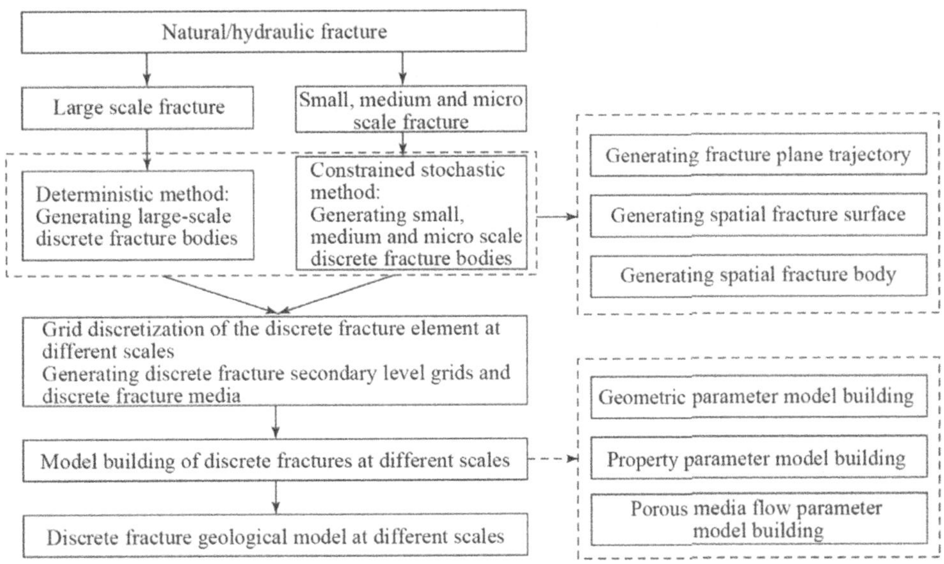

Fig. 5.5 Flow chart of discrete fracture modeling at different scale

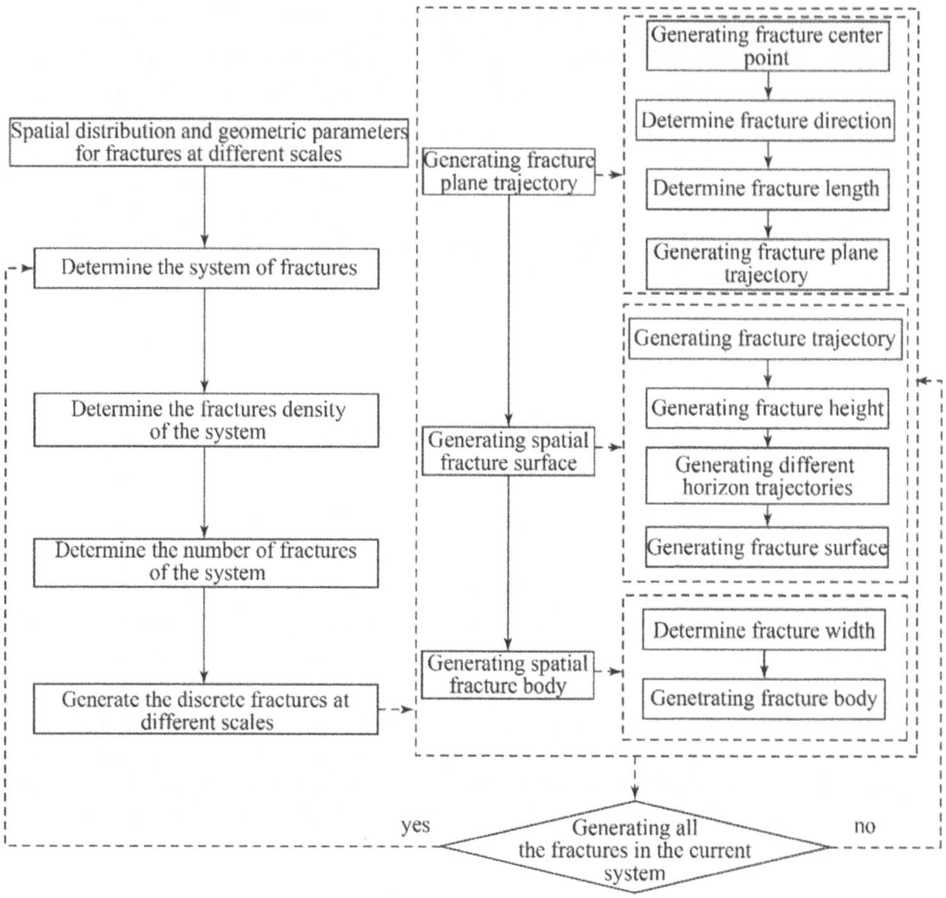

Fig. 5.6 Flow chart of generating different scales of natural/hydraulic fractures

The patterns of spatial distribution and the ranges of geometric parameters for small and medium scale fractures can be characterized by data at well locations. But the specific spatial distribution and geometric parameters between the wells are uncertain. Therefore, we use the patterns of spatial distribution and the ranges of geometric parameters obtained from the data at the well locations as deterministic constraint conditions. Then, we use stochastic methods including fractal geometry and Fisher's distribution to generate discrete fractures, in order to form small and medium scale discrete fracture elements (Xu 2009).

According to the core analysis, thin section analysis, scanning electron microscope characterization and other methods, we can obtain a statistical patterns of spatial distribution and the ranges of geometric parameters for the micro and nano-scale fractures. Then we take them as the constraint conditions. We use stochastic methods to generate discrete fractures, in order to form the micro and nano-scale discrete fracture elements (Li et al. 2016).

The methods and steps for generating different scales of natural/hydraulic fractures are as follows (Fig. 5.6):

(1) The obtainment of the spatial distribution patterns of fractures and geometric parameters

Based on the description results for fractures at different scales, we obtain such data as the spatial distribution of fractures (including the fracture system settings, densities/numbers of fractures, directions/inclinations and trajectories) and geometric parameters (length, width, and height, etc.).

(2) The determination of the fracture system

We divide the fracture systems according to the fracture types, geometry scales, directions and other characteristics.

(3) The determination of the fracture densities in different fracture systems

For macroscopic large, medium and small scale fractures, we use such methods as outcrop and core observation and imaging logging and fracture interpretation using seismic to describe and analyze these macroscopic fractures, which is used to determine the fracture densities of the fracture systems (Zhang et al. 2009).

For the micro and nano scale fractures, we can use the box method to describe and analyze the cores and thin sections in order to obtain the fracture densities and fractal dimensions. Then, we establish the equation for the correlation between fracture densities and fractal dimensions. In the situation of no micro and nano-fracture density data between wells, we can assume that the spatial distribution of cross-well micro and nano-scale fractures follows the fractal dimension distribution. Thus, according to the equation for the correlation between fracture densities and fractal dimensions, we determine the densities of cross-well micro and nano-scale fractures (Kim and Schechter 2009).

① According to the core and thin section data of different well locations, we use the box method to describe and count the fractal dimensions of fracture distribution and fracture densities.
② We establish the relationship model between the fractal dimension of fracture distribution and fracture density.

$$\ln(N_{\mathrm{n}}) = \ln m - F_{\mathrm{v}} \ln(r_{\mathrm{n}}) \tag{5.1}$$

where, r_{n} is the side length of the grid, F_{v} is the fractal dimension, and m is a linear constant.

③ Under the constraint of fractal dimensions of different well locations, we use the Kriging method to calculate the fractal dimensions of different spatial locations between the wells.
④ According to the fractal dimensions at different locations between wells, we use the fractal dimension and fracture density relationship model to determine the fracture densities between the wells.

(4) The determination of the number of fractures in the fracture system

For macroscopic large and medium scale fractures, we can obtain the deterministic number of fractures through the reservoir characterization. For small scale and nano/micro-fractures, we can obtain the number of fractures using the already acquired fracture densities. In addition, we use the stochastic method of the Poisson process to determine the number of fractures.

(5) The generation of fractures at different scales

The spatial distribution of large-scale fractures in the regions (number of fractures, directions/inclinations and trajectories) and geometric parameters (length, width and height, etc.) are deterministic. We use spatial three-dimensional geometry deterministic algorithms to generate discrete fractures.

On the other hand, the medium, small and nano/micro-fractures have deterministic spatial distribution patterns (fracture densities, the range of directions, and inclinations) and range of geometric parameters (length, width and height). Also, we use them as constraints. We use stochastic methods including fractal geometry and Fisher's distribution to generate the discrete fractures.

The steps of the constrained stochastic method for generating discrete fractures are as follows:

First Step: The Generation of the Plane Trajectories of Fractures

We use the constrained stochastic method to determine the specific center point coordinates, directions and lengths of the fractures, which generates the fracture plane trajectories.

(1) The generation of the fracture center points: the method of variable location probability for non-poisson mode

Under the influence of macroscopic geological principles, faults/large-scale fractures have control of certain degree over the distribution densities, directions, and scales of their surrounding fractures. We can use the method of variable location probability to characterize the effects of the faults (large scale fractures) on distribution densities, directions and scales of the surrounding fractures (Zheng and Yao 2009).

① In the regions, we use an evenly distributed stochastic algorithm to generate the coordinates (x, y) of the fracture center points.
② According to the distances of the center points (x, y) to the nearest faults (large-scale fractures), after considering the effects of faults/large-scale fractures on the surrounding fractures, we calculate the possibility of its existence P:

$$P(x, y, z) = e^{-kl} \tag{5.2}$$

where, $P(x, y, z)$ represents the probability of generating a fracture at point (x, y, z). l represents the distance from the point to faults/large-scale fractures. k is the influence coefficient, where $k = 0.5$ here. This coefficient can be obtained through outcrop data.

③ A random number Pr is generated in the range of (0, 1). We compare the values of Pr and P. When $Pr < P$, we keep this point. Otherwise, we again re-generate the coordinates of anew center point and determine whether it is preserved. The distribution of fracture center points generated by the spatially variable probability method for a non-Poisson mode is shown in Fig. 5.7.

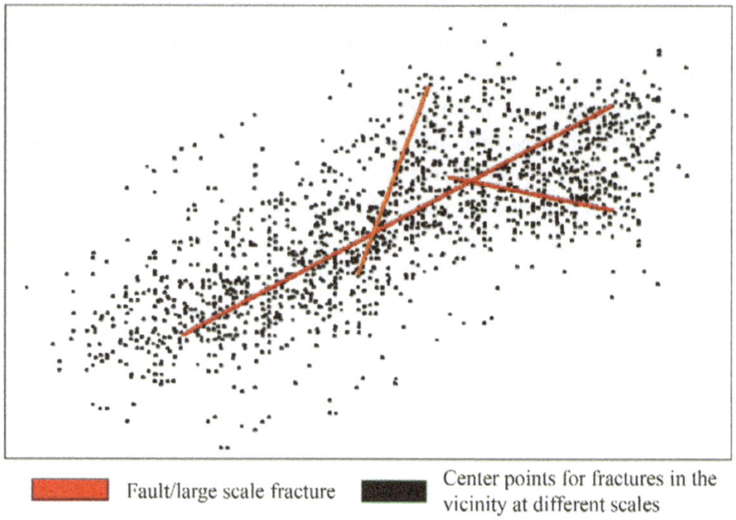

Fig. 5.7 Distribution of fracture center points generated by spatially variable probability method for non-Poisson mode

(2) Determination of fracture directions

We take the range of fracture directions at different scales as a constraint and use an evenly distributed stochastic algorithm to generate the fracture directions (Kim and Schechter 2009). The evenly distributed density function is as follows:

$$f(x) = \begin{cases} \frac{1}{b-a}, a \leq x \leq b \\ 0, \text{others} \end{cases} \quad (5.3)$$

where, $f(x)$ is the distribution density function of the direction angle; x is the direction angle; a is the minimum angle; and b is the maximum angle.

(3) Determination of fracture lengths

We take the range of fracture lengths at different scales as a constraint and assume that the stochastic distribution of fracture lengths follows the characteristics of fractal distribution. Therefore, we can use the equation for the correlation between fracture length and fractal dimension to calculate the fracture lengths.

① In the range of (0, 1), an evenly distributed random number p is generated.
② According to the range of fracture lengths, we use the equation for the correlation between fracture length L_{lf} and fractal dimension F_v (Eq. 5.4) to determine the fracture lengths (Zheng and Yao 2009).

$$l_{f_p} = [(1-p)l_{f_{min}}^{-F_v} + pl_{f_{max}}^{-F_v}]^{-\frac{1}{F_v}} \quad (5.4)$$

where, l_{fp} is the fracture length; l_{fmax} is the maximum length within the fracture length range; l_{fmin} is the minimum length within the fracture length range; F_v is the fractal dimension; p is a random number evenly distributed on the range of (0, 1).

(4) Generation of fracture plane trajectories

According to the location of center points, direction and length of fractures, we use the plane geometry algorithm to generate the plane trajectory of the fractures. By using the above method, we generate two-dimensional distribution maps for fractures of different scales and of different types (natural/hydraulic) and of different methods (deterministic and stochastic) (Fig. 5.8, 5.9, 5.10 and 5.11).

Second Step: The Generation of Spatial Fracture Planes

On the basis of fracture plane trajectories, according to the dip angle and height of fractures at different scales, we use spatial a three-dimensional geometry algorithm to generate spatial fracture planes.

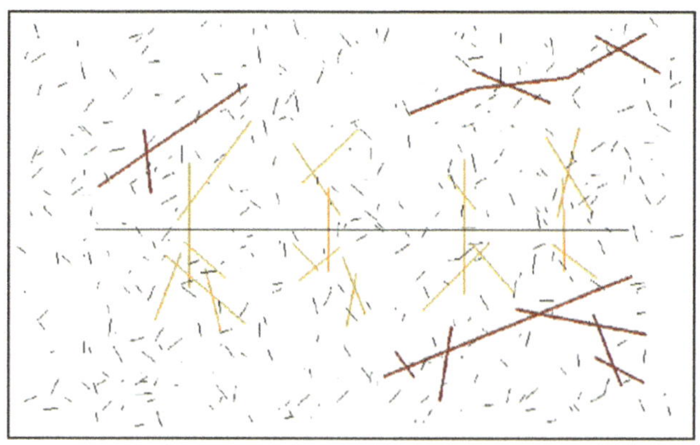

Fig. 5.8 Natural fractures and hydraulic fractures at different scales

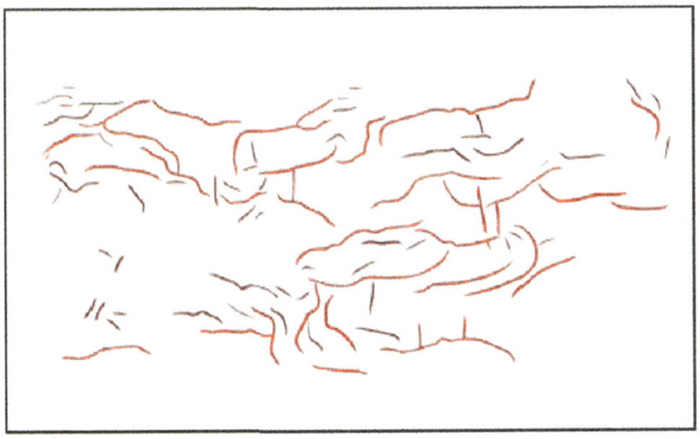

Fig. 5.9 Large and medium scale natural fractures (deterministic)

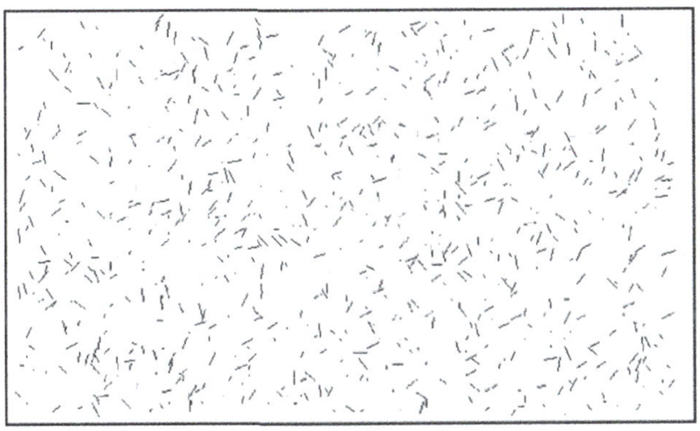

Fig. 5.10 Small and micro scale natural fractures (stochastic)

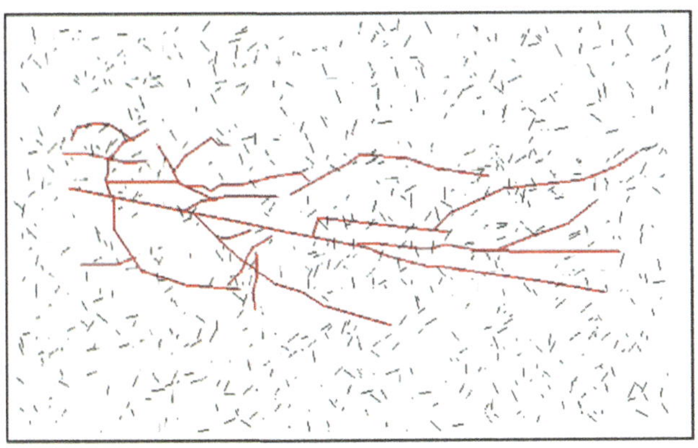

Fig. 5.11 Natural fractures at different scales (deterministic +stochastic)

(1) Generation of fracture dip angle

According to the distribution of normal vectors of fractures, which complies the symmetrically distributed geological rule on the spherical surface, and by taking the range of fracture dip angle as a constraint, we use Fisher's stochastic method to generate the dip angle of fractures at different spatial locations.

① Based on the theory of Fisher distribution, we build the probability density equation of dip angles (Kim and Schechter 2009).

$$f(\theta) = \frac{F_{\text{ish}} \sin \theta e^{F_{\text{ish}} \cos \theta}}{e^{F_{\text{ish}}} - e^{-F_{\text{ish}}}} (0 < \theta < \frac{\pi}{2}) \tag{5.5}$$

where, θ is the angular deviation value of the dip angle compared with the average dip angle; F_{ish} is the Fisher constant, which is obtained by fitting the actual data.

② According to the density equation of the Fisher distribution, by generating a Gaussian random number, we use the dip angle generation formula (Formula (5.6)) to determine the dip angles of the fractures (Kim and Schechter 2009).

$$R_{\text{F,K}}^{i} = \cos^{-1}(\frac{\ln(1 - R_{\text{G,1}}^{i})}{K} + 1) \tag{5.6}$$

where, $R_{\text{G,l}}^{i}$ is a Gaussian random number in the range of (0, 1); $R_{\text{F,K}}^{i}$ is the generated dip angle.

(2) Generation of fracture height

We take the fracture height range as a constraint. The fracture height and fracture length are in direct proportion. Based on the fracture length, we use the proportional coefficient method to determine the fracture height (Kim and Schechter 2009).

$$H_i = K \times R_i \tag{5.7}$$

where, K is the proportional constant obtained from the rock sample experiment; R_i is the fracture length; H_i is the fracture height.

(3) Generation of fracture trajectories of different strata

According to the plane trajectories of fractures, considering the inclination of fractures, we use a three-dimensional geometry algorithm to calculate the fracture trajectories (directions, inclinations, and coordinates) on different strata.

(4) Generation of spatial fracture planes

Based on fracture trajectories on different strata, we use a three-dimensional geometry algorithm to connect lines into planes, in order to form the fracture planes.

Third Step: The Generation of Spatial Three-Dimensional Fracture Bodies.

(1) Generation of fracture widths

According to the constraint of the range of fracture width, based on the assumption that fracture width and fracture length are in direct proportion, in combination with the fracture length values, we use the proportional coefficient method to calculate the fracture width (Zheng and Yao 2009).

$$W_i = K \times R_i \tag{5.8}$$

where, K is the proportional constant obtained by the rock sample experiment; R_i is the fracture length; W_i is the fracture width.

(2) Generation of spatial fracture bodies

According to the spatial distribution of fracture planes at different scales, considering changes in the width of the fractures, we use a three-dimensional geometry algorithm to generate spatially distributed discrete fracture bodies at different scales (Fig. 5.12).

5.2.3 Division of Discrete Fracture Elements and Grid Discretization

The above-mentioned discrete fracture bodies at different scales are regarded as discrete fracture elements at different scales. By using the unstructured grid partitioning technology, we divide the different discrete fracture elements into the first level grids. Then, according to the geometry of the fracture elements at different scales, we select different unstructured grid types and sizes. Next, we discretize each discrete fracture element to generate the second grids for the discrete fractures, which form the media at different scales discrete fractures (Fig. 5.13).

5.2.4 Geological Modeling of Discrete Fracture at Different Scales

Fractures on different scales are discontinuously discretely distributed in space. A traditional continuous modeling method cannot be used here. Only a discontinuous discrete modeling method can be used. There exists big discrepancy of geometric shapes and filling/supporting conditions of natural and hydraulic fractures on different scales. Taking into consideration such aspects as the roughness and filling conditions of natural fractures, the proppant concentrations, sizes and combinations, plus the supporting methods for hydraulic fractures, we have developed the discrete medium modeling method of natural/hydraulic fractures. We establish a model of geometric parameters, property parameters, and flow parameters of flow in pore media in different grids, which forms the discrete fracture geological model.

5.2.4.1 Geometric Parameters for Second-Level Grid

In the gridding process of a discrete fracture body, the discrete fracture body is divided into several fracture grids. Each fracture grid approximates the geometric characteristics of that fracture segment. The length, aperture, height, area, and volume of fractures follow the law of conservation of total quantities. The length of fractures should be equal to the sum of the lengths of the fracture grids along the fracture trajectory. The fracture aperture at different spatial positions should be consistent with the aperture of the fracture grid at that location. The height of the fractures should be the same as the total height of fracture grids in the vertical direction.

According to the media type of the discrete fracture second-level grid and other grids, we extract the geometric

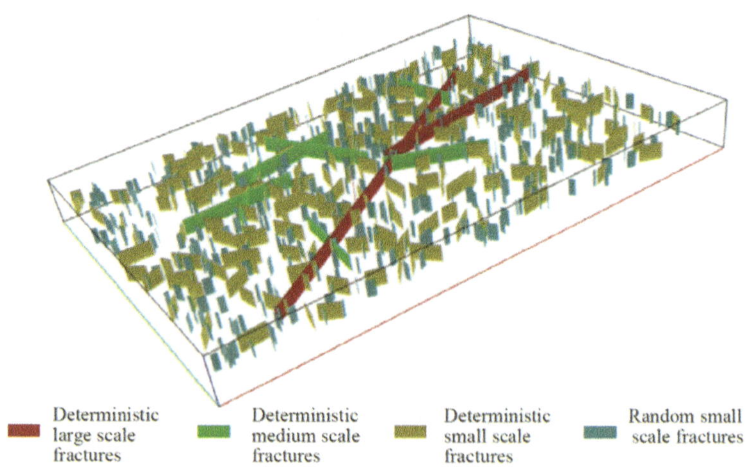

Fig. 5.12 Distribution of discrete fracture bodies at different scales

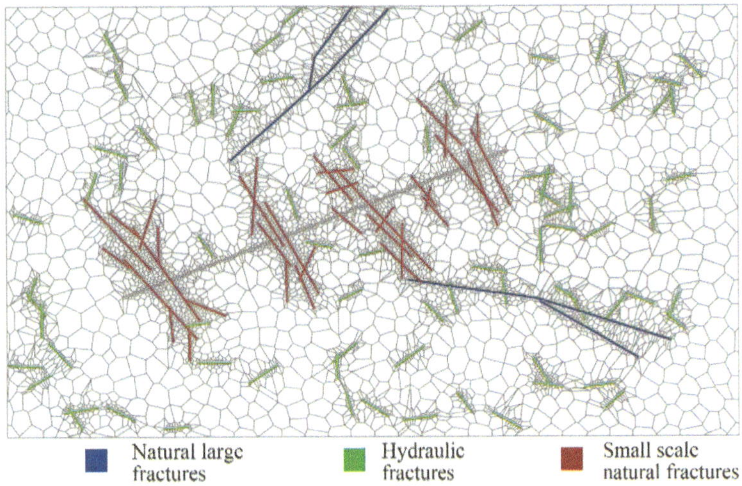

Fig. 5.13 Media with discrete fracture and grid partitioning at different scales

data such as the vertex coordinates of each grid. Then, we can obtain the geometric parameters such as the length, aperture, height, area and volume of the fracture grid by 3D geometric algorithms.

5.2.4.2 Property Parameters of Natural Fracture Media

Physical properties and flow characteristics through the pore and fracture media of tight reservoirs of different natural fractures are quite different, due to differences in their roughness and filling conditions. Natural fractures can be divided into conventional natural fractures, rough natural fractures and mineral-filled natural fractures by the roughness and filling conditions. Different discrete modeling methods are used to build aperture, porosity, permeability and flow conductivity parameter models of different grids and different media with discrete fractures.

1. The aperture of natural fractures

The aperture of natural fractures refers to the geometric aperture of the corresponding point on a fracture surface. The size of the aperture reflects the flowable space of a fluid in the fracture. It is also an important indicator to measure the properties of the fracture. The fracture aperture can be divided into hydraulic aperture and mechanical aperture based on the fracture initiation mode of natural fractures (McClure et al. 2016).

The hydraulic aperture is defined as such aperture when an upper fracture surface and the lower are contacting with each other and the fracture not opened significantly. Due to the roughness of the fracture wall, the fluid can still pass through the fracture. At this time, the fluid pressure in the fracture is less than the normal stress that the fracture surface bears. The degree of fracture deformation is related to the amount of fluid pressure inside the fracture (Fig. 5.14).

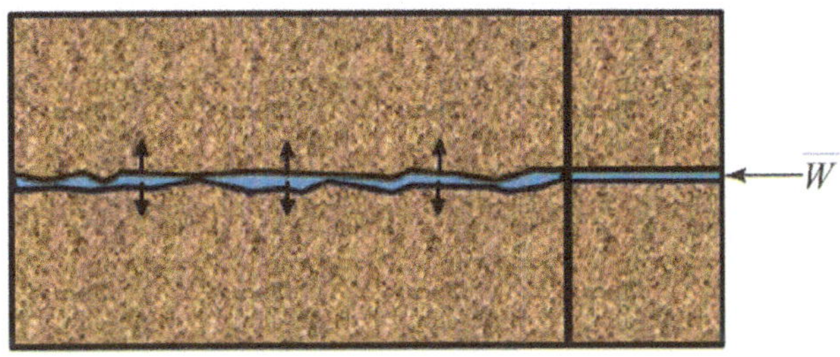

Fig. 5.14 Facture with hydraulic aperture

The mechanical aperture refers to such fracture aperture in the situation when an upper fracture surface is completely separated with a lower one and the fracture is opened clearly. At this time, the fluid pressure in the fracture is greater than the normal stress that the fracture surface bears. The total aperture of the fracture W_t consists of hydraulic aperture $\overline{W}$ and mechanical aperture W (Fig. 5.15).

When the fracture aperture is lower than the minimum hydraulic aperture, fracture surfaces are in close contact. Since a fluid is difficult to pass through such a fracture, we call this fracture as a closed fracture. Under the circumstances that the fracture surfaces are still in contact and the hydraulic aperture can be maintained for fluid to flow through, we call such a fracture as a hydraulic initiation fracture. When the fluid pressure in the fracture is higher

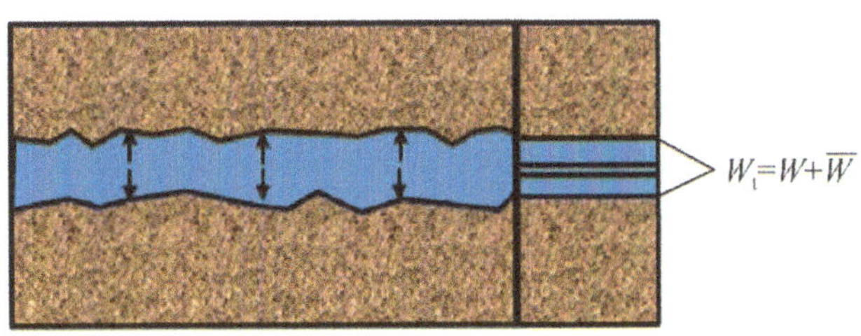

Fig. 5.15 Fracture with mechanical aperture

Fig. 5.16 Conventional natural fracture

than the normal stress that the fracture surface bears, fracture surfaces completely separate, which forms a clear flow channel. We name such a fracture as a mechanical initiation fracture.

For the conventional natural fractures with smooth surfaces and no mineral filling (Fig. 5.16), the geometric and physical parameters in a grid can be calculated by a planer model.

(1) Aperture

For conventional natural fractures with the same aperture, the fracture aperture in the grid can be calculated through the equivalent planer model (Yang and Wei 2004):

$$W_{fs} = \frac{A_f \phi_f}{l_f} \tag{5.9}$$

where, W_{fs} is the fracture aperture of the planer model, m; A_f is the end surface area of the fracture, m^2; ϕ_f is the porosity of the fracture, decimal; l_f is the length of the fracture.

(2) Porosity

The natural fracture porosity is determined by the ratio of the pore volume of a natural fracture to the volume of the fracture grid. In the conventional planer model, the volume of the fracture is the volume of the rectangular block planer fracture, which is the product of length, width and height of the fracture.

$$\phi_{fs} = \frac{V_f}{V_r} \tag{5.10}$$

where, ϕ_{fs} is the fracture porosity of the planer model, decimal; V_f is the volume of the fracture pore, m^3; V_r is the volume of the fracture grid, m^3.When the entire volume of a fracture grid is the fracture pore, the fracture porosity of the grid is 100%.

(3) Permeability

The flow in a fracture is regarded as flow between parallel planer. The flow rate through a planer fracture is calculated by the Boussinesq equation. On the other hand, it is assumed that a fracture is a porous medium. The flow rate passing through a porous medium is calculated by Darcy's law. Then we calculate the fracture permeability based on the principle of equivalent resistance in pore media flow.

$$K_{fs} = \phi_{fs} \frac{W_{fs}^2}{12} \tag{5.11}$$

where, K_{fs} is the fracture permeability of the planer model, mD; ϕ_{fs} is the fracture porosity of the planer model, decimal; W_{fs} is the fracture aperture of the planer model, m.

(4) Flow conductivity

Natural fracture conductivity is the product of natural fracture aperture and fracture permeability.

$$C_{fs} = K_{fs} \times W_{fs} = \phi_{fs} \frac{W_{fs}^3}{12} \tag{5.12}$$

where, C_{fs} is the fracture flow conductivity of the planer model, mD·m; K_{fs} is the fracture permeability of the planer model, mD; W_{fs} is the fracture aperture of the planer model, m; ϕ_{fs} is the fracture porosity of the planer model, decimal.

2. Natural fracture parameters with roughness

In fact, a natural fracture surface is not smooth. Additionally, the roughness can be measured (Fig. 5.17). For a natural fracture with roughness, the calculation of geometric and physical parameters in a grid must employ a calculation model that takes into account the fracture roughness.

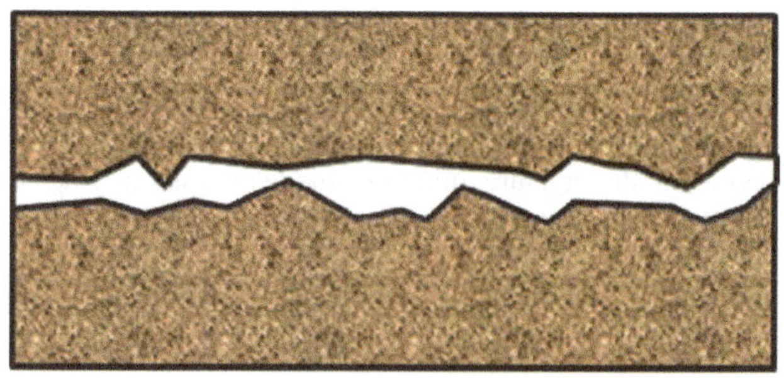

Fig. 5.17 Natural fracture with roughness

(1) Aperture

We can build a natural fracture aperture model considering the roughness within the grid by introducing the fracture roughness coefficient D_r.

$$W_f = (1 - D_r)W_{fs} \quad (5.13)$$

where, W_f is the natural fracture aperture with roughness, m; D_r is the fracture roughness coefficient, dimensionless; W_{fs} is the fracture aperture of the planer model, m. The fracture roughness coefficient D_r value is between 0 and 1. The rougher the fracture surface is, the larger the fracture roughness coefficient is, and the smaller the fracture aperture is. When $D_r = 1$, the fracture aperture is 0. The smoother the fracture surface is, the smaller the fracture roughness coefficient D_r is, and the greater the fracture aperture is. When $D_r = 0$, the fracture aperture is just the conventional natural fracture aperture.

(2) Porosity

The fracture porosity considering the roughness is still the ratio of the pore volume of a natural fracture to the volume of the fracture grid. However, the calculation of the fracture volume requires the introduction of the roughness coefficient D_r. The fracture volume considering the roughness is the product of the volume of a planer fracture and $(1 - D_r)$.

$$\phi_f = (1 - D_r)\phi_{fs} \quad (5.14)$$

where, ϕ_f is the natural fracture porosity with roughness, decimal; D_r is the fracture roughness coefficient, dimensionless; ϕ_{fs} is the fracture porosity of the planer model, decimal. The rougher the fracture surface is, the smaller the porosity is. When $D_r = 1$, the fracture porosity is 0. The smoother the fracture surface is, the greater the porosity is. When $D_r = 0$, the fracture porosity is 100%.

(3) Permeability

From the perspective of practical natural fractures, the fracture surface tortuosity τ, roughness ε and dip angle θ. Then, the equivalent relation between actual pressure gradient and pseudo pressure gradient of the flow in the fracture as well as the equivalent relation between an actual cross-sectional area and a pseudo cross-sectional area of the fluid flow profile are concerned. The calculation model for the natural fracture permeability considering the roughness within a grid was established (Qu et al. 2016).

$$K_f = \frac{10^9 W_f^2/12}{\tau^2 \cos\theta\left[1 + A(\varepsilon/W_f)^B\right]} \quad (5.15)$$

where, K_f is the natural fracture permeability with roughness, mD; W_f is the natural fracture porosity with roughness, decimal; τ is the surface tortuosity of a natural fracture with roughness, dimensionless; ε is the surface tortuosity of a natural fracture, dimensionless; θ is the dip angle of the fracture surface, (°); A and B are the coefficients to be determined.

(4) Flow conductivity

The flow conductivity of a natural fracture is the product of natural fracture aperture and fracture permeability.

$$C_f = K_f \times W_f = \frac{10^9 W_f^3/12}{\tau^2 \cos\theta\left[1 + A(\varepsilon/W_f)^B\right]} \quad (5.16)$$

where, C_f is the flow conductivity of a natural fracture with roughness, mD·m; K_f is the permeability of the natural fracture with roughness, mD; W_f is the porosity of the natural fracture with roughness, decimal;τ is the surface tortuosity of a natural fracture with roughness, ε is the surface tortuosity of the natural fracture, dimensionless; θ is the dip angle of the fracture surface, (°); A and B are the coefficients to be determined.

3. Mineral-filled natural fracture parameters

Parts of natural fractures in a reservoir are filled with minerals (Fig. 5.18). Regarding to these natural fractures filled with minerals, the calculation of geometric and physical parameters for a grid must employ a calculation model considering the degree of mineral filling.

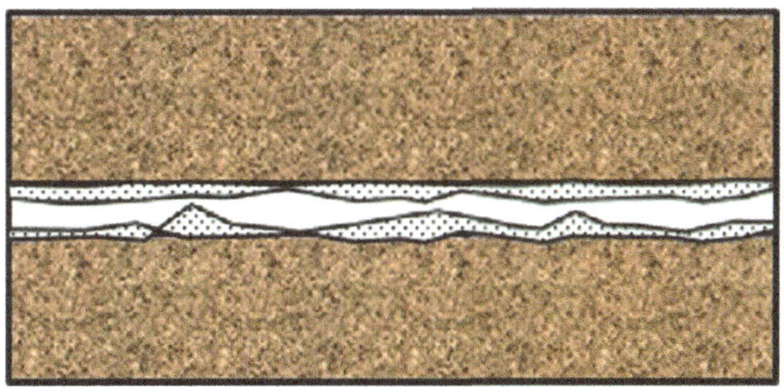

Fig. 5.18 Natural fracture with mineral filling

(1) Aperture

For natural fractures with mineral filling, the fuller the mineral filling is, the smaller the effective fracture aperture is. Here we introduce the mineral filling coefficient to calculate the effective aperture of natural fractures. The calculation model of fracture aperture is as follows (Feng et al. 2011):

$$W_{\mathrm{f}} = (1 - C_{\mathrm{c}})W_{\mathrm{fs}} \tag{5.17}$$

where, W_{f} is the aperture of natural fractures with mineral filling, m; C_{c} is the mineral filling coefficient, dimensionless; W_{fs} is the fracture aperture of the planer model, m. The mineral filling coefficient C_{c} indicates the degree that the mineral fills the fractures. It is the ratio of the mineral-filled fracture volume to the total fracture volume, with its value varying between 0 and 1. When a fracture is full of mineral, $C_{\mathrm{c}} = 1$ and the fracture aperture is 0. When the fracture is half-filled with mineral, $C_{\mathrm{c}} = 0.5$. When the fracture is not filled with minerals, $C_{\mathrm{c}} = 0$, with the fracture aperture the same as that of a normal natural fracture.

(2) Porosity

The fracture porosity considering the roughness is still the ratio of the pore volume of a natural fracture to the volume of the fracture grid. The calculation of the mineral-filled fracture volume requires the introduction of the mineral filling coefficient C_{c}. The mineral-filled fracture volume with roughness is the product of the volume of a planer fracture and $(1 - C_{\mathrm{c}})$.

$$\phi_f = (1 - C_{\mathrm{c}})\phi_{fs} \tag{5.18}$$

where, ϕ_{f} is the mineral-filled natural fracture porosity, decimal; C_{c} is the mineral filling coefficient, dimensionless; ϕ_{fs} is the fracture porosity of the planer model, decimal. The higher the degree of a mineral filling fracture is, the greater C_{c} is. When the fracture is completely filled and cemented, C_{c} is set to 1 and the fracture porosity is 0. When the fracture is not filled and is of entire initiation, C_{c} is set to 0 and the fracture porosity is 100%.

(3) Permeability

According to Poiseuille's law, Darcy's law and the principle of an equal flow rate in fractures,the permeability of the fractures in a grid can be obtained when the aperture of the mineral filling fracture is known.

$$K_{\mathrm{f}} = \phi_{\mathrm{f}}\frac{W_{\mathrm{f}}^2}{12} = \phi_{\mathrm{fs}}\frac{(1 - C_{\mathrm{c}})^3 W_{\mathrm{fs}}^2}{12} \tag{5.19}$$

where, K_{f} is the permeability of a mineral filling natural fracture, mD; ϕ_{f} is the natural fracture porosity, decimal; ϕ_{fs} is the porosity of the planer model, decimal; C_{c} is the mineral filling coefficient, dimensionless; W_{f} is the aperture of the mineral filling natural fracture, m; W_{fs} is the fracture aperture of the planer model, m.

(4) Flow conductivity

The flow conductivity of natural fractures is the product of natural fracture aperture and fracture permeability.

$$C_{\mathrm{f}} = K_{\mathrm{f}} \times W_{\mathrm{f}} = \phi_{\mathrm{fs}}\frac{(1 - C_{\mathrm{c}})^3 W_{\mathrm{fs}}^3}{12} \tag{5.20}$$

where, C_{f} is the flow conductivity of a mineral filling natural fracture, mD·m; K_{f} is the permeability of the mineral filling natural fracture, mD; W_{f} is the porosity of the mineral filling natural fracture, m; C_{c} is the mineral filling coefficient, dimensionless; ϕ_{fs} is the fracture porosity of the planer model, decimal; W_{fs} is the fracture aperture of the planer model, m.

5.2.4.3 Property Parameters of Media with Hydraulic Fracture

Physical properties and flow characteristics through pore and fracture media of tight reservoirs of hydraulic fractures are different because of the existence of proppants and different arrangement of proppants. According to proppants and their arrangement in hydraulic fractures, hydraulic fractures can be divided into hydraulic fractures without proppants and hydraulic fractures with proppants. Different discrete modeling methods are used to parameter models for creating aperture, porosity, permeability and flow conductivity with different grids and different media with discrete fractures.

1. Parameters of hydraulic fractures without proppants

The geometric and physical parameters of hydraulic fractures without proppants (Fig. 5.19) that are filled with a fracturing fluid can be calculated by a planer model.

(1) Aperture

According to the law of mass conservation, the aperture of hydraulic fractures can be calculated from the amount of a fracturing fluid, the number of hydraulic fractures, the half-length of fractures, and the height of fractures (Sun and Schechter 2015).

$$W_{\mathrm{Fl}} = \frac{V_{\mathrm{Fl}}}{2nX_{\mathrm{F}}h_{\mathrm{F}}} \tag{5.21}$$

Fig. 5.19 Hydraulic fracture without proppant

where, W_{Fl} is the hydraulic fracture aperture with the fracturing fluid injected, m; V_{Fl} is the amount of the fracturing fluid, m^3; n is the number of the hydraulic fractures, integer; X_{F} is the half-length of the fractures, m; h_{F} is the height of the fractures, m.

(2) Porosity

The porosity of hydraulic fractures is determined by the ratio of the pore volume of the fractures with a fracturing fluid injected to the volume of the hydraulic fractures.

$$\phi_{\mathrm{F}} = \frac{V_{\mathrm{F}\phi}}{V_{\mathrm{F}}} \tag{5.22}$$

where, ϕ_{F} is the porosity of hydraulic fractures, decimal; V_{F} is the volume of the hydraulic fractures, m^3; $V_{\mathrm{F}\phi}$ is the pore volume of the hydraulic fractures, m^3.

(3) Permeability

According to Poiseuille's law, Darcy's law and the principle of an equal flow rate in fractures, the fracture permeability can be calculated when the aperture is known (Qin and Li 2006).

$$K_{\mathrm{F}} = \phi_{\mathrm{F}} \frac{W_{\mathrm{Fl}}^2}{12} \tag{5.23}$$

where, K_{F} is the permeability of hydraulic fractures, mD; ϕ_{F} is the porosity of the hydraulic fractures, decimal; W_{Fl} is the hydraulic fracture aperture with a fracturing fluid injected, m.

(4) Flow conductivity

The flow conductivity of hydraulic fractures is the product of hydraulic fracture aperture and hydraulic fracture permeability.

$$C_{\mathrm{F}} = K_{\mathrm{F}} \times W_{\mathrm{Fl}} = \phi_{\mathrm{F}} \frac{W_{\mathrm{Fl}}^3}{12} \tag{5.24}$$

where, C_{F} is the flow conductivity of hydraulic fractures, mD·m; K_{F} is the permeability of the hydraulic fractures, mD; W_{Fl} is the hydraulic fracture aperture with a fracturing fluid injected, m; ϕ_{F} is the porosity of the hydraulic fractures, decimal.

2. Parameters of hydraulic fractures with proppants

(1) Unsupported stage

In the initial stage of proppant injection, the proppants in hydraulic fractures are suspended (Fig. 5.20). At this time, the geometric and physical parameters of the hydraulic fractures can still be obtained by a planer model.

1) Aperture

The added aperture due to proppant injection can be acquired by the mass of the proppants injected, the density of the proppants, the number of hydraulic fractures, the half-length and the height of the fractures.

$$W_{\mathrm{FT}} = \frac{V_{\mathrm{Fl}} + M_{\mathrm{prop}}/\rho_{\mathrm{prop}}}{2nX_{\mathrm{F}}h_{\mathrm{F}}} \tag{5.25}$$

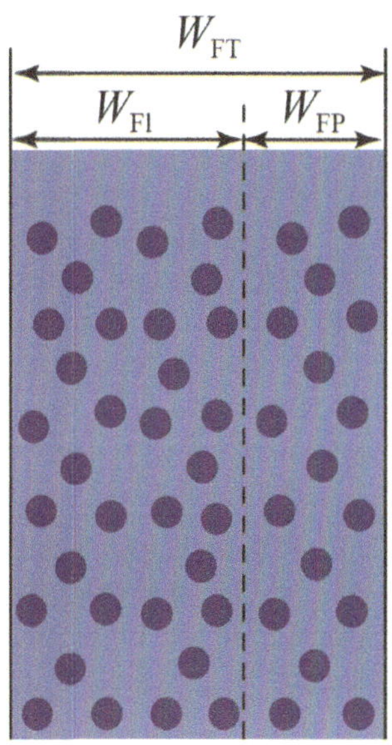

Fig. 5.20 Hydraulic fracture with proppant suspended

where, W_{FT} is the hydraulic fracture aperture with proppants injected, m; V_{FI} is the cumulative volume of a fracturing fluid, m^3; M_{prop} is the gross mass of proppants, t; ρ_{prop} is the proppant density, t/m^3; n is the number of the hydraulic fractures, integer; X_F is the half-length of the fractures, m; h_F is the height of the fractures, m.

2) Porosity

The porosity of hydraulic fractures can be obtained by the ratio of the porosity volume with proppants injected to the volume of the hydraulic fractures.

$$\phi_F = \frac{V_{F\phi}}{V_F} \tag{5.26}$$

where, ϕ_F is the porosity of hydraulic fractures, decimal; V_F is the volume of the hydraulic fractures, m^3. $V_{F\phi}$ is the pore volume of hydraulic fractures. When the fractures are completely filled and cemented, $V_{F\phi}$ is set to 0. When the fractures are not filled and are of entire initiation, $V_{F\phi}$ is set to 100%. When the fractures are partly filled or supported, the value of $V_{F\phi}$ is between 0 and 100%, m^3.

3) Permeability

According to Poiseuille's law, Darcy's law and the principle of an equal flow rate in fractures, the fracture permeability can be calculated when the fracture aperture is known at the stage for proppant injection.

$$K_F = \phi_F \frac{W_{FT}^2}{12} \tag{5.27}$$

where, K_F is the permeability of hydraulic fractures, mD; ϕ_F is the porosity of the hydraulic fractures, decimal; W_{FT} is the hydraulic fracture aperture with proppants injected, m.

4) Flow conductivity

The flow conductivity of hydraulic fractures is the product of hydraulic fracture aperture and hydraulic fracture permeability.

$$C_F = K_F \times W_{FT} = \phi_F \frac{W_{FT}^3}{12} \tag{5.28}$$

where, C_F is the flow conductivity of hydraulic fractures, mD·m; K_F is the permeability of the hydraulic fractures, mD; W_{FT} is the hydraulic fracture aperture with proppants injected, m; ϕ_F is the porosity of the hydraulic fractures, decimal.

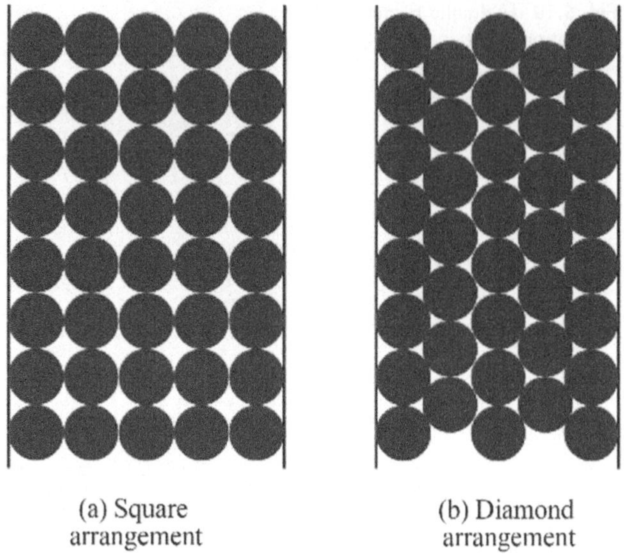

Fig. 5.21 Hydraulic fracture of different propping way

(2) Supported stage

After the fracturing fluid flows back, the proppants in hydraulic fractures are in the supporting state (Fig. 5.21). According to the arrangement of the proppants, a calculation model of the geometric and physical parameters is different.

1) Aperture

After a fracturing fluid flows back, there are two ways to obtain the fracture aperture at the initial stage of proppant support: i. The total volume of hydraulic fractures is the difference between the total volume for the fracturing fluid and proppants and the flow back volume for the fracturing fluid. According to the volume equivalent, the aperture of hydraulic fractures can be calculated on the flow back volume of the fracturing fluid. ii. The total volume of hydraulic fractures is equal to the sum of the pore volume supported by the proppants in the fractures (related to the proppant arrangement) and the volume of the injected proppants. According to the volume equivalent, the aperture of hydraulic fractures can be calculated on the volume of the injected proppants and the proppant arrangement.

$$W_F = \frac{V_{FI} + M_{prop}/\rho_{prop} - V_{PI}}{2nX_Fh_F} \tag{5.29}$$

where, W_F is the aperture of hydraulic fractures, m; V_{FI} is the cumulative volume of the injected fracturing fluid, m^3; V_{PI} is the flow back total volume of the fracturing fluid, m^3; M_{prop} is the gross mass of proppants, t; ρ_{prop} is the proppant density, t/m^3; n is the number of the hydraulic fractures, integer; X_F is the half-length of the fractures, m; h_F is the height of the fractures, m.

$$W_F = \frac{M_{prop}}{2nX_F h_F (1 - \phi_{prop})\rho_{prop}} \quad (5.30)$$

where, W_F is the aperture of the hydraulic fractures, m; M_{prop} is the gross mass of proppants, t; ρ_{prop} is the proppant density, t/m^3; n is the number of the hydraulic fractures, integer; X_F is the half-length of the hydraulic fractures, m; h_F is the height of the hydraulic fractures, m; ϕ_{prop} is the porosity of the hydraulic fractures (proppant), decimal.

2) Porosity

Hydraulic fractures are filled with proppants. The porosity can be obtained by the particle size and arrangement of the proppants. When the proppants are regularly arranged in equal diameter spheres, the porosity of the hydraulic fractures is only related to the arrangement of proppants. It has nothing to do with a particle size of proppants (Guo 1994).

When the proppants are in a positive array, the porosity of the hydraulic fractures is

$$\phi_F = 1 - \frac{\pi}{6} = 47.6\% \quad (5.31)$$

When the proppants are in a diamond array, the porosity of the hydraulic fractures is

$$\phi_F = 1 - \frac{\pi}{3\sqrt{2}} = 25.95\% \quad (5.32)$$

3) Permeability

According to Kozeny's capillary model, the ideal proppant alignment model was transformed into a diversion beam model. The permeability of hydraulic fractures was calculated based on the relationship between the capillary number and the number of proppant particles on the cross-sections of the two models.

When the proppants are in a positive array, the permeability of the hydraulic fractures is

$$K_F = \partial \frac{1}{8\pi} \phi_F^2 D^2 \quad (5.33)$$

When the proppants are in a diamond array, the permeability of the hydraulic fractures is

$$K_F = \partial \frac{1}{16\pi} \phi_F^2 D^2 \quad (5.34)$$

where, K_F is the fracture permeability, mD; ϕ_F is the fracture porosity, decimal; ∂ is the filling coefficient, which can be obtained by a linear regression of experimental data; D is the particle size of proppants, m.

4) Fracture conductivity

A calculation model for the permeability of hydraulic fractures is different when proppants are arranged in different ways. Hydraulic fracture conductivity is the product of hydraulic fracture aperture and fracture permeability. Therefore, the calculation model for the fracture conductivity is different.

When the proppants are in an orthogonal arrangement (positive array), the fracture conductivity of the hydraulic fractures is

$$C_F = K_F \times W_F = \partial \frac{1}{8\pi} \phi_F^2 D^2 W_F \quad (5.35)$$

When the proppants are in a staggered arrangement (diamond array), the fracture conductivity of the hydraulic fractures is

$$C_F = K_F \times W_F = \partial \frac{1}{16\pi} \phi_F^2 D^2 W_F \quad (5.36)$$

where, C_F is the fracture conductivity of the hydraulic fractures, mD·m; K_F is the permeability of the hydraulic fractures, mD; ϕ_F is the porosity of the hydraulic fractures, decimal form; ∂ is the proppant filling coefficient, which can be obtained by a linear regression of experimental data; D is the particle size of the proppants, m.

5.2.4.4 Flow Parameters for Fracture Media

The flow parameters of natural/hydraulic fracture media with at different scales generally include capillary pressure, relative permeability curves and compressibility coefficients. The calculation of model parameters varies because of the distinctions of natural fractures and hydraulic fractures. The calculation of model parameters for natural fractures varies with the difference between closed fractures and opened fractures. The calculation of model parameters for hydraulic fractures varies with the differences between supporting conditions and filling levels.

1. Capillary pressure

The natural fracture aperture is generally greater than 10 μm. Thus, the capillary pressure in the fractures can be ignored as it is approximately 0. The hydraulic fractures with support of proppants are equivalent to high pore media. The capillary pressure is greater than 0 but less than the capillary pressure in reservoir matrix pores.

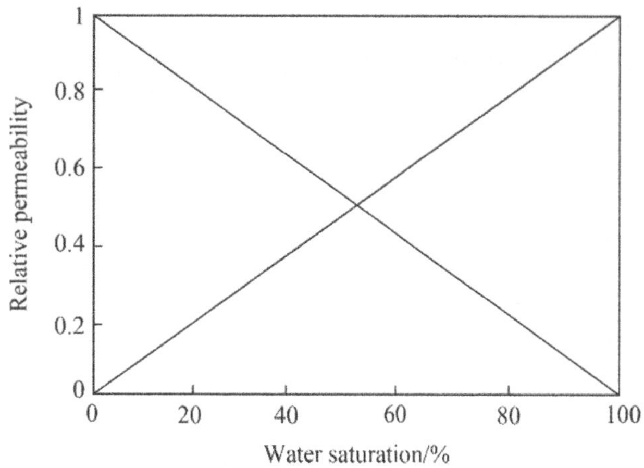

Fig. 5.22 Relative permeability curve of large-scale natural fracture

2. Relative permeability curves

There are great differences in the relative permeability curves between natural fractures and hydraulic fractures at different scales. A relative permeability curve in large-scale natural fractures can be represented by a cross-type curve (Fig. 5.22). The large-scale hydraulic fractures with support of proppants are equivalent to high pore media. Therefore, the relative permeability for oil-water in hydraulic fractures is less than that in natural fractures, but it is greater than the relative permeability in reservoir matrix pores (Fig. 5.23).

3. Compressibility coefficients

Because of the strong compressibility of natural fractures, the compressibility is generally 10 times that of the reservoir matrix compressibility. The compression coefficient of hydraulic fractures with proppants supporting is greater than the compressibility coefficient of reservoir matrix and less than the compressibility coefficient of natural fractures.

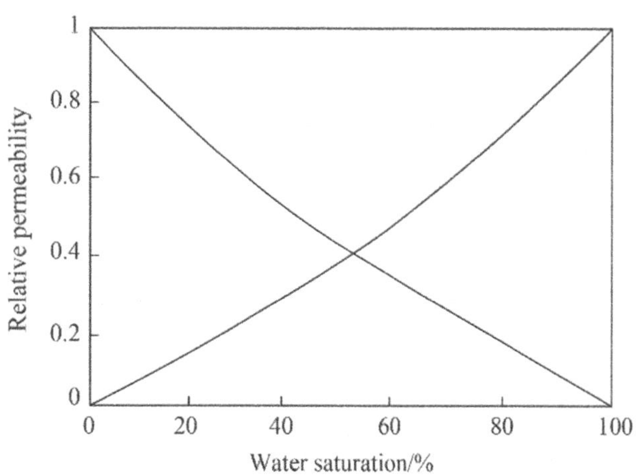

Fig. 5.23 Relative permeability curve of large-scale hydraulic fracture

5.3 Geological Modeling Technology for Discrete Multiple Media at Different Scales

The pore distribution for conventional reservoirs is relatively concentrated, which is characterized by a single type of pore media and continuous distribution. The property parameters of pore media change continuously. Therefore, a single type of pore media continuous modeling method is generally used to establish are reservoir matrix porous model. The pores in unconventional tight reservoirs are smaller, with nano-nano/micro-micro scale pore media at different scales. The composition and quantity distribution patterns of pores at different scales are quite different. Moreover, pores at different scales show discrete discontinuous distribution in space. The spatial distribution patterns are quite different. The geometric and physical parameters of pore media at different scales are quite different. At the same time, natural/hydraulic nano/micro-fractures also exist in tight reservoirs. These nano/micro-fractures are of large quantity and small sizes, which cannot be treated as deterministic discrete fractures. But on the microscopic scale they still have certain storage capacity and flow capability to communicate pores of different scales. In addition, they show discrete and discontinuous distribution in space. Therefore, the pores and nano/micro-fractures at different scales are regarded as discretely distributed micro scale discrete media. The concept of discrete multiple media is proposed for the first time. The spatial distribution and quantity distribution of pores and fractures at different scales are used as constraints to generate discrete multiple media. It is a leap forward for the continuous modeling method of the property parameters of a single type pore media. The modeling technique of the property parameters for discontinuous discrete multiple media has been established.

The discrete multiple media modeling process of pores and fractures at different scales is shown in Fig. 5.24. i. According to the magnitude and strength of the macroscopic heterogeneity of a reservoir, the first level regions based on large scale heterogeneity are partitioned. ii. Within the same regions, the second level representative elements are partitioned based on small scale heterogeneity. Also, we choose the appropriate unstructured grids to partition the regions in order to form the first level discrete grids that correspond to representative elements. iii. Within a representative element, the third levels of multiple media are partitioned according to the quantity and spatial distribution of microscopic pores and fractures at different scales. For the

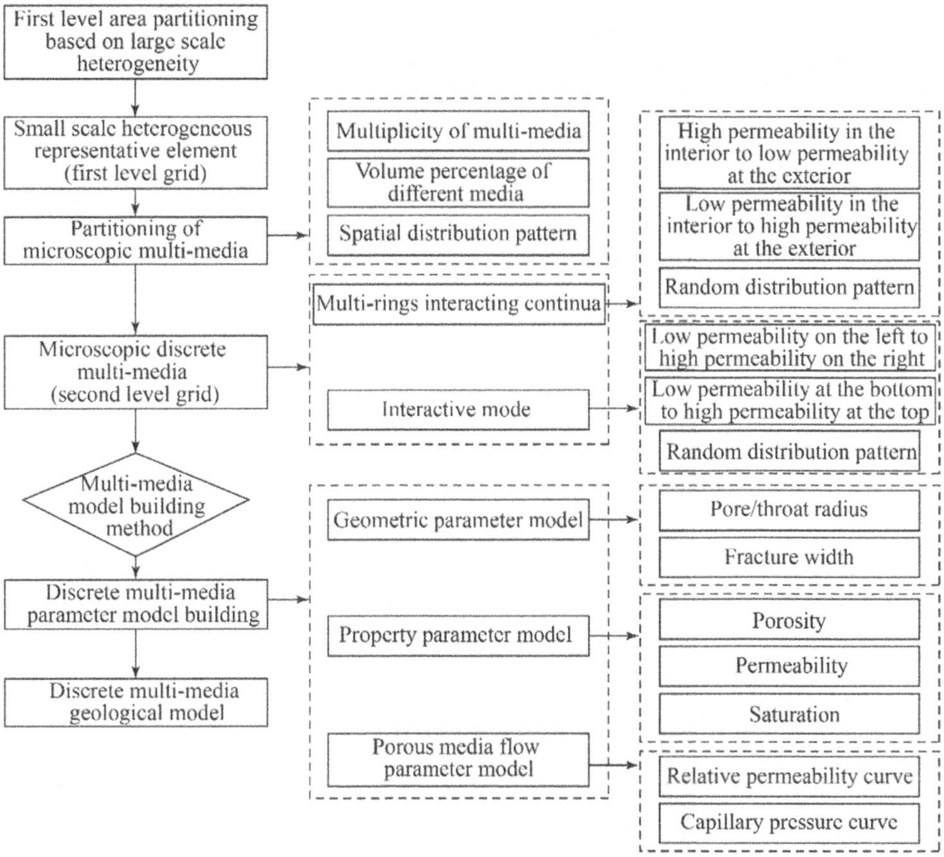

Fig. 5.24 Modeling flowchart for discrete pore and fracture multiple media and at different scales

situations where the micro and small scale fractures and pores at different scales are extensive, where there are big differences in the quantity and spatial distribution characteristics of pores and fractures at different scales, and where the micro scale heterogeneity and the multiple media characteristics are prominent, we decide the number of medium types of multiple media according to the quantitative distribution and the volume percentage of multiple media at different scales. According to the spatial distribution law of multiple media at different scales, they are divided into two kinds of multiple media modes: nested distribution type (in the following text, it is referred as NDT) media and randomly interacting distribution type (in the following text, it is referred as RIDT) media. For the NDT media, the fluid only flows from one single medium to the other single medium. For the RIDT media, the fluid can flow from one single medium to server media surrounding that single medium. By selecting appropriate unstructured grids, representative elements are discretized into nested and interactive second level grids. Moreover, we generate discrete multiple media grids and discrete multiple media accordingly. Using the discrete multiple media modeling methods, we establish the geometric parameters (pore/pore throat radius, etc.), property parameters (porosity and permeability, etc.), flow parameters and other models of the discrete multiple media. Eventually, we form a discrete multiple media geological model.

5.3.1 Partitioning of Macroscopic Heterogeneous Regions and Representative Elements

According to the macroscopic heterogeneity for spatial distribution of such geological conditions as lithology, lithofacies, reservoir types, degree and scales of natural fractures, magnitude and direction of groud stress, and fluid properties, the first level regions are partitioned with macroscopic large-scale heterogeneity. There are relatively large differences in the geological conditions and heterogeneities between different regions. And there are relatively small differences in the geological conditions and heterogeneities within the same regions. As shown in Fig. 5.25, there are many types of lithologies within the study domain. According to the spatial distribution of lithology, it is divided into several different first level lithological regions.

According to the differences in small-scale geological conditions and heterogeneity within the same region, it is partitioned into several second level representative elements. Between different representative elements, the small-scale heterogeneity varies greatly. Within the same representative element, there exists micro/small scale fractures and pores at different scales. But the distribution and property characteristics of pores and fractures at different scales are quite different. According to the distribution characteristics of

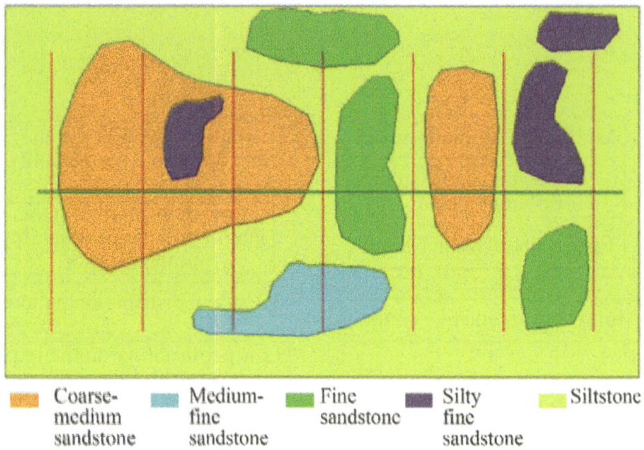

Fig. 5.25 First level area partitioning based on lithology distribution

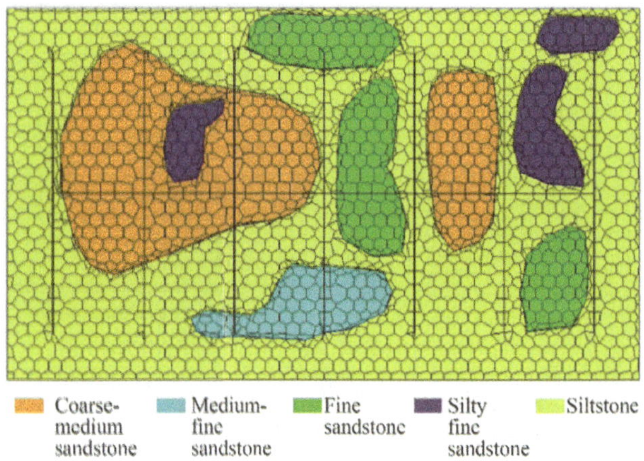

Fig. 5.26 Second level representative element partitioning within the same area

different representative elements, a suitable unstructured grid is used to partition this area, which forms the first level discrete grids that correspond to representative elements (Fig. 5.26).

5.3.2 Division of Discrete Multiple Media and Generation of the Second Level Discrete Grid

In a representative element, three levels of multiple media are partitioned according to the quantity and spatial distributions of microscopic porous and fractures at different scales.

5.3.2.1 Partitioning of Discrete Multiple Media

1. Establishing the mode of quantity distribution of pore and fracture media of different scales

According to such experimental methods as constant rate mercury injection and high pressure mercury injection, the composition and quantity distribution of pores at different scales can be obtained as shown in Fig. 5.27. There are large differences in composition and quantity distribution of pores at different scales for different lithologies, lithofacies, and reservoir types as shown in Figs. 5.28, 5.29 and 5.30.

Unconventional tight reservoirs not only have nano-nano/micro-micro scale pore media, but also have nano/micro-scale natural/hydraulic fractures. Described and calculated through core analysis and thin section analysis, we can get the composition and quantity distribution of nano/micro-fractures at different scales (shown in Fig. 5.31). Pores and nano/micro-fractures at different scales can be regarded as discretely distributed micro scale discrete media.

2. Partitioning of multiple media

Within the same representative element, both pores of different scales and nano/micro-fractures of different scales exist within the same representative element. According to the composition and quantity distribution of multiple media at different scales, based on the fluid properties, flow mechanism, and production conditions, we can decide the

Fig. 5.27 Distribution plot for the sizes and volume percentages of pores at different scales

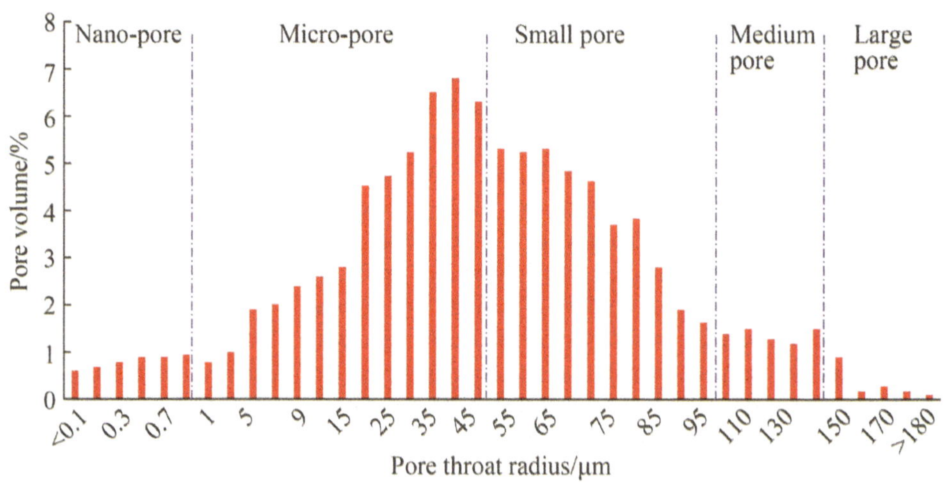

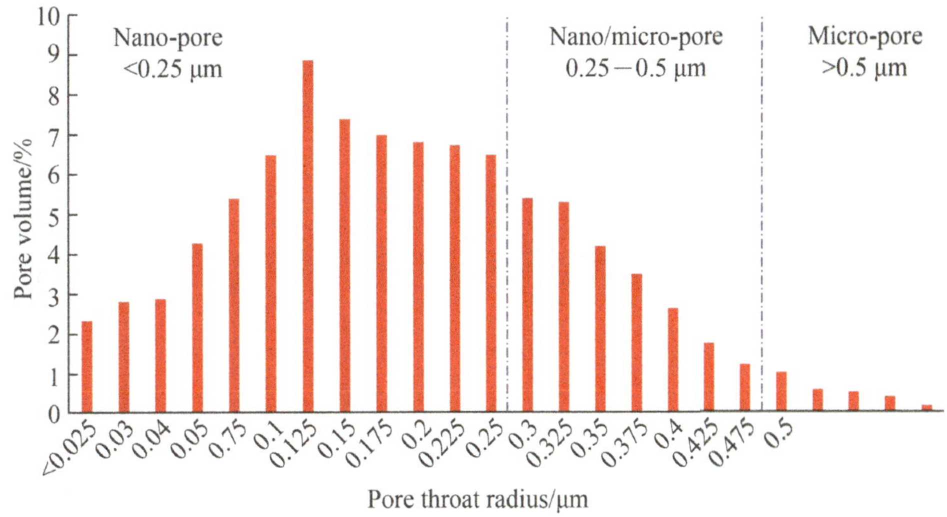

Fig. 5.28 Sizes and volume percentages of pores with nano-pores dominant

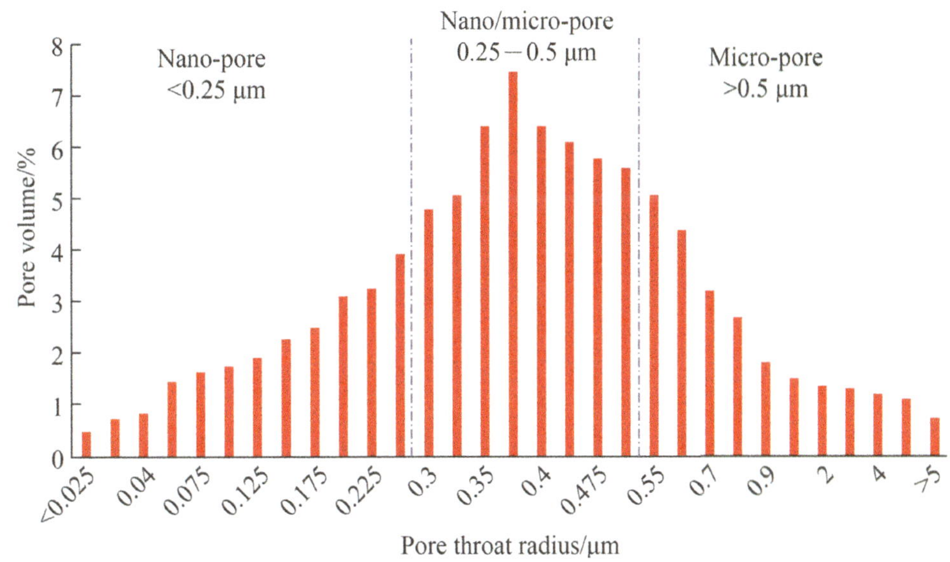

Fig. 5.29 Sizes and volume percentages of pores with nano/micro-pores dominant

number of media types, and determine the boundary and volume percentage of different media (Fig. 5.32).

5.3.2.2 Generation of the Second Level Discrete Grids

1. Establishing the spatial distribution pattern for pores and fractures at different scales

In the representative element of a tight reservoir, the porous medium with fractures at different scales show discontinuous and discrete distribution in space. Controlled by different geological principles, they show different spatial distribution modes, which can be divided into two distribution modes, NDT distribution and the RIDT distribution.

(1) NDT distribution mode

According to the spatial distribution behavior of pores and fractures at different scales in the representative element, the NDT distribution mode can generally be divided into three categories: scale of the media becoming smaller from inside to outside, scale of the media becoming larger from inside to outside, and randomly distributed mode (Table 5.3).

(2) RIDT distribution mode

According to the spatial distribution behavior of pores and fractures at different scales within the representative element, the RIDT distribution mode can be generally divided into five categories, the scale of the media becoming larger from left to right, the scale of the media becoming smaller from

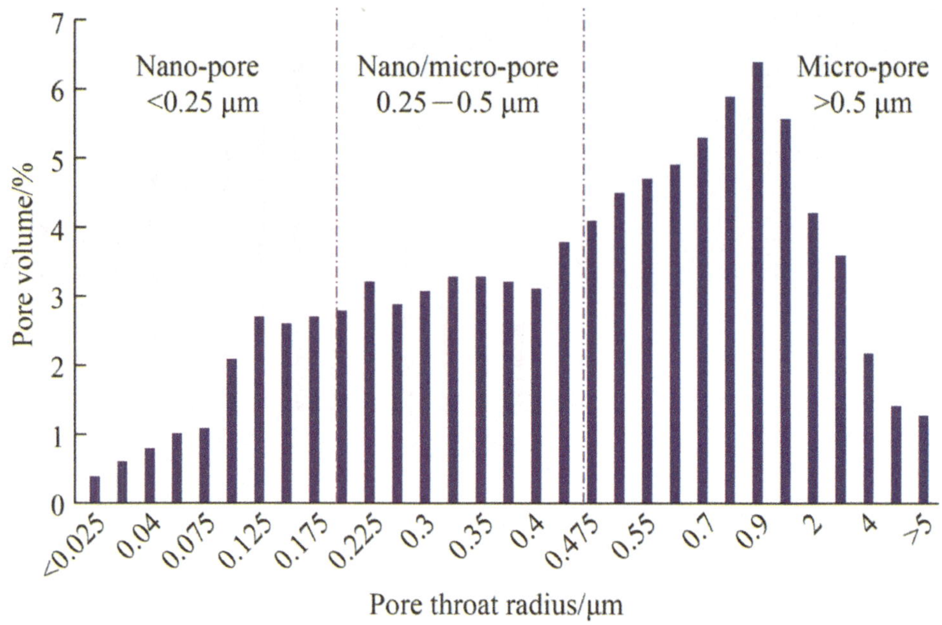

Fig. 5.30 Sizes and volume percentages of pores with micro-pores dominant

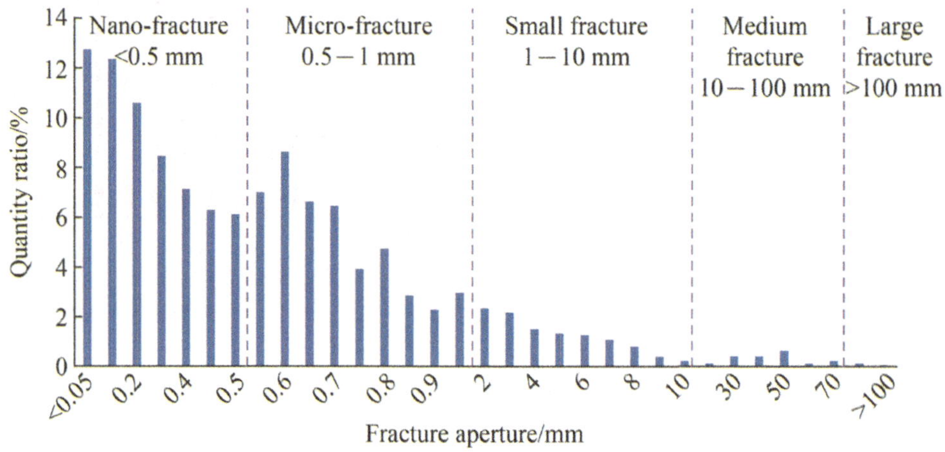

Fig. 5.31 Apertures and quantity percentages for fractures at different scales

left to right, the scale of the media becoming smaller from top to bottom, the scale of the media becoming larger from top to bottom, and randomly distributed mode. Three typical modes are shown in Table 5.4.

2. Generation of second level discrete grids for multiple media

① Establishing the composition and quantity distribution mode for pores and fractures at different scales with different lithologies, lithofacies, and reservoir types. According to the quantity composition and distribution of pore and fracture media at different scales, pores and fractures at different scales are partitioned into different multiple media. Also, the number of medium types and the volume percentage corresponding to each medium are identified.

② The spatial distribution patterns of pores and fractures at different scales for different lithologies, lithofacies, and reservoir types were established. In addition, the spatial distribution patterns of NDT/RIDT distribution for porous medium with fractures at different scales in different elements are determined.

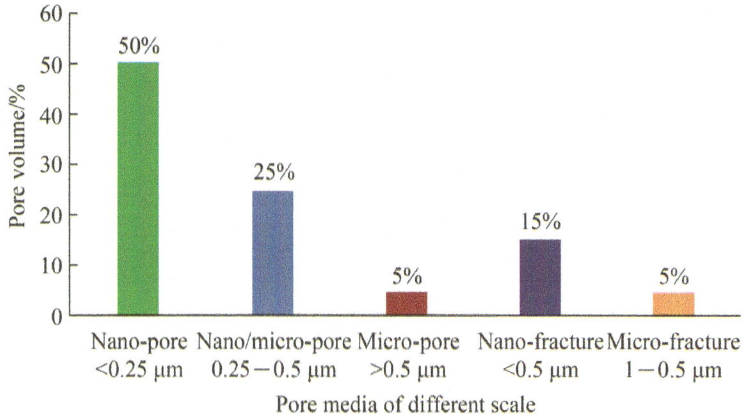

Fig. 5.32 Paritioning and volume percentage of multiple media at different scales within the same element

Table 5.3 NDT distribution mode

NDT distribution	Description of distribution of pores and fractures within the element	Schematic of pores and fractures distribution within the element	Schematic of the second level grids for different media within the element
Scale of the media becoming smaller from the inside to the outside	Pore media at different scales appear in the space as follows: There are micro-pores in the middle, gradually transitioning outwards into nano/micro-pores and eventually nanopores on the exterior. The physical properties of the pore media gradually deteriorate, from high permeability at the center gradually to low permeability at the exterior	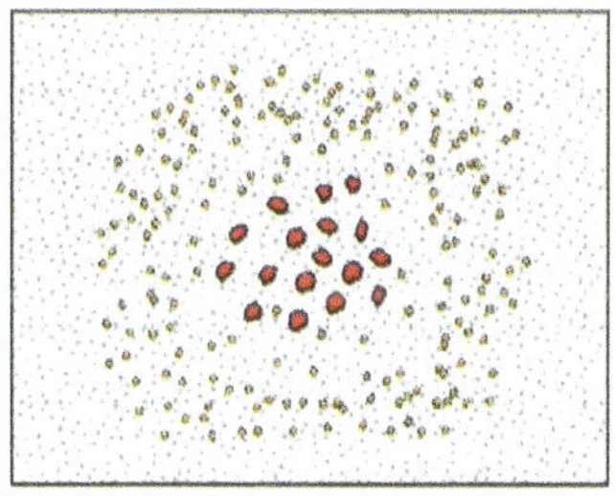	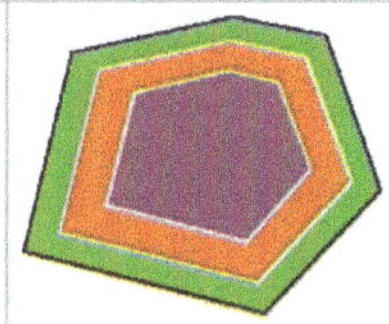
Scale of the media becoming larger from the inside to the outside	Pore media at different scales appear in space as follows: There are micro-pores from the exterior, gradually transitioning into nano/micro-pores towards the middle, and eventually nano-pores at the center. The physical properties of the pore media gradually deteriorate, from high permeability at the exterior to low permeability at the center	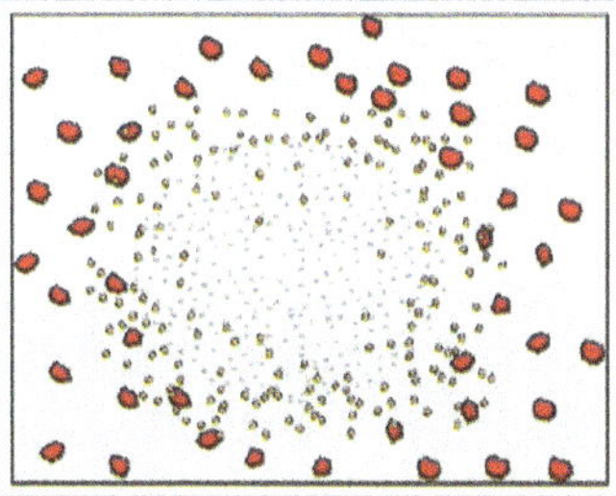	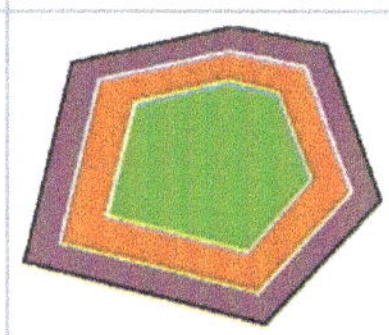
Randomly distributed mode	The micro-pores and nano-pores are randomly distributed in space. And the physical properties of the pore media vary randomly and discontinuously	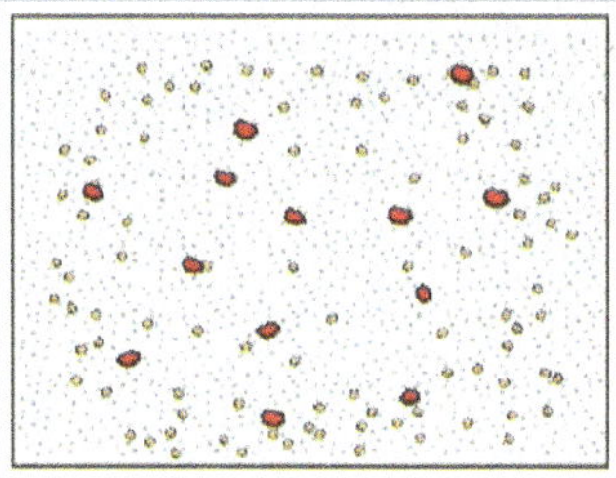	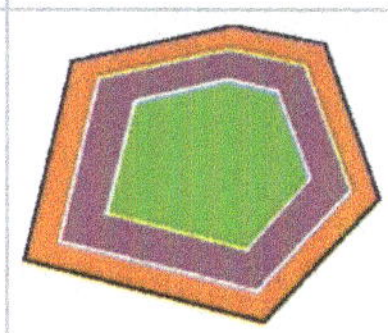

③ According to the number of media types, volume percentage, and spatial distribution pattern for pore and fracture media of different scales in different elements, an appropriate unstructured grid is selected to discretize the representative element into the second level grids, which have a certain number of media types,volume percentage, and nested/interactive spatial distribution patterns. Furthermore, we generate the discrete multiple media grids and discrete multiple media (Figs. 5.33, 5.34 and 5.35).

5.3.3 Geological Modeling for Discrete Multiple Media

Pores and nano/micro-fractures at different scales are regarded as discretely distributed micro scale discrete media. According to the composition, quantity, and spatial distribution of media at different scales pores and fractures in different elements, we go beyond the concept of the continuous modeling method for property parameters for single porosity media. Using the property parameter modeling method for discontinuous and discrete

Table 5.4 RIDT distribution mode

RIDT distribution mode	Description of distribution of pores and fractures within the element	Schematic of distribution of pores and fractures in the element	Schematic of the second level grids for different media in the element
Scale of the media becoming larger from the left to the right	Pore media of different scales appear in space as nano-pores from the left, gradually transitioning to nano/micro-pores towards the right, and micro-pores on the right. The physical properties of the pore media gradually improve from low permeability on the left to high permeability on the right	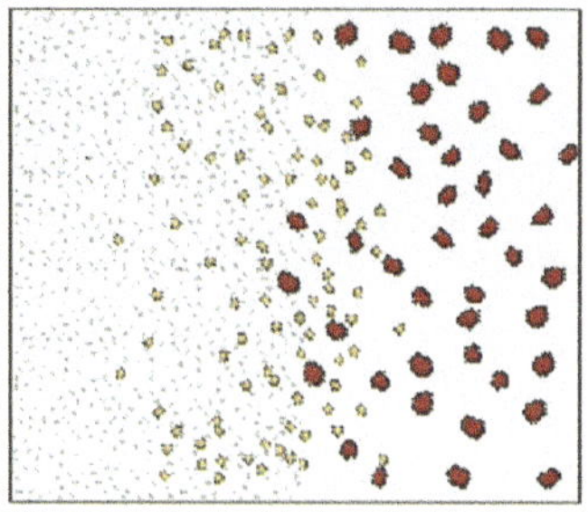	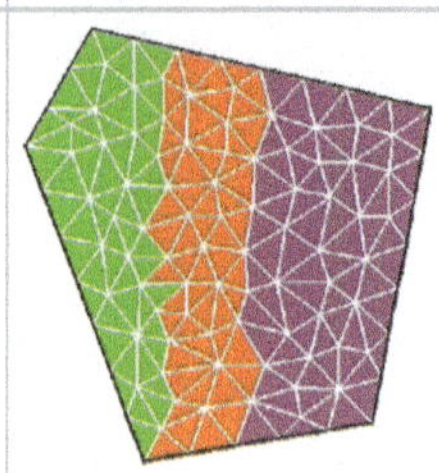
Scale of the media becoming smaller from the top to the bottom	Pore media of different scales appear in space as nano-pores from top, gradually transitioning downwards to micro and nano-pores, and to micro-pores on the bottom. The physical properties of the pore media gradually improve from the low permeability on the top to high permeability on the bottom, which in turn becomes better	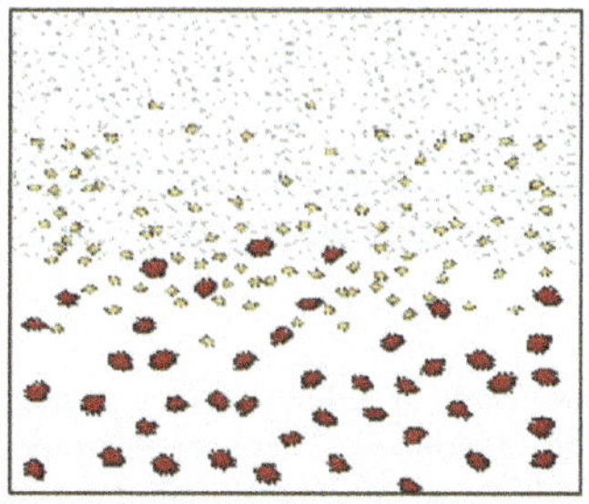	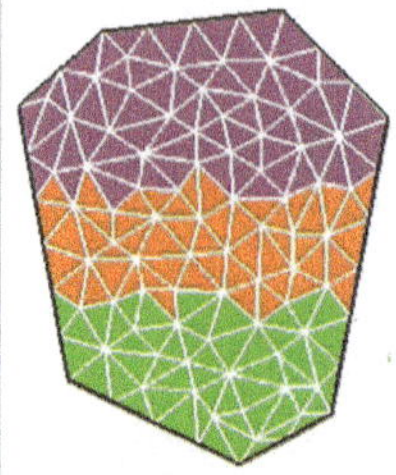
Randomly distributed mode	The stripped regularity of the distribution of the pore media at different scales is not obvious, with strong randomness. The locations and proportions of micro and nano-pores are randomly distributed. And the physical properties of the pore media randomly change	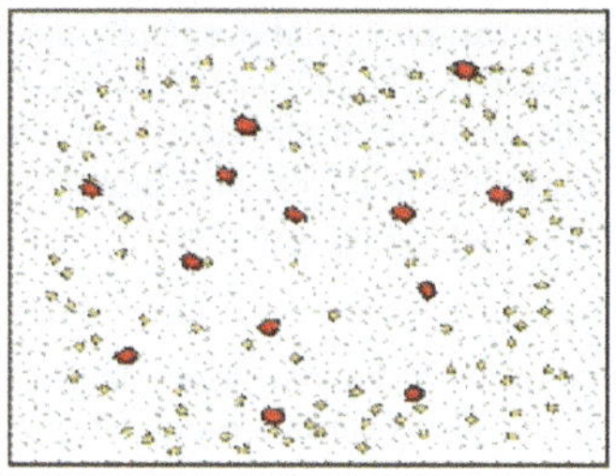	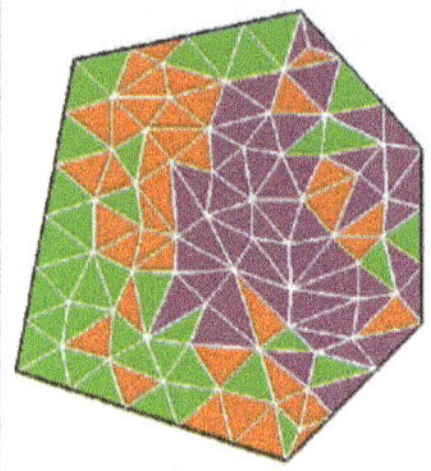

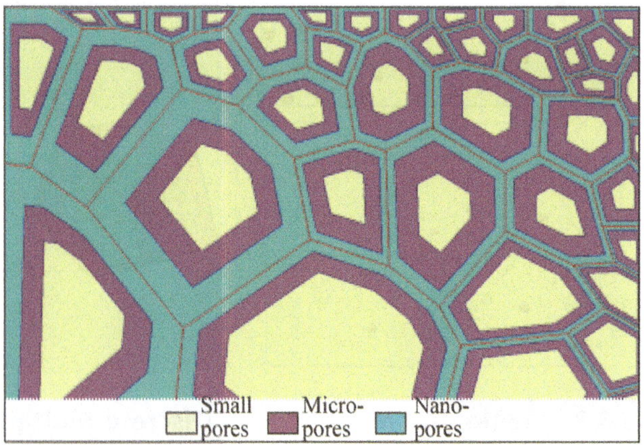

Fig. 5.33 NDT multiple media

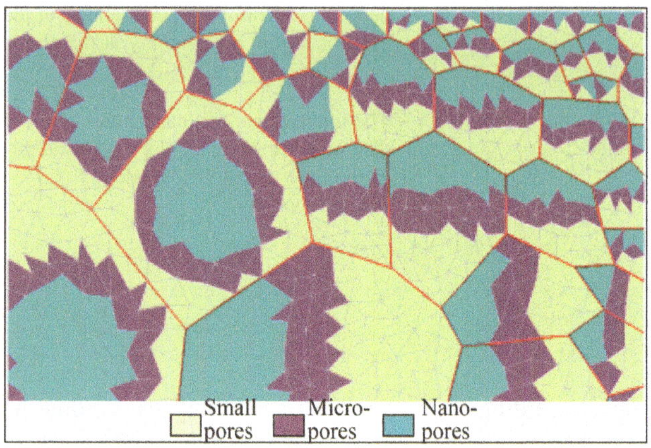

Fig. 5.34 RIDT multiple media

multiple media, based on a discrete multiple media the second level grid (Fig. 5.36), we establish the geometric parameters model for the second level grids (pore radius, pore throat radius, and fracture apertures), property parameter model (porosity, permeability, and oil saturation) and parameter model of flow in pore media(relative permeability curve, capillary force curve).

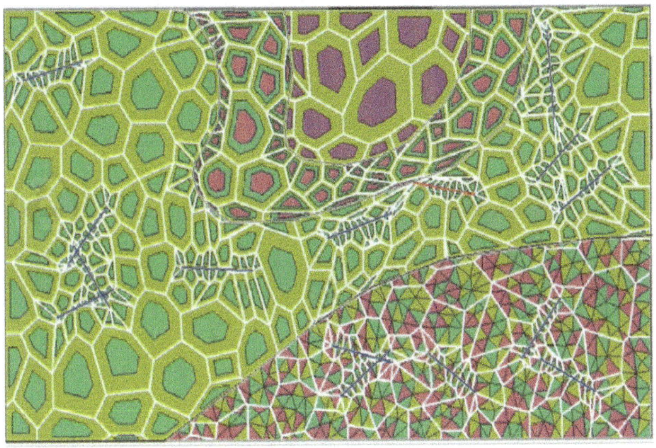

Fig. 5.35 Multiple media for different areas and elements

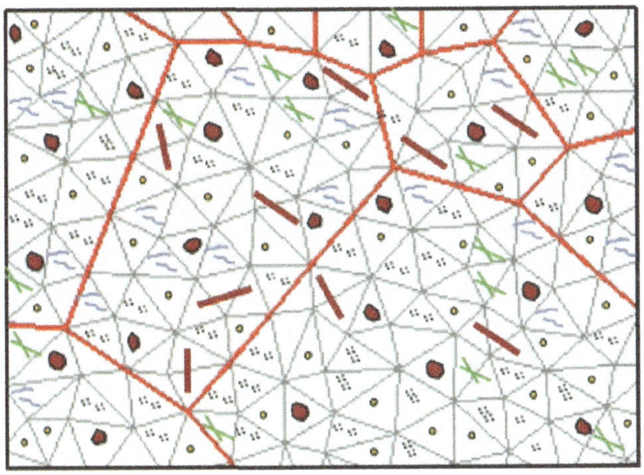

Fig. 5.36 Second level grid for different representative element and discrete multiple media modeling for geometric parameter

5.3.3.1 Modeling for Geometric Parameters

1. Grid geometry parameters for pore and fracture media of different scales

According to the geometric characteristics of unstructured grids of different types, we can calculate the geometric parameters for each second level grid (including grid length, width, height, area, and volume).

2. Geometric parameters for pore media at different scales

(1) Pore radius

According to a constant rate mercury injection experiment, the pore radius distribution and volume percentage of pores at different scales in a representative element can be obtained (Fig. 5.37). From this, the pore radius of different pore media can be calculated: i. According to the pore radius distribution for pores at different scales and their volume percentage, the total average pore radius can be determined. ii. According to the scale boundaries of different pore media and the volume percentage of pores at different scales in this range, the average pore radius and the peak pore radius of the pore media are calculated.

(2) Pore throat radius

According to a high pressure mercury injection experiment, the pore throat radius distribution and volume percentage of pores at different scales in a representative element can be obtained (Fig. 5.37). In addition, the pore throat radius of different pore media can be calculated: i. According to the pore throat radius distribution for pores at different scales and their volume percentage, the total average pore throat radius can be determined. ii. According to the scale boundaries of different pore media and the volume percentage of pores at different scales in this range, the average pore throat radius and the mainstream pore throat radius of the pore media can be calculated.

3. Geometry parameters for fracture media of different scales

Through quantitative characterization of cores and thin section fractures, we can obtain the distribution and quantity percentage of nano/micro-fracture apertures at different scales in representative elements (Fig. 5.38). Moreover, we can calculate the fracture widths for fracture media of different scales. i. According to the aperture distribution for fractures at different scales and its quantitative percentage, we can determine the total average fracture aperture; ii. According to the scale boundaries of fractures at different scales and the quantity percentage of fractures at different scales in this range, we can calculate the average fracture width and the peak value of fracture width.

5.3.3.2 Modeling for Physical Parameters

1. Pore media at different scales

(1) Porosity

According to high pressure/constant rate mercury injection experiments, the total porosity of a representative element and the pore volume percentage of pore media at different scales can be obtained. According to the principle of pore volume conservation, by using the total porosity of the representative element and the pore volume percentage of the pore media at different scales, the porosity of the pore media at different scales is determined.

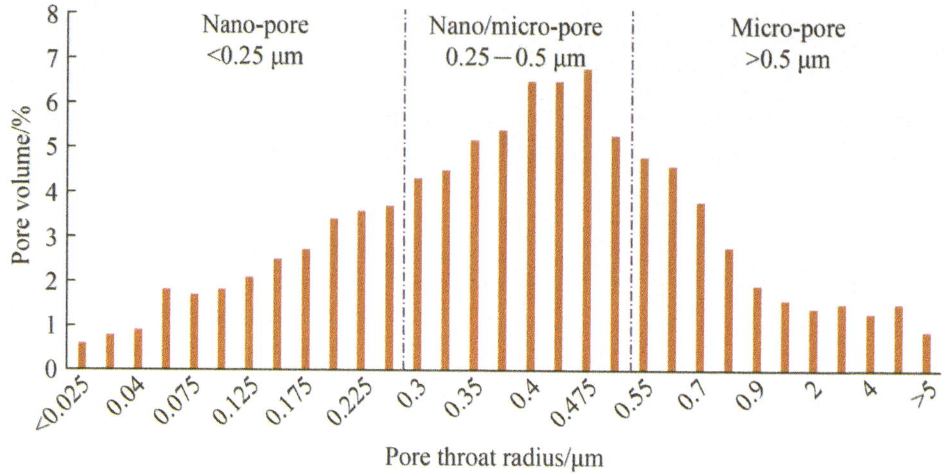

Fig. 5.37 Pore throat radius and volume percentage at different scales

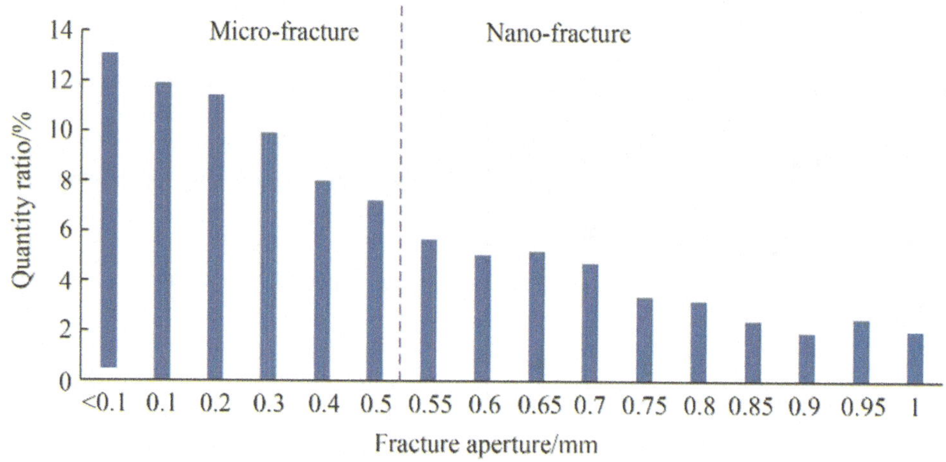

Fig. 5.38 Fracture apertures and quantity percentages at different scales

(2) Permeability

According to high pressure/constant rate mercury injection experiments, the total permeability of a representative element can be obtained. Moreover, the relationship between the permeability and the mainstream pore throat radius (Fig. 5.39) can be established. The corresponding permeability can be calculated by the mainstream pore throat radius for the pore media at different scales which is obtained from above.

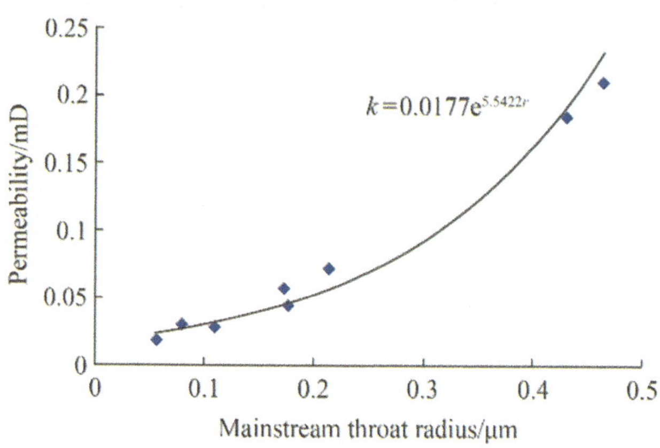

Fig. 5.39 Relationship between permeability and mainstream pore throat radius

(3) Oil saturation

The oil-bearing condition of tight oil reservoirs are greatly influenced by their lithology and physical properties. We can establish the relationship between oil-bearing properties and physical properties (the relationship between oil saturation and throat radius and permeability). Moreover, we can use the throat radius and permeability of the pore media at different scales to calculate the oil saturation.

2. Fracture media of different scales

(1) Porosity

Through the apertures and length of fractures at different scales, as well as the density/number and face-porosity ratio of fractures, which are obtained by the quantitative characterization of core and thin section fractures, the porosity of fractures at different scales is calculated.

(2) Permeability

For natural nano/micro-fractures in tight reservoirs, the roughness and filling conditions are not considered. For the nano/micro-fractures generated by hydraulic fracturing, the filling and supporting of proppants are not considered. Therefore, based on the nano/micro-fracture width obtained previously, a planer model is used to calculate the permeability of the fractures. Also, the fracture conductivity is calculated accordingly.

(3) Oil saturation

Referring to the oil saturation of pore media with similar diameter to fracture width, the oil saturation of media with different scales fractures is determined.

5.3.3.3 Modeling for Flow Parameter

1. Pore media at different scales

The relative permeability curves and capillary pressure curves for pore media at different scales are obtained through pore media flow experiments or digital rock pore media flow theoretical models.

2. Fractured media at different scales

Referring to the relative permeability curves and capillary pressure of pore media with similar pore's size to fracture width, the relative permeability curves and capillary pressure curves of the media with different scales fractures are determined.

5.4 Equivalent Modeling Technology for Porous Medium with Fractures at Different Scales

In tight reservoirs, the lithology, lithofacies, thickness, physical properties and oil-related properties change rapidly at the macroscopic scale and in large differences. In addition, they have strong macroscopic heterogeneity. In contrast, on the microscopic scale, there exist nano-nano/micro-micro scale pore media and fracture media of different scales. Moreover, the microscopic heterogeneity and multiple media characteristics are prominent. The heterogeneity and multiple media characteristics at different scales from micro scale to macro scale have great influence on the flow and production behaviors of oil and gas fluids in tight reservoirs. Conventional geological modeling methods are difficult to achieve quantitative description and characterization of macroscopic heterogeneity, multiple-scale characteristics, and microscopic heterogeneity, multiple media characteristics of pore media at different scales and nano/micro-fractures, with clear reflection of these factors in geological models. We have innovatively developed equivalent modeling technologies for multiple media at different scales, including reservoir-scale equivalent modeling technology and unconventional upscaling equivalent modeling technology. Geological models established by these technologies can fully reflect the macroscopic heterogeneity, microscopic heterogeneity, and the influence of number and spatial distribution of multiple media at different scales on the flow behavior and production dynamics.

5.4.1 Equivalent Modeling Method for Representative Elements at Different Scales

A tight reservoir has strong macroscopic heterogeneity. The multi-scale characteristics of microscopic multiple media are prominent. Within the same representative element, there exist both the pore media at different scales and the media with different scales fractures. The composition, quantity, and spatial distribution pattern of pore and fracture media of different scales have a great influence on the physical properties and flow characteristics through pore and fracture media of tight reservoirs of the representative element. According to the composition, number and spatial distribution of porous medium with fractures at different scales, the fractures at different scales are discretized into discrete fracture media by unstructured grid technology. Additionally, the pore media at different scales are discretized into discrete multiple media, accordingly. Based on the principle of pore volume equivalence and flux equivalence, we generate such a single medium that is equivalent to the porous medium with fractures at different scales in the representative element. Moreover, an equivalent medium property parameters model of the representative element is established.

5.4.1.1 Equivalent Modeling for Representative Elements

The steps for equivalent modeling for the representative element are as follows (Fig. 5.40).

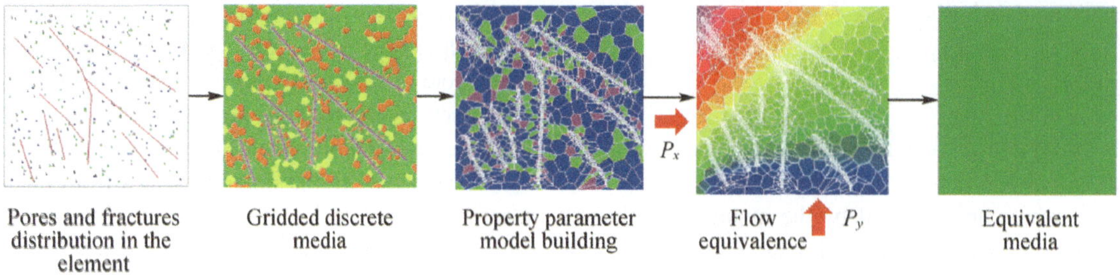

Fig. 5.40 Equivalent modeling diagram of media with different scale pores and fractures of the representative element

① We determine the number and spatial distribution of fractures at different scales, based on the result of fracture descriptions. Then, we determine the composition, quantity, and spatial distribution patterns of pore media at different scales based on high-pressure/constant-rate mercury injection experiments, thin section analysis, and other data. At the same time, we generate the distribution of pores and fractures at different scales within the element.

② By using the unstructured grid technology, the fractures at different scales are discretized into discrete fracture media. Additionally, the pore media at different sizes are discretized into discrete multiple media. Furthermore, we establish geometric, physical, and parameter models of flow in pore media for media with pores and fractures in different grid blocks.

③ By using pore volume equivalence and flux equivalence methods, we obtain the geometric, physical properties and flow parameters in the equivalent medium of are presentative element. Then, we establish the equivalent medium model for the representative element.

By establishing the equivalent medium model for the representative element, we can analyze the influence of the composition, quantity and spatial distribution patterns of the porous and fractured media at different scales on the physical properties and flow characteristics of the representative element.

5.4.1.2 Influence of the Composition and Distribution of Porous Medium with Fractures at Different Scales on the Physical Properties and Pore Media Flow Capabilities of Representative Elements

1. Influence of pore media at different scales

The differences in the geometric scales and physical properties of pore media at different scales result in significant differences in the physical properties and capabilities of flow in pore media of the representative element. At the same time, the different quantitative composition and spatial distributions of pore media at different scales also have a great influence on the physical properties and capabilities of flow in pore media of the representative element.

(1) Effect of quantitative composition of pore media at different scales on the equivalent permeability of a representative element

Due to the influence of microscopic heterogeneity of tight reservoirs, the quantity distributions of micro-pores, nano/micro-pores, and nano-pores at different scales in the same representative element vary to a large extent. In addition, the differences in sizes and quantitative

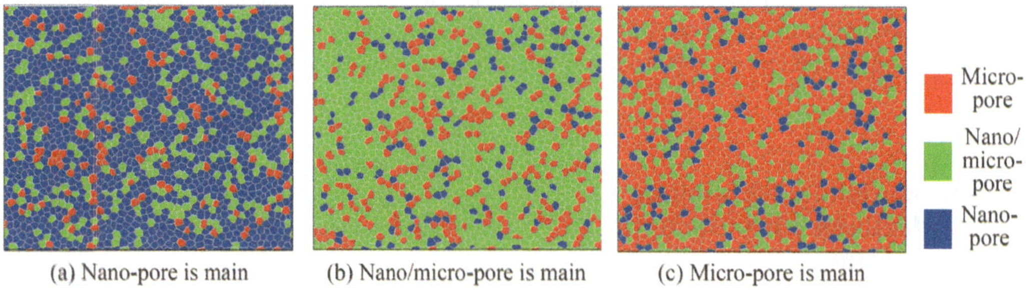

Fig. 5.41 Schematic diagram of pore distributions at different scales

Table 5.5 Equivalent permeability of pore media at different scales

Characteristics for the composition of pore media	Quantity composition for pore media at different scales		
	Micro-pores/%	Nano/micro-pores/%	Nano-pores/%
Nano-pores dominant	10	20	70
Nano/micro-pores and nano-pores dominant	10	40	50
Nano/micro-pores dominant	20	70	10
Micro-pores and nano/micro-pores dominant	50	40	10
Micro-pores dominant	70	20	10

Micro-pores: porosity 15%, permeability 0.2 mD; Nano/micro-pores: porosity 10%, permeability 0.05 mD; Nano-pores: porosity 5%, permeability 0.01 mD

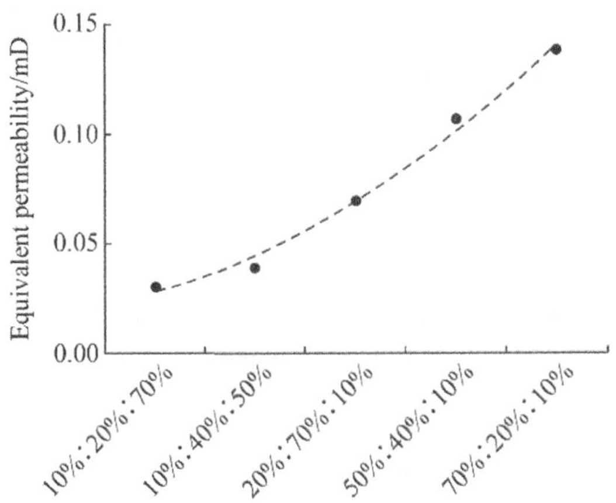

Fig. 5.42 Effect of quantitative composition of pore media at different scales on the equivalent permeability of representative elements

composition of pore media at different scales affect greatly the pore media flow capabilities of the representative element.

According to the equivalent modeling method for the representative element, we compare the effects of the size and composition of pore media at different scales on the equivalent permeability of representative elements (Fig. 5.41). Furthermore, the quantitative composition of pore media at different scales within the representative element is shown in Table 5.5.

It can be seen from Fig. 5.42 that the quantitative composition of pore media at different scales has a great influence on the equivalent permeability of representative elements. As the proportion of micro-pores increasing, the proportion of nano-pores decreases and the equivalent permeability increases. When the pore media are dominated by small-scale nano-pores (the ratio of nano-pores is 70%), the equivalent permeability is only 0.03 md. In addition, when the pore media are dominated by medium-scale nano/micro-pores (the ratio of nano/micro-pores is 70%), the equivalent permeability is 0.07 mD. When a porous medium is dominated by large-scale micro-pores (the ratio of micro-pores is 70%), the equivalent permeability would reach 0.14 mD.

(2) Effect of spatial distribution models for pore media at different scales on the equivalent permeability of representative elements

Under different geological conditions, the spatial distribution characteristics of pore media at different scales existing in tight formations are quite different. The variation of spatial distribution of the media within the representative element influences the contacts and flow relationships between the media, thus affects the equivalent permeability of the representative element.

1) Effect of different NDT modes on the equivalent permeability of representative elements

Pore media at different scales have NDT distribution patterns. In other words, pore media at different scales are proportionally in a ring-shaped distribution within representative elements (the ratio of micro-pores: nano/micro-pores: nano-pores are 30%:40%:30%, respectively). According to the media distribution rules for different scales, such as micro-pores, micro- and nano-pores, and nano-pores, at different scales, we can divide these patterns into three categories. First, the scales of the media become larger from inside to outside. Second, the scales of the media vary randomly from inside to outside. Third, the scales of the media become smaller from inside to outside (Fig. 5.43). According to the equivalent method for representative elements, we compare the effect of the arrangement of media at different scales on the equivalent permeability of representative elements.

From the result of equivalence, it can be seen that under the NDT mode the ordering sequence of pore media at different scales is different, which affects the value of equivalent permeability. Under the NDT flow relationship, the flow relationship is a single serial relationship. Then, the media successively is involved in the flow from inside to outside. For situations where the media scales become larger and larger from inside to outside, the medium with the best physical properties is located outside, which is preferentially involved in flow first. Moreover, the equivalent permeability

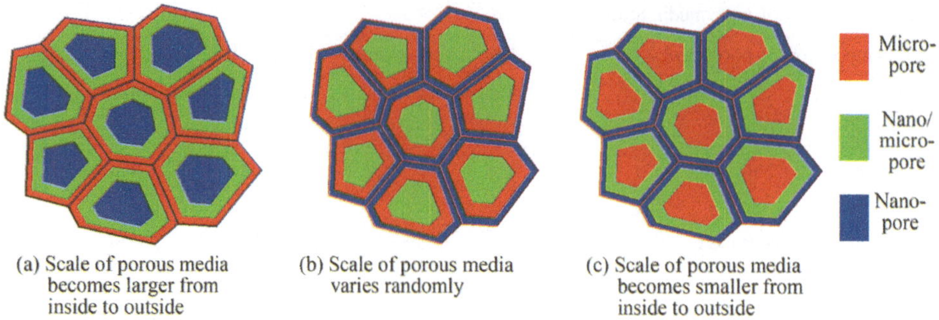

Fig. 5.43 Distribution patterns of pore media at different scales in NDT mode

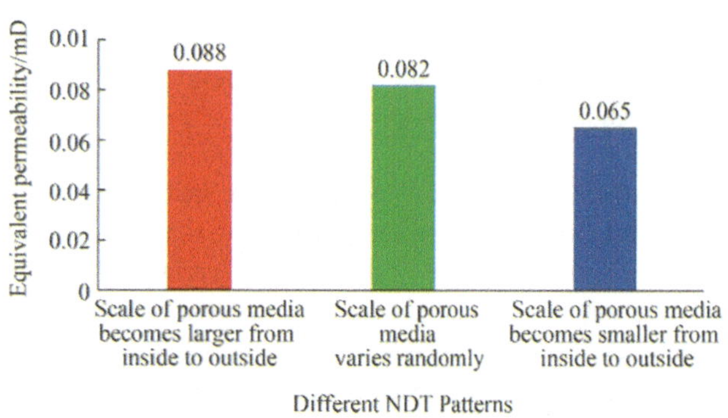

Fig. 5.44 Effect of different NDT patterns on the equivalent permeability

of representative element is relatively large. In contrast, for situations where the media scales become smaller and smaller from inside to outside, the medium with the worst physical properties is located in the outermost. At the same time, the medium with the best physical properties is at the center. In addition, the peripheral nano-pores restrict the flow of the micro-pores at the center. Therefore, the equivalent permeability of such representative element becomes relatively small. The equivalent permeability value for the randomly distributed mode is between the first two (Fig. 5.44).

2) Effect of different RIDT distribution modes on the equivalent permeability of representative elements

Due to the influence of geological principles, pore media at different scales can exhibit different distribution patterns within the representative element. The RIDT distribution model is classified into five categories (Fig. 5.45). In the case of certain proportions of pore media at different scales (the ratio of micro-pores: nano/micro-pores: nano-pores are 30%: 40%: 30%, respectively), we conduct the respective studies on the effect of different distribution modes on the equivalent permeability.

In the case of certain proportions of pore media at different scales, under the RIDT flow relationship, the spatial distribution modes of the media in the representative element are different. Furthermore, the contacts and flow relationships between the media are different. Therefore, the equivalent permeabilities of representative elements are different. However, under the RIDT flow relationship, the contacts and flow relationships between the media are staggered and complicated. Moreover, the equivalent permeability is high when the micro-pores with better physical properties are distributed continuously (Fig. 5.46).

2. Effect of media with different scales fractures

The extent, quantity, spatial distribution, connectivity and the fracture network complexity of the natural fractures at different scales affect the flow of the reservoir matrix rock, which greatly influences the internal connectivity and the equivalent permeability of the representative element.

(1) Effect of natural fractures at different scales on the equivalent permeability of representative elements

Tight reservoirs usually have fractures at different scales. The geometric characteristics and properties of fractures at

Fig. 5.45 Spatial distribution patterns of pore media at different scales under RIDT flow patterns

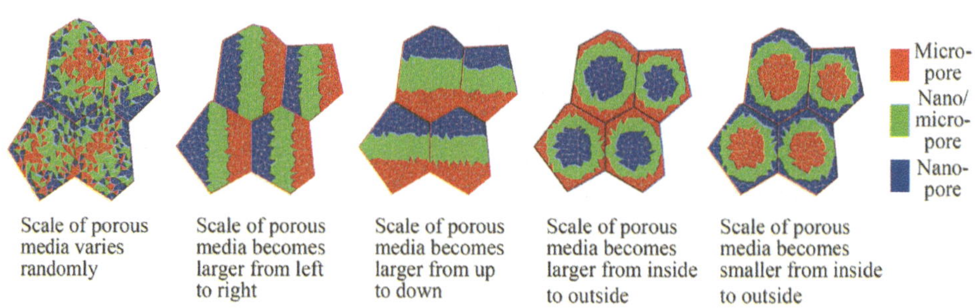

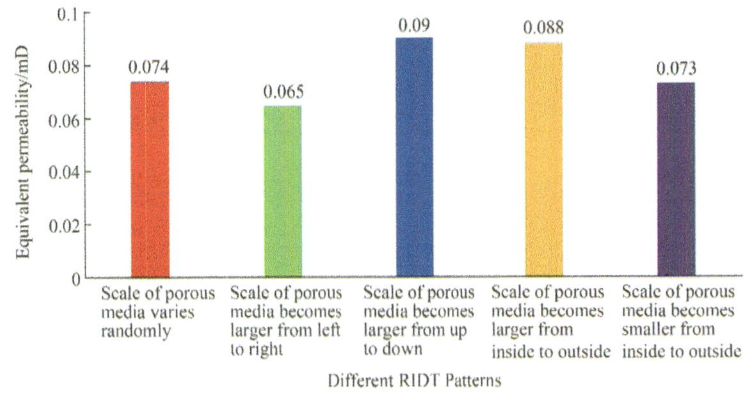

Fig. 5.46 Effect of different RIDT distribution modes on the equivalent permeability

different scales are different, which affects the size of the reservoir matrix rock to be formed. Furthermore, this eventually has a great influence on the equivalent permeability of representative elements.

1) Influence of the random distribution of natural fractures at different scales on the equivalent permeability for representative elements

According to the equivalent modeling method of representative elements, the influences of natural fractures at different scales, including large-scale, medium-scale, small-scale and micro-scale, on the equivalent permeability of representative elements are compared here (Fig. 5.47).

It can be seen that natural fracture combinations at different scales have a relatively large influence on the equivalent permeability of representative elements (Fig. 5.48). When only a single scale of fractures exists, the larger the fracture size is, the higher the equivalent permeability of representative elements is. This is because the larger the fractures size is, the higher the fracture conductivity is. Also, the longer the extension distance is, the greater the communication range is, and the better the communication effect is, which also leads to higher equivalent permeability of representative elements. When the fractures at different scales exist in a reservoir, including large-scale, medium-scale, small-scale and micro-scale fractures, the equivalent permeability of representative elements is the highest.

It can be seen that when natural and hydraulic fractures of different scales are in full coupling with the reservoir matrix to truly enhance the communication with the reservoir and improve field development results.

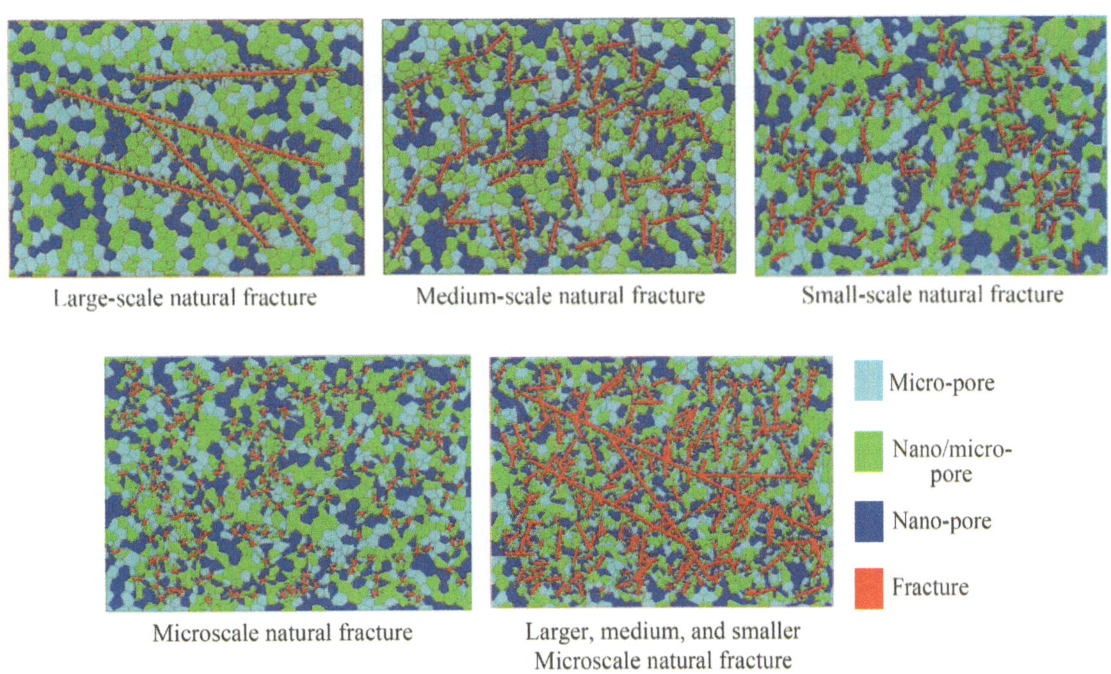

Fig. 5.47 Planar distribution of fractures at different scales within the representative element

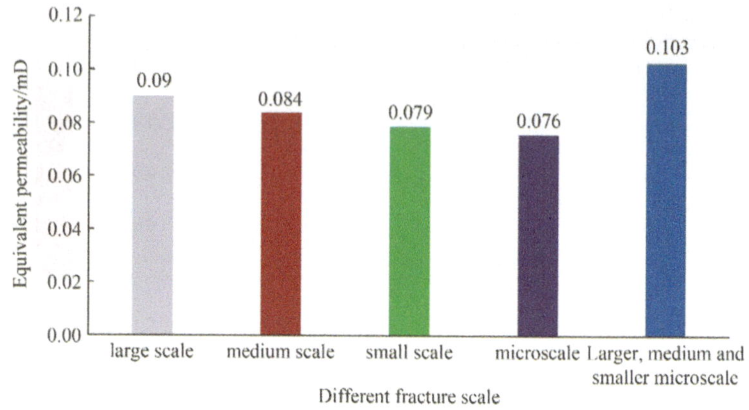

Fig. 5.48 Effect of fractures at different scales on the equivalent permeability

2) Effect of parallel distribution of natural fractures at different scales on the equivalent permeability

When the total internal fracture lengths with in representative elements are the same, the corresponding numbers of fractures with different scales, including large-scale, medium-scale, small-scale, and micro-scale are different. Under the parallel distribution condition for natural fractures at different scales (Fig. 5.49), based on the equivalent modeling method of representative elements, we compare the effects of fractures with different scales on the equivalent permeability of representative elements.

It can be seen that in the parallel distribution mode, when the total lengths of the fractures are the same, the larger the scale of the fractures is, the higher the equivalent permeability is (Fig. 5.50). Because the scale of the fractures is larger, the fracture conductivity is higher; the extension distance is higher; the communication effect is better. Thus, the equivalent permeability of representative elements is high.

(2) Effect of spatial distribution pattern of natural fractures on the equivalent permeability of representative elements

The spatial distribution pattern of natural fractures mainly refers to the density and communication of fractures. The density (quantity) of fractures and the communication between fractures (number of intersections) reflect the development degree and complexity of the fracture net work. This determines the size of the reservoir matrix rock, which affects the equivalent permeability for the representative elements.

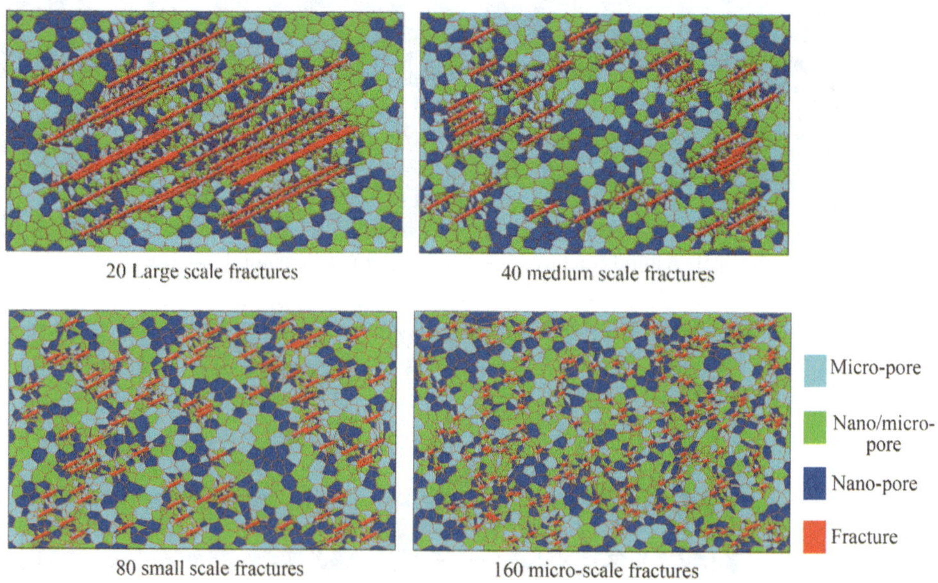

Fig. 5.49 Parallel distribution of fractures at different scales

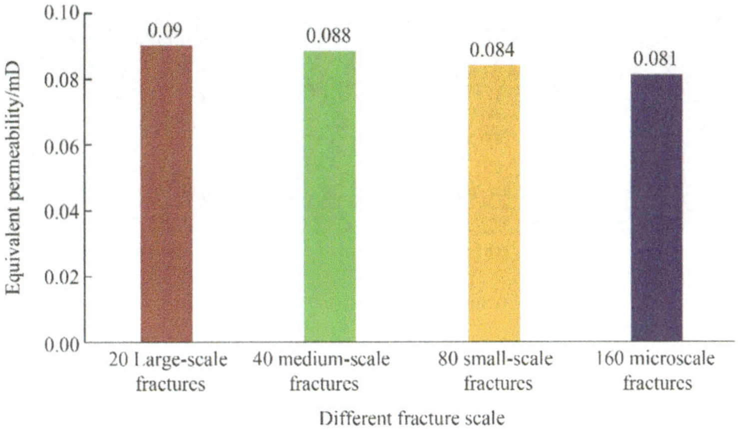

Fig. 5.50 Effect of natural fractures at different scales on the equivalent permeability

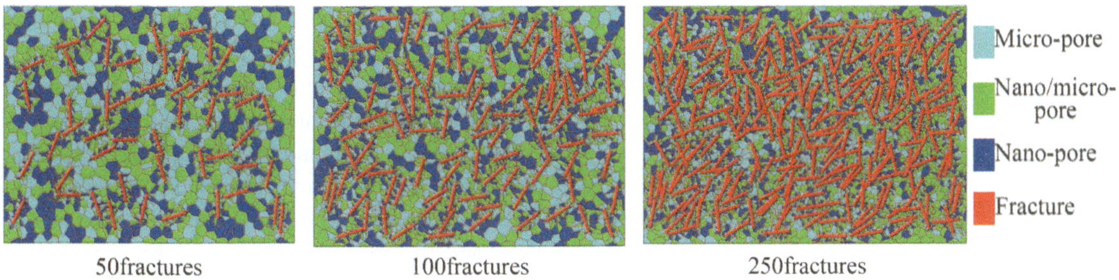

Fig. 5.51 Planar distribution of different numbers of fractures

1) Effect of natural fractures with the different numbers of fractures and the same numbers of intersections on the equivalent permeability of representative elements

According to the representative element equivalent modeling method, the effects of the different numbers of fractures on the equivalent permeability are compared, under the condition when the numbers of fracture intersections are the same (Fig. 5.51).

It can be seen that when the numbers of fracture intersections are the same, the numbers of fractures affect the equivalent permeability of representative elements (Fig. 5.52). The more the fractures are, the higher the fracture density is, the more extended the fracture network is, the stronger its communication ability is, and the greater the equivalent permeability of representative elements is. However, after the number of fractures reaches a certain number, the reservoir matrix rock has been fully partitioned. Additionally, the increase effect in the equivalent permeability of representative elements due to higher number of fractures is reduced.

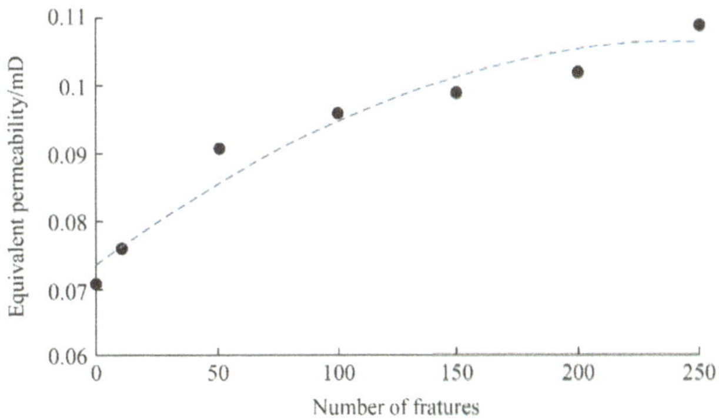

Fig. 5.52 Effect of different numbers of fractures on the equivalent permeability

2) Effect of the nature fractures with the same numbers of fracture and the different number of intersections on the equivalent permeability of representative elements

According to the equivalent modeling method of representative elements, the effects of different numbers of fracture

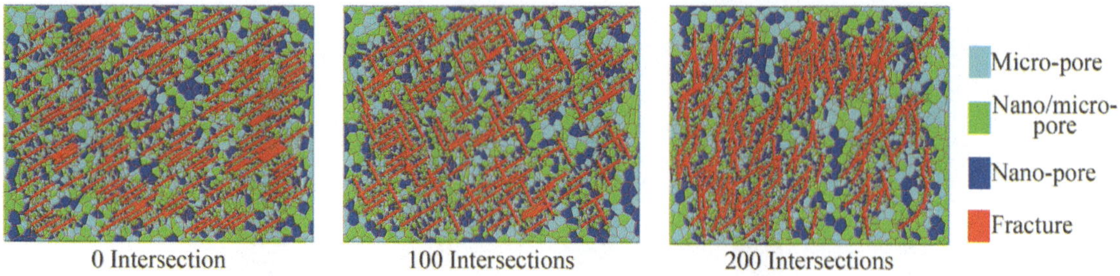

Fig. 5.53 Planar distribution of different numbers of fracture intersections

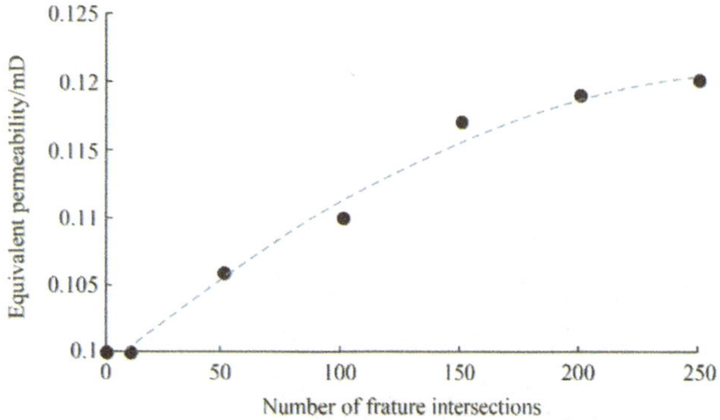

Fig. 5.54 Effect of different numbers of fracture intersections on the equivalent permeability

intersections on the equivalent permeability of representative elements are compared, under the condition when the numbers of fractures are the same (Fig. 5.53).

It can be seen that when the numbers of fractures are the same, the number of intersection points has a large influence on the equivalent permeability of representative elements (Fig. 5.54). The more the intersection points are, the higher the complexity of the fracture network is, the shorter the distance from the reservoir matrix to the fractures is, in other words, the stronger the communication ability within representative elements is, the higher the equivalent permeability of representative elements is.

5.4.2 Equivalent Modeling Techniques at Reservoir Scale

Unconventional tight reservoirs are characterized by strong macroscopic heterogeneity and microscopic heterogeneity. According to the modeling strategy of unconventional reservoirs, we establish a geological model, at different scales, that was divided into different regions based on large-scale heterogeneity, further into simulation element based on small-scale heterogeneity, and finally into different pore and fracture media based on micro-scale heterogeneity. This geological model has high degree of resolution. But the number of grids is large and the calculation speed is slow. In order to reduce the number of grids and increase the computational speed, these fine-grid geological models that can reflect the micro-scale heterogeneous multiple media are upscaled. In addition, the second level refined grids of the discrete multiple media are upscaled into coarse grids of representative elements, realizing the upscaling equivalence for the second grids at the same scale in the horizontal plane. After making the equivalence, the parameters of the first level grids of representative elements reflect the influence of the compositions, quantities, and spatial distribution patterns of the porous and fractured media at different scales on the grid block physical properties and flow parameters through pore and fracture media of tight reservoir of representative elements.

The steps for the equivalent modeling method at the reservoir scale are as follows (flow chart in Fig. 5.55):

① According to the description results of tight reservoirs and fractures, the spatial distribution of reservoir sand bodies and physical properties, and the number and spatial distribution of natural/hydraulic fractures and pore media at different scales are determined.

② According to the macroscopic heterogeneity of the reservoir, the first level regions based on the large-scale heterogeneity are partitioned. Moreover, the second level representative elements (grids in the first level regions) are partitioned according to the differences in small-scale geological conditions and the heterogeneity within a certain area.

③ According to the distribution characteristics of fractures and pores at different scales in different representative elements, the unstructured grid technology is used to discretize the fractures at different scales into discrete fracture media and discretize the pores at different scales into discrete multiple media, in order to form the second level discretized grids corresponding to discrete fracture media and discrete multiple media. Also, the parameter models for geometric attributes, physical properties, and flow of different multiple media within the second level discretized grids are established.

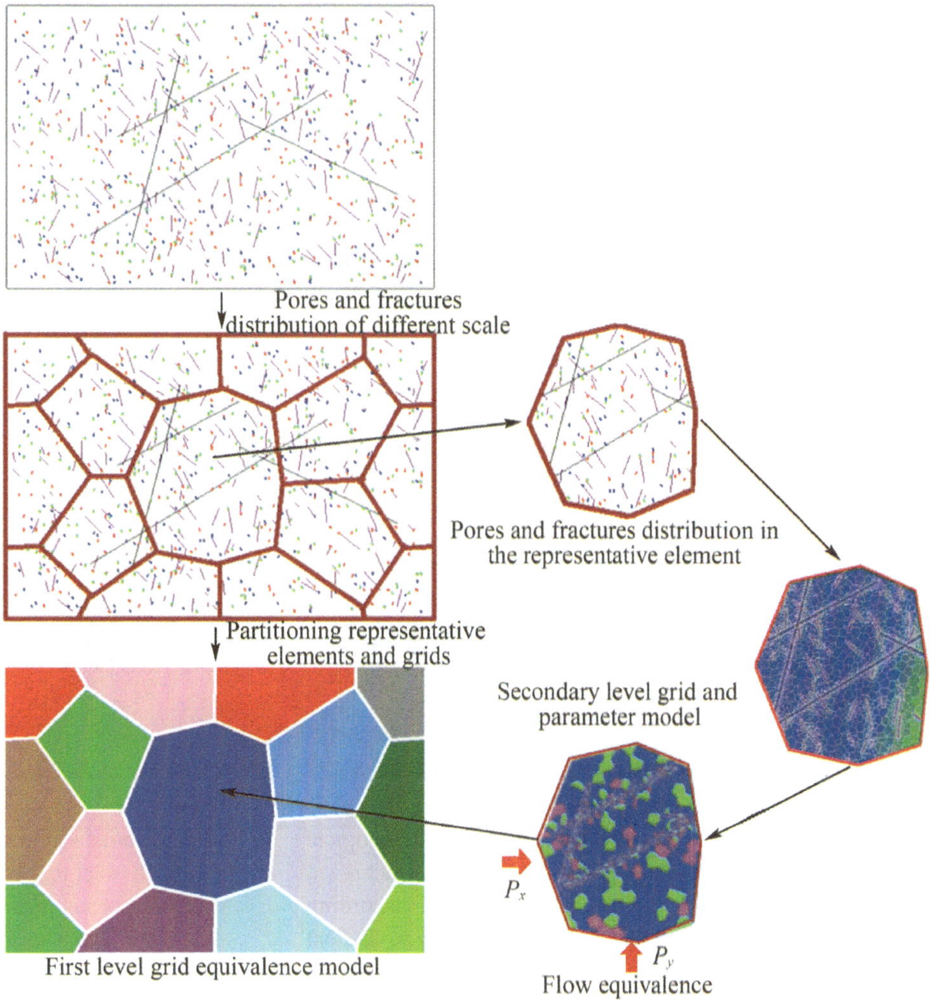

Fig. 5.55 Schematics for the equivalent modeling of multiple media at different scales

④ According to the porosity and oil saturation of different pore and fracture media with the second level discretized grids, the porosity and oil saturation of the equivalent media within representative elements are obtained by the principle and method of volume conservation. According to the permeability, relative permeability curve and capillary pressure curve of different porous medium with fractures within the second level discretized grids, through the principle and method of equivalence of single-phase and multi-phase flow, the permeability, relative permeability curves and capillary pressure curves of the equivalent media of representative elements are obtained. Then the equivalent media model of representative elements is established.

⑤ Through the equivalent modeling processing for each representative element, an equivalent media geological model for dividing different regions and different elements is established.

5.4.3 Equivalent Modeling Technology for Upscaling of Unconventional Reservoirs

Tight reservoirs are distributed over a large area. But the lithology, lithofacies, thickness, physical properties, and oil-bearing conditions vary dramatically with large differences and strong macroscopic heterogeneity between different planes, different wells, and different well sections from the same well. However, on the microscopic scale, there exist nano-nano/micron-micro scale pore media and fracture media. The micro scale heterogeneity and multiple medium features are prominent. The macroscopic and microscopic heterogeneity and multiple media characteristics at different scales have great influence on the flow and recovery behaviors of tight oil and gas fluids. How to describe and quantitatively characterize the macro scale heterogeneity and its multi-scale characteristics, and micro

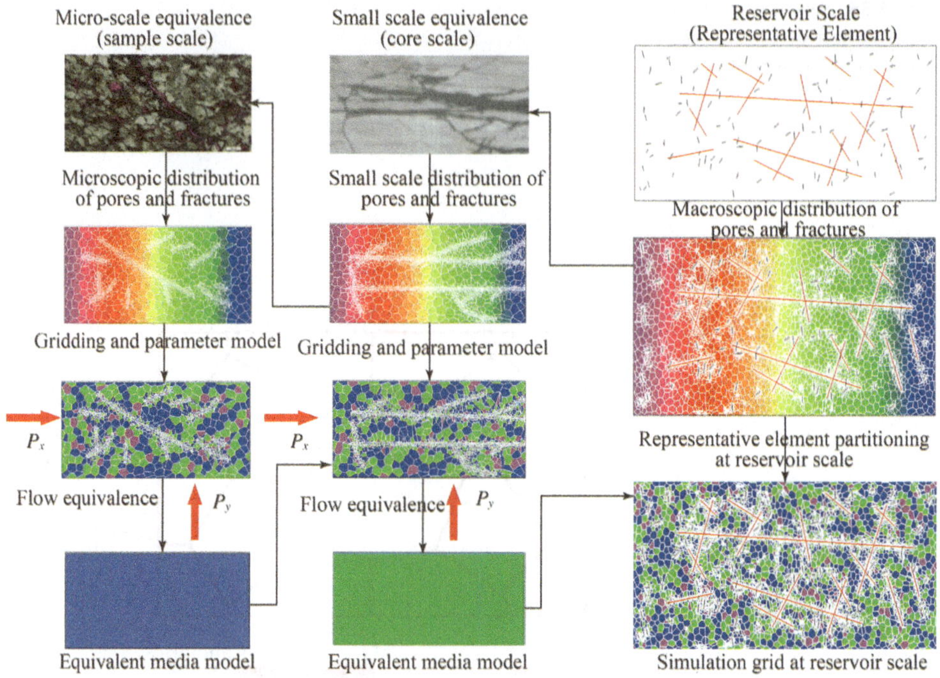

Fig. 5.56 Upscaling equivalent modeling workflow for unconventional reservoirs at different scales

scale heterogeneity and its multiple medium characteristics of pore media and nano/micro-fractures at different-scales, together with how to represent them in the geological models and numerical simulations, is difficult to be achieved in conventional geological modeling. Therefore, according to the upscaling theory, a step-by-step upscaling equivalent modeling technique from the micro scale to the macro scale has been developed. The modeling technique can achieve equivalent upscaling from the micro scale, small scale to the reservoir scale, and upscale the micro scale, small scale heterogeneity into the heterogeneity in the reservoir scale geological model, which fully represents the influence of heterogeneity at different scales on the flow behavior and production dynamics.

The steps of upscaling equivalent modeling technique is as follows (Fig. 5.56).

5.4.3.1 Establishment of Equivalent Media Models for the Sample Units

① The composition, quantity and spatial distribution patterns of nano/micro-fractures and pores on the microscopic pore scale were obtained by performing high-pressure/constant-rate mercury injection analysis, thin section analysis, scanning electron microscopy, and digital rock analysis.

② Using the unstructured grid technology, fractures of different scales and pore media of different scales are discretized into discrete fractures media and discrete multiple media, in order to form discretized grids corresponding to different porous medium with fractures. In addition, the parametric models of different discrete grids are established. The equivalent media model of the sample unit is established by considering the method and principle of volume conservation and flow rate equivalence.

5.4.3.2 Establishment of Equivalent Media Models for Core Units

① Based on the heterogeneity of cores, the heterogeneity at the micro-porosity scale is analyzed at different locations of the sample. Using the unstructured grid technology and combining with the core heterogeneity and micro scale heterogeneity of different samples, we partition the cores into grid blocks of different samples. The equivalent media model of different sample units at different locations of a core is established.

② According to the property parameters of the equivalent media of different sample units at different locations of a core, by considering the principle and method of volume conservation and flow equivalence, several sample units are upscaled to the equivalent media model of the core unit.

5.4.3.3 Establishment of Equivalent Media Models of Reservoir Elements

According to the macroscopic heterogeneity of a reservoir, the reservoir is partitioned into several representative elements. For such representative elements with coring data, the property parameters of the equivalent media of the sample units are established through sampling and analysis. Then they are upscaled to the equivalent media model of the core units. For representative elements without coring data, by establishing the correlation model between the property parameters of the equivalent media of the core units and different lithology, lithofacies and reservoir types, and also by using the lithology, lithofacies, and reservoir types of representative elements, we determine the corresponding element's property parameters.

5.4.3.4 Through the Processing of Upscaling Equivalent Modeling for Each Reservoir Element, Establishment of an Equivalent Media Geological Model of Different Reservoir Element

5.4.4 Equivalent Modeling Method for Porous Medium with Fractures at Different Scales

Fractures and pores at different scales exist in tight reservoirs. The geometric, physical and flow properties of pore and fracture media of different scales are very different. The heterogeneity and multiple media characteristics between different media are prominent, which greatly influences the flow behavior and production dynamics. Through the equivalent modeling method of pore and fracture media of different scales, the geometric, property, and flow parameters of small-scale pore and fracture media of different scales are upscaled into larger-scale elements, which we can establish an equivalent media model. In geological models and numerical simulations, the effects of heterogeneity and multiple media characteristics between different media on the flow behavior and production dynamics are fully reflected.

According to the differences in the equivalence principle and method, there are mainly two methods. One is a porosity and saturation equivalence method based on volume conservation. The second is a permeability, relative permeability curves, and capillary pressure curves equivalence method based on the flow rate conservation.

5.4.4.1 Equivalence Modeling Method Based on Volume Conservation

1. Equivalence modeling method for porosity

In the equivalent elements of tight reservoirs, there are large differences in the geometric sizes of the fractures and pores at different scales. Also, the quantity and spatial distribution of the fractures and pores are different. Therefore, the porosities and the proportions of the occupied pore volume of pore and fracture media with fractures at different scales are quite different. According to the principle of conservation of pore volume, the equivalent porosity of the equivalent elements is established by using the volume equivalence method, according to the porosity of multiple media at different scales. The equivalence method is as follows:

① According to the quantity and spatial distribution of pores and fractures at different scales, the fractures and pores at different scales are discretized through gridding into discrete fracture media and discrete multiple media, respectively. We generate the discrete media grids and determine the porosity parameters for different discrete media grids.
② According to the geometric characteristics of different discrete media grids, we calculate the volumetric parameters for each grid.
③ According to the porosity and volume for different discrete media grids, the volume-weighted equivalence modeling method is used to calculate the equivalent porosity of the equivalent element (Xu et al. 2010).

$$\phi_{\mathrm{eq}} = \frac{V_\phi}{V_{\mathrm{eq}}} = \frac{\sum_{i=1}^{n} \phi_i v_i}{\sum_{i=1}^{n} v_i} \tag{5.37}$$

where, ϕ_i is the porosity of the discrete media grid i; v_i is the block volume of the discrete media grid i, V_{eq} is the total volume of the equivalent element, and ϕ_{eq} is the equivalent porosity.

2. Equivalence modeling method for saturation

In the equivalent elements of tight reservoirs, the oil saturation of the multiple media at different scales is controlled by the characteristics of geometrical features and physical parameters. Thus, the oil saturations of the multiple media at different scales vary greatly. According to the principle of oil volume conservation, the equivalent oil saturation of the equivalent element is established by the volume equivalence method, according to the oil saturation of the pore and fracture media with fractures at different scales. The equivalence method is as follows:

① According to the quantity and spatial distribution of pores and fractures at different scales, the fractures and pores at different scales are discretized through gridding into media with discrete fractures and discrete multiple media, respectively. We generate the discrete media grids and determine the porosity and oil saturation parameters for different discrete media grids.
② According to the geometric characteristics of different discrete media grids, we calculate the volumetric parameters of each grid.
③ According to the oil saturation, porosity, and grid volume for different discrete media grids, the equivalent oil saturation for the equivalent element is calculated by using the volume-weighted equivalence modeling method (Xu et al. 2010):

$$S_{\mathrm{oeq}} = \frac{\sum_{i=1}^{n} S_{\mathrm{o}i}\phi_i v_i}{\sum_{i=1}^{n} \phi_i v_i} \quad (5.38)$$

where, $S_{\mathrm{o}i}$ is the oil saturation of the discrete media grid i, S_{oeq} is the equivalent oil saturation.

5.4.4.2 Equivalence Modeling Method for Permeability Based on Single-Phase Flow and Flow Rate Conservation

Because the geometric scales of pores and fractures at different scales have large differences, their permeability values also vary greatly. At the same time, the quantity and spatial distribution of pore and fracture media at different scales also vary to a large extent. Therefore, it has a great influence on the capability and characteristics of flow in pore media of the equivalent elements. According to the single-phase fluid theory of flow in pore media and flow rate equivalent principle, based on the permeability for the multiple media at different scales, the equivalent permeability values of the equivalent elements in different directions are established by using the numerical simulation method. The equivalence method is as follows.

① According to the quantity and spatial distribution of pores and fractures at different scales, the fractures and pores at different scales are discretized through gridding into media with discrete fractures and discrete multiple media, respectively, to generate the discrete media grids.
② The geometric parameters of different discrete media grids and their physical parameters, such as porosity, saturation, and permeability, are determined.
③ Since the absolute permeability is based on the single-phase fluid theory of flow in pore media, it is necessary to determine the parameters of the single-phase fluid, including fluid parameters such as fluid viscosity μ and density.
④ For a certain direction of the equivalent element, we determine the flowing pressure difference ($P_{\mathrm{b}}^{\mathrm{wi}}$, $P_{\mathrm{b}}^{\mathrm{wp}}$) along that element (the distance between the two ends is L and the flowing area at the two ends is A), and get the fluid flow rate Q through the single-phase flow numerical simulation.
⑤ According to the principle of flow rate conservation, Darcy's law formula of single-phase flow is used to calculate the effective permeability of the equivalent element based on the differential pressure at the two ends of the element and the flow rate.

$$K_{\mathrm{eq}} = \frac{Q\mu L}{A\Delta P} = \frac{Q\mu L}{A(P_{\mathrm{b}}^{\mathrm{wi}} - P_{\mathrm{b}}^{\mathrm{wp}})} \quad (5.39)$$

⑥ For other different directions, we repeat steps ④ and ⑤, and use the same method to calculate the equivalent permeability of the equivalent element in other different directions.

5.4.4.3 Equivalence Modeling Method Based on Multiphase Flow and Flow Rate Conservation

1. Equivalence modeling method for relative permeability curve

Due to the large differences in the geometric scales and physical parameters of pores and fractures at different scale, there are large differences in their relative permeability curves. The quantity and spatial distribution of the porous medium with fractures at different scales in different elements vary greatly, which has a great influence on the relative permeability curves of the equivalent elements. According to the two-phase fluid flow theory, based on the flow rate equivalence principle, and considering the relative permeability curves of porous medium with fractures at different scales, the equivalent relative permeability curves of the equivalent elements are established through numerical simulation.

① According to the quantity and spatial distribution of pores and fractures at different scales, the fractures and pores at different scales are discretized through gridding into media with discrete fractures and discrete multiple media, respectively, to generate discrete media grids.
② The geometric parameters for different discrete media grids and physical parameters, such as porosity, permeability, and relative permeability curves, are determined.
③ The relative permeability equivalence is based on the two-phase fluid flow theory. Thus, it is necessary to determine such fluid parameters as the viscosity μ and density of the two-phase fluids.
④ Because a relative permeability curve is the reflection of the laws of relative permeability of different fluids varying with the changing of saturations, the relative permeability of different fluids under different saturation conditions should be required. Therefore, saturation can be equally spaced in order to calculate the equivalent relative permeability of different fluids under the conditions that the different saturation is S_l.
⑤ Given a certain saturation value, we determine the flowing differential pressure (P_b^{wi}, P_b^{wp}) at the two ends of the element (the distance between the two ends is L and the flowing area at the two ends is A). Through two-phase flow numerical simulation, the total fluid flow rate Q and the flow rate for each phase are obtained.
⑥ According to the principle of flow rate conservation, Darcy's law formula based on two-phase flow is used to calculate the equivalent relative permeability $K_{req}(S_l)$ of a certain phase in the equivalent element at a saturation value of S_l, according to the pressure difference at the two ends of the element and the flow rate of a certain phase.

$$K_{rleq}(S_l) = \frac{Q_r \mu_r L}{A(P_b^{wi} - P_b^{wp}) \cdot K_{eq}} \tag{5.40}$$

⑦ For different saturation values, we repeat steps ⑤ and ⑥. Using the same method, we calculate the equivalent relative permeability of different fluids at different saturations within the equivalent element, in order to obtain the equivalent relative permeability curve for the equivalent element.

2. Equivalence modeling method for capillary pressure curve

There are large differences in the capillary pressure curves of multiple media at different scales. The quantity and spatial distribution of multiple media at different scales in different elements have a great influence on the capillary pressure curves of the equivalent element. According to the two-phase fluid theory of flow in pore media and flow rate equivalence principle, based on the capillary pressure curves of multiple media at different scales, the equivalent capillary pressure curves of the equivalent elements are established through numerical simulation.

① According to the quantity and spatial distribution of pores and fractures at different scales, the fractures and pores at different scales are discretized through gridding into media with discrete fractures and discrete multiple media to generate discrete media grids.
② The geometric parameters and other parameters, such as porosity and permeability of different discrete media grids, are determined.
③ The capillary pressure curve equivalence is based on the two-phase fluid flow theory. Hence such fluid parameters as viscosity μ and density of the two-phase fluid need to be determined.
④ Since the capillary pressure curve is the reflection of the laws of the capillary pressure values between the two phases varying with the change of saturations, the values of the capillary pressure under different saturation conditions are required. Therefore, saturations can be equally spaced to calculate the equivalent capillary pressures between the two-phase fluids at different saturation conditions.
⑤ Given a certain saturation value, we determine the flowing differential pressure (P_b^{wi},P_b^{wp}) between the two ends of the element (the distance between the two ends is L and the flowing area at the two ends is A). Through the two-phase flow numerical simulation, the pressure values and capillary pressures of different fluids in each grid are obtained.
⑥ According to the grid block volume v_i, porosity ϕ_i, capillary pressure $p_{ci}(S_l)$ of different media grids, the equivalent capillary pressure $p_{cleq}(S_l)$ of the equivalent element at saturation value S_l is calculated using the pore volume weighted method (Eq. 5.41).

$$p_{cleq}(S_l) = \frac{\sum_{i=1}^{n} p_{ci}(S_l)\phi_i v_i}{\sum_{i=1}^{n} \phi_i v_i} \tag{5.41}$$

⑦ For different saturation values, we repeat steps ⑤ and ⑥. Moreover, we use the same method to calculate the equivalent capillary pressure at different saturations for the equivalent element, in order to obtain the equivalent capillary pressure curve for the equivalent element.

References

Feng JW, Dai JS, Ma ZR, Zhang YJ, Wang ZK (2011) Theoretical model for relationship between fracture parameters and stress field in low permeability sandstones. J Pet 32(4):664–671

Guo RB (1994) Porosity of regular pellet accumulations. Earth Sci 19 (4):503–508

Kim TH, Schechter DS (2009) Estimation of fracture porosity of naturally fractured reservoirs with no matrix porosity using fractal discrete fracture networks. SPE Reserv Eval Eng 110720:232–242

Lang XL, Guo ZJ (2013) Fractured reservoir modeling method based on DFN discrete fracture network model. J Peking Univ (Natural Science Edition) 49(6):964–972

Li Y, Hou JG, Li YQ (2016) Reservoir characteristics and classification geological modeling of fractured reservoirs in carbonate rock. Pet Explor Dev 41(4):1–7

Liang YT, Liu PC, Feng GC (2014) Modeling method of fractured reservoir based on ant tracinging technology. Complex Oil Gas Reserv 7(3):11–15

McClure MW, Babazadeh M, Shiozawa S, Huang J (2016) Fully coupled fluid dynamics simulation of hydraulic fracturing based on three-dimensional discrete fracture network. Pet Technol News 9:40–59

Qin JS, Li AF (2006) Reservoir physics. China University of Petroleum Press, Shandong, pp 138–139

Qu GZ, Qu ZQ, Dolye HR, David F, Rahman M (2016) Geometry description and permeability calculation of shale tensile micro fractures. Pet Explor Dev 43(1):115–120

Sun J, Schechter D (2015) Investigating the Effect of improved fracture conductivity on production performance of hydraulically fractured wells: field case studies and numerical simulations. J Can Pet Technol 25(11):442–449

Xu ZH (2009) Research on discrete fracture network geological modeling technology. Master's thesis of China University of Petroleum

Xu X, Yang ZM, Zu LK, Liu SB (2010) Equivalent continuum model and numerical simulation of seepage in multiple media reservoirs 17 (6):733–737

Yang SL, Wei JZ (2004) Reservior physics. Petroleum Industry Press, Beijing, pp 152–154

Zhang M, Li JM, Zhu WM (2009) Observations and detection methods of reservoir fractures. Fault Block Oil Gas Fields 16(5):40–42

Zheng SQ, Yao ZL (2009) Stochastic modeling method for discrete fracture networks. Oil Gas Sci 31(4):106–110

6 Numerical Simulation of Multiple Media at Different Scales

Unconventional tight oil and gas reservoirs contain multiple media with different scale pores and fractures. There are huge differences in geometry, property and flow characteristics between these media. The mechanisms of multiple phase flow through these multiple media are complex and diverse. Based on simulation of conventional continuous dual-porosity media, in this study, pore media and fracture media were further subdivided into different types of multiple pore media and multiple fracture media, respectively, and, finally, the simulation of continuous multiple media based on dual-porosity models was developed.

Because multiple media with different scale pores and fractures are discretely distributed in space, their properties do not change continuously, causing these media difficult to be dealt with like conventional continuous media in terms of simulation. As a result, in this study, the numerical simulation of discontinuous-discrete-multiple media was innovatively developed. For large-scale natural/hydraulic fractures, the dynamic simulation of multiple-scale discrete fractures was formed. For nano/micro-scale pores and fractures at different scales, the numerical simulations of discontinuous discrete multiple media were innovatively developed by proposing two grid patterns. Each grid represents one medium but the interacting relationships of these two grid patterns are different. The interacting relationship of the first grid pattern is the nested distribution type (in the following, it is referred as NDT) to represent the flow behavior of relay supply and drainage. The interacting relationship of the second grid pattern is the randomly interacting distribution type (in the following, it is referred as RIDT). For the NDT, the fluid only flows from one single medium to the other single medium. For the RIDT, the fluid can flow from one single medium to server media surrounding that single medium. For complex multiple media containing large-scale discrete fractures, micro-scale fractures and pores, the numerical simulation of discontinuous discrete hybrid multiple media was developed to achieve a numerical simulation transition from continuous dual-porosity media to discontinuous discrete multiple media (summarized in Table 6.1).

6.1 Numerical Simulation of Continuous Multiple Media at Different Scales

Unconventional tight reservoirs contain multiple-scale pores and fractures. According to the seepage theory of continuous media, multiple media at different scales can be divided into continuous pore media and continuous fracture media with different geometric and property parameters. The simulation of continuous dual-porosity media has been developed, which includes the simulation of dual-porosity single-permeability media and the simulation of dual-porosity dual-permeability media (Barrenblatt et al. 1960; Warren and Root 1963).

Meanwhile, there are obvious differences in geometry, property and flow characteristics between multi-scale pores and fractures so that the pore system and fracture system can be further subdivided into continuous multiple-scale pore media and multiple-scale fracture media. Based on conventional simulation and models of dual-porosity media, the simulation of multiple media (Table 6.2) was developed and the corresponding simulation process was formed.

6.1.1 Numerical Simulation of Dual-Porosity Single-Permeability Media

For dual-porosity media with poor pore properties in matrix, the fluid is difficult to flow within matrix and fails to directly flow from matrix to wellbore, while the fluid inside matrix can only flow into fractures and then directly flow from fractures into wellbore. For the flow characteristics of this kind of dual-porosity media, the simulation of dual-porosity single-permeability was developed (Barrenblatt et al. 1960; Warren and Root 1963).

Q. Ran, *Unconventional Tight Reservoir Simulation: Theory, Technology and Practice*,
https://doi.org/10.1007/978-981-32-9848-4_6

Table 6.1 Numerical simulation for different scale multiple media

Numerical simulation			Method description	Application condition
Numerical simulation for continuous multiple media	Numerical simulation for dual-porosity media	Numerical simulation for dual-porosity single-permeability media	Two individual grid systems are composed of structured grids represent single scale pore media and single scale fracture media respectively, and they separately distribute in space. The fluid flow fails to occur between adjacent grids representing pore media while the fluid flow occurs between adjacent grids representing fracture media. The interporosity flow occurs between pores and fractures. The fluid flow fails to occur between pores and wells while the fluid flow occurs between fractures and wells	Matrix pores continuously distribute and their permeability can be ignored. Fractures are well developed and forms complex fracture networks
		Numerical simulation for dual-porosity dual-permeability media	Two individual grid systems are composed of structured grids represent single scale pore media and single scale fracture media, respectively and they separately distribute in space. The fluid in pores and fractures can flow through adjacent grids. The interporosity flow occurs between pores and fractures. The fluid flow occurs between pores/fractures and wells	Matrix pores continuously distribute and their permeability cannot be ignored. Fractures are well developed and forms complex fracture networks
	Numerical simulation for multiple media based on model for dual-porosity media		Two individual grid systems are composed of structured grids represent multiple-scale pore media and multiple-scale fracture media, respectively and they separately distribute in space. The fluid in pores and fractures can flow through adjacent grids. The interporosity flow occurs between pores and fractures. The fluid flow occurs between pores/fractures and wells	Multiple scale pores continuously distribute. Multiple-scale fractures are well developed and their density is high
Numerical simulation for discontinuous multiple media	Numerical simulation for discrete fracture		The multiple-scale discrete natural/hydraulic fractures are meshed by unstructured grids and identified as different grid systems of discrete fractures. The parameters of different scale discrete fracture change nonlinearly. The fluid flow occurs between adjacent and connected fractures. The fluid flow occurs between different scale natural/hydraulic fractures and wells	Mese/large scale hydraulic/natural fractures discretely distribute in reservoirs and their density is low and connectivity is poor
	Numerical simulation for multiple media based on the grid pattern of NDT		Unstructured grids are used to mesh different scale discrete microscopic fractures/pores to be different microscopic scale elements. According to NDT grid pattern, those elements form a nested grid representing different discrete pores and fractures. The parameters of those pores and fractures change nonlinearly. The fluid flow occurs between adjacent and connected	The density of microscopic scale fractures is high and distribution of such fractures is discrete. The pore distribution is subjected to the NDT grid pattern and, is intensive and discrete in space

(continued)

Table 6.1 (continued)

Numerical simulation		Method description	Application condition
		elements. The fluid flow between pores and fractures is subjected to behavior of flow through NDTs. The fluid flow can occurs between different pores/fractures and wells	
	Numerical simulation for multiple media based on the grid pattern of RIDT	According to RIDT grid pattern, unstructured grids are used to mesh different scale discrete microscopic fractures and discrete pores to be different microscopic scale elements. The parameters of those pores and fractures change nonlinearly. The fluid flow occurs between adjacent and connected pores/fractures. The fluid flow occurs between different pores/fractures and wells	The density of microscopic scale fractures is high and distribution of such fractures is discrete. The pore distribution is subjected to the RIDT grid pattern and, is intensive and discrete in space
	Numerical simulation for hybrid discrete multiple media	Using unstructured grids generates discrete fracture elements to represent meso/large-scale natural/hydraulic fractures, and generates microscopic scale pore/fracture element in terms of both NDTs and RIDTs to represent microscopic pores/fractures. The fluid flow occurs between adjacent and connected pores/fractures. The fluid flow occurs between different pores/fractures and wells	The distribution of meso/large-scale fractures is discrete. The density of microscopic scale fractures is high and distribution of such fractures is discrete. The pore distribution is subjected to the NDT or the RIDT grid pattern and, is intensive and discrete in space

6.1.1.1 Grid System

A tight reservoir contains pores and fractures so it can be divided into pore media and fracture media. Because pore media and fracture media continuously distribute in space and their property parameters continuously change, using structured grids, these media can be represented by two individual grid systems, i.e., a pore grid system and a fracture grid system. The pore grid system represents the single pore media while the fracture grid system represents the single fracture media. Different grids can have different properties (Fig. 6.1).

In these two grid systems, natural/hydraulic fractures and horizontal wells are separately meshed to form complete pore and fracture grid systems.

6.1.1.2 Behavior of Fluid Flow Through Different Media in Reservoirs

In modeling dual-porosity single-permeability, due to poor pore properties in reservoir matrix, the fluid does not flow between adjacent grids in the matrix system. For the fracture system, the fluid can flow between adjacent grids. Inter-porosity flow occurs between corresponding grids of pore and fracture systems (Fig. 6.2).

According to the above grid systems, the pore and fracture grids are respectively ordered and numbered. According to the adjacency relationship and flow relationship between grids of different media, the connectivity list between different grids is established (taking the figures in Table 6.2 as an example), as shown in Table 6.3.

6.1.1.3 Flow Behavior Between Reservoir Media and Wellbores

In modeling dual-porosity single-permeability, the fluid in matrix pores can only flow into wellbores through fractures (as shown in Fig. 6.2). Based on the meshing of natural/hydraulic fractures and horizontal wells, well grids are ordered and numbered. According to the adjacency relationship and the flow behavior between fractures and well grids, the connectivity list between fractures and well grids is established (taking the figures in Table 6.2 as an example), as shown in Table 6.4.

Table 6.2 Simulation for continuous multiple media

Classification	Simulation for continuous dual-porosity media		Simulation for continuous multiple media
	Simulation for dual-porosity single-permeability media	Simulation for dual-porosity dual-permeability media	Simulation for multiple media based on model for dual-porosity media
Conceptual model	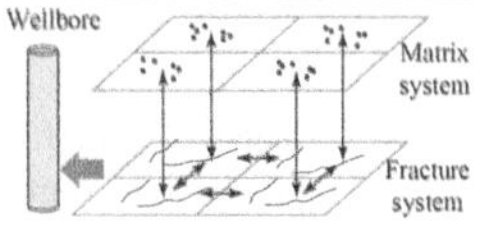	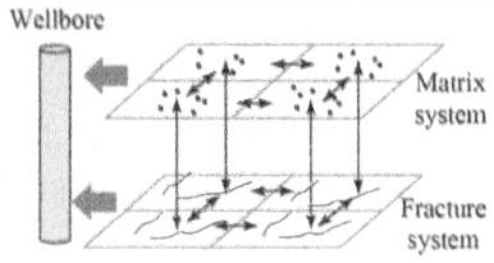	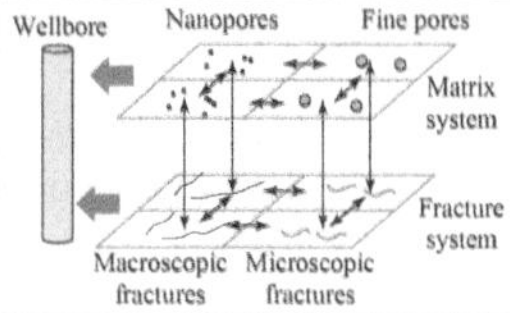
Nested grid number	Grid for matrix and fracture, respectively	Grid for matrix and fracture, respectively	Grid for matrix and fracture, respectively
Medium type	Single pore media Single fracture media	Single pore media Single fracture media	Multiple pore media Multiple fracture media
Spatial distribution and parameter change of different media	Two individual grid systems represent pore media and fracture media, respectively. The distributions of single pore media and single fracture media are continuous in corresponding grid system and their parameters continuously change		Two individual grid systems represent pore media and fracture media, respectively. The distributions of multiple-scale pore media and multiple-scale fracture media are continuous in corresponding grid system and their parameters continuously change
Relationship of fluid flow between different media	① Fluid flow fails to occur in grid system of pore media ② Fluid flow occurs between adjacent grids in grid system of fracture media ③ The interporosity flow occurs between grids representing pores and fractures	① Fluid flow occurs between adjacent grids in grid system of pore media ② Fluid flow occurs between adjacent grids in grid system of fracture media ③ The interporosity flow occurs between grids representing pores and fractures	① Fluid flow occurs between adjacent grids in grid system of pore media ② Fluid flow occurs between adjacent grids in grid system of fracture media ③ The interporosity flow occurs between grids representing pores and fractures
Relationship of fluid flow between reservoirs and wellbores	① Fluid flow fails to occur between pores and wells ② Fluid flow occurs between fractures and wells	① Fluid flow occurs between pores and wells ② Fluid flow occurs between fractures and wells	① Fluid flow occurs between different scale pores and wells ② Fluid flow occurs between different scale fractures and wells
Application conditions	① Matrix pores are intensively developed and continuously distribute but their permeability is quite low so it can be ignored ② Fractures are well developed and their density is high to generate complex fracture networks. When fractures are less developed and their density is low, the simulation method is inapplicable ③ Because the simulation method is based on structured grid, the calculation is relatively simple and fast	① Matrix pores are intensively developed, continuously distribute and their permeability cannot be ignored ② Fractures are well developed and their density is high to generate complex fracture networks. When fractures are less developed and their density is low, the simulation method is inapplicable ③ Because the simulation method is based on structured grid, the calculation is relatively simple and fast	① Multiple-scale pores are intensively developed, continuously distribute and the simulation method can deal with fluid flow between different scale pores ② Fractures are well developed and their density is high. The simulation method can deal with fluid flow between different scale fractures. When fractures are less developed and their density is low, the simulation method is inapplicable ③ Because the simulation method is based on structured grid, the calculation is relatively simple and fast

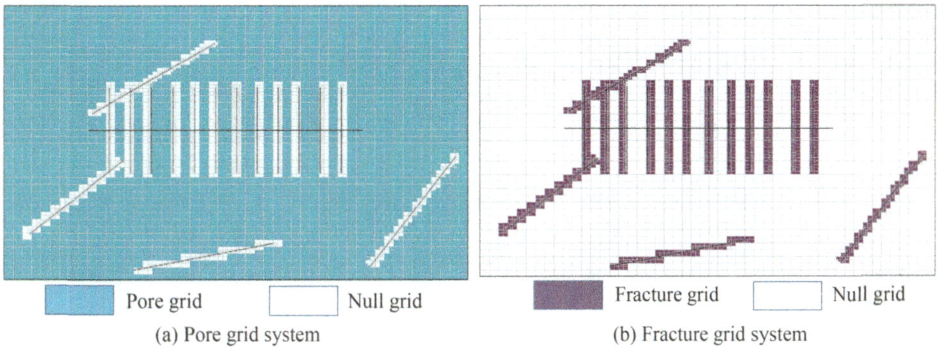

Fig. 6.1 Grid illustration of simulation of dual-porosity single-permeability media

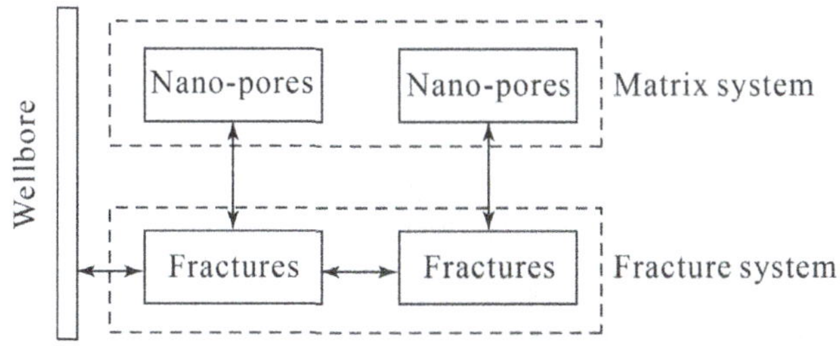

Fig. 6.2 Illustration of flow relationship between different media for dual-porosity single-permeability model

6.1.1.4 Numerical Simulation Model

1. Mathematical model of flow between different media

The transmissibility between structured grids in a fracture system is

$$T_{\mathrm{f}i,j} = \alpha_{\mathrm{f}}\left(\frac{2K_iK_j}{K_i + K_j}\right)_{\mathrm{f}} \tag{6.1}$$

A multiple-phase flow model in dual-porosity single-permeability media is

(1) Pore system

$$-\alpha_{\mathrm{f,m}}K_{\mathrm{m}}\sum_{\mathrm{p}}\left[\rho_{\mathrm{p}}X_{\mathrm{cp}}\frac{K_{\mathrm{rp,m}}}{\mu_{\mathrm{p}}}\left(\phi_{\mathrm{p,m}} - \phi_{\mathrm{p,f}}\right)\right]^{n+1} = \frac{V}{\Delta t}\left\{\left[\phi\sum_{\mathrm{p}}\left(S_{\mathrm{p}}\rho_{\mathrm{p}}X_{\mathrm{cp}}\right)\right]^{n+1} - \left[\phi\sum_{\mathrm{p}}\left(S_{\mathrm{p}}\rho_{\mathrm{p}}X_{\mathrm{cp}}\right)\right]^{n}\right\}_{i,m} \tag{6.2}$$

(2) Fracture system

$$\left\{\sum_{j}\left[T_{i,j}\sum_{\mathrm{p}}\left(\rho_{\mathrm{p}}X_{\mathrm{cp}}\frac{K_{\mathrm{rp}}}{\mu_{\mathrm{p}}}\Delta\phi_{\mathrm{p}}\right)\right]_{i,\mathrm{f}} + \alpha_{\mathrm{f,m}}K_{\mathrm{m}}\sum_{\mathrm{p}}\left[\begin{array}{c}\rho_{\mathrm{p}}X_{\mathrm{cp}}\frac{K_{\mathrm{rp,m}}}{\mu_{\mathrm{p}}}\\ \left(\phi_{\mathrm{p,m}} - \phi_{\mathrm{p,f}}\right)\end{array}\right] + q_{\mathrm{p,f}}^{\mathrm{W}}\right\}^{n+1} = \frac{V}{\Delta t}\left\{\left[\phi\sum_{\mathrm{p}}\left(S_{\mathrm{p}}\rho_{\mathrm{p}}X_{\mathrm{cp}}\right)\right]^{n+1} - \left[\phi\sum_{\mathrm{p}}\left(S_{\mathrm{p}}\rho_{\mathrm{p}}X_{\mathrm{cp}}\right)\right]^{n}\right\}_{i,\mathrm{f}} \tag{6.3}$$

2. Mathematical model of flow between fractures and wellbores

A well index between structured grids of fractures and wellbores is

$$\mathrm{WI_f} = \frac{2\pi K_{\mathrm{f}}\Delta x}{\ln\left(\frac{r_{\mathrm{e}}}{r_{\mathrm{w}}}\right) + s} \tag{6.4}$$

Based on the flow behavior between dual-porosity single-permeability media and wellbores, a model of flow between fractures and wellbores is

$$q_{\mathrm{c,f}}^{\mathrm{W}} = \mathrm{WI_f}\sum_{\mathrm{p}}\left\{\left(\frac{K_{\mathrm{rp}}\rho_{\mathrm{p}}}{\mu_{\mathrm{p}}}\right)_{\mathrm{f}}X_{\mathrm{cp}}\left[\left(P_{\mathrm{p,f}} - \rho_{\mathrm{p}}gD_{\mathrm{f}}\right) - \left(P^{\mathrm{W}} - \rho_{\mathrm{p}}gD^{\mathrm{W}}\right)\right]\right\} \tag{6.5}$$

6.1.2 Numerical Simulation of Dual-Porosity Dual-Permeability Media

For dual-porosity media with good pore properties in reservoir matrix, the fluid can flow inside matrix so that the fluid can flow not only from matrix into wells through fractures but also directly from matrix into wells. The

Table 6.3 Grid ordering of simulation for dual-porosity single-permeability media and corresponding conductivity list

Pore system	Grid *i*	Grid *j*
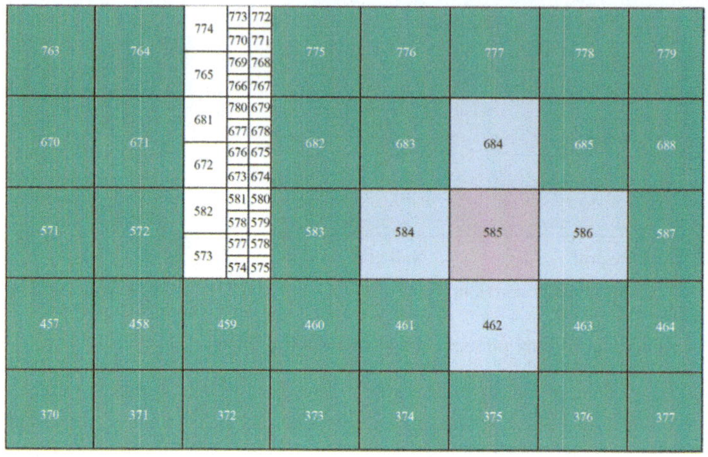	…	…
	585	462
	585	584
	585	586
	585	684
	…	…

Fracture system	Grid *i*	Grid *j*
	…	…
	1027	1022
	1027	1024
	1027	1026
	1027	1028
	…	…

Table 6.4 Connectivity list between fracture and well grids in simulation of dual-porosity single-permeability media

Fracture system-Wellbore system	Reservoir grid	Wellbore grid
	…	…
	833	755
	834	754
	835	753
	662	746
	661	747
	660	748
	…	…

following simulation of dual-porosity dual-permeability was developed to deal with this case.

6.1.2.1 Grid System

Structured grids used for tight reservoirs can be classified into two systems: a pore system and a fracture system. Those two systems represent single pore media and single fracture media, respectively, and have different parameters (Fig. 6.3).

After the grid generation for natural/hydraulic fractures and horizontal wells, the grid systems of pores and fractures can be completed.

6.1.2.2 Flow Behavior Between Different Media in Reservoirs

In a dual-porosity dual-permeability model, fluid flow can occur between adjacent grids in grid systems of pores and fractures. The interporosity flow occurs between grids of pores and fractures (Fig. 6.4).

According to the above grid systems, the pore and fracture grids are respectively ordered and numbered. According to the adjacency relationship and flow behavior between grids of different media, the connectivity list between different grids is established (taking the figures in Table 6.2 as an example), as shown in Table 6.5.

6.1.2.3 Flow Behavior Between Reservoir Media and Wellbores

In the model of dual-porosity dual-permeability, both the fluid flow between pores and wellbores and the one between fractures and wellbores can occur (as shown in Fig. 6.4). Based on the mesh generation of natural/hydraulic fractures and horizontal wells, the well grids are ordered and numbered. According to the adjacency relationship and flow behavior between fracture and well grids, the connectivity list between pores, fractures and well grids is established (taking the figures in Table 6.2 as an example), as shown in Table 6.6.

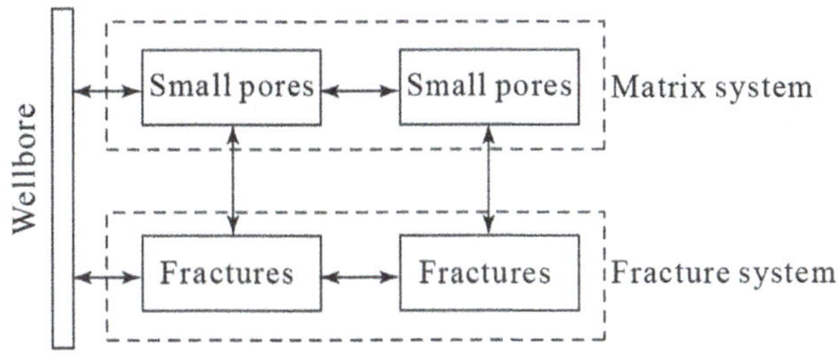

Fig. 6.4 Illustration of flow relationship between different media in dual-porosity dual-permeability model

6.1.2.4 Numerical Simulation Model

1. Mathematical model of flow between different media

The transmissibility between structured grids of a pore system is

$$T_{\mathrm{m}i,j} = \alpha_{\mathrm{m}}\left(\frac{2K_iK_j}{K_i + K_j}\right)_{\mathrm{m}} \tag{6.6}$$

The transmissibility between structured grids of a fracture system is

$$T_{\mathrm{f}i,j} = \alpha_{\mathrm{f}}\left(\frac{2K_iK_j}{K_i + K_j}\right)_{\mathrm{f}} \tag{6.7}$$

A Multiple-phase (oil/gas/water) flow model of dual-porosity dual-permeability media is

(1) Pore system

$$\left\{\sum_j\left[T_{i,j}\sum_{\mathrm{p}}\left(\rho_{\mathrm{p}}X_{\mathrm{cp}}\frac{K_{\mathrm{rp}}}{\mu_{\mathrm{p}}}\Delta\phi_{\mathrm{p}}\right)\right]_{i,\mathrm{m}} - \alpha_{\mathrm{f,m}}K_{\mathrm{m}}\sum_{\mathrm{p}}\left[\begin{matrix}\rho_{\mathrm{p}}X_{\mathrm{cp}}\frac{K_{\mathrm{rp,m}}}{\mu_{\mathrm{p}}}\\(\phi_{\mathrm{p,m}} - \phi_{\mathrm{p,f}})\end{matrix}\right] + q_{\mathrm{p,m}}^{\mathrm{W}}\right\}^{n+1} = \frac{V}{\Delta t}\left\{\left[\phi\sum_{\mathrm{p}}(S_{\mathrm{p}}\rho_{\mathrm{p}}X_{\mathrm{cp}})\right]^{n+1} - \left[\phi\sum_{\mathrm{p}}(S_{\mathrm{p}}\rho_{\mathrm{p}}X_{\mathrm{cp}})\right]^{n}\right\}_{i,\mathrm{m}} \tag{6.8}$$

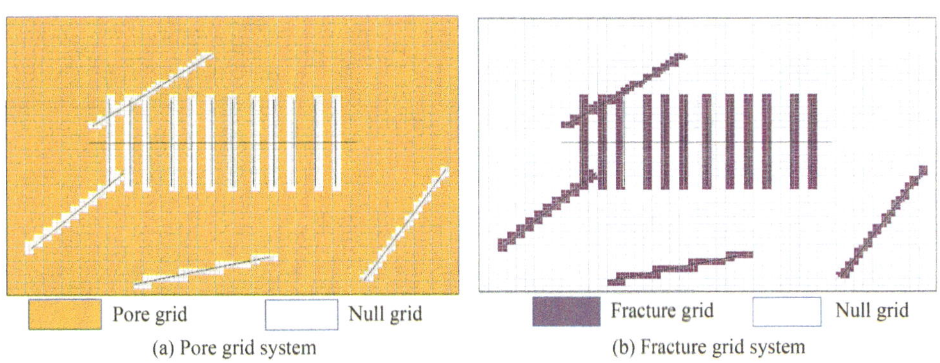

Fig. 6.3 Illustration of grid for dual-porosity dual-permeability simulation

Table 6.5 Grid ordering of simulation of dual-porosity dual-permeability media and corresponding conductivity list

Pore system

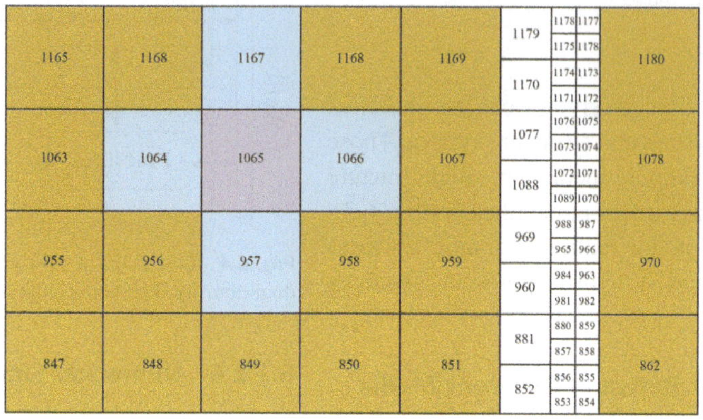

Grid i	Grid j
…	…
1065	957
1065	1064
1065	1066
1065	1167
…	…

Fracture system

Grid i	Grid j
…	…
1027	1022
1027	1024
1027	1026
1027	1028
…	…

(2) Fracture system

$$\left\{\sum_j\left[T_{i,j}\sum_p\left(\rho_p X_{cp}\frac{K_{rp}}{\mu_p}\Delta\phi_p\right)\right]_{i,f}+\alpha_{f,m}K_m\sum_p\left[\frac{\rho_p X_{cp}\frac{K_{rp,m}}{\mu_p}}{(\phi_{p,m}-\phi_{p,f})}\right]+q^W_{p,f}\right\}^{n+1}=\frac{V}{\Delta t}\left\{\left[\phi\sum_p(S_p\rho_p X_{cp})\right]^{n+1}-\left[\phi\sum_p(S_p\rho_p X_{cp})\right]^n\right\}_{i,f} \tag{6.9}$$

2. Mathematical model of flow between fractures and wellbores

A well index between structured grids of pores and wellbores is

$$\mathrm{WI_m}=\frac{2\pi K_m\Delta x}{\ln\left(\frac{r_e}{r_w}\right)+s} \tag{6.10}$$

A well index between structured grids of fractures and wellbores is

$$\mathrm{WI_f}=\frac{2\pi K_f\Delta x}{\ln\left(\frac{r_e}{r_w}\right)+s} \tag{6.11}$$

Based on the flow behavior between dual-porosity dual-permeability media and wellbores, a model of flow between pore media and wellbores is

$$q^W_{c,m}=\mathrm{WI_m}\sum_p\left\{\left(\frac{K_{rp}\rho_p}{\mu_p}\right)_m X_{cp}\left[(P_{p,m}-\rho_p g D_m)-(P^W-\rho_p g D^W)\right]\right\} \tag{6.12}$$

THE model of flow between fracture media and wellbores is

Table 6.6 Connectivity list between fracture and well grids in simulation of dual-porosity dual-permeability media

Pore system-Wellbore system

Reservoir grid	Wellbore grid
…	…
956	848
761	848
957	849
762	849
958	850
763	850
……	……

Fracture system-Wellbore system

Reservoir grid	Wellbore grid
…	…
960	861
961	860
835	753
962	859
774	852
773	853
…	…

$$q_{c,f}^{W} = WI_{f}\sum_{p}\left\{\left(\frac{K_{rp}\rho_{p}}{\mu_{p}}\right)_{f}X_{cp}\left[(P_{p,f}-\rho_{p}gD_{f})-(P^{W}-\rho_{p}gD^{W})\right]\right\} \quad (6.13)$$

6.1.3 Numerical Simulation of Multiple Media Based on Dual-Porosity Model

For tight reservoirs with multiple-scale pores (such as nano/micro-pores and macro/micro-fractures) and natural/hydraulic fractures, fluid can flow inside multiple-scale pore media and fracture media, and can also flow from matrix into fractures, and directly from matrix and fractures into wellbores. In terms of numbers of multiple media, distribution and flow characteristics of such tight reservoirs, the simulation of multiple media is innovatively developed based on a model of dual-porosity media.

6.1.3.1 Grid System

Tight reservoirs contain multiple-scale media with pores and fractures can be classified into two media: a pore medium and a fracture medium. Because those media are continuously distributed in space, their properties also continuously change in space. Structured grids can be classified into two systems: a pore system and a fracture system. The pore grid system contains different scale pore media and fracture grid system contains different scale fracture media (Fig. 6.5).

The grids of natural/hydraulic fractures and horizontal wellbores are generated to form the complete grid systems of pores and fractures.

6.1.3.2 Flow Behavior Between Different Media in Reservoirs

In a continuous multiple media model, fluid flow can occur between adjacent grids of a pore system for multiple-scale pore media. Fluid flow can occur between adjacent grids of a fracture system for multiple-scale fracture media. The interporosity flow between corresponding grids occurs between pore and fracture systems (Fig. 6.6).

According to the above grid systems, the pore and fracture grids are respectively ordered and numbered. According to the adjacency relationship and flow behavior between

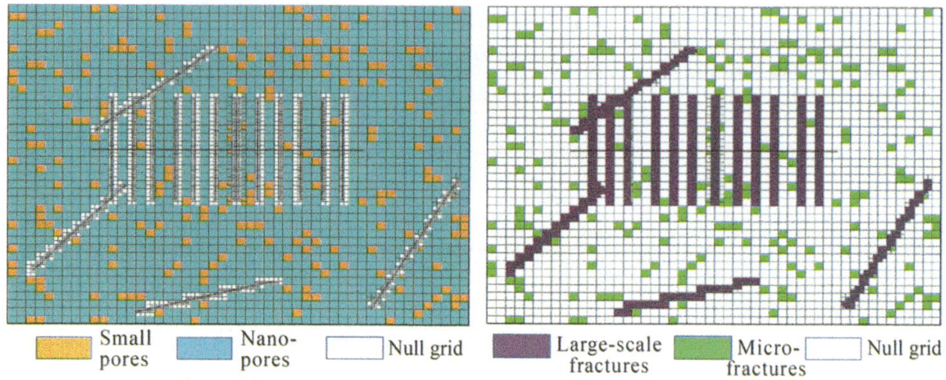

(a) Pore grid system (b) Fracture grid system

Fig. 6.5 Illustration of simulation for multiple media based on dual-porosity model

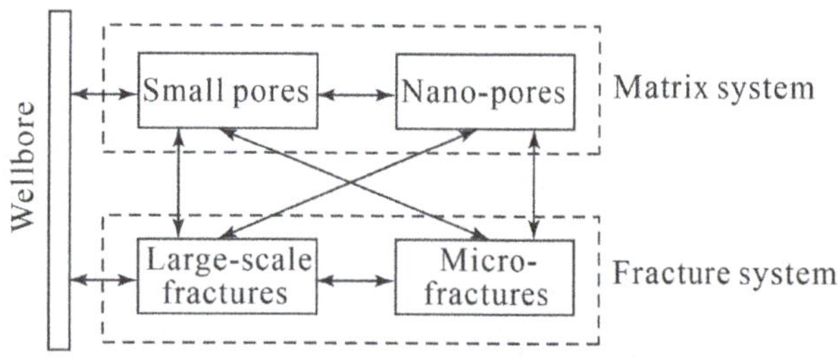

Fig. 6.6 Illustration of relationship of flow between multiple media based on dual-porosity model

grids of pores and fractures the connectivity list between different grids is established (taking the figures in Table 6.2 as an example), as shown in Table 6.7.

6.1.3.3 Flow Behavior Between Reservoir Media and Wellbores

In a model for multiple media, wellbores can obtain fluids from both a pore system and a fracture system (Fig. 6.6). Based on the mesh generations of natural/hydraulic fractures and horizontal wells, the well grids are ordered and numbered. According to the adjacency relationship and flow behavior between fracture and well grids, the connectivity list between pore, fracture and well grids is established (taking the figures in Table 6.2 as an example), as shown in Table 6.8.

6.1.3.4 Numerical Simulation Model

1. Mathematical model of flow between different media

The transmissibility between structured grids of a pore system is

$$T_{\mathrm{m1,m2}} = \alpha_{\mathrm{m1m2}} \frac{2K_{\mathrm{m1}}K_{\mathrm{m2}}}{K_{\mathrm{m1}} + K_{\mathrm{m2}}} \tag{6.14}$$

The transmissibility between structured grids of a fracture system is

$$T_{\mathrm{f,F}} = \alpha_{\mathrm{fF}} \frac{2K_{\mathrm{f}}K_{\mathrm{F}}}{K_{\mathrm{f}} + K_{\mathrm{F}}} \tag{6.15}$$

Taking the flow model shown in Fig. 6.6 as an example, the pore system includes micro-pores and nano-pores, and the fracture system includes large fractures and micro-fractures. In this case, a multiple-phase (oil/gas/water) flow model of multiple media is

(1) Pore system

1) Media with microscopic pores (m1)

$$\left\{\sum_j\left[T_{i,j}\sum_{\mathrm{p}}\left(\rho_{\mathrm{p}}X_{\mathrm{cp}}\frac{K_{\mathrm{rp}}}{\mu_{\mathrm{p}}}\Delta\phi_p\right)\right]_{\mathrm{m1}} - \alpha_{\mathrm{f,m1}}K_{\mathrm{m1}}\sum_{\mathrm{p}}\left[\begin{matrix}\rho_{\mathrm{p}}X_{\mathrm{cp}}\frac{K_{\mathrm{rp,m1}}}{\mu_{\mathrm{p}}}\\(\phi_{\mathrm{p,m1}}-\phi_{\mathrm{p,f}})\end{matrix}\right] + q^{\mathrm{W}}_{\mathrm{p,m1}}\right\}^{n+1} = \frac{V}{\Delta t}\left\{\left[\phi\sum_{\mathrm{p}}(S_{\mathrm{p}}\rho_{\mathrm{p}}X_{\mathrm{cp}})\right]^{n+1} - \left[\phi\sum_{\mathrm{p}}(S_{\mathrm{p}}\rho_{\mathrm{p}}X_{\mathrm{cp}})\right]^{n}\right\}_{\mathrm{m1}} \tag{6.16}$$

2) Media with nano-pores (m2)

$$\left\{\sum_j\left[T_{i,j}\sum_{\mathrm{p}}\left(\rho_{\mathrm{p}}X_{\mathrm{cp}}\frac{K_{\mathrm{rp}}}{\mu_{\mathrm{p}}}\Delta\phi_p\right)\right]_{\mathrm{m2}} - \alpha_{\mathrm{f,m2}}K_{\mathrm{m2}}\sum_{\mathrm{p}}\left[\begin{matrix}\rho_{\mathrm{p}}X_{\mathrm{cp}}\frac{K_{\mathrm{rp,m2}}}{\mu_{\mathrm{p}}}\\(\phi_{\mathrm{p,m2}}-\phi_{\mathrm{p,f}})\end{matrix}\right] + q^{\mathrm{W}}_{\mathrm{p,m2}}\right\}^{n+1} = \frac{V}{\Delta t}\left\{\left[\phi\sum_{\mathrm{p}}(S_{\mathrm{p}}\rho_{\mathrm{p}}X_{\mathrm{cp}})\right]^{n+1} - \left[\phi\sum_{\mathrm{p}}(S_{\mathrm{p}}\rho_{\mathrm{p}}X_{\mathrm{cp}})\right]^{n}\right\}_{\mathrm{m2}} \tag{6.17}$$

Table 6.7 Grid ordering of simulation of multiple media based on dual-porosity model and corresponding conductivity list

Pore system	Grid i	Grid j
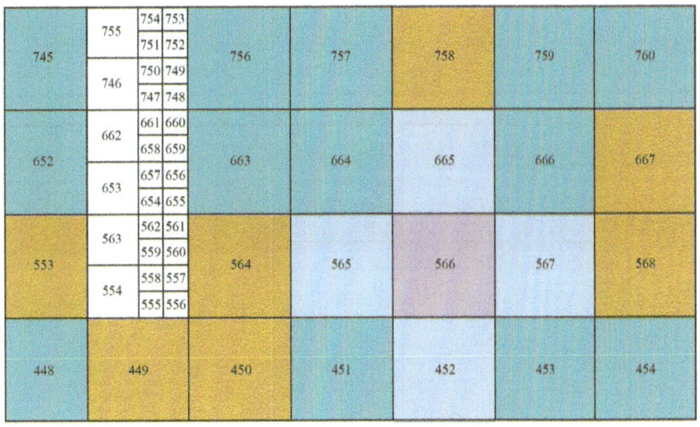	… 566 566 566 566 …	… 452 565 567 665 …
Fracture system	**Grid i**	**Grid j**
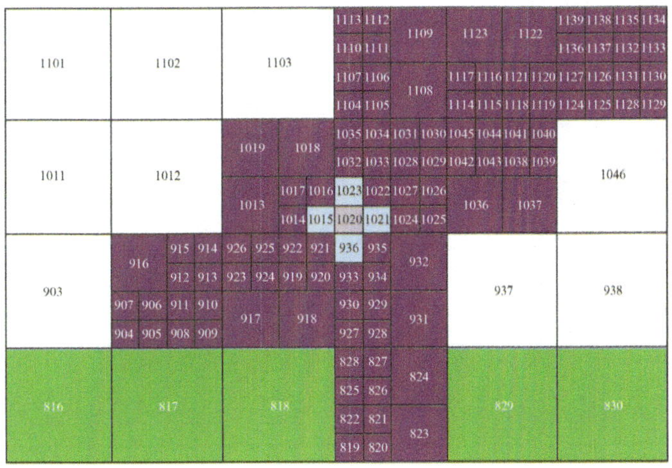	… 1020 1020 1020 1020 …	… 936 1015 1021 1023 …

(2) Fracture system

1) Large-scale fractures (F)

$$\left\{\sum_j\left[T_{i,j}\sum_p\left(\rho_p X_{cp}\frac{K_{rp}}{\mu_p}\Delta\phi_p\right)\right]_{i,F}+\alpha_{f,m2}K_{m2}\sum_p\left[\rho_p X_{cp}\frac{K_{rp,m2}}{\mu_p}(\phi_{p,m2}-\phi_{p,F})\right]+q_{p,F}^{W}\right\}^{n+1}$$
$$=\frac{V}{\Delta t}\left\{\left[\phi\sum_p(S_p\rho_p X_{cp})\right]^{n+1}-\left[\phi\sum_p(S_p\rho_p X_{cp})\right]^{n}\right\}_{i,F} \tag{6.18}$$

2) Microscopic fractures media (f)

$$\left\{\sum_j\left[T_{i,j}\sum_p\left(\rho_p X_{cp}\frac{K_{rp}}{\mu_p}\Delta\phi_p\right)\right]_{i,f}+\alpha_{f,m1}K_{m1}\sum_p\left[\rho_p X_{cp}\frac{K_{rp,m1}}{\mu_p}(\phi_{p,m1}-\phi_{p,f})\right]+q_{p,f}^{W}\right\}^{n+1}$$
$$=\frac{V}{\Delta t}\left\{\left[\phi\sum_p(S_p\rho_p X_{cp})\right]^{n+1}-\left[\phi\sum_p(S_p\rho_p X_{cp})\right]^{n}\right\}_{i,f} \tag{6.19}$$

2. Mathmatical model of flow between fractures and wellbores

This model contains models of flow between multiple-scale pore media, multiple-scale fracture media and wellbores:

(1) Model of flow between microscopic pores and wellbores

A well index between microscopic pores and wellbores is

$$\mathrm{WI}_{m1}=\frac{2\pi K_{m1}\Delta x}{\ln\left(\frac{r_e}{r_w}\right)+s} \tag{6.20}$$

A model of flow between microscopic pores and wellbores is

$$q_{c,m1}^{W}=\mathrm{WI}\sum_p\left\{\left(\frac{K_{rp}\rho_p}{\mu_p}\right)_{m1}X_{cp}\left[(P_{p,m1}-\rho_p g D_{m1})-(P^{W}-\rho_p g D^{W})\right]\right\} \tag{6.21}$$

Table 6.8 Connectivity list between fracture and well grids in simulation of multiple media based on model for dual-porosity media

Pore system-Wellbore system

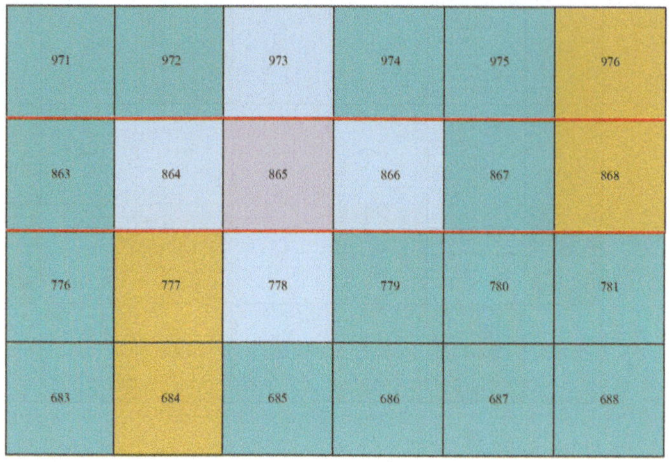

Reservoir grid	Well grid
…	…
777	864
972	864
778	865
973	865
779	866
974	866
…	…

Fracture system-Wellbore system

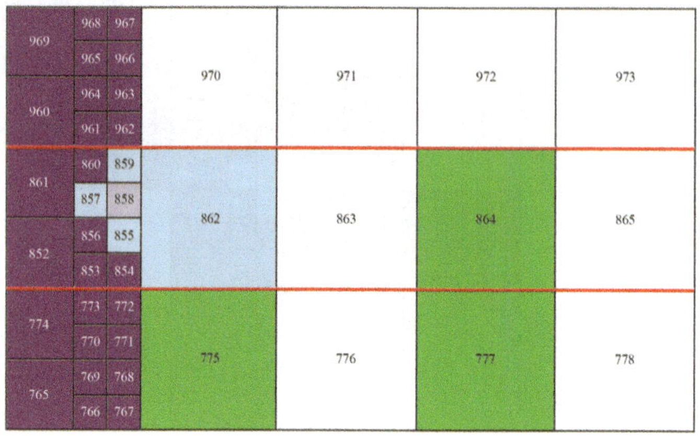

Reservoir grid	Well grid
…	…
960	861
961	860
962	859
774	852
773	853
…	…

(2) Model of flow between nano-pores and wellbores

A well index between nano-pores and wellbores is

$$\mathrm{WI} = \frac{2\pi K_{\mathrm{m2}} \Delta x}{\ln\left(\frac{r_e}{r_w}\right) + s} \tag{6.22}$$

A model of flow between nano-pores and wellbores is

$$q_{\mathrm{c,m2}}^{\mathrm{W}} = \mathrm{WI} \sum_{\mathrm{p}} \left\{ \left(\frac{K_{\mathrm{rp}}\rho_{\mathrm{p}}}{\mu_{\mathrm{p}}}\right)_{\mathrm{m2}} X_{\mathrm{cp}} \left[(P_{\mathrm{p,m2}} - \rho_{\mathrm{p}} g D_{\mathrm{m2}}) - (P^{\mathrm{W}} - \rho_{\mathrm{p}} g D^{\mathrm{W}}) \right] \right\} \tag{6.23}$$

(3) Model of flow between large-scale fracture and wellbores

A well index between large-scale fracture and wellbores is

$$\mathrm{WI} = \frac{2\pi K_{\mathrm{F}} \Delta x}{\ln\left(\frac{r_e}{r_w}\right) + s} \tag{6.24}$$

A model of flow between large-scale fracture and wellbores is

$$q_{\mathrm{c,F}}^{\mathrm{W}} = \mathrm{WI} \sum_{\mathrm{p}} \left\{ \left(\frac{K_{\mathrm{rp}}\rho_{\mathrm{p}}}{\mu_{\mathrm{p}}}\right)_{\mathrm{F}} X_{\mathrm{cp}} \left[(P_{\mathrm{p,F}} - \rho_{\mathrm{p}} g D_{\mathrm{F}}) - (P^{\mathrm{W}} - \rho_{\mathrm{p}} g D^{\mathrm{W}}) \right] \right\} \tag{6.25}$$

(4) Model of flow between microscopic fracture and wellbores

A well index between microscopic fracture and wellbores is

$$\mathrm{WI}_{\mathrm{f}} = \frac{2\pi K_{\mathrm{f}} \Delta x}{\ln\left(\frac{r_e}{r_w}\right) + s} \tag{6.26}$$

A model of flow between microscopic fracture and wellbores is

$$q_{\mathrm{c,f}}^{\mathrm{W}} = \mathrm{WI}\sum_{\mathrm{p}}\left\{(\frac{K_{\mathrm{rp}}\rho_{\mathrm{p}}}{\mu_{\mathrm{p}}})_{\mathrm{f}}X_{\mathrm{cp}}\left[(P_{\mathrm{p,f}} - \rho_{\mathrm{p}}gD_{\mathrm{f}}) - (P^{\mathrm{W}} - \rho_{\mathrm{p}}gD^{\mathrm{W}})\right]\right\} \tag{6.27}$$

6.1.4 Numerical Simulation Process of Continuous Multiple Media

In numerical simulation of continuous dual-porosity media, it is assumed that tight reservoirs are composed of two kinds of media with continuous geometric parameters and properties, i.e., a pore medium and a fracture medium. According to the fluid exchanging behavior between media and wells, this simulation is extended to a numerical simulation of dual-porosity single-permeability media and a numerical simulation of dual-porosity dual-permeability media. However, in reality, both different scales pores and different scales fractures in tight reservoirs have a relatively huge difference in characteristics of geometry, property and fluid flow. In this case, the pores and fractures are further classified into the multiple-scale pore media and the multiple-scale fracture media, respectively. In this chapter, based on conventional simulation of dual-porosity media, the simulation of continuous multiple media is developed. Its process and detailed steps are as follows:

① According to a spatial distribution of multiple-scale pore media and fracture media, a tight reservoir is classified into a pore system and a fracture system. In the simulation of dual-porosity media, the pore system has a single pore medium, and the fracture system has a single fracture medium. However, in the simulation of dual-porosity media, the pore system has several pore media at different scales, and the fracture system has several fracture media at different scales.

② According to the information of domain boundaries, well location, well trajectory, spatial distributions of pores and fractures, and volume percentages of those pores, using structured grids, grids of a tight reservoir are divided into a pore grid system and a fracture grid system (Fig. 6.7).

③ The grids of a pore system and a fracture system are ordered and numbered. According to flow behaviors between different media and the ones between those media and wells, the connectivity list between adjacent grids is established.

④ Through geology modeling of continuous dual-porosity/multiple media, the media type of each grid for the pore system and fracture system is determined and then each grid is assigned values of corresponding parameters which include physical properties (porosity and permeability), fluid parameters (saturation, viscosity, density, and capillary pressure) and mechanism types (Darcy flow). Therefore, the numerical simulation model of continuous dual-porosity/multiple media is built.

⑤ According to the connectivity list and the numerical simulation model of dual-porosity/multiple media, the conductivities between different grids and the well indices between different media and wells are calculated.

⑥ Running dynamic simulations of oil and gas reservoirs. According to the numerical simulation model, the flow between pore systems, the one between fracture systems and the one between systems, the dynamic changes in pressure and saturation of oil and gas reservoirs are obtained.

⑦ Forecasting development index. According to different well indices and the flow between multiple media and wells obtained by the simulations in the step ⑥, the productions of oil, gas and water and their dynamic changes can be obtained.

6.2 Numerical Simulation of Discontinuous Multiple Media at Different Scales

Unconventional tight oil and gas reservoirs contain multiple media with different scales pores and fractures but those media discontinuously distribute in space so their properties also change discontinuously in space. There are huge differences in geometry, property and flow characteristics between those media. Conventional simulation for dual-porosity media (Wu 2016; Chen et al. 2006) is difficult to explicitly represent characteristics of those multiple media in terms of discontinuous discrete distribution, multiple-scales and medium components and consider the effects of flow regimes and complex flow mechanisms. Therefore, in order to overcome these limits of conventional simulation for dual-porosity media, the innovative simulation for multiple-regime flow through discontinuous multiple media at multiple-scales is developed, including the dynamic simulation of multiple-scale natural/hydraulic discrete fractures, simulations for multiple-scale discrete media based on the NDT and the RIDT grid patterns, and the simulation for discontinuous-discrete-hybrid multiple media (Table 6.9).

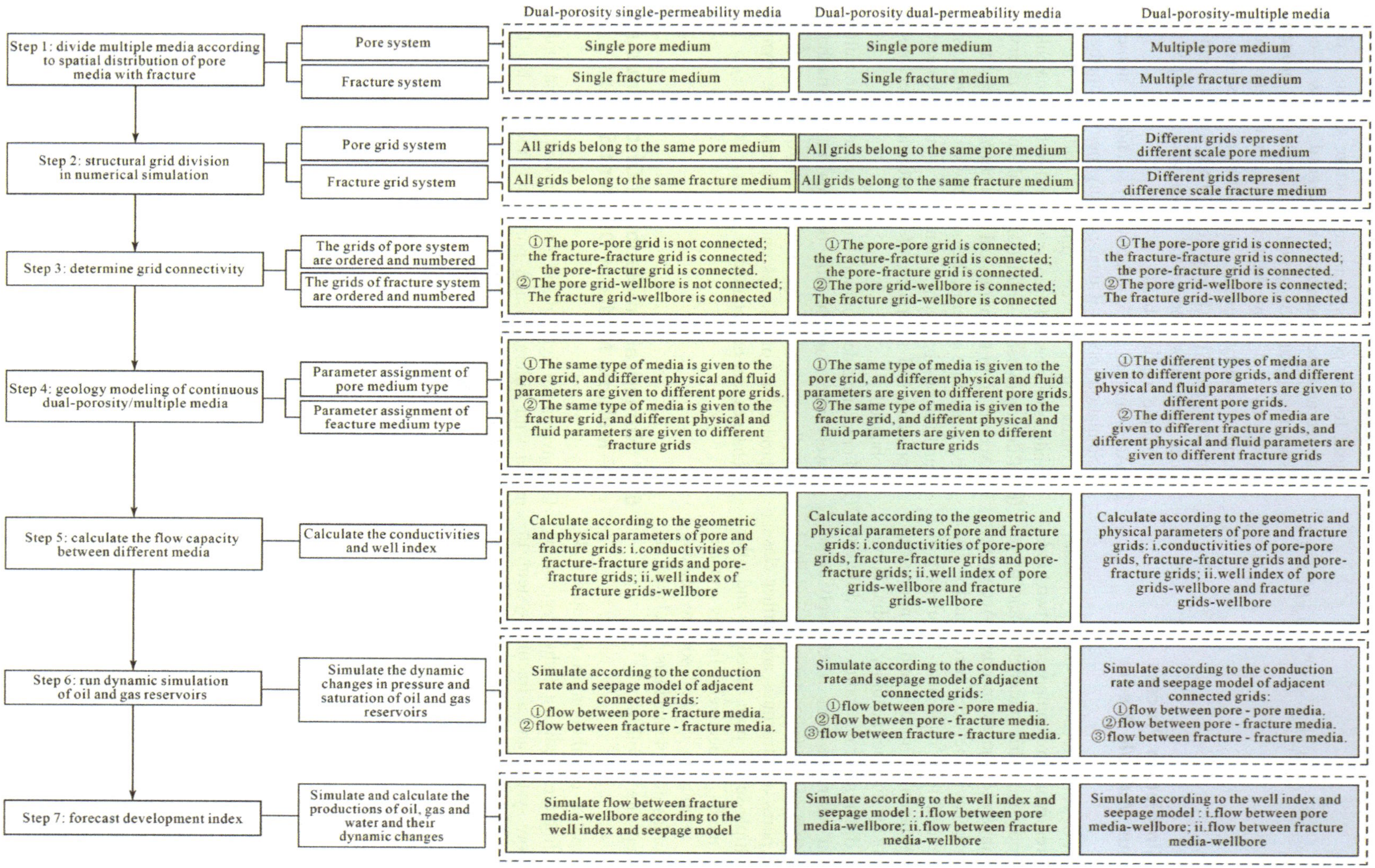

Fig. 6.7 Technique process of simulation for continuous multiple media

Table 6.9 Comparison between numerical simulations for discontinuous multiple media

classification	Numerical simulation for discrete fractures	Numerical simulation for discrete multiple media		Numerical simulation for hybrid discrete multiple media
		Numerical simulation for multiple media based on the grid pattern of NDT	Numerical simulation for multiple media based on the grid pattern of RIDT	
Grid pattern	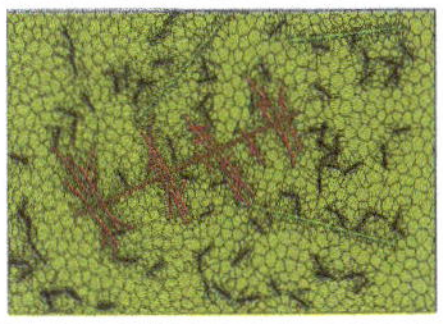		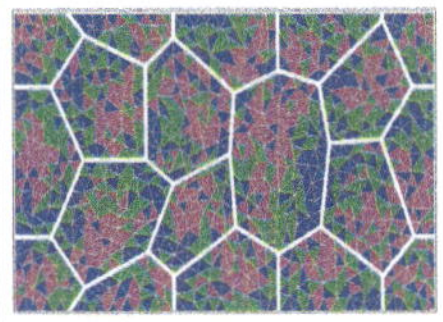	
Meshing by different scale grids	① Area partitioning based on macroscopic heterogeneity ② Assigning different discrete fracture elements to large/meso-scale natural/hydraulic fractures ③ Meshing the domain and using discrete fracture elements represents discrete fractures ④ Forming a set of grid system composed of different discrete fracture grids	① Area partitioning based on macroscopic heterogeneity ② Assigning different microscopic scale pore/fracture element to pore/fractures according to microscopic heterogeneity ③ According to quantity and spatial distribution of microscopic scale pores/fractures, using NDT grid pattern represents different discrete pores/fractures ④ Forming a set of grid system composed of different discrete pore/fracture grids	① Area partitioning based on macroscopic heterogeneity ② According to quantity and spatial distribution of different scale pores/fractures, using RIDT grid pattern represents different discrete pores/fractures ③ Forming a set of grid system composed of different discrete pore/fracture grids	① Area partitioning based on macroscopic heterogeneity ② Assigning different discrete fracture element to large/meso-scale fractures; Assigning different microscopic scale pore/fracture element to pore/fractures according to microscopic heterogeneity ③ According to scale magnitude, using discrete fracture elements represents discrete fractures; According to quantity and spatial distribution of different scale pores/fractures, using MMTG or RIDT grid pattern represents different discrete pores/fractures ④ Forming a set of grid system composed of different discrete pore/fracture grids
Media type	① Different scale discrete natural fracture ② Different scale discrete hydraulic fracture	① Different scale discrete microscopic fracture ② Different scale discrete pore	① Different scale discrete microscopic fracture ② Different scale discrete pore	① Different scale discrete natural/hydraulic fracture ② Different scale discrete microscopic fracture ③ Different scale discrete pore
Spatial distribution and parameter change of different media	① The discrete distribution of different scale natural/hydraulic fractures ② Discontinuous change of properties of different scale discrete fractures	① The spatial distribution of different scale microscopic fractures and different scale pores is subjected to the NDT grid pattern ② Discontinuous change of properties of the above pores and fractures	① The spatial distribution of different scale microscopic fractures and different scale pores is subjected to the RIDT grid pattern ② Discontinuous change of properties of the above pores and fractures	① The spatial distribution of different scale natural/hydraulic fractures is discrete ② The spatial distribution of different scale microscopic fractures and different scale pores is subjected to the NDT or the RIDT grid pattern ③ Discontinuous change of properties of the above pores and fractures
Relationship of fluid flow between different media	① Fluid flow occurs between adjacent and connected multi-scale fractures ② Fluid flow fails to occur between nonadjacent and disconnected multi-scale fractures	① Fluid flow occurs between adjacent and connected multi-scale fractures ② The flow between adjacent and connected pores/fractures obeys the mechanism of flow through media with matryoshka-doll-type distribution	① Fluid flow occurs between adjacent and connected multi-scale pores/fractures ② Fluid flow fails to occur between nonadjacent and disconnected multi-scale pores/fractures	① Fluid flow occurs between adjacent and connected multi-scale pores/fractures ② Fluid flow occurs between adjacent and connected multi-scale pores/fractures elements ③ The flow between adjacent and connected pores/fractures obeys the mechanism of flow through media with matryoshka-doll-type distribution

(continued)

Table 6.9 (continued)

classification	Numerical simulation for discrete fractures	Numerical simulation for discrete multiple media		Numerical simulation for hybrid discrete multiple media
		Numerical simulation for multiple media based on the grid pattern of NDT	Numerical simulation for multiple media based on the grid pattern of RIDT	
Relationship of fluid flow between reservoirs and wellbores	Fluid flow between different scale natural/hydraulic fractures and wellbores	① Fluid flow between different scale hydraulic fractures and wellbores ② Fluid flow between different scale pores/fractures and wellbores	① Fluid flow between different scale hydraulic fractures and wellbores ② Fluid flow between different scale pores/fractures and wellbores	① Fluid flow between different scale hydraulic fractures and wellbores ② Fluid flow between different scale pores/fractures and wellbores ③ Fluid flow between different scale pores/fractures elements and wellbores
Application conditions	① The distribution of large/meso-scale hydraulic/natural fractures is discrete, the density of those fractures is low and their connectivity is poor ② Using unstructured grid can actually approach the spatial distribution and the geometry of fractures, therefore enhances the calculation accuracy	① The microscopic scale fractures are well developed, their density is high, and their distribution is discrete ② Different scale pores are controlled by geological laws. The pore distribution subjected to the matryoshka-doll-type is intensive and discrete in space ③ Using unstructured nested grid can actually approach the discrete distribution of different scale pores and the relationship of flow between those pores	① The microscopic scale fractures are well developed, their density is high, and their distribution is discrete ② Different scale pores are controlled by geological laws. The pore distribution subjected to the puzzle-piece-type distribution is intensive and discrete in space ③ Using unstructured interactive refined grid can actually approach the discrete distribution of different scale pores/fractures and the relationship of flow between those pores/fractures	① The distribution of large/meso-scale hydraulic/natural fractures is discrete, their density is low, and connectivity is poor ② The microscopic scale fractures are well developed, their density is high and their distribution is discrete ③ Different scale pores are controlled by geological laws. The pore distribution subjected to the matryoshka-doll-type or puzzle-piece-type is intensive and discrete ③ Using unstructured discrete fracture grid, nested grid and interactive refined grid can actually approach the discrete distribution of different scale pores/fractures and the relationship of flow between those pores/fractures

6.2.1 Numerical Simulation of Discrete Natural/Hydraulic Fractures at Large-Scale

For large/meso-scale natural fractures/hydraulic fractures, they discretely distribute in space and their parameters discontinuously change in space as well, so that the conventional simulation for continuous fracture media is inapplicable for those media. According to spatial distribution characteristics of large/meso-scale natural fractures/hydraulic fractures, the fractures elements can be classified into different discrete fracture elements. Through grid techniques, those different discrete fracture elements can be assigned to different discrete fracture media, forming a grid system for discrete fractures. Using the modeling of discrete natural fractures based on filling and roughness characteristics, and the modeling of discrete hydraulic fractures with different proppant concentrations and proppant modes, a numerical simulation model is built to represent geometry and property characteristics of multiple-scale discrete fractures, complex flow regimes and flow mechanisms, and the dynamic simulation of multiple-scale discrete natural/hydraulic fractures is formed (Yuan et al. 2004; Lv et al. 2013; Yao and Huang 2014).

6.2.1.1 Grid System

For the multiple-scale natural/hydraulic fractures, they have the following features: i. Their macroscopic heterogeneity in spatial distribution is extremely strong. ii. Their characteristics of discontinuous discrete distribution and multiple-scales are obvious. iii. Their fracture density is low, their connectivity is poor. iv. Their property parameters and flow characteristics change discontinuously. It is difficult to use structured grids to deal with those fractures. As a result, in this section, the unstructured grids are used to deal with the discretization of those fractures.

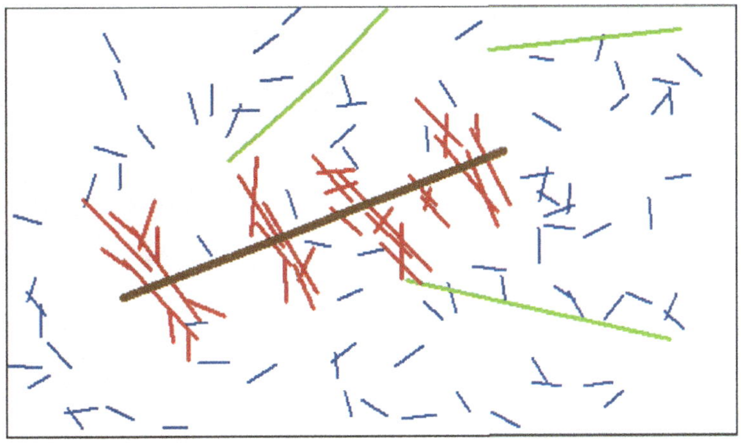

Fig. 6.8 Illustration of distribution of discrete fractures

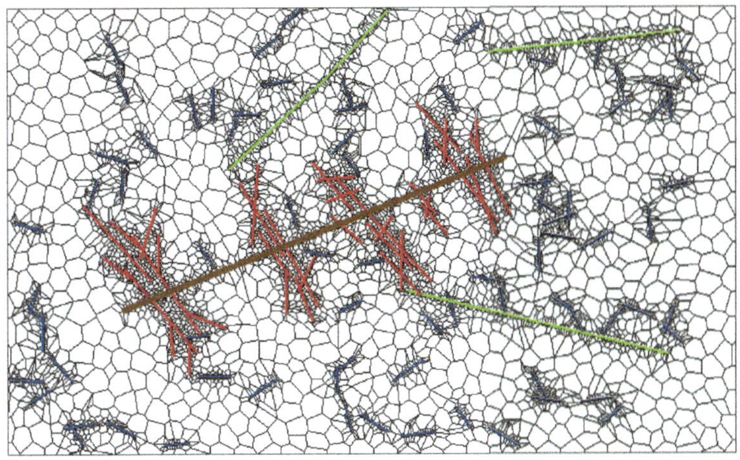

Fig. 6.9 Illustration of discrete fracture grids

① According to macroscopic heterogeneity in lithology, lithofacies and reservoir category, a tight reservoir is partitioned into several first-level areas. Different areas have a huge difference in quantity and spatial distributions of multiple-scale pores and natural/hydraulic fractures (Fig. 6.8).
② Large/meso-scale natural/hydraulic fractures are classified into different representative elements of discrete fractures.
③ Using unstructured grids, the above representative elements of discrete fracture are assigned to different discrete fracture media through mesh generation (Gong et al. 2006, 2010) and different grids represent different discrete fracture media.
④ Different discrete fracture grids generate a grid system (Fig. 6.9).
⑤ Using the modeling of discrete natural fractures based on filling and roughness characteristics, and the modeling of discrete hydraulic fractures based on proppant concentrations and proppant modes, the geometry and property models of discrete fracture media in different grids are built. Those property parameters of discrete fracture media for different grids change discontinuously.

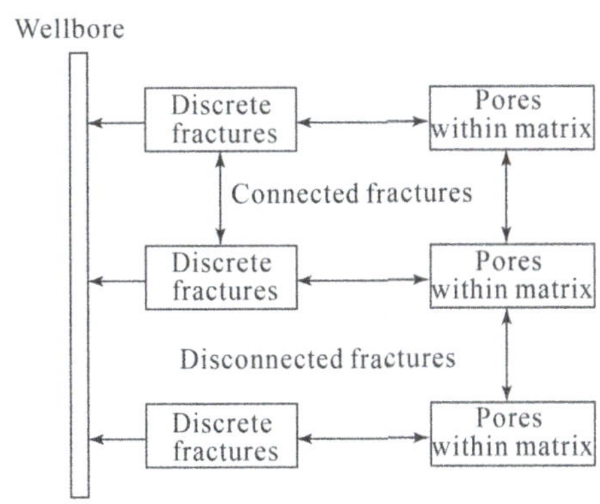

Fig. 6.10 Illustration of flow relationship between different media

6.2.1.2 Flow Behavior Between Different Media in Reservoirs

In a discrete fracture system, fractures discontinuously distribute in matrix blocks. Some fractures connect with each other while other fractures do not. Therefore, fluid flow can occur between matrix and fractures as well as between adjacent and connected fractures, while fluid flow fails to occur between nonadjacent and disconnected fractures (Fig. 6.10).

Table 6.10 Ordering and connectivity list of discrete fracture grids

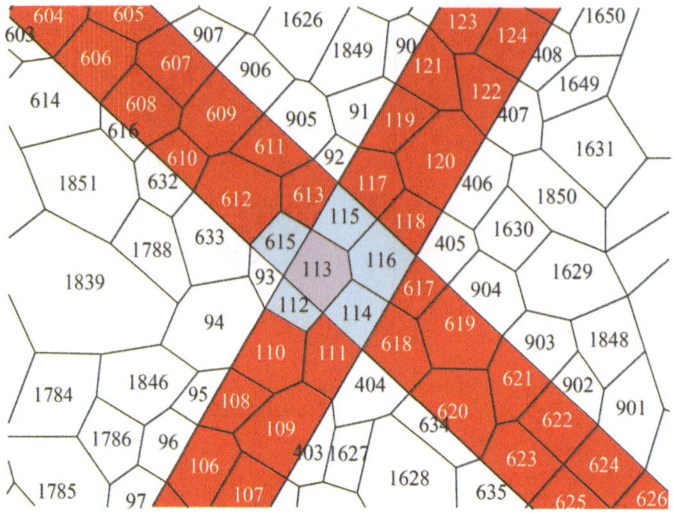

Grid *i*	Grid *j*
…	…
113	112
113	114
113	115
113	116
113	615
…	…

Table 6.11 Connectivity list between reservoir and wellbore grids in simulation with discrete fractures

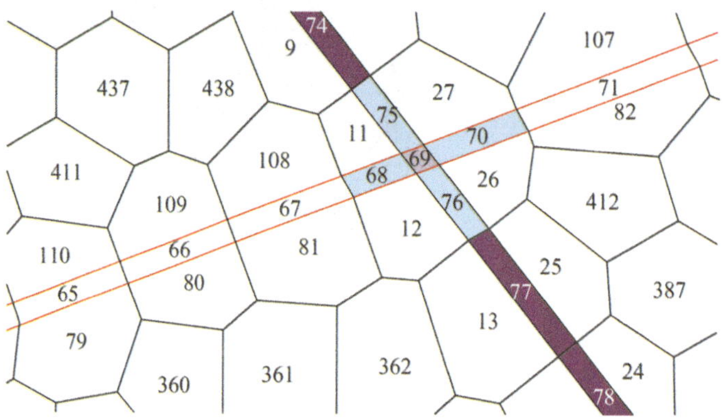

Reservoir grid	Wellbore grid
…	…
75	69
76	69
11	68
12	68
27	70
…	…

According to the above grid systems, the discrete fracture grids are respectively ordered and numbered. According to the adjacency relationship and flow behavior between grids of different media, the connectivity list between different grids is established (Table 6.10).

6.2.1.3 Flow Behavior Between Reservoir Media and Wellbores

In a discrete fracture system, the fluid in different multiple-scale natural/hydraulic fractures can flow into wellbores. Based on the mesh generation of natural/hydraulic fractures and horizontal wells, the wellbore grids are respectively ordered and numbered. According to the adjacency and flow behavior between reservoir grids and wellbore grids, the connectivity list between discrete fracture grids and wellbore grids is established (Table 6.11).

6.2.1.4 Numerical Simulation Model

1. Numerical simulation model of flow between different media

The transmissibility between fracture media at different scales (Karimi-Fard et al. 2003) is

$$T_{\mathrm{F}i,\mathrm{F}j} = \frac{\alpha_{\mathrm{F}i}\alpha_{\mathrm{F}j}}{\alpha_{\mathrm{F}i} + \alpha_{\mathrm{F}j}} \tag{6.28}$$

$$\alpha_{\mathrm{F}i} = A_{\mathrm{F}i,\mathrm{F}j}\frac{K_{\mathrm{F}i}}{L_{\mathrm{F}i}}\boldsymbol{n}_{\mathrm{F}i}\cdot\boldsymbol{f}_{\mathrm{F}i} \tag{6.29}$$

The transmissibility between matrix and fractures is

$$T_{\mathrm{F,m}} = \frac{\alpha_{\mathrm{F}}\alpha_{\mathrm{m}}}{\alpha_{\mathrm{F}} + \alpha_{\mathrm{m}}} \tag{6.30}$$

$$\alpha_{\mathrm{m}} = A_{\mathrm{F,m}}\frac{K_{\mathrm{m}}}{L_{\mathrm{m}}}\boldsymbol{n}_{\mathrm{m}}\cdot\boldsymbol{f}_{\mathrm{m}} \tag{6.31}$$

For oil and gas flow through discrete fractures,it is affected by complex flow mechanisms and has several flow regimes. According to oil/gas flow behavior between discrete fractures at different scales, the numerical simulation model with multiple flow regimes and flow mechanisms for discrete fractures is established.

(1) Numerical simulation model of high-speed nonlinear flow

When the pressure gradient of oil/gas flow through discrete fractures is higher than the critical pressure gradient of high-speed nonlinear flow, the flow regime of oil/gas flow is the high-speed nonlinear flow, and its numerical simulation model is

$$\sum_{\mathrm{F}j}\left\{T_{\mathrm{F}i,\mathrm{F}j}\sum_{\mathrm{p}}F_{\mathrm{ND}}\left(\frac{K_{\mathrm{rp}}\rho_{\mathrm{p}}}{\mu_{\mathrm{p}}}\right)_{\mathrm{F}ij}X_{\mathrm{cp}}\left[(P_{\mathrm{p,F}j}-\rho_{\mathrm{p}}gD_{\mathrm{F}j})-(P_{\mathrm{p,F}i}-\rho_{\mathrm{p}}gD_{\mathrm{F}i})\right]\right\}^{n+1} + (q_{\mathrm{pF}i}^{\mathrm{W}})^{n+1} = \frac{V}{\Delta t}\left[\left(\phi\sum_{\mathrm{p}}\rho_{\mathrm{p}}S_{\mathrm{p}}X_{\mathrm{cp}}\right)^{n+1}-\left(\phi\sum_{\mathrm{p}}\rho_{\mathrm{p}}S_{\mathrm{p}}X_{\mathrm{cp}}\right)^{n}\right]_{\mathrm{F}i} \tag{6.32}$$

(2) Numerical simulation model of quasi linear flow

When the pressure gradient of oil/gas flow through discrete fractures falls in between the critical pressure gradient of quasi linear flow and that of high-speed nonlinear flow, the flow regime of oil/gas flow is the quasi linear flow, and its numerical simulation model is

$$\sum_{\mathrm{F}j}\left\{T_{\mathrm{F}i,\mathrm{F}j}\sum_{\mathrm{p}}\left(\frac{K_{\mathrm{rp}}\rho_{\mathrm{p}}}{\mu_{\mathrm{p}}}\right)_{\mathrm{F}ij}X_{\mathrm{cp}}\left[(P_{\mathrm{p,F}j}-\rho_{\mathrm{p}}gD_{\mathrm{F}j})-(P_{\mathrm{p,F}i}-\rho_{\mathrm{p}}gD_{\mathrm{F}i})-G_{\mathrm{c,F}i,\mathrm{F}j}\right]\right\}^{n+1} + (q_{\mathrm{pF}i}^{\mathrm{W}})^{n+1} = \frac{V}{\Delta t}\left[\left(\phi\sum_{\mathrm{p}}\rho_{\mathrm{p}}S_{\mathrm{p}}X_{\mathrm{cp}}\right)^{n+1}-\left(\phi\sum_{\mathrm{p}}\rho_{\mathrm{p}}S_{\mathrm{p}}X_{\mathrm{cp}}\right)^{n}\right]_{\mathrm{F}i} \tag{6.33}$$

(3) Numerical simulation model of low-speed nonlinear flow

When the pressure gradient of oil/gas flow through discrete fractures is higher than the start-up pressure gradient and lower than the critical pressure gradient of quasi linear flow, the flow regime of oil/gas flow is the low-speed nonlinear flow affected by the start-up pressure, and its numerical simulation model is

$$\sum_{\mathrm{F}j}\left\{T_{\mathrm{F}i,\mathrm{F}j}\sum_{\mathrm{p}}\left(\frac{K_{\mathrm{rp}}\rho_{\mathrm{p}}}{\mu_{\mathrm{p}}}\right)_{\mathrm{F}ij}X_{\mathrm{cp}}\left[(P_{\mathrm{p,F}j}-\rho_{\mathrm{p}}gD_{\mathrm{F}j})-(P_{\mathrm{p,F}i}-\rho_{\mathrm{p}}gD_{\mathrm{F}i})-G_{\mathrm{a,F}i,\mathrm{F}j}\right]^{n'}\right\}^{n+1} + (q_{\mathrm{pF}i}^{\mathrm{W}})^{n+1} = \frac{V}{\Delta t}\left[\left(\phi\sum_{\mathrm{p}}\rho_{\mathrm{p}}S_{\mathrm{p}}X_{\mathrm{cp}}\right)^{n+1}-\left(\phi\sum_{\mathrm{p}}\rho_{\mathrm{p}}S_{\mathrm{p}}X_{\mathrm{cp}}\right)^{n}\right]_{\mathrm{F}i} \tag{6.34}$$

2. Numerical simulation model of flow between reservoirs and wellbores

The well index between fracture media and wellbores is

$$\mathrm{WI} = \frac{2\pi K_{\mathrm{F}}L_{\mathrm{eff,F}}}{\ln\left(\frac{r_{\mathrm{e}}}{r_{\mathrm{w}}}\right)+s} \tag{6.35}$$

The regimes of oil/gas flow from discrete fractures to wellbores change with flow mechanisms. According to flow behavior between discrete fractures as different scales and wellbores, the numerical simulation model of flow between discrete fractures and wellbores is built.

(1) Numerical simulation model of high-speed nonlinear flow between discrete fractures and wellbores

When the pressure gradient of oil/gas flow from discrete fractures to wellbores is higher than the critical pressure gradient of high-speed nonlinear flow, the flow regime of oil/gas flow is the high-speed nonlinear flow, and its numerical simulation model is

$$(q_{\mathrm{hs,c}})^{\mathrm{W}} = \mathrm{WI}\sum_{\mathrm{p}}\left\{F_{\mathrm{ND}}\left(\frac{K_{\mathrm{rp}}\rho_{\mathrm{p}}}{\mu_{\mathrm{p}}}\right)_{\mathrm{F}i}X_{\mathrm{cp}}\begin{bmatrix}(P_{\mathrm{p,F}i}-\rho_{\mathrm{p}}gD_{\mathrm{F}i})\\-(P^{\mathrm{W}}-\rho_{\mathrm{p}}gD^{\mathrm{W}})\end{bmatrix}\right\} \tag{6.36}$$

(2) Numerical simulation model of quasi linear flow between discrete fractures and wellbores

When the pressure gradient of oil/gas flow from discrete fractures to wellbores falls in between the critical pressure gradient of quasi linear flow and that of high-speed nonlinear flow, the flow regime of oil/gas flow is the quasi linear flow, and its numerical simulation model is

$$(q_{\mathrm{quasi,c}})^{\mathrm{W}} = \mathrm{WI}\sum_{\mathrm{p}}\left\{\left(\frac{K_{\mathrm{rp}}\rho_{\mathrm{p}}}{\mu_{\mathrm{p}}}\right)_{\mathrm{F}i}X_{\mathrm{cp}}\begin{bmatrix}(P_{\mathrm{p,F}i}-\rho_{\mathrm{p}}gD_{\mathrm{F}i})\\-(P^{\mathrm{W}}-\rho_{\mathrm{p}}gD^{\mathrm{W}})-G_{\mathrm{c,F}i}^{\mathrm{W}}\end{bmatrix}\right\} \tag{6.37}$$

(3) Numerical simulation model of low-speed nonlinear flow between discrete fractures and wellbores

When the pressure gradient of oil/gas flow from discrete fractures to wellbores is higher than the start-up pressure gradient and lower than the critical pressure gradient of quasi linear flow, the flow regime of oil/gas flow is the low-speed nonlinear flow affected by the start-up pressure, and its numerical simulation model is

$$(q_{\text{start-up,c}})^{\text{W}} = \text{WI}\sum_{\text{p}}\left\{(\frac{K_{\text{rp}}\rho_{\text{p}}}{\mu_{\text{p}}})_{\text{F}i}X_{\text{cp}}\begin{bmatrix}(P_{\text{p,F}i} - \rho_{\text{p}}gD_{\text{F}i}) \\ -(P^{\text{W}} - \rho_{\text{p}}gD^{\text{W}}) - G^{\text{W}}_{\text{a,F}i}\end{bmatrix}^{n^*}\right\} \quad (6.38)$$

6.2.2 Numerical Simulation of Discrete Multiple Media with Different Scale Pores and Micro-Fractures

Tight reservoirs contain discontinuous discrete pore media (as shown in Fig. 6.11). For different scale pores, their quantitative distributions and spatial distributions have huge differences (as shown in Figs. 6.13 and 6.14). Meanwhile, the discrete nano/micro-fractures (as shown in Fig. 6.12) are massive and their sizes are small. They cannot be treated as discrete fractures in simulations. In conventional simulations, different scale pores are treated as homogeneous continuous single pore media and different scale fractures are treated as homogeneous continuous single fracture media. The above conventional simulation is the simulation for continuous dual-porosity media. This kind of simulation cannot deal with production mechanisms and fluid flows between different scale pores and fractures, occurrence states of different scale media, and the impacts of difference in fluid property and fluid flow on production performance. Therefore, the different scale pores and nano/micro-fractures in tight reservoirs are treated as microscopic scale discrete multiple media in this study. The simulation method for discontinuous discrete multiple media is used to innovatively build the simulation for different scale discrete multiple media (Table 6.12).

6.2.2.1 Numerical Simulation for Multiple Media Based on the Grid Pattern of NDT

Pruess (1983, 1985) proposed the MINC (Multiple Interaction Continua) model where the matrix in reservoirs is further divided into several parts and the non-transient changes of pressure and saturation in matrix can be described. Based on the MINC model, Wu and Pruess (1988) Wu et al. (2009) developed the simulation of triple scale and multiple-scale media for reservoirs with fractures and caves, and, meanwhile, used a refined mesh to obtain the simulation of quasi-steady flow.

Due to the existence of discontinuous discrete pores and nano/micro-fractures in tight reservoirs, there is a huge difference in quantitative and spatial distributions of different scale pores and nano/micro-fractures between different elements. Therefore, according to modes of composition and quantitative distributions of different scale pores and fractures, the volume percentages of different pores and fractures are determined. According to the spatial discrete distribution of matryoshka-doll-type, different discrete porous and fracture media are generated, the flow behaviors between different media are determined, and the numerical simulation model is built to describe the geometry and property characteristics, and different flow mechanisms of multiple media with different scale pores and fractures. Finally, the simulation for discrete multiple media based on the grid pattern of NDT is innovatively developed.

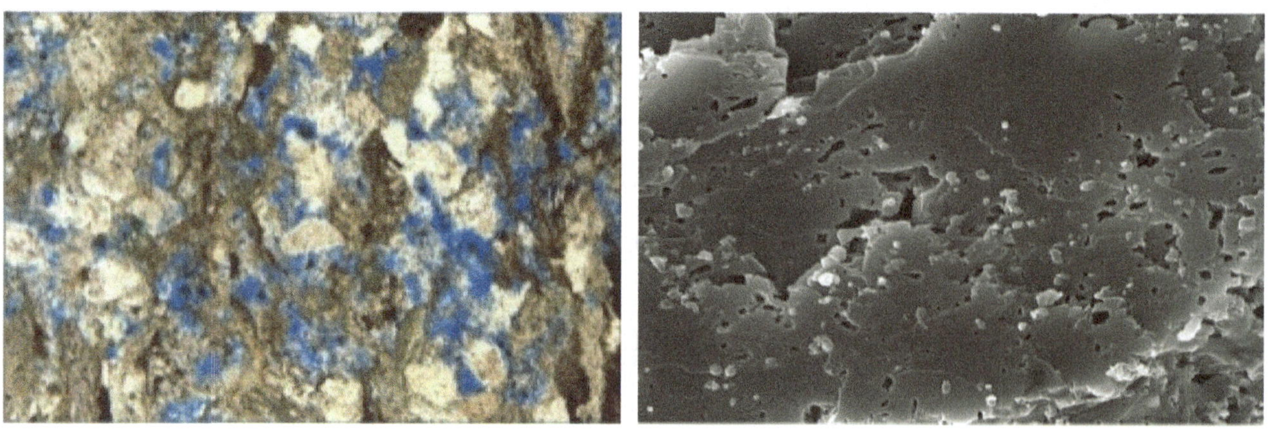

Fig. 6.11 Interactive discrete distribution of different scale pores

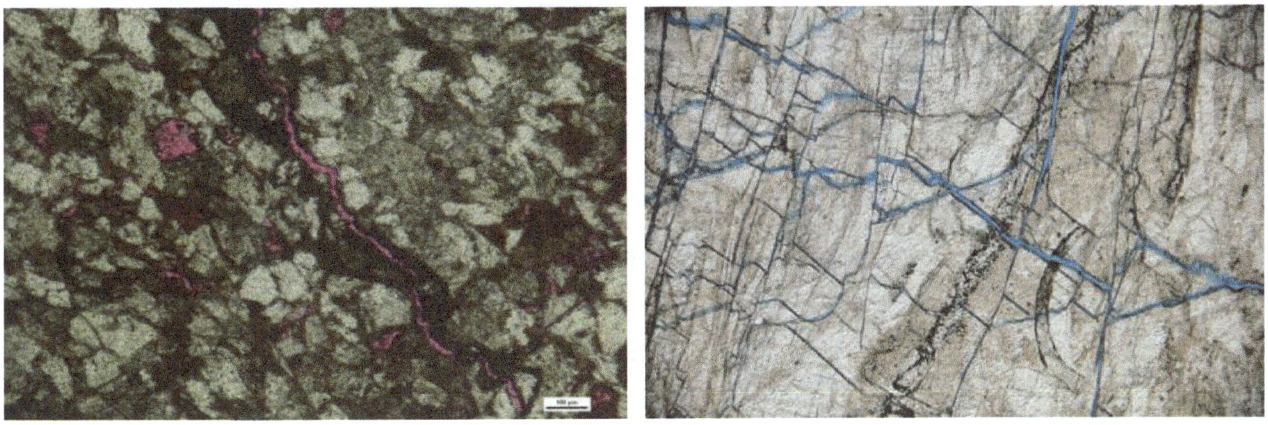

Fig. 6.12 Interactive discrete distribution of different scale fractures

Fig. 6.13 Spatial distribution of different scale pores

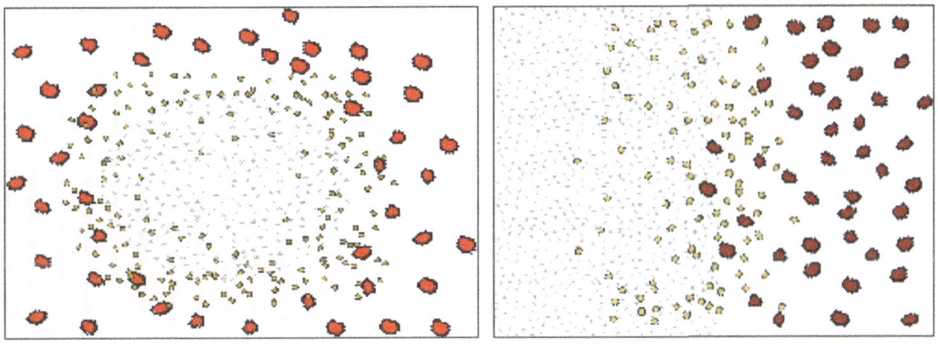

(a) Microscopic pores located in element center, macroscopic pores located in the external

(b) Microscopic pores located in the left, macroscopic pores located in the right

1. Grid system

The spatial distributions of nano/micro-fractures and different scale pores in tight reservoirs obey the matryoshka-doll-type distribution. Meanwhile, the property parameters of nano/micro-scale fractures and pores discontinuously change. Therefore, using unstructured grids transforms the discontinuous discrete pores and nano/micro-fractures into discrete multiple media.

① According to the difference in macroscopic heterogeneities in lithology, lithofacies and reservoir category of tight reservoirs, the reservoir is partitioned into several areas. There are great differences in quantitative and spatial distributions of different scale pores and natural/hydraulic fractures between different domains (as shown in Fig. 6.15).

② According to microscopic scale heterogeneity, the elements are classified into different types to represent microscopic scale pores and fractures. The element shape can be unstructured grids with arbitrary shapes (as shown in Fig. 6.16).

③ According to quantitative (as shown in Fig. 6.17) and spatial (as shown in Fig. 6.18) distributions of different scale pores and fractures in different elements, the number of media in elements representing microscopic pores and fractures is determined. After that, the volume percentages of different scale media are determined. According to the grid pattern of NDT, those media are classified as different discrete pore media and fracture media. The shape of those media is represented by the nest unstructured grids. Each nest unstructured grid represents one kind of medium. The number of nest grids equals the media number in this element. The volume of nest element should obey the characteristics of a pore volume percentage of media in the element.

④ The grids representing different discrete pores and fractures media constitute a complete set of grids used in the simulation for multiple media based on the NDT grid pattern, as shown in Fig. 6.19.

Fig. 6.14 Components and quantitative distribution of different scale pores

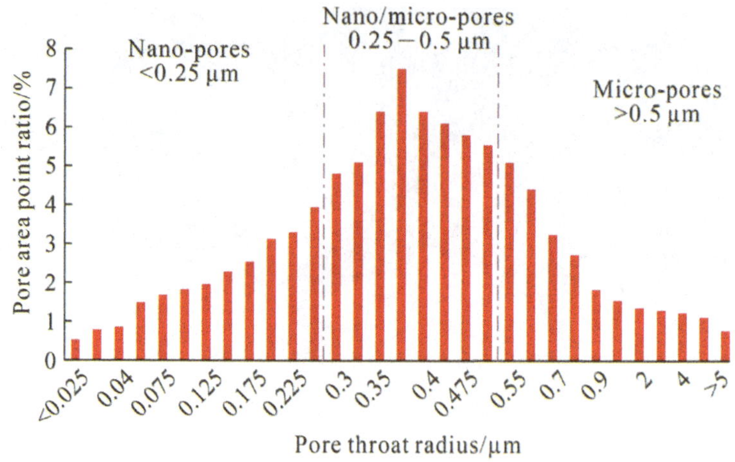

(a) Nano/micro-pores are dominant

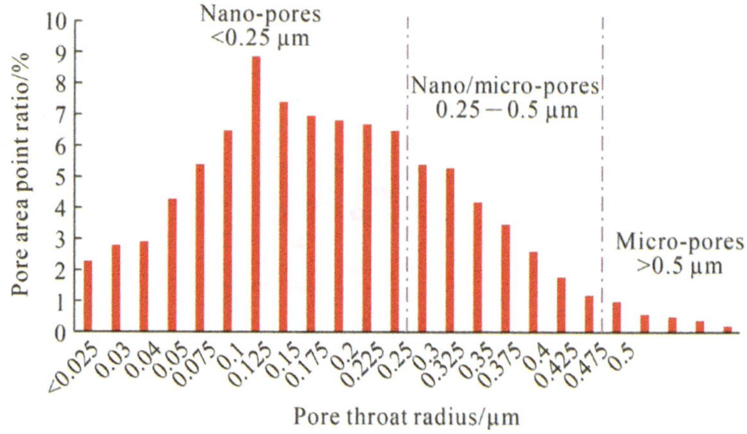

(b) Nano-pores are dominant

Table 6.12 Comparison between numerical simulation for conventional dual-porosity media and simulation for discrete multiple media

	Conventional dual-porosity media	Discrete multiple media
Pore/fracture type	Single pore media, Single fracture media	Different scale pores and nano/micro-scale fractures are identified as microscopic scale discrete multiple media
Geological laws	Pores and fractures continuously distribute and their physical parameters continuously change	The distribution of different scale pores and nano/micro-fractures fractures is discrete is space and their physical parameters discretely change. The spatial and quantitative distribution of those pores/fractures and their physical parameters in one element have certain patterns
Simulation Method	Simulation for continua media	Simulation for discontinuous discrete media

⑤ Using the spatial and quantitative distributions of different scale pores and fractures as a constraint condition, the modeling of property parameters for a continuous single pore medium is broken through. Using modeling of property parameters of discontinuous discrete multiple media, the model of geometry and property parameters for multiple media based on NDT grid pattern is built.

2. Flow behavior between different media in reservoirs

For a model of multiple media based on the NDT grid pattern, fluid flow can interactively occur between adjacent and connected elements representing different scale pores and fractures. In one element, according to the NDT grid pattern, sequential flow of fluid can occur between adjacent and

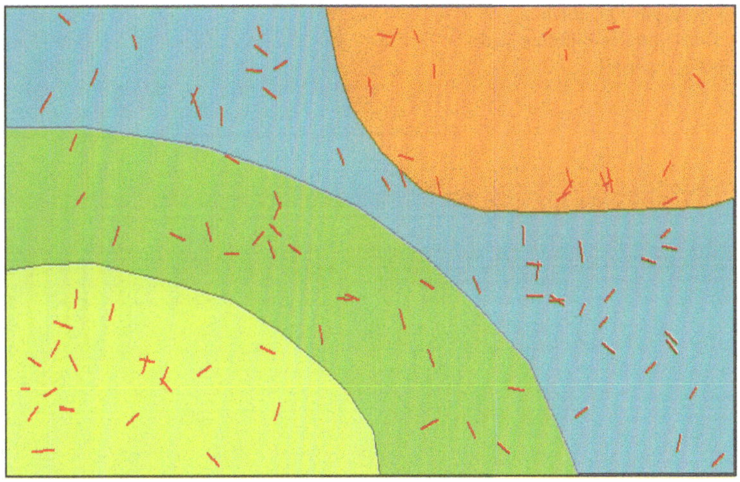

Fig. 6.15 Illustration of reservoir partition

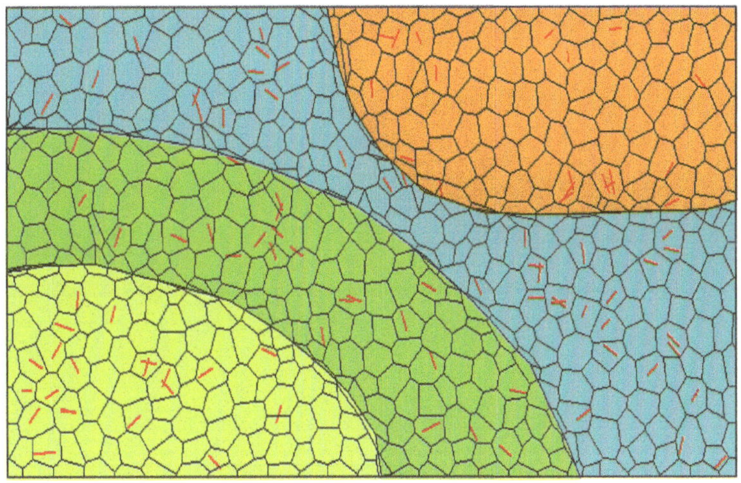

Fig. 6.16 Illustration of microscopic scale pore element

connected multi-scale pore media with fractures, as shown in Fig. 6.20.

According to the above grid systems, the element representing pores and fractures and the grids representing pores and fractures in one element are respectively ordered and numbered. According to the adjacency relationship and the flow behavior between different elements, and between grids of different media, the connectivity lists between different elements/grids are established (Table 6.13).

3. Flow behavior between reservoir media and wellbores

In a model for multiple media based on the NDT grid pattern, fluid flow can occur between elements/media representing different scale pores/fractures and wellbores. Based on the mesh generation of horizontal wells, the wellbore grids are respectively ordered and numbered. According to the adjacency relationship and the flow behavior between reservoir media grids and wellbore grids, the connectivity list between elements/media representing different scale pores/fractures and wellbore grids is established (as shown in Table 6.14).

4. Numerical simulation model

(1) Numerical simulation model of flow between different media

For oil and gas flow through discrete multiple media, it is affected by complex flow mechanisms so it has several flow regimes. According to flow behaviors both between elements representing pores/fractures and between media with different scale pores and fractures in one element, the numerical simulation model with multiple flow regimes and mechanisms for discrete multiple media is established.

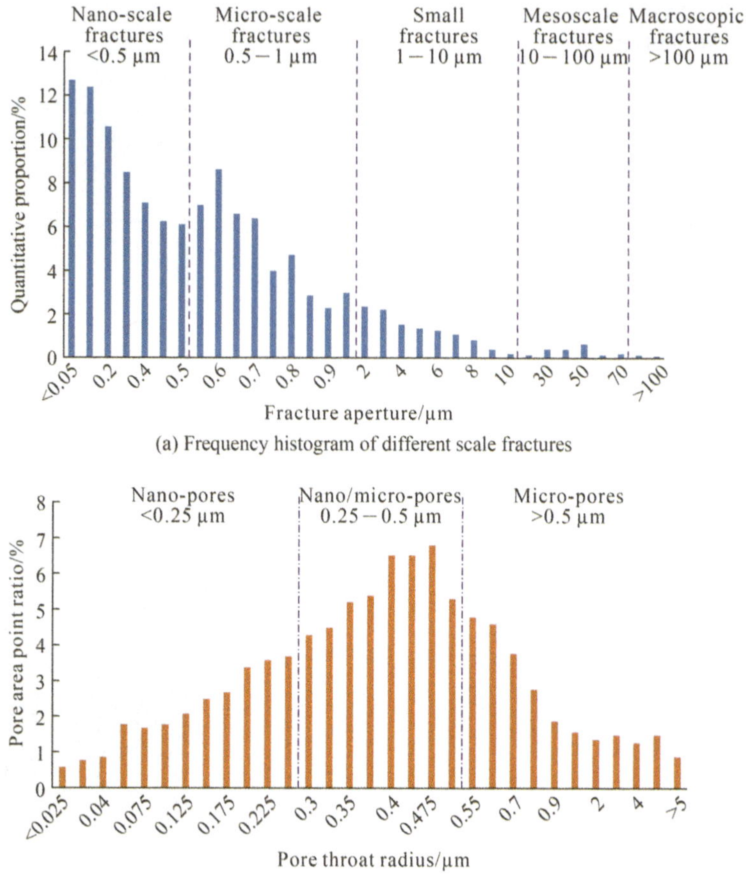

(a) Frequency histogram of different scale fractures

(b) Frequency histogram of different scale pores

Fig. 6.17 Quantitative distribution of different scale pores/fractures

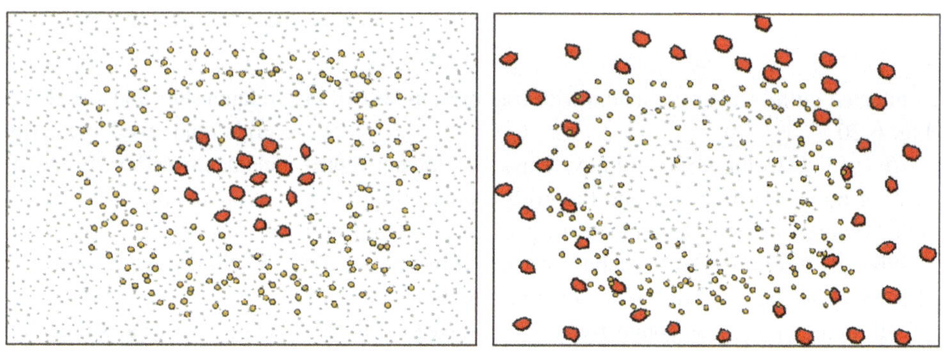

(a) Macroscopic pores located in element center, micro-pores located in the external

(b) Micro-pores located in element center, macroscopic pores located in the external

Fig. 6.18 Spatial distribution of different scale pores/fractures

1) Numerical simulation model of flow between different elements of pores/fractures

The transmissibility between different elements of pores/fractures is

$$T_{ei,ej} = \frac{\alpha_{lei\_J_N}\alpha_{lej\_J_N}}{\alpha_{lei\_J_N} + \alpha_{lej\_J_N}} \quad (6.39)$$

$$\alpha_{lei\_J_N} = A_{ei,ej}\frac{K_{ei\_J_N}}{L_{lei\_J_N}}\boldsymbol{n}\cdot\boldsymbol{f} \quad (6.40)$$

$$a_{lej\_J_N} = A_{ei,ej}\frac{K_{ej\_J_N}}{L_{lej\_J_N}}\boldsymbol{n}\cdot\boldsymbol{f} \quad (6.41)$$

where $T_{ei,ej}$ is the transmissibility between two adjacent elements i and j; $A_{ei,ej}$ is the contact area of two adjacent

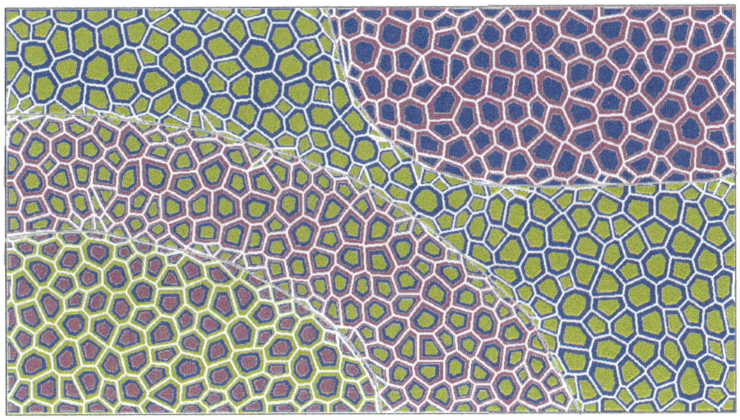

Fig. 6.19 Illustration of simulation for multiple media based on the NDT grid pattern

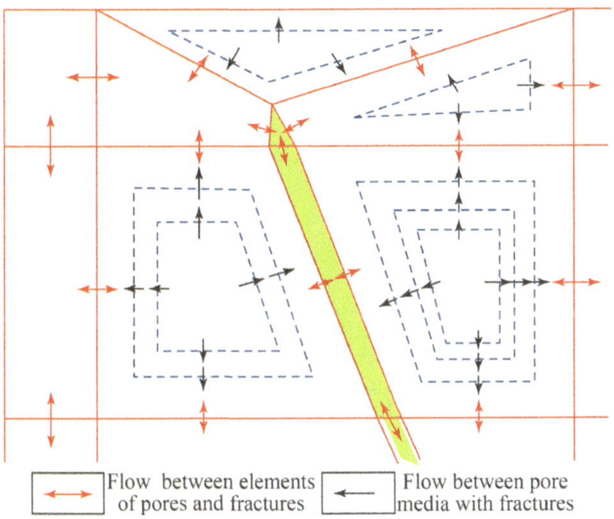

Fig. 6.20 Illustration of different scale pores/fractures element and relationship of flow between pores and fractures

elements i and j, and its unit is m^2; $K_{ei\_J_N}$, $K_{ej\_J_N}$ is the permeability of the medium at the outermost layer of elements i and j, respectively, and their unit is mD; $\alpha_{lei\_J_N}$, $\alpha_{lej\_J_N}$ are the shape factors of the critical local element of the medium J_N at the outmost layer of elements i and j, respectively; $L_{lei\_J_N}$, $L_{lej\_J_N}$ are the actual distance from the gravity center of the local element (i or j) to the contact area center of its adjacent element; $\boldsymbol{n}\cdot\boldsymbol{f}$ is the correction of orthogonality normal of one local element.

2) Numerical simulation model of fluid flow between different elements of pores/fractures

$$\sum_{ej}\left\{T_{ei,ej}\sum_{p}\left(\frac{K_{rp}\rho_p}{\mu_p}\right)_{ei\_J_N|j\_J_N}X_{cp}\left[\begin{array}{c}(P_{p,ej\_J_N}-\rho_p g D_{ej\_J_N})\\-(P_{p,ei\_J_N}-\rho_p g D_{ei\_J_N})\end{array}\right]\right\}^{n+1}+(q^{W}_{pei\_J_N})^{n+1}=\frac{V}{\Delta t}\left[\left(\phi\sum_{p}\rho_p S_p X_{cp}\right)^{n+1}-\left(\phi\sum_{p}\rho_p S_p X_{cp}\right)^{n}\right]_{ei\_J_N} \tag{6.42}$$

This model is a general model. Because the media of two connecting elements are different, the corresponding flow regimes and mechanisms of those media are also different. Therefore, the specific expression of the above model should be determined by flow regimes and mechanisms.

(2) Numerical simulation model of flow between media with different scale pores and fractures in one element

In one element, the transmissibility between media with different scale pores and fractures is

$$T_{mi,mi+1}=\frac{4A_{mi,mi+1}}{d_{mi}+d_{mi+1}}\left(\frac{K_{mi}K_{mi+1}}{K_{mi}+K_{mi+1}}\right) \tag{6.43}$$

where, $T_{mi,mi+1}$ is the transmissibility between two adjacent media i and i + 1 in one element; $A_{mi,mi+1}$is the contact area between two adjacent media i and i + 1, and the unit is m^2; K_{mi}, K_{mi+1} is the permeability of medium i and i + 1, respectively, and the unit is mD; d_{mi}, $d_{mi+1}$is the width of the nest grid of medium i and i + 1, respectively, and the unit is m.

Because the flow regimes and flow mechanisms of media with different scale pores and fractures in one element are significantly various, the corresponding numerical simulation models are different between different flow regimes and flow mechanisms.

1) Numerical simulation model of quasi linear flow

When the pressure gradient of oil/gas flow through media with different scale pores and fractures falls between the critical pressure gradient of quasi linear flow and that of high-speed nonlinear flow, the flow regime of oil/gas flow is the quasi linear flow, and its numerical simulation model is

$$\sum_{mi+1}\left\{T_{mi,mi+1}\sum_{p}\left(\frac{K_{rp}\rho_p}{\mu_p}\right)_{mi|i+1}X_{cp}\left[\begin{array}{c}(P_{p,mi+1}-\rho_p g D_{mi+1})\\-(P_{p,mi}-\rho_p g D_{mi})-G_{c,mi,mi+1}\end{array}\right]\right\}^{n+1}+(q^{W}_{pmi})^{n+1}=\frac{V}{\Delta t}\left[\left(\phi\sum_{p}\rho_p S_p X_{cp}\right)^{n+1}-\left(\phi\sum_{p}\rho_p S_p X_{cp}\right)^{n}\right]_{mi} \tag{6.44}$$

Table 6.13 Grid ordering of simulation for multiple media based on the NDT grid pattern and corresponding conductivity list

<table>
<tr><td>Different pores/fractures elements
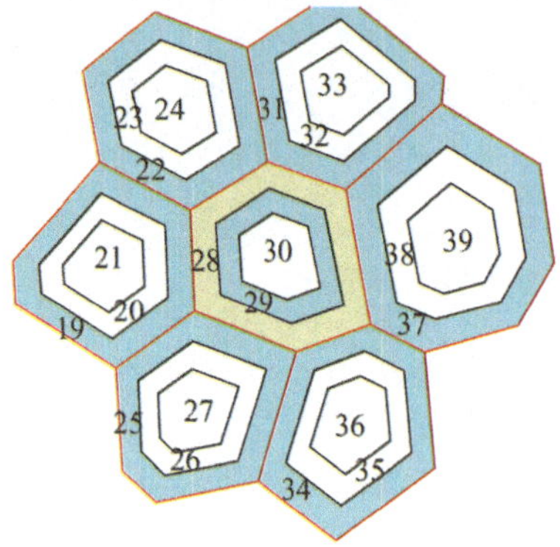
</td><td>Element i
…
28
28
28
28
28
…</td><td>Element j
…
19
22
25
31
34
…</td></tr>
<tr><td>Different pores and fractures in one element
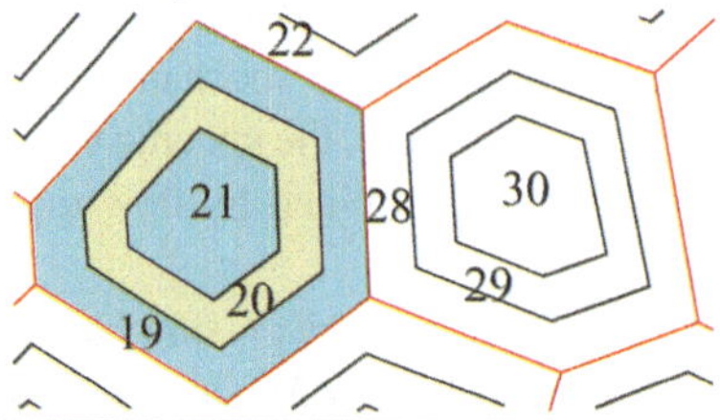
</td><td>Grid i
…
20
29
…</td><td>Grid j
…
21
30
…</td></tr>
</table>

Table 6.14 Connectivity list between pore/fracture and well grids in simulation

<table>
<tr><td>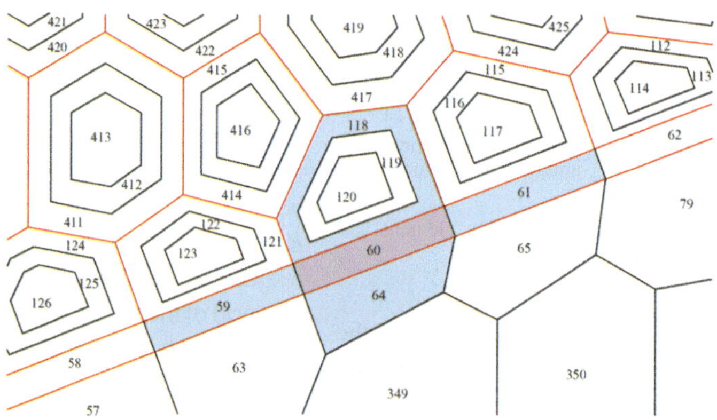
</td><td>Reservoir grid i
…
121
118
115
112
…</td><td>Wellbore j
…
59
60
61
62
…</td></tr>
</table>

2) Numerical simulation model of low-speed nonlinear flow

When the pressure gradient of oil/gas flow through media with different scale pores and fractures is higher than the start-up pressure gradient and is lower than the critical pressure gradient of quasi linear flow, the flow regime of oil/gas flow is the low-speed nonlinear flow affected by the start-up pressure, and its numerical simulation model is

$$\sum_{\mathrm{m}i+1}\left\{T_{\mathrm{m}i,\mathrm{m}i+1}\sum_{\mathrm{p}}\left(\frac{K_{\mathrm{rp}}\rho_{\mathrm{p}}}{\mu_{\mathrm{p}}}\right)_{\mathrm{m}i|i+1}X_{\mathrm{cp}}\begin{bmatrix}(P_{\mathrm{p,m}i+1}-\rho_{\mathrm{p}}gD_{\mathrm{m}i+1})\\-(P_{\mathrm{p,m}i}-\rho_{\mathrm{p}}gD_{\mathrm{m}i})-G_{\mathrm{a,m}i,\mathrm{m}i+1}\end{bmatrix}^{n*}\right\}^{n+1} + (q_{\mathrm{pm}i}^{\mathrm{W}})^{n+1}=\frac{V}{\Delta t}\left[\left(\phi\sum_{\mathrm{p}}\rho_{\mathrm{p}}S_{\mathrm{p}}X_{\mathrm{cp}}\right)^{n+1}-\left(\phi\sum_{\mathrm{p}}\rho_{\mathrm{p}}S_{\mathrm{p}}X_{\mathrm{cp}}\right)^{n}\right]_{\mathrm{m}i} \tag{6.45}$$

3) Numerical simulation model with slippage effect

Under low pressure, gas flow through media with nano/micro-scale pores and fractures is affected by the slippage effect. The numerical simulation model (the slippage factor (b) between two adjacent grids equals the critical pressure (P_{mi})) is

$$\sum_{mi+1}\left\{T_{mi,mi+1}\sum_{p}\left(\frac{K_{rp}\rho_{p}}{\mu_{p}}\right)_{mi|i+1}X_{cp}\left[\begin{array}{c}\left(1+\frac{b}{P_{mi+1}}\right)(P_{p,mi+1}-\rho_{p}gD_{mi+1})\\-\left(1+\frac{b}{P_{mi}}\right)(P_{p,mi}-\rho_{p}gD_{mi})\end{array}\right]\right\}^{n+1}+(q_{pmi}^{W})^{n+1}=\frac{V}{\Delta t}\left[\left(\phi\sum_{p}\rho_{p}S_{p}X_{cp}\right)^{n+1}-\left(\phi\sum_{p}\rho_{p}S_{p}X_{cp}\right)^{n}\right]_{mi} \quad (6.46)$$

4) Numerical simulation model with diffusion effect

Under low pressure, gas flow through media with nano/micro-scale pores and fractures is affected by the diffusion effect. The numerical simulation model (the diffusion coefficients of two adjacent grids are equal) is

$$\sum_{mi+1}\left\{T_{mi,mi+1}\sum_{p}\left(\frac{K_{rp}\rho_{p}}{\mu_{p}}\right)_{mi|i+1}X_{cp}\left[\begin{array}{c}\left(1+\frac{32\sqrt{2}\sqrt{RT}\mu_{g}}{3r\sqrt{\pi M}P_{mi+1}}\right)(P_{p,mi+1}-\rho_{p}gD_{mi+1})\\-\left(1+\frac{32\sqrt{2}\sqrt{RT}\mu_{g}}{3r\sqrt{\pi M}P_{mi}}\right)(P_{p,mi}-\rho_{p}gD_{mi})\end{array}\right]\right\}^{n+1}+(q_{pmi}^{W})^{n+1}=\frac{V}{\Delta t}\left[\left(\phi\sum_{p}\rho_{p}S_{p}X_{cp}\right)^{n+1}-\left(\phi\sum_{p}\rho_{p}S_{p}X_{cp}\right)^{n}\right]_{mi} \quad (6.47)$$

5) Numerical simulation model with desorption effect

When the reservoir pressure decreases to the desorption pressure, the gas in media with pores and fractures has desorption. The numerical simulation model with the gas desorption effect is

$$\sum_{mi+1}\left\{T_{mi,mi+1}\sum_{p}\left(\frac{K_{rp}\rho_{p}}{\mu_{p}}\right)_{mi|i+1}X_{cp}\left[\begin{array}{c}(P_{p,mi+1}-\rho_{p}gD_{mi+1})\\-(P_{p,mi}-\rho_{p}gD_{mi})\end{array}\right]\right\}^{n+1}+(q_{pmi}^{W})^{n+1}=\frac{V}{\Delta t}\left[\begin{array}{c}\left(\phi\sum_{p}\rho_{p}S_{p}X_{cp}+\rho_{g}\rho_{R}\cdot\frac{V_{L}}{1+bP_{g}}\right)^{n+1}\\-\left(\phi\sum_{p}\rho_{p}S_{p}X_{cp}+\rho_{g}\rho_{R}\cdot\frac{V_{L}}{1+bP_{g}}\right)^{n}\end{array}\right]_{mi} \quad (6.48)$$

6) Numerical simulation model with imbibition effect

In tight reservoirs, fluid flow between media with different scale pores and fractures is affected by an imbibition effect. The numerical simulation model of water-phase fluid flow is

$$\sum_{mi+1}\left\{T_{mi,mi+1}\left(\frac{K_{rw}\rho_{w}}{\mu_{w}}\right)_{mi|i+1}\left[\begin{array}{c}(P_{o,mi+1}-P_{o,mi})-\rho_{w}g(D_{mi+1}\\-D_{mi})-(P_{cow,mi+1}-P_{cow,mi})\end{array}\right]\right\}^{n+1}+(q_{w,mi}^{W})^{n+1}=\frac{V}{\Delta t}\left[(\phi\rho_{w}S_{w})^{n+1}-(\phi\rho_{w}S_{w})^{n}\right]_{mi} \quad (6.49)$$

(3) Numerical simulation model of fluid flow between different elements/media with pores/fractures and wellbores

The well index between elements of pores/fractures and wellbores is

$$\mathrm{WI}=\frac{2\pi K_{ei\_J_N}L_{eff,ei\_J_N}}{\ln\left(\frac{r_e}{r_w}\right)+s} \quad (6.50)$$

Because the flow regimes and flow mechanisms of different elements/media with pores and fractures are significantly various, the corresponding numerical simulation models are different for different flow regimes and mechanisms.

1) Numerical simulation model of quasi linear flow between elements with pores/fractures and wellbores

When the pressure gradient of oil/gas flow from elements with pores and fractures to wellbores falls between the critical pressure gradient of quasi linear flow and that of high-speed nonlinear flow, the flow regime of oil/gas flow is the quasi linear flow, and its numerical simulation model is

$$(q_{quasi,c})^{W}=\mathrm{WI}\sum_{p}\left\{\left(\frac{K_{rp}\rho_{p}}{\mu_{p}}\right)_{ei\_J_N}X_{cp}\left[\begin{array}{c}(P_{p,ei\_J_N}-\rho_{p}gD_{ei\_J_N})\\-(P^{W}-\rho_{p}gD^{W})-G_{c,ei\_J_N}^{W}\end{array}\right]\right\} \quad (6.51)$$

2) Numerical simulation model of low-speed nonlinear flow between elements with pores/fractures and wellbores

When the pressure gradient of oil/gas flow from elements with pores and fractures to wellbores is higher than the start-up pressure gradient, and lower than the critical pressure gradient of quasi linear flow, the flow regime of oil/gas flow is the low-speed nonlinear flow affected by the start-up pressure, and its numerical simulation model is

$$(q_{start-up,c})^{W}=\mathrm{WI}\sum_{p}\left\{\left(\frac{K_{rp}\rho_{p}}{\mu_{p}}\right)_{ei\_J_N}X_{cp}\left[\begin{array}{c}(P_{p,ei\_J_N}-\rho_{p}gD_{ei\_J_N})\\-(P^{W}-\rho_{p}gD^{W})-G_{a,ei\_J_N}^{W}\end{array}\right]^{n*}\right\} \quad (6.52)$$

3) Numerical simulation model with slippage effect between elements with pores/fractures and wellbores

Under low pressure, gas flow through elements with nano/micro-scale pores and fractures is affected by the slippage effect. The numerical simulation model (the slippage factor (b) between two adjacent grids equal to the critical pressure (P_{mi})) is

$$(q_{\mathrm{slippage,c}})^{\mathrm{W}} = \mathrm{WI}\sum_{\mathrm{p}}\left\{\begin{array}{c}\left(\frac{K_{\mathrm{rp}}\rho_{\mathrm{p}}}{\mu_{\mathrm{p}}}\right)_{ei\_J_{\mathrm{N}}} X_{\mathrm{cp}}\left(1+\frac{b}{P_{\mathrm{e}i\_J_{\mathrm{N}}}}\right)\\ \left[(P_{\mathrm{p,e}i\_J_{\mathrm{N}}} - \rho_{\mathrm{p}} g D_{\mathrm{e}i\_J_{\mathrm{N}}}) - (P^{\mathrm{W}} - \rho_{\mathrm{p}} g D^{\mathrm{W}})\right]\end{array}\right\} \tag{6.53}$$

4) Numerical simulation model with diffusion effect between elements with pores/fractures and wellbores

Under low pressure, gas flow through elements with nano/micro-scale pores and fractures is affected by diffusion effect. The numerical simulation model (the diffusion coefficients of two adjacent grids are equal) is

$$(q_{\mathrm{diffusion,c}})^{\mathrm{W}} = \mathrm{WI}\sum_{\mathrm{p}}\left\{\begin{array}{c}\left(\frac{K_{\mathrm{rp}}\rho_{\mathrm{p}}}{\mu_{\mathrm{p}}}\right)_{ei\_J_{\mathrm{N}}} X_{\mathrm{cp}}\left(1+\frac{32\sqrt{2}\sqrt{RT}\mu g}{3r\sqrt{\pi \mathrm{M}}P_{\mathrm{e}i\_J_{\mathrm{N}}}}\right)\\ \left[(P_{\mathrm{p,e}i\_J_{\mathrm{N}}} - \rho_{\mathrm{p}} g D_{\mathrm{e}i\_J_{\mathrm{N}}}) - (P^{\mathrm{W}} - \rho_{\mathrm{p}} g D^{\mathrm{W}})\right]\end{array}\right\} \tag{6.54}$$

6.2.2.2 Numerical Simulation for Multiple Media Based on the Grid Pattern of RIDT

There is a huge difference in quantitative and spatial distributions of media with different scale pores and nano/micro-fractures. Therefore, according to modes of composition and quantitative distributions of different scale pores and fractures, the volume percentages of different pores and fractures are determined. According to the discrete puzzle-piece-type distribution in space, different discrete porous and fracture media are generated, the flow behavior between different media are determined, and the numerical simulation model is built to describe the geometry and property characteristics, and different flow mechanisms of media with different scale pores and fractures. Finally, the simulation for discrete multiple media based on the grid pattern of RIDT is innovatively developed.

1. Grid system

The spatial distributions of media with nano/micro-fractures and different scale pores in tight reservoirs obey the puzzle-piece-type distribution. Meanwhile, the property parameters of media with nano/micro-scale fractures and pores discontinuously change. Therefore, by using unstructured grids, the discontinuous discrete pores and nano/micro-fractures are treated as discrete multiple media.

① According to the macroscopic heterogeneities in lithology, lithofacies and reservoir category of tight reservoirs, the reservoir is partitioned into several domains. There are great differences in quantitative and spatial distributions of different scale pores and nano/micro-fractures between different domains (as shown in Fig. 6.21).

② According to microscopic heterogeneity, the elements with microscopic pores and fractures are created. The elements with different modalities can be represented by unstructured grids with arbitrary shapes (as shown in Fig. 6.22).

③ According to quantitative and spatial distributions of media with different scale pores and fractures (as shown in Fig. 6.23), the media number of elements with microscopic scale pores and fractures is determined. After that, the volume percentages of different scale media are determined. According to the grid pattern of RIDT, those media are classified as different discrete pore media with fractures.

④ The grids of media with different scale discrete pores and fractures constitute a complete set of grids used in the simulation for multiple media based on the grid pattern of RIDT, as shown in Fig. 6.24.

⑤ Using the spatial and quantitative distributions of different scale pores and fractures as a constraint conditions, the modeling of property parameters for a continuous single pore medium is broken through. Using modeling of property parameters of discontinuous discrete multiple media, the model of geometry and property parameters for multiple media based on the RIDT grid pattern is built.

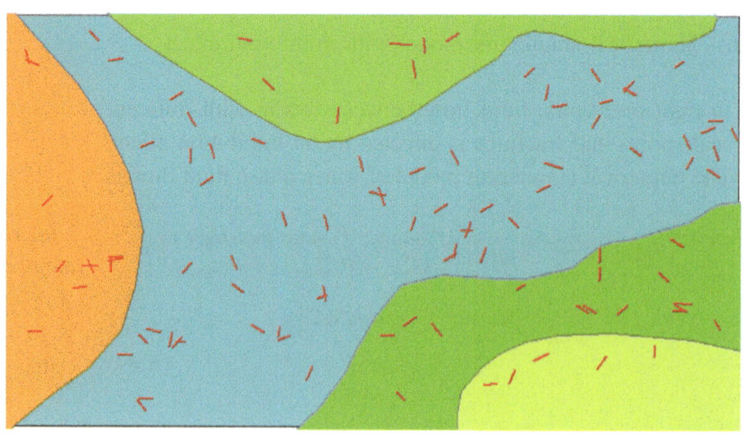

Fig. 6.21 Illustration of reservoir partition

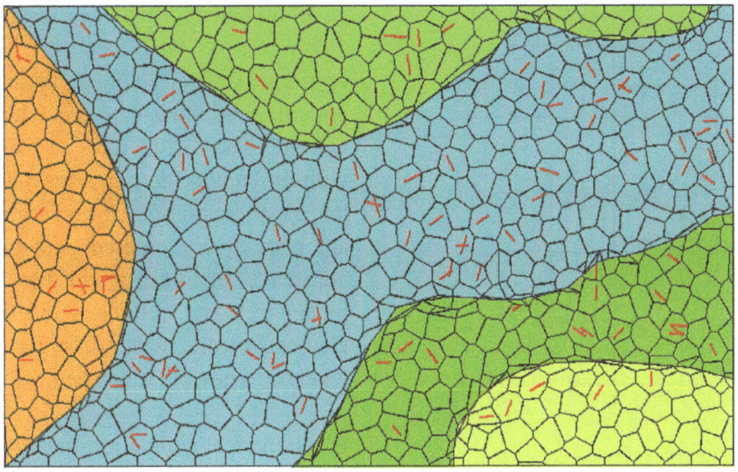

Fig. 6.22 Illustration of interactive element of different scale pores/fractures

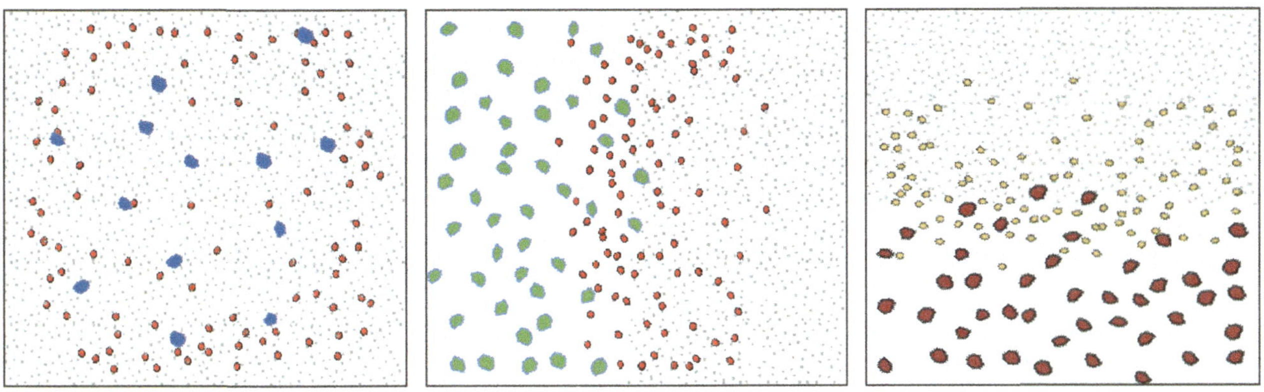

(a) Random distribution of different scale media

(b) Microscopic pores located in the left, macroscopic pores located in the right

(c) Microscopic pores located in the above, macroscopic pores located in the bottom

Fig. 6.23 Spatial distribution of different scale pores/fractures

Fig. 6.24 Illustration of Grid used in simulation for interactive multiple media

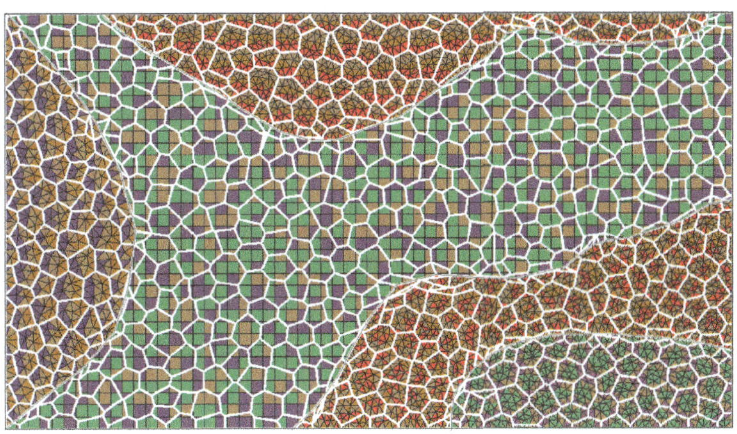

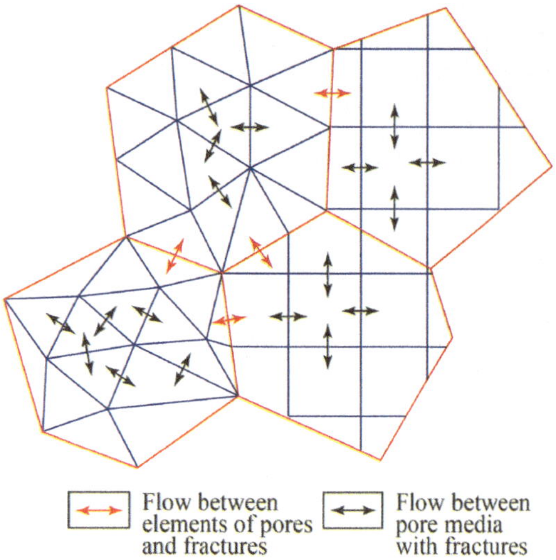

Fig. 6.25 Illustration of relationship of flow between different scale pores/fractures

2. Flow behavior between different media in reservoirs

For a model for multiple media based on the RIDT grid pattern, fluid flow can occur between adjacent and connected elements representing different scale pores and fractures. Fluid flow cannot occur between nonadjacent and disconnected elements representing different scale pores and fractures, as shown in Fig. 6.25.

According to the above grid systems, the grids representing media with pores/fractures are respectively ordered and numbered. According to the adjacency relationship and flow behavior between different elements, and grids of different media, the connectivity list between different elements/grids is established (Table 6.15).

3. Flow behavior between reservoir media and wellbores

In the model for multiple media based on the RIDT grid pattern, fluid flow can occur both between different scale natural/hydraulic fractures and wellbores, and between media with different scale pores/fractures and wellbores. Based on the mesh generation of horizontal wells, the wellbore grids are respectively ordered and numbered. According to the adjacency relationship and flow behavior between reservoir media grids and wellbore grids, the connectivity list between elements/media with different scale pores/fractures and wellbore grids is established (Table 6.16).

4. Numerical simulation model

(1) Numerical simulation model of flow between different media

For oil and gas flow through discrete multiple media,it is affected by complex flow mechanisms and has several flow regimes. According to flow behavior between media with different scale pores/fractures, the numerical simulation model with multiple flow regimes and flow mechanisms for discrete multiple media is established.

The transmissibility between media with different scale pores/fractures (Karimi-Fard et al. 2003) is

$$T_{\mathrm{m(f)}i,\mathrm{m(f)}j} = \frac{\alpha_{\mathrm{m(f)}i}\alpha_{\mathrm{m(f)}j}}{\alpha_{\mathrm{m(f)}i} + \alpha_{\mathrm{m(f)}j}} \tag{6.55}$$

$$\alpha_{\mathrm{m(f)}i} = A_{\mathrm{m(f)}i,\mathrm{m(f)}j}\frac{K_{\mathrm{m(f)}i}}{L_{\mathrm{m(f)}i}}\boldsymbol{n}\cdot\boldsymbol{f} \tag{6.56}$$

$$\alpha_{\mathrm{m(f)}j} = A_{\mathrm{m(f)}i,\mathrm{m(f)}j}\frac{K_{\mathrm{m(f)}j}}{L_{\mathrm{m(f)}j}}\boldsymbol{n}\cdot\boldsymbol{f} \tag{6.57}$$

Table 6.15 Grid ordering of simulation for interactive multiple media and corresponding conductivity list

	Grid i	Grid j
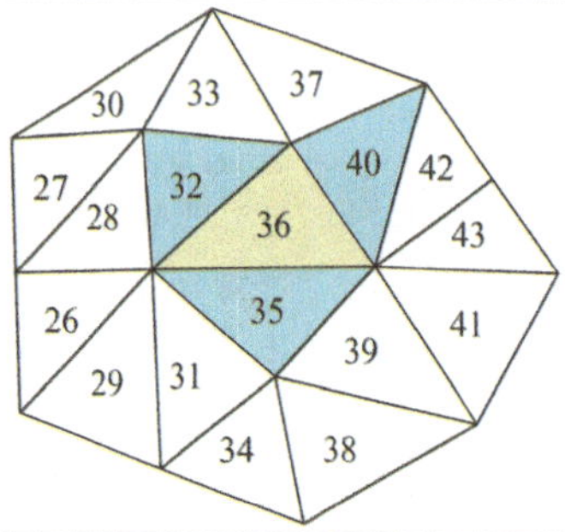	…	…
	36	32
	36	35
	36	40
	…	…

Table 6.16 Connectivity list between interactive multiple media and well grids

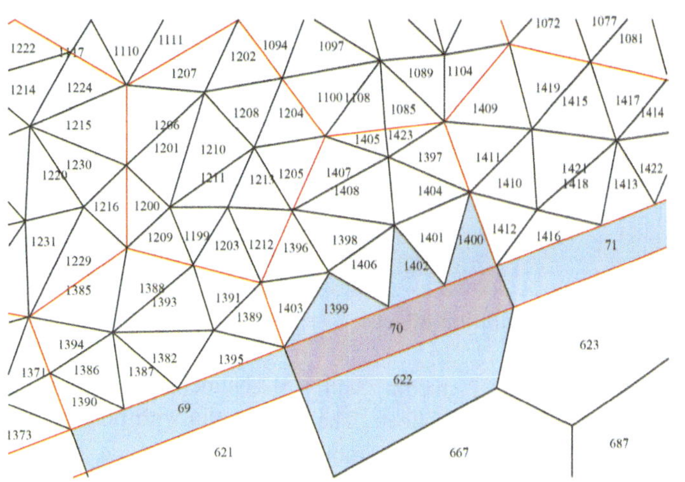

Reservoir grid i	Well grid j
…	…
622	70
1399	70
1400	70
1402	70
…	…

1) Numerical simulation model of quasi linear flow

When the pressure gradient of oil/gas flow through media with different scale pores/fractures falls between the critical pressure gradient of quasi linear flow and that of high-speed nonlinear flow, the flow regime of oil/gas flow is the quasi linear flow, and its numerical simulation model is

$$\sum_{m(f)j}\left\{T_{m(f)i,m(f)j}\sum_{p}\left(\frac{K_{rp}\rho_p}{\mu_p}\right)_{m(f)i|m(f)j}X_{cp}\left[\begin{array}{c}(P_{p,m(f)j}-\rho_p g D_{m(f)j})\\ -(P_{p,m(f)i}-\rho_p g D_{m(f)i})-G_{c,m(f)i,m(f)j}\end{array}\right]\right\}^{n+1}+(q^{W}_{pm(f)i})^{n+1}=\frac{V}{\Delta t}\left[\begin{array}{c}\left(\phi\sum_{p}\rho_p S_p X_{cp}\right)^{n+1}\\ -\left(\phi\sum_{p}\rho_p S_p X_{cp}\right)^{n}\end{array}\right]_{pm(f)i} \tag{6.58}$$

2) Numerical simulation model of low-speed nonlinear flow

When the pressure gradient of oil/gas flow through media with different scale pores/fractures is higher than the start-up pressure gradient and is lower than the critical pressure gradient of quasi linear flow, the flow regime of oil/gas flow is the low-speed nonlinear flow affected by the start-up pressure, and its numerical simulation model is

$$\sum_{m(f)j}\left\{T_{m(f)i,m(f)j}\sum_{p}\left(\frac{K_{rp}\rho_p}{\mu_p}\right)_{m(f)i|m(f)j}X_{cp}\left[\begin{array}{c}(P_{p,m(f)j}-\rho_p g D_{m(f)j})\\ -(P_{p,m(f)i}-\rho_p g D_{m(f)i})-G_{a,m(f)i,m(f)j}\end{array}\right]^{n'}\right\}^{n+1}+(q^{W}_{pm(f)i})^{n+1}=\frac{V}{\Delta t}\left[\left(\phi\sum_{p}\rho_p S_p X_{cp}\right)^{n+1}-\left(\phi\sum_{p}\rho_p S_p X_{cp}\right)^{n}\right]_{pm(f)i} \tag{6.59}$$

3) Numerical simulation model with slippage effect

Under low pressure, gas flow through media with nano/micro-scale pores and fractures is affected by the slippage effect. The numerical simulation model (the slippage factor (b) between two adjacent grids equals to the critical pressure (P_{mi})) is

$$\sum_{m(f)j}\left\{T_{m(f)i,m(f)j}\sum_{p}\left(\frac{K_{rp}\rho_p}{\mu_p}\right)_{m(f)i|m(f)j}X_{cp}\left[\begin{array}{c}\left(1+\frac{b}{P_{m(f)j}}\right)(P_{p,m(f)j}-\rho_p g D_{m(f)j})\\ -\left(1+\frac{b}{P_{m(f)i}}\right)(P_{p,m(f)i}-\rho_p g D_{m(f)i})\end{array}\right]\right\}^{n+1}+(q^{W}_{pm(f)i})^{n+1}=\frac{V}{\Delta t}\left[\left(\phi\sum_{p}\rho_p S_p X_{cp}\right)^{n+1}-\left(\phi\sum_{p}\rho_p S_p X_{cp}\right)^{n}\right]_{pm(f)i} \tag{6.60}$$

4) Numerical simulation model with diffusion effect

Under low pressure, gas flow through media with nano/micro-scale pores and fractures is affected by the diffusion effect. The numerical simulation model (the diffusion coefficients of two adjacent grids are equal) is

$$\sum_{m(f)j}\left\{T_{m(f)i,m(f)j}\sum_{p}\left(\frac{K_{rp}\rho_p}{\mu_p}\right)_{m(f)i|m(f)j}X_{cp}\left[\begin{array}{c}\left(1+\frac{32\sqrt{2}\sqrt{RT}\mu_g}{3r\sqrt{\pi M}P_{m(f)j}}\right)(P_{p,m(f)j}-\rho_p g D_{m(f)j})\\ -\left(1+\frac{32\sqrt{2}\sqrt{RT}\mu_g}{3r\sqrt{\pi M}P_{m(f)i}}\right)(P_{p,m(f)i}-\rho_p g D_{m(f)i})\end{array}\right]\right\}^{n+1}+(q^{W}_{pm(f)i})^{n+1}=\frac{V}{\Delta t}\left[\left(\phi\sum_{p}\rho_p S_p X_{cp}\right)^{n+1}-\left(\phi\sum_{p}\rho_p S_p X_{cp}\right)^{n}\right]_{pm(f)i} \tag{6.61}$$

5) Numerical simulation model with desorption effect

When reservoir pressure decreases to the desorption pressure, the gas in pores and fractures has desorption. The numerical simulation model with the gas desorption effect is:

$$\sum_{\mathrm{m(f)}j}\left\{T_{\mathrm{m(f)}i,\mathrm{m(f)}j}\sum_{p}\left(\frac{K_{\mathrm{rp}}\rho_{\mathrm{p}}}{\mu_{\mathrm{p}}}\right)_{\mathrm{m(f)}i|\mathrm{m(f)}j}X_{\mathrm{cp}}\begin{bmatrix}(P_{\mathrm{p,m(f)}j}-\rho_{\mathrm{p}}gD_{\mathrm{m(f)}j})\\-(P_{\mathrm{p,m(f)}i}-\rho_{\mathrm{p}}gD_{\mathrm{m(f)}i})\end{bmatrix}\right\}^{n+1}$$
$$+(q^{\mathrm{W}}_{\mathrm{pm(f)}i})^{n+1}=\frac{V}{\Delta t}\begin{bmatrix}\left(\phi\sum_{\mathrm{p}}\rho_{\mathrm{p}}S_{\mathrm{p}}X_{\mathrm{cp}}+\rho_{\mathrm{g}}\rho_{\mathrm{R}}\cdot\frac{V_{\mathrm{L}}}{1+bP_{\mathrm{g}}}\right)^{n+1}\\-\left(\phi\sum_{\mathrm{p}}\rho_{\mathrm{p}}S_{\mathrm{p}}X_{\mathrm{cp}}+\rho_{\mathrm{g}}\rho_{\mathrm{R}}\cdot\frac{V_{\mathrm{L}}}{1+b_{\mathrm{L}}P_{\mathrm{g}}}\right)^{n}\end{bmatrix}_{\mathrm{pm(f)}i} \tag{6.62}$$

6) Numerical simulation model with imbibition effect

In tight reservoirs, fluid flow between media with pores/fractures is affected by an imbibition effect. The numerical simulation model considering water-phase imbibition is

$$\sum_{\mathrm{m(f)}j}\left\{T_{\mathrm{m(f)}i,\mathrm{m(f)}j}\left(\frac{K_{\mathrm{rw}}\rho_{\mathrm{w}}}{\mu_{\mathrm{w}}}\right)_{\mathrm{m(f)}i|\mathrm{m(f)}j}\begin{bmatrix}(P_{\mathrm{o,m(f)}j}-P_{\mathrm{o,m(f)}i})+\rho_{\mathrm{w}}g(D_{\mathrm{m(f)}j}\\-D_{\mathrm{m(f)}i})-(P_{\mathrm{cow,m(f)}j}-P_{\mathrm{cow,m(f)}i})\end{bmatrix}\right\}^{n+1}$$
$$+(q^{\mathrm{W}}_{\mathrm{w,m(f)}i})^{n+1}=\frac{V}{\Delta t}\left[(\phi\rho_{\mathrm{w}}S_{\mathrm{w}})^{n+1}-(\phi\rho_{\mathrm{w}}S_{\mathrm{w}})^{n}\right]_{\mathrm{pm(f)}i} \tag{6.63}$$

(2) Numerical simulation model of fluid flow between reservoir media and wellbores

The well index between media with pores/fractures and wellbores is

$$\mathrm{WI}=\frac{2\pi K_{\mathrm{m(f)}}L_{\mathrm{eff,m(f)}}}{\ln\left(\frac{r_{\mathrm{e}}}{r_{\mathrm{w}}}\right)+s} \tag{6.64}$$

Because the flow regimes and flow mechanisms of media with different pores/fractures and wellbores are obviously different, the corresponding numerical simulation models are different for different flow regimes and flow mechanisms.

1) Numerical simulation model of quasi linear flow between media with pores/fractures and wellbores

When the pressure gradient of oil/gas flow from media with pores/ fractures to wellbores falls between the critical pressure gradient of quasi linear flow and that of high-speed nonlinear flow, the flow regime of oil/gas flow is the quasi linear flow, and its numerical simulation model is

$$(q_{\mathrm{quasi,c}})^{\mathrm{W}}=\mathrm{WI}\sum_{\mathrm{p}}\left\{\left(\frac{K_{\mathrm{rp}}\rho_{\mathrm{p}}}{\mu_{\mathrm{p}}}\right)_{\mathrm{m(f)}i}X_{\mathrm{cp}}\begin{bmatrix}(P_{\mathrm{p,m(f)}i}-\rho_{\mathrm{p}}gD_{\mathrm{m(f)}i})\\-(P^{\mathrm{W}}-\rho_{\mathrm{p}}gD^{\mathrm{W}})-G^{\mathrm{W}}_{\mathrm{c,m(f)}i}\end{bmatrix}\right\} \tag{6.65}$$

2) Numerical simulation model of low-speed nonlinear flow between media with pores/fractures and wellbores

When the pressure gradient of oil/gas flow from media with pores/fractures to wellbores is higher than the start-up pressure gradient, and lower than the critical pressure gradient of quasi linear flow, the flow regime of oil/gas flow is the low-speed nonlinear flow affected by the start-up pressure, and its numerical simulation model is

$$(q_{\mathrm{start-up,c}})^{\mathrm{W}}=\mathrm{WI}\sum_{\mathrm{p}}\left\{\left(\frac{K_{\mathrm{rp}}\rho_{\mathrm{p}}}{\mu_{\mathrm{p}}}\right)_{\mathrm{m(f)}i}X_{\mathrm{cp}}\begin{bmatrix}(P_{\mathrm{p,m(f)}i}-\rho_{\mathrm{p}}gD_{\mathrm{m(f)}i})\\-(P^{\mathrm{W}}-\rho_{\mathrm{p}}gD^{\mathrm{W}})-G^{\mathrm{W}}_{\mathrm{a,m(f)}i}\end{bmatrix}^{n^{*}}\right\} \tag{6.66}$$

3) Numerical simulation model with slippage effect between media with pores/fractures and wellbores

Under low pressure, gas flow through media with nano/micro-scale pores and fractures is affected by the slippage effect. The numerical simulation model (the slippage factor (b) between two adjacent grids equals the critical pressure ($P_{\mathrm{m}i}$)) is

$$(q_{\mathrm{slippage,c}})^{\mathrm{W}}=\mathrm{WI}\sum_{\mathrm{p}}\left\{\begin{matrix}\left(\frac{K_{\mathrm{rp}}\rho_{\mathrm{p}}}{\mu_{\mathrm{p}}}\right)_{\mathrm{m(f)}i}X_{\mathrm{cp}}\left(1+\frac{b}{P_{\mathrm{m(f)}i}}\right)\\\left[(P_{\mathrm{p,m(f)}i}-\rho_{\mathrm{p}}gD_{\mathrm{m(f)}i})-(P^{\mathrm{W}}-\rho_{\mathrm{p}}gD^{\mathrm{W}})\right]\end{matrix}\right\} \tag{6.67}$$

4) Numerical simulation model with diffusion effect between media with pores/fractures and wellbores

Under low pressure, gas flow through media with nano/micro-scale pores and fractures is affected by the diffusion effect. The numerical simulation model (the diffusion coefficients of two adjacent grids are equal) is

$$(q_{\mathrm{diffusion,c}})^{\mathrm{W}}=\mathrm{WI}\sum_{\mathrm{p}}\left\{\begin{matrix}\left(\frac{K_{\mathrm{rp}}\rho_{\mathrm{p}}}{\mu_{\mathrm{p}}}\right)_{\mathrm{m(f)}i}X_{\mathrm{cp}}\left(1+\frac{32\sqrt{2}\sqrt{RT}\mu_{\mathrm{g}}}{3r\sqrt{\pi M}P_{\mathrm{m(f)}i}}\right)\\\left[(P_{\mathrm{p,m(f)}i}-\rho_{\mathrm{p}}gD_{\mathrm{m(f)}i})-(P^{\mathrm{W}}-\rho_{\mathrm{p}}gD^{\mathrm{W}})\right]\end{matrix}\right\} \tag{6.68}$$

6.2.2.3 Numerical Simulation of Hybrid Discrete Multiple Media with Large-Scale Scale Fractures, Different Size Pores and Micro-Fractures

In the simulation for practical development of tight reservoirs, large/meso-scale natural/hydraulic fractures, micro-fractures and nano/micro-pores coexist in reservoirs, they have huge difference in geometry and property characteristics and their flow regimes and mechanisms are complex. Therefore, through the integration of media with large-scale discrete fractures and discrete media with pores/fractures with matryoshka-doll-type and puzzle-piece-type distribution, the simulation for

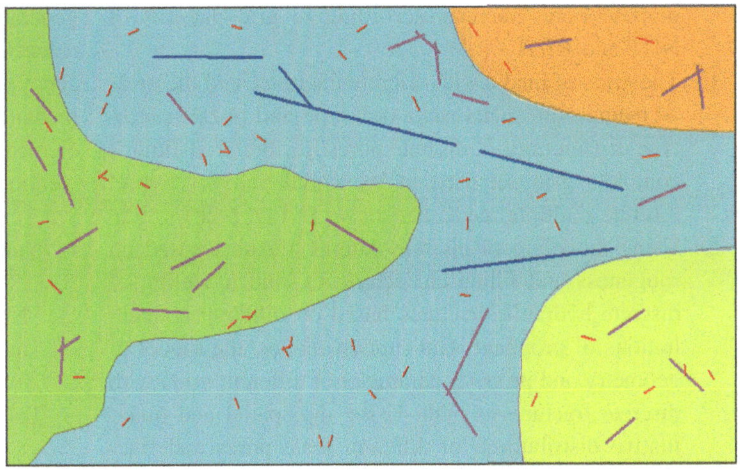

Fig. 6.26 Illustration of reservoir partition

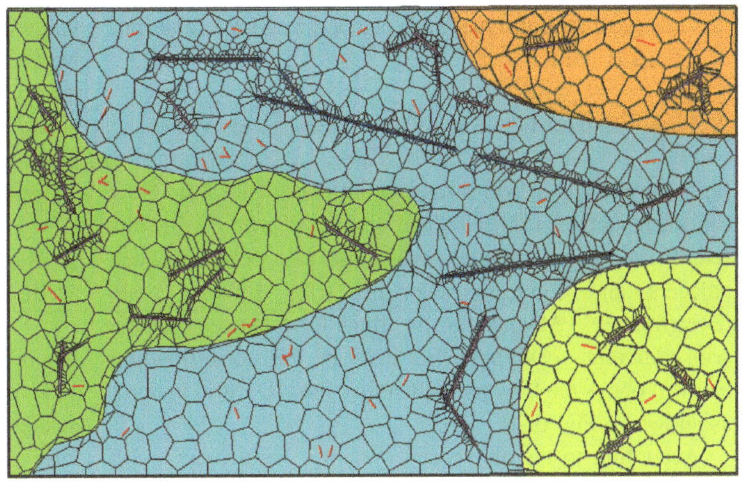

Fig. 6.27 Illustration of element partition of discrete fractures and microscopic pores/fractures

discontinuous discrete hybrid multiple media is developed (Lee et al. 2001; Sarda et al. 2002).

1. Grid system

The distributions of large-scale natural/hydraulic fractures, nano/micro-fractures and different scale pores in tight reservoirs obey the discrete distribution in space, and their property parameters discontinuously change. The difference in geometric characteristics and distribution pattern between different large-scale natural/hydraulic fractures is huge. The spatial distributions of media with nano/micro-fractures and different scale pores obey the matryoshka-doll-type and puzzle-piece-type discrete distributions. Therefore, using unstructured grids transforms the media with discontinuous discrete natural/hydraulic fractures, nano/micro-fractures and different scale pores into the hybrid discrete multiple media which integrates media with discrete fractures and media with matryoshka-doll-type and puzzle-piece-type distributions of discrete pores/fractures.

① According to the macroscopic heterogeneities in lithology, lithofacies and reservoir category of tight reservoirs, a reservoir is partitioned into several domains. Between different domains, the spatial and quantitative characteristics of different pores and natural/hydraulic fractures are different (as shown in Fig. 6.26).

② The large/meso-scale natural/hydraulic fractures are classified as elements with different discrete fractures; meanwhile, according to microscopic heterogeneity, the nano/micro-scale fractures and pores are assigned to elements with different microscopic pores and fractures (as shown in Fig. 6.27).

③ Using unstructured grids, through meshing, the elements with discrete fractures are assigned to media with different discrete fractures. Meanwhile, according to quantitative and spatial distributions of different scale pores and fractures, the elements with microscopic pores/fractures are assigned to media with different discrete pores/fractures according to grid patterns of NDT and RIDT.

④ The grids of media with discrete fractures, and the grids of media with matryoshka-doll-type and puzzle-piece-type distributions of discrete pores/fractures constitute a complete grid set used in the simulation for hybrid discrete multiple media (as shown in Fig. 6.28).

⑤ Using modeling of discrete natural fractures based on roughness and filling characteristics, and modeling of discrete hydraulic fractures based on different concentrations of proppants and support modes, the model of geometry and property parameter of different grids with discrete fractures is built. Using the spatial and quantitative distributions of different scale pores and fractures as a constraint conditions, the model of geometry and property parameters for multiple media with matryoshka-doll-type and puzzle-piece-type distributions is built.

2. Flow behavior between different media in reservoirs

For a model for hybrid discrete multiple media, fluid flow can occur between adjacent and connected media with different scale pores/fractures, and it can also occur between adjacent and connected elements with different scale pores/fractures. Meanwhile, the fluid flow between media with different scale pores/fractures in the corresponding element obeys the flow mechanism for the NDT grid pattern, as shown in Fig. 6.29.

According to the above grid systems, the elements with discrete fractures and the elements with pores/fractures, and the grids for media with pores/fractures in one element are respectively ordered and numbered. According to the adjacency relationship and flow behavior between different elements and between grids of different media, the connectivity list between different elements/grids is established (Table 6.17).

3. Flow behavior between reservoir media and wellbores

In a model for hybrid discrete multiple media, there are three flow behaviors and contact types between reservoir media and wellbores. These three behaviors include the fluid flow between different scale natural/hydraulic fractures and wellbores, the fluid flow between media with different scale pores/fractures and wellbores, and the fluid flow between single medium with different scale pores/fractures near wellbores and wellbores. Different scale based on the mesh generation of horizontal wells, the wellbore grids are respectively ordered and numbered. According to the adjacency relationship and flow behavior between reservoir media grids and wellbore grids, the connectivity list between elements/media with different scale pores/fractures and wellbore grids is established (Table 6.18).

4. Numerical simulation model

(1) Numerical simulation model of flow between different media

1) Transmissibility between different reservoir media

The transmissibility between different reservoir media includes the transmissibility between media with large/meso-scale discrete fractures, the transmissibility between media with fractures and media with pores, the transmissibility between matryoshka-doll-type media with pores/fractures, and the transmissibility between puzzle-piece-type media with pores/fractures. Their specific expressions are the same as the above ones.

2) Numerical simulation model of flow through hybrid discrete multiple media

For oil and gas flow through media with discrete fractures, and matryoshka-doll-type/puzzle-piece-type media with discrete pores/fractures, it is affected by complex flow mechanisms and it has several flow regimes. According to flow behavior between media with pores/fractures, the numerical simulation models with multiple flow regimes and flow mechanisms of hybrid discrete multiple media are established. These models include the numerical simulation model for large/meso-scale discrete fractures and the one for matryoshke-doll-type/puzzle-piece-type discrete multiple media. Their specific expressions are as same as the above ones so that there is no more detailed description in this section.

(2) Numerical simulation model of fluid flow between reservoir media and wellbores

1) Well index between reservoir media and wellbores

The well index between reservoir media and wellbores include the well index between media with large/meso-scale discrete fractures, the matryoshka-doll-type/puzzle-piece-

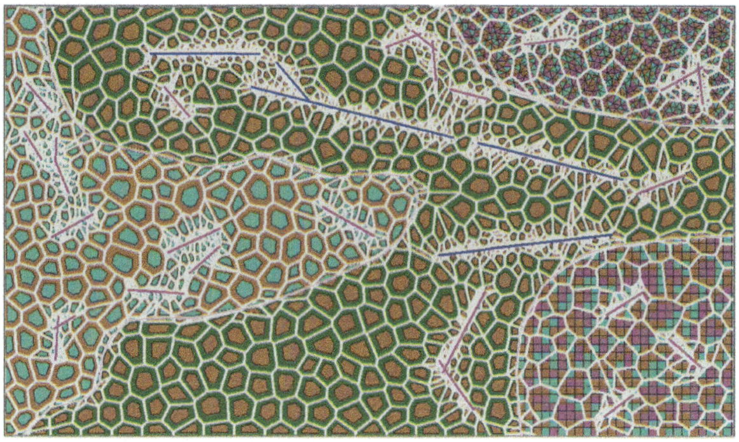

Fig. 6.28 Illustration of grid used in simulation for hybrid discrete multiple media

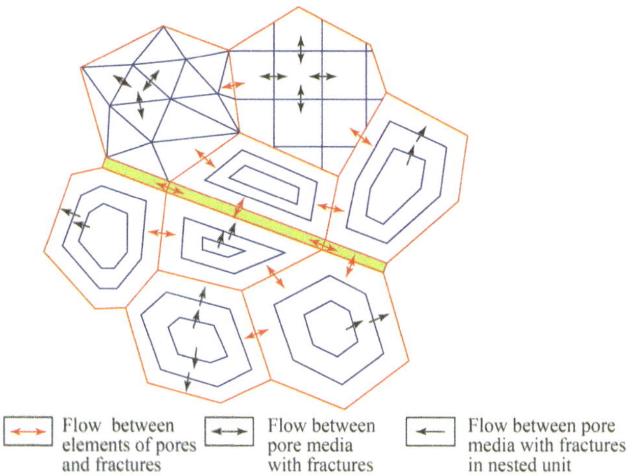

Fig. 6.29 Illustration of relationship of flow between hybrid discrete multiple media

type media with pores/fractures and wellbores. Their specific expressions are the same as the above ones.

2) Numerical simulation model of fluid flow between hybrid discrete multiple media and wellbores

For oil and gas flow from hybrid discrete multiple media to wellbores, the corresponding flow regimes change with flow mechanisms. According to flow behavior between different scale reservoir media and wellbores, the numerical simulation model of flow between discrete multiple media and wellbores is established and it includes the numerical simulation model of flow between media with large/meso-scale discrete fractures, the matryoshka-doll-type/puzzle-piece-type media with pores/fractures and wellbores. Their specific expressions are the same as the above ones so that there is no more detailed description in this section.

5. Numerical simulation process for hybrid discrete multiple media

① According to quantitative and spatial distributions of different scale pores and natural/hydraulic fractures, the reservoir is partitioned into different elements based on the macroscopic heterogeneities. The large/meso-scale natural/hydraulic fractures are assigned to elements with different discrete fractures. Meanwhile, according to the microscopic heterogeneities, micro-scale fractures and different scale pores are assigned to elements with nano/micro-scale pores and fractures.

② Using unstructured grids, the elements with discrete fractures are assigned to grids for media with different discrete fractures; meanwhile, according to quantitative and spatial distributions of different scale pores and fractures, based on the matryoshka-doll-type and puzzle-piece-type distributions, the elements with nano/micro-scale pores and fractures are assigned to nest grids of media with discrete pores/fractures, and interactive grids of media with discrete pores/fractures.

③ The above three types of grids as a whole are ordered and numbered. According to the flow behavior between different reservoir media, and between different media and wellbores, the connectivity list between adjacent grids is established.

④ Through the simulation for discontinuous discrete multiple media, the media types of grids for media with

Table 6.17 Grid ordering of simulation for hybrid discrete multiple media and corresponding conductivity list

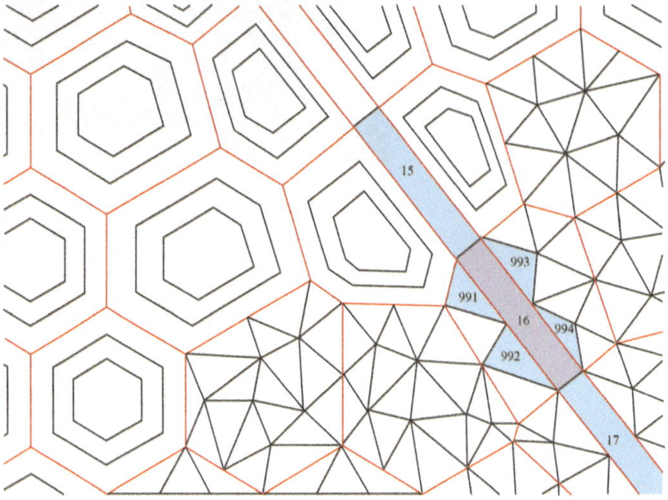

Grid i	Grid j
…	…
16	991
16	992
16	993
16	994
…	…

Table 6.18 Connectivity list between pore/fracture and well grids in simulation for hybrid discrete multiple media

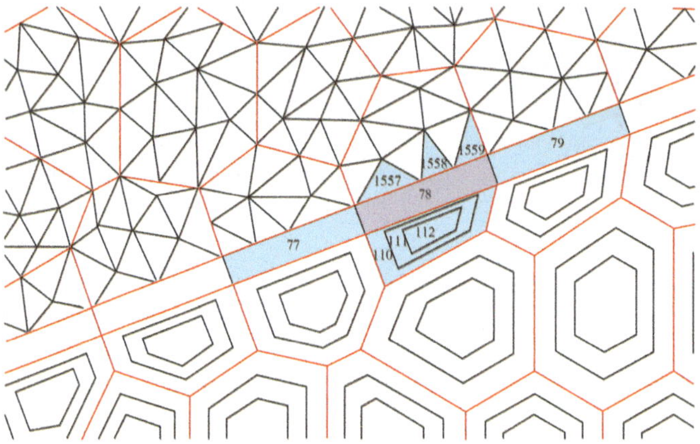

Reservoir grid i	Well grid j
…	…
110	78
1557	78
1558	78
1559	78
…	…

discrete fractures, nest grids of media with discrete pores/fractures, and interactive grids of media with discrete pores/fractures are determined and are assigned proper parameters. Those parameters include property parameters (porosity, permeability), fluid parameters (saturation, viscosity, density, relative permeability, capillary pressure, and high pressure property), mechanism parameters (stress sensitivity, high-speed nonlinear flow, start-up pressure gradient, slippage effect, diffusion effect, and imbibition effect). The simulation for hybrid discrete multiple media is built.

⑤ Through a conductivity list and simulation for hybrid discrete multiple media, the transmissibility between grids of different media and a well index between different media and wellbores are calculated.

⑥ The simulation for oil/gas reservoirs is conducted. According to the simulation of fluid flow between media with discrete fractures, and between nest/interactive media with discrete pores/fractures, the dynamic changes in pressure and saturation of oil/gas reservoirs are obtained.

⑦ The forecast of development index is conducted. According to different well indexes and the above simulation results, the production of oil/gas/water and its dynamic changes are calculated (Fig. 6.30).

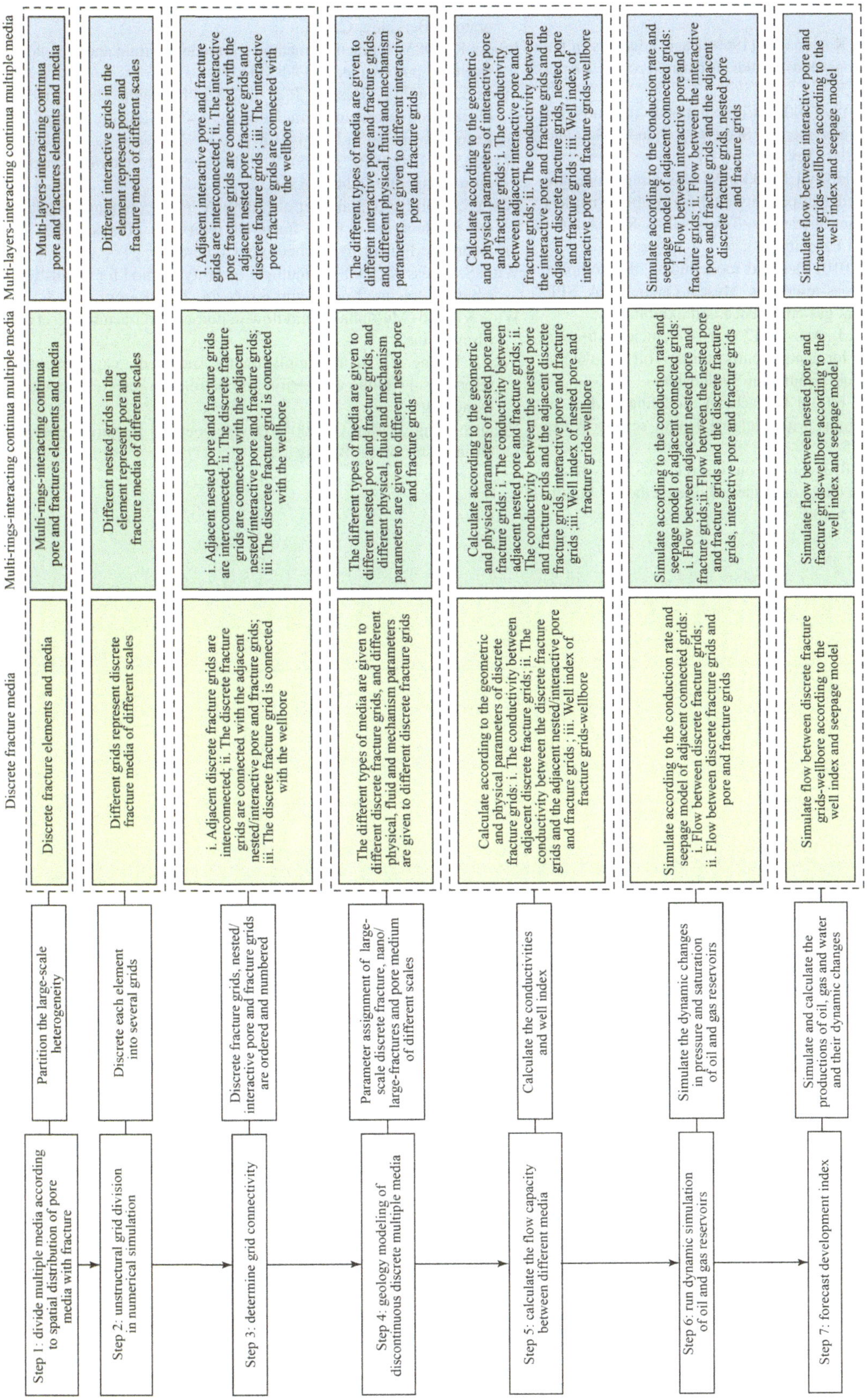

Fig. 6.30 Technique process of simulation for hybrid discrete multiple media

References

Barrenblatt GI, Zehltov YuP, Kochina IN (1960) Basic concepts in the theory of seepage of homogeneous liquids in fissured rocks. J Appl Math Mech 24(5):1286–1303

Chen ZX, Huan GR, Ma YL (2006) Computational methods for multiphase flows in porous media. Society for Industrial and Applied Mathematics, Dallas, Texas

Gong B, Karimi-Fard M, Durlofsky LJ (2006) An Upscaling procedure for constructing generalized dual-porosity/dual-permeability models from discrete fracture characterizations. Texas, USA, SPE Annual Technical Conference and Exhibition

Gong B, Qin G, Shiyi Y (2010) Detailed modeling of the complex fracture network of shale gas reservoirs. Muscat Oman, USA, SPE middle east unconventional gas conference and exhibition

Karimi-Fard M, Durlofsky LJ, Aziz K (2003) An efficient discrete fracture model applicable for general purpose reservoir simulators. Texas, USA, SPE Reservoir Simulation Symposium

Lee S, Lough M, Jensen C (2001) Hierachical modeling of floe in naturally fractured formations with multiple length scales. Water Resour Res 37(3):443–445

Lv X, Yao J, Huang C (2013) Study on discrete fracture model two-phase flow simulation based on finite volume method. J Southwest Petrol Univ 34(6):123–130

Pruess K (1983) GMINC-a mesh generator for flow simulation in fractured reservoirs. Report LBL-15227, Lawrence Berkeley Laboratory, Berkeley, CA

Pruess K (1985) A practical method for modeling fluid and heat flow in fractured porous media. Old SPE J

Sarda S, Jeannin L, Basquet R et al (2002) Hydraulic characterization of fractured reservoirs: Simulation on discrete fracture models. SPE Reserv Eval Eng 5(2):154–162

Warren JE, Root PJ (1963) The behavior of naturally fractured reservoirs. SPE J, 245–255

Wu YS, Moridis G, Bai B et al (2009) A multi-continuum model for gas production in tight fractured reservoirs. Texas, USA, SPE Hydraulic Fracturing Technology Conference

Wu YS, Pruess K (1988) A multiple-porosity method for simulation of naturally fractured petroleum reservoirs. SPE Reserv Eng

Wu YS (2016) Multiphase fluid flow in porous and fractured reservoirs. Elsevier Inc

Yao J, Huang C (2014) Simulation for fractured—vuggy carbonate reservoirs. China university of petroleum press, Shangdong, Dongying

Yuan S, Song X, Ran Q (2004) Development of fractured reservoirs. Oil industry press, Beijing

7 Coupled Multiphase Flow-Geomechanics Simulation for Multiple Media with Different-Size Pores and Natural/Hydraulic Fractures in Fracturing-Injection-Production Process

During the processes of fracturing, injection, and production of unconventional tight reservoirs, significant changes in effective stress are caused by a large range of pore pressure changes, resulting in deformation of multiple media such as matrix pores, natural fractures, and hydraulic fractures at different scales. The geometric characteristics and physical parameters of media with pore/fracutres change dynamically. Changes in pore pressure and physical parameters will in turn lead to changes in the conductivity and well index between media, which will greatly affect the reservoir dynamics performance and production characteristics. In order to solve the dynamic simulation problem of the above processes, dynamic models of geometry, properties, conductivity, and well indices for multiple media are established by investigating the deformation mechanism and laws of natural/hydraulic fractures and matrix pores, during the above processes, and finally the coupled flow-geomechanics simulation for multiple media during the fracturing-injection-production process is built.

7.1 Coupled Flow-Geomechanics Deformation Mechanism of Multiple Media with Different Scales Pores and Fractures

In order to understand the coupled flow-geomechanics deformation mechanism of media with multiple scale pores and fractures during the fracturing-injection-production process, based on the effective stress principle, the deformation mechanism of matrix pores, natural/hydraulic fractures with increasing pressure and decreasing pressure is described in detail, and the dynamic characteristics of the media property during the fracturing-injection-production process are demonstrated by laboratory experiments.

7.1.1 The Principle of Effective Stress in Multiple Media with Different Scales Pores and Fractures

7.1.1.1 The Basic Principle of Effective Stress in a Single Medium

Oil and gas reservoirs are porous media composed of rock grains and saturated with oil, gas, and water. For such a porous medium in reservoirs, its total stress partially acts on the fluids in the porous medium and this part of total stress is the pore pressure. Meanwhile the other part is applied to the rock's grain skeleton and this part is the effective stress (Xu et al. 2001).

The total stress refers to the vector sum of the external stress generated by overlying rock formations and lateral surrounding rocks. Generally, the vertical stress generated by the overlying rock is greater than the horizontal stress generated by the surrounding rock. Pore pressure is the pressure generated by fluids inside pores. The effective stress is the stress applied to the rock skeleton of a porous medium. It is an equivalent stress and is the combination result of total stress and pore pressure (Fig. 7.1). For different reservoir media and a different number of fluid phases, their expressions of effective stress principles are different.

Because the reservoir rock is mostly at a compressed state under its initial conditions, the direction of compressive stress is set to be positive. The total stress σ^{T} and initial effective stress of the rock σ' are generally positive, while the pore fluid pressure P is tensile and negative. Based on this, the general mathematical expression of the effective stress principle for a single medium is (Yang 2004):

$$\sigma' = \sigma^{\mathrm{T}} + \alpha_{\mathrm{e}} P \delta_{ij} \tag{7.1}$$

Q. Ran, *Unconventional Tight Reservoir Simulation: Theory, Technology and Practice*,
https://doi.org/10.1007/978-981-32-9848-4_7

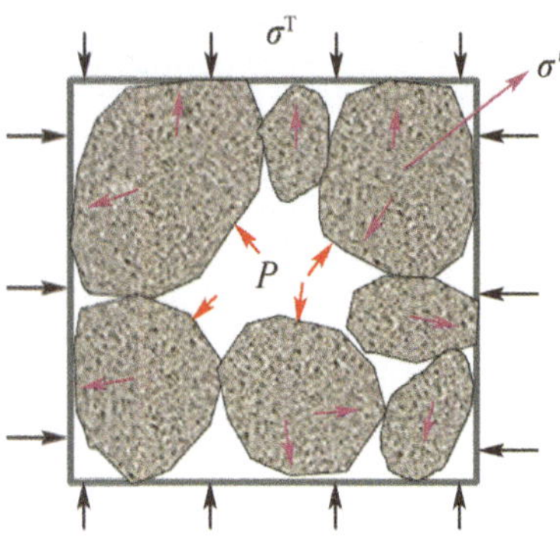

Fig. 7.1 Illustration of principle of effective stress

where, α_e is the effective stress correction coefficient ($0 < \alpha \leq 1$) and δ_{ij} is the Kronecker constant.

7.1.1.2 Effective Stress Principle of Multiple Media

Unconventional tight reservoirs contain matrix pores and natural/hydraulic fractures at different scales. There are significant differences in the flow state of different media with pores/fractures. The traditional effective stress principle of a single medium has been difficult to characterize the impact of pore pressure of different media with pores/fractures on the effective stress of reservoirs. This impact should be described by the effective stress principle for multiple media.

The effective stress principle of multiple media mainly considers the effect of different flow states in multiple media such as pores, fractures, and vugs at different scales on the effective stress. The mathematical expression is as follows (Cai et al. 2009):

$$\sigma = \sigma^T + \alpha_{ij}^1 P_1 + \alpha_{ij}^2 P_2 + \alpha_{ij}^3 P_3 + \cdots + \alpha_{ij}^{n-1} P_{n-1} + \alpha_{ij}^n P_n \tag{7.2}$$

where, α_{ij}^1, α_{ij}^1, $\alpha_{ij}^1, \ldots, \alpha_{ij}^{n-1}$, α_{ij}^n are the effective stress coefficient tensors of various media, which are related to the elastic properties and direction of the media. P_1, P_1, P_2, P_3, …, P_n are, respectively, referred to as the first pore pressure, the second pore pressure, and the nth pore pressure.

When only the flow difference between reservoir matrix pores, natural fractures, and hydraulic fractures is considered, the effective stress of multiple media can be simplified as the following effective stress expression of triple media (Chen and Chen 1999):

$$\sigma' = \sigma^T + (\alpha_m P_m + \alpha_f P_f + \alpha_F P_F)\delta_{ij} \tag{7.3}$$

where, α_m, α_f, and α_F are the effective stress factors for matrix pores, natural fractures, and hydraulic fractures; P_m is the fluid pressure inside matrix pores, P_f is the fluid pressure in natural fractures, and P_F is the fluid pressure in hydraulic fractures.

7.1.1.3 Flow-Geomechanics Coupled Dynamic Change of Multiple Media with Different Scales Pores/Fractures

During the development of oil and gas reservoirs, under flow-geomechanics coupled impacts, the change of pore pressure will cause the change of reservoir effective stress, which further leads to the change of geometric and physical parameters of multiple media with pores/fractures. Therefore, dynamic changing media with pores/fractures at different scales are formed in different types of oil reservoirs. The impacts of dynamic changes of different media on dynamic production cannot be ignored. These media include porous media in conventional reservoirs, fractured media in low permeability/fractured reservoirs, and multiple media in unconventional reservoirs.

① Dynamic change of porous media: during the depletion development of conventional reservoirs, with the flow and production of oil and gas, the pore pressure and effective stress of reservoirs dynamically change, resulting in the dynamic changes of the geometric and physical parameters of a porous medium (see Fig. 7.2 a). Dynamic changes in the parameters of a porous medium ultimately affect the dynamics production of the reservoir. Therefore, the impact of dynamic changes of the porous medium on production must be taken into account in simulating the development of conventional reservoirs.

② Dynamic change of a fracture medium: during the waterflooding development of a low-permeability/fractured reservoir, the initiation, expansion, and propagation of natural fractures lead to dynamic changes in geometric and physical properties of a fractured medium, and thereby dynamic fractures between injection and production wells are formed (Fan et al. 2015; Wang et al. 2015; Xie et al. 2015; Kyunghaeng et al. 2011; Ji et al. 2004; Van den Hoek et al. 2008) (as shown in Fig. 7.2b). Dynamic fractures can cause flooding, water channeling, and other development problems. Therefore, the impact of dynamic fractures on dynamic production must be considered when simulating the oil/gas development in these types of reservoirs.

③ Dynamic change of multiple media: during the fracturing-injection-production process of unconventional tight reservoirs, the geometric characteristics and

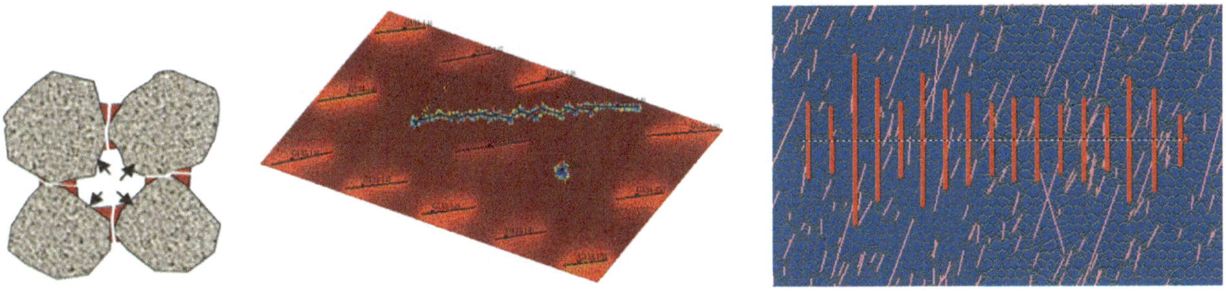

(a) Porous media deformation (b) Dynamic fractures between injection and production wells (c) Dynamic multiple media during fracturing-injection-production

Fig. 7.2 Schematic diagram of porous medium with fractures deformation during oil and gas production

attribute parameters of multiple media such as different scales matrix pores and natural/hydraulic fractures will change with the change of pore pressure and effective stress (as shown in Fig. 7.2c). These changes will lead to changes in conductivity between these media and well indices, and further affect the fluid exchange capacity between media and the one between media and wellbores. As a result, the impact of dynamic changes of multiple media on dynamic characteristics of reservoir properties and production has to be considered in the simulation of development of tight reservoirs.

7.1.1.4 Critical Pressure of Dynamic Changes of Different Media During Fracturing-Injection-Production Process

Unconventional oil and gas reservoirs are stimulated by a hydraulic fracturing treatment. During the fracturing and injection processes, the pore pressure increases and there are three key critical pressures, namely the rock breakdown pressure, the initiation pressure and the extension pressure. The initiation pressure of natural fractures has two types: initiation pressure caused by fluid flow (IPFF) and initiation pressure caused by rock mechanics (IPRM). During the processes of flowback and production, the pore pressure decreases and there are three critical pressures: fracture closure pressure, proppant crushing pressure, and hydraulic fracture failure pressure see Table 7.1.

1. Key critical pressures during pore pressure increasing

During fracturing/injection of tight reservoirs, fractures undergo rock breakdown, natural fracture reopening, and fracture extension. Therefore, key critical pressures during the pore pressure increasing include: breakdown pressure, initiation pressure, and extension pressure.

(1) Breakdown pressure

The breakdown pressure is the pressure generated by the liquid column in a wellbore and sufficient to generate new hydraulic fractures in reservoirs (Zhou et al. 2013). Its calculation expression is

$$P_{\mathrm{f}} = 3\sigma_{\mathrm{h}} - \sigma_{\mathrm{H}} - P + S_{\mathrm{rt}} \tag{7.4}$$

The breakdown pressure is mainly affected by the magnitude of the in-situ stress and pore pressure, and the reservoir lithology. The greater minimum horizontal principal stress and the less horizontal principal stress difference leads to the larger fracture pressure; and the smaller pore pressure of the reservoir leads to the greater fracture pressure. Among the rocks with different lithologies, the breakdown pressure of dolomite is relatively large while the breakdown pressure of shale is small (as shown in Fig. 7.3). For the tight sandstone reservoir with a depth of 2000 m, the breakdown pressure is usually 40–70 MPa.

(2) Initiation pressure

Initiation pressure refers to the external pressure required to reopen the existing closed natural fractures in reservoirs under current stress state (Zhu et al. 2016). According to the initiation mode of natural fractures, initiation pressure has two types: initiation pressure caused by fluid flow (IPFF) and initiation pressure caused by rock mechanics (IPRM) (McClure et al. 2016).

1) Initiation pressure caused by fluid flow (IPFF)

When the fluid pressure in fractures is less than the normal stress applied to fractures, the wall surfaces of fractures contact each other. However, due to the rough wall surfaces

Table 7.1 Key critical pressures and related concepts during fracturing-injection-production

Fracturing-injection-production	Critical pressures	Fracture type	Descriptions of critical pressures
Injection process	Breakdown pressure	Hydraulic fracture	The breakdown pressure is the pressure generated by the liquid column in the wellbore and sufficient to cause fractures in reservoirs without fractures to form hydraulic fractures
	Initiation pressure	Natural fracture	Initiation pressure refers to the external pressure required to reopen the existing closed natural fractures in reservoirs under current stress state
	Extension pressure	Natural fracture/ Hydraulic fracture	Fracture extension pressure refers to the initial fluid pressure required for reopened or broken natural/hydraulic fractures to expand in three directions of length, width, and height
Production process	Fracture closure pressure	Natural fracture/ Hydraulic fracture	Closure pressure refers to the fluid pressure needed to keep the natural/hydraulic fractures almost closed
	Proppant crushing pressure	Hydraulic fracture	The proppant crushing pressure is the minimum fluid pressure that avoids plastic deformation of proppant within hydraulic fractures
	Hydraulic fracture failure pressure	Hydraulic fracture	The hydraulic fracture failure pressure refers to the fluid pressure when the hydraulic fracture completely fails due to crushing and embedding of the proppant

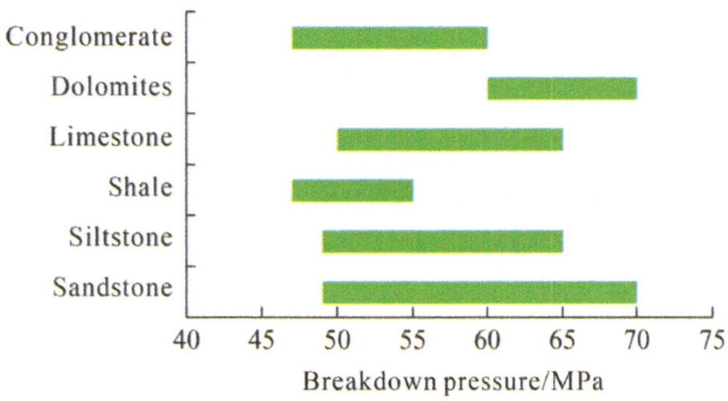

Fig. 7.3 Breakdown pressures of different type of rocks with the maximum and minimum horizontal principal stresses are respectively 42 MPa and 37 MPa

of fractures, the fluid can also flow through natural fractures. The IPFF refers to the minimum fluid pressure in natural fractures which makes the fluid flow through closed natural fractures with rough fracture surfaces. Calculation models of IPFF are different for the different filling conditions of natural fractures.

IPFF of natural fractures without filling is

$$P_i = \left(\frac{v}{1-v}\sigma_v \cos\theta + \sigma_v \sin\theta - P\right) + \sigma_H \cos\theta \cos\psi + \sigma_h \cos\theta \sin\psi \tag{7.5}$$

Hydraulic opening pressure of natural fractures with filling is

$$P_i = \left(\frac{v}{1-v}\sigma_v \cos\theta + \sigma_v \sin\theta - P\right) + \sigma_H \cos\theta \cos\psi + \sigma_h \cos\theta \sin\psi + S_{ft} \tag{7.6}$$

2) Initiation pressure caused by rock mechanics (IPRM)

The IPRM refers to the minimum fluid pressure in fractures when the fracture wall surfaces no longer contact because the fluid pressure in fractures is greater than the normal stress. For different filling conditions of natural fractures, calculation models of the IPRM are different.

The IPRM of natural fractures without filling is

$$P_i = \left(\frac{v}{1-v}\sigma_v \sin\theta + \sigma_v \cos\theta - P\right) + \sigma_H \sin\theta \sin\psi + \sigma_h \sin\theta \cos\psi \tag{7.7}$$

The fracturing pressure caused by rock mechanics of natural fractures with filling is

$$P_i = \left(\frac{\nu}{1-\nu}\sigma_v \sin\theta + \sigma_v \cos\theta - P\right) + \sigma_H \sin\theta \sin\psi + \sigma_h \sin\theta \cos\psi + S_{ft} \tag{7.8}$$

Fracture initiation pressure is mainly affected by the magnitude of the in-situ stress, reservoir lithology and fracture angles. The smaller normal stress on the fracture surface makes fractures easily open, so the initiation pressure is lower. In different types of rocks, the initiation pressure of dolomite is relatively large while the initiation pressure of shale is small (Ye et al. 1991). The larger the angle between the natural fracture and the maximum horizontal principal stress, the greater the fracture initiation pressure (as shown in Fig. 7.4). The initiation pressure of natural fractures in tight sandstone reservoirs with a depth of 2000 m is usually 30–60 MPa.

(3) Fracture extension pressure

Fracture extension pressure refers to the initial fluid pressure required for reopened or broken natural/hydraulic fractures to expand in three directions of length, width, and height. According to the theory of fracture mechanics, the extension pressure at fracture tips (Micar et al. 2002; Na 2009; Irwin 1957) is

$$P_{tip} = \frac{K_{Ic}}{\sqrt{\pi L}} + \sigma_3 \tag{7.9}$$

The fracture extension pressure is mainly affected by the magnitude of the in-situ stress, reservoir lithology and a fracture size. The smaller minimum horizontal principal stress leads to the smaller fracture extension pressure. In different types of rocks, the fracture extension pressure of dolomite is relatively large while the fracture extension pressure of shale is small; a larger fracture size leads to the smaller extension pressure. The fracture extension pressure in a tight sandstone reservoir with a depth of 2000 m is usually 30–60 MPa.

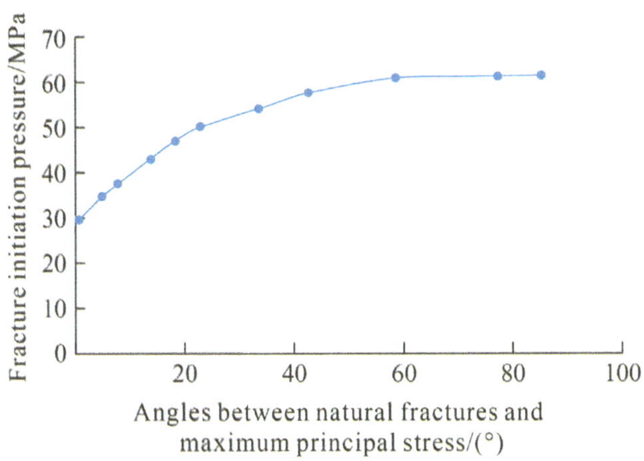

Fig. 7.4 Initiation pressure of natural fractures with different angles of Chang 63 reservoir in Huaqing exploration area

2. Key critical pressures during pore pressure decreasing

In the production process of tight reservoirs, as the pore pressure decreases, the fracture width gradually narrows until it closes, and the proppants within the hydraulic fractures will also break and lose efficiency. Therefore, the key pressures in the production process include: closure pressure, crushing pressure and failure pressure.

(1) Closure pressure

Closure pressure refers to the fluid pressure needed to keep the natural/hydraulic fractures almost closed (Sun et al. 2010). The natural fracture closure pressure is mainly affected by the magnitude of the in-situ stress, and its calculation model is

$$P_{fC} = \sigma_h \sin\theta + \sigma_v \cos\theta \tag{7.10}$$

In addition, the closure pressure of the hydraulic fractures is also related to the fracture opening and the mechanical properties of fractures, and its calculation expression is

$$P_{FC} = \frac{W_F E}{2h_F(1-\nu^2)} + P_{min} \tag{7.11}$$

In general, the larger minimum horizontal principal stress leads to the greater closure pressure of fractures. The closure pressure of fractures in tight sandstone reservoirs with a depth of 2000 m is usually 30–50 MPa.

(2) Proppant crushing pressure

The proppant crushing pressure is the minimum fluid pressure that avoids plastic deformation of proppants within hydraulic fractures, and its calculation expression is:

$$P_{PC} = \sigma_h \sin\theta + \sigma_v \cos\theta - R_{Pro} \tag{7.12}$$

The proppant crushing pressure has a close relationship with in-situ stress and proppant strength (Gao et al. 2006). Ceramic and resin sands have high strength, so they are difficultly broken. In a tight sandstone reservoir with a depth of 2000 m, the fluid pressure when proppants are crushed is usually less than 8–10 MPa. Because the strength of quartz sands is relatively low, in a tight sandstone reservoir with a depth of 2000 m, the fluid pressure when proppants are crushed is usually less than 8–39 Mpa.

(3) Hydraulic fracture failure pressure

The hydraulic fracture failure pressure refers to the fluid pressure when the hydraulic fracture completely fails due to crushing and embedding of proppants. Its calculation expression is:

$$P_{\mathrm{FL}} = \sigma_{\mathrm{h}} \sin\theta + \sigma_{\mathrm{v}} \cos\theta - \alpha_{\mathrm{Pro}} R_{\mathrm{Pro}} \tag{7.13}$$

The failure of hydraulic fractures is related to the crushing/embedding of proppants and Young's modulus of reservoirs. The mechanism of hydraulic fracture failure is complex and it is difficult to be described by a simple formula. In the practical application, the hydraulic fractures failure usually occurs after 5–6 years of production (Ikonnikova et al. 2014) and the hydraulic fractures propped by quartz sands are more prone to failure.

7.1.2 Mechanisms of Matrix Pore Expansion, Hydraulic/Natural Fracture Propagation During Pore Pressure Increasing

7.1.2.1 Mechanism of Matrix Pore Expansion Deformation

At the original state of reservoirs, the rock grains subjected to the pore pressure in matrix and confining pressure are in equilibrium. During the fracturing and injection processes, the effective stress of rock grains gradually decreases with the increase of pore pressure. With the increase of the pore pressure, the matrix pores experience linear or nonlinear elastic expansion deformation; when the pore pressure is larger than the breakdown pressure of the rock, the cement breaks and opens. Affected by shear stress, the rock grains and cement dislocate, and the matrix breaks down (as shown in Fig. 7.5).

The most widely used failure criteria in the development of oil and gas are the maximum tensile stress (strain) criterion, the Mohr-Coulomb criterion and the Drucker-Prager criterion. The method of combining the maximum tensile stress criterion with the Mohr-Coulomb criterion is used in this study.

A reservoir medium is affected by the in situ stress, and the magnitude and direction of the in-situ stress have differences in spatial distribution. However, in-situ stresses at different directions are normally different. Affected, the deformation of porous medium is affected by the in-situ stresses with different magnitudes from different directions and pore pressure. The magnitude of the deformation depends on the effective normal stress and shear stress in different directions. The change in stress state can be characterized by the Mohr circle (Chen et al. 2011; Wang et al. 2015; Shen 1980; Li and Wang 2006) (as shown in Fig. 7.6). In this figure, the abscissa is the effective normal stress, the ordinate is the shear stress, and the oblique line and the left vertical line are critical failure lines representing the Mohr-Coulomb and the maximum tensile stress criteria, respectively. The Mohr circle reflects the stress state at a certain point of the reservoir rock. The diameter of Mohr circle is the difference between the maximum principal stress and the minimum principal stress at that point. The 1/2 of the angle between the radius of one point on the Mohr circle and the abscissa is equal to the angle between the oblique section represented by that point on the Mohr circle and the surface of the maximum principal stress. The right endpoint of the Mohr circle represents the stress state along the direction of the maximum principal stress of this specific location in a reservoir, and the left endpoint represents the stress state along the direction of minimum principal stress of that specific location. At these two points, only the normal stress acts, and the shear stress is zero. The stress value at the right endpoint of the circle is the maximum effective principal stress, and the stress at the left endpoint is the minimum effective principal stress.

The intercept of a critical failure line representing the Mohr-Coulomb criterion and the coordinate is rock cohesion, and the slope of that line is the tangent value of a rock frictional angle. Different types of reservoir rocks or media have different values of tensile strength, a cohesive force and an internal friction angle, so the damage lines are different. In general, the matrix is the most difficult to be broken so its tensile stress is the largest, and the slope and intercept of the skew line are the largest. The extension of the fracture line follows the damage line of matrix. In addition, the reopening

Fig. 7.5 Schematic diagram of matrix expansion deformation

(a) Initial state (b) Tensile failure (c) Shear failure

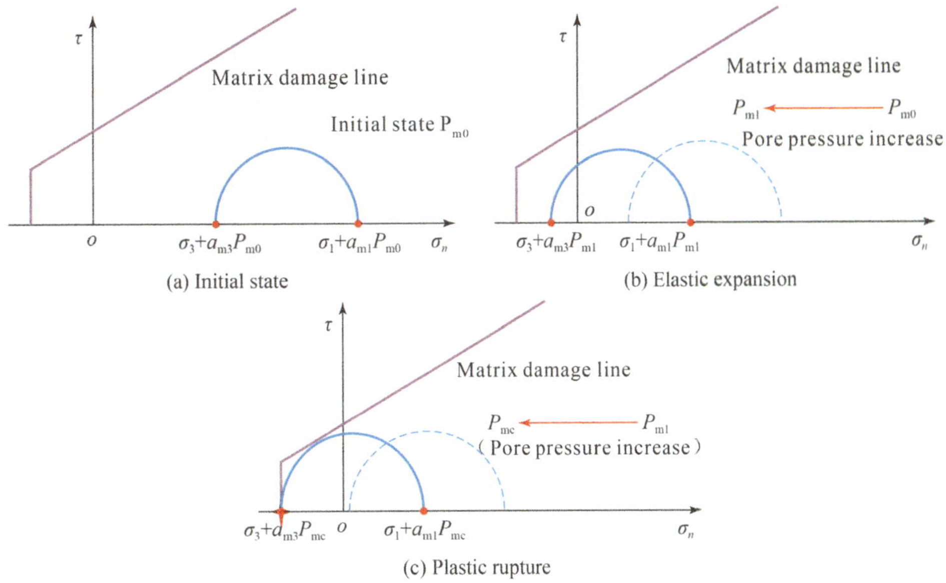

Fig. 7.6 The expansion deformation mechanism of pores inside matrix explained by the criterions of maximum tensile stress and Mohr-Coulomb

of fractures generally only needs to overcome the influence of in-situ stress, so the maximum tensile stress, oblique line slope and intercept of the line representing natural fracture reopening are the minimum.

Figure 7.6 shows the expansion deformation mechanism of pores inside matrix explained by the criteria of maximum tensile stress and Mohr-Coulomb. There are mainly three stages:

① Initial state: Before a fracturing fluid is injected, the pores inside matrix are in equilibrium and the pore fluid pressure is P_{m0}.

② Elastic expansion: As the large amounts of the fracturing fluid and proppants are injected, the pressure in the matrix pores rises from the initial pressure P_{m0} to P_{m1}, which causes the Mohr circle to move to the left. At this time, the Mohr circle is still on the right side of the fracture line. The deformation is elastic and reversible.

③ Plastic rupture: As the formation pressure increases further, the pressure in the pores inside matrix increases from P_{m1} to the fracture pressure P_{mc}. Currently, the Mohr circle is tangent to the matrix maximum tensile stress failure line or the Mohr-Coulomb oblique line, indicating that there is an irreversible tensile fracture or shear fracture deformation of the matrix pores under the influence of tension stress and shear stress.

7.1.2.2 Mechanisms of Natural Fracture Initiation and Expansion

At the original state of reservoirs, natural fractures are closed under confining pressure. In the fracturing injection process, as a fracturing fluid enters natural fractures, the pore pressure in fractures gradually increases, and the effective stress on natural fracture walls gradually decreases. When the pore pressure of natural fractures reaches the initiation pressure of natural fractures, natural fractures reopen; with the increase of pore pressure in natural fractures, the width of open natural fractures continues to increase; when the pore pressure at a natural fracture tip reaches the fracture extension pressure, the rock around the natural fracture tips is broken and natural fractures extend forward (as shown in Fig. 7.7).

Figure 7.8 shows the mechanism of hydraulic fractures generated from natural fractures explained by the criteria of maximum tensile stress and Mohr-Coulomb. The movement of the Mohr circle mainly has the following four stages:

① Initial closed state: under the initial conditions, the natural fractures are at a closed state and the fluid pressure in fractures is P_{f0}.

② Natural fracture initiation: as the pore pressure rises, the Mohr circle moves to the left, and the pressure in fractures gradually increases from the initial P_{f0} to the fracture initiation pressure P_{f0}. At this moment, the Mohr circle is tangent to the natural fracture initiation line, and natural fractures reopen.

③ Natural fractures expansion: after natural fractures are initiated, a natural fracture initiation line is replaced by a natural fracture extension line. As the pressure in fractures gradually increases from P_{f1} to P_{f2}, the Mohr circle will move to the left and correspondingly, the fracture width will increase according to the elasticity theory.

④ Natural fracture expansion: when the pressure in fractures continues to increase to the natural fracture expansion pressure P_{fe}, the Mohr circle is tangent to the

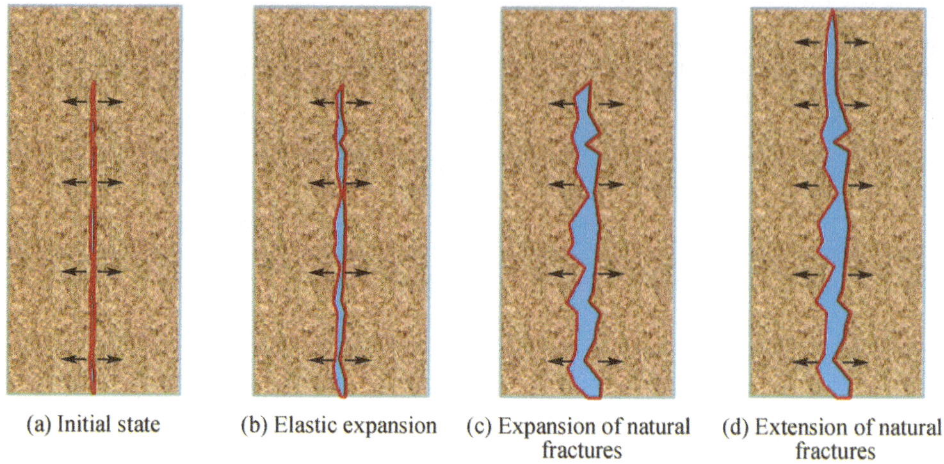

Fig. 7.7 Schematic diagram of their opening and expansion of natural fractures

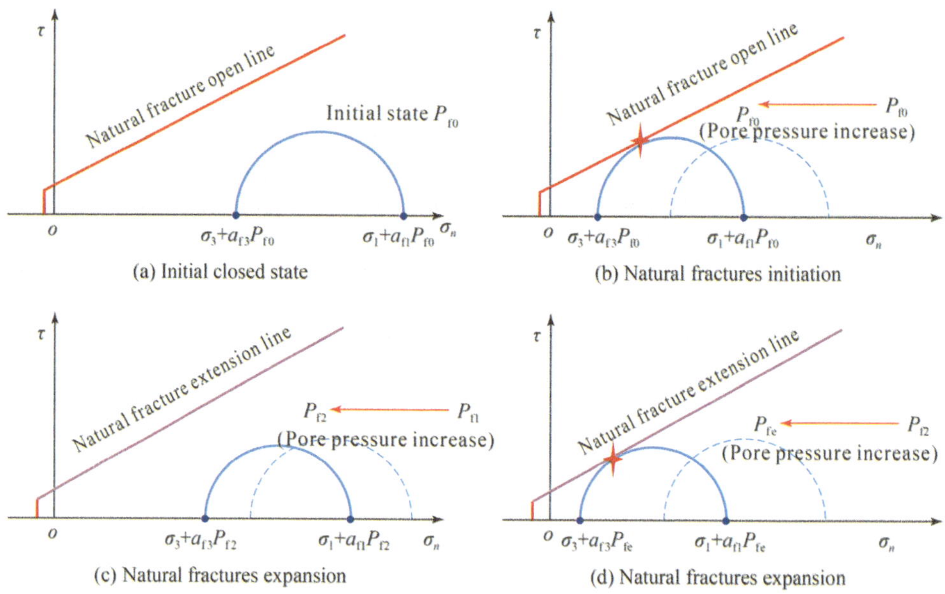

Fig. 7.8 The mechanism of hydraulic fracture generated from natural fractures explained by the criterions of maximum tensile stress and Mohr-Coulomb

extension line of natural fractures. At this point, the rock at natural fracture tips ruptures and the fractures extend forward.

7.1.2.3 Mechanisms of Hydraulic Fracture Generation and Expansion

For homogeneous reservoirs, there is no natural fracture developed, and the direction of the minimum horizontal principal stress is shown in Fig. 7.9. During the fracturing and injection process, with the injection of a fracturing fluid or water, the pore pressure in matrix increases, which leads to the expansion of matrix and pore throat. When the treating pressure reaches the breakdown pressure of the reservoir rock, the rock around a wellbore is first broken to generate symmetrical hydraulic bi-wing fractures perpendicular to the direction of minimum horizontal principal stress. With the injection of fracturing fluid and proppant, the pore pressure in hydraulic fractures increases, leading to a gradual increase in fracture opening. When the pressure reaches the fracture extension pressure, the rock around fracture tips is broken, and the hydraulic fracture controlled by in-situ stress continue to extend along the direction perpendicular to the minimum principal stress.

For reservoirs with a large number of natural fractures, the tension strength of natural fractures is much lower than the one of matrix. During fracturing and injection processes, when the pore pressure reaches the initiation pressure of natural fractures, the originally closed natural fractures are first activated and reopened. With the injection of the fracturing fluid, proppants or water, the pore pressure in the fracture increases, causing gradual increase of fracture width. When the pressure at the tip of a initiated fracture reaches the initiation pressure of the other natural fracture, this natural fracture is reopened. In this case, the extension of hydraulic fractures is achieved, and the expansion direction is controlled by natural fractures.

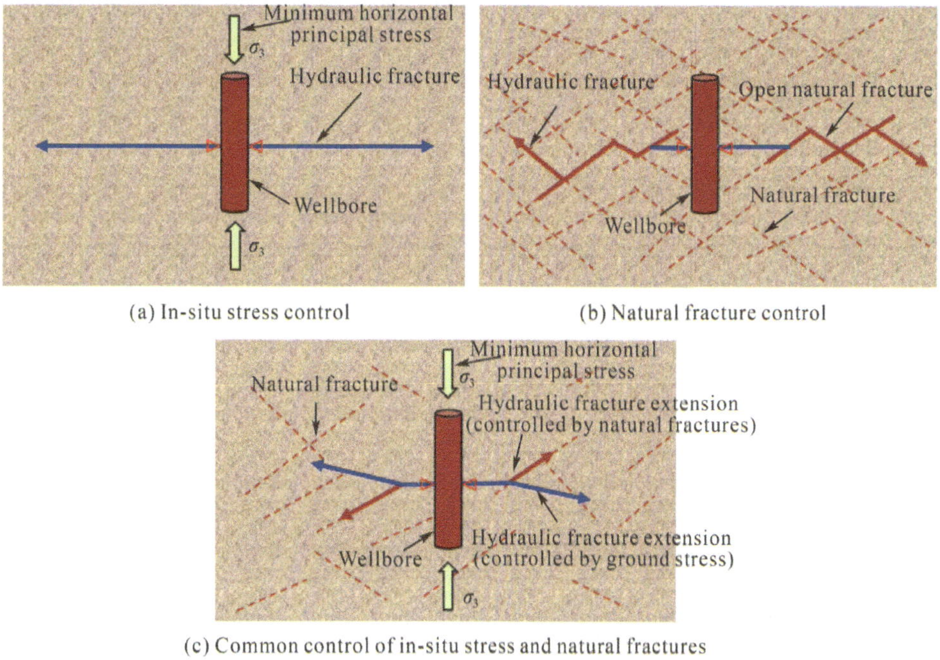

Fig. 7.9 Expansion mechanism of hydraulic fractures in reservoirs without natural fracture

In reservoirs with few natural fractures, the expansion of hydraulic fractures generated by the matrix crushing is controlled by natural fractures and hydraulic fractures. A part of the fracturing fluid or water enters the natural fractures and initiates the natural fractures to extend forward; the other part of the fracturing fluid under the control of local stress makes the matrix to be broken to generate new fractures. In this case, the extension of hydraulic fractures is achieved.

The mechanisms of initiation, expansion and extension of natural fractures are discussed in the above sections. In the following, the mechanisms of generation, expansion and extension of hydraulic fractures controlled by in-situ stress are explained by criteria of maximum tensile stress and Mohr-Coulomb. The movement of the Mohr circle mainly has the following four stages (as shown in Fig. 7.10).

① Expansion of matrix pores: with the injection of the large amounts of fracturing fluids and proppants, the formation pressure rises sharply, and the pore pressure rises from the initial pressure P_{m0} to P_{m1}. In this case, the Mohr circle moves left within the area on the right hand of a damage line and the deformation of the matrix at this stage is elastic and reversible.

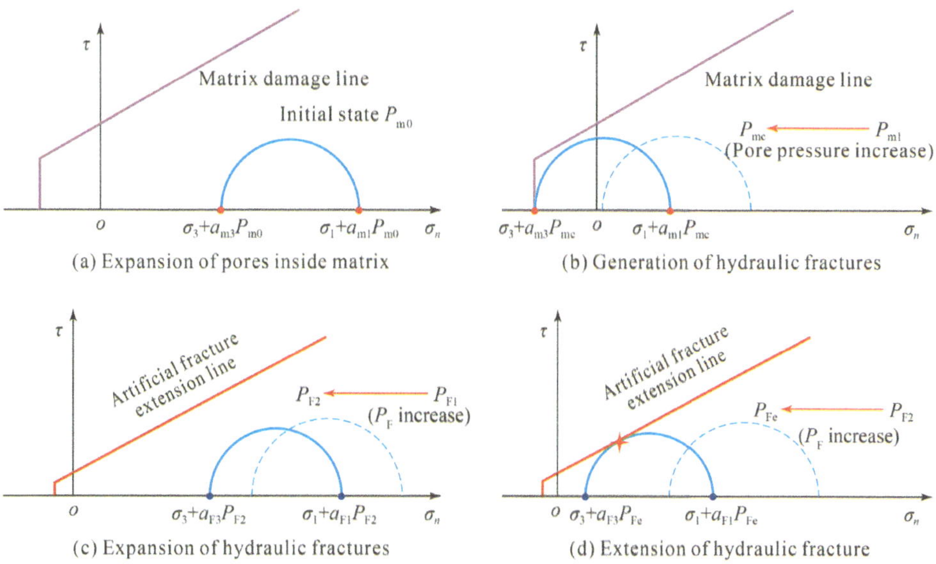

Fig. 7.10 The mechanisms of generation, expansion and extension of hydraulic fractures controlled by in-situ stress are explained by criteria of maximum tensile stress and Mohr-Coulomb

② Generation of hydraulic fractures: when the Mohr circle moves left to the location tangent to the damage line, the pore pressure rises from P_{m1} to the matrix fracture pressure (P_{mc}). In this case, the matrix pores experience deformation caused by plastic elongation and shear failure, resulting in hydraulic fractures.

③ Expansion of hydraulic fractures: after the generation of hydraulic fractures, the damage line of matrix is replaced by the fracture extension line. As the pore pressure in the fractures increases, the Mohr circle moves left, as shown in Fig. 7.10c. The pressure in the hydraulic fractures gradually increases from P_{F1} to P_{F2}. In this case, the Mohr circle is still on the right side of the damage line, and the elastic deformation of fractures leads to an increase in fracture width.

④ Extension of hydraulic fractures: when the pressure in the hydraulic fractures rises from P_{F2} to the fracture extension pressure (P_{Fe}), the Mohr circle is tangent to the extension line of the hydraulic fractures. In this case, the fracture tips have plastic damage and the fractures extend forward.

7.1.3 Mechanisms of Matrix Pore Compression, Hydraulic/Natural Fractures Closure Deformation During Pore Pressure Decreasing

7.1.3.1 Mechanism of Matrix Pore Compression Deformation

During the production process, the effective stress applied on the reservoir rock increases with recovery of oil and gas. Under the influence of compressive stress, pores are elastically compressed and deformed. If the reservoir pore pressure further reduces, under the influence of shear stress, rock grains and cement can experience shear damage (as shown in Fig. 7.11).

As shown in Fig. 7.12, the law of the Mohr circle movement during the compression of matrix pores can be divided into three stages:

① Initial state: prior to flow back and production, the matrix pores are in equilibrium and the pore pressure is P_{m0}.

② Elastic compression: with the flowback of a fracturing fluid and recovery of oil/gas/water, the pore pressure decreases from the initial pressure P_{m0} to P_{m1}, and the Mohr circle moves right. Due to the anisotropy of matrix properties, the Mohr circle radius increases. In this case the whole Mohr circle is still within the area on the right side of the damage line, and the pore deformation is elastic and reversible.

③ Plastic failure: as the pore pressure further reduces, the pore pressure decreases from P_{m1} to the failure pressure (P_{mb}). In this case, due to the influence of anisotropy, the Mohr circle radius increases to be tangent to the matrix damage line, indicating that the matrix pores experience irreversible shear failure.

7.1.3.2 Mechanism of Natural Fracture Closure Deformation

During the production of oil and gas, the pore pressure in natural fractures gradually decreases while the effective compressive stress on fracture walls increases, which makes the natural fracture aperture gradually decrease. Initially, a natural fracture surface has a small contact area and low stiffness, so natural fractures close quickly and experience large decreases in fracture apertures. In the later period, with the increase of the contact area of fractures, their bearing capacity increases. Although the value of fracture aperture continues to decrease, the closing speed of fractures decreases.

When the pressure within natural fractures is less than the closure pressure of fractures, the natural fractures can experience closure deformation. Compared with the matrix, the natural fractures have a lower Young's Modulus and the higher deformation under the same stress. As the pressure drop near a production well is funnel-shaped, the natural fractures near the wellbore are closed first. With production, the pressure in natural fractures away from the wellbore continues to decrease. When it reaches the closure pressure

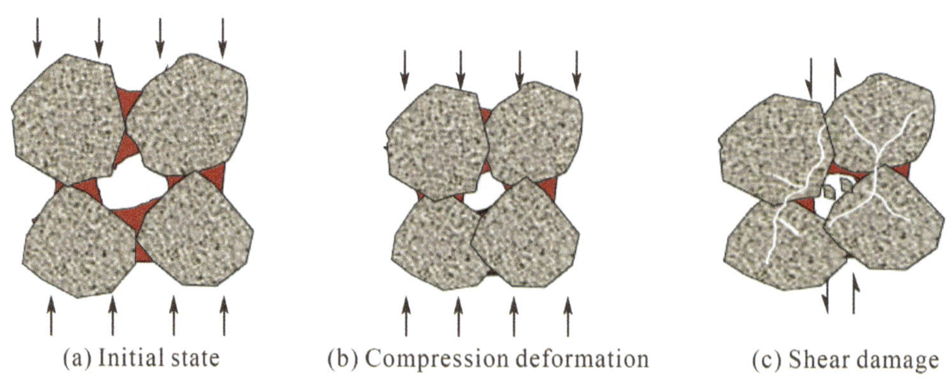

Fig. 7.11 Matrix pore deformation mechanism during the production process

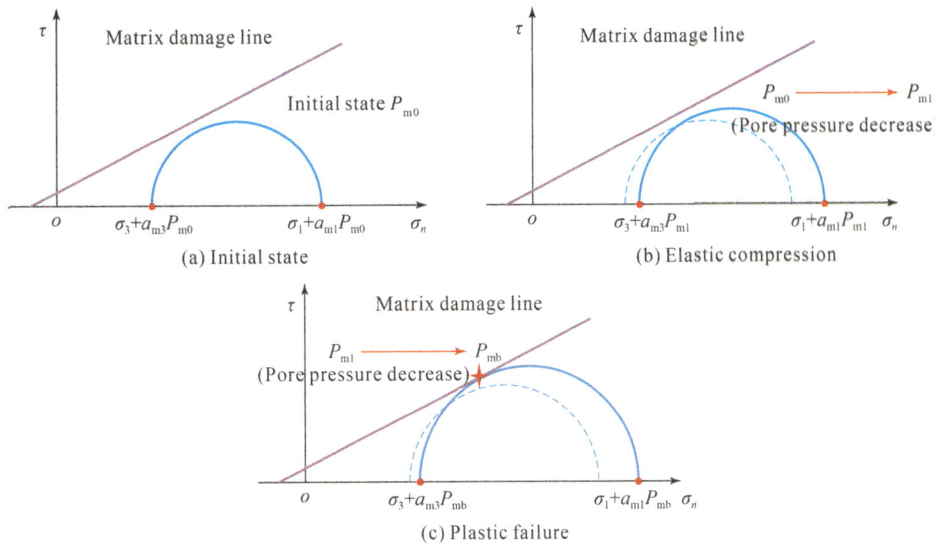

Fig. 7.12 The law of Mohr circle's movement during the compression of matrix pores

of fractures, these fractures also gradually close (as shown in Fig. 7.13).

Figure 7.14 shows the mechanism of closure deformation of natural fractures explained by the Mohr-Coulomb criterion. The movement of the Mohr circle mainly has the following three stages:

① Initial opening state: under initial conditions, the natural fractures are open and the fluid pressure in fractures is P_{f0}. During production, the stress applied to natural fracture walls is less than the fracture closure pressure, and the natural fractures are open. In this case, the Mohr circle is tangent to the natural fracture initiation line.

② Decrease in natural fracture aperture: as the flowback and recovery progress, the pressure in fractures gradually decreases from P_{f0} to P_{f1}, the Mohr circle moves right to go away from the natural fracture initiation line. In this case, the natural fractures are still open while their aperture decreases.

③ Closure of natural fractures: as the pressure further reduces, the pressure in natural fractures gradually decreases from P_{f1} to the pressure lower than the closure pressure of natural fractures, P_{fc}. The Mohr circle has obviously moved away from the natural fracture initiation line, and the natural fractures close.

7.1.3.3 Mechanism of Hydraulic Fracture Closure Deformation

During the flowback and recovery processes, the hydraulic fractures also dynamically change with the formation pressure. Because hydraulic fractures are propped by proppants, the dynamic change of hydraulic fractures is closely related to proppant properties. The mechanism of hydraulic fractures closure deformation mainly has the following four stages, as shown in Fig. 7.15.

① Suspension of proppants in hydraulic fractures: at the early stage of fracturing fluid flowback, due to the high pressure in hydraulic fractures, the hydraulic fracture aperture is large, and the proppants suspend in a fracturing fluid.

② Force acting on the proppant in hydraulic fractures: during the fracturing fluid backflow, the pressure in hydraulic fractures and fracture width gradually decrease, and the proppants originally suspending in fractures become compressed. During the production, hydraulic fractures are propped by proppants to keep open.

③ Compressive deformation of proppants in hydraulic fractures: with the production of oil and gas, the pressure in hydraulic fractures gradually decreases. The proppant-filled hydraulic fractures are compressed and experience closure deformation, resulting in the shrinkage of proppant particles and a decrease in fracture volume and fracture width. Due to the support of proppants, the stiffness of fractures significantly increases. As a result, with the same closure pressure, the closure deformation of hydraulic fractures with proppants is much lower than that of natural fractures without proppants.

④ Breakdown and embedding of proppants in hydraulic fractures: during the long-term production, proppants can be embedded into fracture walls under the impact of closure pressure. When the closure pressure is large, especially for tight reservoirs with large Young's Modulus, proppants can be crushed during proppant

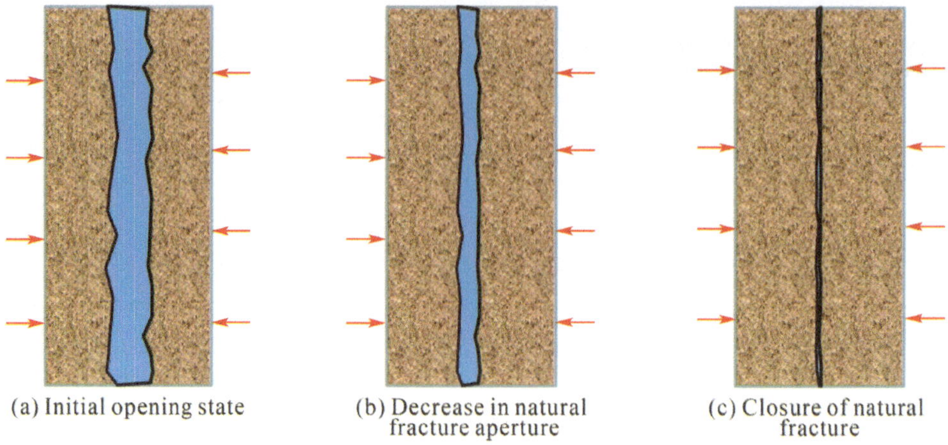

Fig. 7.13 Mechanism of natural fracture closure during the production process

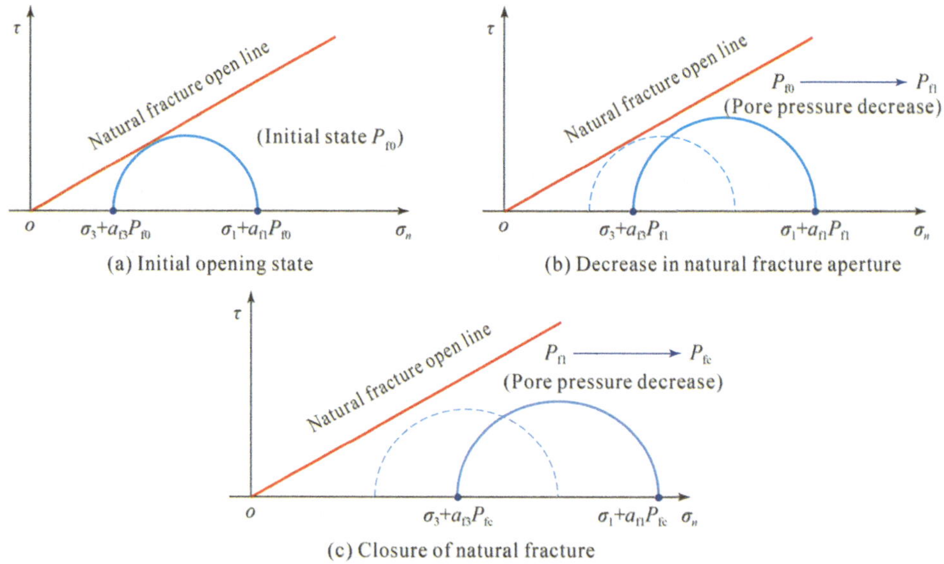

Fig. 7.14 The mechanism of closure deformation process of natural fractures explained by the Mohr-Coulomb criterion

embedment. Under the influence of proppant embedment and crushing, the hydraulic fractures with proppants can shrink and even close.

Figure 7.16 shows the closure deformation process of hydraulic fractures explained by the Mohr-Coulomb criterion. There are mainly three stages in the movement of the Mohr circle:

① Initial suspension stage of proppants: proppant particles are in suspension before returning and production. The fluid pressure in hydraulic fractures is P_{F0}. Under the influence of hydrostatic pressure, the maximum principal stress σ_1 and the minimum principal stress σ_3 applied to proppant particles are equal, so the Mohr circle appears as a point.

② Elastic support stage of proppants: with the flowback of a fracturing fluid and production of oil/gas/water, the pressure in hydraulic fractures reduces from the initial pressure P_{F0} to P_{FC} (hydraulic fracture closure pressure), and the Mohr circle moves to the right. Under the compressive impact of fracture walls, the maximum principal stress σ_1 and the minimum principal stress σ_3 of applied to proppant particles begin to be different, and the radius of the Mohr circle gradually increases. In this case, the whole Mohr circle is still located in the right side of the failure line, indicating that proppants under compressive conditions experience elastic compression deformation.

③ Plastic fracture stage of proppants: as the formation pressure further reduces, the pressure in hydraulic fractures reduces from P_{FC} to the proppant breakdown pressure P_{PC}. In this case, due to the influence of anisotropy, the radius of the Mohr circle increases and the Mohr circle can be tangent to the damage line, indicating that the proppant particles experience deformation caused by shear failure.

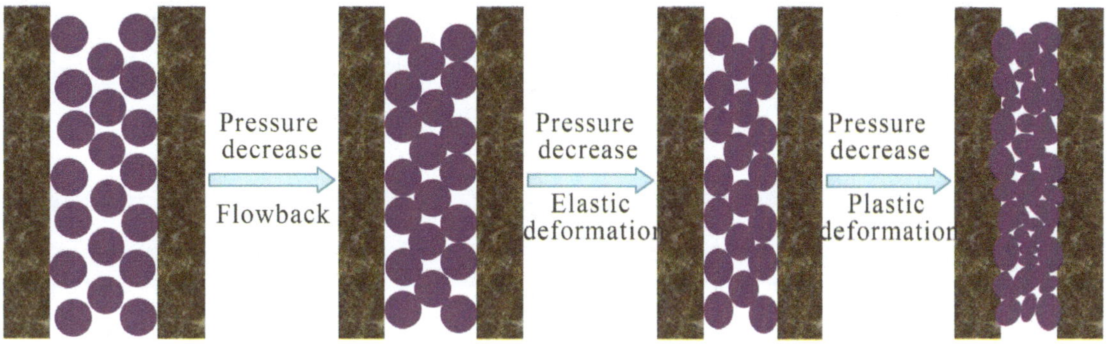

(a) Suspension of proppantin hydraulicfractures (b) Force acting on the in hydraulic fracture (c) Compressive deformation of proppants in hydraulic fractures (d) Breakdown and embedding of proppants in hydraulic fractures

Fig. 7.15 Compression, breakdown, and embedding of proppants in hydraulic fractures

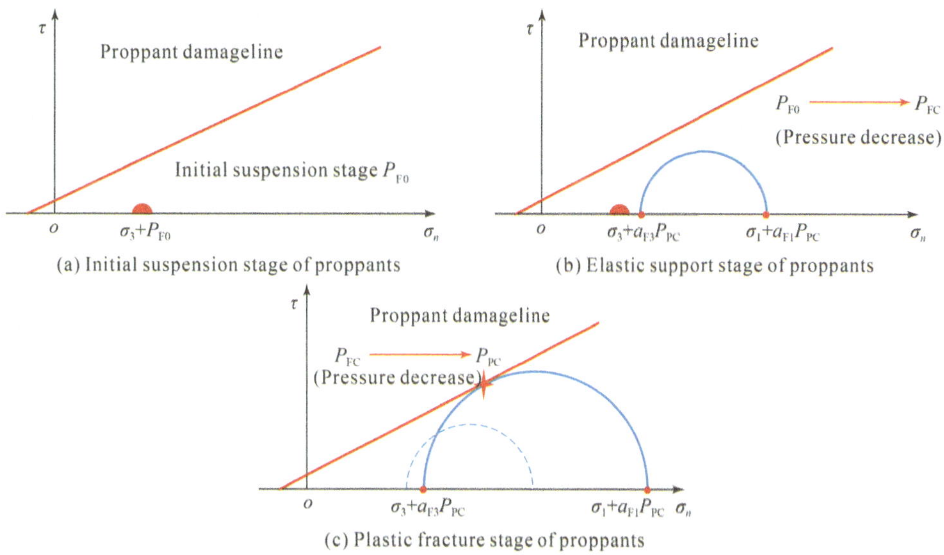

Fig. 7.16 The closures deformation process of hydraulic fractures explained by the Mohr-Coulomb criterion

7.1.4 Characteristics Analysis of Dynamic Change of Multiple Media with Different Scales Pores and Fractures

During the processes of injection and production in tight reservoirs, with the change of pore pressure, media deform, resulting in changes of geometric characteristics and physical properties parameters of these media. The dynamic change characteristics of different media with pores/fractures are different, and the rules of these dynamic change characteristics are complex.

7.1.4.1 Dynamic Change Characteristics of Different Media with Pores/Fractures During Pore Pressure Increasing

During the processes of fracturing and injection in tight reservoirs, the pore pressure changes with fluid flow, and the effective stresses applied to different media change. Matrix pores expand, pore throats widen, and natural/hydraulic fractures experience expansion deformation, resulting in dynamic changes in the geometry, size, and physical properties of different media.

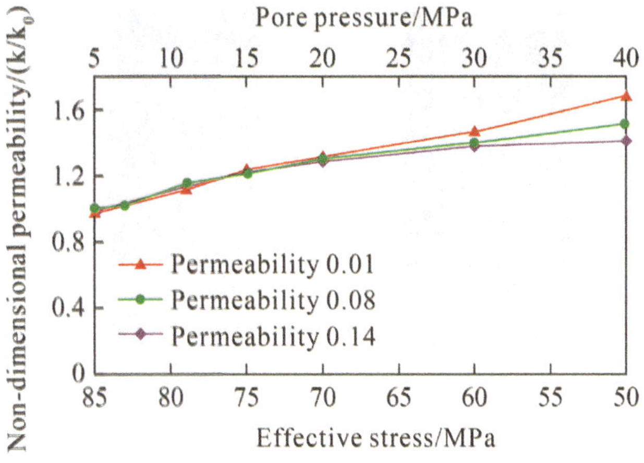

Fig. 7.17 Relationship between permeability and effective stress during injection process

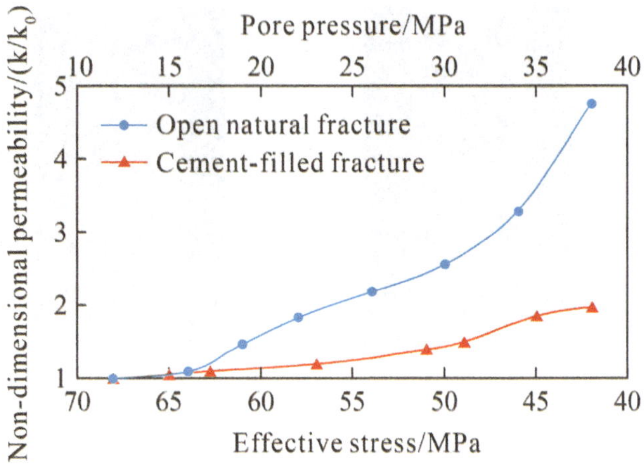

Fig. 7.18 Relationship between permeability of natural fracture and effective stress during injection process

1. Dynamic change characteristics of matrix pore during pore pressure increasing

During the processes of fracturing and injection in tight sandstone reservoirs, as the pore pressure increases, the effective stress of the reservoir gradually decreases, resulting in expansion of pores/throats and increase of matrix permeability. The experiments under a condition of constant confining pressure and variable pore pressure were conducted to investigate the impact of stress on matrix pores in thigh reservoirs. The experimental results show as Fig. 7.17:

① As the pore pressure increases, the effective stress decreases while the matrix permeability increases. When the pore pressure increases to 40 MPa, the matrix permeability will increase by 30% to 70%.
② The magnitudes of permeability changing with pore pressure for matrix with different permeability values are different. The smaller the matrix permeability is, the stronger the stress sensitivity of matrix permeability is.

2. Dynamic change characteristics of natural fractures during pore pressure increasing

During the processes of fracturing and injection of tight sandstone reservoirs, as the reservoir pressure increases, the normal effective stress of natural fractures decreases. The open natural fractures expand, and the cement-filled fractures deform, making the fracture permeability increase dramatically. The impact of stress sensitivity on natural fractures under different conditions was analyzed by stress sensitive experiments of filled and open natural fractures. Experiment results show as Fig. 7.18:

① As the pore pressure increases, the effective stress decreases, and the permeability of the open natural fractures natural fracture increases. When the pore pressure increases to 38 MPa, the permeability of open natural fractures will increase by 4.8 times.
② As the pore pressure increases, the effective stress decreases and the permeability of cement-filled natural fractures obviously increases. When the pore pressure increases to 38 MPa, the permeability of cement-filled natural fractures increases by 2 times.
③ The change magnitudes of the permeability of natural fractures at different states are different. The change magnitude of the permeability of open natural fractures with the effective stress is significantly greater than that of the cement-filled fractures. The stress sensitivity of cement-filled fractures is weaker than that of open natural fractures.

3. Comparison of dynamic change characteristics between different media with pores/fracture during pore pressure increasing

For different media with pores/fractures, the permeability changes with increasing pore pressure during fracturing and injection processes were compared and the results are shown in Fig. 7.19. Experimental results illustrate:

① The stress sensitivity of matrix pores in tight reservoirs is the weakest, and the magnitude of permeability changes with pore pressure is minor.
② For cement-filled natural fractures in tight sandstone reservoirs, with the increase of pore pressure, the normal effective stress on fracture wall surfaces reduces. The magnitude of fracture deformation is relatively

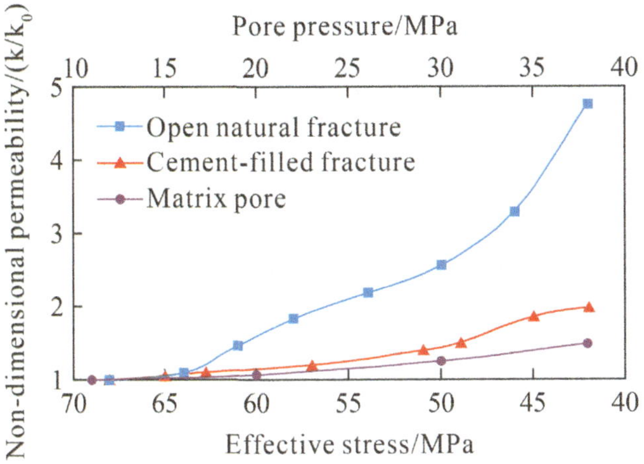

Fig. 7.19 Comparison of stress sensitivity in different media with pores/fractures during injection

weak, and the stress sensitivity of cement-filled natural fractures is slightly greater than that of the matrix.

③ For open natural fractures in tight sandstone reservoirs, with the pore pressure increasing, the effective stress applied to fracture walls decreases, and the magnitude of deformation is the largest and the stress sensitivity is the strongest among these three cases.

7.1.4.2 Dynamic Change Characteristics of Different Media with Pores/Fractures During Pore Pressure Decreasing

Tight sandstone reservoirs contain nano/micro/millimeter scale media with pores/fractures. Natural/hydraulic fractures form a complex fracture network. During production, as the reservoir pressure changes, the effective stress applied to different media changes. The pore throats and fractures inside matrix change dynamically. The pores shrink, the pore throats narrow, and natural/hydraulic fractures deform and even close, resulting in dynamic changes in the geometry, size, and physical properties of different media (as shown in Fig. 7.20).

1. Dynamic change characteristics of matrix pores during pressure decreasing

During the production of tight sandstone reservoirs, the decrease of pore pressure causes the change of effective stress, resulting in shrinkage of pores and pore throats and changes in porosity, permeability and other parameters of reservoirs. The experiments under a condition of constant confining pressure and variable pore pressure were conducted to investigate the impact of stress on porosity and permeability of matrix pores in tight reservoirs.

(1) The experimental results about the impact of stress sensitivity on porosity of matrix pores indicate

① As the pore pressure decreases, the effective stress increases and the matrix porosity decreases. When the pore pressure drops to 5 MPa, the matrix porosity will decrease by 1.5% to 2.5%.

② The impacts of stress sensitivity on porosity of matrix pores with different initial permeability are different. The matrix pore with lower initial permeability has stronger stress sensitivity.

(2) The experimental results about the impact of stress sensitivity on permeability of matrix pores indicate

① As the pore pressure decreases, the effective stress increases and the matrix permeability decreases. When the pore pressure drops to 5 MPa, the matrix permeability will drop by 10% to 35%.

② The impacts of stress sensitivity on permeability of matrix pores with different initial permeability are different. The matrix pores with lower initial permeability have stronger stress sensitivity.

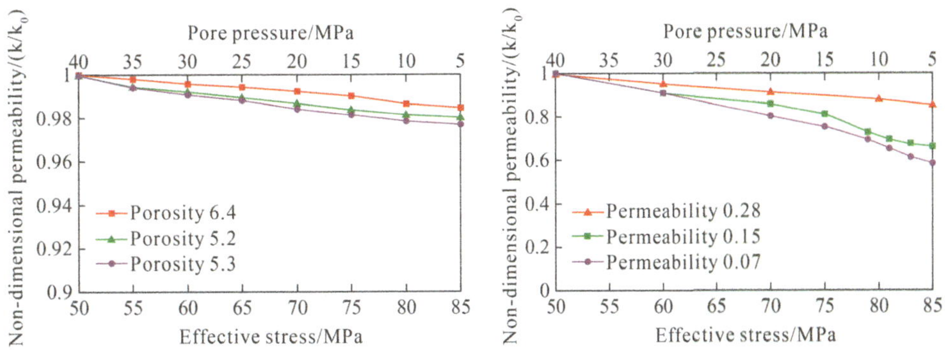

(a) Relationship between porosity and effective stress

(b) Relationship between permeability and effective stress

Fig. 7.20 The relationship between porosity and permeability with effective stress during pore pressure decrease

Fig. 7.21 The relationship between the permeability of natural fracture and the effective stress during the pressure decrease process

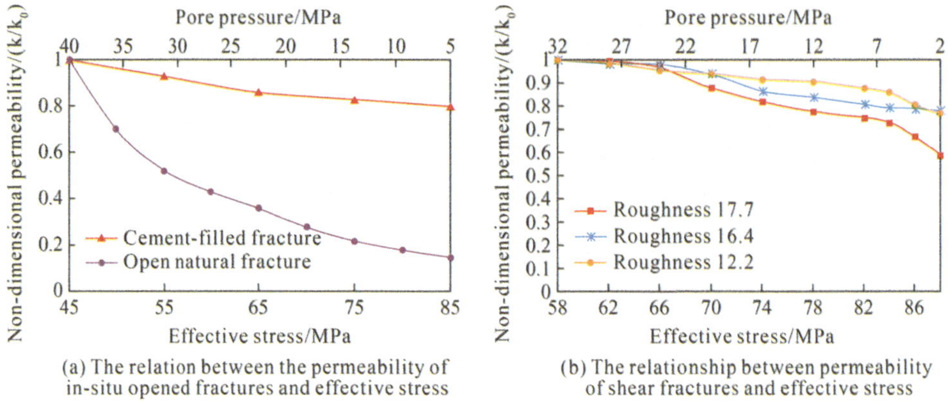

2. Dynamic change characteristics of natural fractures during pore pressure decreasing

Tight sandstone reservoirs contain multiple scale natural fractures. The open natural fractures, cement-filled natural fractures and closure natural fractures coexist in reservoirs. Under the impact of the normal stress applied to wall surfaces, natural fractures deform, affecting the characteristics of fluid flow in natural fractures.

During the production of oil and gas in tight reservoirs, as the formation pressure decreases, the normal effective stress of natural fractures increases. In this case, the open natural fractures narrow down and close, and cement-filled natural fractures deform, resulting in a sharp decrease of fracture permeability (as shown in Fig. 7.21). Through the experiments of stress sensitivity for different types of natural fractures, the impacts of stress sensitivity on permeability of natural fractures under different conditions were analyzed (Zhang et al. 2015; Zhou et al. 2016).

(1) The results of stress sensitivity experiments for originally open and cement-filled natural fractures indicate

① As the pore pressure decreases, the effective stress increases and the permeability of cement-filled natural fracture decreases. When the pore pressure drops to 5 MPa, the permeability of cement-filled natural fractures will decrease by 22%.
② As the pore pressure decreases, the effective stress increases, and the permeability of open natural fractures decreases significantly. When the pore pressure drops to 5 MPa, the permeability of open natural fractures will decrease by 87%.
③ The change magnitudes of permeability of different types of natural fractures are different. The change magnitude of permeability of open natural fractures with effective stress is significantly greater than that of cement-filled natural fractures.

(2) The results of stress sensitivity experiments for shear fractures with different surface roughness show

① As the pore pressure decreases, the effective stress increases and the permeability of shear fractures decreases. When the pore pressure decreases to 2 MPa, the permeability of shear fractures will decrease ranging from 22% to 41%.
② The stress sensitivity magnitudes of shear fractures with different fracture surface roughness are different. The greater roughness of fracture wall surfaces leads to the greater residual flow space after shear deformation. With the increase of effective stress, the permeability decreases obviously, and the stress sensitivity is strong.

3 Dynamic change characteristics of hydraulic fracture during pore pressure decreasing

There are proppants in hydraulic fractures in tight sandstone reservoirs. Under the normal stress applied to fracture surfaces, the hydraulic fractures deform. Meanwhile, the proppant particles are easily broken, which causes the proppant failure and hydraulic fractures further deform and close. Due to the difference of dynamic change laws between natural fractures and hydraulic fractures, their fluid flow characteristics are different.

(1) Dynamic changes of fractures with different types of proppants during pore pressure decreasing

Different types of proppants have different compressive strengths and different embedment magnitudes, resulting in different laws of changes of fracture widths, porosity, permeability and fracture flow conductivity with different closing pressures (Guo and Zhang 2011; Zou et al. 2012; Wen et al. 2005).

Three types of proppants, like ceramist, quartz sand and resin sand, were used to test their conductivity at different

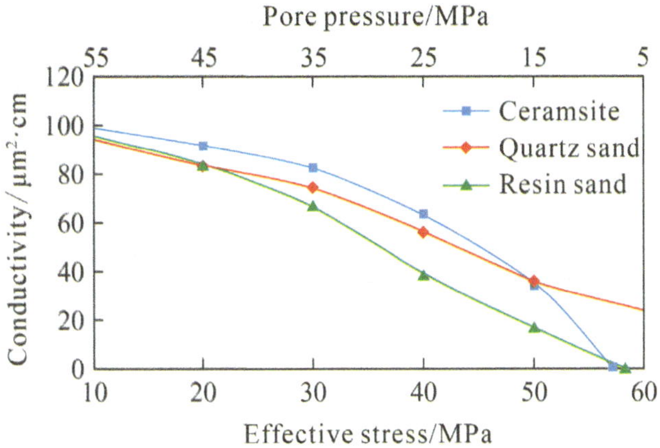

Fig. 7.22 Change of conductivity with different types of proppant at closing pressure

closing pressures (as shown in Fig. 7.22). The experimental results show that:

① The change laws of proppant conductivity changing with closure pressure are different between different types of proppants. Among these three types of proppants, the compressive strength of ceramics is the best, followed by resin sands. The quartz sands have the lowest compressive strength and are the most easily to be broken. The proppant crushing causes the support ability of proppants to become worse, and hydraulic fractures deform to close. Therefore, the magnitude of quartz sands conductivity changing with the closure pressure is highest.

② After the proppants are broken, affected by compaction, the proppants are compacted more than before. The ceramic grains (rigidity, almost no crushing) have a strong compressive ability so they deform slightly. However, the quartz sands have a weak compressive ability so their plastic deformation is large (broken more), resulting in decrease of their porosity and permeability.

③ The different magnitudes of proppants being embedded into fracture walls are different between different types of proppants. Among these three types of proppants, ceramic garnishment is easily to be embedded into a fracture wall, followed by quartz sands. Resin sands are the most difficult to be embedded into a fracture wall. The reason of this phenomenon is that the proppants with higher compressive strength is more difficulty to be broken and easier to be embedded into surface walls and cause fractures to close. The decreases in fracture width, the porosity and permeability finally result in decrease in fracture conductivity.

④ In the early stage (low closure pressure), proppants are more easily broken, and decreases of their porosity, permeability and conductivity are higher. In the later stage (high closure pressure), the proppants with stronger anti-embedding ability have a higher ultimate conductivity.

(2) Dynamic changes of fractures propped by proppants with different particle sizes during pore pressure decreasing

The proppants with different mesh sizes have different particle sizes. At the initial stage (under a condition of low closure pressure), the propped fractures with those proppants have a huge difference in their flow conductivities, and their crushing rates under closure pressure are also different, resulting in different laws of these fractures changing with closure pressure (Jin et al. 2007; Wang et al. 2005). Experiments were conducted under different closure pressures to measure conductivities of fractures with the same width. Those fractures were propped by the same type of proppants with three different mesh sizes. The experimental results are shown in Fig. 7.23.

It can be seen from Fig. 7.23 that the above fractures have the same porosity but different permeability, so their conductivities are also different. The laws of conductivity changing with closure pressure are different between fractures with different mesh size proppants:

① Under the effect of closure pressure, the proppant particles are broken and fail to support fractures, causing fractures to close. Large size proppants are easily to be broken and their particle sizes after proppant crushing are similar, causing a fracture size decrease.

② After the proppant crushing, the crushed particles at different scales are compacted, resulting in decreases in fracture width, porosity, permeability, and conductivity.

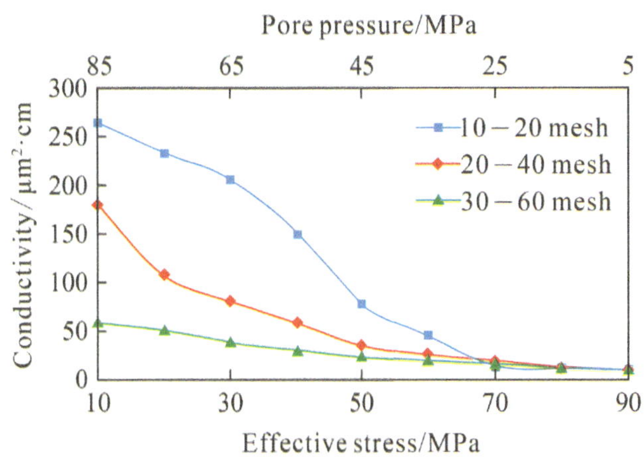

Fig. 7.23 Change of conductivity of different mesh size proppant under different closure stress

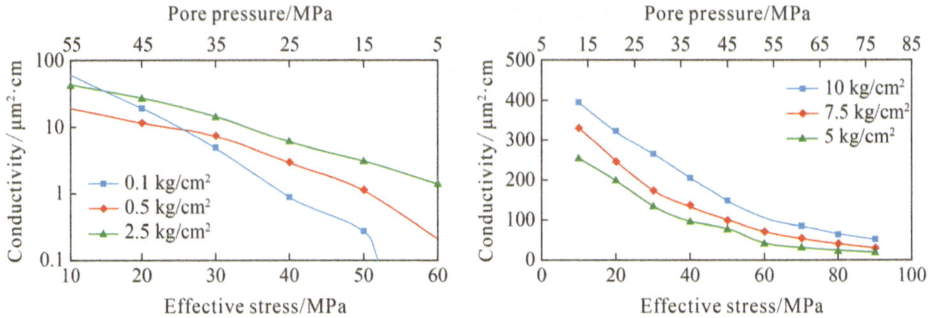

(a) Variation of conductivity with different sanding methods and sand concentration

(b) Variation of conductivity with the same sanding method and different sand concentrations

Fig. 7.24 Change of fracture conductivity with different sand concentration

With the increase of closure pressure, proppants are gradually broken and pores are filled by fragments of crushed proppants. In this case, the advantages of large size proppant gradually disappear. The conductivity gap between large size proppants and small size proppants gradually reduces and tends to be disappeared.

(3) Dynamic changes of fractures propped by proppants with different concentrations during pore pressure decreasing

Different proppant concentrations affect the sanding method and the number of sanding layers of proppants in fractures, and then affect width, porosity, permeability, and conductivity of fractures (Zhang et al. 2008).

Experiments were conducted under different closure pressures to measure conductivities of propped fractures. Those fractures were propped by the same proppants with three different concentrations. The experimental results are shown in Fig. 7.24.

From Fig. 7.24, it can be seen that fractures propped by proppants with different concentrations have different porosity and permeability, so their conductivities are also different. The laws of conductivity changing with closure pressure are different between fractures propped by proppants with different concentrations:

① The partial monolayer sanding props fractures at partial surfaces. Fracture surfaces without proppants have large porosity and permeability so the conductivity of the whole fractures is also large in the early stage (low closure pressure). However, the partial monolayer sanding has few support sites in fractures, resulting in low compressive strength of fractures. In the later stage (high closure pressure), magnitudes of proppant crushing, embedment and fracture closure deformation are strong, so these fractures close first.

② The full monolayer sanding props the whole fractures and has a lot of support sites in fractures. It makes the compressive strength of fractures high, resulting in a low magnitude of fracture closure deformation. In the early stage (low closure pressure), porosity, permeability and conductivity of fractures propped by the full monolayer sanding is lower than that of fractures propped by the spatial monolayer sanding. However, in the later stage (high closure pressure), the conductivity of fractures under full monolayer sanding is higher than that of fractures under partial monolayer sanding.

③ The full multilayer sanding has the same supporting ability, compressive strength, porosity and permeability with the full monolayer sanding. However, the width and conductivity of fractures propped by full multilayer sanding is higher than those of fractures propped by full monolayer sanding. The deformation law of fractures propped by full multilayer sanding is similar to that of fractures proppants by full monolayer sanding.

④ When all the fractures propped by full multilayer sanding, their porosity and permeability are the same while their widths are different so their conductivity are also different. In the full multilayer sanding pattern, the higher proppant concentration leads to the more numbers of sanding layers, and higher conductivity of the propped fractures. However, with the increase of closure pressure, laws of fracture closure deformation and conductivity change of all those fractures are similar.

4. Comparison of dynamic change characteristics between different media with pores/fractures during pore pressure decreasing

During production of tight reservoirs, for different media with pores/fractures, their permeability changes with pore pressure. These permeability changes with pore pressure were compared and illustrated in Fig. 7.25. Through this figure, it can be seen that:

① The stress sensitivity of matrix pores in tight reservoirs is the lowest, and with pore pressure decrease, the change magnitude of permeability is low.

② For natural fractures filled with cement in tight sandstone reservoirs, under the effect of effective overburden pressure, with pore pressure decreasing, the

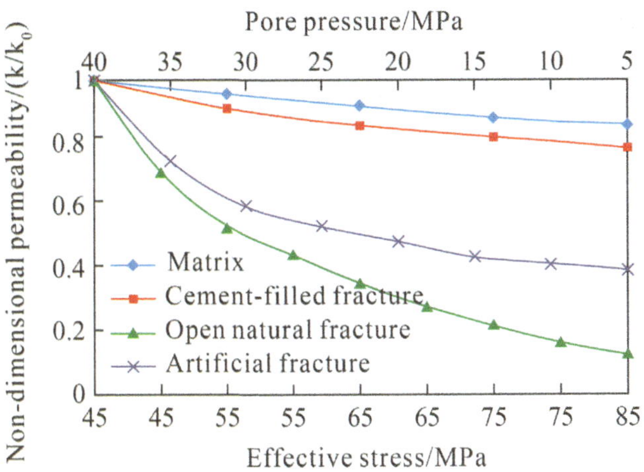

Fig. 7.25 Comparison of stress sensitivity in different media with pores/fractures during pressure decrease

deformation magnitudes are low and the stress sensitivity is relatively low.

③ The hydraulic fractures are propped by proppants. Applied by lateral closure stress, the proppants deform and are easily broken to fail to support fractures, further enhancing the deformation or closure of hydraulic fractures. Overall, the deformation magnitude of hydraulic fractures is strong. Its magnitude of stress sensitivity falls between ones of cement-filled fractures and open natural fractures.

④ For open natural fractures in tight sandstone reservoirs, when they are subjected to vertical closure stress to deform or close, the magnitudes of deformation and stress sensitivity are the largest among all these four kinds of media.

7.1.4.3 Dynamic Changes of Different Media During Pore Pressure Increasing and Decreasing

For different ways of tight reservoir development, the sequences of pore pressure increasing and decreasing are different, and the dynamic changes of media in reservoirs are quite different.

When water injection or gas injection is used to supplement energy after oil well depletion production, a flow medium in the reservoir experiences a process of pore pressure decreasing followed by a process of pore pressure increasing. The dynamic change of a flow medium is shown in Fig. 7.26a. During depletion production, the pore pressure decreases while the effective stress increases, and, therefore, the permeability of the flow medium decreases. After depletion production, the water/gas injection is used to supplement energy. In this case, the pore pressure increases while the effective stress decreases, which causes the matrix permeability to gradually recover. However, due to the generation of plastic deformation of a flow medium during a pore pressure decrease, the permeability cannot recover its initial value when the pore pressure recovers.

When hydraulic fracturing or early water/gas injection is used for production, a flow medium in the reservoir experiences a process of a pore pressure increase followed by a process of a pore pressure decrease. The dynamic change of the flow medium is shown in Fig. 7.26b. During fracturing or early water/gas injection, the pore pressure increases while the effective stress decreases, and, therefore, the permeability of the flow medium increases. During the production, the pore pressure decreases while the effective stress increases, and, therefore, the permeability of the flow medium gradually recovers. There are both plastic deformation and elastic deformation during the process of increasing and then decreasing in pore pressure in tight reservoirs.

7.2 Coupled Flow-Geomechanics Dynamic Simulation for Multiple Media with Different Scales Pores and Fractures

According to the effective stress principle of multiple media and the deformation mechanism and laws of multiple media with different scales pores and fractures, dynamic models of geometry and properties of matrix pores and hydraulic/natural fractures during a fracturing-injection-production process were built, and models of transmissibility between different media and well indices between wellbores and media were also established. The flow-geomechanics coupled dynamic simulation for this multiple media was developed.

7.2.1 Pressure-Deformation Law of Multiple Media with Different Scales Pores and Fractures

7.2.1.1 Pressure-Strain Law of Multiple Media During Fracturing and Injection Process

The traditional stress-strain law describes the relationship between the effective stress applied at rock skeleton and the deformation of rock skeleton. During the process of fracturing, injection and production in tight oil reservoirs, the magnitude and thresholds of dynamic changes of geometric characteristics and physical parameters of different media with pores/fractures are directly related to the formation pressure. Based on the stress-strain relationship and the effective stress principle of unconventional tight reservoirs, the relationship between formation pressure and media deformation was established (as shown in Fig. 7.27).

Fig. 7.26 Dynamic change of flow medium during different development processes

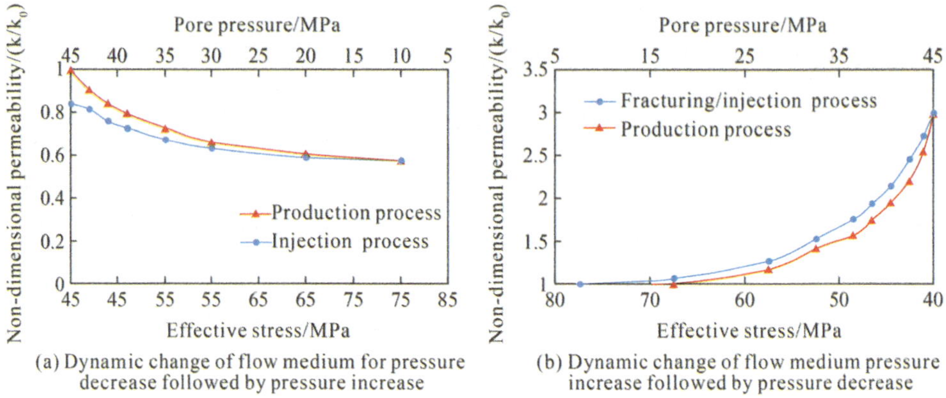

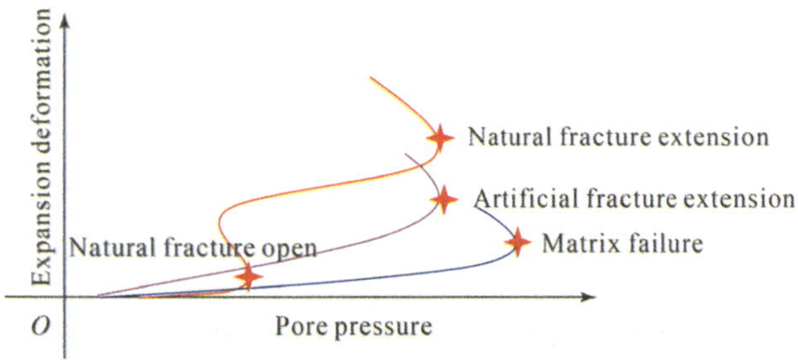

Fig. 7.27 Law between stress and deformation of media with pores/fractures during pressure increase process

During processes of fracturing and injection, because of large amounts of a fracturing fluid, proppants, or water flow into reservoirs, pore pressure increases while the effective stress applied at the rock skeleton decreases. The volume of matrix pore increases in the original state. Before the pore pressure reaches the breakdown pressure of matrix, the deformation of matrix pores is elastic. When the breakdown pressure is reached, matrix pores generate plastic failure. Although the pore pressure decreases, the pore deformation keeps increasing. After hydraulic fractures are generated, the fracture deformation further increases with the increase of pressure in fractures. When the extension pressure is reached, the rock at the tips of hydraulic fractures is broken, causing hydraulic fractures to extend forward. In this case, the pressure in hydraulic fractures decreases while the deformation obviously increases.

When the pressure in natural fractures is low, the fractures are closed, and the expansion deformation is minor. When the pressure increases to the initiation pressure of natural fractures, the fractures reopen. Although the pressure in natural fractures decreases, the natural fracture deformation increases sharply. With the increase of pressure in natural fractures, both the width and expansion deformation of natural fractures increases. When the extension pressure of natural fracture is reached, the rock at the tips of natural fractures is broken and natural fractures extend forward. The pressure in fractures decreases while the deformation significantly increases.

7.2.1.2 Pressure-Strain Law of Multiple Media During Production

During flowback and production, with the production of fracturing fluids and oil/gas, the pore pressure in matrix decreases while the effective stress applied to the rock skeleton increases. Initially, the pores inside matrix generate compression deformation and a volume decrease. After the pore pressure reaches the yield pressure of the matrix, the pores inside the matrix generate plastic compression deformation, and the compression deformation rate accelerates significantly.

Because there is no proppant to prop natural fractures, when the pressure decreases, the fracture closure speed is quite fast, and the volumetric compression magnitude is large. When the pressure decreases to the closure pressure, the natural fractures close and start to be compacted. The deformation magnitude of natural fractures decreases with a pressure decrease (as shown in Fig. 7.28).

Initially, the proppants in hydraulic fractures are in suspension. When the pressure decreases, the magnitude of fracture closure deformation is large. When the closure

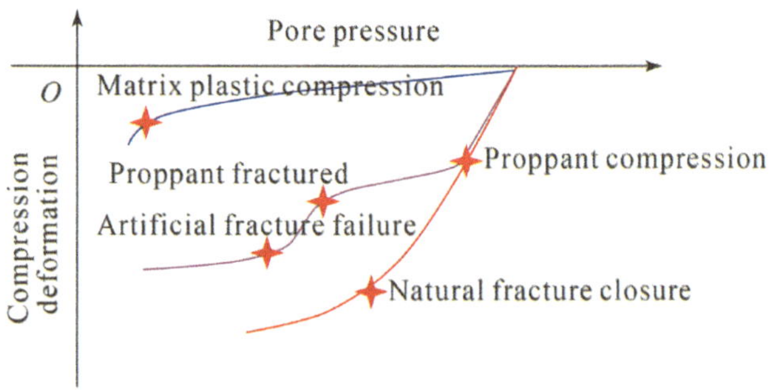

Fig. 7.28 Law between stress and deformation of media with pores/fractures during pressure decrease process

pressure of hydraulic fracture is reached, the proppants start to be compacted and generate elastic deformation, leading to a decrease in fracture closure speed. When the breakdown pressure of proppants is reached, the proppants begin to generate plastic deformation. In this case the fractures close with a high speed and the magnitude of their compression deformation obviously increases. When the pressure further decreases to reach the failure pressure of hydraulic fractures, the proppants embed into fractures and fail to prop fractures, causing the generation of closure deformation of hydraulic fractures.

7.2.2 Dynamic Models of Geometric and Physical Parameters for Multiple Media in Fracturing-Injection-Production Process

7.2.2.1 Fracturing

1. Dynamic model of matrix porosity
(1) Dynamic change of matrix pores

During fracturing, the pore pressure in matrix increases rapidly with the injection of a large amount of fracturing fluids. Under the influence of fluid expansion, matrix pores generate expansion deformation, and pore throat radii gradually increase. When the pore pressure is lower than the yield pressure, the matrix pores generate elastic expansion deformation, and the radius of pore throats inside matrix linearly changes with the pore pressure. This deformation process is reversible. When the pore pressure is greater than the yield pressure, the rock grains or cements are broken, the matrix pores begin to generate plastic deformation, and the radius of pore throats inside matrix changes with the pore pressure. This deformation process is irreversible (Fig. 7.29a).

(2) Dynamic models of geometric and physical parameters

According to the dynamic change mechanism of matrix pores during fracturing, dynamic models of a pore radius, porosity and permeability were built by piecewise functions to represent their elastic changes and plastic changes, respectively. The above models can describe the yield pressure of matrix pores, and mechanisms and behaviors of their elastic and plastic deformation.

Elastic deformation: During fracturing, the amount of fracturing fluids injected into reservoirs is large, the matrix pores expand rapidly, the elastic deformation is fast, and the time to reach the yield pressure is short.

Plastic deformation: During fracturing, the construction pressure is high, the matrix pores generate obvious deformation, and the matrix pores mainly generate plastic deformation.

The dynamic models of a pore throat radius, porosity and permeability during fracturing are shown in Table 7.2.

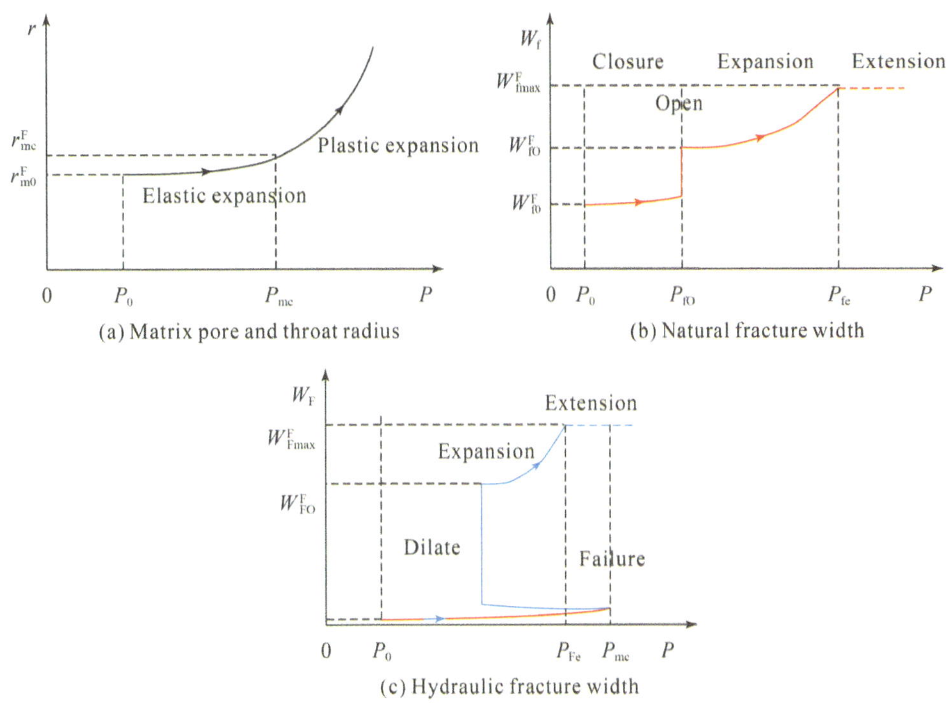

Fig. 7.29 Variations of properties of media with pores/fractures during fracturing

Table 7.2 Dynamic change model of parameters during the fracturing process

	Matrix pores	Natural fractures	Hydraulic fractures
Pore and throat radius/fracture width	$\begin{cases} r_m^F = r_{m0}^F & P = P_0 \\ r_m^F = r_{m0}^F e^{a_E^F(P-P_0)} & P_0 < P < P_{mc} \\ r_m^F = r_{mc}^F & P = P_{mc} \\ r_m^F = r_{mc}^F e^{a_P^F(P-P_{mc})} & P_{mc} < P \end{cases}$	$\begin{cases} W_f^F = W_{f0}^F e^{a_f^F(P-P_0)} & P_0 < P < P_{fO} \\ W_f^F = W_{fO}^F & P = P_{fO} \\ W_f^F = W_{fO}^F e^{a_f^F(P-P_{fO})} & P_{fO} < P < P_{fe} \\ W_f^F = W_{fmax}^F & P > P_{fe} \end{cases}$	$\begin{cases} W_F^F = r_{m0}^F e^{a_E^F(P-P_0)} & P_0 < P < P_{mc} \\ W_F^F = W_{FO}^F & P = P_{mc} \\ W_F^F = W_{FO}^F e^{a_F^F(P-P_{FO})} & P_{mc} < P < P_{Fe} \\ W_F^F = W_{Fmax}^F & P > P_{Fe} \end{cases}$
Porosity	$\begin{cases} \phi_m^F = \phi_{m0}^F & P = P_0 \\ \phi_m^F = \phi_{m0}^F e^{b_E^F(P-P_0)} & P_0 < P < P_{mc} \\ \phi_m^F = \phi_{mc}^F & P = P_{mc} \\ \phi_m^F = \phi_{mc}^F e^{b_P^F(P-P_{mc})} & P_{mc} < P \end{cases}$	$\begin{cases} \phi_f^F = \phi_{f0}^F e^{b_f^F(P-P_0)} & P_0 < P < P_{fO} \\ \phi_f^F = \phi_{fO}^F & P = P_{fO} \\ \phi_f^F = \phi_{fO}^F e^{b_f^F(P-P_{fO})} & P_{fO} < P < P_{fe} \\ \phi_f^F = \phi_{fmax}^F & P > P_{fe} \end{cases}$	$\begin{cases} \phi_F^F = \phi_{m0}^F e^{b_E^F(P-P_0)} & P_0 < P < P_{mc} \\ \phi_F^F = \phi_{FO}^F & P = P_{mc} \\ \phi_F^F = \phi_{FO}^F e^{b_F^F(P-P_{FO})} & P_{mc} < P < P_{Fe} \\ \phi_F^F = \phi_{Fmax}^F & P > P_{Fe} \end{cases}$
Permeability	$\begin{cases} K_m^F = K_{m0}^F & P = P_0 \\ K_m^F = K_{m0}^F e^{c_E^F(P-P_0)} & P_0 < P < P_{mc} \\ K_m^F = K_{mc}^F & P = P_{mc} \\ K_m^F = K_{mc}^F e^{c_P^F(P-P_{mc})} & P_{mc} < P \end{cases}$	$\begin{cases} K_f^F = K_{f0}^F e^{c_f^F(P-P_0)} & P_0 < P < P_{fO} \\ K_f^F = K_{fO}^F & P = P_{fO} \\ K_f^F = K_{fO}^F e^{c_f^F(P-P_{fO})} & P_{fO} < P < P_{fe} \\ K_f^F = K_{fmax}^F & P > P_{fe} \end{cases}$	$\begin{cases} K_F^F = K_{m0}^F e^{c_E^F(P-P_0)} & P_0 < P < P_{mc} \\ K_F^F = K_{FO}^F & P = P_{mc} \\ K_F^F = K_{FO}^F e^{c_F^F(P-P_{FO})} & P_{mc} < P < P_{Fe} \\ K_F^F = K_{Fmax}^F & P > P_{Fe} \end{cases}$

2. Dynamic model of natural fractures

(1) Dynamic change of natural fractures

During fracturing, with a large amount of fracturing fluids injected into formations within a short period, the deformation of natural fractures experiences the following four stages: the initial closure stage, the natural fractures are closed under original conditions of the formation; when the pressure in natural fractures rises to the natural fracture initiation pressure, the natural fractures reopens, and their width increases instantaneously; with the pressure in the natural fractures continues to increase, the natural fractures expand and their width gradually increases; when the pressure in these fractures reaches the extension pressure of their tips, they extend forward, and their length increases (Fig. 7.29b).

(2) Dynamic models of geometric and physical parameters

Based on the dynamic change mechanism of natural fractures during fracturing, dynamic models of the width, porosity and permeability were established for natural fractures in stages of closure, initiation, expansion and extension, respectively (as shown in Table 7.2). These models describe two critical pressures of natural fractures during fracturing: initiation pressure and extension pressure. Meanwhile, these models also describe nonlinear changes of geometric and physical parameters of natural fractures in these four stages.

Initial closure stage: for natural fractures with different cementation degree they have different initial opening.

Initiation stage: during fracturing, a large amount of fracturing fluids is injected into reservoirs in a short period, and the natural fractures are initiated quite fast.

Expansion stage: the viscosity of fracturing fluids is large, natural fractures are rapidly expanded, and the fracture width is large.

Extension stage: during fracturing, the amount of fracturing fluids injected into reservoirs is large, the fracture width increases rapidly, and the extension distance is quite long.

3. Dynamic model of hydraulic fractures

(1) Dynamic change of hydraulic fractures

During fracturing, changes of hydraulic fractures mainly experience in the following four stages: in the stage of pre-crushing of hydraulic fractures, the matrix pores expand rapidly with the injection of a fracturing fluid; in the stage of hydraulic fracture initiation, when the pore pressure increases to the rock breakdown pressure, rocks are broken to generate hydraulic fractures; in the stage of hydraulic fracture expansion, as the pressure in hydraulic fractures keeps increasing, the width of hydraulic fractures rapidly increases; in the stage of hydraulic fracture extension, when the pressure in hydraulic fractures reaches the extension pressure, the rock at tips of hydraulic fractures are broken, resulting in growth of hydraulic fractures and an increase in their length (as shown in Fig. 7.29c).

(2) Dynamic models of geometric and physical parameters

According to the dynamic change mechanism of hydraulic fractures during fracturing, dynamic models of the width, porosity and permeability were established for hydraulic fractures in stages of pre-crushing, initiation, expansion and extension, respectively (as shown in Table 7.2). These models describe two critical pressures of hydraulic fractures during fracturing: initiation pressure and extension pressure. Meanwhile, these models also describe change of geometric and physical parameters of hydraulic fractures in stages of pre-crushing, generation, expansion and extension.

The stage of pre-crushing: before hydraulic fractures being generated, the matrix pores experience expansion deformation, and hydraulic fractures do not have the initial opening.

The stage of initiation: after initiation of hydraulic fractures, the fracture width and permeability increase rapidly.

In stage of expansion: because the viscosity of fracturing fluids is quite large, hydraulic fractures rapidly expand. The width of hydraulic fractures is large, which is usually larger than that of natural fractures.

In stage of extension: a large amount of fracturing fluids is quickly injected into reservoirs. The width of hydraulic fractures increases quickly, and their length is large. The extension distance of hydraulic fractures is greater than that of natural fractures.

7.2.2.2 Injection

1. Dynamic model of matrix pores

(1) Dynamic change of matrix pores

During injection, the pore pressure in matrix increases slowly with the fluid injection. Matrix pores first generate elastic expansion deformation, and pore throat radii gradually increase. When the pore pressure is larger than the yield pressure, the rock grains and cements are broken so the matrix pores generate plastic deformation, and the change magnitude of matrix pore throat radius increasing with the pore pressure enlarges a little bit (Fig. 7.30a).

(2) Dynamic models of geometric and physical parameters

According to the dynamic change mechanism of matrix pores during injection, dynamic models of a pore radius, porosity and permeability were built for stages of elastic and plastic

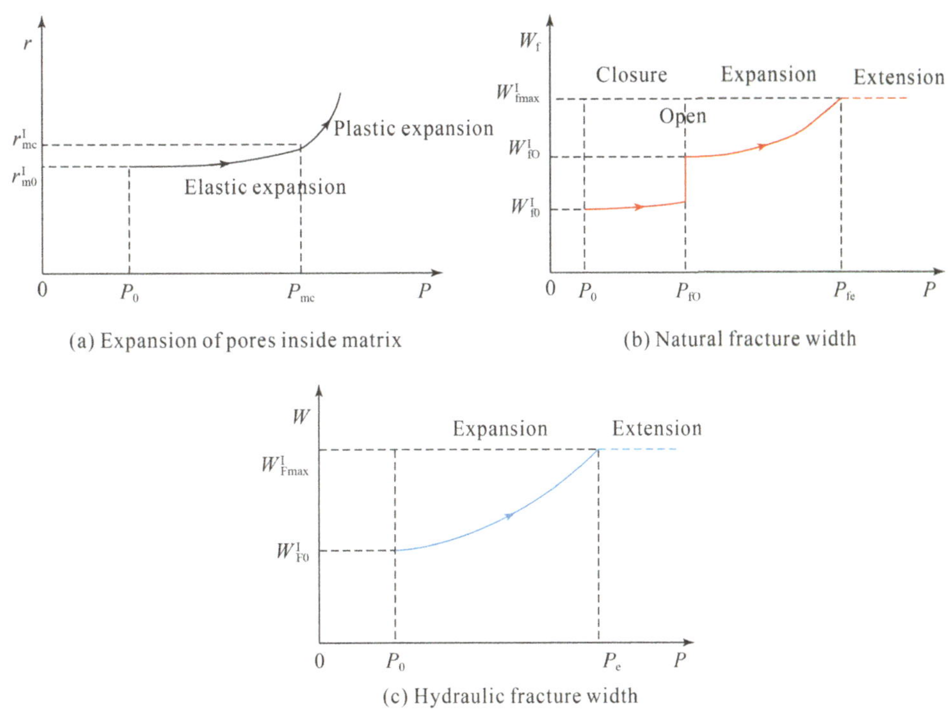

Fig. 7.30 Variations of the property parameters of the media with pores/fractures during the injection process

expansion of matrix pores, respectively. The above models can describe the yield pressure of matrix pores, and mechanisms and behaviors of their elastic and plastic deformation.

Elastic deformation: compared with the fluid amount used in fracturing, during injection, the amount of fluids injected into reservoirs are relatively low, the matrix pores expand gradually, the elastic deformation is slow, and the time to reach the yield pressure is long.

Plastic deformation: compared with the treatment pressure during fracturing, during injection, the treatment pressure is low, the deformation of matrix pores is quite small, and the injection mainly generates elastic deformation of matrix pores (as shown in Table 7.3).

The dynamic models of a pore throat radius, porosity, and permeability during injection are shown in Table 7.3.

2. Dynamic model of natural fracture

(1) Dynamic change of natural fractures

During injection, with gradual injection of fluids into reservoirs, the deformation of natural fractures mainly experiences four stages: the first one is the initial closure stage, and the natural fractures are closed under original reservoir conditions; the second one is the initiation stage. When the pressure in a natural fracture increases to the natural fracture initiation pressure, the natural fracture reopens, and its width increases instantaneously; the third one is the expansion stage. With the pressure in the natural fractures continues to increase, the natural fractures expand, and their width gradually increases; the fourth stage is the extension stage. When the pressure in these fractures reaches the extension pressure of their tips, they extend forward, and their length increases (as shown in Fig. 7.30b).

(2) Dynamic models of geometric and physical parameters

Based on the dynamic change mechanism of natural fractures during injection, dynamic models of the width, porosity and permeability were established for natural fractures in stages of closure, initiation, expansion and extension, respectively (as shown in Table 7.2). These models describe two critical pressures of natural fractures during injection: initiation pressure and extension pressure. Meanwhile, these models also describe nonlinear changes of geometric and physical parameters of natural fractures in these four stages:

Initiation stage: according to cementation degrees, natural fractures have different initial opening.

Initiation stage: compared with characteristics of natural fractures during fracturing, during injection, the injection period is long, the injection rate is quite low and the natural fractures are initiated quite slowly.

Expansion stage: compared with characteristics of natural fractures during fracturing, during injection, the viscosity of injection fluids is quite low, natural fractures are slowly expanded, and the fracture width is minor.

Extension stage: compared with characteristics of natural fractures during fracturing, during injection, the amount of fluids injected into reservoirs is low, the fracture width increases slowly, and the extension distance is quite short.

3. Dynamic model of hydraulic fracture

(1) Dynamic change of hydraulic fractures

During injection, changes of hydraulic fractures mainly occur in the following two stages: in the stage of hydraulic fracture expansion, with the gradual fluid flow into hydraulic fractures, the pressure in hydraulic fractures keeps increasing, the width of hydraulic fracture slowly increases; in the stage of hydraulic fracture extension, when the pressure in hydraulic fractures reaches the extension pressure, the rock at tips of hydraulic fractures are broken, resulting in growth of hydraulic fractures and an increase in their length (as shown in Fig. 7.30c).

(2) Dynamic models of geometric and physical parameters

According to the dynamic change mechanism of hydraulic fractures during injection, dynamic models of the width, porosity and permeability were established for hydraulic fractures in stages of expansion and extension, respectively (as shown in Table 7.3). These models describe the extension pressure of hydraulic fractures during injection. Meanwhile, these models also describe changes of geometric and physical parameters of hydraulic fractures in stages of expansion and extension.

In stage of expansion: compared with characteristics of natural fractures during fracturing, the viscosity of fluids injected into reservoirs is low, hydraulic fractures slowly expand, and the width of hydraulic fractures is small, which is usually larger than that of natural fractures.

In stage of extension: compared with characteristics of natural fractures during fracturing, a minor amount of fluids is slowly injected into reservoirs, the width of hydraulic fractures increases slowly and their length is short. However, the extension distance of hydraulic fractures is greater than that of natural fractures.

Table 7.3 Dynamic change models of properties parameters of injection process

	Matrix pores	Natural fractures	Hydraulic fractures
Pore and throat radius/fracture width	$\begin{cases} r_{\mathrm{m}}^{\mathrm{I}} = r_{\mathrm{m0}}^{\mathrm{I}} & P = P_0 \\ r_{\mathrm{m}}^{\mathrm{I}} = r_{\mathrm{m0}}^{\mathrm{I}} \mathrm{e}^{a_{\mathrm{E}}^{\mathrm{I}}(P-P_0)} & P_0 < P < P_{\mathrm{mc}} \\ r_{\mathrm{m}}^{\mathrm{I}} = r_{\mathrm{mc}}^{\mathrm{I}} & P = P_{\mathrm{mc}} \\ r_{\mathrm{m}}^{\mathrm{I}} = r_{\mathrm{mc}}^{\mathrm{I}} \mathrm{e}^{a_{\mathrm{P}}^{\mathrm{I}}(P-P_{\mathrm{mc}})} & P_{\mathrm{mc}} < P \end{cases}$	$\begin{cases} W_{\mathrm{f}}^{\mathrm{I}} = W_{\mathrm{f0}}^{\mathrm{I}} \mathrm{e}^{a_{\mathrm{f}}^{\mathrm{I}}(P-P_0)} & P_0 \le P < P_{\mathrm{fO}} \\ W_{\mathrm{f}}^{\mathrm{I}} = W_{\mathrm{fO}}^{\mathrm{I}} & P = P_{\mathrm{fO}} \\ W_{\mathrm{f}}^{\mathrm{I}} = W_{\mathrm{fO}}^{\mathrm{I}} \mathrm{e}^{a_{\mathrm{f}}^{\mathrm{I}}(P-P_{\mathrm{fO}})} & P_{\mathrm{fO}} < P < P_{\mathrm{e}} \\ W_{\mathrm{f}}^{\mathrm{I}} = W_{\mathrm{fmax}}^{\mathrm{I}} & P \ge P_{\mathrm{e}} \end{cases}$	$\begin{cases} W_{\mathrm{F}}^{\mathrm{I}} = W_{\mathrm{F0}}^{\mathrm{I}} & P = P_0 \\ W_{\mathrm{F}}^{\mathrm{I}} = W_{\mathrm{FO}}^{\mathrm{I}} \mathrm{e}^{a_{\mathrm{F}}^{\mathrm{I}}(P-P_0)} & P_0 < P < P_{\mathrm{e}} \\ W_{\mathrm{F}}^{\mathrm{I}} = W_{\mathrm{Fmax}}^{\mathrm{I}} & P \ge P_{\mathrm{e}} \end{cases}$
Porosity	$\begin{cases} \phi_{\mathrm{m}}^{\mathrm{I}} = \phi_{\mathrm{m0}}^{\mathrm{I}} & P = P_0 \\ \phi_{\mathrm{m}}^{\mathrm{I}} = \phi_{\mathrm{m0}}^{\mathrm{I}} \mathrm{e}^{b_{\mathrm{E}}^{\mathrm{I}}(P-P_0)} & P_0 < P < P_{\mathrm{mc}} \\ \phi_{\mathrm{m}}^{\mathrm{I}} = \phi_{\mathrm{mc}}^{\mathrm{I}} & P = P_{\mathrm{mc}} \\ \phi_{\mathrm{m}}^{\mathrm{I}} = \phi_{\mathrm{mc}}^{\mathrm{I}} \mathrm{e}^{b_{\mathrm{P}}^{\mathrm{I}}(P-P_{\mathrm{mc}})} & P_{\mathrm{mc}} < P \end{cases}$	$\begin{cases} \phi_{\mathrm{f}}^{\mathrm{I}} = \phi_{\mathrm{f0}}^{\mathrm{I}} \mathrm{e}^{b_{\mathrm{f}}^{\mathrm{I}}(P-P_0)} & P_0 < P < P_{\mathrm{fO}} \\ \phi_{\mathrm{f}}^{\mathrm{I}} = \phi_{\mathrm{fO}}^{\mathrm{I}} & P = P_{\mathrm{fO}} \\ \phi_{\mathrm{f}}^{\mathrm{I}} = \phi_{\mathrm{fO}}^{\mathrm{I}} \mathrm{e}^{b_{\mathrm{f}}^{\mathrm{I}}(P-P_{\mathrm{fO}})} & P_{\mathrm{fO}} < P < P_{\mathrm{e}} \\ \phi_{\mathrm{f}}^{\mathrm{I}} = \phi_{\mathrm{fmax}}^{\mathrm{I}} & P > P_{\mathrm{e}} \end{cases}$	$\begin{cases} \phi_{\mathrm{F}}^{\mathrm{I}} = \phi_{\mathrm{F0}}^{\mathrm{I}} & P = P_0 \\ \phi_{\mathrm{F}}^{\mathrm{I}} = \phi_{\mathrm{FO}}^{\mathrm{I}} \mathrm{e}^{b_{\mathrm{F}}^{\mathrm{I}}(P-P_{\mathrm{F0}})} & P_0 < P < P_{\mathrm{e}} \\ \phi_{\mathrm{F}}^{\mathrm{I}} = \phi_{\mathrm{Fmax}}^{\mathrm{I}} & P > P_{\mathrm{e}} \end{cases}$
Permeability	$\begin{cases} K_{\mathrm{m}}^{\mathrm{I}} = K_{\mathrm{m0}}^{\mathrm{I}} & P = P_0 \\ K_{\mathrm{m}}^{\mathrm{I}} = K_{\mathrm{m0}}^{\mathrm{I}} \mathrm{e}^{c_{\mathrm{E}}^{\mathrm{I}}(P-P_0)} & P_0 < P < P_{\mathrm{mc}} \\ K_{\mathrm{m}}^{\mathrm{I}} = K_{\mathrm{mc}}^{\mathrm{I}} & P = P_{\mathrm{mc}} \\ K_{\mathrm{m}}^{\mathrm{I}} = K_{\mathrm{mc}}^{\mathrm{I}} \mathrm{e}^{c_{\mathrm{P}}^{\mathrm{I}}(P-P_{\mathrm{mc}})} & P_{\mathrm{mc}} < P \end{cases}$	$\begin{cases} K_{\mathrm{f}}^{\mathrm{I}} = K_{\mathrm{f0}}^{\mathrm{I}} \mathrm{e}^{c_{\mathrm{f}}^{\mathrm{I}}(P-P_0)} & P_0 \le P < P_{\mathrm{fO}} \\ K_{\mathrm{f}}^{\mathrm{I}} = K_{\mathrm{fO}}^{\mathrm{I}} & P = P_{\mathrm{fO}} \\ K_{\mathrm{f}}^{\mathrm{I}} = K_{\mathrm{fO}}^{\mathrm{I}} \mathrm{e}^{c_{\mathrm{f}}^{\mathrm{I}}(P-P_{\mathrm{fO}})} & P_{\mathrm{fO}} < P < P_{\mathrm{e}} \\ K_{\mathrm{f}}^{\mathrm{I}} = K_{\mathrm{fmax}}^{\mathrm{I}} & P \ge P_{\mathrm{e}} \end{cases}$	$\begin{cases} K_{\mathrm{F}}^{\mathrm{I}} = K_{\mathrm{F0}}^{\mathrm{I}} & P = P_0 \\ K_{\mathrm{F}}^{\mathrm{I}} = K_{\mathrm{FO}}^{\mathrm{I}} \mathrm{e}^{c_{\mathrm{F}}^{\mathrm{I}}(P-P_0)} & P_0 < P < P_{\mathrm{e}} \\ K_{\mathrm{F}}^{\mathrm{I}} = K_{\mathrm{Fmax}}^{\mathrm{I}} & P \ge P_{\mathrm{e}} \end{cases}$

7.2.2.3 Production

1. Dynamic model of matrix pore throat

(1) Dynamic change of matrix pores

During production, the matrix deformation in reservoirs mainly occurs in the following two stages: the stage of elastic compression deformation and the stage of plastic compression deformation. The pore pressure in matrix decreases with the production of oil and gas. Under the influence of overburden stress, the rock skeleton gradually generates compression deformation, and the radius of a pore throat decreases with decrease of pore pressure. However, in this stage, the deformation could recover with recovery of pore pressure. When the pore pressure is reached at the yield pressure, the rock is broken and generates plastic deformation due to shear failure. In this stage of plastic deformation, the rate of matrix pore throat radius decreasing with the pore pressures is larger than that in the stage of elastic deformation, and the deformation fails to totally recover with recovery of pore pressure (as shown in Fig. 7.31a).

(2) Dynamic models of geometric and physical parameters

According to the dynamic change mechanism of matrix pores during production, dynamic models of pore radius, porosity and permeability were built for stages of elastic and plastic compression of matrix pores, respectively. The above models can describe the yield pressure of matrix pores, and mechanisms and behaviors of their elastic and plastic deformation.

Elastic deformation: with a decrease in pore pressure, the matrix pores shirk and generate elastic deformation. During production, dynamic changes of matrix pores mainly include elastic compression deformation.

Plastic deformation: with a further decrease in pore pressure, the matrix pores further shrink and generate plastic deformation.

The theoretical models for the pore throat radius, porosity, and permeability of the matrix during production are shown in Table 7.4.

2. Dynamic model of natural fractures

(1) Dynamic change of natural fractures

During production of oil and gas, the dynamic change of natural fractures mainly experiences two stages: the first stage is the one where natural fractures narrow down. In this stage, the pressure in natural fractures is higher than the closure pressure, and the width of natural fractures decreases with pressure decreasing. The second stage is the one where natural fractures close. In this stage, the pressure in natural fractures is less than the closure pressure, and natural fractures close to be inefficient for fluids to flow through (as shown in Fig. 7.31b).

(2) Dynamic models of geometric and physical parameters

Based on the dynamic change mechanism of natural fractures during production, dynamic models of the width, porosity and permeability were established for natural fractures in stages of shrinkage and closure (as shown in Table 7.4). These models describe the closure pressure of natural fractures and deformation mechanisms of natural fractures before and after they close.

Shrinkage stage: with a decrease in pressure in natural fractures, natural fractures narrow down, and their porosity and permeability significantly decrease.

Closure stage: when the pressure in natural fractures is lower than the closure pressure, natural fractures close, and their porosity and permeability slowly decrease.

3. Dynamic model of hydraulic fractures

(1) Dynamic change of hydraulic fractures

During production of oil and gas, the width of hydraulic fractures rapidly decreases. When the pressure in hydraulic fractures is lower than the closure pressure, hydraulic fractures will close. Changes of hydraulic fractures mainly experience in the following four stages: in the stage of hydraulic fracture closure, with the pressure decrease, the width of hydraulic fractures decreases. In this case, the proppants in hydraulic fractures are in suspension; in the stage of elastic compression of proppants, proppants contact with fracture surface walls and the fracture width gradually decreases with the decrease of pressure in fractures. In this case, proppants generate elastic deformation; in the stage of plastic deformation of proppants, the pressure in fractures keeps decreasing. When the effective stress applied to proppant particles is higher than the strength of proppants, proppants will generate plastic deformation; in the stage of hydraulic fracture failure, majority of proppants are broken and embedded into fractures, causing failure of hydraulic fractures (as shown in Fig. 7.31c).

(2) Dynamic models of geometric and physical parameters

According to the dynamic change mechanism of hydraulic fractures during production, dynamic models of the width, porosity and permeability were established for hydraulic fractures in four stages, respectively (as shown in Table 7.3). These models describe the closure pressure of hydraulic fractures, breakdown pressure of proppants and deformation mechanisms of hydraulic fractures before and after closing.

Stage of hydraulic fracture closure: as the pressure in hydraulic fractures decreases, hydraulic fractures without proppants rapidly narrows down, and the porosity and permeability of hydraulic fractures obviously decrease.

Stage of elastic compression of proppants: with the pressure in hydraulic pressure, the proppants in fractures generate elastic deformation, causing the porosity and permeability of fractures to slowly decrease.

Stage of plastic compression of proppants: as the pressure in hydraulic fractures decreases, the proppants in fractures generate plastic deformation, causing the porosity and permeability of fractures to rapidly decrease.

In stage of hydraulic fracture failure: due to failure and embedment of proppants, hydraulic fractures are no longer efficient, and their porosity and permeability slowly decrease.

7.2.2.4 Changes of Physical Properties Under Different Development Modes

The mechanisms and models of changes in physical parameters during fracturing, injection and production processes were introduced in detail in the above sections. However, these processes of fracturing, injection and production are an integrated system rather than three individual processes to make the development of a tight reservoir success. In the following, taking permeability as an example, the dynamic change of permeability at each of the above three processes was introduced and illustrated as a linear change; however, the real (linear or nonlinear) changes of permeability requires experiments to be determined.

1. Changes of physical properties during fracturing and production

The dynamic changes of permeability during fracturing, soaking, flowback, and production are shown in Fig. 7.32a. In this figure, p_i is the initial pressure of the reservoir, p_{mc} is the breakdown pressure of the swelling rock, p'_{mc} means the breakdown pressure of the compressive rock and p_{max} is the maximum treatment pressure during fracturing. The figure shows that at the initial stage of fracturing, with the injection of a fracturing fluid, the reservoir pressure increases, the matrix generates expansion deformation, causing the matrix pore throat to expand. In this case, the matrix permeability slowly increases; when the reservoir pressure reaches the matrix breakdown pressure, the matrix is broken to generate hydraulic fractures.

In this case, the permeability increases rapidly, and it further increases with opening of hydraulic fractures increasing. Followed by the completion of fracturing, the soaking begins and the fracturing fluids in hydraulic fractures are gradually imbibed into the matrix. It causes the decreases of pressure in the hydraulic fractures, the hydraulic fracture width, and the permeability; after the soaking, the flowback of a fracturing fluid begins.

The permeability gradually decreases with the fracture opening decrease; at the early stage of the production, the opening of hydraulic fractures decreases due to elastic fracture deformation, and the permeability of hydraulic fractures further decreases; at the later stage of the production, with the massive production of oil and gas, the proppants are compressed to be broken and embedded, the decrease magnitude of permeability increases, and this decrease is irreversible.

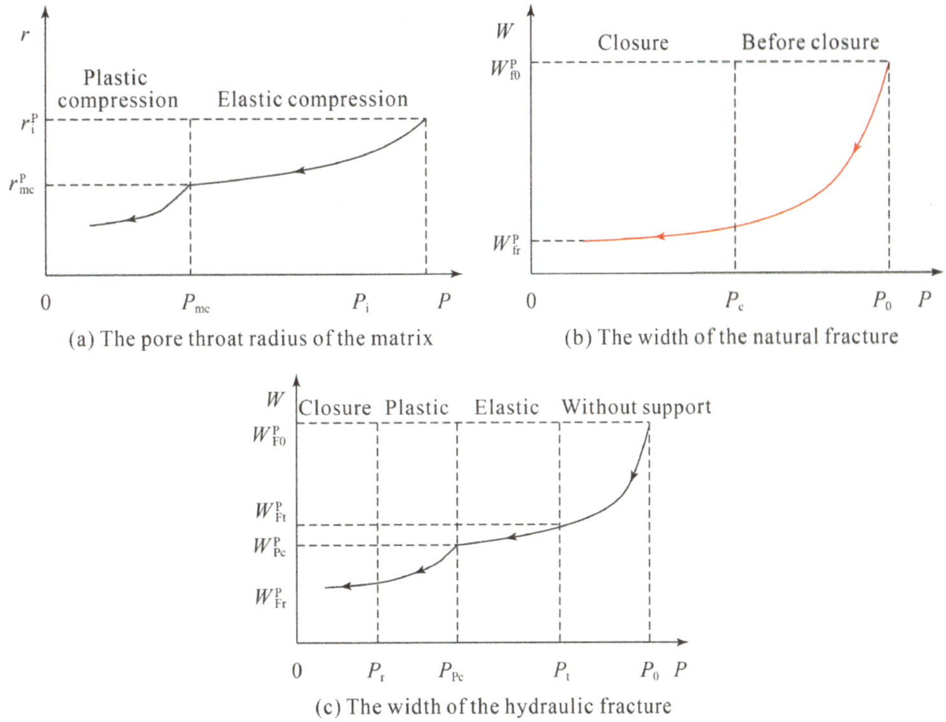

Fig. 7.31 Change of the property parameters of the media with pores/fractures during the production process

2. Changes of physical properties during fracturing, injection, and production processes

The dynamic change of permeability during fracturing, injection and production processes is shown in Fig. 7.32b. It is mainly used to simulate the actual development treatment of water/gas injection to supplement formation energy after the flowback of fracturing fluids. From this figure, it can be seen that after the first stage of fracturing, flowback and production, the permeability of hydraulic fractures decreases significantly; after water/gas injection, the pressure in fractures recovers, the elastic deformation of compressive proppants in the above stage recovers, and the opening of hydraulic fractures increases, causing the elastic recovery of permeability; during the production following the above injection, as the pressure in hydraulic fractures decreases, the fracture opening decreases, causing the permeability to decrease.

3. Changes of physical properties during fracturing and huff-n-puff

Figure 7.32c shows the physical property parameter change after fracturing and huff-n-puff. The change of physical parameters during fracturing is consistent with the previous one. In fact, the huff-n-puff is a process of multiple circles of injection and production. The change in physical parameters during each circle of injection and production is the same as the one during the development mode 2.

7.2.3 Dynamic Model of Transmissibility and Well Index

During fracturing, injection and production processes, fluid injection and production result in the increase or decrease of reservoir pressure, and then multiple media with pores/fractures experience different magnitude of deformation under influence of stress, leading to their physical parameters (porosity, permeability) change dynamically. This change affects the dynamic change of reservoir transmissibility and well indices, and finally affects the dynamic evolutions of fluid flow through porous media and production.

7.2.3.1 Dynamic Model of Transmissibility Between Multiple Media with Pores/Fractures

The dynamic changes of geometric and physical parameters of different media lead to dynamic changes in the transmissibility between media. After discretization, different grids represent different media, and the transmissibility changes dynamically between the following media (as shown in Fig. 7.33): a. transmissibility between matrix grids; b. transmissibility between hydraulic fractures grids; c. transmissibility between natural fractures grids; d. transmissibility between hydraulic fractures and matrix grids; e. transmissibility between natural fractures and matrix grids; f. transmissibility between hydraulic fractures and natural fractures grids.

Table 7.4 The theoretical model for the pore throat radius, porosity, and permeability of the matrix during production

	Matrix pores	Natural fractures	Hydraulic fractures
Pore throat radius/width	$\begin{cases} r_m^P = r_{m0}^P & P = P_0 \\ r_m^P = r_{m0}^P e^{a_E^P(P-P_0)} & P_{mc} < P < P_0 \\ r_m^P = r_{mc}^P & P = P_{mc} \\ r_m^P = r_{mc}^P e^{a_P^P(P-P_{mc})} & P < P_{mc} \end{cases}$	$\begin{cases} W_f^P = W_{f0}^P & P = P_0 \\ W_f^P = W_{f0}^P e^{a_o^P(P-P_0)} & P_c < P < P_0 \\ W_f^P = W_{fr}^P & P = P_c \\ W_f^P = W_{fr}^P e^{a_c^P(P-P_c)} & P < P_c \end{cases}$	$\begin{cases} W_F^P = W_{F0}^P & P = P_0 \\ W_F^P = W_{Fo}^P e^{a_F^P(P-P_0)} & P_t < P < P_0 \\ W_F^P = W_{Ft}^P & P = P_t \\ W_F^P = W_{Ft}^P e^{a_{ProE}^P(P-P_t)} & P_{Pc} < P < P_t \\ W_F^P = W_{Pc}^P & P = P_{Pc} \\ W_F^P = W_{Pc}^P e^{a_{ProP}^P(P-P_{Pc})} & P_r < P < P_{Pc} \\ W_F^P = W_{Fr}^P & P = P_r \\ W_F^P = W_{Fr}^P e^{a_r^P(P-P_r)} & P < P_r \end{cases}$
Porosity	$\begin{cases} \phi_m^P = \phi_{m0}^P & P = P_0 \\ \phi_m^P = \phi_{m0}^P e^{b_E^P(P-P_0)} & P_{mc} < P < P_0 \\ \phi_m^P = \phi_{mc}^P & P = P_{mc} \\ \phi_m^P = \phi_{mc}^P e^{b_P^P(P-P_{mc})} & P < P_{mc} \end{cases}$	$\begin{cases} \phi_f^P = \phi_{f0}^P & P = P_0 \\ \phi_f^P = \phi_{f0}^P e^{b_o^P(P-P_0)} & P_c < P < P_0 \\ \phi_f^P = \phi_{fr}^P & P = P_c \\ \phi_f^P = \phi_{fr}^P e^{b_c^P(P-P_0)} & P < P_c \end{cases}$	$\begin{cases} \phi_F^P = \phi_{F0}^P & P = P_0 \\ \phi_F^P = \phi_{Fo}^P e^{b_F^P(P-P_0)} & P_t < P < P_0 \\ \phi_F^P = \phi_{Ft}^P & P = P_t \\ \phi_F^P = \phi_{Ft}^P e^{b_{ProE}^P(P-P_t)} & P_{Pc} < P < P_t \\ \phi_F^P = \phi_{Pc}^P & P = P_{Pc} \\ \phi_F^P = \phi_{Pc}^P e^{b_{ProP}^P(P-P_{Pc})} & P_r < P < P_{Pc} \\ \phi_F^P = \phi_{Fr}^P & P = P_r \\ \phi_F^P = \phi_{Fr}^P e^{b_r^P(P-P_r)} & P < P_r \end{cases}$
Permeability	$\begin{cases} K_m^P = K_{m0}^P & P = P_0 \\ K_m^P = K_{m0}^P e^{c_E^P(P-P_0)} & P_{mc} < P < P_0 \\ K_m^P = K_{mc}^P & P = P_{mc} \\ K_m^P = K_{mc}^P e^{c_P^P(P-P_{mc})} & P < P_{mc} \end{cases}$	$\begin{cases} K_f^P = K_{f0}^P & P = P_0 \\ K_f^P = K_{f0}^P e^{c_o^P(P-P_0)} & P_c < P < P_0 \\ K_f^P = K_{fr}^P & P = P_c \\ K_f^P = K_{fr}^P e^{c_c^P(P-P_0)} & P < P_c \end{cases}$	$\begin{cases} K_F^P = K_{F0}^P & P = P_0 \\ K_F^P = K_{Fo}^P e^{c_F^P(P-P_0)} & P_t < P < P_0 \\ K_F^P = K_{Ft}^P & P = P_t \\ K_F^P = K_{Ft}^P e^{c_{ProE}^P(P-P_t)} & P_{Pc} < P < P_t \\ K_F^P = K_{Pc}^P & P = P_{Pc} \\ K_F^P = K_{Pc}^P e^{c_{ProP}^P(P-P_{Pc})} & P_r < P < P_{Pc} \\ K_F^P = K_{Fr}^P & P = P_r \\ K_F^P = K_{Fr}^P e^{c_r^P(P-P_r)} & P = P_r \end{cases}$

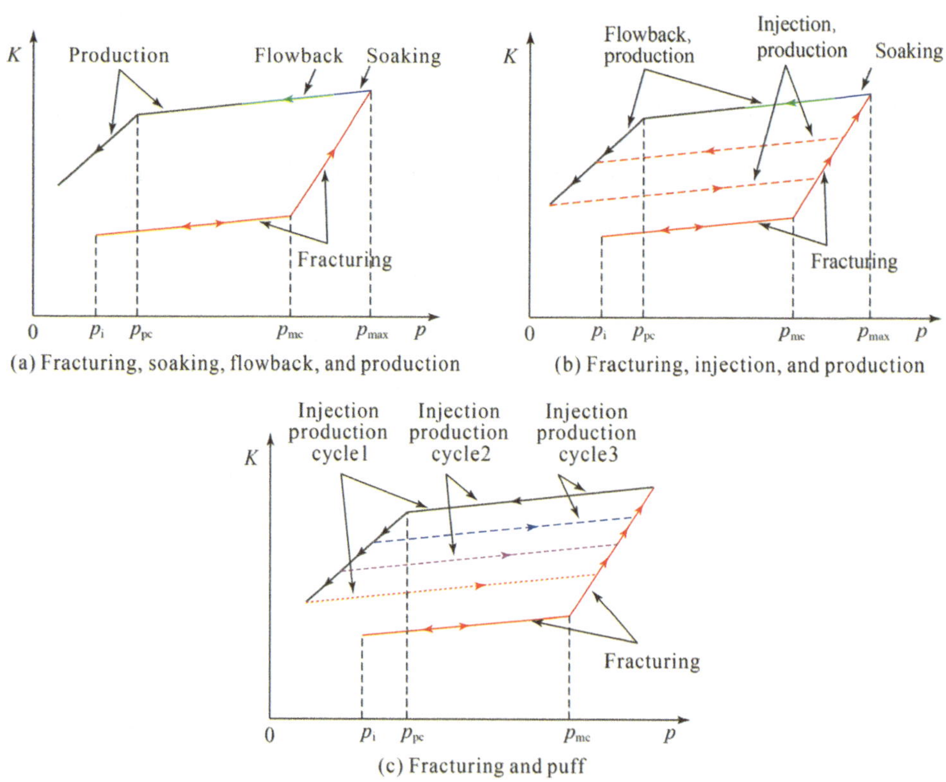

Fig. 7.32 The change of physical properties under different development modes

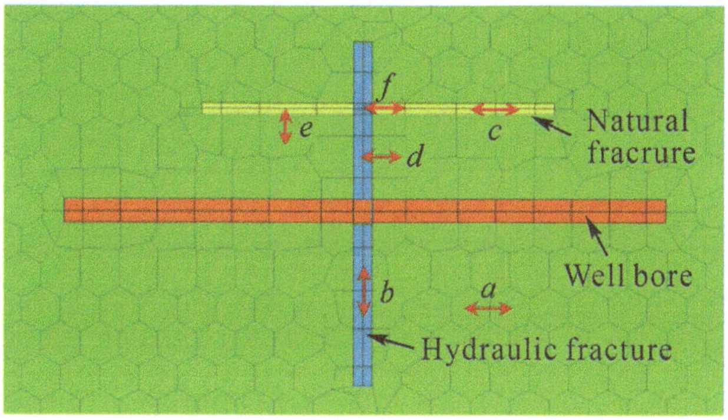

Fig. 7.33 Change in transmissibility between different media grids

Without considering the dynamic changes of media with pores/fractures, the calculation model of transmissibility between adjacent grids is as follows:

$$T_{m,n} = \frac{\alpha_m(K_m) \cdot \alpha_n(K_n)}{\alpha_m(K_m) + \alpha_n(K_n)} \tag{7.14}$$

where, $T_{m,n}$ is the transmissibility between adjacent grids; α_m is the shape factor of the grid m; α_n is the shape factor of the grid n; α_m and α_n are functions of K_m and K_n representing the permeability of the corresponding grid, respectively; m and n are subscripts representing two adjacent grids, and they can be matrix grids, natural fracture grids, or hydraulic fracture grids.

Considering the dynamic changes of media with pores/fractures during the fracturing, injection and production processes, the transmissibility between adjacent grids also changes. In this case, the dynamic model of transmissibility is required to calculate the change in transmissibility:

$$T_{m,n}[K(p)] = \frac{\alpha_m[K(p)]\alpha_n[K(p)]}{\alpha_m[K(p)] + \alpha_n[K(p)]} \tag{7.15}$$

where, $\alpha_m[K(p)]$, $\alpha_n[K(p)]$ denote the change of a shale factor of two adjacent grids m and n with the corresponding permeability of these two grids.

7.2.3.2 Dynamic Model of Well Index Between Different Multiple Media with Pores and Fracures and Wellbore

The magnitude of the well index between the different media with pores/fractures and wellbore reflects the fluid exchange capacity between the reservoir medium and the wellbore. The original calculation model is as follows:

$$\mathrm{WI} = \frac{2\pi\sqrt{K_x K_y}L}{\mathrm{In}\left(\frac{r_e}{r_w}\right) + s} \tag{7.16}$$

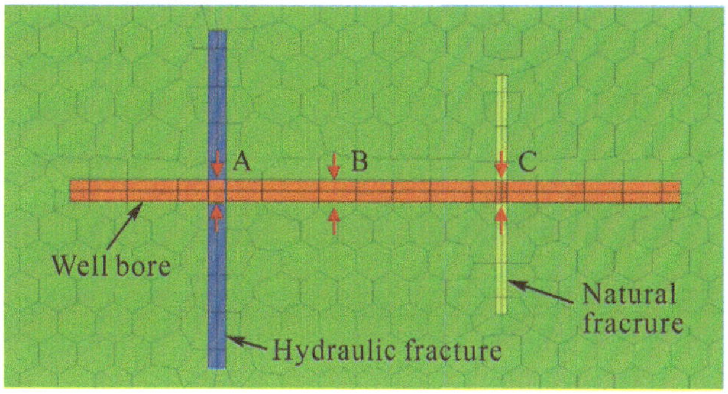

Fig. 7.34 Change in well index

where, WI is the well index; K_x is the permeability in the x direction; K_y is the permeability in the y direction; L is the length of the wellbore involved in the calculation; r_e is the radius of the area controlled by wellbore; r_w is the radius of the wellbore; s is the well skin factor.

In the depletion production, the pressure around the wellbore decreases sharply, whereas the pressure rises sharply during the fracturing and injection, leading to a significantly sharp change of fractures and matrix parameters. Due to the property changes of reservoir media around wellbore, such as matrix and natural/hydraulic fractures, the fluid exchange capacity between the media and the wellbore changes as well, finally leading to the change of the well index (as shown in Fig. 7.34, A. the well index between hydraulic fractures and wellbores, B. the well index between matrix and wellbores, C. the well index between natural fractures and wellbores). In this case, the calculation of the change in the well index requires the dynamic model of the well index.

$$\mathrm{WI}(p) = \frac{2\pi\sqrt{K_x(p)K_y(p)}L}{\mathrm{In}\left(\frac{r_e}{r_w}\right) + s} \tag{7.17}$$

7.2.4 Process of Coupled Flow-Geomechanics Simulation for Multiple Media with Different Scales Pores and Fractures

Based on the effective stress principle of multiple media, the deformation mechanism and law of multiple media with pores/fractures, and the dynamic models of matrix pores and natural/hydraulic fractures during fracturing, injection and production, the coupled flow-geomechanics simulation of multiple media with pores/fractures was formed. The specific steps of coupled simulation are as follows (as shown in Fig. 7.35):

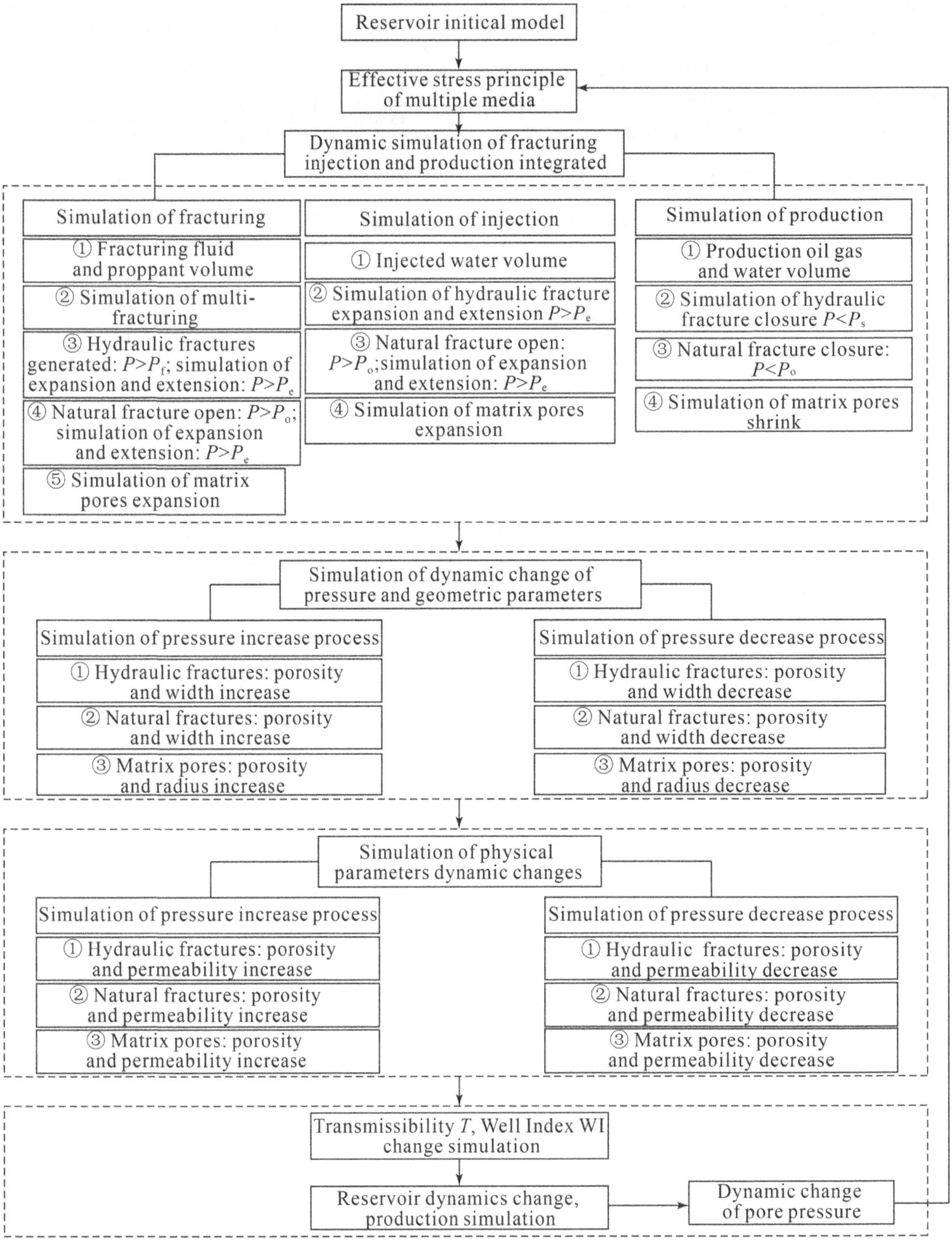

Fig. 7.35 Flowchart of multi-media dynamic simulation of coupled flow-geomechanics during fracturing injection production injection

(1) The establishment of initial reservoir model

The unconventional tight reservoirs are meshed and divided into several media types. The initial geometric, properties and fluid parameters of different media with pores/fractures are given. Through these settings an initial model of the integration simulation for fracturing, injection and production is established.

(2) Distribution of pore pressure and effective stress of multiple media

According to the flow state and the effective stress principle of different media with pores/fractures, such as matrix pores and natural/hydraulic fractures, the initial distributions of fluid pressure and the effective stress of different media with pores/fractures are determined.

(3) Integration simulation for fracturing, injection and production

Using the established integration simulation model for fracturing, injection and production, the corresponding simulation for these processes is conducted.

The simulation for fracturing includes: i. the simulation of injection of a fracturing fluid and proppants: ii. the simulation of multi-stage fracturing; iii. the simulation of hydraulic fracture generation, expansion and extension; iv. the simulation of natural fracture initiation, expansion and extension; v. the simulation of matrix pore expansion.

The simulation for injection includes: i. the simulation of water injection; ii. the simulation of hydraulic fracture expansion and extension; iii. the simulation of natural fracture initiation, expansion and extension; iv. the simulation of matrix pore expansion.

The simulation for production includes: i. the simulation of oil, gas and water production; ii. the simulation of hydraulic fracture closure; iii. the simulation of natural fracture closure; iv. the simulation of matrix pore shrinkage.

(4) Simulation of dynamic changes of pressure and geometric parameters

The simulation of pore pressure during the fracturing injection production is used to numerically calculate the increases of hydraulic/natural fracture width and a matrix pore radius during pore pressure increasing, and the decreases of hydraulic/natural fracture width and a matrix pore radius during pore pressure decreasing.

(5) Simulation of dynamic change of physical parameters

According to the changes of the fluid pressure and the geometric parameters of different media with pores/fractures, the increases of physical parameters of the hydraulic/natural fractures and matrix pores during pore pressure increasing, and the decreases of these physical parameters during pore pressure decreasing are simulated.

(6) Simulation of change in transmissibility and well index

Based on the changes of the fluid pressure and the geometric parameters of different media with pores/fractures, the dynamic changes of the transmissibility between different media with pores/fractures and the well index between the media and the wellbore are simulated.

(7) Simulation of reservoir performance and productivity forecast

Based on the comprehensive consideration of the dynamic changes of fluid pressure, physical parameters, transmissibility between different media, and well index, for unconventional tight oil reservoirs, the simulation of reservoir performance and productivity is conducted to match production data and predict their future performance and productivity.

References

Cai XS, Chen M et al (2009) Effective stress law of anisotropic multiple porous media. Eng Mech 26(4):57–60

Chen M, Chen ZD (1999) The effective stress law of multiple porous media. Appl Math Mech 20(in Chinese)

Chen M, Jin Y, Zhang GQ (2011) Fundamentals of rock mechanics in petroleum engineering. Petroleum Industry Press

Fan TY, Song XM, Wu SH, Li QY, Wang BH, Li XB, Li H, Liu HL (2015) Mathematical model and numerical simulation of water-driven fractures in low-permeability reservoirs. Pet Explor Develop 42(4):496–501

Gao WL, Jie JL, Zhang WM (2006) Research and field application of a resin coated sand proppant. Oilfield Chem 23(1):39–41

Guo TK, Zhang SC (2011) Research on the factors affecting proppant embedding. Fault Block Oil Gas Fields 18(4):527–529

Ikonnikova S, Browning J, Horvath S, Tinker SW (2014) Well recovery, drainage area, and future drill-well inventory: Empirical study of the Barnett Shale gas play, 171552-PA

Irwin GR (1957) Analysis of stresses and strain near the end of afracture traversing a plate. JAPP Mech 24

Ji LJ, Settari AT, Sullivan RB, Orr D (2004) Methods for modeling dynamic fractures in coupled reservoir and geomechanics simulation. SPE 90874, SPE Annual Technical Conference and Exhibition

Jin ZR, Guo JC, Zhao JZ, Zhou CL, Qin Y, Luo W, Wang Y (2007) Experimental study and analysis of influencing factors in supporting fracture diversion capability. J Drill Prod 30(5):36–38

Kyunghaeng L, Chun H, Mukul MS (2011) Impact of fracture growth on well injectivity and reservoir sweep during waterflood and chemical EOR processes. SPE 146778, SPE Annual Technical Conference and Exhibition

Li HY, Wang ZR (2006) Analysis of stress moiré circle variation and its application in typical hydraulic plastic molding. J Appl Basic Eng Sci 14(1):59–68

McClure MW, Babazadeh M, Shiozawa S, Huang J (2016) Hydraulic fracturing full coupling fluid dynamics simulation based on three-dimensional discrete fracture network. Pet Sci Technol 9:40–59

Micar J, Ekron M, Kenneth GN, Zhang BP (2002) Reservoir stimulation measures. Petroleum Industry Press

Na ZQ (2009) Research on fracture initiation mechanism and fracture extension model for horizontal wells. China University of Petroleum (East China)

Shen ZJ (1980) The yield theory of Mohr-Cullen materials. Water Res Water Transp Res 1:1–9

Sun CR, Wang NT, Zhang WC, Liu YC, Fang SJ (2010) Study on determination method of hydraulic fracturing closure pressure. J Chongqing Univ Sci Technol Nat Sci Ed, 12(2)

Van den Hoek PJ, Hustedt B, Sobera M, Mahani H, Masfry RA, Snippe J, Zwarts D (2008) Dynamic induced fractures in waterfloods and EOR. SPE 115204, SPE Russia Oil & Gas Technical Conference and Exhibition

Wang L, Zhang SC, Zhang WZ, Wen QZ (2005) Experimental study on long-term conductivity of composite fracturing proppant with different particle sizes. Nat Gas Ind 25(9):64–66

Wang KT, Zhao N, Lu G (2015) Movement law of effective moiré circle caused by seepage in soil. Shanxi Archit 41(26):81–83

Wang YJ, Song XM, Tian CB, Shi CF, Li JH, Hui G, Hou JF, Gao CN, Wang XJ, Liu P (2015) Dynamic fractures are new developmental geological attributes in waterflooding development of ultra-low permeability reservoirs[J]. Pet Explor Develop 42(2):222–228

Wen QZ, Zhang SC, Wang L, Liu YS (2005) Effect of proppant embedding on long-term conductivity of fractures. Nat Gas Ind 25 (5):65–68

Xie JB, Long GQ, Tian CB, Hou JF, Li JS, Wang YJ (2015) Genesis of dynamic fractures in ultra-low permeability sandstone reservoirs and its impact on waterflood development. Pet Geol Recovery 22 (3):106–110

Xu XZ, Li PC, Li CL (2001) Study on the principle of effective stress in porous media. Mech Eng 23:42–45

Yang MP (2004) Deformation theory and application of porous media in oil and gas reservoirs. Southwest Pet Inst, Sichuan

Ye JH et al (1991) Rock mechanics parameters manual. Water Power Press, Beijing

Zhang SC, Yan SB, Zhang J, Wang L (2008) Experimental study on the influence of coal rock on long-term conductivity of fracturing fractures. Acta Geol Sinica 82(10):1444–1448

Zhang Y, Pan LH, Zhou T, Zou YS, Li N, Xu ZH (2015) Evaluation of stress sensitivity of longmaxi formation shale. Sci Technol Eng 15 (8):37–41

Zhou XG, Zhang LY, Huang CJ (2013) Reservoir fracture pressure and fracture opening pressure estimation and development proposal for Chang 63 in Huaqing exploration area. J Central South Univ Nat Sci 44(7):2814–2718

Zhou T, Zhang SC, Zou Y, Li N, Hao SY (2016) Experimental study on permeability characteristics of natural fractures filled with shale gas reservoir. Journal of Xi'an Shiyou Univ Nat Sci 31(1):73–78

Zhu SG, Zhao XY, Zhang XS, Wang JH, Zhang YY, Liu P, Jiao J, Zeng LB (2016) Opening pressure of natural fractures and influencing factors in low-permeability sandstone reservoirs. J Northwest Univ Nat Sci Ed 46(44)

Zou YS, Zhang SC, Ma XF (2012) Evaluation of the effectiveness of fracturing support fractures in shale gas reservoirs. Nat Gas Ind 32 (9):52–55

8 Identification of Flow Regimes and Self-adaption Simulation of Complex Flow Mechanisms in Multiple Media with Different-Scale Pores and Fractures

Tight reservoirs have nano/micro pores and natural/hydraulic fracture networks, so they have features of multiple media at different scales. For different media at different scales, under different production stages, the flow regimes of fluid flow through these media are different and the affecting factors are also various, therefore, the flow mechanisms and behaviors of flow through these media are different. For conventional oil/gas reservoirs, the media are almost homogeneous and simple, and the flow mechanisms and flow regimes are relatively simple. However, between different stages of tight oil/gas production, production and flow mechanisms are different. The coupled flow behaviors between different-scale media are complex. Therefore, the flow characteristics of oil/gas in tight reservoirs are that media have several kinds of pores and fractures at different scales and the flow through these media have several flow regimes. According to factors influencing production mechanisms of tight reservoirs, an index system of flow regime identification for tight reservoirs was determined. The criterion of flow regime identification of oil/gas flow through multiple-scale multiple media in tight reservoirs was built. The flow regime identification for oil/gas flow through multiple-scale multiple media and the self-adaption simulation of complex flow mechanisms in tight reservoirs were established. Those achievements lay a foundation for mathematical models of fluid flow through tight reservoirs, productivity evaluation and numerical simulation of tight reservoirs.

8.1 Identification Index System of Flow Regimes in Multiple Media for Tight Oil and Gas

Nano/micro- pores and millimeter/micro-scale fractures are developed in tight reservoirs. The fluid flow through multiple-scale media and has different flow regimes. Meanwhile, the flow regimes of flow through matrix pores/fractures are affected by the reservoir pressure and the water/gas occurrence states. The flow regimes of fluid flow in tight reservoirs can generally be classified into three kinds of flow regimes, which are high-speed nonlinear flow, quasi linear flow and low-speed nonlinear flow. However, the flow mechanisms of different flow regimes have quite huge differences, affecting the numerical simulation for multiple-scale multiple media. Therefore, how to identify the fluid flow regimes is the theoretical basis for building a numerical simulation model.

Tight oil/gas reservoirs contain multiple-scale multiple media between which flow regimes are different. According to factors influencing complex flow features and production mechanisms of tight reservoirs, an index system of flow regime identification was determined (as shown in Fig. 8.1). This system includes geometric parameters, physical parameters, pressure gradients and kinetic parameters.

The patterns of oil/gas flow in tight reservoirs have two types which are macroscopic flow and microscopic flow. The pattern of oil flow in tight reservoirs is the macroscopic flow where the flow regime can be identified by the pressure gradients and the Reynolds number. The gas flow in tight reservoirs has the above two flow patterns at the same time. The flow regime of macroscopic gas flow is identified by the pressure gradients and the Reynolds number while the flow regime of microscopic gas flow is identified only by the Knudsen number.

The Reynolds number is used to identify flow regimes and the critical Reynolds number is a constant value which is inapplicable for the macroscopic flow in tight reservoirs. However, a pressure gradient reflects the comprehensive influence of geometry, physical properties and fluid properties on flow regimes and has the closest relationship with flow regimes, so it is more practical in real reservoir developments. Therefore, it is usually used to identify flow regimes. The flow curves are used to determine the inflection points of flow regimes. Through the flow velocity under different start-up pressures, the Reynolds numbers of those

Q. Ran, *Unconventional Tight Reservoir Simulation: Theory, Technology and Practice*,
https://doi.org/10.1007/978-981-32-9848-4_8

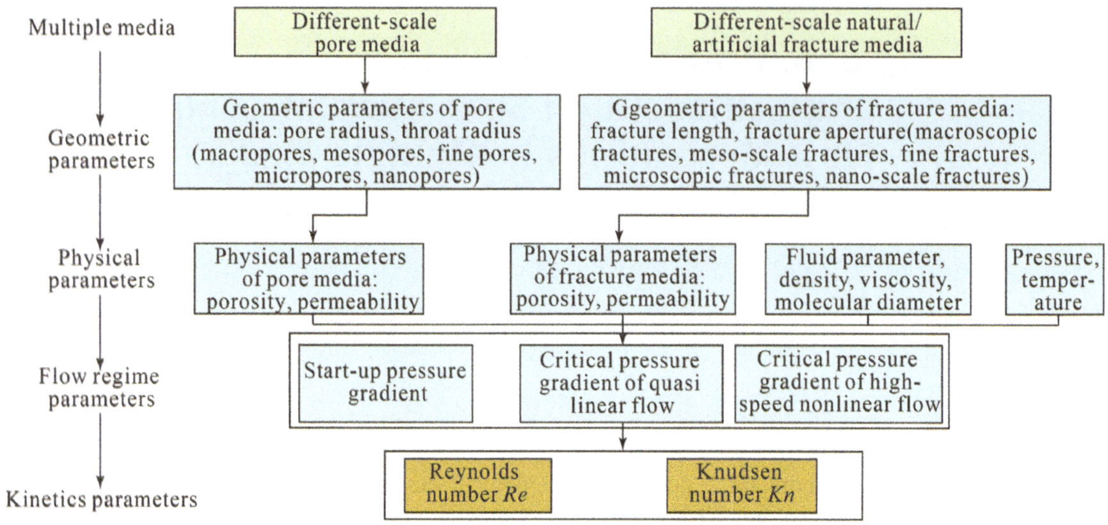

Fig. 8.1 An index system for various flow regimes

inflection points are calculated. A Reynolds number is an indirect parameter which indirectly represents a pressure gradient, and it can be selected to identify the macroscopic flow regimes.

8.1.1 Determination of Critical Parameters of Macroscopic Flow Regimes Based on Flow Characteristic Curves

8.1.1.1 Pressure Gradients

Tight sandstone reservoirs have different-scale media, and flows through different media have different flow curves. The experiments of microscopic flow show that i. flow regimes of oil flow through different pore media in tight reservoirs have two patterns which are low-speed nonlinear flow and quasi linear flow; ii. flow regimes of flow in different fractured media have three patterns which are low-speed nonlinear flow, quasi linear flow and high-speed nonlinear flow. The characteristics curves of those flow patterns are shown in Figs. 8.2 and 8.3. According to those curves, the start-up pressure gradient, the pressure gradient of quasi linear flow, and the pressure gradient of high-speed nonlinear flow can be determined. The flow regimes of flow in different media can be identified by these three pressure gradients.

1. The flow characteristics curves

During fluid flow, the curves representing relationships between a flux and a pressure gradient are the flow characteristic curves. The flow characteristics curves of pore media and fractured media are different.

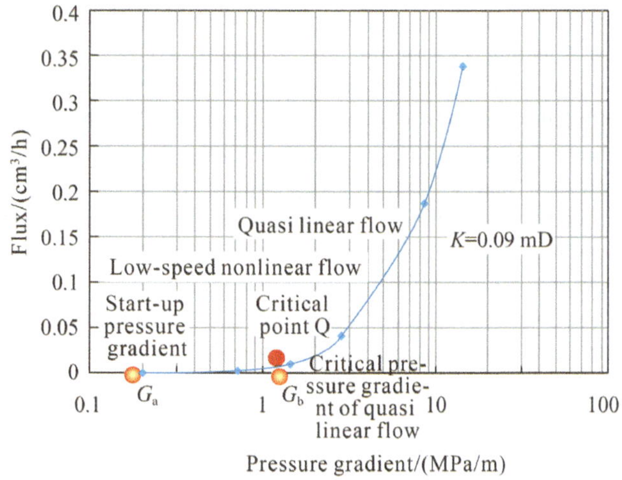

Fig. 8.2 Flow curves of pore media

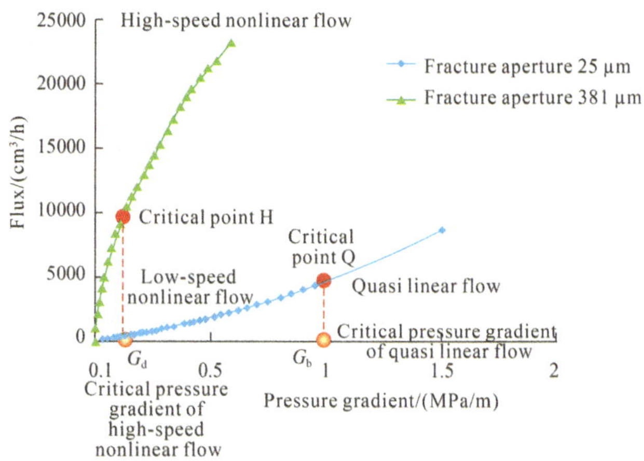

Fig. 8.3 Flow curves of fracture media

(1) Flow characteristics curve of pore media

The experiments of nonlinear flow were conducted on pore media of tight sandstones (32 cores from Chongqing and 26 cores from Sichuan). The experimental results show that flow regimes of flow through pore media are low-speed nonlinear flow and quasi linear flow, and the impacts of geometry, properties and fluid viscosity on flow regime of flow through pore media are obvious.

1) Flow characteristics curve of flow through media with different pore throat radii

Cores of tight sandstones contain different-scale pore media which have different pore structures. For oil samples with the same viscosity, the flow regimes of oil flow through pore media with various throat radii are different, ranging from low-speed nonlinear flow to quasi linear flow. In pore media with large throat radius, a fluid starts to flow even if the pressure gradient is quite low, and its flow velocity increases rapidly, resulting in steep flow characteristics curves. In the pore media with small throat radius, a fluid starts to flow when the pressure gradient is large, and its flow velocity increases slowly, resulting in gentle flow characteristics curves. As the throat radius decrease, the critical pressure gradient of the quasi linear flow increases, and the critical flow velocity decreases (as shown in Fig. 8.4).

2) Flow characteristics curve of flow through media with different permeability

Because of the difference of pore structures of tight reservoirs, the media with different pore structures have different properties. For the oil samples with the same viscosity, the flow regimes of flow through pore media with different properties are different. In the pore media with large permeability, a fluid begins to flow even if the pressure gradient is small. With the increase of a pressure gradient, the flow velocity increases rapidly, and the flow characteristics curves are steep. In the pore media with low permeability, the fluid starts to flow when the pressure gradient is large, and the flow velocity increases slowly, resulting in gentle flow characteristics curves. As the permeability decreases, the critical pressure gradient of quasi linear flow increases while the critical flow velocity decreases. Meanwhile, the flow regime of flow through pore media switches from coexistence of two flow patterns which are low-speed nonlinear flow and quasi linear flow to single low-speed nonlinear flow, as shown in Fig. 8.5.

3) Flow characteristics curve of flow through media with different mobility

The flow regime of tight oil flow through pore media is affected by fluid viscosity. The viscosity of a fluid is different, resulting in different mobility and different flow regimes. The flow regimes of tight oil flow through the same media are different with different mobility, ranging from low-speed nonlinear flow to quasi linear flow. When the mobility is large, the fluid starts to flow even if the pressure gradient is very small, and the flow velocity increases rapidly, resulting in steep flow curves. When the mobility is small, the fluid starts to flow when the pressure gradient is large, the flow velocity increases slowly, and the flow curves are gentle. As the mobility decreases, the critical pressure gradient of the quasi linear increases and the critical flow velocity decreases, as shown in Fig. 8.6.

(2) Flow characteristics curve of fractured media

A plate model was used to simulate the microscopic flow in fractured media. The experimental results show that affected by combined effects of fractures' geometry and properties and fluid viscosity, the flow through nano/micro-fractures obeys behavior of the low-speed nonlinear flow or quasi linear flow, whereas the flow through micro/millimeter

Fig. 8.4 Flow curves of pore media with different throat radii

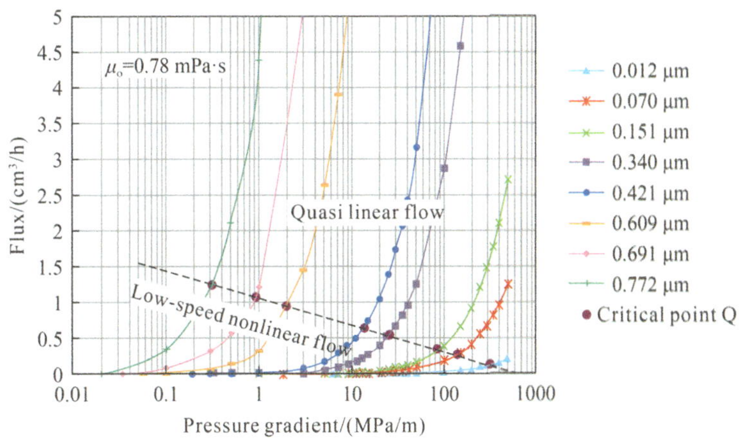

Fig. 8.5 Flow curves of pore media with different permeability

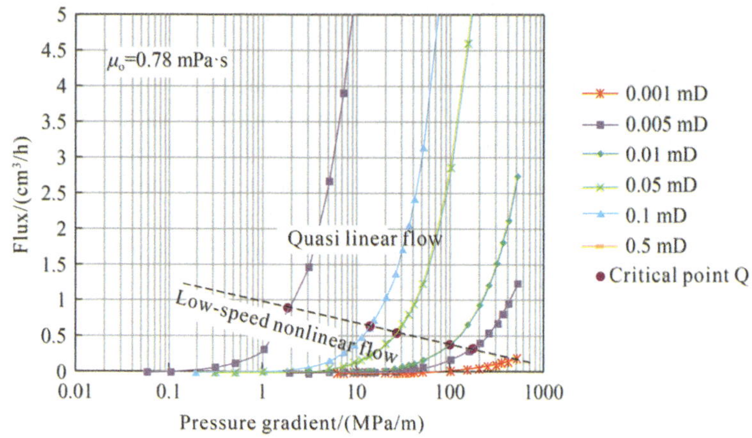

Fig. 8.6 Flow curves of pore media with different mobility

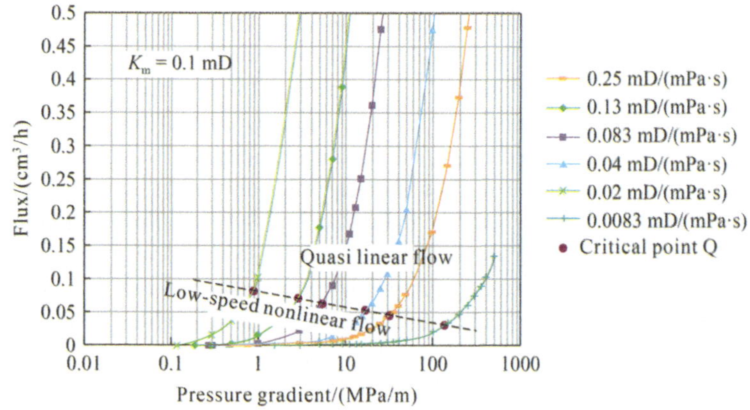

fractures obeys behavior of the quasi linear flow, linear flow or high-speed nonlinear flow.

1) Flow characteristics curve of flow through media with different fracture apertures

Tight sandstones contain different-scale fractures. For crude oil samples with the same viscosity, flow regimes of oil flow through different-scale fractures are different, including high-speed nonlinear flow, linear flow, quasi linear flow and low-speed nonlinear flow. When the fracture aperture is very small, the fluid velocity is low, and the flow characteristics curves are gentle, which indicates that the corresponding flow regimes are low-speed nonlinear flow and quasi linear flow. With the fracture aperture increases, when the fracture aperture is quite large, the changes of flow velocity and pressure gradient are approximately linear so the corresponding flow regime is linear flow. Meanwhile, with the increase of a pressure gradient, the magnitude of flow velocity increases with pressure gradient decreasing so the corresponding flow regime switches from linear flow into high-speed nonlinear flow. In this case, the larger fracture aperture causes the flow velocity to become higher; therefore, the high-speed nonlinear flow is more easy to occur (as shown in Fig. 8.7).

2) Flow characteristics curve of flow through media with different permeability

For the oil samples with the same viscosity, the flow regimes of flow through fractures with different properties are different. When the fracture permeability is large while the pressure gradient is low, the flow velocity increases rapidly and the flow characteristics curve is steep. In this case, the fluid flow obeys the high-speed nonlinear flow and the increase of flow velocity with a pressure gradient is nonlinear. As the fracture permeability decreases, the flow resistance increases, and the increase of flow velocity with the pressure gradient is almost linear. When the fracture permeability is low, the flow velocity increases slowly and the flow characteristics is gentle. In this case, with a small pressure gradient, the corresponding flow regime is the low-speed nonlinear flow. As the displacement pressure gradient decreases, the flow regime switches from the low-speed nonlinear flow into quasi linear flow (as shown in Fig. 8.8).

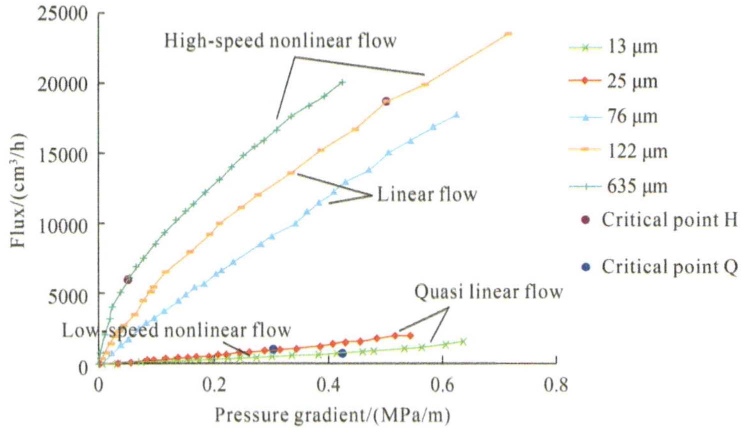

Fig. 8.7 Flow curves of fracture media with different fracture aperture

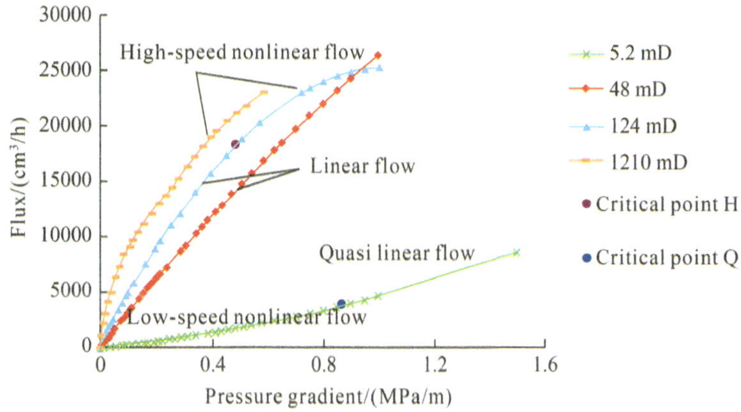

Fig. 8.8 Flow curves of fracture media with different permeability

3) Flow characteristics curve of flow through media with different mobility

The flow regime of tight oil flow through fractures is affected by fluid viscosity. The higher viscosity of the fluid causes the worse fluid mobility. In this case, the difference in flow regimes is obvious. The flow regimes of tight oil flow through the fractures are different with different flow mobility. When the mobility of fluid in fractures is large, the flow velocity increases rapidly even if the pressure gradient is very small, resulting in steep flow characteristics curves, and the corresponding flow regime is the high-speed nonlinear flow. As the flow mobility decreases, the flow regime switches from high-speed nonlinear flow into linear flow and low-speed nonlinear flow. When the mobility is small, with the increase of a displacement pressure gradient, the flow velocity increases slowly, and the flow characteristics curve is gentle. When the pressure gradient increases to a certain value, the increase of flow velocity with a pressure gradient is almost linear, and the corresponding flow regime switches from low-speed flow into quasi linear flow. When the mobility is extremely low, the increase of flow velocity with the pressure gradient is also extremely slow and only low-speed nonlinear flow occurs in this case (as shown in Fig. 8.9).

Compared with the flow through pore media, the resistance of flow through fractures is lower, the flow velocity is faster, the flow flux is higher, and the flow characteristics curve is steeper. In addition, the flow regimes switch more rapidly and have several complex flow regimes such as high-speed nonlinear flow, linear flow, quasi linear flow and low-speed nonlinear flow. The resistance of flow through pore media is large, the flow velocity increases slowly, and the flow characteristics curve is gentle. The flow regimes of flow through macrospores are mainly low-speed nonlinear flow and quasi linear flow. With the decrease in the geometry size of pores, the flow regime of flow through micropores is low-speed nonlinear flow (as shown in Fig. 8.10).

2. The start-up pressure gradient

When the displacement pressure gradient is higher than a certain pressure gradient, the fluid starts to flow. That certain pressure gradient is called as the start-up pressure gradient. In the flow characteristics curve, the start-up pressure gradient is the pressure gradient making the flow flux be zero. The start-up pressure gradients for different-scale pores (as shown in Fig. 8.2) and fractures (as shown in Fig. 8.3) are different.

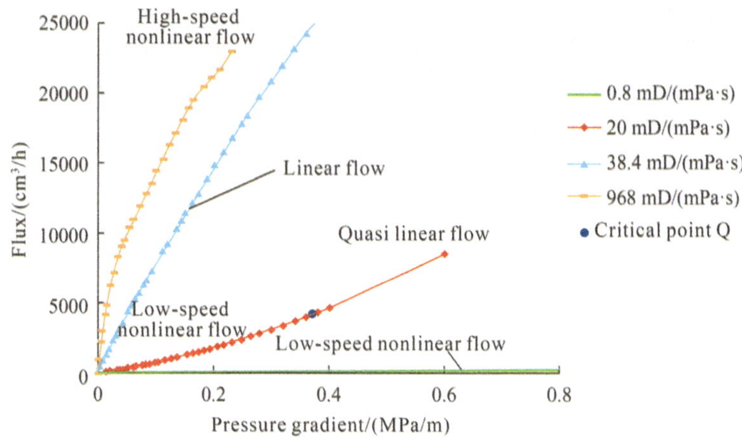

Fig. 8.9 Flow curves of fracture media with different mobility

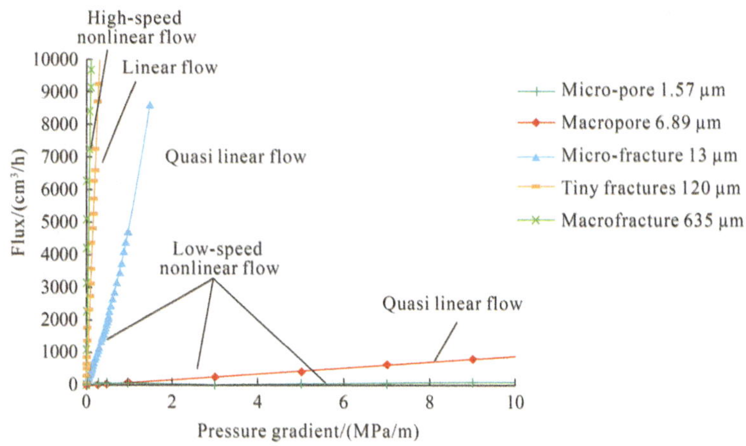

Fig. 8.10 Comparison of flow curves of fracture and pore media

(1) Start-up pressure gradient of pore media

Through the curve obtained from experiments of nonlinear flow through tight sandstone cores, the start-up pressure gradients of different-scale pore media can be determined, building the relationships between start-up pressure gradients of different-scale pore media and pore throat radii (as shown in Fig. 8.11a). As the radius of pore throat decreases, the start-up pressure gradient increases. When the pore throat radius is large, the start-up pressure gradient is quite low. When the pore throat radius decreases to a critical value, the start-up pressure gradient will increase sharply with the decrease of pore throat radius.

The physical properties of different pore media and fluid properties have great influences on the start-up pressure gradient. As the permeability of pore media increases, the flow resistance decreases, and the start-up pressure gradient decreases as well (as shown in Fig. 8.11b). Meanwhile, due to the difference of fluid properties, as the viscosity of crude oil increases, the mobility deteriorates and the start-up pressure gradient increases (as shown in Fig. 8.11c).

It can be seen that the magnitude of start-up pressure gradient is affected by the pore throat radius (permeability) and fluid property (viscosity) of different-scale pore media. Therefore, the start-up pressure gradient should be determined by its correlation with mobility.

(2) Start-up pressure gradient of fractured media

Through the curve obtained from experiments of nonlinear flow through tight sandstone parallel plates, the start-up pressure gradients of different-scale fractured media can be determined, building the relationships between start-up pressure gradients of different-scale fractured media and fracture apertures (as shown in Fig. 8.12a). As the fracture aperture in nano/micro fractured media decreases, the start-up pressure gradient increases. When the fracture aperture reaches a certain critical value, the start-up pressure gradient will increase sharply. Overall, the start-up pressure gradient in fractured media is quite low.

The physical properties of different fractured media and fluid properties have great influences on the start-up pressure

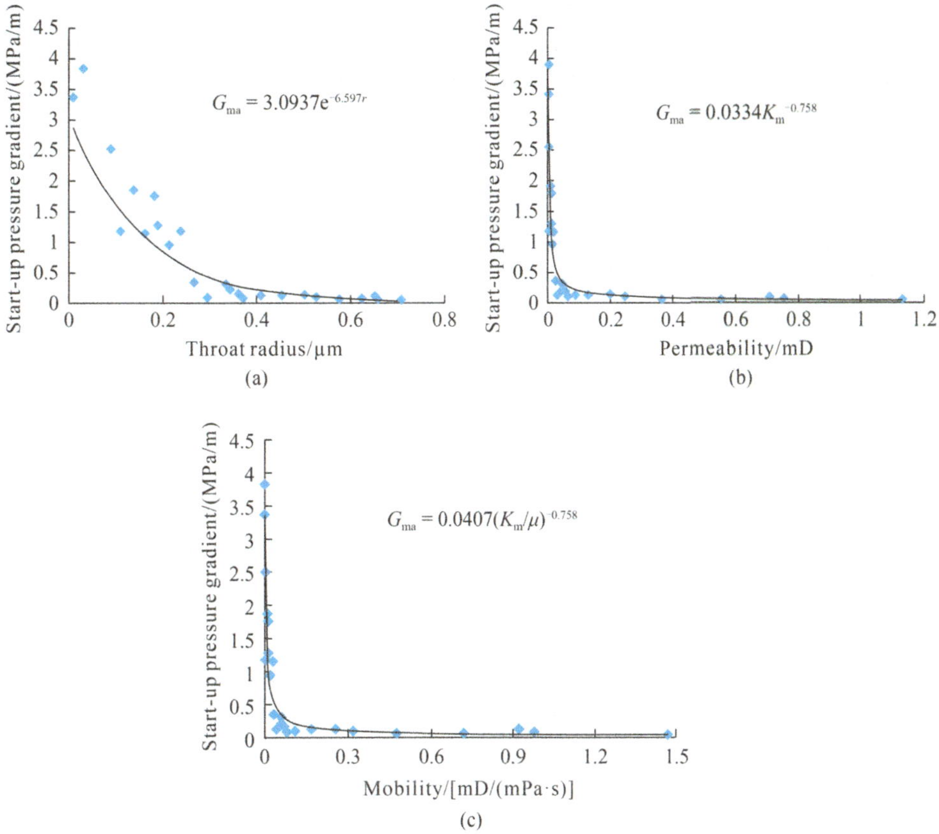

Fig. 8.11 The variation laws of the threshold pressure gradient under different throat radii, permeability and mobility

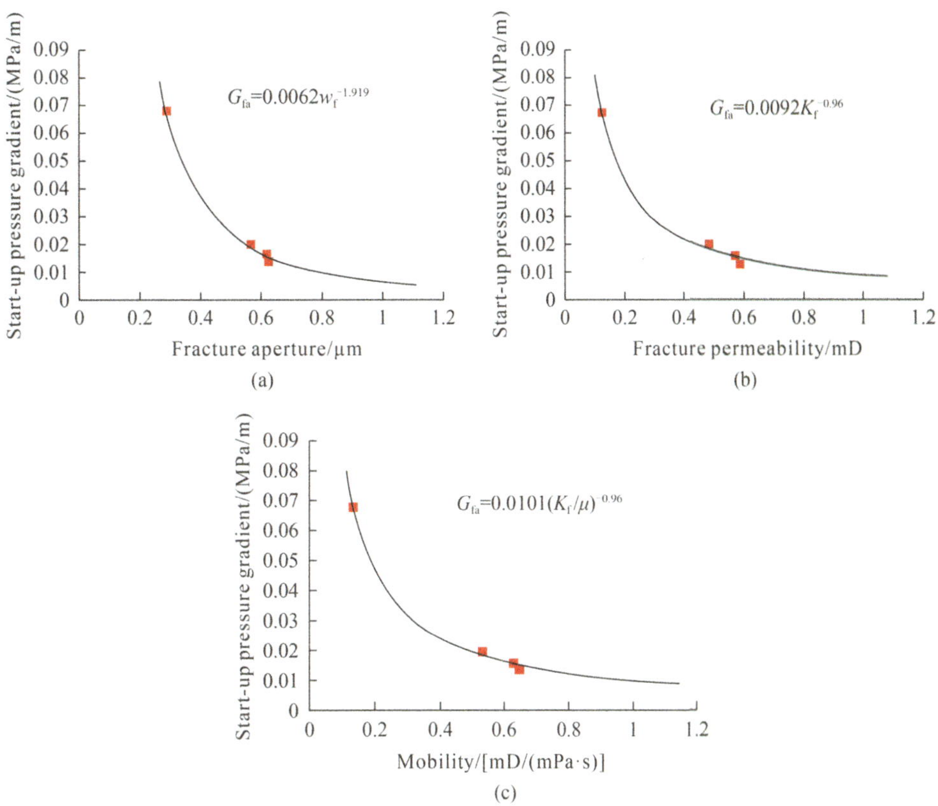

Fig. 8.12 The variation laws of the threshold pressure gradient under different fracture apertures, permeability and mobility

gradient. As the permeability of fractured media increases, the flow resistance and the start-up pressure gradient decrease (as shown in Fig. 8.12b). Meanwhile, due to the difference of fluid properties, as the viscosity of crude oil increases, the mobility deteriorates and the start-up pressure gradient increases (as shown in Fig. 8.12c).

The magnitude of a start-up pressure gradient is affected by the fracture aperture, permeability and fluid properties. Meanwhile, the mobility closely relates with fracture properties and fluid properties. Therefore, the start-up pressure gradient should be determined by its relationship with mobility.

Through the comparison of characteristics curves of fractured media and pore media (Fig. 8.10), it is can be seen that nano/micro-fractures are similar with large-scale pores in terms of flow characteristics. The flow characteristics of nano/micro fractured media are similar with those of large/micro-scale pore media and they include low-speed nonlinear flow, quasi linear flow and linear flow. Generally, the start-up pressure gradient of fractured media is low while that of pore media is high. The flow through large-scale fractures does not require the start-up pressure gradient. When the fracture permeability is lower than a certain critical value, the flow through fractured media is affected by the start-up pressure gradient. However, this pressure gradient is quite low and obviously lower than that of pore media which has the same magnitude of permeability with fractured media (as shown in Fig. 8.13). With the decrease of permeability, the start-up pressure gradient increases.

3. The critical pressure gradient of quasilinear flow

The critical pressure gradient of quasilinear flow is the pressure gradient used to estimate the switch of flow regime from low-speed nonlinear flow into quasilinear flow (Zheng et al. 2016). In a flow characteristics curve, the critical pressure gradient of quasilinear flow is the pressure gradient

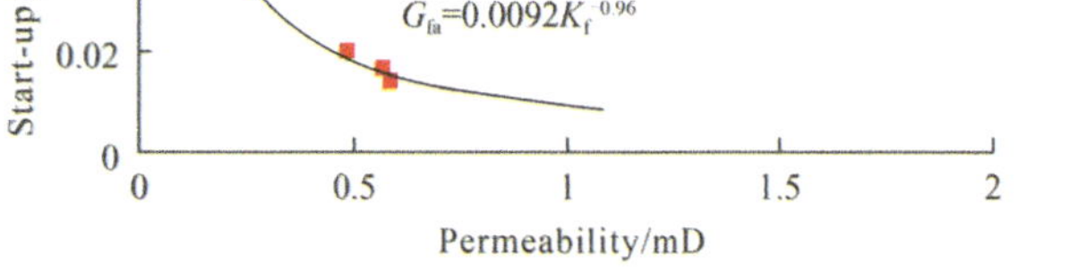

Fig. 8.13 Comparison of the threshold pressure gradient of fracture and pore media

where the nonlinear relationship of the pressure gradient with flux switches into the linear relationship.

(1) The critical pressure gradient of quasilinear flow of pore media

Through the curve obtained from experiments of nonlinear flow through tight sandstone cores, the critical pressure gradients of quasilinear flow of different-scale pore media can be determined, building the relationships between the critical pressure gradients of quasilinear flow of different-scale pore media and pore throat radii (as shown in Fig. 8.14a). As the radius of pore throat decreases, this critical pressure gradient increases. When the pore throat radius is large, this critical pressure gradient will be quite low. When the pore throat radius decreases to a certain critical value, this critical pressure gradient will increase sharply with the decrease of the pore throat radius.

The physical properties of pore media and fluid properties have great influences on this critical pressure gradient. As the permeability of pore media increases, this critical pressure gradient decreases (as shown in Fig. 8.14b). Meanwhile, as the viscosity of crude oil increases, the mobility deteriorates and this critical pressure gradient increases (as shown in Fig. 8.14c).

The magnitude of the critical pressure gradient of quasilinear flow is affected by the pore structure (pore throat radius and permeability) and fluid property (viscosity). Therefore, the critical pressure gradient should be determined by its correlation with mobility.

(2) The critical pressure gradient of quasilinear flow of fractured media

The displacement pressure gradient in fractured media is quite small. When the velocity of flow through fractured media is low, the flow regime is low-speed nonlinear flow. With the increase of a pressure gradient, the flow velocity increases. As the pressure gradient increases to a certain value, the flow regime of flow through fracture media will switch from low-speed nonlinear flow into quasilinear flow. This critical value of the pressure gradient is called the critical pressure gradient of quasilinear flow in fractured media.

Through the curve obtained from experiments of nonlinear flow through tight sandstone parallel plates, the critical pressure gradients of quasilinear flow of different-scale fractured media can be determined, building relationships between this critical pressure gradients of different-scale fractured media and fracture apertures (as shown in Fig. 8.15a). As the fracture aperture decreases, this critical pressure gradient increases. Generally, this critical pressure gradient in fractured media is quite small.

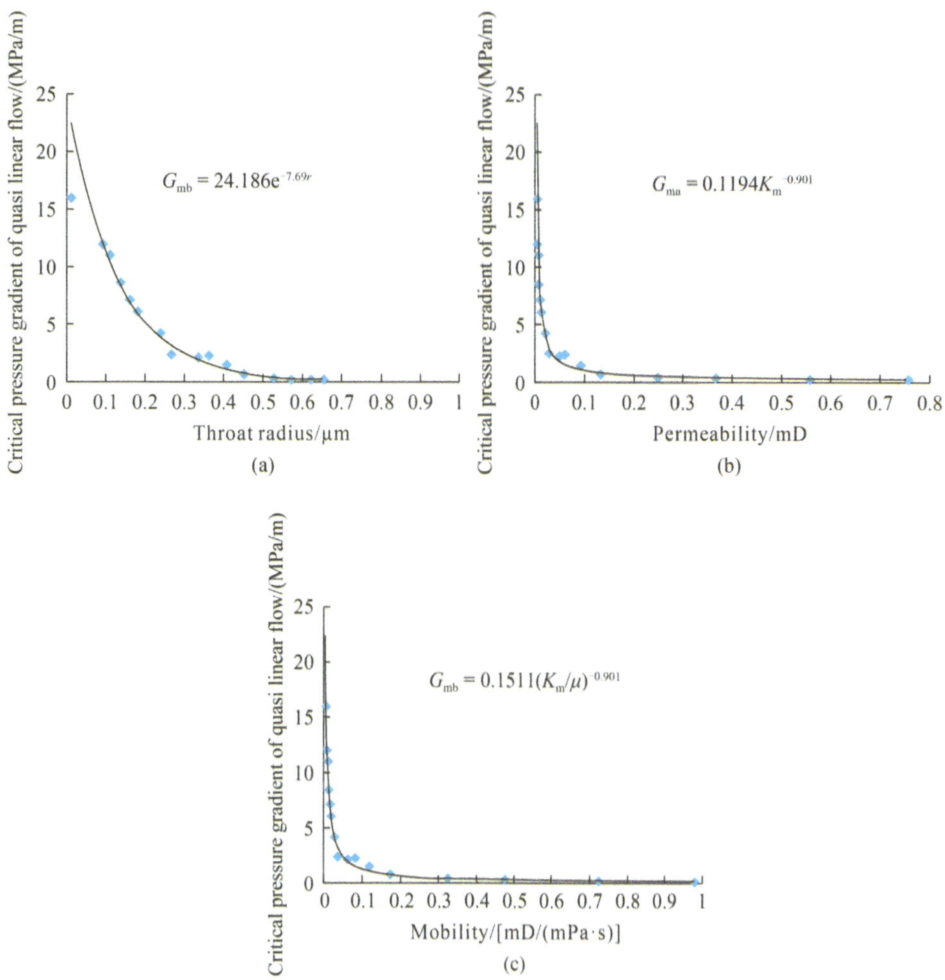

Fig. 8.14 The variation laws of the pressure gradient of quasilinear under different throat radii, permeability and mobility

As the permeability of fractured media increases, the critical pressure gradient of quasilinear flow decreases (as shown in Fig. 8.15b). As the viscosity of crude oil increases, the flow resistance decreases while the critical pressure gradient increases (as shown in Fig. 8.15c). The magnitude of this critical pressure gradient is affected by the fracture aperture, permeability and fluid properties. Meanwhile, the mobility results from fracture properties and fluid properties. Therefore, this critical pressure gradient should be determined by its correlation with mobility.

Tight reservoirs contain microscopic fractures and matrix pores at different scales. The conditions where the flow regime becomes quasilinear flow are different between different-scale media. Compared with flows through pore throats, the resistance of flow through fractures is lower, causing a lower critical pressure gradient of quasilinear flow, and, therefore, the quasilinear flow tends to appear in the flow through fractures (as shown in Fig. 8.16). If magnitudes of media sizes are the same, this pressure gradient of fractured media will be obviously lower than that of pore media. With the increase of the magnitude of media sizes (pore throat radius and fracture aperture), the critical pressure gradient of quasilinear flow decreases.

4. The critical pressure gradient of high speed nonlinear flow

The critical pressure gradient of high-speed nonlinear flow is the pressure gradient used to estimate the switch of flow regime from quasilinear flow or linear flow into high-speed nonlinear flow. In a flow characteristics curve, the critical pressure gradient of high-speed nonlinear flow is the pressure gradient where the linear relationship of flux with a pressure gradient switches into the nonlinear relationship.

(1) The critical pressure gradient of high-speed nonlinear flow of pore media

The flow resistance of oil flow through pore media in tight reservoirs is large so the flow velocity is low and the high-speed nonlinear flow is difficult to occur. Compared with tight oil flow, the flow resistance of tight gas flow is

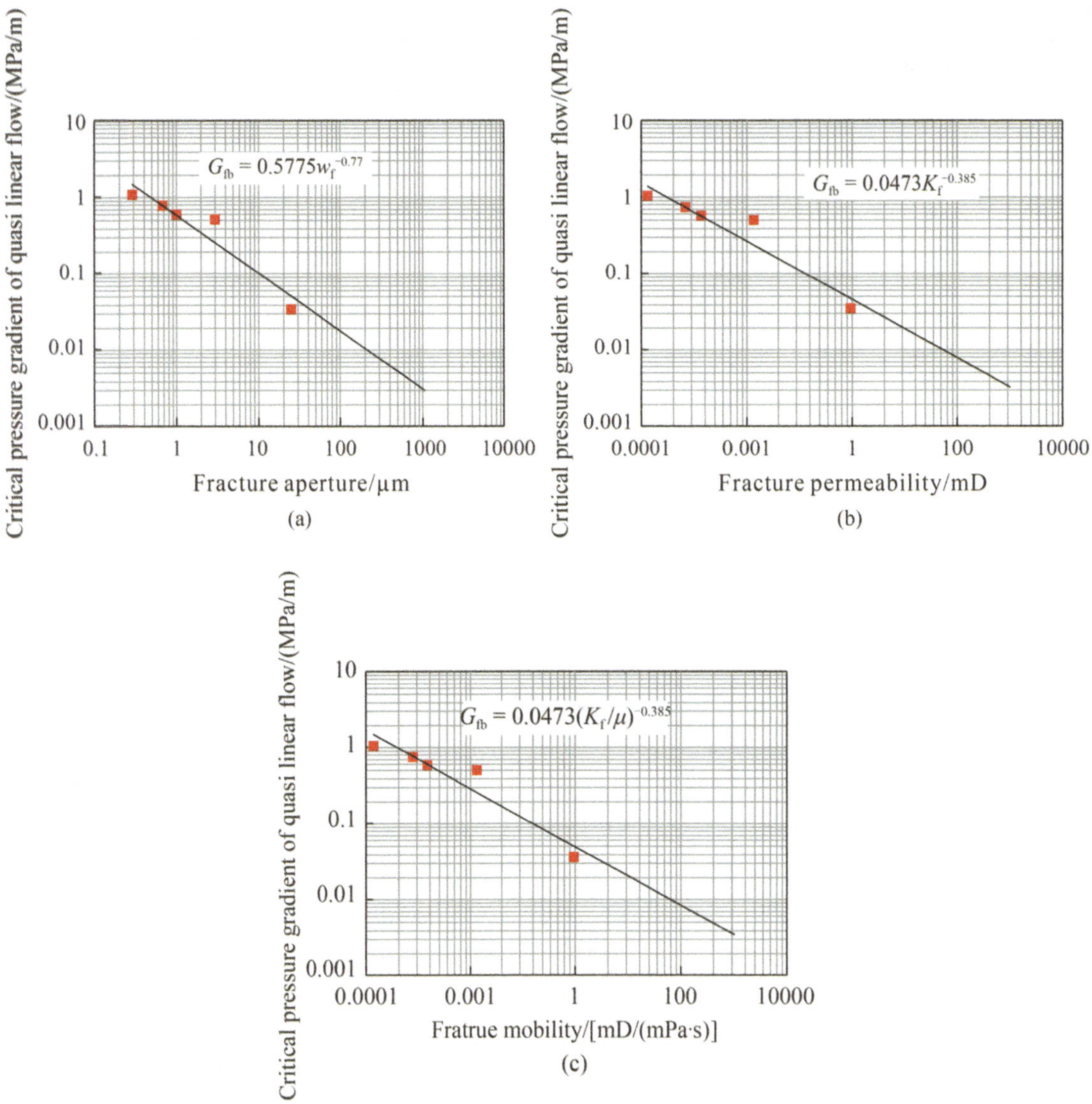

Fig. 8.15 The variation laws of the pressure gradient of quasilinear under different fracture apertures, permeability and mobility

lower so the flow velocity of tight gas is larger and the high-speed nonlinear flow is easy to occur in large pores. According to experiments of nonlinear tight gas flow through pore media, the critical pressure gradients of high-speed nonlinear flow of different-scale pore media can be determined, building the relationships between the critical pressure gradients of high-speed nonlinear flow of different-scale pore media and pore throat radii (as shown in Fig. 8.17a). As the radius of pore throat decreases, this critical pressure gradient increases. When the pore throat radius is large, this critical pressure gradient will be low. When the pore throat radius decreases to a certain critical value, this critical pressure gradient will increase sharply.

The viscosity difference between different gases is quite small so the impact of viscosity on the critical pressure gradient of high-speed nonlinear can be ignored. With the increase of the permeability of different pore media, the flow resistance decreases, the flow velocity increases, and the critical pressure gradient of high-speed nonlinear flow decreases (as shown in Fig. 8.17b).

(2) The critical pressure gradient of high-speed nonlinear flow of fractured media

The flow resistance of tight oil flow through fractured media at large-scale is low. In this case, the high-speed nonlinear flow can occur under a low pressure gradient. According to experiments of nonlinear tight gas flow through fractured media, the critical pressure gradients of high-speed nonlinear flow of different-scale fractured media can be determined (as shown in Fig. 8.18a). As the fracture aperture increases, this critical pressure gradient decreases. Generally, the critical

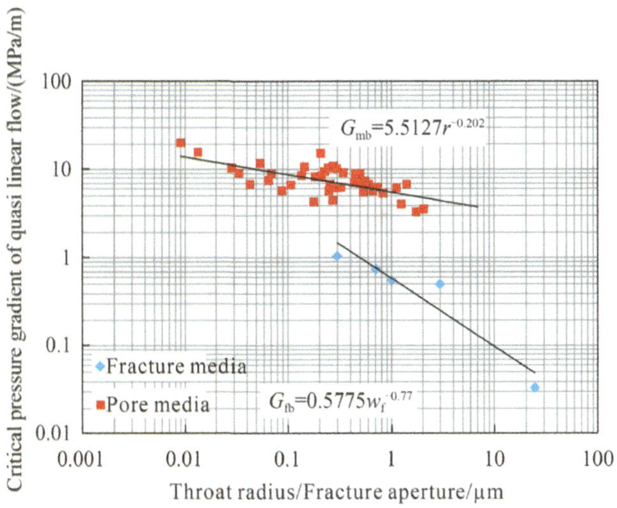

Fig. 8.16 Comparison of the pressure gradient of quasilinear of fracture and pore media

pressure gradient of high-speed nonlinear flow through fractured media is low.

As the permeability of fractured media increases, the critical pressure gradient of high-speed nonlinear flow decreases (as shown in Fig. 8.18b). As the viscosity of crude oil increases, the flow resistance and the flow mobility decrease (as shown in Fig. 8.18c). The magnitude of this critical pressure gradient is affected by the fracture aperture, permeability and fluid properties. Meanwhile, the mobility results from fracture properties and fluid properties. Therefore, this critical pressure gradient should be determined by its correlation with mobility.

8.1.1.2 The Reynolds Number

Tight reservoirs contain nano/micro/millimeter-scale pores and fractures. For tight oil, the Reynolds number can be used to determine the flow regime. For tight gas, the flow through large-scale pores and fractures is macroscopic flow where the flow regime can be determined by the Reynolds number. With the decrease in sizes of pores and fractures, under low pressure conditions, the gas flow is microscopic flow where the flow regime can be determined by the Knudsen number.

The Reynolds number is a ratio of viscous resistance to inertial resistance and is the basic dynamic parameter for determining the flow regime. According to the flow characteristics curve, the Reynolds number curve can be calculated and then the critical Reynolds number can be determined.

1. The Reynolds number curve obtained from the flow characteristics curve

(1) Reynolds number curve of pore media

According to the flow characteristics curve, for one pore medium with a pore throat radius of d, the flow velocity of flow through this medium can be determined under a certain pressure gradient. The following equation is used to calculate the Reynolds number of the pore medium:

$$R_e = \frac{\rho_{o,g} \boldsymbol{v}_{o,g} d}{\mu_{o,g}} \tag{8.1}$$

where, $\boldsymbol{v}_{o,g}$ is the flow velocity of oil/gas flow through the pore medium, m/s,the subscripts o refers to oil, g refers to gas, $\boldsymbol{v}_{o,g} = \frac{K_m}{\mu_{o,g}} \cdot \nabla P_{o,g}$; K_m is the permeability of the pore medium, mD; d is the pore throat diameter, μm; $\rho_{o,g}$ is the density of oil or gas, g/cm^3; $\mu_{o,g}$ is the viscosity of oil or gas, mPa·s; $\nabla P_{o,g}$ is the pressure gradient of oil or gas, MPa/m.

For pore media with different pore throat diameters, the variations of the Reynolds number with different pressure

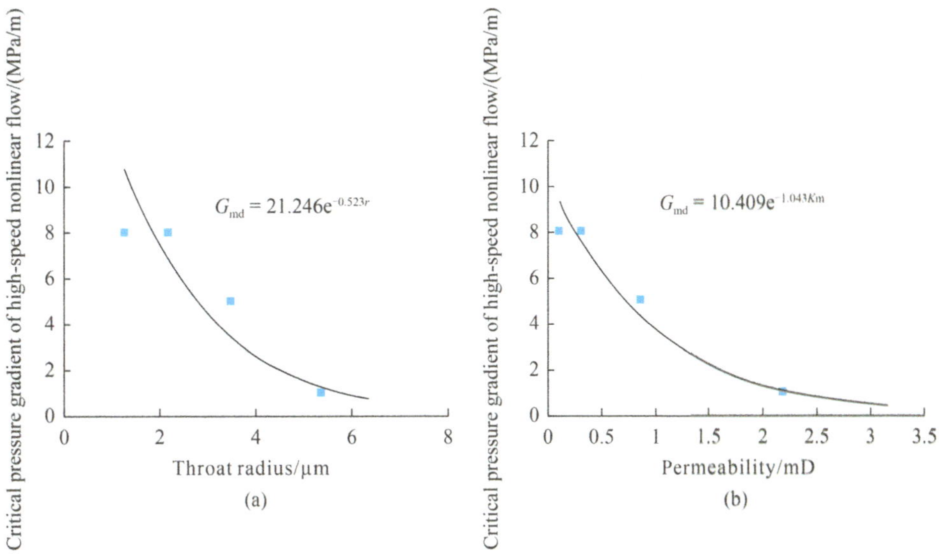

Fig. 8.17 The variation laws of the pressure gradient of nonlinear at high flow velocity under different throat radii and permeability

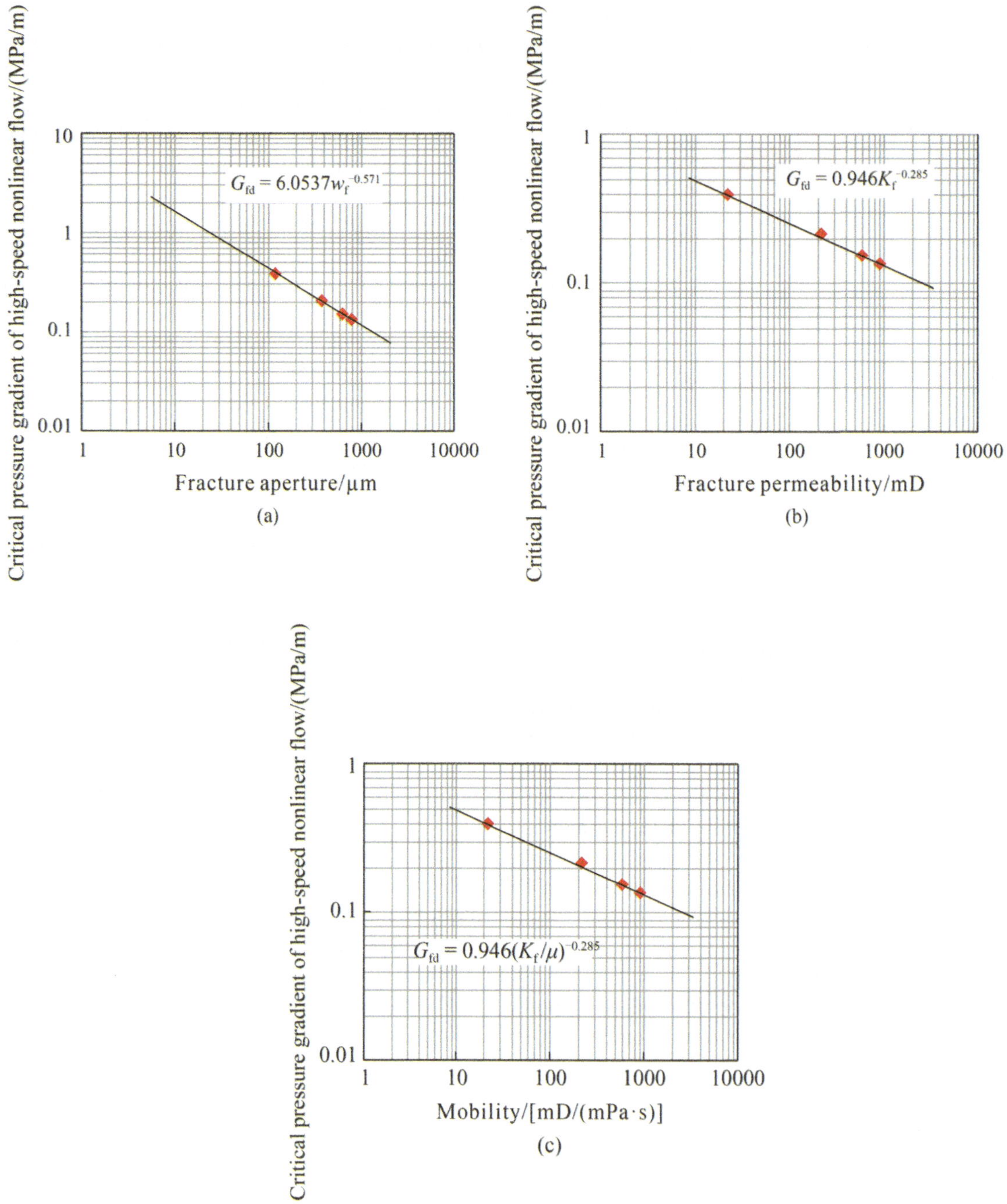

Fig. 8.18 The variation laws of the pressure gradient of nonlinear at high flow velocity under different fracture apertures, permeability and mobility

gradients are shown in Fig. 8.19. Under the same pressure gradient, the flow velocity of flow through pore media with large pore throat diameters is high, and the corresponding Reynolds number is large. However, under the same pressure gradient, the flow velocity of flow through pore media with small pore throat diameters is low and the corresponding Reynolds number is small.

(2) Reynolds number curve of fracture media

According to the flow characteristics curves, for fractured media with the fracture aperture of w_f, the flow velocity of flow through the media can be determined under a certain pressure gradient. The following equation is used to calculate the Reynolds number of the fracture media:

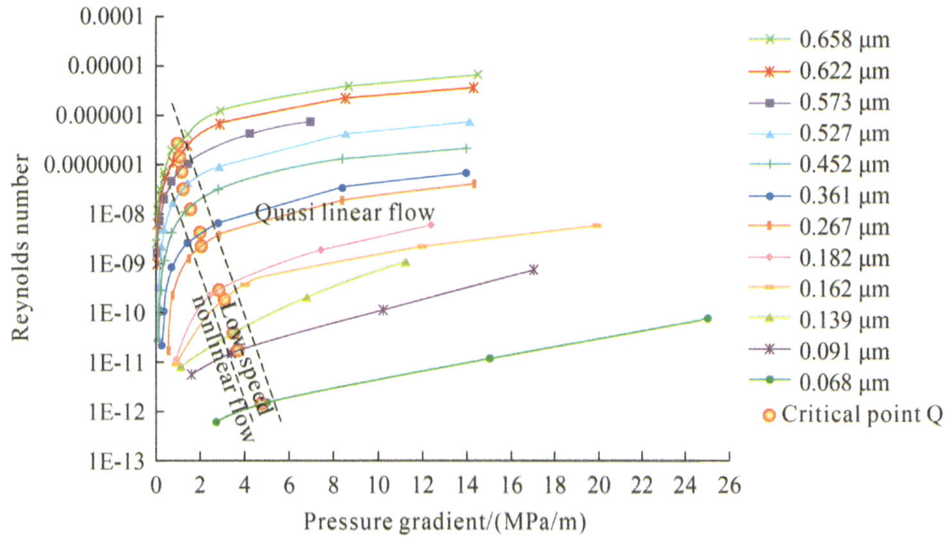

Fig. 8.19 The Reynolds number curves of pore media

$$R_e = \frac{\rho_{o,g} v_{o,g} w_f}{2\mu_{o,g}} \tag{8.2}$$

where, $v_{o,g}$ is the flow velocity of the oil phase or gas phase in the fracture media, m/s, $v_{o,g} = \frac{K_f}{\mu_{o,g}} \cdot \nabla P_{o,g}$; w_f is the fracture width, μm; K_f is the fracture permeability, mD.

For fracture media with different fracture apertures, the variations of the Reynolds number with different pressure gradients are shown in Fig. 8.20. Under the same pressure gradient, the flow velocity of flow through fracture media with large fracture aperture is high, and the corresponding Reynolds number is big. However, under the same pressure gradient, the flow velocity of flow through fracture media with small fracture aperture is low and the corresponding Reynolds number is small.

As the fracture apertures are small, there are two flow regimes which are low-speed nonlinear and quasilinear flow existing in fracture media. With the increase of fracture apertures, the flow velocity when the low-speed nonlinear flow in fracture media switches into quasilinear flow increases, so the Reynolds number limit increases. When the fracture apertures increase to a certain value, the high-speed nonlinear flow will occur in fracture media. In this case, with the increase of fracture apertures, the flow velocity when the quasilinear flow switches into high-speed nonlinear flow decreases, so the Reynolds number limit decreases.

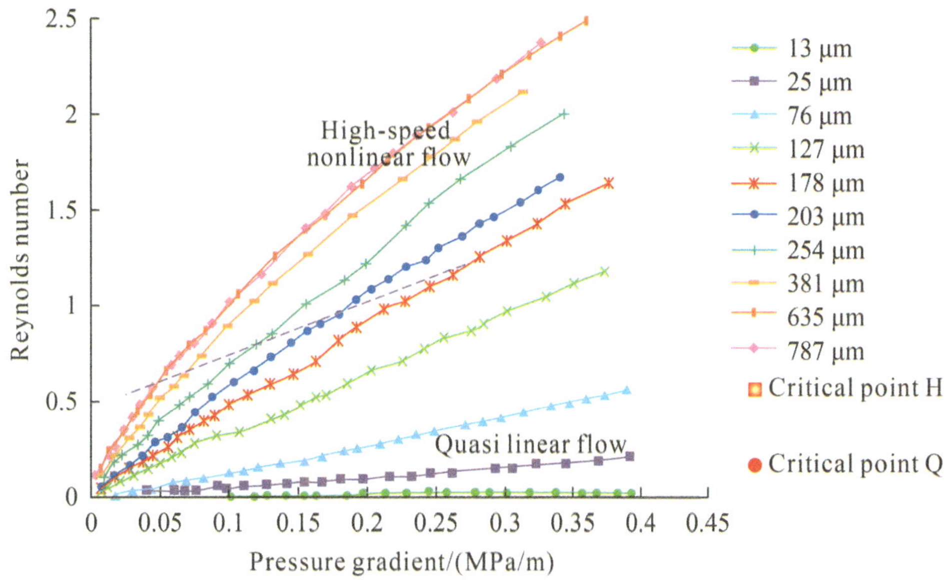

Fig. 8.20 The Reynolds number curves of fracture media

2. The Reynolds number curve

(1) The minimum reynolds numbers

On the Reynolds number curve, the minimum Reynolds number is the Reynolds number when a fluid starts to flow. According to the Reynolds number curves of pore media (as shown in Fig. 8.19), the minimum Reynolds number is determined, building the relationship of the minimum Reynolds number with a pore throat radius (as shown in Fig. 8.21a). With the increase of the pore throat radius, the minimum Reynolds number in pore media approximately linearly increases. A smaller pore throat radius induces a smaller minimum Reynolds number.

The physical properties of pore media and fluid properties have great influences on the minimum Reynolds number. As the permeability of pore media increases, the flow resistance decreases and the flow velocity of fluid starting to flow increases, so the minimum Reynolds number increases as well. When the permeability is low, the increase of the minimum Reynolds number with permeability is significant. After a certain value, the increase of the minimum Reynolds number with permeability is slow (as shown in Fig. 8.21b). Meanwhile, as the viscosity of crude oil increases, the mobility decreases and the flow velocity of the fluid starting to flow decreases so the minimum Reynolds number decreases as well (as shown in Fig. 8.21c). The magnitude of the minimum Reynolds number is affected by the pore throat radius (permeability) and fluid property (viscosity). Therefore, the minimum Reynolds number should be determined by its correlation with mobility.

(2) The critical Reynolds number of quasilinear flow

The critical Reynolds number of quasilinear flow is the Reynolds number corresponding to the critical pressure gradient of quasilinear flow. The critical Reynolds numbers of quasilinear flow of different-scale pore media (as shown in Fig. 8.19) and fracture media (as shown in Fig. 8.20) are different.

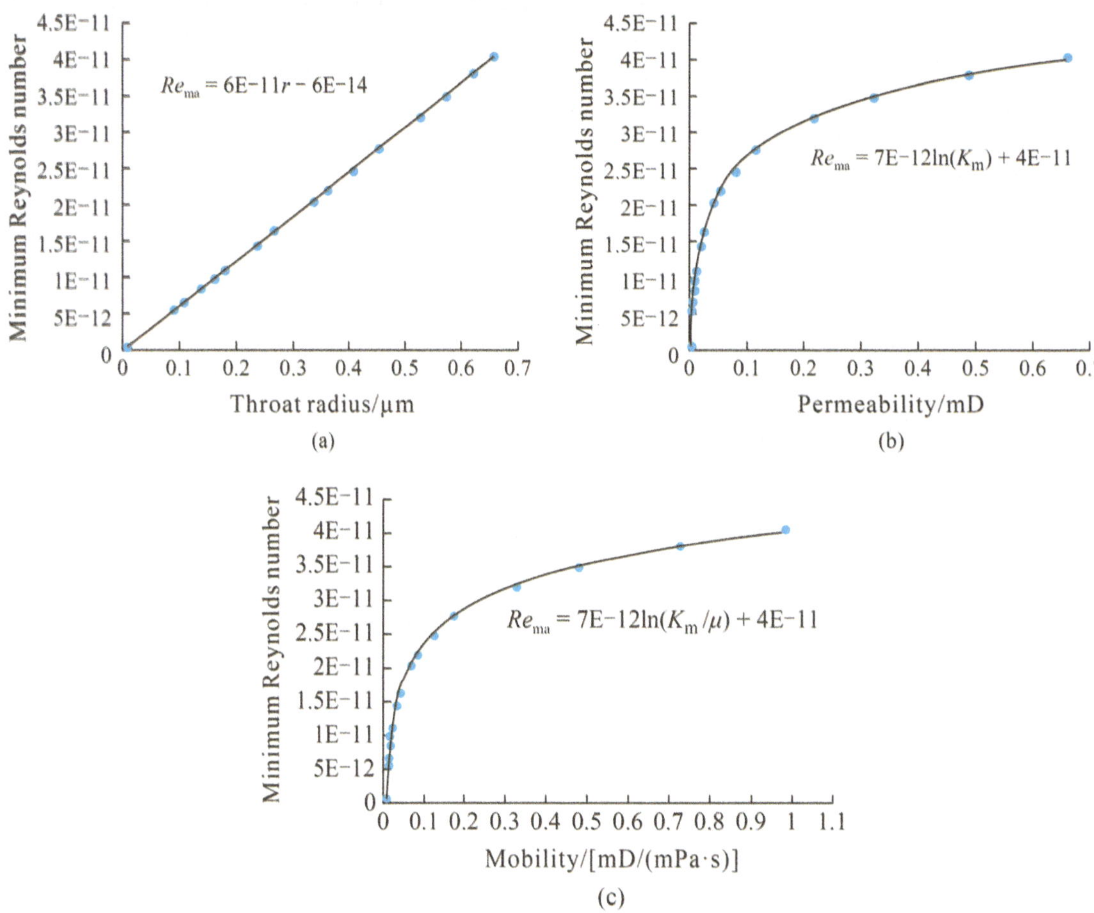

Fig. 8.21 The variation laws of the minimum Reynolds number under different throat radii, permeability and mobility

1) The critical Reynolds number of quasilinear flow of pore media

According to the Reynolds number curves of pore media, the critical Reynolds numbers of quasilinear flow through different-scale pore media are determined, building the relationships of this critical Reynolds numbers with pore throat radii of different-scale pore media (as shown in Fig. 8.22a). With the increase of a pore throat radius, this critical Reynolds number of pore media increases. A smaller pore throat radius induces a smaller critical Reynolds number.

The physical properties of pore media and fluid properties have great influences on the critical Reynolds number. As the permeability of pore media increases, the critical Reynolds number sharply increases. After a certain value, the increase of the critical Reynolds number becomes slow (as shown in Fig. 8.22b). Meanwhile, as the viscosity of crude oil increases, the mobility decreases and the critical Reynolds number decreases (as shown in Fig. 8.22c).

The magnitude of the critical Reynolds number is affected by the pore throat radius (permeability) and fluid property (viscosity). Therefore, the critical Reynolds number should be determined by its correlation with mobility.

2) The critical Reynolds number of quasilinear flow of fracture media

According to the Reynolds number curves of fracture media, the critical Reynolds numbers of quasilinear flow through different-scale fracture media are determined, building the relationships of these critical Reynolds numbers with fracture apertures of different-scale fracture media (as shown in Fig. 8.23a). When the fracture scale is quite small, the

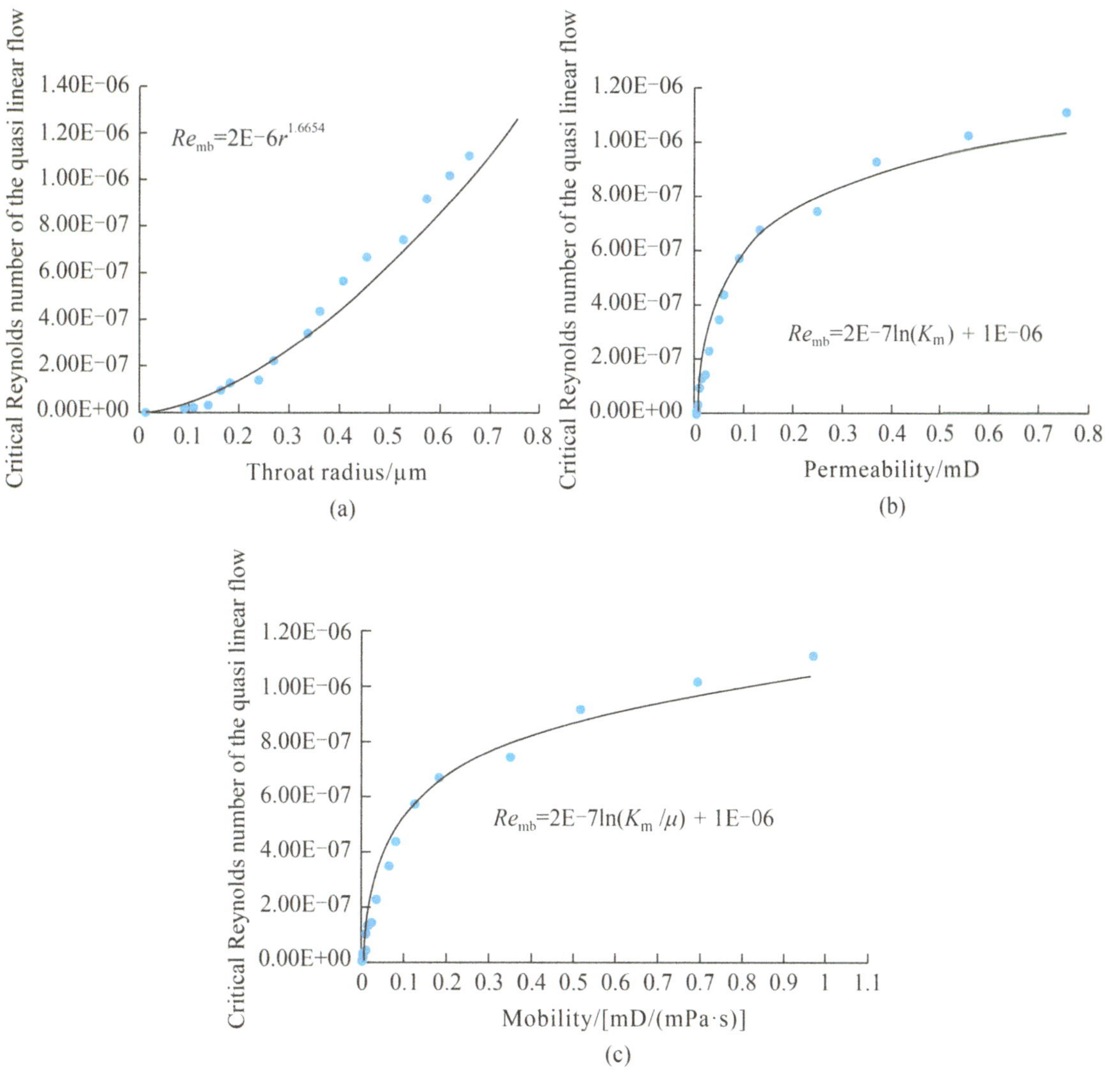

Fig. 8.22 The variation laws of the critical Reynolds number of quasilinear under different throat radii, permeability and mobility

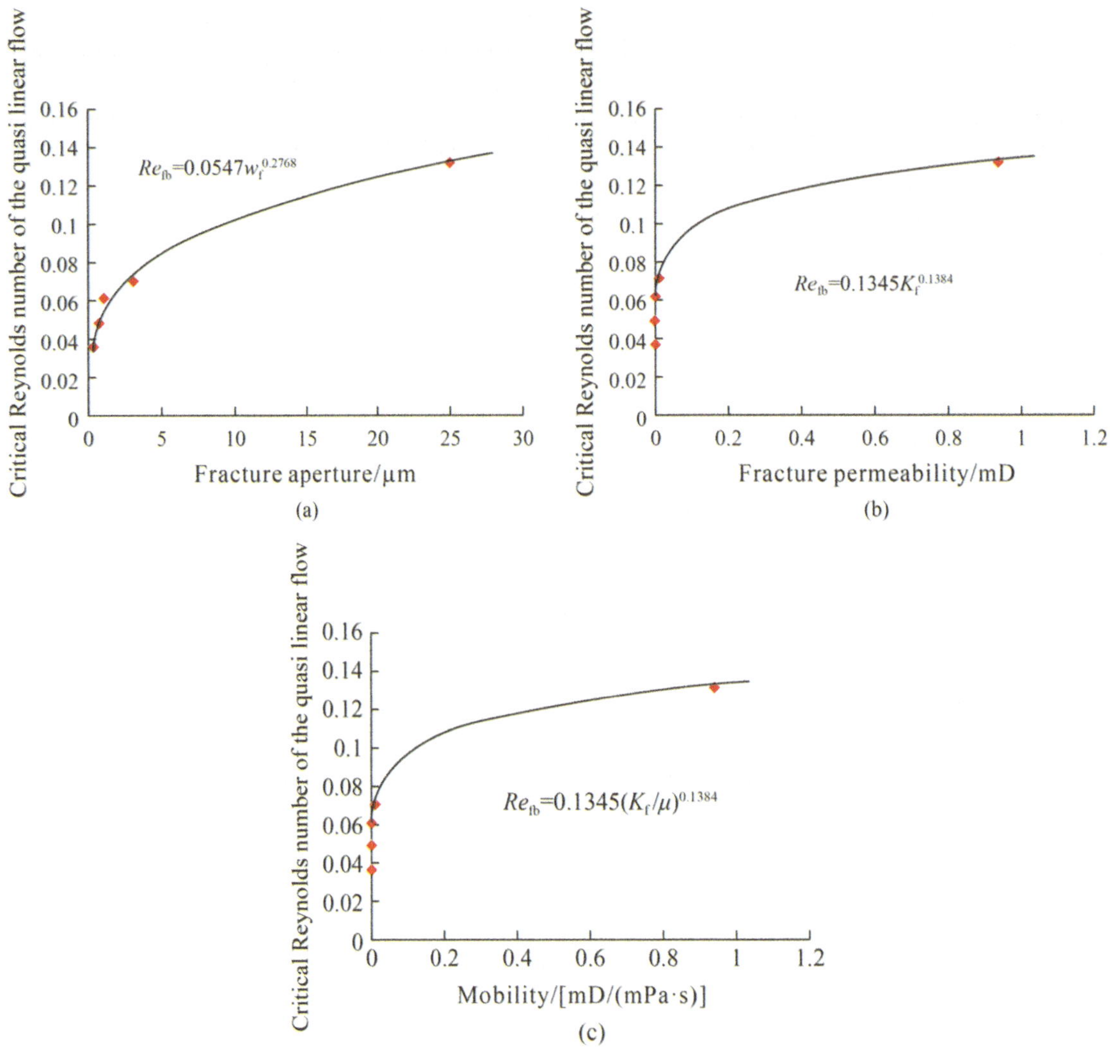

Fig. 8.23 The variation laws of the critical Reynolds number of quasilinear under different throat radii, permeability and mobility

quasilinear flow exists in these fractures. With the increase of a fracture aperture, this critical Reynolds number increases. A smaller fracture aperture induces a smaller critical Reynolds number.

The physical properties of fracture media and fluid properties have great influences on the critical Reynolds number. As the fracture permeability increases, the critical Reynolds number sharply increases. When the permeability increases to a certain value, the increase rate of the critical Reynolds number becomes slow (as shown in Fig. 8.23b). Meanwhile, as the viscosity of crude oil increases, the mobility decreases and the critical Reynolds number decreases (as shown in Fig. 8.23c).

The magnitude of the critical Reynolds number is affected by the fracture aperture, permeability and fluid properties. Therefore, the critical Reynolds number should be determined by its correlation with mobility.

The conditions where the flow regime becomes the quasilinear flow between pore media and fracture media are different. With the same permeability, compared with pore throats, the flow resistance of flow through fractures is lower and the critical Reynolds number of quasilinear flow is easier to be reached so the quasilinear flow in fractures more easily occurs (as shown in Fig. 8.24). With the same permeability, the critical Reynolds number of quasilinear flow in fracture media is more obviously lower than that in pore media. With a fracture permeability increase, the critical Reynolds number of quasilinear flow increases.

3) The critical Reynolds number of high-speed nonlinear flow

The critical Reynolds number of high-speed nonlinear flow is the Reynolds number corresponding to the critical

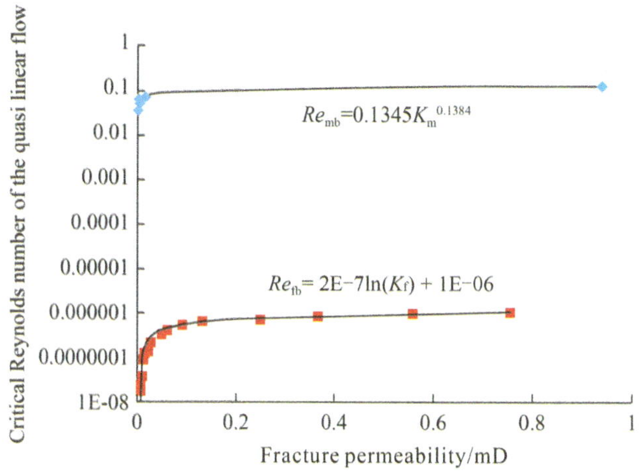

Fig. 8.24 Comparison of the critical Reynolds number of quasilinear of fracture and pore media

pressure gradient of high-speed nonlinear flow. The pore throats of tight reservoirs are small, and the flow resistance is large, so the high-speed nonlinear flow is difficult to occur. However, the flow resistance in fractures is small, and the flow velocity is large, so the high-speed nonlinear flow tends to occur. According to experiments of nonlinear flow through fracture media, the relationships of critical Reynolds numbers of different-scale fracture media with fracture apertures are determined (as shown in Fig. 8.25a). Fluids in large-scale fracture media are prone to have the high-speed nonlinear flow. With the increase of fracture apertures, the critical Reynolds number of high-speed nonlinear flow decreases.

As the fracture permeability increases, the critical Reynolds number of high-speed nonlinear flow decreases (as shown in Fig. 8.25b). As the viscosity of crude oil increases, both of the mobility and the flow velocity of crude oil flow through fracture media decrease, so the critical Reynolds number of high-speed nonlinear flow increases (as shown in Fig. 8.25c). This critical Reynolds number is affected by fracture apertures, permeability and fluid properties. Meanwhile, the mobility results from properties of fractures and fluids. Therefore, this critical Reynolds number should be determined by its correlation with mobility.

8.1.2 Critical Parameters of Microscopic Flow of Tight Gas

Under a high pressure condition, the tight gas flow through large-scale pore media is affected by the viscous force so the flow pattern is macroscopic flow. In this case, the flow regime is determined by a pressure gradient and Reynolds number. With a decrease in pore sizes, under the low pressure condition, the gas slippage and diffusion occur so the gas flow pattern is no longer the macroscopic flow but becomes the microscopic flow where the flow regime is determined by the Knudsen number.

The Knudsen number is the ratio of the free path of gas molecules to the characteristic length of the media and is proposed for the discontinuous gas flow. The flow regime of tight gas flow through nano/micro-scale media under low pressure can be determined by the Knudsen number (Wang et al. 2016). The detailed expression of the Knudsen number is

$$Kn = \frac{\bar{\lambda}}{d} \tag{8.3}$$

where, $\bar{\lambda}$ is the mean free path of gas molecules, m.

$$\bar{\lambda} = \frac{\kappa T}{\sqrt{2}\pi D_g^2 \bar{P}} = \sqrt{\frac{\pi R T}{2M}}\frac{\mu_g}{\bar{P}_g} \tag{8.4}$$

where, κ is the Boltzmann's constant, $\kappa = 1.38 \times 10^{-23}$, J/K; μ_g is the gas viscosity, mPa·s; M is the gas molecular mass; D_g is the effective diameter of gas molecules (mean spacing of molecules), m; $\bar{P}_g$ is the mean reservoir pressure of tight gas reservoirs, MPa; T is the reservoir temperature, K.

When the above parameters cannot be determined, the Knudsen number can be calculated by the following expression proposed by Freeman (2011a, b):

$$Kn = \frac{\bar{\lambda}}{d} = \frac{\sqrt{\frac{\pi}{2}}\frac{1}{P}\mu_g\sqrt{\frac{RT}{M}}}{2.81708\sqrt{\frac{K_m}{\phi_m}}} \tag{8.5}$$

where, d is the diameter of a pore throat, m; μ_g is the gas viscosity, mPa·s; M is the gas molar mass, g/mol; R is the universal gas constant, J/mol.K; K_m is the permeability of pore media, mD; ϕ_m is the porosity of pore media.

For gas flow through different pore media, a great number of experiments were conducted by a lot of scholars. Based on these experiment results, the flow regimes of microscopic gas flow can be classified into four groups which are macroscopic flow, slippage flow, transition flow and molecular diffusion. The Knudsen numbers for these four flow regimes are different. When Kn is smaller than 0.001, the flow regime of gas flow through pore media will be the macroscopic flow which is affected by the viscous force and obeys Darcy's law (Wu et al. 2015a, b; Wang et al. 2016). When Kn is higher than 0.001 and smaller than 0.1, the gas will slip on a pore wall so the gas flow regime will be the slippage flow (Javadpour et al. 2007; Sondergeld 2010).

When Kn is higher than 0.1 and smaller than 10, the flow regime of gas flow through pore media falls between slippage flow and diffusion flow. In this case, the solution accuracy of

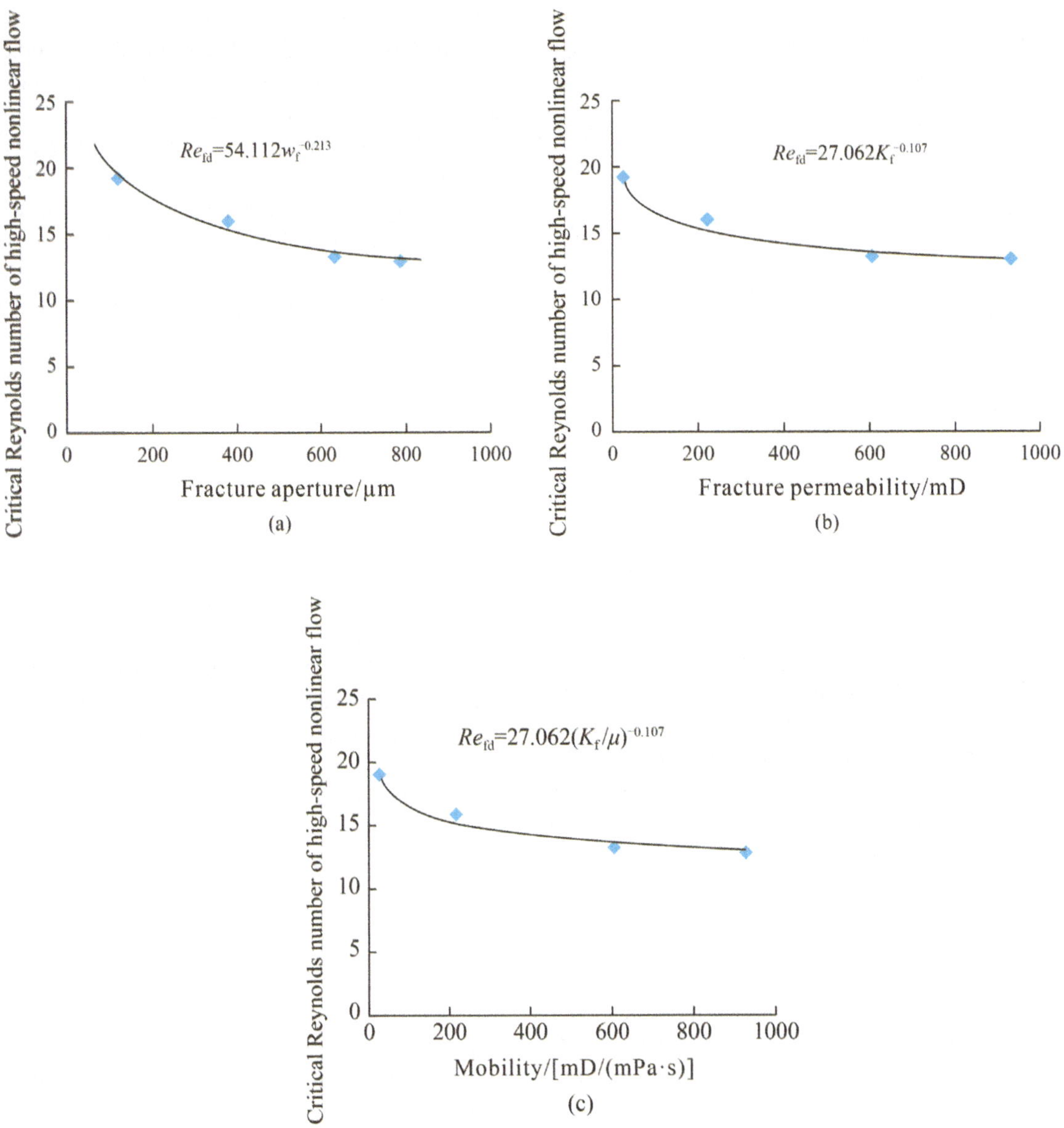

Fig. 8.25 The variation laws of the critical Reynolds number of nonlinear at high flow velocity under different fracture apertures, permeability and mobility

Table 8.1 Dividing the flow regimes by the Knudsen number

Kn	Flow regime	Mathematical model
<0.001	Macroscopic flow	Model of linear or nonlinear flow
0.001–0.1	Slippage flow	Klingenberg equation
>0.1	Diffusion	Knudsen diffusion model

gas flow can increase by using the Knudsen diffusion method (Civan 2002; Javadpour 2009; Wang et al. 2016). When *Kn* is higher than 10, the flow regime of gas flow through pore media will be the diffusion and obey the Knudsen diffusion law. Tight gas reservoirs contain nano/micro-scale pores. The Knudsen number of gas flow through nano/micro-scale pore media under low pressure is small and the majority of the corresponding Knudsen number is smaller than 10 (Wang et al. 2016). In order to simplify calculations and improve calculation accuracy, in the following, the classification of gas flow regimes was simplified according to the Knudsen number, as shown in Table 8.1.

8.2 Identification Criterion of Flow Regimes in Multiple Media for Tight Oil and Gas

8.2.1 Identification Method of Flow Regimes in Multiple Media for Tight Oil and Gas

8.2.1.1 Identification Method of Flow Regimes of Tight Oil

Tight oil reservoirs contain different-scale pores and fractures and flow regimes of fluid flow through them are different. In the early stage of production, the viscous flow is dominant and the dominant flow regimes of flow through pore and fracture media are high-speed nonlinear flow, linear flow, quasilinear flow and low-speed nonlinear flow. As the production progresses, in the later stage of production the reservoir pressure decreases, and the flow energy and flow velocity decreases as well so the dominant flow regime becomes the low-speed nonlinear flow. The flow regimes in different production stages can be determined by the critical pressure gradient or the critical Reynolds number.

1. Flow regimes determined by the critical pressure gradient

The flow regime of oil flow through tight reservoirs can be determined by the pressure gradient and mainly has three different types which are high-speed nonlinear flow, quasilinear flow and low-speed nonlinear flow. The pressure gradient criterion used to determine the high-speed nonlinear flow is the critical pressure gradient of high-speed nonlinear flow. The pressure gradient criterion used to determine the quasilinear flow is the critical pressure gradient of quasilinear flow. The pressure gradient criterion used to determine the low-speed nonlinear flow is the critical pressure gradient of low-speed nonlinear flow.

Through experiments of nonlinear flow through different-scale pore and fracture media, the models of critical pressure gradients for different-scale pore and fracture media were built (as shown in Table 8.2). Those models include models of a start-up pressure gradient, the critical pressure gradient of quasilinear flow and the critical pressure gradient of high-speed nonlinear flow. The mobility relates to properties of pore media and fluids. Therefore, these parameters can be obtained by models of correlations of those critical pressure gradients with mobility.

Where, G_a is the start-up pressure gradient, MPa/m; G_b is the critical pressure gradient of quasilinear flow, MPa/m; G_d is the critical pressure gradient of high-speed nonlinear flow, MPa/m; r is the throat radius of pore media, μm; K_m is the permeability of pore media, mD; μ is the viscosity of crude oil, mPa·s; w_f is the fracture apertures of fracture media, μm; K_f is the fracture permeability of fracture media, mD; A_r、B_r、A_k、B_k、A_λ、B_λ、C_r、D_r、C_k、D_k、C_λ、D_λ、C_w、D_w、C_{kf}、D_{kf}、$C_{\lambda f}$、$D_{\lambda f}$、E_w、F_w、E_{kf}、F_{kf}、$E_{\lambda f}$、$F_{\lambda f}$ are empirical parameters.

2. Flow regimes determined by the reynolds number

Currently, the critical Reynolds number to differentiate the quasilinear flow and low-speed nonlinear flow is 10^{-6}. When the Reynolds number is smaller than 10^{-6}, the corresponding flow regime will be low-speed nonlinear flow (Wang 2008). When the Reynolds number is higher than 10^{-6}, the corresponding flow regime will be quasilinear flow. Based on the experimental results of nonlinear flow through tight oil reservoirs, this critical value is inapplicable for majority cases.

Through experiments of nonlinear flow through different-scale pore and fracture media, the models of critical Reynolds numbers for different-scale pore and fracture media were built (as shown in Table 8.3). Those models include models of the minimum Reynolds number, critical Reynolds number of quasilinear flow and critical Reynolds number of high-speed nonlinear flow. The mobility relates to pore throat radius (permeability) and fluid properties. Therefore, these parameters can be obtained by models of correlations of those critical Reynolds numbers with mobility.

Where, Re_a is the minimum Reynolds number; Re_b is the critical Reynolds number of quasilinear flow; Re_d is the critical Reynolds number of high-speed nonlinear flow; a_r、b_r、a_k、b_k、a_λ、b_λ、c_r、d_r、c_k、d_k、c_λ、d_λ、c_w、d_w、c_{kf}、d_{kf}、$c_{\lambda f}$、$d_{\lambda f}$、e_w、f_w、e_{kf}、f_{kf}、$e_{\lambda f}$、$f_{\lambda f}$ are empirical parameters.

8.2.1.2 Identification Method of Flow Regimes of Tight Gas

Tight gas reservoirs have two types which are water-bearing tight gas reservoirs and dry tight gas reservoirs. No matter what types of tight gas reservoirs are, gas flow regimes mainly have four types which are high-speed nonlinear flow, linear flow, quasilinear flow and low-speed nonlinear flow. In the early stage of production, the viscous flow is dominant and the dominant flow regimes of gas flow through pore and fracture media are high-speed nonlinear flow, linear flow, quasilinear flow and low-speed nonlinear flow. As the production progresses, in the later stage of production, the reservoir pressure further decreases, and impacts of gas slippage and diffusion on flow increase. The flow regimes in different production stages can be determined by the combination of the critical pressure gradient or the critical Reynolds number.

Table 8.2 Computational model of the critical pressure gradient

Parameters to identify flow regimes	Pore media		Fracture media	
	Mathematical model	Parameter value	Mathematical model	Parameter value
Start-up pressure gradient	$G_{ma} = A_r \cdot r^{-B_r}$ $G_{ma} = A_k \cdot K_m^{-B_k}$ $G_{ma} = A_\lambda \cdot (K_m/\mu)^{-B_\lambda}$	$A_r = 0.087$, $B_r = 0.927$; $A_k = 0.034$, $B_k = 0.658$; $A_\lambda = 0.0296$, $B_\lambda = 0.704$;	–	–
Critical pressure gradient of quasi linear flow	$G_{mb} = C_r \cdot r^{-D_r}$ $G_{mb} = C_k \cdot K_m^{-D_k}$ $G_{mb} = C_\lambda \cdot (K_m/\mu)^{-D_\lambda}$	$C_r = 2.9623$, $D_r = 0.491$; $C_k = 1.6838$, $D_k = 0.347$; $C_\lambda = 1.5556$, $D_\lambda = 0.386$;	$G_{fb} = C_w \cdot w_f^{-D_w}$ $G_{fb} = C_{kf} \cdot K_f^{-D_{kf}}$ $G_{fb} = C_{\lambda f} \cdot (K_f/\mu)^{-D_{\lambda f}}$	$C_w = 0.5775$, $D_w = 0.77$; $C_{kf} = 0.0473$, $D_{kf} = 0.385$; $C_{\lambda f} = 0.0473$, $D_{\lambda f} = 0.385$;
Critical pressure gradient of high-speed nonlinear flow	–	–	$G_{fd} = E_w \cdot w_f^{-F_w}$ $G_{fd} = E_{kf} \cdot K_f^{-F_{kf}}$ $G_{fd} = E_{\lambda f} \cdot (K_f/\mu)^{-F_{\lambda f}}$	$E_w = 6.0537$, $F_w = 0.571$; $E_{kf} = 0.946$, $F_{kf} = 0.285$; $E_{\lambda f} = 0.946$, $F_{\lambda f} = 0.285$;

Table 8.3 Computational model of the critical Reynolds number

Parameters to identify flow regimes	Pore media		Fracture media	
	Mathematical model	Parameter value	Mathematical model	Parameter value
Minimum Reynolds number	$Re_{ma} = a_r \cdot r^{b_r}$ $Re_{ma} = a_k \cdot \ln(K_m) + b_k$ $Re_{ma} = a_\lambda \cdot \ln(K_m/\mu) + b_\lambda$	$a_r = 6\times10^{-11}$, $b_r = 0.9972$; $a_k = 7\times10^{-12}$, $b_k = 4\times10^{-11}$; $a_\lambda = 7\times10^{-12}$, $b_\lambda = 4\times10^{-11}$;	–	–
Critical Reynolds number of quasi linear flow	$Re_{mb} = c_r \cdot r^{d_r}$ $Re_{mb} = c_k \cdot K_m^{d_k}$ $Re_{mb} = c_\lambda \cdot (K_m/\mu)^{d_\lambda}$	$c_r = 1\times10^{-7}$, $d_r = 1.6287$; $c_k = 7\times10^{-7}$, $d_k = 1.2369$; $c_\lambda = 8\times10^{-7}$, $d_\lambda = 1.2432$;	$Re_{fb} = c_w \cdot w_f^{d_w}$ $Re_{fb} = c_{kf} \cdot K_f^{d_{kf}}$ $Re_{fb} = c_{\lambda f} \cdot (K_f/\mu)^{d_{\lambda f}}$	$c_w = 0.0547$, $d_w = 0.2768$; $c_{kf} = 0.1345$, $d_{kf} = 0.1384$; $c_{\lambda f} = 0.1345$, $d_{\lambda f} = 0.1384$;
Critical Reynolds number of high-speed nonlinear flow	–	–	$Re_{fd} = e_w \cdot w_f^{-f_w}$ $Re_{fd} = e_{kf} \cdot K_f^{-f_{kf}}$ $Re_{fd} = e_{\lambda f} \cdot (K_f/\mu)^{-f_{\lambda f}}$	$e_w = 54.112$, $f_w = 0.213$; $e_{kf} = 27.062$, $f_{kf} = 0.107$; $e_{\lambda f} = 27.062$, $f_{\lambda f} = 0.107$;

1. Identification method of macroscopic flow regimes of tight gas

(1) Macroscopic flow regimes determined by the critical pressure gradient

The flow regimes of gas flow through tight reservoirs mainly have three types which are high-speed nonlinear, quasilinear flow and low-speed nonlinear flow. The critical pressure gradient of high-speed nonlinear flow is used to determine the flow regime of high-speed nonlinear flow. The critical pressure gradient of quasilinear flow is used to determine the flow regime of quasilinear flow. The start-up pressure gradient is used to determine the flow regime of low-speed nonlinear flow.

Through experiments of nonlinear flow through different-scale pore and fracture media, the models of critical pressure gradients for different-scale pore and fracture media were built (as shown in Table 8.4). Those models include models of a start-up pressure gradient, the critical pressure gradient of quasilinear flow and the critical pressure gradient

Table 8.4 Computational model of the critical pressure gradient

Parameters to identify flow regimes	Pore media	Fitting coefficient	Fracture media	Fitting coefficient
Start-up pressure gradient	$G_{mga} = A_{gr} \cdot r^{-B_{gr}}$ $G_{mga} = A_{gk} \cdot K_m^{-B_{gk}}$	$A_{gr} = 0.0206$, $B_{gr} = 2.595$; $A_{gk} = 0.0007$, $B_{gk} = 1.205$;	–	–
Critical pressure gradient of quasi linear flow	$G_{mgb} = C_{gr} \cdot r^{-D_{gr}}$ $G_{mgb} = C_{gk} \cdot K_m^{-D_{gk}}$	$C_{gr} = 1.5959$, $D_{gr} = 1.737$; $C_{gk} = 0.1605$, $D_{gk} = 0.807$;	–	–
Critical pressure gradient of high-speed nonlinear flow	$G_{mgd} = E_{gr} \cdot r^{-F_{gr}}$ $G_{mgd} = E_{gk} \cdot K_f^{-F_{gk}}$	$E_{gw} = 16.508$, $F_{gw} = 1.372$; $E_{gkf} = 2.6905$, $F_{gkf} = 0.637$;	$G_{fgd} = E_{gw} \cdot e^{F_{gw} \cdot w_f}$ $G_{fgd} = E_{gkf} \cdot e^{F_{gkf} \cdot K_f}$	$E_{gw} = 0.0632$, $F_{gw} = 0.001$; $E_{gkf} = 0.0634$, $F_{gkf} = 0.0003$;

of high-speed nonlinear flow. The permeability relates to pore structure characteristics and geometry sizes of pore media. However, the gas viscosity values of different gases are close so they have less impact on gas flow. Therefore, these parameters can be obtained by models of correlations of those critical pressure gradients with permeability.

Where, G_{ga} is the start-up pressure gradient, MPa/m; G_{gb} is the critical pressure gradient of quasilinear flow, MPa/m; G_{gd} is the critical pressure gradient of high-speed nonlinear flow, MPa/m; r is the throat radius of pore media, μm; K_m is the permeability of pore media, mD; w_f is the fracture apertures, μm; K_f is the fracture permeability, mD; A_{gr} 、B_{gr} 、A_{gk} 、B_{gk} 、C_{gr} 、D_{gr} 、C_{gk} 、D_{gk} 、E_{gr} 、F_{gr} 、E_{gk} 、F_{gk} 、E_{gw} 、F_{gw} 、E_{gkf} 、F_{gkf} are empirical parameters.

(2) Macroscopic flow regimes determined by the Reynolds number

At present, there are many experiments about the flow regimes of gas flow. However, the transition condition of gas flow regimes has not been clearly defined, especially the transition conditions where the low-speed nonlinear flow regime of gas flow switches into the quasilinear flow, and the transition condition where the quasilinear flow switches into the high-speed nonlinear flow.

Through experiments of nonlinear flow through different-scale pore and fracture media, the models of critical Reynolds numbers for different-scale pore and fracture media were built (as shown in Table 8.5). Those models include models of the minimum Reynolds number, the critical Reynolds number of quasilinear flow and critical Reynolds number of high-speed nonlinear flow. The permeability relates to pore structure characteristics and geometry sizes of pore media. Therefore, these parameters can be obtained by models of correlations of those critical Reynolds numbers with permeability.

Where, Re_{ga} is the minimum Reynolds number; Re_{gb} is the critical Reynolds number of quasilinear flow; Re_{gd} is the critical Reynolds number of high-speed nonlinear flow; a_{gr} 、b_{gr} 、a_{gk} 、b_{gk} 、c_{gr} 、d_{gr} 、c_{gk} 、d_{gk} 、e_{gr} 、f_{gr} 、e_{gk} 、f_{gk} 、e_{gw} 、f_{gw} 、e_{gkf} 、f_{gkf} are empirical parameters.

2. Microscopic flow regimes identified by the Knusen number

Tight gas reservoirs contain nano/micro-scale pore media. When the size of pores in pore media is large, the flow pattern of gas flow through pore media will be macroscopic flow and the flow regime could be identified by the critical pressure gradient or the critical Reynolds number. With a decrease in pore size, the mean free path of gas molecules and the size of pore throat are almost of the same order of magnitude, and, therefore, the flow pattern of gas flow through such pores is microscopic flow and the Knudsen number is used to identify the flow regime. The identification criterion based on a Knudsen number is used to identify the flow regime of tight gas flow through nano/micro-scale pore media under a low-pressure condition and is shown in Table 8.6.

When Kn is smaller than 0.001, the flow pattern of gas flow through pore media willbe macroscopic flow, and the critical pressure gradient or the critical Reynolds number will be used to identify the corresponding flow regime. When $0.001 < Kn$ is higher than 0.001 but smaller than 0.1, the flow regime of gas flow through pore media will be the slippage flow. When Kn is higher than 0.1, the flow regime of gas flow through pore media will be Knudsen diffusion.

Table 8.5 Computational model of the critical Reynolds number

Parameters to identify flow regimes	Pore media		Fracture media	
	Mathematical model	Parameter value	Mathematical model	Parameter value
Minimum Reynolds number	$\mathrm{Re}_{\mathrm{mga}} = a_{\mathrm{gr}} \cdot r^{b_{\mathrm{gr}}}$ $\mathrm{Re}_{\mathrm{mga}} = a_{\mathrm{gk}} \cdot K_{\mathrm{m}}^{b_{\mathrm{gk}}}$	$a_r = 7\times10^{-14}$, $b_r = 0.8255$; $a_k = 2\times10^{-13}$, $b_k = 0.3833$;	–	–
Critical Reynolds number of quasi linear flow	$\mathrm{Re}_{\mathrm{mgb}} = c_{\mathrm{gr}} \cdot r^{d_{\mathrm{gr}}}$ $\mathrm{Re}_{\mathrm{mgb}} = c_{\mathrm{gk}} \cdot K_{\mathrm{m}}^{d_{\mathrm{gk}}}$	$c_{gr} = 4\times10^{-10}$, $d_{gr} = 1.1013$; $c_{gk} = 2\times10^{-9}$, $d_{gk} = 0.5113$;	–	–
Critical Reynolds number of high-speed nonlinear flow	$\mathrm{Re}_{\mathrm{mgd}} = e_{\mathrm{gr}} \cdot r^{f_{\mathrm{gr}}}$ $\mathrm{Re}_{\mathrm{mgd}} = e_{\mathrm{gk}} \cdot K_{\mathrm{m}}^{f_{\mathrm{gk}}}$	$e_{gr} = 4\times10^{-6}$, $f_{gr} = 2.1628$; $e_{gk} = 7\times10^{-5}$, $f_{gk} = 1.0042$;	$\mathrm{Re}_{\mathrm{fgd}} = e_{\mathrm{gw}} \cdot w_{\mathrm{f}}^{f_{\mathrm{gw}}}$ $\mathrm{Re}_{\mathrm{fgd}} = e_{\mathrm{gkf}} \cdot K_{\mathrm{f}}^{f_{\mathrm{gkf}}}$	$e_{gw} = 0.0005$, $f_{gw} = 2.0485$; $e_{gkf} = 0.0001$, $f_{gkf} = 1.8794$;

Table 8.6 Identification criteria based on the Knudsen number

Macroscopic flow	$Kn < 0.001$
Slippage flow	$0.001 < Kn < 0.1$
Knudsen diffusion	$Kn > 0.1$

8.2.2 Identification Criterion of Flow Regime in Multiple Media for Tight Oil and Gas

8.2.2.1 Identification Chart of Flow Regime of Tight Oil Flow

1. Identification chart of flow regime based on critical pressure gradient

Based on models of critical pressure gradients, the charts used to identify flow regimes of fluid flow through pore media and fracture media were built, including charts illustrating relationships of a pressure gradient with a throat radius (fracture aperture), permeability and mobility.

(1) Identification chart of flow regime of fluid flow through pore media

Based on models of critical pressure gradients of fluid flow through pore media, the charts illustrating relationships of a critical pressure gradient with a throat radius, permeability and mobility were built (as shown in Fig. 8.26).

This chart illustrates that both the start-up pressure gradient and the critical pressure gradient of quasilinear flow decrease with the increase of throat radius, permeability and mobility. The region where the pressure gradient is smaller than the start-up pressure gradient is the region where no flow occurs. The region where the pressure gradient falls between the start-up pressure gradient and the critical pressure gradient of quasilinear flow is the region where the low-speed nonlinear flow occurs. The region where the pressure gradient is larger than the critical pressure gradient of quasilinear flow is the region where the quasilinear flow occurs. The mobility relates to properties of pore media and fluids. Therefore, the flow regime is identified by the chart illustrating the relationship of a pressure gradient and mobility.

(2) Identification chart of flow regime of fluid flow through fracture media

Based on models of critical pressure gradients of fluid flow through fracture media, the charts illustrating relationships of a critical pressure gradient with fracture aperture, fracture permeability and mobility were built (as shown in Fig. 8.27).

The identification charts for flow regimes in fracture media (Fig. 8.27) illustrate that both the critical pressure gradient of quasilinear flow and the critical pressure gradient of high-speed nonlinear flow decrease with the increase of fracture aperture, permeability and mobility. The region where the pressure gradient is smaller than the critical pressure gradient of quasilinear flow is the region where the low-speed nonlinear flow occurs. The region where the pressure gradient falls between the critical pressure gradient of quasilinear flow and the critical pressure gradient of high-speed nonlinear flow is the region where the quasilinear flow occurs. The region where the pressure gradient is larger than the critical pressure gradient of high-speed nonlinear flow is the region where the high-speed nonlinear flow occurs. The mobility relates to properties of fracture media

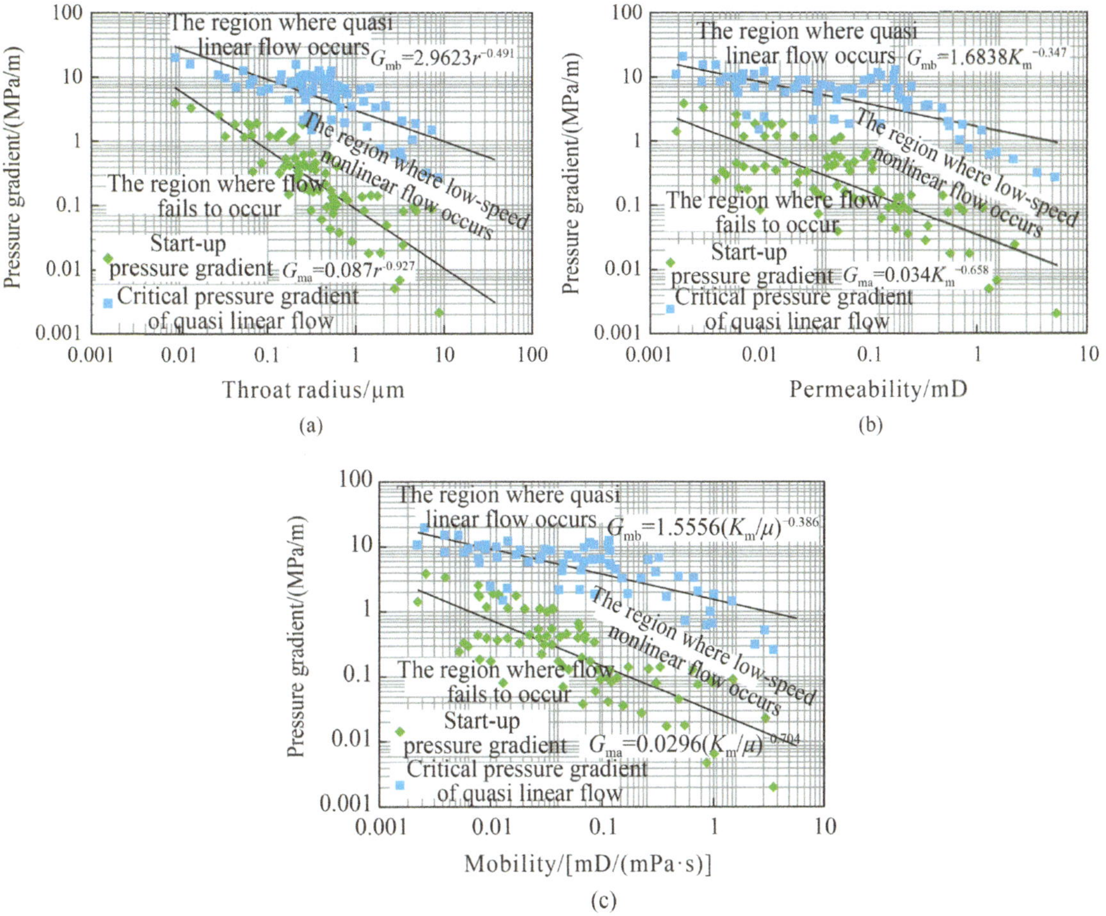

Fig. 8.26 Identification charts for flow regimes in pore media

and fluids. Therefore, the flow regime is identified by the chart illustrating the relationship of a pressure gradient and mobility.

(3) Identification chart of flow regime of fluid flow through different locations around wellbores

Based on the above models of critical pressure gradients of fluid flow, and the models of a pressure distribution around locations near wellbores, the identification chart of flow regimes of fluid flow through different locations around wellbores was built (as shown in Fig. 8.28a, b).

According to the chart illustrating the relationship of pressure gradient with wellbores, the following conclusions can be obtained. The pressure gradients at locations near wellbores are relatively high and higher than the critical pressure gradient of quasilinear flow, therefore, the quasilinear flow occurs in this region. At the region away from the wellbores, the pressure gradient falls between the start-up pressure gradient and the critical pressure gradient of quasilinear flow, therefore, the low-speed nonlinear flow occurs in this region. At the region far away from the wellbores, the pressure gradient is low and smaller than the start-up pressure gradient, therefore, the fluid flow in this region fails to occur (as shown in Fig. 8.28).

With the increase of distance from wellbore, the pressure gradient decreases, and the radius of throat through which a fluid can flow and the throat radius where the quasilinear flow can occur increase as well. Under the fixed distance from the wellbore and the value of the pressure gradient, if the pressure gradient in pores is smaller than the start-up pressure gradient, the flow in such pores will keep motionless; if the pressure gradient in pores falls between the start-up pressure gradient and the critical pressure gradient of quasilinear flow, the flow regime of fluid flow through such pores will be the low-speed nonlinear flow; if the pressure gradient in pores is higher than the critical pressure gradient of quasilinear flow, the flow regime of fluid flow through such pores will be the quasilinear flow.

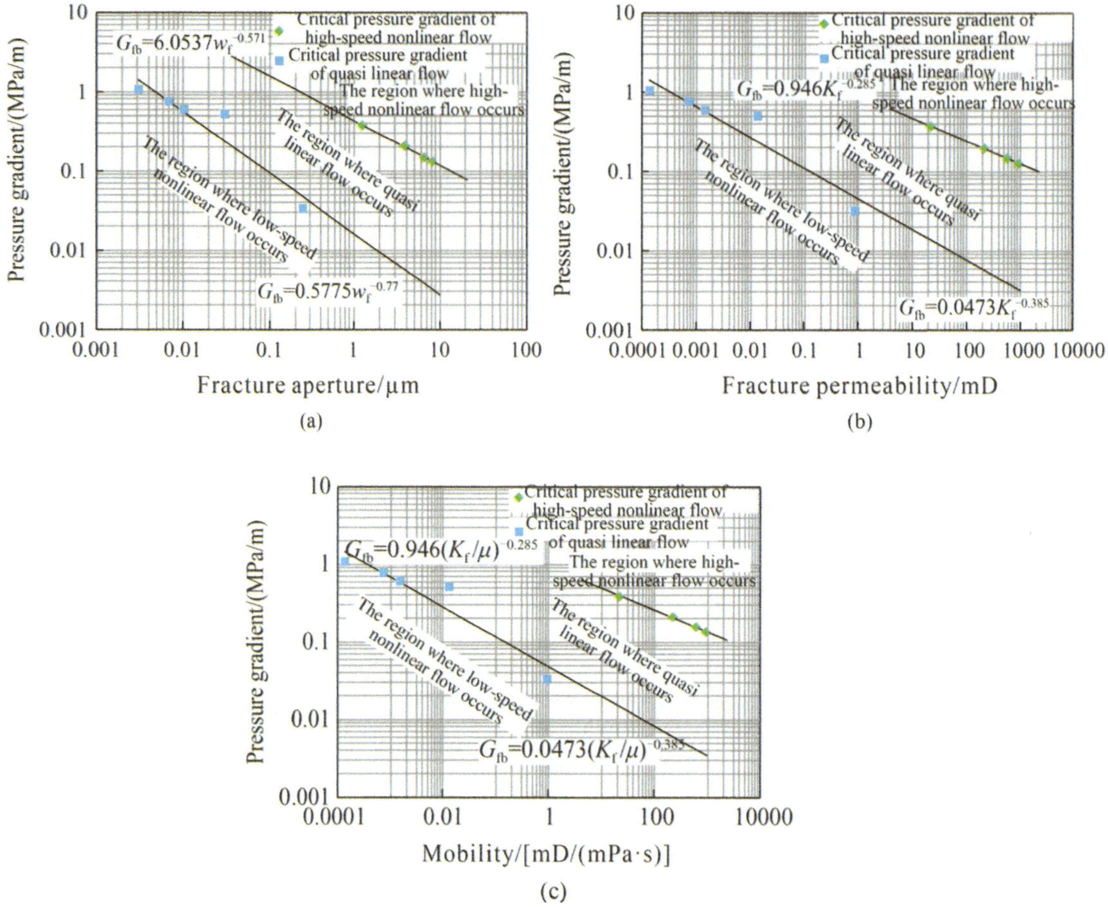

Fig. 8.27 Identification charts for flow regimes in fracture media

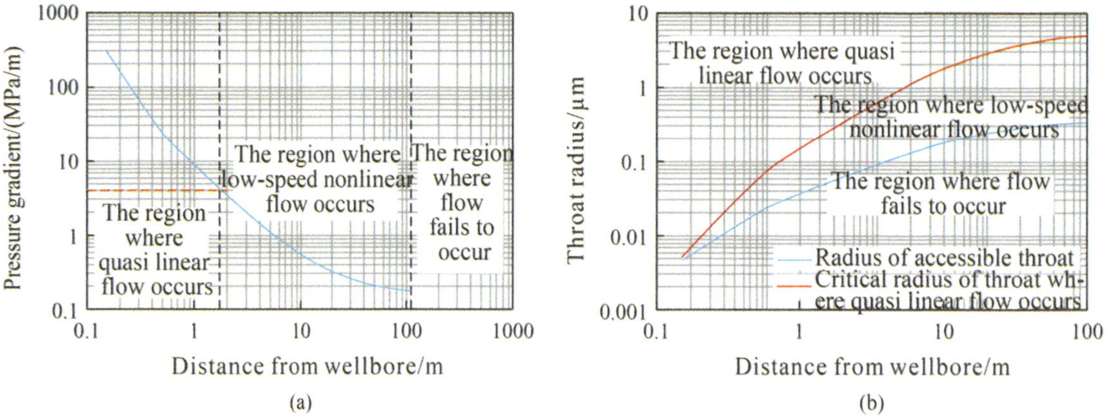

Fig. 8.28 Identification charts for flow regimes at different positions near the wellbore

2. Identification chart of flow regime based on the Reynolds number

Based on models of Reynolds numbers, the charts used to identify flow regimes of fluid flow through pore media and fracture media were built, including charts illustrating relationships of the Reynolds number with throat radius (fracture apertures), permeability and mobility.

(1) Identification chart of flow regimes of fluid flow through pore media based on Reynolds numbers

Based on models of critical Reynolds numbers of fluid flow through pore media, the charts illustrating relationships of critical Reynolds numbers with throat radius, permeability and mobility were built (as shown in Fig. 8.29).

Identification charts for flow regimes in pore media (Fig. 8.29) illustrate that both the minimum Reynolds number and the critical Reynolds number of quasilinear flow increase with the increase of throat radius, permeability and mobility. The region where the Reynolds number is smaller than the minimum Reynolds number is the region where no flow occurs. The region where the Reynolds number falls between the minimum Reynolds number and the critical Reynolds number of quasilinear flow is the region where the low-speed nonlinear flow occurs. The region where the Reynolds number is larger than the critical Reynolds number of quasilinear flow is the region where the quasilinear flow occurs. The mobility relates to pore structures (throat radius and permeability) and fluid properties. Therefore, the flow regime is identified by the chart illustrating the relationship of a Reynolds number and mobility.

(2) Identification chart of flow regimes of fluid flow through fracture media based on Reynolds number

Based on models of critical Reynolds numbers of fluid flow through fracture media, the charts illustrating relationships

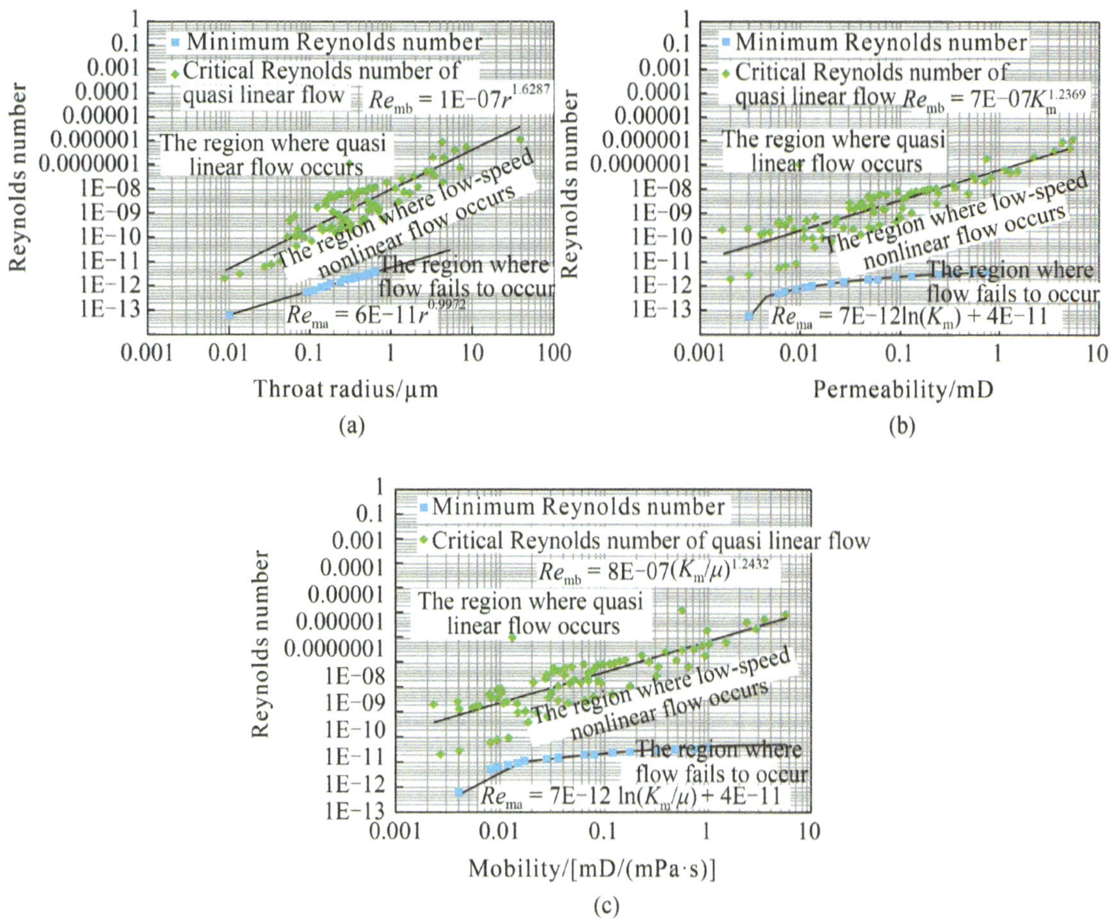

Fig. 8.29 Identification charts for flow regimes in pore media

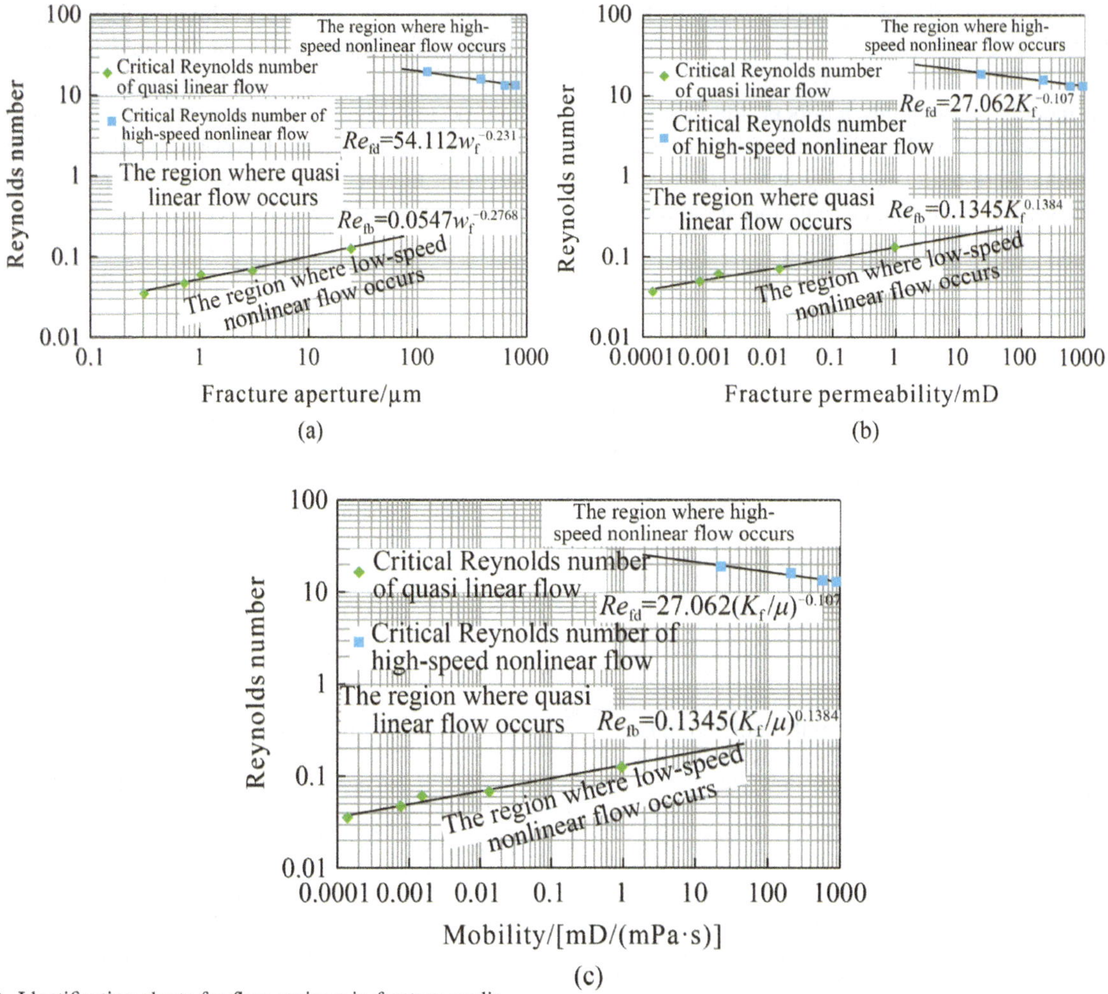

Fig. 8.30 Identification charts for flow regimes in fracture media

of the critical Reynolds number with fracture aperture, fracture permeability and mobility were built (as shown in Fig. 8.30).

Identification charts for flow regimes in fracture media (Fig. 8.30) illustrate that the critical Reynolds number of quasilinear flow increases while the critical Reynolds number of high-speed nonlinear flow decreases with the increase of fracture aperture, permeability and mobility. The region where the Reynolds number is smaller than the critical Reynolds number of quasilinear flow is the region where the low-speed nonlinear flow occurs. The region where the Reynolds number falls between the critical Reynolds number of quasilinear flow and the critical Reynolds number of high-speed nonlinear flow is the region where the quasilinear flow occurs. The region where the Reynolds number is larger than the critical Reynolds number of high-speed nonlinear flow is the region where the high-speed nonlinear flow occurs. The mobility relates to properties of fracture media and fluids. Therefore, the flow regime is identified by the chart illustrating the relationship of Reynolds number and mobility.

8.2.2.2 Identification Chart of Flow Regime of Tight Gas Flow

1. Identification chart of macroscopic flow regime based on critical pressure gradient

Based on models of critical pressure gradients of tight gas flow, the charts used to identify flow regimes of gas flow through pore media and fracture media were built, including charts illustrating relationships of pressure gradient with throat radius (fracture aperture), permeability and mobility.

(1) Identification chart of flow regime of gas flow through pore media

Based on models of critical pressure gradients of gas flow through pore media, the charts illustrating relationships of a critical pressure gradient with a throat radius, permeability and mobility were built (as shown in Fig. 8.31).

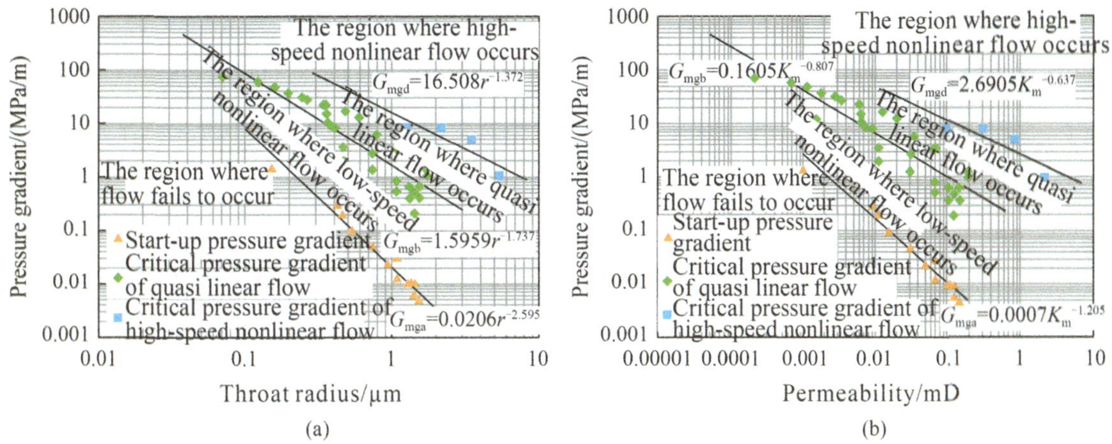

Fig. 8.31 Identification charts for flow regimes in pore media

Identification charts for flow regimes in pore media (Fig. 8.31) illustrate that the start-up pressure gradient, critical pressure gradient of quasilinear flow and the critical pressure gradient of high-speed nonlinear flow decrease with the increase of throat radius and permeability. The region where the pressure gradient is smaller than the start-up pressure gradient is the region where no flow occurs. The region where the pressure gradient falls between the start-up pressure gradient and the critical pressure gradient of quasilinear flow is the region where the low-speed nonlinear flow occurs. The region where the pressure gradient falls in between the critical pressure gradient of quasilinear flow and the critical pressure gradient of high-speed nonlinear flow is the region where the quasilinear flow occurs. The region where the pressure gradient is larger than the critical pressure gradient of high-speed nonlinear flow is the region where the high-speed nonlinear flow occurs. The permeability relates to pore structures and geometric dimensioning of pore media. Therefore, the flow regime is identified by the chart illustrating the relationship of pressure gradient and permeability.

(2) Identification chart of flow regime of gas flow through fracture media

Based on models of critical pressure gradients of gas flow through fracture media, the charts illustrating relationships of critical pressure gradients with fracture aperture and fracture permeability were built (as shown in Fig. 8.32).

The critical pressure gradient of high-speed nonlinear flow increases with the increase of fracture aperture and permeability. The region where the pressure gradient is smaller than the critical pressure gradient of high-speed nonlinear flow is the region where the quasilinear flow occurs. The region where the pressure gradient is larger than the critical pressure gradient of high-speed nonlinear flow is the region where the high-speed nonlinear flow occurs. The fracture permeability relates to the fracture size and structure

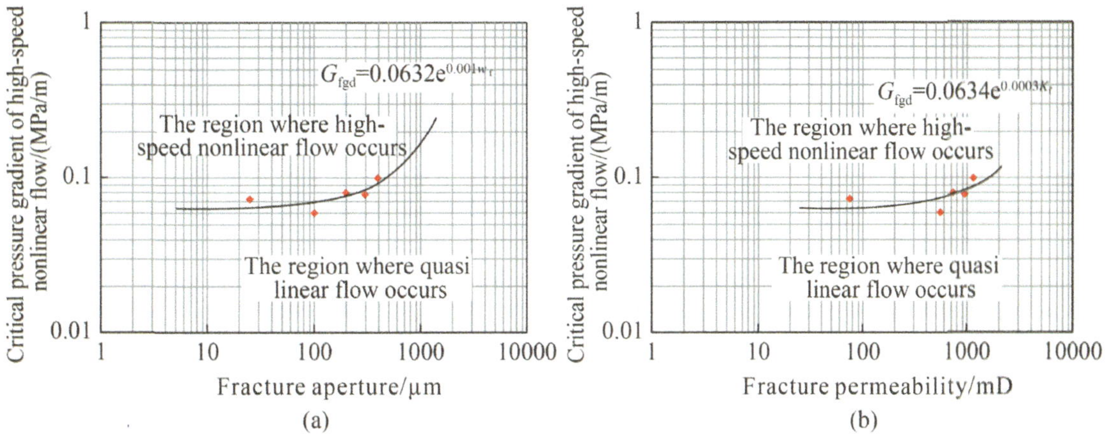

Fig. 8.32 Identification charts for flow regimes in fracture media

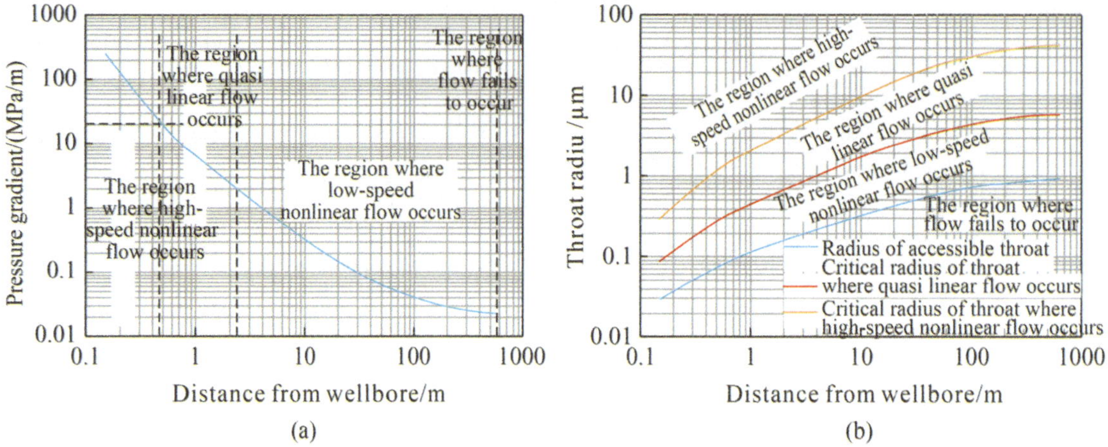

Fig. 8.33 Identification charts for flow regimes at different positions near the wellbore

characteristics of fracture media. Therefore, the flow regime is identified by the chart illustrating the relationship of pressure gradient and permeability.

(3) Identification chart of flow regime of gas flow through different locations around wellbores

Based on the above models of critical pressure gradients of gas flow and the models of pressure distribution around locations near wellbores, the identification chart of flow regimes of gas flow through different locations around wellbores was built (as shown in Fig. 8.33).

According to the chart illustrating relationship of pressure gradient with wellbores, the following conclusions can be obtained. The pressure gradients at locations near wellbores are relatively high and are higher than the critical pressure gradient of high-speed nonlinear flow, therefore, the high-speed nonlinear flow occurs in this region. The region where the pressure gradient falls between the critical pressure gradient of quasilinear flow and the critical pressure gradient of high-speed nonlinear flow is the region where the quasilinear flow occurs. At the region away from the wellbores, the pressure gradient falls between the start-up pressure gradient and the critical pressure gradient of quasilinear flow, therefore, the low-speed nonlinear flow occurs in this region. At the region far away from the wellbores, the pressure gradient is low and smaller than the start-up pressure gradient, therefore, the fluid flow in this region fails to occur (as shown in Fig. 8.33).

With the increase of distance from the wellbore, the pressure gradient decreases, the radius of throat through which gas can flow and the throat radius where the quasilinear flow and high-speed nonlinear can occur increase as well. Under the fixed distance from wellbore and the value of the pressure gradient, if the pressure gradient in pores is smaller than the start-up pressure gradient, the flow in such pores will keep motionless; if the pressure gradient in pores falls between the start-up pressure gradient and the critical pressure gradient of quasilinear flow, the flow regime of gas flow through such pores will be the low-speed nonlinear flow; if the pressure gradient in pores falls between the critical pressure gradient of quasilinear flow and the critical pressure gradient of high-speed nonlinear flow, the flow regime of gas flow through such pores will be the quasilinear flow; if the pressure gradient in pores is higher than the critical pressure gradient of high-speed nonlinear flow, the flow regime of gas flow through such pores will be the high-speed nonlinear flow.

2. Identification chart of flow regime based on the Reynolds number

Based on models of Reynolds numbers, the charts used to identify flow regimes of gas flow through pore media and fracture media were built, including charts illustrating relationships of a Reynolds number with throat radius (fracture aperture) and permeability.

(1) Identification chart of flow regime of gas flow through pore media based on Reynolds number

Based on models of critical Reynolds numbers of gas flow through pore media, the charts illustrating relationships of a critical Reynolds number with throat radius and permeability were built (as shown in Fig. 8.34).

Identification charts for flow regimes in pore media (Fig. 8.34) illustrate that the minimum Reynolds number, the critical Reynolds number of quasilinear flow and the critical Reynolds number of high-speed nonlinear flow increase with the increase of throat radius and permeability. The region where the Reynolds number is smaller than the minimum Reynolds number is the region where no flow

Fig. 8.34 Identification charts for flow regimes in pore media

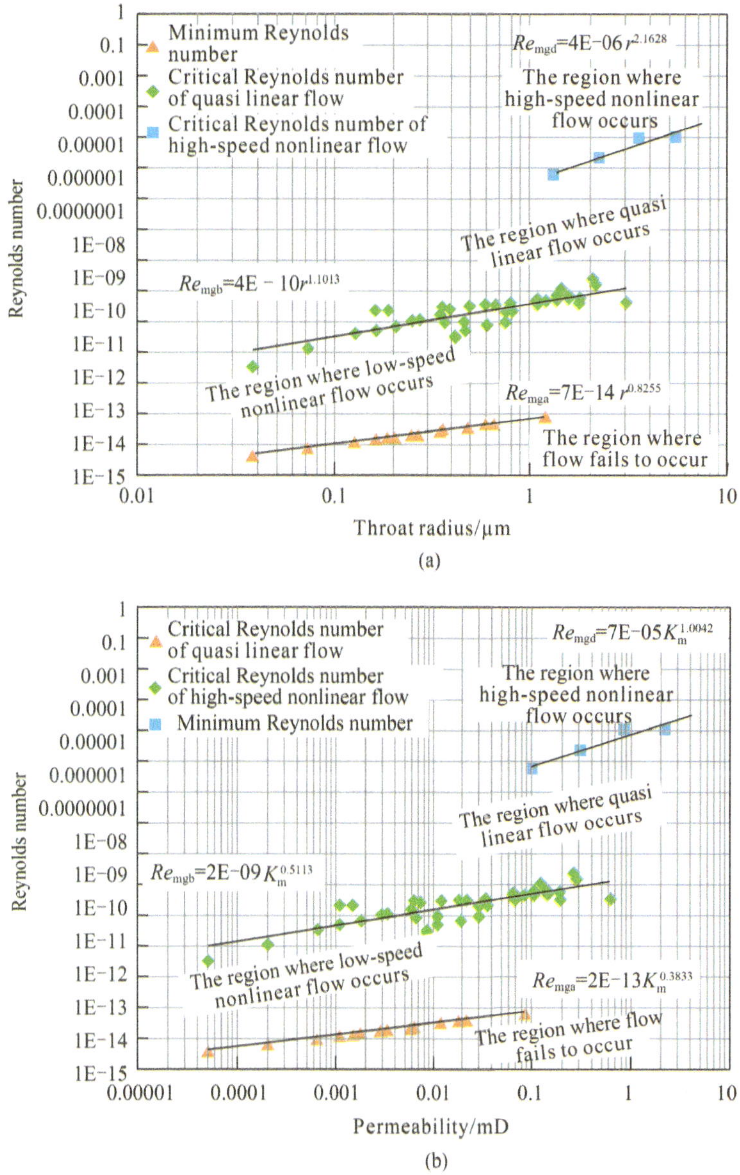

occurs. The region where the Reynolds number falls between the minimum Reynolds number and the critical Reynolds number of quasilinear flow is the region where the low-speed nonlinear flow occurs. The region where the Reynolds number falls between the critical Reynolds number of quasilinear flow and the critical Reynolds number of high-speed nonlinear flow is the region where the quasilinear flow occurs. The region where the Reynolds number is larger than the critical Reynolds number of high-speed nonlinear flow is the region where the high-speed nonlinear flow occurs. The permeability relates to pore structures and geometric dimensioning of pores. Therefore, the flow regime is identified by the chart illustrating the relationship of Reynolds number and permeability.

(2) Identification chart of flow regime of gas flow through fracture media based on Reynolds number

Based on models of critical Reynolds numbers of gas flow through fracture media, the charts illustrating relationships of critical Reynolds number with fracture aperture and fracture permeability were built (as shown in Fig. 8.35).

Identification charts for flow regimes in fracture media (Fig. 8.35) illustrate that the critical Reynolds number of high-speed nonlinear flow increases with the increase of fracture aperture and permeability. The region where the Reynolds number is smaller than the critical Reynolds number of high-speed nonlinear flow is the region where the quasilinear flow occurs. The region where the Reynolds

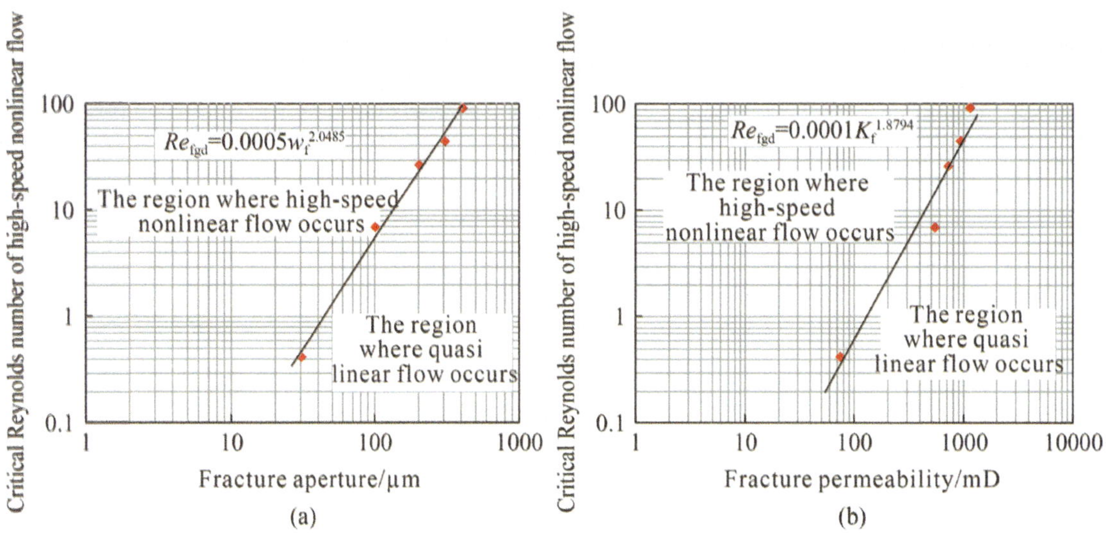

Fig. 8.35 Identification charts for flow regimes in fracture media

number is larger than the critical Reynolds number of high-speed nonlinear flow is the region where the high-speed nonlinear flow occurs. The permeability relates to structures and geometric dimensioning of fractures. Therefore, the flow regime is identified by the chart illustrating the relationship of Reynolds number and fracture permeability.

3. Identification chart of macroscopic flow regime based on Knudsen number

A large number of microscopic flow experiments show that under the same pore pressure, the Knudsen number decreases with the increase of throat radius, and the flow regime of gas flow in pore media transitions from discontinuous Knudsen diffusion and slip flow to continuous flow. The smaller throat increases the difficulty degree of gas flow through pore media and decreases the flow velocity. In this case, the phenomena of slippage flow and diffusion tends to occur so the Knudsen number becomes larger. With the pressure decrease, the Knudsen number in the same pore medium increases, and the microscopic flow tends to occur. A smaller throat diameter causes a higher Knudsen number. As the pressure increases, the Knudsen number decreases and the gas flow transitions from discontinuous flow to continuous flow (as shown in Fig. 8.36).

8.2.2.3 Identification Criteria of Flow Regime of Tight Oil and Gas

The identification criteria of flow regimes of tight oil and gas include four criteria based on geometric parameters, permeability, pressure gradient parameters and kinetics parameters, respectively. Each criterion has different adaptability to identification of flow regimes (as shown in Table 8.7).

Through the comparison between different identification criteria of flow regimes, the following summaries can be obtained. The identification criterion based on geometrical parameters identifies the flow regime through geometric dimensioning of media with different-scale pores/fractures and ignores the impacts of pore structures and fluid properties on fluid flow. Therefore, there is deviation in flow regime identification by this identification criterion. The permeability relates to pore structures and geometric parameters rather than fluid properties, so it also has disadvantages in identification of flow regimes. The pressure gradient relates to geometry, rock properties and fluid properties so it is the most direct dynamic parameter to identify flow regimes. Therefore, the pressure gradient is generally used to identify flow regimes. In addition, the kinetic parameters are calculated by the flow velocities under different pressure gradients, which indirectly represent the pressure gradients. Therefore, the kinetic parameter can also be used to identify flow regimes.

Using parameter models of flow regime identification of fluid flow through tight reservoirs, for the certain reservoir conditions, the flow regimes of gas/oil flow through different-scale pore and fracture media under the same production conditions were analyzed, and the identification criterion of flow regimes of flow of fluids with different viscosity through pore and fracture media was built (as shown in Figs. 8.37 and 8.38).

Because identification criteria of flow regimes of flow through different media are affected by the pore/fracture sizes, fluid properties, production pressure gradients and spatial location, any change of these factors results in a change of switch conditions of flow regimes. Therefore, it is necessary to use the self-adaptive identification of flow

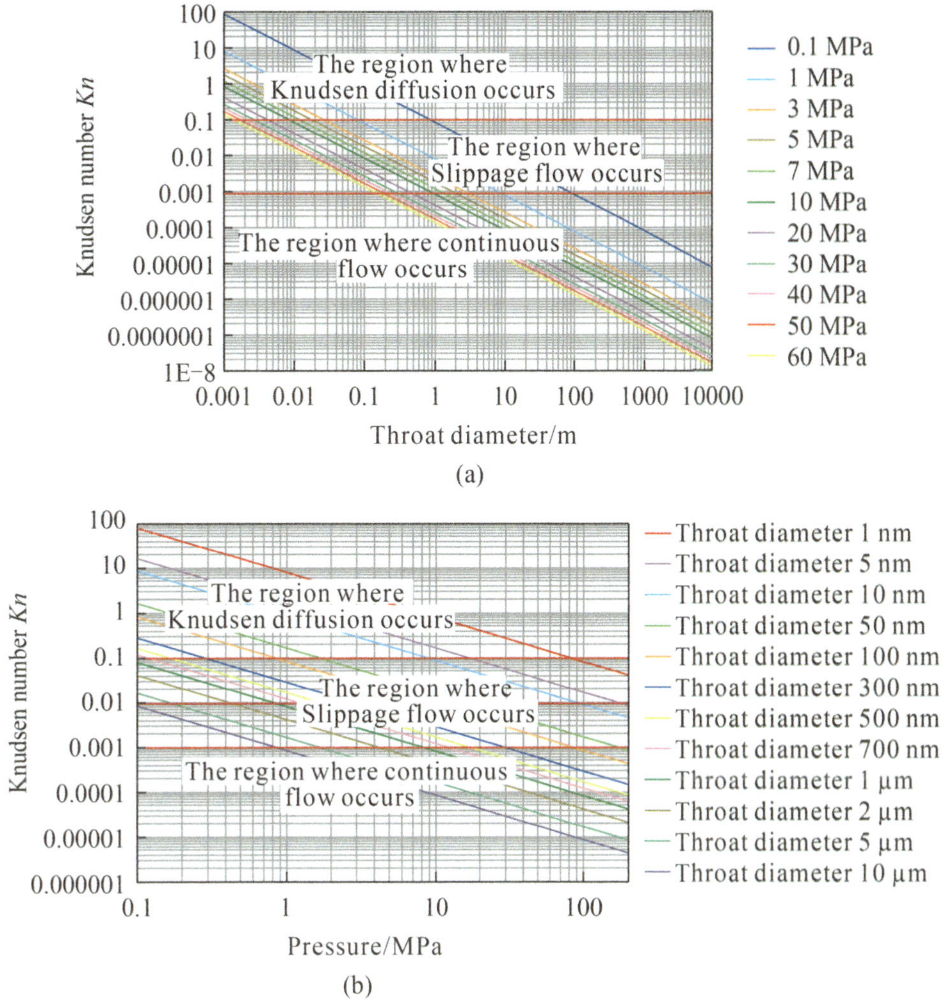

Fig. 8.36 Identification charts for flow regimes

Table 8.7 Comparison of various identification criteria of flow regimes

Identification criteria of tight oil and gas	Advantage	Disadvantage
Geometric parameter	The geometric scales of media with different-scale pores and fractures are considered in geometric parameters	The impacts of pore structures and fluid properties are not considered so the assumptions about pore structures and fluid properties deviate from the actual situation
Permeability	The pore structures and geometric parameters of media with different-scale pores and fractures are considered in permeability model	The impact of fluid properties is not considered so this method has certain limitations
Pressure gradient	The combined influences of geometric scales, physical parameters and fluid properties are considered in calculation of pressure gradient. The pressure gradient is the most direct dynamic parameter to reflect flow regimes	–
Kinetic parameter	The combined influences of geometric scales, physical parameters and fluid properties are considered in calculation of kinetic parameter	The kinetic parameter is calculated from the flow velocities under different pressure gradients, and it is an indirect characterization of pressure gradient

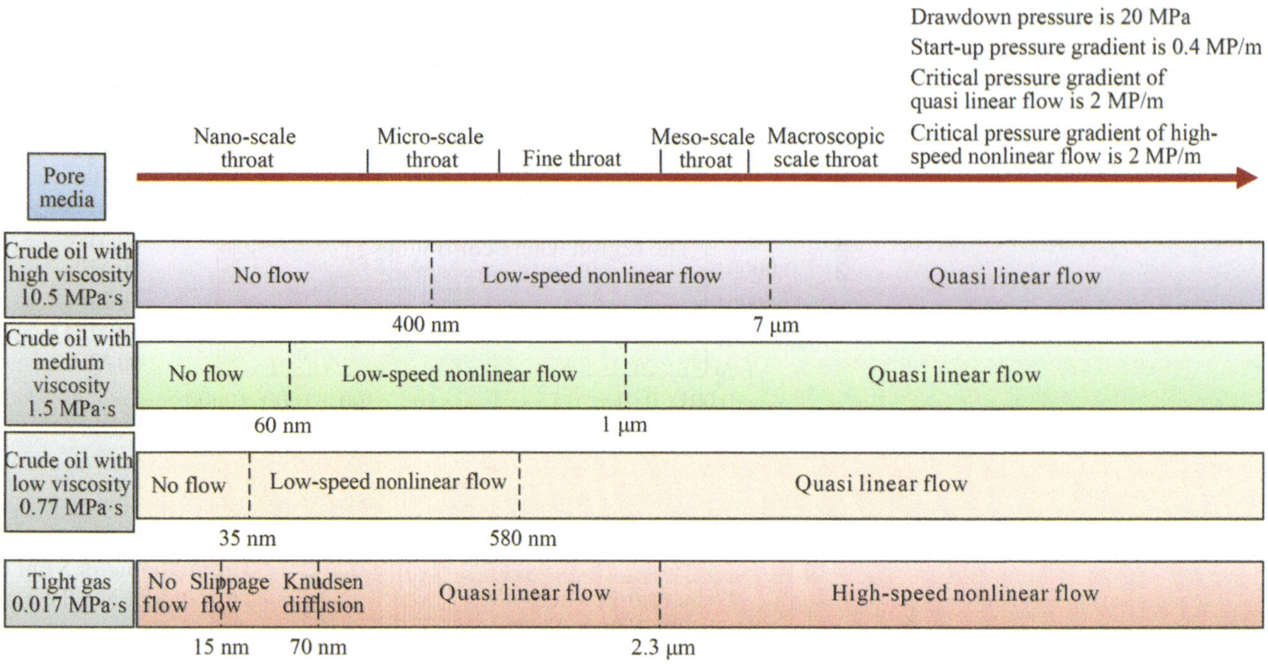

Fig. 8.37 Identification criteria for flow regimes under different fluid viscosity in pore media

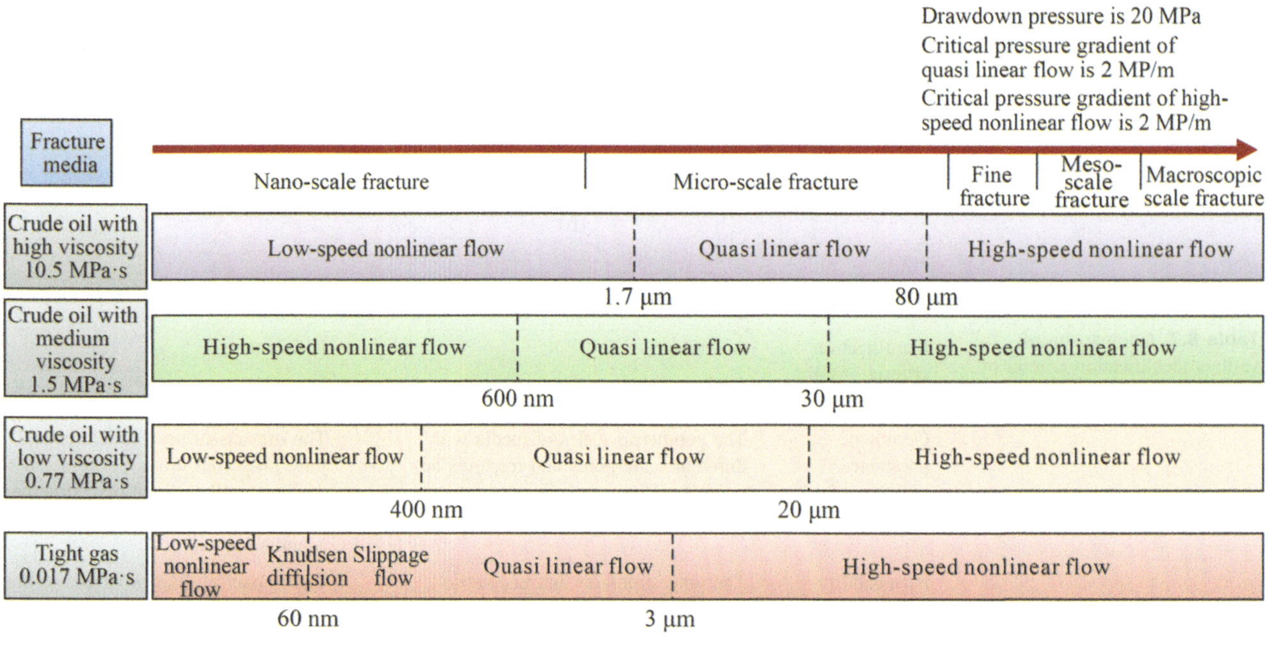

Fig. 8.38 Identification criteria for flow regimes different fluid viscosity in fracture media

regimes to identify the flow regimes of different fluids through different media under different production conditions.

Tight reservoirs contain different-scale pore and fracture media. For different fluids, the identification criteria of flow regimes of flow through different media are different. In the same pore medium, the switch conditions of flow regimes for tight oil/gas with different properties are different (as shown in Fig. 8.39). The start-up pressure gradient of tight gas is obviously smaller than the critical pressure gradient of quasilinear flow of tight oil so the flow regime of tight gas tends to switch from the low-speed nonlinear flow into the

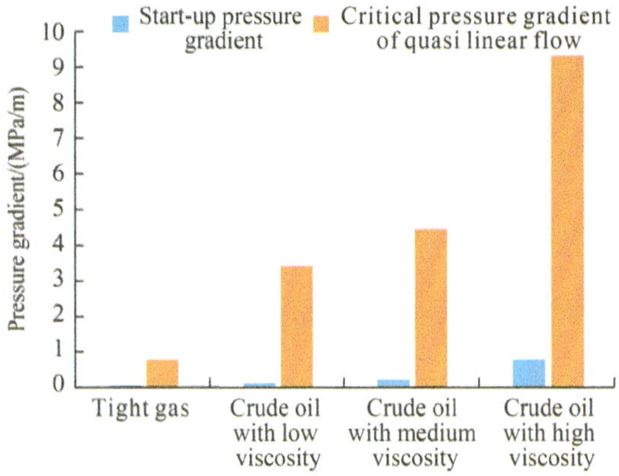

Fig. 8.39 Comparison of the threshold pressure gradient and the critical pressure gradient of quasilinear

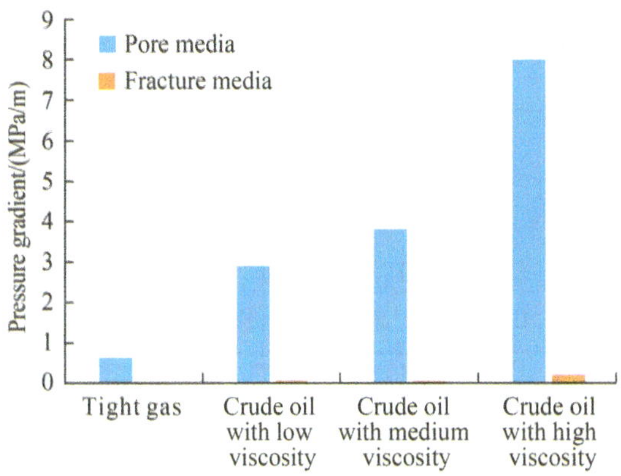

Fig. 8.40 Comparison of the critical pressure gradient of quasilinear of pore and fracture media

quasilinear flow. As the oil viscosity increases, the switch conditions of a flow regime of crude oil in the same medium become tough and the flow regime switch becomes difficult.

The start-up pressure gradient of fluid flow through fracture media is extremely low and can be negligible. Through the comparison of the critical pressure gradient of quasilinear flow of oil and gas with different properties in fracture media (as shown in Fig. 8.40), it can be seen that the critical pressure gradient of the tight gas quasilinear flow is obviously lower than that of the tight oil quasilinear flow under the same permeability, and the quite low critical pressure gradient of quasilinear flow makes the flow regime of gas flow tend to be the quasilinear flow or linear flow. However, with the increase of crude oil viscosity, the critical pressure gradient of quasilinear flow increases and the quasilinear flow of tight oil is difficult to occur.

8.3 Flow Regimes Identification and Self-Adaption Simulation of Complex Flow Mechanisms in Multiple Media with Different-Scale Pores and Fractures

Tight reservoirs have nano/micro-pores, micro-fractures, and complex networks of natural-hydraulic fractures. The geometric and attribute characteristics of those media are different, inducing obvious differences of flow mechanisms and flow regimes between those media (Singh and Azom 2013; Swami et al. 2013). Because identification criteria of flow regimes of flow through different media are affected by pore/fracture sizes, fluid properties, production pressure gradients and spatial location, any change of these factors will lead to changes in switch conditions of flow regimes. Therefore, it is necessary to use the self-adaptive identification of flow regimes to identify the flow regimes of different fluids through different media under different production conditions. Based on the laboratory experiments and production data, the flow regime of fluid flow through media was identified by using the identification criterion of flow regimes of tight oil/gas flow through different-scale media. After that, a proper model of flow through media was selected based on the flow regime identification, and the techniques of identification of flow regimes in different-scale media and the self-adaption simulation of complex flow mechanisms were built as well.

8.3.1 Identification of Flow Regimes and Self-Adaption Simulation of Complex Flow Mechanisms in Multi-Media for Tight Oil Reservoirs

Tight oil reservoirs contain different-scale pore and fracture media which have different geometric and attribute characteristics. Based on the difference of geometric, attribute characteristics and flow regimes between different-scale media, the flow regime of flow through different media is identified by a flow regime identification criterion of tight oil. Based on the identification results, a proper model of flow through media was selected and the identification of flow regimes in different-scale media and the self-adaption simulation of complex flow mechanisms were built (as shown in Fig. 8.41).

Technical process and detailed steps are as follows:

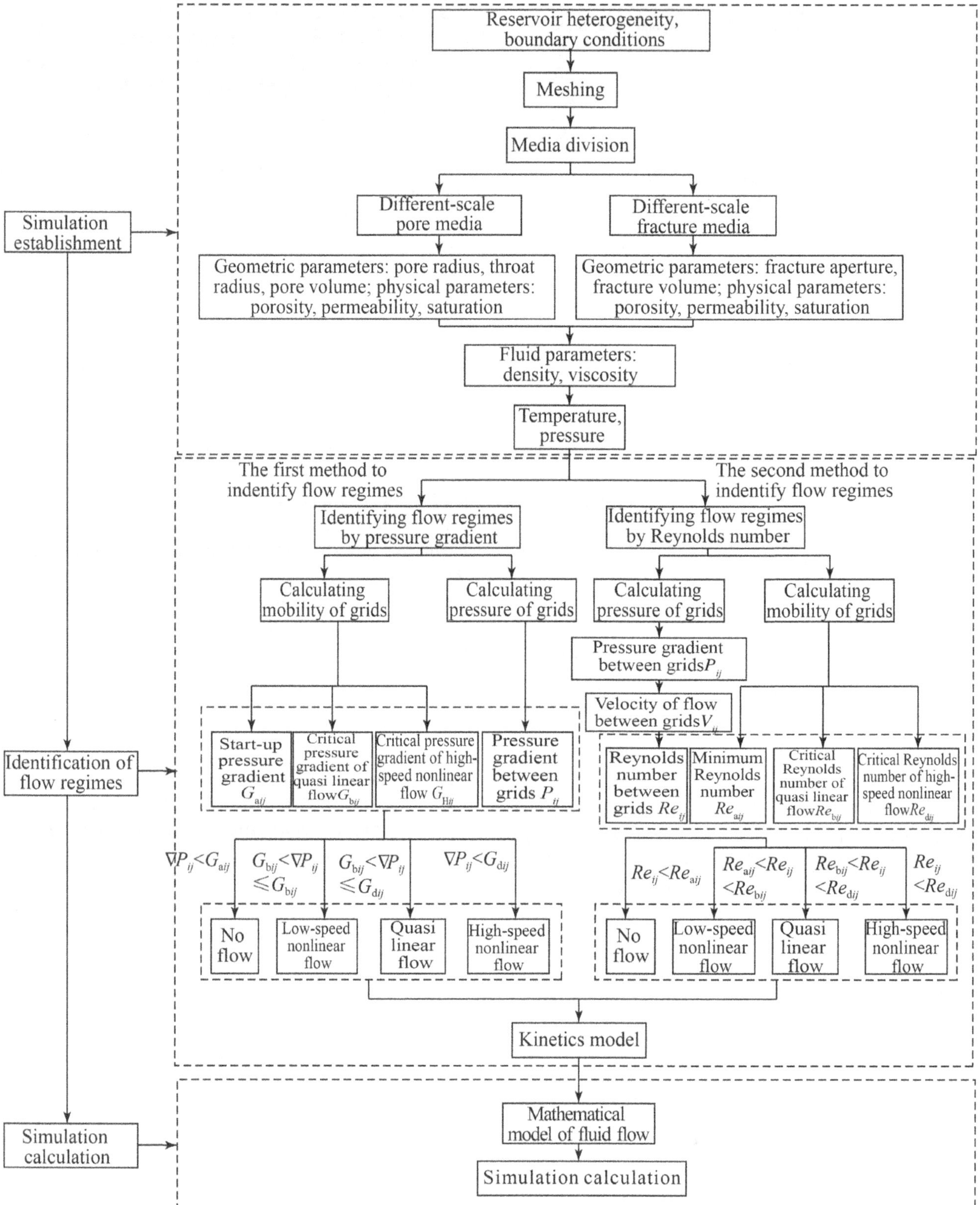

Fig. 8.41 Adaptive simulation for oil flow in tight pore media using identification of flow regimes

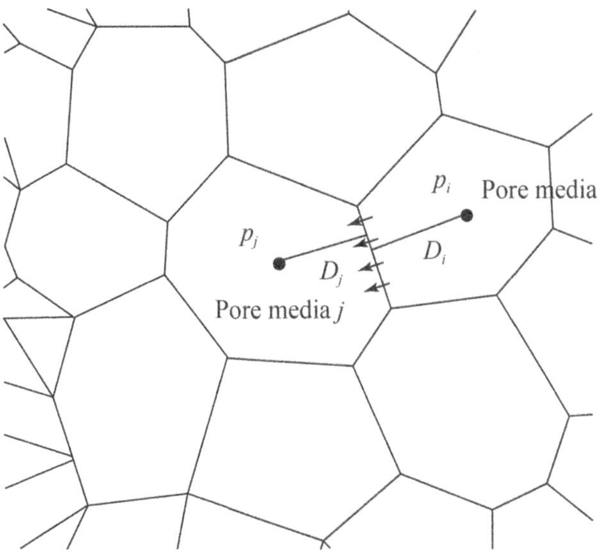

Fig. 8.42 The relationship between two adjacent grids

8.3.1.1 A Numerical Model

1. Meshing

The meshing is generated based on reservoirs heterogeneity and conditions of internal and external boundaries (as shown in Fig. 8.42).

2. Assignment of media with different-scale pores and fractures to grids

First, according to characteristics of media with different-scale pores and fractures, these media is classified into several different types of media.

Secondly, each grid is assigned with a certain type of medium (different-scale pores and fractures).

3. A parametric model for grids

① Assigning geometric parameters to each grid. A pore medium is assigned with the following parameters which are the pore radius r_p, the throat radius r, and the pore volume V_p. The fracture medium is assigned with the following parameters which are the fracture aperture w_f and the fracture volume V_f.
② Assigning physical parameters (porosity ϕ, permeability K, and saturation S_o) to each grid.
③ Assigning fluid parameters (density ρ_o, viscosity μ_o) to each grid.
④ Assigning the temperature T and the reservoir pressure P to each grid.

8.3.1.2 Identification of Flow Regimes

The flow regime of tight oil flow is the macroscopic flow which can be identified by the pressure gradient or the Reynolds number. The main flow regimes of fluid flow through pore media are low-speed nonlinear flow and quasilinear flow. The main flow regimes of fluid flow through fracture media are low-speed nonlinear flow, quasilinear flow and high-speed nonlinear flow. Based on the flow regimes, a proper dynamic model of fluid flow is selected. The detailed information is shown in Table 8.8.

1. Flow regime identification using a pressure gradient

The specific process of flow regime identification is as follows:

① Calculating the mobility K/μ_o in each grid.
② Calculating the following parameters in each grid: the start-up pressure gradient G_a, the critical pressure gradient of quasilinear flow G_b and the critical pressure gradient of high-speed nonlinear flow G_d.
③ Calculating the pressure P in each grid.

Table 8.8 Flow regimes and kinetic characteristics within different media

Media	Pressure gradient	Reynolds number	Flow regime	Kinetic model
Pore media	$\lvert\nabla P_{ij}\rvert < G_{aij}$	$Re_{ij} < Re_{aij}$	No flow	$v_{ij} = 0$
	$G_{aij} < \lvert\nabla P_{ij}\rvert \le G_{bij}$	$Re_{aij} < Re_{ij} < Re_{bij}$	Low-speed nonlinear flow	$v_{ij} = -\frac{K_{ij}}{\mu_o}(\nabla P_{ij} - G_{aij})^n$
	$G_{bij} < \lvert\nabla P_{ij}\rvert < G_{dij}$	$Re_{bij} < Re_{ij} < Re_{dij}$	Quasi linear flow	$v_{ij} = -\frac{K_{ij}}{\mu_o}(\nabla P_{ij} - G_{cij})$
Fracture media	$\lvert\nabla P_{ij}\rvert < G_{aij}$	$Re_{ij} < Re_{aij}$	No flow	$v_{ij} = 0$
	$G_{aij} < \lvert\nabla P_{ij}\rvert \le G_{bij}$	$Re_{aij} < Re_{ij} < Re_{bij}$	Low-speed nonlinear flow	$v_{ij} = -\frac{K_{ij}}{\mu_o}(\nabla P_{ij} - G_{aij})^n$
	$G_{bij} < \lvert\nabla P_{ij}\rvert < G_{dij}$	$Re_{bij} < Re_{ij} < Re_{dij}$	Quasi linear flow	$v_{ij} = -\frac{K_{ij}}{\mu_o}(\nabla P_{ij} - G_{cij})$
	$\lvert\nabla P_{ij}\rvert > G_{dij}$	$Re_{ij} > Re_{dij}$	High-speed nonlinear flow	$-\nabla P_{ij} = \frac{\mu_o v_{ij}}{K_{ij}} + \beta_{ij}\rho_o v_{ij}\lvert v_{ij}\rvert$

④ Calculating the pressure gradient ∇P_{ij} between two adjacent grids, $\nabla P_{ij} = \frac{P_i - P_j}{D_{ij}}$.

⑤ Flow regime identification: i. When the pressure gradient is smaller than the threshold pressure gradient, $|\nabla P_{ij}| < G_{aij}$, the fluid flow will fail to occur; ii. When the pressure gradient is larger than the start-up pressure gradient and smaller than the critical pressure gradient of quasilinear flow, $G_{aij} < |\nabla P_{ij}| \le G_{bij}$, the flow regime will be the low-speed nonlinear flow; iii. When the pressure gradient falls between the critical pressure gradient of quasilinear flow and the critical pressure gradient of high-speed nonlinear flow, $G_{bij} < |\nabla P_{ij}| < G_{dij}$, the flow regime will be quasilinear flow; iv. When the pressure gradient is larger than the critical pressure gradient of high-speed nonlinear flow, $|\nabla P_{ij}| > G_{dij}$, the flow regime will be the high-speed nonlinear flow.

⑥ Selection of a kinetic model: i. When the pressure gradient is smaller than the start-up pressure gradient, the kinetic model will be $\boldsymbol{v}_{ij} = 0$; ii. When the pressure gradient is larger than the start-up pressure gradient and smaller than the critical pressure gradient of quasilinear flow, the kinetic model will be $\boldsymbol{v}_{ij} = -\frac{K_{ij}}{\mu_o}\left(\nabla P_{ij} - G_{aij}\right)^n$, where n ranges from 0.9 to 1.2; iii. When the pressure gradient falls between the critical pressure gradient of quasilinear flow and the critical pressure gradient of high-speed nonlinear flow, the kinetic model will be $\boldsymbol{v}_{ij} = -\frac{K_{ij}}{\mu_o}\left(\nabla P_{ij} - G_{cij}\right)$; iv. When the pressure gradient is larger than the critical pressure gradient of high-speed nonlinear flow, the kinetic model will be $-\nabla P_{ij} = \frac{\mu_o \boldsymbol{v}_{ij}}{K_{ij}} + \beta_{ij}\rho_o \boldsymbol{v}_{ij}|\boldsymbol{v}_{ij}|$.

2. Flow regime identification using the reynolds number

The specific process of flow regime identification using the Reynolds number is as follows:

① Calculating the following parameters in each grid: i. the mobility K/μ_o; ii. the minimum Reynolds number Re_a, the critical Reynolds number of quasilinear flow Re_b and the critical Reynolds number of high-speed nonlinear flow Re_d; iii. the pressure P; iv. the pressure gradient ∇P_{ij} between two adjacent grids, $\nabla P_{ij} = \frac{P_i - P_j}{D_{ij}}$; v. the flow velocity between two adjacent grids, $\boldsymbol{v}_{ij} = -\frac{K_{ij}}{\mu_o}\left(\nabla P_{ij} - G_{aij}\right)^n$; vi. the Reynolds number between two adjacent grids, $\mathrm{Re}_{ij} = \frac{\rho_o \cdot \boldsymbol{v}_{ij} \cdot d}{\mu_o}$.

② Flow regime identification: i. When the Reynolds number is smaller than the minimum Reynolds number, $Re_{ij} < Re_{aij}$, the fluid flow will fail to occur; ii. When the Reynolds number is larger than the minimum Reynolds number and smaller than the critical Reynolds number of quasilinear flow, $Re_{aij} < Re_{ij} < Re_{bij}$, the flow regime will be the low-speed nonlinear flow; iii. When the Reynolds number falls between the critical Reynolds number of quasilinear flow and the critical Reynolds number of high-speed nonlinear flow, $Re_{bij} < Re_{ij} < Re_{dij}$, the flow regime will be quasilinear flow; iv. When the Reynolds number is larger than the critical Reynolds number of high-speed nonlinear flow, $Re_{ij} > Re_{dij}$, the flow regime will be the high-speed nonlinear flow.

③ Selection of kinetic model: i. When the Reynolds number is smaller than the minimum Reynolds number, the kinetic model will be $\boldsymbol{v}_{ij} = 0$; ii. When the Reynolds number is larger than the minimum Reynolds number and smaller than the critical Reynolds number of quasilinear flow, the kinetic model will be $\boldsymbol{v}_{ij} = -\frac{K_{ij}}{\mu_o}\left(\nabla P_{ij} - G_{aij}\right)^n$, where n ranges from 0.9 to1.2; iii. When the Reynolds number falls between the critical Reynolds number of quasilinear flow and the critical Reynolds number of high-speed nonlinear flow, the kinetic model will be $\boldsymbol{v}_{ij} = -\frac{K_{ij}}{\mu_o}\left(\nabla P_{ij} - G_{cij}\right)$; iv. When the Reynolds number is larger than the critical Reynolds number of high-speed nonlinear flow, the kinetic model will be $-\nabla P_{ij} = \frac{\mu_o \boldsymbol{v}_{ij}}{K_{ij}} + \beta_{ij}\rho_o \boldsymbol{v}_{ij}|\boldsymbol{v}_{ij}|$.

8.3.1.3 Numerical Simulation

According to the flow regime identification results in different grids, the corresponding mathematical model was selected, and numerical simulation was conducted. In the above the subscripts i,j refer to the ith and jth elements, respectively; K_{ij} is the permeability between the ith and jth elements, mD; D_{ij} is the distance between two centroids of the ith and jth elements, $D_{ij} = D_i + D_j$, m; G_{cij} is the quasi-start-up pressure gradient between the ith and jth elements, $G_{cij} = 0.1518 \cdot \left(k_{ij}/\mu_{ij}\right)^{-0.659}$, MPa/m.

8.3.2 Flow Regimes Identification and Self-Adaption Simulation of Complex Flow Mechanisms in Multiple Media for Tight Gas Reservoirs

The main difference between tight gas flow and tight oil flow is that the slippage flow and Knusen diffusion can occur when the tight gas flows through micro/nano pores under low pressure. Based on the differences of geometric, attribute characteristics and flow regimes between different-scale media, a flow regime of gas through different media was identified by the flow regime identification criterion of tight gas. Based on the identification results, a proper model of gas flow through media was selected and the identification of flow regimes in different-scale media and the self-adaption simulation of complex flow mechanisms were built (as shown in Fig. 8.43).

The main difference of flow regime identification between tight gas and tight oil is that the flow regime of gas flow can be identified by a pressure gradient, the Reynolds number as well as the Knudsen number. The Reynolds number and the pressure gradient are mainly used to identify the macroscopic flow through different-scale multiple media. Under the low-pressure condition, the microscopic flow through nano/micro-pores, such as slippage flow and Knudsen diffusion, cannot be identified by the Reynolds number and the pressure gradient. Therefore, this kind of microscopic flow is identified by the Knudsen number. The technical process and detailed steps of identification of flow regimes of tight gas flow through multiple media and self-adaption simulation of complex flow mechanisms are as follows:

8.3.2.1 A Numerical Model

1. Meshing

The meshing is generated based on reservoirs heterogeneity and conditions of internal and external boundaries (as shown in Fig. 8.42).

2. Assignment of media with different-scale pores and fractures to grids

First, according to characteristics of media with different-scale pores and fractures, these media is classified into several different types of media.

Secondly, each grid is assigned to a certain type of medium (different-scale pores and fractures).

3. A parametric model

① Assigning geometric parameters to each grid.

A pore medium is assigned with the following parameters which are the pore radius r_p, the throat radius r, and the pore volume V_p. A fracture medium is assigned with the following parameters which are the fracture aperture w_f and the fracture volume V_f.

② Assigning physical parameters (porosity ϕ, permeability K, and saturation S_g) to each grid.

③ Assigning fluid parameters (density ρ_g, viscosity μ_g) to each grid.

④ Assigning the temperature T and the reservoir pressure P to each grid.

8.3.2.2 Identification of Flow Regimes

The flow pattern of tight gas flow contains macroscopic flow and microscopic flow. The flow regime of macroscopic flow can be identified by a pressure gradient and the Reynolds number. The flow regime of microscopic flow can be identified by the Knudsen number. First, the Knudsen number is used to identify whether the gas flow pattern is the macroscopic flow or the microscopic flow. Secondly, the detailed flow regime of macroscopic flow can be identified by the pressure gradient or the Reynolds number. Finally, the flow regime of microscopic flow can be identified by the Knudsen number. The main flow regimes of tight gas flow through pore media are the Knudsen diffusion, the slippage flow, the low-speed nonlinear flow, the quasilinear flow and the high-speed nonlinear flow. The main flow regimes of tight gas flow through fracture media are the low-speed nonlinear flow, the quasilinear flow and the high-speed nonlinear flow. Based on the identified flow regime, a proper model of gas flow is selected. The detailed information is shown in Table 8.9.

1. Identification of macroscopic and microscopic tight gas flow using the Knudsen number

① Calculating the pressure P in each grid.

② Calculating the throat diameter d in each grid, $d = 2 \cdot r$.

③ Calculating the slippage factor b in each grid, $b = \frac{2\sqrt{2}c\kappa T}{\pi} D_g^2 \frac{1}{r}$

④ Calculating the Knudsen number Kn in each grid, $Kn = \frac{\sqrt{\frac{\pi RT}{2M} \frac{\mu_g}{P}}}{d}$.

⑤ When the Knudsen number is smaller than 0.001, $Kn < 0.001$, the gas flow pattern will be the macroscopic flow that can be identified by the pressure gradient or the Reynolds number.

⑥ When the Knudsen number is larger than 0.001, $Kn > 0.001$, the gas flow pattern will be the microscopic flow that can be identified by the Knudsen number.

2. Flow regime identification of macroscopic tight gas flow

(1) Flow regime identification of macroscopic tight gas flow using a pressure gradient

Detailed steps of the flow regime identification are as follows:

① Calculating the following parameters in each grid: the start-up pressure gradient G_{ga}, the critical pressure

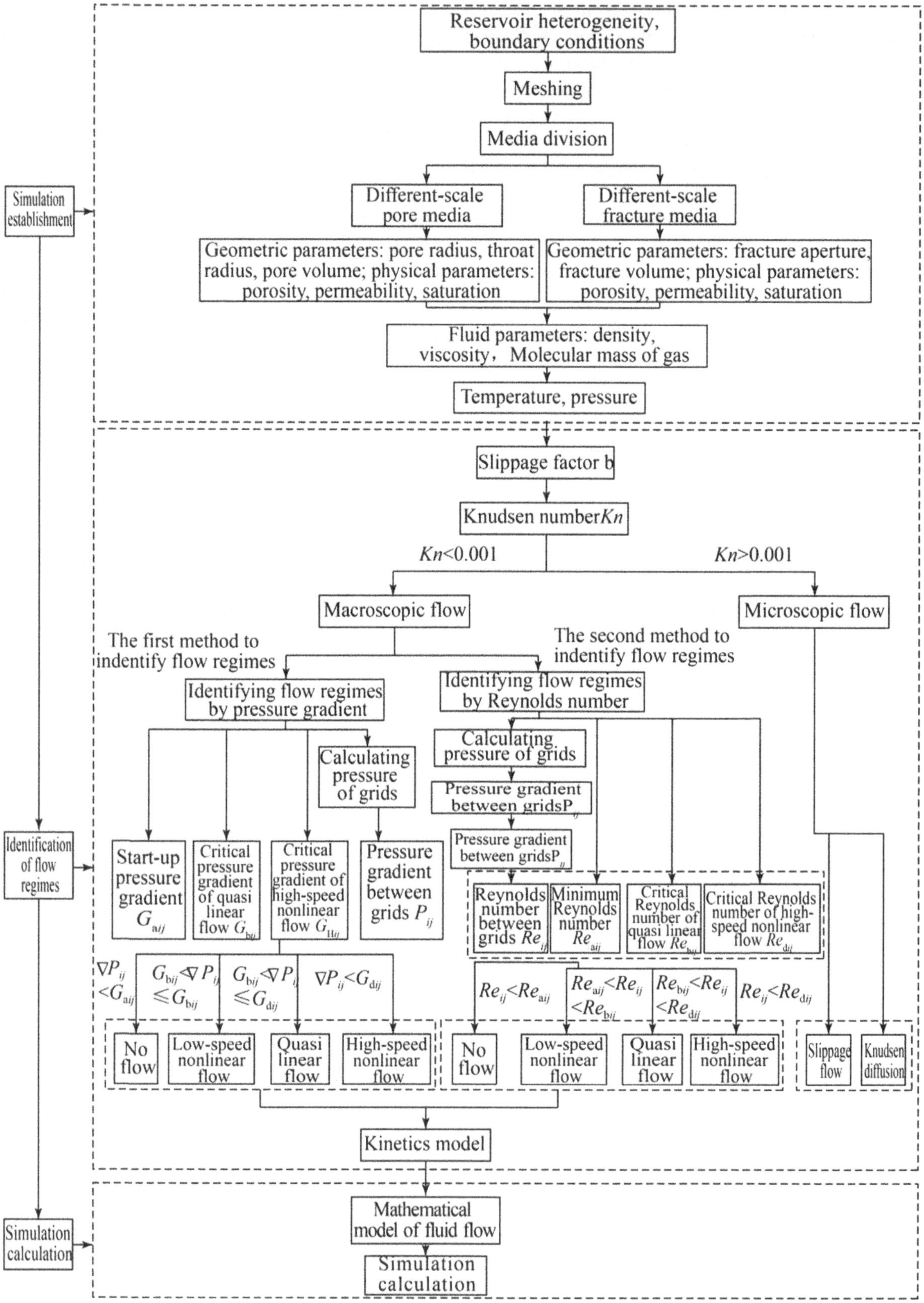

Fig. 8.43 Adaptive simulation for gas flow in tight pore media using identification of flow regimes

Table 8.9 Flow regimes and kinetic characteristics within different media

Media	Knudsen number	Pressure gradient	Reynolds number	Flow regime	Kinetic model
Pore media	$0.001 < Kn < 0.1$	$\lvert\nabla P_{ij}\rvert > G_{gaij}$	$Re_{ij} > Re_{gaij}$	Slippage flow	$v_{ij} = -\frac{K_{ij}}{\mu_g}\frac{b_{ij}}{P_{ij}}\nabla P_{ij}$
	$Kn > 0.1$			Knudsen diffusion	$v_{ij} = -\frac{32\sqrt{2RT_{ij}}}{3r\sqrt{\pi M_{ij}}P_{ij}}K_{ij}\nabla P_{ij}$
	$Kn < 0.001$	$G_{gaij} < \lvert\nabla P_{ij}\rvert \le G_{gbij}$	$Re_{gaij} < Re_{ij} < Re_{gbij}$	Low-speed nonlinear flow	$v_{ij} = -\frac{K_{ij}}{\mu_g}\left(\nabla P_{ij} - G_{gaij}\right)^n$
		$G_{gbij} < \lvert\nabla P_{ij}\rvert < G_{gdij}$	$Re_{gbij} < Re_{ij} < Re_{gdij}$	Quasi linear flow	$v_{ij} = -\frac{K_{ij}}{\mu_g}\left(\nabla P_{ij} - G_{gcij}\right)$
		$\lvert\nabla P_{ij}\rvert > G_{gdij}$	$Re_{ij} > Re_{gdij}$	High-speed nonlinear flow	$-\nabla P_{ij} = \frac{\mu_g v_{ij}}{K_{ij}} + \beta_{ij}\rho_g v_{ij}\lvert v_{ij}\rvert$
Fracture media	$Kn < 0.001$	$G_{gaij} < \lvert\nabla P_{ij}\rvert \le G_{gbij}$	$Re_{gaij} < Re_{ij} < Re_{gbij}$	Low-speed nonlinear flow	$v_{ij} = -\frac{K_{ij}}{\mu_g}\left(\nabla P_{ij} - G_{gaij}\right)^n$
		$G_{gbij} < \lvert\nabla P_{ij}\rvert < G_{gdij}$	$Re_{gbij} < Re_{ij} < Re_{gdij}$	Quasi linear flow	$v_{ij} = -\frac{K_{ij}}{\mu_g}\left(\nabla P_{ij} - G_{gcij}\right)$
		$\lvert\nabla P_{ij}\rvert > G_{gdij}$	$Re_{ij} > Re_{gdij}$	High-speed nonlinear flow	$-\nabla P_{ij} = \frac{\mu_g v_{ij}}{K_{ij}} + \beta_{ij}\rho_g v_{ij}\lvert v_{ij}\rvert$

gradient of quasilinear flow G_{gb} and the critical pressure gradient of high-speed nonlinear flow G_{gd}.

② Calculating the pressure P in each grid.

③ Calculating the pressure gradient ∇P_{ij} between two adjacent grids, $\nabla P_{ij} = \frac{P_i - P_j}{D_{ij}}$.

④ Flow regime identification: i. When the pressure gradient is smaller than the start-up pressure gradient, $\lvert\nabla P_{ij}\rvert < G_{gaij}$, the fluid flow will fail to occur ;ii. When the pressure gradient is larger than the start-up pressure gradient and smaller than the critical pressure gradient of quasilinear flow, $G_{gaij} < \lvert\nabla P_{ij}\rvert \le G_{gbij}$, the flow regime will be the low-speed nonlinear flow; iii. When the pressure gradient falls between the critical pressure gradient of quasilinear flow and the critical pressure gradient of high-speed nonlinear flow, $G_{gbij} < \lvert\nabla P_{ij}\rvert < G_{gdij}$, the flow regime will be the quasilinear flow; iv. When the pressure gradient is larger than the critical pressure gradient of high-speed nonlinear flow, $\lvert\nabla P_{ij}\rvert > G_{gdij}$, the flow regime will be the high-speed nonlinear flow.

⑤ Selection of kinetic model: i. When the pressure gradient is smaller than the threshold pressure gradient, the kinetic model will be $v_{ij} = 0$; ii. When the pressure gradient is larger than the start-up pressure gradient and smaller than the critical pressure gradient of quasilinear flow, the kinetic model will be $v_{ij} = -\frac{K_{ij}}{\mu_g}\left(\nabla P_{ij} - G_{gaij}\right)^n$, where n ranges from 0.9 to1.2; iii. When the pressure gradient falls between the critical pressure gradient of quasilinear flow and the critical pressure gradient of high-speed nonlinear flow, the kinetic model will $v_{ij} = -\frac{K_{ij}}{\mu_g}\left(\nabla P_{ij} - G_{gcij}\right)$; iv. When the pressure gradient is larger than the critical pressure gradient of high-speed nonlinear flow, the kinetic model will be $-\nabla P_{ij} = \frac{\mu_g v_{ij}}{K_{ij}} + \beta_{ij}\rho_g v_{ij}\lvert v_{ij}\rvert$.

(2) Flow regime identification of macroscopic tight gas flow using the Reynolds number

Detailed steps of identification of macroscopic tight gas flow using the Reynolds number are as follows:

① Calculating the parameters in each grid: i. the minimum Reynolds number Re_{ga}, the critical Reynolds number of quasilinear flow Re_{gb} and the critical Reynolds number of high-speed nonlinear flow Re_{gd}; ii. the pressure P; iii. the pressure gradient ∇P_{ij} between two adjacent grids, $\nabla P_{ij} = \frac{P_i - P_j}{D_{ij}}$; iv. the flow velocity between two adjacent grids, $v_{ij} = -\frac{K_{ij}}{\mu_g}\left(\nabla P_{ij} - G_{gaij}\right)^n$; v. the Reynolds number between two adjacent grids, $Re_{ij} = \frac{\rho_g \cdot v_{ij} \cdot d}{\mu_g}$;

② Flow regime identification: i. When the Reynolds number is smaller than the minimum Reynolds number, $Re_{ij} < Re_{gaij}$, the fluid flow will fail to occur; ii. When the Reynolds number is larger than the minimum Reynolds number and smaller than the critical Reynolds number of quasilinear flow, $Re_{gaij} < Re_{ij} < Re_{gbij}$, the flow regime will be the low-speed nonlinear flow; iii. When the Reynolds number fails between the critical Reynolds number of quasilinear flow and the critical Reynolds number of high-speed nonlinear flow, $Re_{gbij} < Re_{ij} < Re_{gdij}$, the flow regime will be the quasilinear flow; iv. When the Reynolds number is larger than the critical Reynolds number of high-speed nonlinear flow, $Re_{ij} > Re_{gdij}$, the flow regime will be the high-speed nonlinear flow.

③ Selection of kinetic model:

i When the Reynolds number is smaller than the minimum Reynolds number, the kinetic model will be $v_{ij} = 0$;

ii When the Reynolds number is larger than the minimum Reynolds number and smaller than the critical Reynolds number of quasilinear flow, the kinetic model will be $v_{ij} = -\frac{K_{ij}}{\mu_g}(\nabla P_{ij} - G_{gaij})^n$, where n ranges from 0.9 to 1.2.

iii When the Reynolds number fails between the critical Reynolds number of quasilinear flow and the critical Reynolds number of high-speed nonlinear flow, the kinetic model will be $v_{ij} = -\frac{K_{ij}}{\mu_g}(\nabla P_{ij} - G_{gcij})$;

iv When the Reynolds number is larger than the critical Reynolds number of high-speed nonlinear flow, the kinetic model will be $-\nabla P_{ij} = \frac{\mu_g v_{ij}}{K_{ij}} + \beta_{ij}\rho_g v_{ij}|v_{ij}|$.

3. Flow regime identification of microscopic tight gas flow

(1) Flow regime identification

① When the Knudsen number is larger than 0.001 and smaller than 0.1, $0.001 < Kn < 0.1$, the flow regime of gas flow will be the slippage flow.

② When the Knudsen number is larger than 0.1, $Kn > 0.1$, the flow regime of gas flow will be the Knudsen diffusion.

(2) Selection of kinetic model

① When the Knudsen number is larger than 0.001 and smaller than 0.1, the kinetic model will be $v_{ij} = -\frac{K_{ij}}{\mu_g}\frac{b_{ij}}{P_{ij}}\nabla P_{ij}$.

② When the Knudsen number is larger than 0.1, the kinetic model will be $v_{ij} = -\frac{32\sqrt{2RT_{ij}}}{3r\sqrt{\pi M_{ij} P_{ij}}} K_{ij}\nabla P_{ij}$.

8.3.2.3 Numerical simulation

According to the results of flow regime identification of different grids, the corresponding mathematical model of fluid flow is selected, and then the numerical simulation is conducted.

References

Civan F (2002) A triple-mechnanism fractal model with hydraulic dispersion for gas permeation in tight reservoirs. SPE74368, In: SPE international petroleum conference and exhibition in Mexico, 2002-02-10-12. Villahermosa, Mexico

Freeman CM, Moridis GJ, Ilk D, Blasingame TA (2011a) A numerical study of tight gas and shale gas reservoir system. SPE paper 124961, SPE annual technical conference and exhibition, New Orleans, Louisiana, 4-9 October

Freeman CM, Moridis GJ, Blasingame TA (2011b) A numerical study of microscale flow behavior in tight gas and shale gas reservoir systems. Transp Pore Media 90(1):253

Javadpour F (2009) Nanopores and apparent permeability of gas flow in mudrocks (shales and siltstone). J Can Pet Technol 48(8):16–21

Javadpour F, Fisher D, Unsworth M (2007) Nanoscale gas flow in shale gas sediments. J Can Pet Technol 46(10):55–61

Singh H, Azom PN (2013) Integration of nonempirical shale permeability model in a dual-continuum reservoir simulator. Presented at the SPE unconventional resources conference Canada, Calgary,5-7 November. SPE-167125-MS

Sondergeld CH (2010) Petrophysical considerations in evaluating and producing shale gas Resources. SPE 131768,In: SPE unconventional gas conference held in Pittsburgh, Pennsylvania, USA, 2010-2-23-25

Swami V, Settari A, Javadpour F (2013) A numerical model for multi-mechanism flow in shale gas reservoirs with application to laboratory scale testing. Presented at the EAGE annual conference & exhibition incorporating SPE Europec, London, UK, 10-13 June. SPE-164840-MS

Wang DF (2008) Ordos basin extra-low permeability oilfield development. Petroleum Industry Press, Beijing

Wang C (2016) Study on productivity calculation methods for fractured gas wells in tight gas reservoirs. China University of Geosciences (Beijing)

Wang HL, Xu WY, Chao ZM et al (2016) Experimental study on slippage effects of gas flow in compact rock. Chin J Geotech Eng 38 (5):185–777

Wu KL, Li XF, Chen ZX et al (2015a) Gas transport behavior through micro fractures of shale and tight gas reservoirs. Chin J Theor Appl Mech 47(6):955–964

Wu KL, Li XF, Wang C, Chen Z et al (2015b) A model for gas transport in microfractures of shale and tight gas reservoirs. AIChE J 61(6):2079–2088

Zheng M, Li JZ, Wu XZ et al (2016) Physical modeling of oil charging in tight reservoirs: a case study of Permian Lucaogou formation in Jimsar Sag, Junggar Basin, NW China, 43(2):219–227

Zhu GY, Liu XG, Yang ZM et al (2016) Study on mechanism of gas seepage characteristics in low permeability gas reservoirs. In: Proceedings of the 1st Oil and gas field development technology conference of China petroleum institute—2005 China science and technology field development science and technology development and refractory reserve mining technology seminar

9 Production Performance Simulation of Horizontal Well with Hydraulic Fracturing

Tight oil/gas reservoirs are developed by horizontal wells and hydraulic fracturing. Different completion methods (perforation and open-hole) lead to different contact types between reservoir matrix and wellbores. Coupled fluid flow exists between different reservoir media and wellbore, and it is especially complicated in the fracture network after hydraulic fracturing. Meanwhile, the pressure varies dramatically near the wellbore, causing strong flow-geomechanics coupling and intense changes in the flow behaviors towards the wellbore. It is necessary to develop a coupled flow model between reservoir media and horizontal well with hydraulic fracturing treatment.

On the basis of coupled flow between reservoir media and horizontal well after hydraulic fracturing, the horizontal well is treated with a line source, discrete grids and a multi-segment well. An innovative dynamic coupled modeling technique is developed for fractured horizontal wells with different wellbore processing modes. It is able to model the flow-geomechanics coupling between reservoirs and wellbores, flow mechanisms in multiple flow regimes, and fluid flow in wellbores, to improve the accuracy of production prediction.

9.1 Coupled Flow Pattern Between Reservoir and Horizontal Well with Hydraulic Fracturing

Tight oil/gas reservoirs are developed by horizontal wells and hydraulic fracturing. In such a reservoir, nano/micro-pores, micro-fractures, natural/artificial fractures are developed. With different completion methods (Fig. 9.1), the contact and flow behavior between multiple media and horizontal wells are distinct. With perforation completion, the multiple media directly contact artificial fractures instead of the wellbore (Table 9.1). With open-hole completion, the multiple media may directly contact either the artificial fractures other wellbore. Reservoir fluids can flow either directly or through the artificial fractures to the wellbore.

During the production process of a tight oil/gas reservoir, the pressure varies greatly near the wellbore, which causes strong flow-geomechanics coupling and an intense change in the flow regime and flow mechanisms. The pore pressure and effective stress in the multiple media at different scales near the wellbore change drastically. Natural/artificial fractures may easily deform or close up, and the pore media at different scales are prone to shrink and deform, which causes the well index to change dynamically, and greatly affects the productivity. In addition, due to the large variations in the production pressure difference (well proration), and the change of the geometric and physical properties of multiple media, the flow patterns and mechanisms of fluid flowing from reservoirs to wellbores affect the simulation of flow dynamics and productivity in different reservoir media. Based on the above description of flow characteristics, a coupled flow model between the reservoir media and the horizontal well with hydraulic fracturing treatment is developed.

9.2 Simulation of Coupled Multiphase Flow Between Reservoir and Horizontal Well with Hydraulic Fracturing

Based on the flow model for the coupled multiphase flow between reservoirs and horizontal wells with hydraulic fracturing, a simulation technology is developed for the coupled multiphase flow between reservoirs and well bores with hydraulic fracturing. Depending on the contact type between reservoirs and wellbores under different completion methods, unstructured grids are used to partition and order the reservoir and the wellbore. The adjacency relationship and the connectivity table between reservoirs and wellbore grids are established. Meanwhile, the geometric features of

Q. Ran, *Unconventional Tight Reservoir Simulation: Theory, Technology and Practice*,
https://doi.org/10.1007/978-981-32-9848-4_9

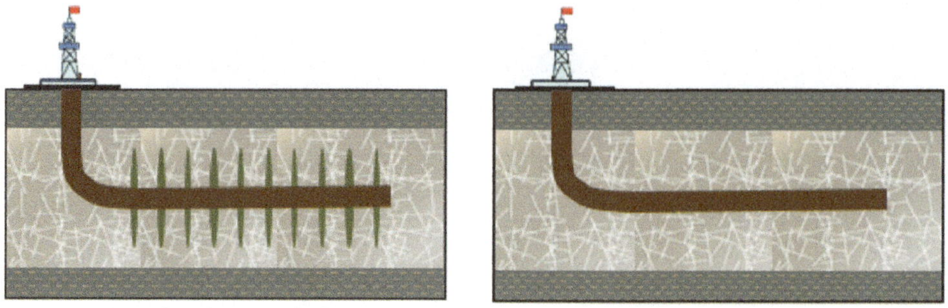

(a) Perforation completion (b) Open-hole completion

Fig. 9.1 Schematics of two completion methods

Table 9.1 Contact and flow behavior between reservoir and wellbore under two completion methods

Completion	Contact between reservoir and wellbore	Flow behavior from reservoir to wellbore	Description of flow path in reservoir media and wellbore
Perforation	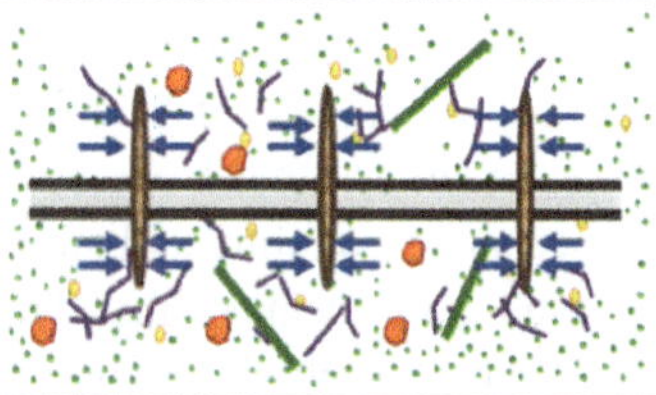	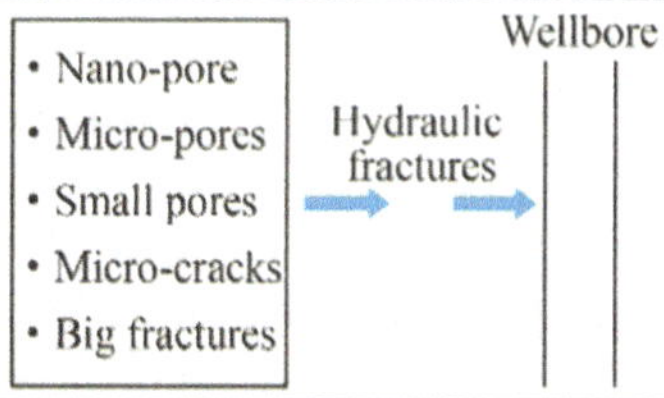	① Fluids flow from pores/natural fractures at different scales to artificial fractures ② Fluids flow from artificial fractures to wellbore
Open-hole	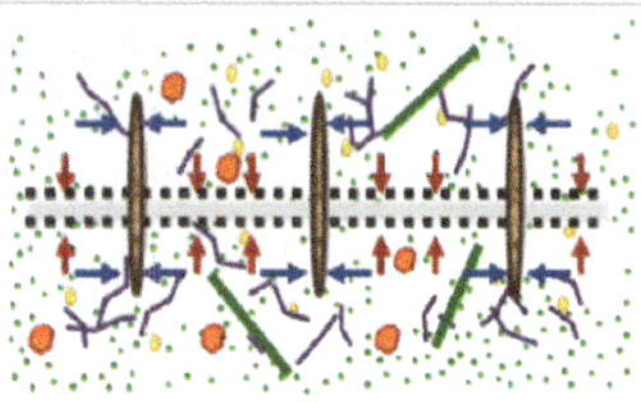	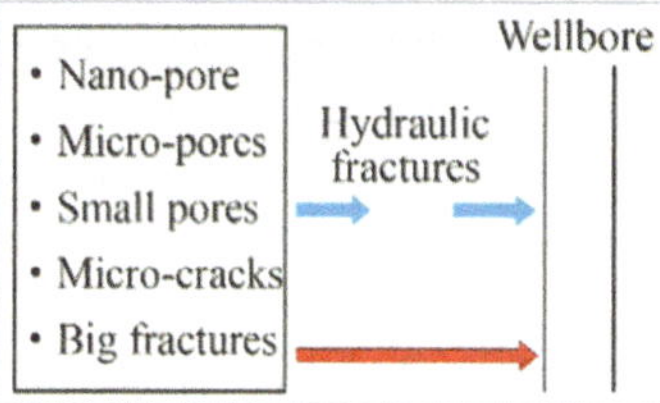	① Fluids flow from pores/natural fractures at different scales to wellbore ② Fluids flow from pore/natural fractures at different scales to wellbore through artificial fractures

reservoir grids, physical properties of multiple media, and flow-geomechanics coupling are taken into account in the development of a dynamic well index calculation model. A line source, discrete gridding, and multi-segment well treatment are applied to treat the horizontal well with hydraulic fracturing under different completion conditions, which lead to three coupled flow simulation techniques for line source, discrete wellbore, multi-segment well, respectively. It is used to simulate the flow behaviors and coupled flow-geomechanics between reservoirs and wellbores, complex flow mechanisms in multiple flow regimes, and the flow dynamics inside the wellbore, in order to improve the accuracy of productivity prediction (Table 9.2).

9.2.1 Coupled Flow Simulation with Line-Source Wellbore Scheme

A line-source wellbore scheme treats a horizontal well as a line source that consists of several point sources. Each point source has fluid exchange with reservoir media. The production of a horizontal well is the sum of the production from all point sources.

According to the contact type and flow paths between reservoir and wellbore under different completion methods (Fig. 9.2), the reservoir grids are partitioned by using the line-source wellbore scheme. The point sources are located, and the grids and point sources are ordered. The grid connectivity table between reservoir grids and wellbore grids, the dynamic well index calculation model with line-source wellbore scheme, the model for fluid exchange between reservoir and wellbore, and the production model are developed. A horizontal well with hydraulic fracturing under different completion methods are treated with the line-source wellbore scheme, and the coupled flow simulation for the line-source wellbore scheme is developed.

9.2.1.1 Grid Partitioning and Ordering

Structured grids are applied for the line-source wellbore scheme. The principal direction of structured grids is set the same as that of a horizontal well. Along the well trajectory, the line-source wellbore is treated as fine grids and the surrounding as coarse grids. The overall ordering is arranged according to the rows and columns, and the grid number increases along the horizontal well direction. The locations of point sources are determined and ordered according to the

Table 9.2 Wellbore treatment techniques

Technique content		Line-source	Discrete grid	Multi-segment well
Completion	Perforation	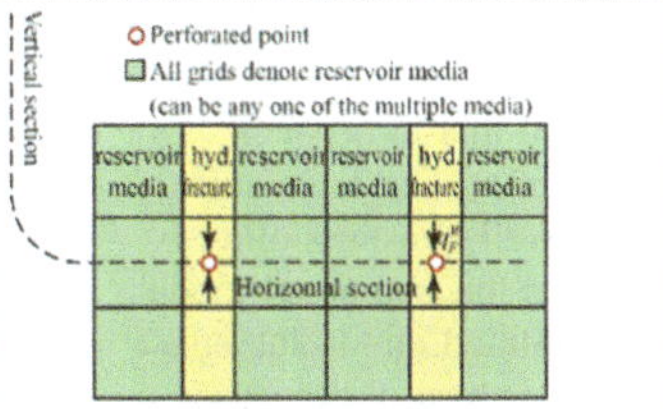	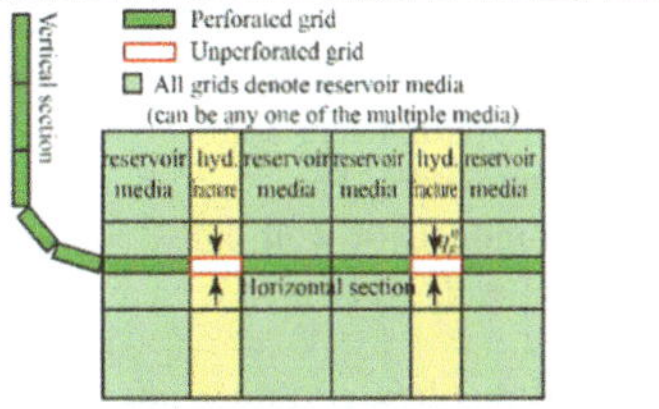	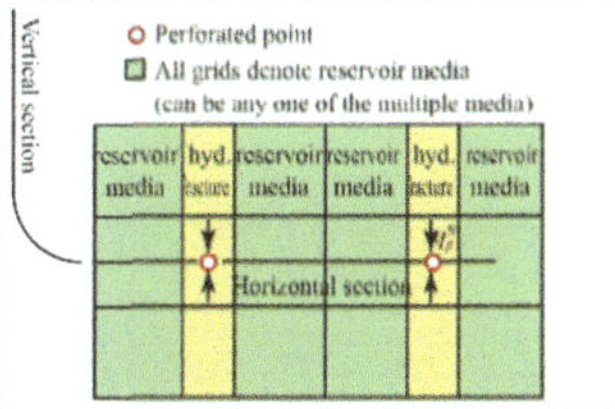
	Open-hole	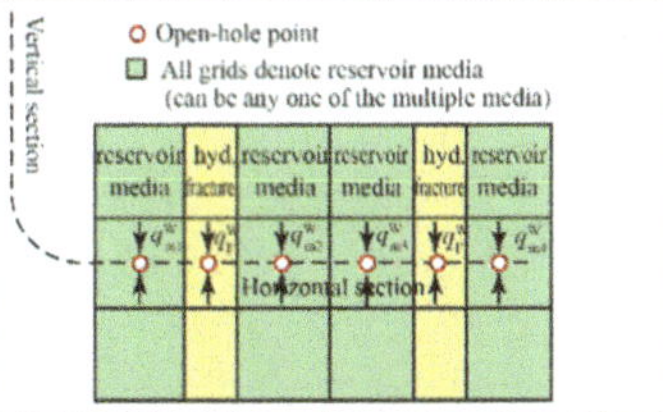	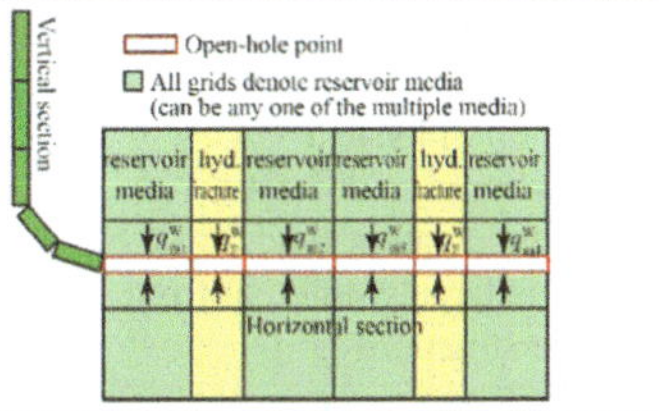	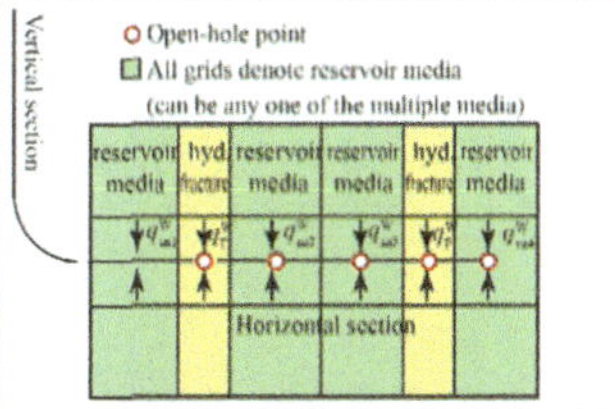
Method description		A horizontal well is treated as a line source that consists of several point sources. Each point source has fluid exchange with the reservoir media. The production of the horizontal well is the sum of that of each point source	A horizontal well is divided into a number of discrete grids. The fluids in the reservoir grids exchange through the adjacent and connected wellbore grids. The fluid exchange occur between wellbore grids. The produciton of the horizontal well is determined by the export flowrate of the discrete wellbore grid	A horizontal well is divided into multiple segments. Each segment contains at most one perforation point. The fluid exchange occurs between the well segment and the adjacent and connected reservoir grids. Inside a well segment, friction, gravity, and multiphase flow are considered; in-between well segments, fluid exchange and pressure loss are considered
Grid partitioning and ordering		① Grid disection: the principal direction of the structural/unstructured grid is set the same as that of the horizontal well. Along the horizontal well trajectory, the source wellbore grid and the surrounding grid are processed respectively ② Grid ordering: order the grids according to the grid shape, the horizontal well direction, and the well trajectory. The number of the source wellbore grid not only represents the number of the reservoir grid, but that of the wellbore	① Grid disection: the principal direction of the structural/unstructured grid is set the same as that of the horizontal well. Along the horizontal well trajectory, the source wellbore grid and the surrounding grid are processed respectively ② Grid ordering: order the grids of discrete well segments and reservoir raccording to the grid shape, the horizontal well direction, and the well trajectory	Based on the line-source wellbore treatment, fluid flows inside wellbore and between wellbore segments are considered. Its grid partitioning and ordering are similar to that of the line-source wellbore
Connectivity table		① Perforation: fluid exchange occurs between reservoir and wellbore girds, also between horizontal well grids and the adjacent and connected surrounding reservoir grids, but does not happen between reservoir and non-performation/fracture grids ② Open-hole: in the grids for fracture, fluid exchange may occur between the artificial fracture and the wellbore; in the grids without fractures, fluid exchange may occur between reservoir media and wellbore; fluid exchange occur between horizontal wellbore grids and the adjacent and connected reservoir grids	① Perforation: fluid exchange may occur between discrete wellbore girds wellbore and the adjacent fracture grids forperforation/fracturing, but does not happen between discrete wellbore grids and multiple grids of different scales for non-perforation/fracturing grids; fluid exchange may occur in-between wellbore grids ② Open-hole: fluid exchange may occur between fractured wellbore discrete grids and the adjacent hydraulic fracture grids, also between non-fractured wellbore grids and the grids for multiple media at different scales, and in-between wellbore grids	Based on the line-source treatment, fluid flow inside wellbores and in-between wellbore segments are considered. Its connectivity table is similar to that of the line-source wellbore

(continued)

Table 9.2 (continued)

Technique content	Line-source	Discrete grid	Multi-segment well
Well index calculation model	① The structured mesh has definite direction, geometric characteristic parameters and physical parameters. The calculation models of well index for different well types are different ② The unstructured grid does not havedifinite direction. There is large difference in geometric parameters and physical parameters between different grids. The calculation model of well index is largely different from that for the structural grid	The fluid exchange between the wellbore grids and reservoir grids is similar to that in-between reservoir grids. The calculation model of well transmissibility is the same as the traditional model, but is simplified due to the effect of reservoir physical properties	The calculation model of well index is the same as that of line-source wellbore
Production prediction model	① The model for fluid exchange between reservoir grids and wellbore grids ② Production calculation model	① The model for fluid exchange between reservoir grids and wellbore grids ② The model for fluid exchange between wellbore grids ③ Production calculation model	① The model for fluid exchange between reservoir grids and wellbore grids ② The model for fluid flow between wellbore segments 3. The model for fluid exchange between wellbore grids 4. Production calculation model
Advantages and disadvantages	Advantages: simple calculation, few variables, and easy to solve	Advantages: consideration of the pressure loss caused by friction, grivity, few variables	Advantages: consideration of the pressure loss caused by friction, grivity, acceleration, high calculation accuracy
	Disadvantages: low accuracy, negligence of pressure loss due to fluid flow in the wellbore, inaccurate production split	Disadvantages: Fast flow in the wellbore grid, bad convergence of the calculation	Disadvantages: complicated calculation, bad convergence due to the incorporation of multiphase flowrate and liquid holdup variable

contact type/flow path between reservoir and wellbore. The serial number of the fine grids stands for not only the serial number of the reservoir media but also those of the wellbore point sources (Fig. 9.3).

Unstructured hybrid grids are applied for the line-source wellbore scheme. The principal direction of the unstructured grids is set the same as that of the horizontal well. Along the well trajectory, a line-source wellbore is set as long-strip rectangular grids and the surrounding as PEBI grids. The overall ordering is: first, the long-strip rectangular grids are ordered along the horizontal well trajectory; then the surrounding PEBI meshes are ordered (as shown in Fig. 9.4). The location of point sources is determined and ordered according to the contact type/flow path between the reservoir and the wellbore. The serial number of the long-strip rectangular grids stands for not only the serial number of the reservoir media but also those of the wellbore point sources (Table 9.3).

9.2.1.2 Flow Path and Connectivity Table

Based on the contact type and flow behavior between a reservoir and a horizontal well under different completion methods, the connectivity table is developed for the reservoir and the horizontal well.

For the perforation completion, in terms of the contact type between reservoir and horizontal well, the multiple media of different scales only contact artificial fractures and cannot directly reach a wellbore; in terms of the flow behavior between reservoir and horizontal well, the artificial fractures may contact the wellbore directly. Fluids in the multiple media of different scales should flow to the artificial fractures first and then the wellbore. A line source consists of several point sources along the horizontal well trajectory. The point source is located in the reservoir grids for perforation/fracturing, in which the fluid exchange happens between reservoir with wellbore and does not exist between non-perforation/fracturing reservoir grids and wellbore. Fluid exchange occurs between the fine grid for the wellbore and the coarse grid for the surrounding area. Based on the contact type and flow behavior between the reservoir and the horizontal well under perforation completion, a connectivity table is established for the reservoir and horizontal well (Table 9.4).

For the open-hole completion, in terms of the contact type between reservoir and horizontal well, the multiple media of

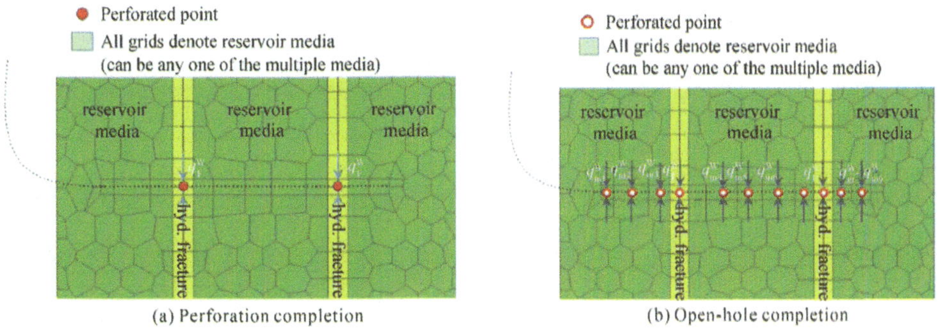

Fig. 9.2 Relationship between reservoir and wellbore for different completion methods under line-source wellbore scheme

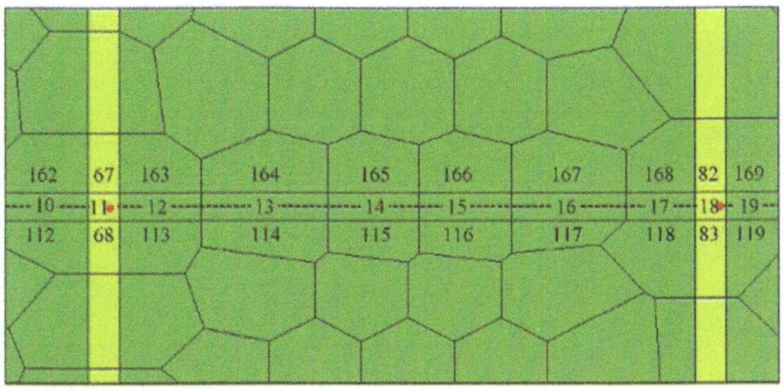

Fig. 9.3 Grid partitioning and ordering for horizontal well under perforation completion

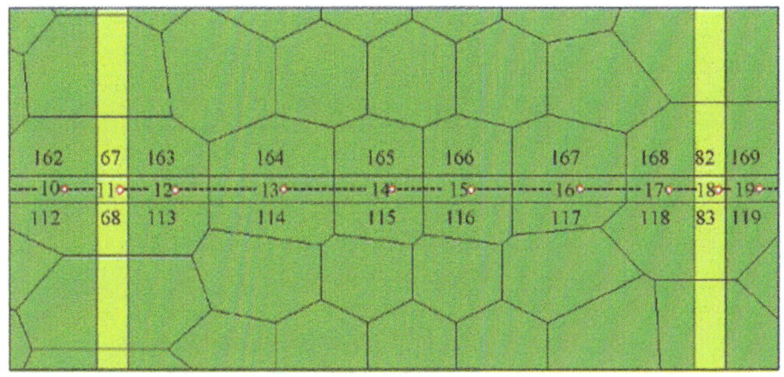

Fig. 9.4 Grid partitioning and ordering for horizontal well under open-hole completion

different scales contacts not only artificial fractures but also a wellbore. in terms of the flow behavior between reservoir and horizontal well, the fluid in the reservoir can flow directly to the wellbore, or indirectly through the artificial fractures. The line source consists of several point sources along the horizontal well trajectory. The point source is located in the long-strip grids. In the long-strip grids for fractures, the fluid exchanges between the fractures and the wellbore. In other long-strip grids, the fluid exchange occurs between the multiple media of different scales and the wellbore. The fluid exchange also happens between the long-strip grids for the wellbore and the PEBI grids for the reservoir matrix. Based on the contact type and flow behavior between the reservoir and the horizontal well under open-hole completion, a connectivity table is established for the reservoir and the horizontal well (Table 9.5).

9.2.1.3 Calculation Model for Dynamic Well Index

For the line-source wellbore scheme, a wellbore is a point source in the center of reservoir grids. The well index represents the geometric characteristics of grids for point sources, physical properties of reservoir, and the effect of wellbore geometric characteristics on productivity. It is a main control factor for the fluid exchange between reservoir and horizontal well.

1. Calculation model for well index with structured grids

With structured grids, the reservoir meshes have definite orientation, geometric characteristics (length, width, and height), and physical parameters (anisotropic permeability). When different well types (vertical well, horizontal well, slant well) are used, the geometric parameters of the point source wellbore are rather different, so are their respective well index calculation models (Table 9.6) (Peaceman 1978, 1982, 1991, 1995; Chen et al. 2006; Chen 2011).

2. Calculation model for well index with unstructured grids

With unstructured grids, reservoir grids have irregular shapes with no definite orientation. The geometric characteristics of different grids and the physical parameters of reservoir media vary greatly, which is rather different from the calculation model for well index with structured grids. The corresponding well index calculation models are shown in Table 9.7.

3. Calculation model for dynamic well index

The pore pressure and effective stress near a wellbore change with time during the fracturing-injection-production process, which leads to the dynamic variation of the geometric

Table 9.3 Grid partitioning and ordering with a line-source wellbore scheme

Grid type	Grid partitioning and ordering	Completion	Schematics of grid partitioning and ordering
Structured grid	The principal direction of the structured grids is set the same as that of the horizontal well. Along the well trajectory, the line-source wellbore is set as fine grids and the surrounding as coarse grids.The serial number of grids increases along the horizontal well	Perforation	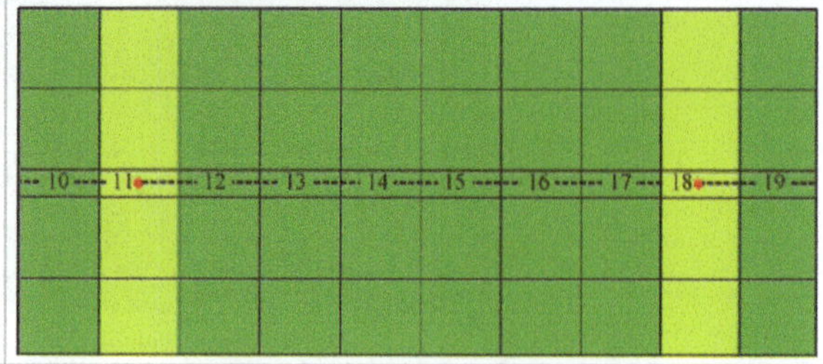
		Open hole	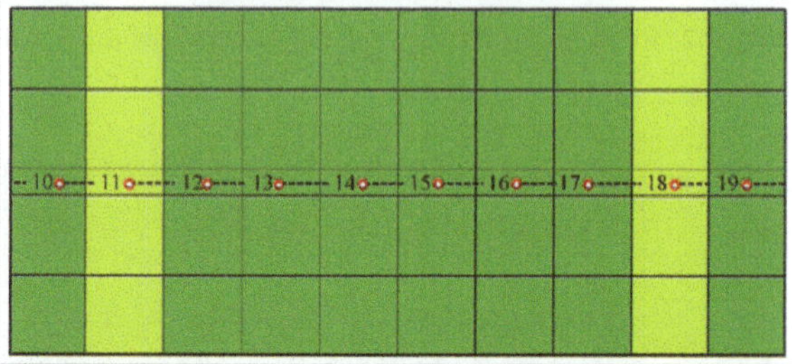
Unstructured grid	The principal direction of the unstructured grid is set the same as that of the horizontal well. Along the well trajectory, the line-source wellbore is set as long-strip rectangular grids and the surrounding as PEBI grids. The long-strip rectangular grids are ordered first along the horizontal well trajectory, and then the surrounding PEBI grids are ordered	Perforation	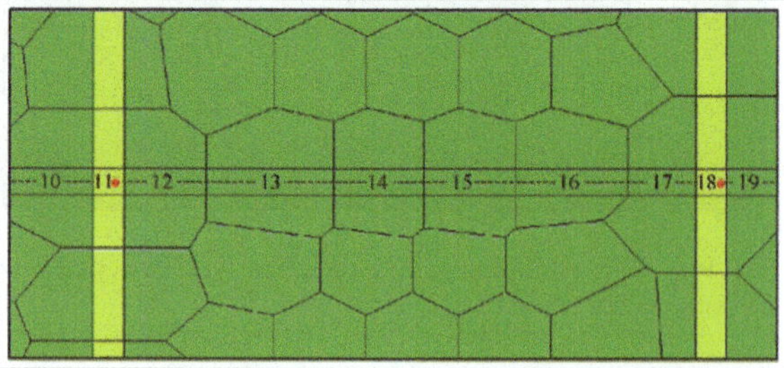
		Open hole	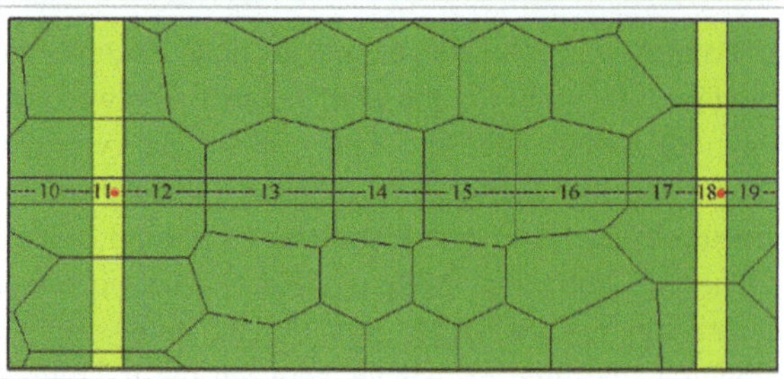

Table 9.4 Connectivity table for reservoir-wellbore grids under perforation completion

Unit	Connectivity table
W_{11}, R_{11}	(W_{11}, R_{11}, $WI_{W11,\ R11}$)
R_{11}, R_{67}	(R_{11}, R_{67}, $T_{R11,\ R67}$)
R_{11}, R_{68}	(R_{11}, R_{68}, $T_{R11,\ R68}$)
R_{11}, R_{10}	(R_{11}, R_{10}, $T_{R11,\ R10}$)
R_{11}, R_{12}	(R_{11}, R_{12}, $T_{R11,\ R12}$)

Note W is the wellbore point source, R is the reservoir media grids; WI is the well index, T is the transmissibility

characteristics of reservoir grids for wellbore and physical parameters of reservoir media, and further affects the well index and productivity.

Table 9.5 Connectivity table for reservoir-wellbore grids under open-hole completion

Unit	Connectivity table
W_{11}, R_{11}	(W_{11}, R_{11}, $WI_{W11,\ R11}$)
R_{11}, R_{67}	(R_{11}, R_{67}, $T_{R11,\ R67}$)
R_{11}, R_{68}	(R_{11}, R_{68}, $T_{R11,\ R68}$)
R_{11}, R_{10}	(R_{11}, R_{10}, $T_{R11,\ R10}$)
R_{11}, R_{12}	(R_{11}, R_{12}, $T_{R11,\ R12}$)
W_{12}, R_{12}	(W_{12}, R_{12}, $WI_{W12,\ R12}$)
R_{12}, R_{10}	(R_{12}, R_{10}, $T_{R12,\ R10}$)
R_{12}, R_{13}	(R_{12}, R_{13}, $T_{R12,\ R13}$)
R_{12}, R_{113}	(R_{12}, R_{113}, $T_{R12,\ R113}$)
R_{12}, R_{163}	(R_{12}, R_{163}, $T_{R12,\ R163}$)

Table 9.6 Well index calculation model for different well types with structured girds

<table>
<tr><th>Well type</th><th>Well index calculation model</th><th>Schematics</th></tr>
<tr><td>Vertical well</td><td>$$\mathrm{WI}=\frac{2\pi\sqrt{K_xK_y\Delta z}}{\ln\left(\frac{r_e}{r_w}\right)+s}\quad r_e=0.28\frac{\left[\left(\frac{K_y}{K_x}\right)^{\frac{1}{2}}\Delta x^2+\left(\frac{K_x}{K_y}\right)^{\frac{1}{2}}\Delta y^2\right]^{\frac{1}{2}}}{\left(\frac{K_y}{K_x}\right)^{\frac{1}{4}}+\left(\frac{K_x}{K_y}\right)^{\frac{1}{4}}}$$</td><td>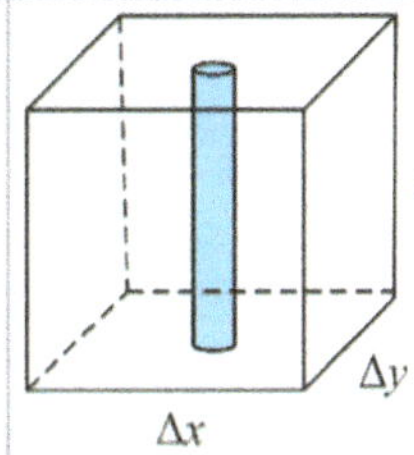
</td></tr>
<tr><td>Horizontal well</td><td>$$\mathrm{WI}=\frac{2\pi\sqrt{K_yK_z}\Delta x}{\ln\left(\frac{r_e}{r_w}\right)+s}\quad r_e=0.28\frac{\left[\left(\frac{K_z}{K_y}\right)^{1/2}\Delta y^2+\left(\frac{K_y}{K_z}\right)^{1/2}\Delta z^2\right]^{1/2}}{\left(\frac{K_z}{K_y}\right)^{1/4}+\left(\frac{K_y}{K_z}\right)^{1/4}}$$</td><td>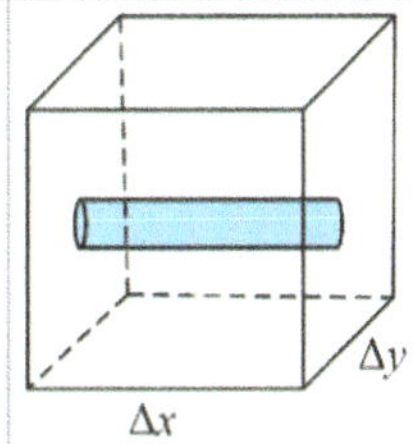
</td></tr>
<tr><td>Slant well</td><td>$$\mathrm{WI}=\sqrt{\mathrm{WI}_x^2+\mathrm{WI}_y^2+\mathrm{WI}_z^2}$$
$$\mathrm{WI}_x=\frac{2\pi\sqrt{K_yK_zL_x}}{\ln\left(\frac{r_{e,x}}{r_w}\right)+s}\quad r_{e,x}=0.28\frac{\left[\left(\frac{K_y}{K_z}\right)^{1/2}\Delta z^2+\left(\frac{K_z}{K_y}\right)^{1/2}\Delta y^2\right]^{1/2}}{\left(\frac{K_y}{K_z}\right)^{1/4}+\left(\frac{K_z}{K_y}\right)^{1/4}}$$
$$\mathrm{WI}_y=\frac{2\pi\sqrt{K_xK_zL_y}}{\ln\left(\frac{r_e}{r_w}\right)+s}\quad r_{e,y}=0.28\frac{\left[\left(\frac{K_z}{K_x}\right)^{\frac{1}{2}}\Delta x^2+\left(\frac{K_x}{K_z}\right)^{\frac{1}{2}}\Delta z^2\right]^{\frac{1}{2}}}{\left(\frac{K_z}{K_x}\right)^{\frac{1}{4}}+\left(\frac{K_x}{K_z}\right)^{\frac{1}{4}}}$$
$$\mathrm{WI}_z=\frac{2\pi\sqrt{K_xK_yL_z}}{\ln\left(\frac{r_{e,z}}{r_w}\right)+s}\quad r_{e,z}=0.28\frac{\left[\left(\frac{K_y}{K_x}\right)^{\frac{1}{2}}\Delta x^2+\left(\frac{K_x}{K_y}\right)^{\frac{1}{2}}\Delta y^2\right]^{\frac{1}{2}}}{\left(\frac{K_y}{K_x}\right)^{\frac{1}{4}}+\left(\frac{K_x}{K_y}\right)^{\frac{1}{4}}}$$</td><td>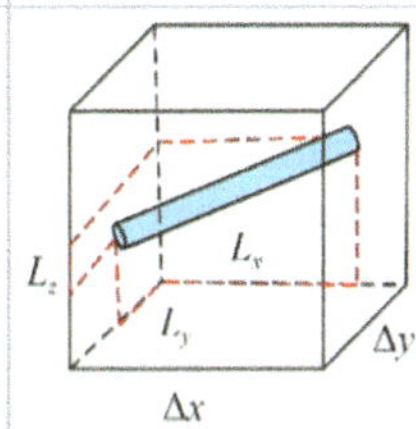
</td></tr>
</table>

Table 9.7 Well index calculation model with unstructured girds

<table>
<tr><th>Well index calculation model</th><th>Schematics</th></tr>
<tr><td>$$\mathrm{WI}=\frac{2\pi KV^{1/3}}{\ln\left(\frac{r_e}{r_w}\right)+s=0.2V^{1/3}}$$</td><td>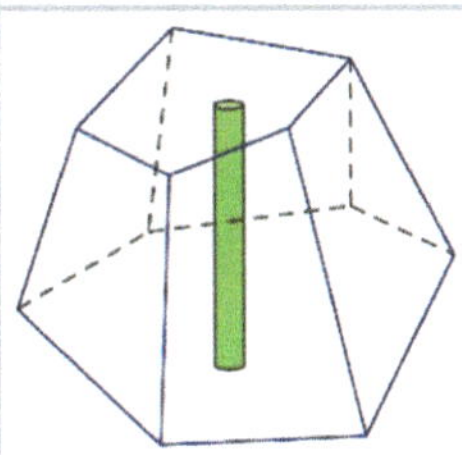</td></tr>
</table>

The pore pressure variation causes the reservoir media to deform and the permeability to shrink, which can be expressed as

$$K_\mathrm{p}=K_0e^{-\alpha_{\mathrm{F(f,m)}}\left(P_\mathrm{e}-P_{\mathrm{F(f,m)}}\right)} \tag{9.1}$$

A well index changes dynamically with the permeability due to the flow-geomechanics coupling:

$$WI_\mathrm{p}(P)=\frac{2\pi K_\mathrm{p}V^{1/3}}{\ln\left(\frac{r_\mathrm{e}}{r_\mathrm{w}}\right)+s} \tag{9.2}$$

9.2.1.4 Calculation Model of Production Rate with Line-Source Wellbore Scheme

A production rate model for the line-source wellbore scheme consists of two parts: a fluid exchange model between the reservoir media and the wellbore, and a production calculation model.

The reservoir media inside a point-source grid include the nano/micro-pores, micro-fractures, and natural/artificial fractures. Within the grid, the fluid exchange occurs between a point-source wellbore and multiple media of different scales, which is constrained by both the geometric properties of grids and physical properties of media, and the effects of flow pattern and mechanisms on the flow capability.

The production from each point source can be calculated according to the fluid exchange model between reservoir and wellbore, which together gives the total production (Fig. 9.5).

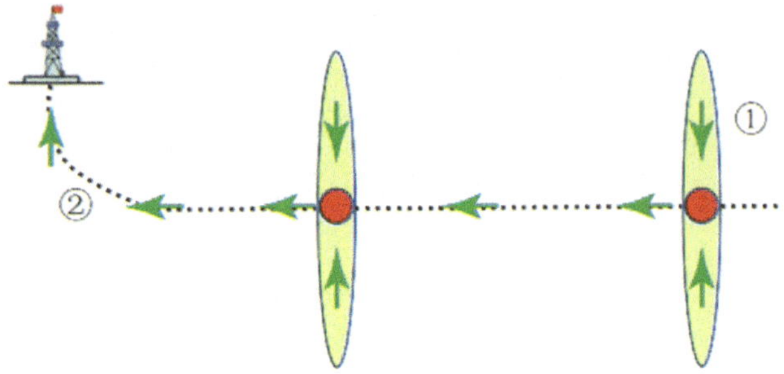

Fig. 9.5 Schematics of formation-wellbore-production with line-source wellbore scheme

1. Calculation model for fluid exchange between reservoir and wellbore

For the line-source wellbore scheme, the fluid exchange between the reservoir grids for every point source and the wellbore is calculated by the following models (Babu et al. 1991a, b; Lee and Miliken 1993; Abou-Kassem and Aziz 1985; Cao 2002):

$$q_{c,i}^{W} = WI_i \sum_{p} \left\{ \left(\frac{K_{rp}\rho_p}{\mu_p}\right) X_{cp}\left[(P_p - \rho_p g D) - (P^W - \rho_p g D^W)\right] \right\}_i^{n+1} \tag{9.3}$$

where $q_{p,i}^{W}$ is the exchange of the Pth phase fluid in the ith point source, g; WI_i is the well index for the ith point source, dimensionless; ρ_p is the density of the Pth phase fluid, g/cm^3; i is the ith point source.

The fluid exchange between reservoir and wellbore is affected by not only the well index and its dynamic change, but also the flow pattern and flow mechanisms. The following section focuses on the effect of the effects of flow pattern and mechanisms on the fluid exchange between reservoir and wellbore.

(1) Calculation model for fluid exchange between pore media of multiple scale and wellbore

In tight oil/gas reservoirs, the fluid exchange between pore media of different scales and a wellbore is mainly influenced by the complex flow mechanisms including the starting pressure gradient, the slippage effect, and the diffusion. The corresponding flow regime is characterized by the quasi-linear flow and the low-speed nonlinear flow (Zhang et al. 2009; Fan et al. 2015; Wang 2016).

1) Quasi-linear flow

When the pressure gradient of oil/gas flow to a wellbore from pore media of different scales is between the pseudo-linear critical pressure gradient and the high-speed nonlinear critical pressure gradient, the flow regime is quasi linear, and the corresponding calculation model is shown as follows:

$$(q_{\text{quasilinear}})_{m}^{W} = WI \cdot \sum (F_{p,i}^{W①} \cdot F_{(\text{linear})p,i}^{W②})_m \tag{9.4}$$

$$F_{p,mi}^{W①} = \left(\frac{K_{rp}\rho_p}{\mu_p}\right)_{mi}^{n+1} \tag{9.5}$$

$$F_{(\text{linear})p,mi}^{W②} = (P_{p,mi}^{n+1} - \rho_p^{n+1} g D_{mi}) - (P_{wfk}^{n+1} - \rho_p^{n+1} g D_k^W) - G_{c,mi}^W \tag{9.6}$$

$$G_{c,mi}^{W} = c^{W}(L_{mi} + r_w) \tag{9.7}$$

2) Low-speed nonlinear flow

When the pressure gradient of oil/gas flow to a wellbore from pore media of different scales is greater than the staring pressure gradient and less than the pseudo-linear critical pressure gradient, the flow regime is the low-speed nonlinear flow and the corresponding calculation model is shown below:

$$(q_g)_{m}^{W} = WI \cdot \sum (F_{p,i}^{W①} \cdot F_{(g)\,p,i}^{W②})_m \tag{9.8}$$

$$F_{p,mi}^{W①} = \left(\frac{K_{rp}\rho_p}{\mu_p}\right)_{mi}^{n+1} \tag{9.9}$$

$$F_{(g)p,mi}^{W②} = \left[(P_{p,mi}^{n+1} - \rho_p^{n+1} g D_{mi}) - (P_{wfk}^{n+1} - \rho_p^{n+1} g D_k^W) - G_{a,mi}^W\right]^{n^*} \tag{9.10}$$

$$G_{a,mi}^{W} = a^{W}(L_{mi} + r_w) \tag{9.11}$$

3) Slippage effect

Under the low-permeability and low-pressure conditions, the gas flow towards wellbore is affected by the slippage effect, and the corresponding calculation model is represented below:

$$(q_{\text{slippage, g}})_{m}^{W} = WI \cdot (F_{g,i}^{W①} \cdot F_{(\text{slippage})g,i}^{W②} + F_{go,i}^{W①} \cdot F_{go,i}^{W②})_m \tag{9.12}$$

$$F_{\mathrm{g,m}i}^{\mathrm{W}①} = \left(\frac{K_{\mathrm{rg}}}{\mu_{\mathrm{g}}} \frac{\rho_{\mathrm{gsc}}}{B_{\mathrm{g}}} \right)_{\mathrm{m}}^{n+1} \tag{9.13}$$

$$F_{(\mathrm{slippage})\,\mathrm{g,m}i}^{\mathrm{W}②} = \left[(1 + \frac{b}{P_{\mathrm{g,m}i}})(P_{\mathrm{g,m}i} - \frac{\rho_{\mathrm{gsc}}}{B_{\mathrm{g}}} g D_{\mathrm{m}i}) - (P_{\mathrm{wfk}} - \frac{\rho_{\mathrm{gsc}}}{B_{\mathrm{g}}} g D_{\mathrm{k}}^{\mathrm{W}}) \right]^{n+1} \tag{9.14}$$

$$F_{\mathrm{go,m}i}^{\mathrm{W}①} = \left(\frac{K_{\mathrm{ro}}}{\mu_{\mathrm{o}}} \frac{\rho_{\mathrm{gsc}} R_{\mathrm{g,o}}}{B_{\mathrm{o}}} \right)_{\mathrm{m}}^{n+1} \tag{9.15}$$

$$F_{\mathrm{go,m}i}^{\mathrm{W}②} = \left[(P_{\mathrm{o,m}i} - P_{\mathrm{wfk}}) - \frac{(\rho_{\mathrm{osc}} + \rho_{\mathrm{gsc}} R_{\mathrm{g,o}})}{B_{\mathrm{o}}} g (D_{\mathrm{m}i} - D_{\mathrm{k}}^{\mathrm{W}}) \right]^{n+1} \tag{9.16}$$

4) Diffusion effect

Under the low-permeability and low-pressure conditions, the gas flow towards wellbore is affected by the diffusion effect, and the corresponding calculation model is represented below:

$$(q_{\mathrm{diffusion},\,\mathrm{g}})_{\mathrm{m}}^{\mathrm{W}} = \mathrm{WI} \cdot (F_{\mathrm{g},i}^{\mathrm{W}①} \cdot F_{(\mathrm{diffusion})\mathrm{g},i}^{\mathrm{W}②} + F_{\mathrm{go},i}^{\mathrm{W}①} \cdot F_{\mathrm{go},i}^{\mathrm{W}②})_{\mathrm{m}} \tag{9.17}$$

$$F_{\mathrm{g,m}i}^{\mathrm{W}①} = \left(\frac{K_{\mathrm{rg}}}{\mu_{\mathrm{g}}} \frac{\rho_{\mathrm{gsc}}}{B_{\mathrm{g}}} \right)_{\mathrm{m}}^{n+1} \tag{9.18}$$

$$F_{(\mathrm{diffusion})\mathrm{g,m}i}^{\mathrm{W}②} = \left[(1 + \frac{32\sqrt{2}\sqrt{RT}\mu g}{3r\sqrt{\pi \mathrm{M}} P_{\mathrm{g,m}i}})(P_{\mathrm{g,m}i} - \frac{\rho_{\mathrm{gsc}}}{B_{\mathrm{g}}} g D_{\mathrm{m}i}) - (P_{\mathrm{wfk}} - \frac{\rho_{\mathrm{gsc}}}{B_{\mathrm{g}}} g D_{\mathrm{k}}^{\mathrm{W}}) \right]^{n+1} \tag{9.19}$$

$$F_{\mathrm{go,m}i}^{\mathrm{W}①} = \left(\frac{K_{\mathrm{ro}}}{\mu_{\mathrm{o}}} \frac{\rho_{\mathrm{gsc}} R_{\mathrm{g,o}}}{B_{\mathrm{o}}} \right)_{\mathrm{m}}^{n+1} \tag{9.20}$$

$$F_{\mathrm{go,m}i}^{\mathrm{W}②} = \left[(P_{\mathrm{o,m}i} - P_{\mathrm{wfk}}) - \frac{(\rho_{\mathrm{osc}} + \rho_{\mathrm{gsc}} R_{\mathrm{g,o}})}{B_{\mathrm{o}}} g (D_{\mathrm{m}i} - D_{\mathrm{k}}^{\mathrm{W}}) \right]^{n+1} \tag{9.21}$$

(2) Calculation model for fluid exchange between small-scale fractures and wellbore

The fluid exchange between a tight reservoir and a wellbore is mainly influenced by the starting pressure gradient. The flow regime is characterized by the quasi-linear flow and the low-speed nonlinear flow.

1) Quasi-linear flow

When the pressure gradient of oil/gas flow from small-scale fractures to a wellbore is between the pseudo-linear critical pressure gradient and the high-speed nonlinear critical pressure gradient, the flow regime is quasi linear and the corresponding calculation model is shown as follows:

$$(q_{\mathrm{quasi\text{-}linear}})_{\mathrm{f}}^{\mathrm{W}} = \mathrm{WI} \cdot \sum (F_{\mathrm{p},i}^{\mathrm{W}①} \cdot F_{(\mathrm{linear})\,\mathrm{p},i}^{\mathrm{W}②})_{\mathrm{f}} \tag{9.22}$$

$$F_{\mathrm{p,f}i}^{\mathrm{W}①} = (\frac{K_{\mathrm{rp}} \rho_{\mathrm{p}}}{\mu_{\mathrm{p}}})_{\mathrm{f}i}^{n+1} \tag{9.23}$$

$$F_{(\mathrm{linear})\mathrm{p,f}i}^{\mathrm{W}②} = (P_{\mathrm{p,f}i}^{n+1} - \rho_{\mathrm{p}}^{n+1} g D_{\mathrm{f}i}) - (P_{\mathrm{wfk}}^{n+1} - \rho_{\mathrm{p}}^{n+1} g D_{\mathrm{k}}^{\mathrm{W}}) - G_{\mathrm{c,f}i}^{\mathrm{W}} \tag{9.24}$$

$$G_{\mathrm{c,f}i}^{\mathrm{W}} = c^{\mathrm{W}} (L_{\mathrm{f}i} + r_{\mathrm{w}}) \tag{9.25}$$

2) Low-speed nonlinear flow

When the pressure gradient of oil/gas flow from small-scale fractures to a wellbore is greater than the staring pressure gradient and less than the pseudo-linear critical pressure gradient, the flow is the low-speed nonlinear flow that is partially dependent on the starting pressure gradient, and the corresponding calculation model is shown below:

$$(q_{\mathrm{G}})_{\mathrm{f}}^{\mathrm{W}} = \mathrm{WI} \cdot \sum (F_{\mathrm{p},i}^{\mathrm{W}①} \cdot F_{(\mathrm{g})\,\mathrm{p},i}^{\mathrm{W}②})_{\mathrm{f}} \tag{9.26}$$

$$F_{\mathrm{p,f}i}^{\mathrm{W}①} = (\frac{K_{\mathrm{rp}} \rho_{\mathrm{p}}}{\mu_{\mathrm{p}}})_{\mathrm{f}i}^{n+1} \tag{9.27}$$

$$F_{(\mathrm{G})\mathrm{p,f}i}^{\mathrm{W}②} = \left[(P_{\mathrm{p,f}i}^{n+1} - \rho_{\mathrm{p}}^{n+1} g D_{\mathrm{f}i}) - (P_{\mathrm{wfk}}^{n+1} - \rho_{\mathrm{p}}^{n+1} g D_{\mathrm{k}}^{\mathrm{W}}) - G_{\mathrm{a,f}i}^{\mathrm{W}} \right]^{n^*} \tag{9.28}$$

$$G_{\mathrm{a,f}i}^{\mathrm{W}} = a^{\mathrm{W}} (L_{\mathrm{f}i} + r_{\mathrm{w}}) \tag{9.29}$$

(3) Calculation model of fluid exchange between large-scale fractures and wellbore

In tight oil/gas reservoirs, the fluid exchange between large-scale fractures and a wellbore is characterized by the high-speed nonlinear flow and quasi-linear flow.

1) High-speed nonlinear flow

When the pressure gradient of oil/gas flow from large-scale fractures to a wellbore is greater than the high-speed non-linear critical pressure gradient, the flow regime is the high-speed nonlinear flow, and the corresponding calculation model is shown as follows:

$$(q_{\text{high-speed}})_{\text{F}}^{\text{W}} = \text{WI} \cdot \sum (F_{\text{ND}}^{n+1} \cdot F_{\text{p},Fi}^{\text{W}①} \cdot F_{\text{p},Fi}^{\text{W}②}) \tag{9.30}$$

$$F_{\text{p},Fi}^{\text{W}①} = (\frac{K_{\text{rp}}\rho_{\text{p}}}{\mu_{\text{p}}})_{Fi}^{n+1} \tag{9.31}$$

$$F_{\text{p},Fi}^{\text{W}②} = (P_{\text{p},Fi}^{n+1} - \rho_{\text{p}}^{n+1} gD_{\text{F}i}) - (P_{\text{wfk}}^{n+1} - \rho_{\text{p}}^{n+1} gD_{\text{k}}^{\text{W}}) \tag{9.32}$$

2) Quasi-linear flow

When the pressure gradient of oil/gas flow from large-scale fractures to a wellbore is between the pseudo-linear critical pressure gradient and the high-speed nonlinear critical pressure gradient, the flow regime is quasi-linear and the corresponding calculation model is shown as follows:

$$(q_{\text{quasi-linear}})_{\text{F}}^{\text{W}} = \text{WI} \cdot \sum (F_{\text{p},i}^{\text{W}①} \cdot F_{(\text{linear})\text{p},i}^{\text{W}②})_{\text{F}} \tag{9.33}$$

$$F_{\text{p},Fi}^{\text{W}①} = (\frac{K_{\text{rp}}\rho_{\text{p}}}{\mu_{\text{p}}})_{Fi}^{n+1} \tag{9.34}$$

$$F_{(\text{linear})\text{p},Fi}^{\text{W}②} = (P_{\text{p},Fi}^{n+1} - \rho_{\text{p}}^{n+1} gD_{\text{F}i}) - (P_{\text{wfk}}^{n+1} - \rho_{\text{p}}^{n+1} gD_{\text{k}}^{\text{W}}) - G_{\text{c},Fi}^{\text{W}} \tag{9.35}$$

$$G_{\text{c},fi}^{\text{W}} = c^{\text{W}}(L_{fi} + r_{\text{w}}) \tag{9.36}$$

In addition, the flow mechanisms in different media at different stages change with the pressure and pressure gradient, so the flow regime can be analyzed and calculated according to an automatic identification process.

2. Calculation model of production with line-source wellbore

For the line-source wellbore treatment scheme, the production of a horizontal well is the sum of productions from all point sources.

$$Q_{\text{p}}^{\text{W}} = \sum_{i=1}^{N} q_{\text{p},i}^{\text{W}} \tag{9.37}$$

9.2.2 Coupled Flow Simulation with Discrete Wellbore Scheme

The discrete wellbore scheme divides a horizontal well into a number of discrete grids. Fluids in reservoir grids exchange through the adjacent and connected wellbore grids. The fluid exchange occurs in-between wellbore grids. The production of the horizontal well is determined by the flowrate of the end grid of the discrete wellbore.

According to the contact type and flow behavior between a reservoir and a wellbore under different completion methods (Fig. 9.6), the reservoir and the wellbore are dissected and ordered with the discrete wellbore scheme. A connectivity table for the reservoir-wellbore grids, a calculation model for the dynamic well index under discrete wellbore scheme, a model for the fluid exchange between the reservoir and wellbore, a model for the fluid exchange between wellbore grids, and a production model are developed, respectively. The hydraulic fractured horizontal well under different completion methods is treated with the discrete wellbore scheme, and the coupled flow simulation technique is established.

9.2.2.1 Grid Partitioning and Ordering

Structured grids are applied for the discrete wellbore scheme. The principal direction of the structured grid is set the same as that of a horizontal well. Along the well trajectory, the horizontal well is set as fine grids and the surrounding formation as coarse grids. Each fine grid stands for a discrete well segment, and each coarse grid stands for a reservoir medium. The overall ordering is arranged according to the row and column of grids. The grid number increases along the horizontal well (Fig. 9.3).

Unstructured hybrid grids are applied for the discrete wellbore scheme. The principal direction of the unstructured grid is set the same as that of a horizontal well. Along the well trajectory, the horizontal well is set as long-strip rectangular meshes and the surrounding formation as PEBI meshes. Each long-strip rectangular mesh stands for a discrete well segment, and each PEBI mesh stands for a reservoir medium. The overall ordering of the unstructured grids is that: the long-strip rectangular meshes are ordered first along the horizontal well trajectory, and the surrounding PEBI meshes are ordered subsequently (Table 9.8).

9.2.2.2 Flow Behavior and Connectivity Table

Based on the contact type and flow behavior between a reservoir and a horizontal well (Figs. 9.7 and 9.8), the connectivity table is developed for the reservoir grids and the horizontal well grids.

With the discrete wellbore scheme, a horizontal well is composed of several discrete long-strip rectangular meshes along the trajectory. The contact type and flow behavior between the horizontal well and the surrounding reservoir can be characterized by the contact type and flow behavior between discrete wellbore grids and reservoir grids. For

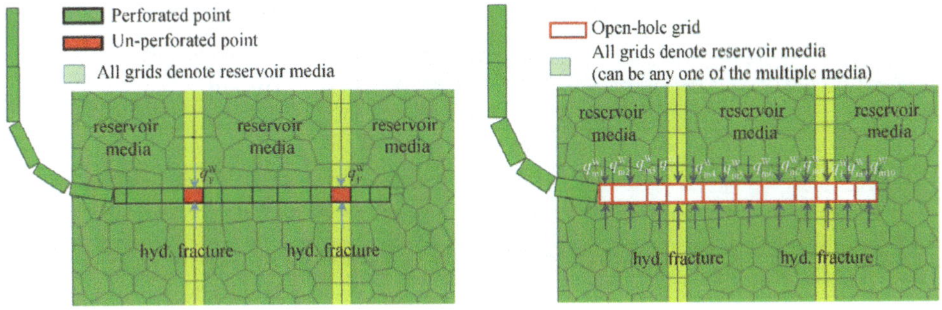

(a) Perforation (b) Openhole

Fig. 9.6 Relation between reservoir and wellbore with different completion under discrete wellbore scheme

Table 9.8 Grid partitioning and ordering with discrete wellbore scheme

Grid type	Grid partitioning and ordering	Completion	Schematics of grid partitioning and ordering
Structured grid	The principal direction of the structured grid is set the same as that of the horizontal well. Along the well trajectory, the horizontal well is set as fine grids and the surrounding formation as coarse grids. The serial number of grids increases along the horizontal well	Perforation	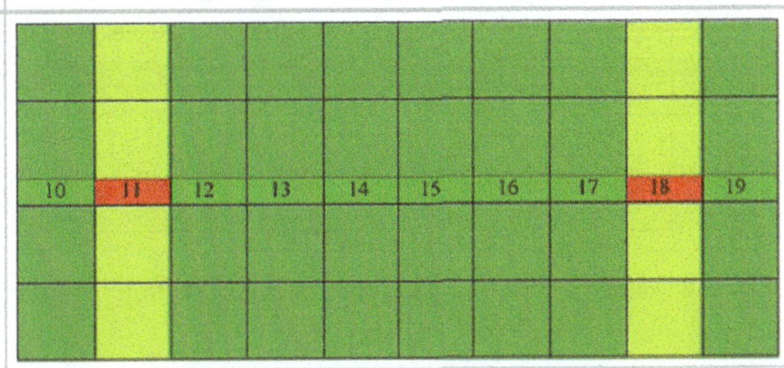
		Open hole	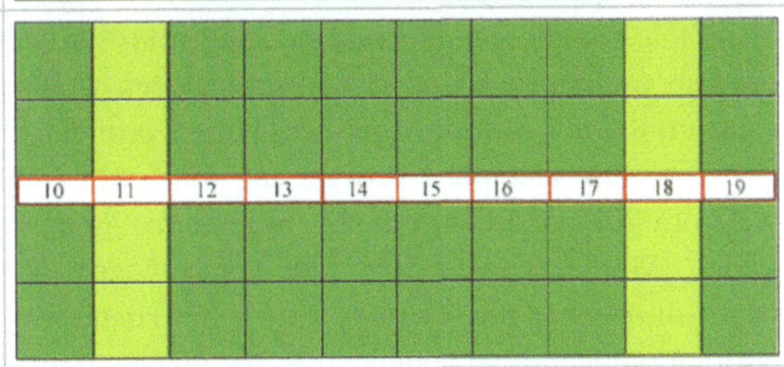
Unstructured grid	The principal direction of the unstructured grid is set the same as that of the horizontal well. Along the well trajectory, the horizontal well is set as long-strip rectangular meshes and the surrounding formation as PEBI meshes. The long-strip rectangular meshes are first ordered along the horizontal well trajectory, then the surrounding PEBI meshes are ordered	Perforation	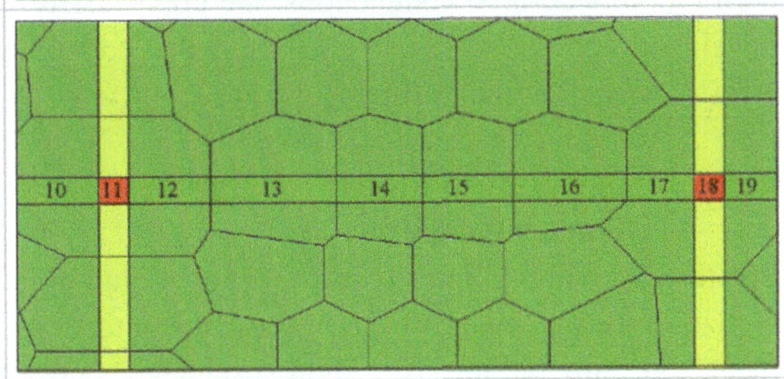
		Open hole	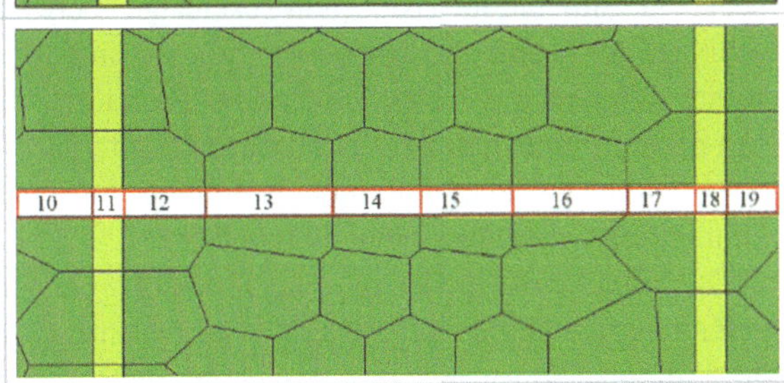

perforation completion, the fluid exchange occurs in-between discrete wellbore grids, may happen between the perforation/fracture discrete wellbore girds and the surrounding artificial fracture girds, but won't take place between the non-perforation/fracture discrete wellbore grids and the girds for multiple media of different scales. For

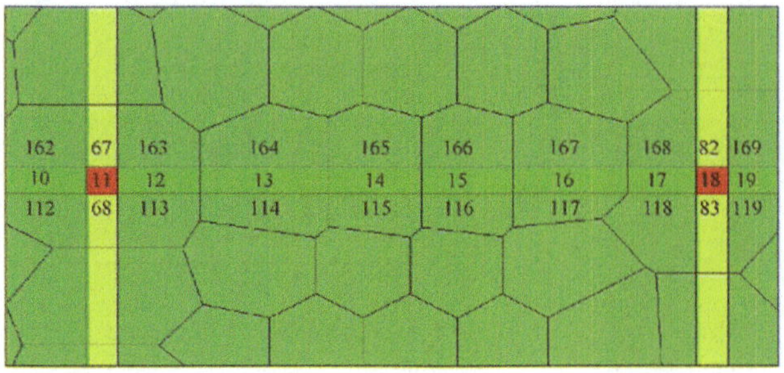

Fig. 9.7 Grid partitioning and ordering for perforation completion

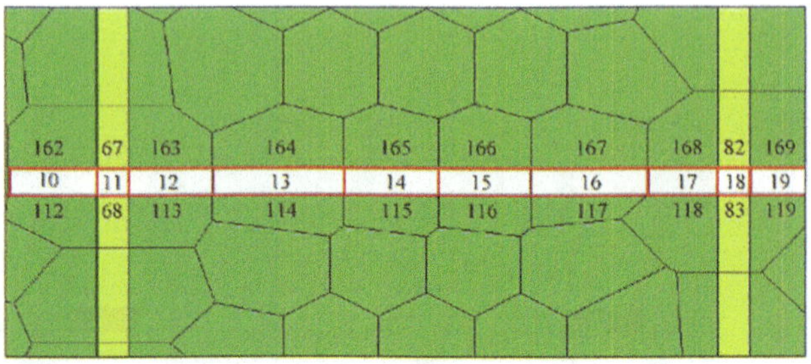

Fig. 9.8 Grid partitioning and ordering for open-hole completion

open-hole completion, the fluid exchange exists between the discrete wellbore grids, may happen between wellbore discrete grids for fracturing and the adjacent artificial fracture grids, can also take place between the non-fractured discrete wellbore grids and the grids for multiple media of different scales. Moreover, based on the contact type and flow behavior between the reservoir and the horizontal well under perforation/open-hole completions, the connectivity table is developed for the reservoir grids and the horizontal well grids (Tables 9.9 and 9.10).

Table 9.9 Connectivity table for perforation completion

Connection unit	Connectivity table
W_{11}, R_{67}	(W_{11}, R_{67}, $T^{R\text{-}W}_{W11,\ R67}$)
W_{11}, R_{68}	(W_{11}, R_{68}, $T^{R\text{-}W}_{W11,\ R68}$)
W_{11}, W_{10}	(W_{11}, W_{10}, $T^{W\text{-}W}_{W11,\ W10}$)
W_{11}, W_{12}	(W_{11}, W_{12}, $T^{W\text{-}W}_{W11,\ W12}$)

Table 9.10 Connectivity table for open-hole completion

Connection unit	Connectivity table
W_{10}, R_{112}	(W_{10}, R_{112}, $T^{R\text{-}W}_{W10,\ R112}$)
W_{10}, R_{162}	(W_{10}, R_{162}, $T^{R\text{-}W}_{W10,\ R162}$)
W_{10}, W_{11}	(W_{10}, W_{11}, $T^{W\text{-}W}_{W10,\ W11}$)
W_{11}, R_{67}	(W_{11}, R_{67}, $T^{R\text{-}W}_{W11,\ R67}$)
W_{11}, R_{68}	(W_{11}, R_{68}, $T^{R\text{-}W}_{W11,\ R68}$)
W_{11}, W_{12}	(W_{11}, W_{12}, $T^{W\text{-}W}_{W11,\ W12}$)
…	…

9.2.2.3 Calculation Model for Dynamic Well Transmissibility

For the discrete wellbore scheme, a horizontal well is discretized into several grids that represent multiple well segments in simulation. The fluid exchange between the wellbore grids and the reservoir grids is similar to that among different reservoir media grids. Therefore, the concept of well transmissibility is introduced to replace a well index, and its calculation formula is different from those for the traditional transmissibility and well index.

The calculation model of transmissibility between the wellbore grid and the reservoir grid is consistent with that of the traditional transmissibility. However, due to the extremely small flow resistance and vitally large permeability, the fluid exchange between the wellbore grid and the reservoir grid is mainly influenced by the physical properties of the reservoir. The calculation model can be simplified as shown in Fig. 9.11 (Table 9.11).

The pore pressure and effective stress near a wellbore vary with time during the fracturing-injection-production process. This leads to the dynamic change of the geometric characteristics of wellbore grids and physical properties of the reservoir media, which causes the well transmissibility to change accordingly. The calculation formula is shown as

$$T^{R-W}_{(\text{pres}-\text{sensivity})i,k} = \alpha^{R}_{(\text{pres}-\text{sensivity})i} = A^{R-W}_{i,k}\frac{K_{(\text{pres}-\text{sensivity})i}}{L_i}\boldsymbol{n}_i\cdot\boldsymbol{f}_i \tag{9.38}$$

where, subscripts i, k are the serial numbers of the reservoir grids and the wellbore grids that are connected to the wellbore, respectively; superscript R-W is there reservoir-wellbore; $T^{R-W}_{i,k}$ is the transmissibility between the adjacent reservoir grid i and wellbore grid k. α^{R}_{i} is the shape factor of the formation grid i that is connected to the wellbore grid k. $A^{R-W}_{i,k}$ is the contact area between the adjacent formation grid i and wellbore grid k, m^2.

Table 9.11 Well transmissibility vs. traditional transmissibility

Name	Well transmissibility	Traditional transmissibility
Grid type	Reservoir grid and wellbore grid	Reservoir grid and reservoir grid
Calculation model	$T^{R-W}_{i,k} = \alpha^{R}_{i}$	$T_{i,j} = \frac{\alpha_i\alpha_j}{\alpha_i+\alpha_j}$

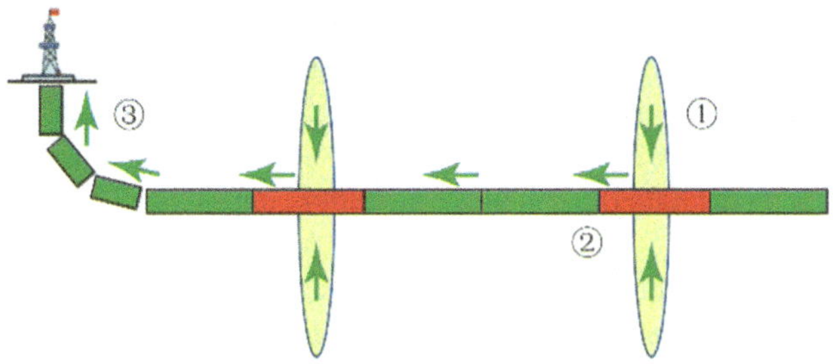

Fig. 9.9 Schematics of formation-wellbore-production with the discrete wellbore scheme

9.2.2.4 Calculation Model for Production Rate with Discrete Wellbore Scheme

A production model for the discrete wellbore scheme consists of three parts: a fluid exchange model between a reservoir and a wellbore, a fluid exchange model between wellbore grids, and a production model (Fig. 9.9). A reservoir grid can be categorized as the micro- and nano-pores, micro fractures, natural/artificial fractures. The fluid exchange may occur between the reservoir and the wellbore, which is affected by not only the geometric/physical properties of reservoir media grids, but also the effects of flow regimes/mechanisms on the flowability. The fluid flow inside the wellbore mostly appeared as nonlinear or even turbulent. The fluid exchange in-between wellbore grids can be modeled as the turbulent flow. The total production of a horizontal well is the flowrate from the wellbore end grid.

1. Fluid exchange between reservoir and wellbore

For the discrete wellbore treatment scheme, the fluid exchange between a wellbore gird and an adjacent reservoir grid is calculated with the following modeled (Wu et al. 1996):

$$\begin{aligned} q_{c,i}^{R-W} &= \sum_i \left\{ T_{i,k}^{W} \sum_p \left(\frac{K_{rp}\rho_p}{\mu_p}\right)_i X_{cp} \left[(P_{p,i} - \rho_p g D_i) - (P_{p,k}^{W} - \rho_p g D_k^{W}) \right] \right\}^{n+1} \\ &= \sum_i \left[T_{i,k}^{W} \cdot \sum \left(F_{p,i}^{W①} \cdot F_{p,i}^{W②} \right) \right] \end{aligned} \quad (9.39)$$

where, $q_{c,i}^{W}$ is the production of the c component from the ith shooting point.

The fluid exchange between a reservoir grid and a wellbore grid is affected by not only the dynamic well transmissibility but also the flow behavior and mechanisms, which shows the features of the high-speed nonlinear flow, quasilinear flow, and low-speed nonlinear flow. They are consistent with the fluid exchange model ($F_{p,i}^{W①}$, $F_{p,i}^{W②}$) between the reservoir media and a line-source wellbore, details will not be shown here.

2 Fluid exchange in-between wellbore grids

Fluids flow fast inside the wellbore, which is usually expressed as nonlinear or even turbulent flow. Thereby, Darcy's law is no longer applicable and it should be described by the turbulent flow model to develop the fluid exchange modeling between wellbore grids (Dikken 1990) (Fig. 9.10):

$$q_{l,k}^{W-W} = \sum_k \left\{ T_{l,k}^{W-W} \sum_p (\zeta_p^{W})_{l|k} X_{cp} \begin{bmatrix} (P_{p,k}^{W} - \rho_p g D_k^{W}) \\ -(P_{p,l}^{W} - \rho_p g D_l^{W}) \end{bmatrix} \right\}^{n+1} \quad (9.40)$$

where, the pseudo-transmissibility formula between adjacent wellbore grids is

$$T_{l,k}^{W-W} = \frac{1.97588 \times d^{5/2}}{d_l + d_k} \quad (9.41)$$

Pseudo-mobility between the adjacent wellbore grids can be calculated as (Wu 2000, 2016):

$$(\zeta_p^{W})_{l|k} = \frac{(S_p\rho_p)_{l|k}}{\bar{\rho}^{0.5}} \left| \frac{(P_{p,k}^{W} - \rho_p g D_k^{W}) - (P_{p,l}^{W} - \rho_p g D_l^{W})}{d_l + d_k} \right|^{-0.5} \quad (9.42)$$

$$\bar{\rho} = \frac{1}{2} \sum_l^k \sum_p (S_p\rho_p)$$

Since the fluid flow inside a wellbore is mostly expressed as the high-speed non-Darcy flow, the for chheimer equation is used:

$$\begin{aligned} &-\frac{(P_{p,k}^{W} - \rho_p g D_k^{W}) - (P_{p,l}^{W} - \rho_p g D_l^{W})}{d_l + d_k} \\ &= \frac{\mu_p}{KK_{rp}} v_p + \beta_p \rho_p v_p |v_p| \end{aligned} \quad (9.43)$$

where, d is the wellbore diameter, cm; d_l and d_k are the distances from the centers of two neighboring wellbore grids to their contact surface, cm; $\bar{\rho}$ is the average fluid density, $\bar{\rho} = \frac{1}{2} \sum_l^k \sum_p (S_p\rho_p)$, g/cm^3; β_p is the coefficient of effective non-Darcy flow, cm^{-1}; μ_p is the viscosity of non-Darcy fluid, cP.

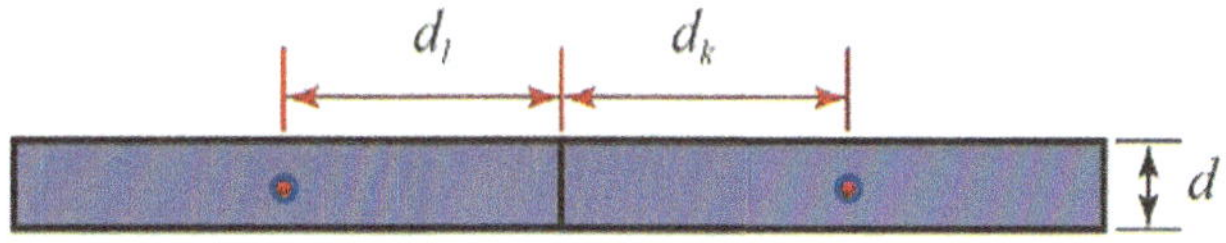

Fig. 9.10 Schematics of adjacent wellbore grids

3. Production model

For the discrete wellbore scheme, the production of a horizontal well is the flowrate from the wellbore end grid:

$$Q_{\mathrm{c}}^{\mathrm{W}} = q_{\mathrm{c,end}}^{\mathrm{W-W}} \tag{9.44}$$

9.3 Coupled Flow Simulation with Multi-Segment Wellbore Scheme

The multi-segment wellbore scheme divides a horizontal well into a number of segments, and each wellbore segment contains one point source at most. Fluid exchange occurs between each point source and the reservoir. Friction, gravity, and multi-phase flow are considered inside the wellbore segment. The flow between adjacent well segments is treated as the pipe flow.

According to the contact type and flow behavior between a reservoir medium and the wellbore under different completion methods (Fig. 9.11), the reservoir is discretized according to the multi-segment wellbore scheme. The location of a point source in a well segment is determined, and the grids and point sources are ordered according to the contact type and flow behavior between the reservoir and the wellbore. A connectivity table for the reservoir-wellbore grids, a calculation model for the dynamic well index under discrete wellbore scheme, a model for the fluid exchange between the reservoir and wellbore, a model for the fluid exchange between wellbore grids, and a production model are developed, respectively. The hydraulic fractured horizontal well under different completion methods is treated with the discrete wellbore scheme, and the corresponding coupled flow simulation technique is established.

9.3.1 Grid Partitioning and Ordering

The grid partitioning and ordering for the multi-segment wellbore scheme are similar to those for the line-source wellbore scheme (Table 9.3).

The principal direction of the structured grid is set the same as that of a horizontal well. Along the well trajectory, the line-source wellbore is set as fine grids and the surrounding as coarse grids. The grid serial number increases along the horizontal well. The location of each point source is determined and ordered according to the contact type and flow behavior between the reservoir and the wellbore. The serial number of the fine grids stands for not only the serial number of the reservoir grids but also the serial number of point source of the multi-segment wellbore (refer to Fig. 9.3).

The principal direction of the unstructured grid is set the same as that of a horizontal well. Along the well trajectory, the wellbore is set as long-strip rectangular meshes and the surrounding as PEBI meshes. The overall ordering is: the long-strip rectangular meshes are first ordered along the horizontal well trajectory, and then the surrounding PEBI meshes are ordered. The location of a point source is determined and ordered according to the contact type and flow behavior between the reservoir and the wellbore. The serial number of the long-strip rectangular meshes stands for not only the serial number of the reservoir grids but also the serial number of the point source of the wellbore (refer to Fig. 9.4).

9.3.2 Flow Behavior and Connectivity Table Between Reservoir and Horizontal Well

Based on the contact type and flow behavior between the reservoir and the horizontal well under the multi-segment wellbore scheme (Figs. 9.3 and 9.4), the connectivity table is developed for the reservoir grids and horizontal well grids (Figs. 9.4 and 9.5).

For perforation completion, the point source is located in a reservoir mesh for perforation/fracturing, where the fluid exchange many happen between the reservoir and the wellbore. The fluid exchange does not take place between non-perforated/fractured grids and the wellbore. The fluid exchange happens between the fine grids for horizontal well and the adjacent coarse grids for surrounding reservoir that connect to the wellbore. For open-hole completion, the point source is located at the long-strip rectangular meshes, where

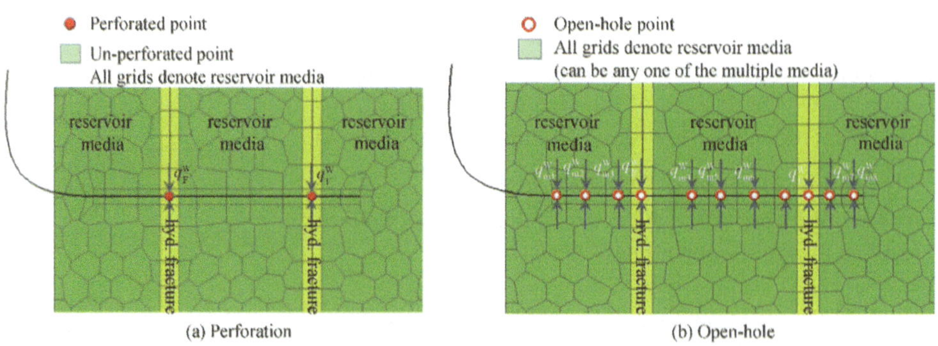

Fig. 9.11 Relation between reservoir and wellbore with different completion under multi-segment wellbore scheme

the fluid exchange may happen between the artificial fractures and the wellbore, between the multiple media and the wellbore in the non-fracture long-strip meshes, and it takes place between the long-strip meshes and the surrounding PEBI meshes connected to the wellbore.

9.3.3 Calculation Model for Dynamic Change of Well Index

For the multi-segment wellbore scheme, the wellbore is a point source in the center of reservoir grids. The well index stands for the effect of the geometric characteristics thereof, formation physical properties, and the effect of wellbore geometric characteristics on the productivity. It is the main control factor for the fluid exchange between the reservoir and the wellbore.

9.3.3.1 Calculation Model for Well Index Model with Structured and Unstructured Grids

With structured grids, the reservoir grids have a definite orientation, geometric characteristic parameters (length, width, and height), physical parameters (anisotropic permeability). Under different well types (vertical well, horizontal well, or slant well), the geometric parameters of the point source wellbore are rather different, and the calculation model is the same as the well index model for the line-source scheme, details are not presented here.

With unstructured grids, the reservoir grids have irregular shapes without definite orientation. The geometric characteristic parameters and physical properties vary greatly, which are largely different from the calculation model for the well index with structured grids, and the calculation model is the same as that for well index with the line-source scheme.

9.3.3.2 Calculation Model for Dynamic Change of Well Index

The pore pressure and effective stress near a wellbore change with time during the fracturing-injection-production process. This leads to the dynamic change of the geometric characteristics of wellbore grids and physical properties of the reservoir media, which causes the well index to change accordingly and affect the well productivity. The calculation model is the same as that for the well index for the line-source scheme.

9.3.4 Calculation Model for Production Rate with Multi-Segment Wellbore Scheme

The production model for the multi-segment wellbore scheme consists of four parts (Christian et al. 2000; Stone et al. 2002; Jiang 2007): A fluid exchange model between the reservoir and the wellbore, a fluid exchange model inside each well segment, a fluid exchange model in-between well segments, and a production calculation model (Fig. 9.12).

9.3.4.1 Fluid Exchange Model Between Reservoir and Wellbore

The reservoir media for a multi-segment wellbore can be categorized as nano/micro-pores, micro-fractures, and natural/artificial fractures. The fluid exchange may occur between the point-source wellbore and the multiple media of different scales, which is constrained by both the geometric/physical properties of meshes and the flow pattern and mechanisms. It is represented by multi flow regimes: the high-speed nonlinear flow, the quasilinear flow, and the low-speed nonlinear flow. The model is the same as the model for fluid exchange between reservoir and wellbore (Eqs. (9.4)–(9.36))with the line-source scheme.

9.3.4.2 Fluid Exchange Model Inside Each Well Segment

The oil-gas-water three-phase flow usually occurs inside the multi-segment wellbore. A flow regime includes the high-speed nonlinear flow and the multi-phase pipe flow (Shi et al. 2005a, b). The distributions of oil-gas-water are influenced by multiple factors. The distribution of different gas/liquid in the flow process is described by the flow behavior, which includes pure oil (liquid) flow, bubble flow, slug flow, circulation, and mist flow. The fluid exchange model inside each well segment is developed, which considers the multiphase pipe flow in a wellbore segment (Jiang et al. 2015; Gao et al. 2015):

$$q_i^{\mathrm{W}} = \left[\left(A \sum_{\mathrm{p}} \rho_{\mathrm{p}} X_{\mathrm{cp}} V_{\mathrm{sp}} \right)_{i,\mathrm{in}} - \left(A \sum_{\mathrm{p}} \rho_{\mathrm{p}} X_{\mathrm{cp}} V_{\mathrm{sp}} \right)_{i,\mathrm{out}} \right]^{n+1} \tag{9.45}$$

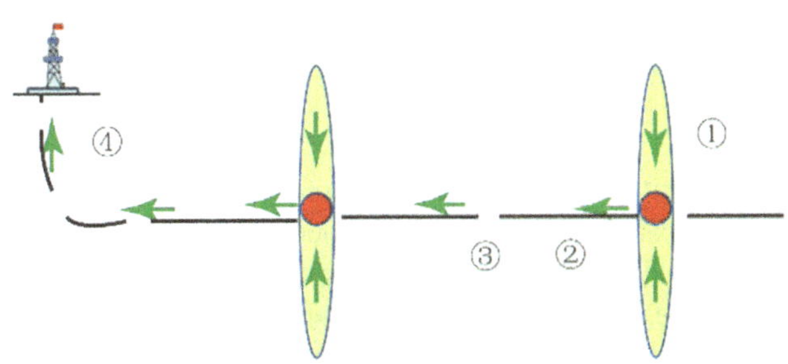

Fig. 9.12 Schematics of formation-wellbore-production for multi-segment wellbore scheme

The Oil-gas-water three-phase flowrate is

$$V_{m} = V_{so} + V_{sw} + V_{sg} \tag{9.46}$$

where, A is the wellbore cross-sectional area, m^2; ρ_o^o is the oil density in the oil phase, shorted as the oil density, g/cm^3; ρ_o^g is the gas density in the oil phase, shorted as the solution gas density, g/cm^3; V_m is the mixture flowrate, m^3/d; V_{sp} is the flowrate of the p phase in the wellbore, m/s; V_{sw} is the flowrate of the water phase, m/s; V_{so} is the flowrate of the oil phase, m/s; V_{sg} is the flowrate of the gas phase, m/s.

According to weather the slippage effect between multi-phases fluids is considered, the flowrates of multiple phases are described as either a homogeneous model or a drift model.

1. Homogeneous model

If the slippage effect does not occur between the multiple phases, all fluids in the wellbore flow at the same rate:

$$V_{sp} = \alpha_p V_m \tag{9.47}$$

The oil-gas-water three-phase model is

$$V_{sw} = \alpha_w V_m \tag{9.48}$$

$$V_{so} = (1 - \alpha_w - \alpha_g) \cdot V_m \tag{9.49}$$

$$V_{sg} = \alpha_g \cdot V_m \tag{9.50}$$

where, α_g is the volume fraction of the gas phase; α_w is the volume fraction of the water phase; α_o is the volume fraction of the oil phase.

2. Drift model

If the slippage effect exists between the multiple phases, all fluids in the wellbore flow at different rates:

$$V_{sg} = \alpha_g C_0 V_m + \alpha_g V_d \tag{9.51}$$

$$V_{so} = \alpha_{ol} C_0^{ol} V_{sl} + \alpha_o V_d^{ol} \tag{9.52}$$

$$\alpha_{ol} = \alpha_o / (\alpha_o + \alpha_w) \tag{9.53}$$

where, V_d is the gas drift rate, m/s; $C_0$is the shape parameter of the drift model measured by experiments, dimensionless.

The parameters in the drift model vary largely under different flow patterns. For the vertical pipe flow, bubble flow appears at a low flowrate V_m and low gas holdup α_g. It transitions to slug flow and eventually mist flow with an increase in the flowrate and gas holdup. Therefore, the gas holdup is the lowest at bubble flow and highest at the mist flow. C_0 is approximately 1.16 at slug flow and 1 at mist flow. Thus, gas slugs and fluid slugs alternately appear at the slug flow with an increased unevenness, so $C_0$is greater than 1; whereas at mist flow, the gas phase is uniformly distributed even though some liquids enter the gas phase, so C_0 is approximately 1. At horizontal pipe flow, the flow stratification is also related to high gas content. At the low average flowrate, laminar flow instead of the mist flow occurs in the vertical pipe flow.

9.3.4.3 Fluid Exchange Model in-Between Well Segments

The fluid exchange in-between well segments depend on three factors: friction, gravity, and acceleration (Li et al. 2013; Li 2012):

$$\Delta P_{i,\text{total}} = \Delta P_{\text{garavity}} + \Delta P_{\text{friction}} + \Delta P_{\text{acceleration}} \tag{9.54}$$

Pressure between well segments is described as

$$\begin{aligned} P_i^W - P_{i-1}^W &= \rho_m g L \sin\theta + \frac{f \rho_m L v_m^2}{D} \\ &+ \left[(0.5\rho_m V_m^2)_i - (0.5\rho_m V_m^2)_{i-1} \right] \end{aligned} \tag{9.55}$$

The density of mixture ρ_m is

$$\rho_m = \alpha_g \rho_g + \alpha_o \rho_o + \alpha_w \rho_w \tag{9.56}$$

The sum of volume fractions of oil-gas-water is

$$\alpha_o + \alpha_g + \alpha_w = 1 \tag{9.57}$$

where, ρ_m is the mixture density, g/cm^3; V_m is the mixture flowrate, m^3/d; θ is the wellbore inclination angle, degree; f_{tp} is the friction coefficient of a tube wall; A is the wellbore cross-sectional area, m^2; ρ_w is the water density, g/cm^3; ρ_o is the oil density, g/cm^3; ρ_g is the gas density, g/cm^3; α_g is the volume fraction of the gas phase; α_w is the volume fraction of the water phase; α_o is the volume fraction of the oil phase.

The flow between well segments is modeled as Fig. 9.13.

$$Q_{i,\text{out}} = q_i + Q_{i,\text{in}} \tag{9.58}$$

9.3.4.4 Production Model

With the multi-segment wellbore scheme, the production of a horizontal well is represented as the flowrate of the end point source:

$$Q_{\text{total}} = Q_{\text{last}} \tag{9.59}$$

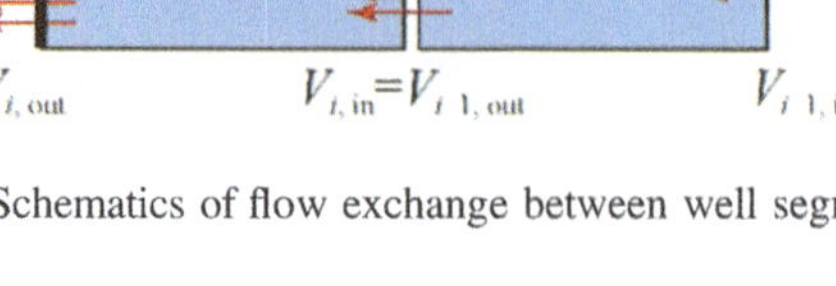

Fig. 9.13 Schematics of flow exchange between well segments

References

Abou-Kassem JH, Aziz K (1985) Analytical well models for reservoir simulation, SPE 11719

Babu DK, Odeh AS, Al-Khalifa AJ, McCann RC (1991a) The relation between wellblock and wellbore pressures in numerical simulation of horizontal wells, SPE 20161

Babu DK, Odeh AS, Al-Khalifa AJ, McCann RC (1991b) Numerical simulation of horizontal wells. SPE 21425

Cao H (2002) Development of techniques for general purpose simulation, PhD thesis, Stanford University

Chen Z (2011) The finite element method. World Scientific

Chen Z, Huan G, Ma Y (2006) Computational methods for multiphase flows in porous media. Society for Industrial and Applied Mathematics

Christian W, Louis JD, Khalid A (2000) Efficient estimation of the effects of wellbore hydraulics and reservoir heterogeneity on the productivity of non-conventional wells, SPE 59399

Dikken BJ (1990) Pressure drop in horizontal wells and its effects on production performance. J Pet Technol 42(11):1426–1433

Fan D et al (2015) Numerical simulation of multi-fractured horizontal well in shale gas reservoir considering multiple gas transport mechanisms. Chinese J Theor Appl Mech 47(6):906–915

Gao D et al (2015) Research progress and development trend of numerical simulation technology for unconventional wells. Appl Mathema and Mech 36(12):1238–1256

Jiang Y (2007) Techniques for modeling complex reservoirs and advanced wells, PhD thesis, 27–78

Jiang R, Xu J, Fu J (2015) Multistage fractured horizontal well numerical simulation and application for tight oil reservoir. J Southwest Petrol Univ (Sci Technol Edn) 3:45–52

Lee SH, Milliken WJ (1993) The productivity index of an inclined well in finite-difference reservoir simulation, paper SPE-25247 presented at 12th SPE Symposium on Reservoir Simulation. Soc of Pet Eng, New Orleans, La

Li L (2012) Study on multi-segment numerical simulation for horizontal well of bottom-water sand reservoir in the oil field. Master Thesis

Li W et al (2013) The policy research of horizontal well segmented utilization in heterogeneous reservoir based on multi-segment well model. Sci Technol Eng 13(33):9935–9939

Peaceman DW (1978) Interpretation of well-block pressures in numerical reservoir simulation. Soc Pet Eng J 253:183–194

Peaceman DW (1982) Interpretation of well-block pressures in numerical reservoir simulation with nonsquare grid blocks and anisotropic permeability, paper SPE-10528 presented at Sixth SPE Symposium on Reservoir Simulation. Soc of Pet Eng, New Orleans, La

Peaceman DW (1991) Representation of a horizontal well in numerical reservoir simulation, paper SPE-21217 presented at 11th SPE Symposium on Reservoir Simulation. Soc of Pet Eng, Anaheim, Calif

Peaceman DW (1995) A new method for representing multiple wells with arbitrary rates in numerical reservoir simulation, paper SPE-29120 presented at 13th SPE Symposium on Reservoir Simulation. Soc of Pet Eng, San Antonio, Tex

Shi H, Holmes JA, Diaz LR, Durlofsky LJ, Aziz K (2005a) Drift-flux parameters for three-phase steady-state flow in wellbores, SPE 89836. SPE J 10(2):130–137

Shi H, Holmes JA, Durlofsky LJ, Aziz K, Diaz LR, Alkaya B, Oddie G (2005b) Drift-flux modeling of two-phase flow in wellbores, SPE 84228. SPE J 10(1):24–33

Stone TW, Bennett J, Law DHS, Holmes JA (2002) Thermal simulation with multi-segment wells, SPE 78131. SPE Reserv Eval Eng 5(3):206–218

Wang W (2016) Numerical modeling of staged cluster fracturing and gas flow in horizontal wells of shale gas reservoirs. Sci Technol Eng 16(14):160–165

Wu Y (2000) A virtual node method for handling wellbore boundary conditions in modeling multiphase flow in porous and fractured media, LBNL-42882. Water Resour Res 36(3):807–814

Wu YS (2016) Multiphase fluid flow in porous and fractured reservoirs. Elsevier, 663–692

Wu Y, Forsyth PA, Jiang H (1996) A consistent approach for applying numerical boundary conditions for multiphase subsurface flow. J Contaminant Hydrol 23:157–184

Zhang X at al (2009) Sensitivity studies of horizontal wells with hydraulic fractures in shales gas reservoirs. IPTC 13338

Generation and Solving Technology of Mathematical Matrix for Multiple Media Based on Unstructured Grids

10

The multiphase flow mathematical model for tight oil/gas reservoirs is highly nonlinear due to multiple media, multiple flow regimes, complex flow mechanisms, and complex-structured wells. In addition, the coefficient matrix based on unstructured grids is large in size and complex in format. To solve these issues, zero and dead nodes are eliminated for point elements, multiple variables are compressed and stored in blocks, and reservoirs and multi-segment wells are stored in partitioned memory. An efficient mathematical matrix generation technology based on unstructured grid is developed for multiple media, which reduces the matrix size and simplifies the matrix format. At the same time, aiming at the poor and slow convergence of the coefficient matrix of the mathematical model, an efficient solving technology for linear algebraic equation set based on unstructured grids for multiple media is developed based on an efficient preconditioning technology, which greatly improves the speed and precision of the solving process.

10.1 Efficient Generation Technology of Mathematical Matrix for Multiple Media Based on Unstructured Grid

Tight sandstone oil/gas reservoirs are characterized by multiple media of different-size pores and fractures, complex-structured wells, and complex geological boundaries, which cause increased invalid grids, a complicated mathematical matrix format, and some issues that conventional grids are unable to handle. On the basis of the connectivity table for unstructured grids and the sparse mathematical matrix derivation rules, the compression and storage technology that eliminates zero elements and dead nodes (reducing storage space and total storage grid number), the mathematical matrix arrangement technology based on blocks (improving the efficiency of data reading), the compactness of data arrangement and the Input/output (I/O) speed are improved, and the computing speed is also accelerated in this section. These together established a mathematical matrix generation technology for multiple media for unstructured grids (Table 10.1).

10.1.1 Mathematical Matrix Generation Technology Based on Unstructured Grid

Tight sandstone reservoirs are characterized by nano/micro-pores, complex natural/artificial fracture networks, complex-structured wells, and complex geological boundaries, which cause complicated mathematical matrix formats, and some issues that conventional grids are unable to describe. On the basis of the connectivity table of unstructured grids, a mathematical matrix generation technology based on unstructured grid for multiple media is developed according to the sparse mathematical matrix derivation rules.

10.1.1.1 Matrix Equation

Tight sandstone reservoirs are composed of multi-scale multiple media of different-size nano/micro-pores, micro fissures, and complex natural/artificial fracture networks. The flow regime corresponding to each medium is different. The distribution of multiple media and the coupled flow mechanism between these media are different, which causes distinctive flow dynamics. Therefore, the differential equation for the nonlinear flow is characterized by multiple scale, multiple media, and multiple flow regimes (Odeh 1981; Han et al. 1999; Cao 2002; Karimi-Fard and Durlofsky 2012).

Directing at the complex characteristics of the differential equation for nonlinear flow in tight sandstone reservoirs, the equations are built according to the procedures below (Fig. 10.1).

The constructed equation has the following format:

$$\underbrace{\begin{pmatrix} J_{\mathrm{RR}} & J_{\mathrm{RW}} \\ J_{\mathrm{WR}} & J_{\mathrm{WW}} \end{pmatrix}}_{\text{coefficient matrix}} \cdot \underbrace{\begin{pmatrix} \delta x_{\mathrm{R}} \\ \delta x_{\mathrm{W}} \end{pmatrix}}_{\text{variable array}} = \underbrace{-\begin{pmatrix} R_{\mathrm{R}} \\ R_{\mathrm{W}} \end{pmatrix}}_{\text{right-hand side}}$$

Q. Ran, *Unconventional Tight Reservoir Simulation: Theory, Technology and Practice*,
https://doi.org/10.1007/978-981-32-9848-4_10

Table 10.1 Technical features with structured versus unstructured grid

Technology	Content	Technical features with structured grid	Technical features with unstructured grid
Mathematical matrix generation and solving of for multiple media based on unstructured grids	Mesh structure	Structure grid (block center, radial, and corner-point grids)	Unstructured grid
	Mesh shape	Positive hexahedron	Polyhedron
	Mesh ordering	Natural ranking	Non-natural ranking (by the narrowest band principle)
	Medium	Single	Multiple
	Flow regime	Pure Darcy flow	Complex high- and low- speed flow
	Wellbore	Vertical well	Complex-structured well
	Coefficient matrix	Banded large sparse matrix	Non-banded large sparse matrix
	Mathematical Matrix generation basis	Fixed pattern (by the inter-hexahedron relation)	Unfixed pattern (by the inter-face relation in the connectivity table)
	Mathematical matrix format	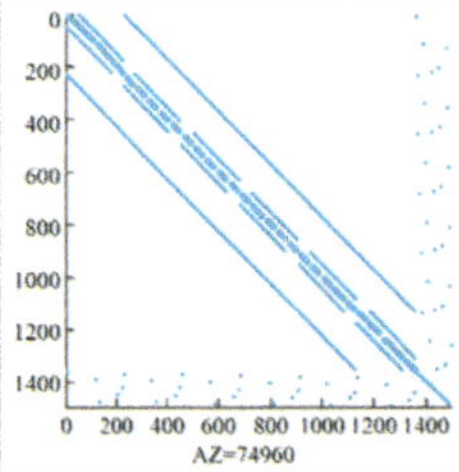	
Compression and storage of complex-structured matrix	Data storage	For a geological model with N grids, seven 1D arrays with a length of $N \times N_p \times N_p$ are constructed to characterize the mathematical matrix, in which N_p is the total fluid component, and $N_p = 3$ for the black oil model. That is, seven 1D arrays (size = 9 N) are required	The arrangement of seven diagonal matrices is no longer followed. In order not to miss the data and save all non-zero data, the zero element and dead nodes are eliminated
	Mathematical matrix format	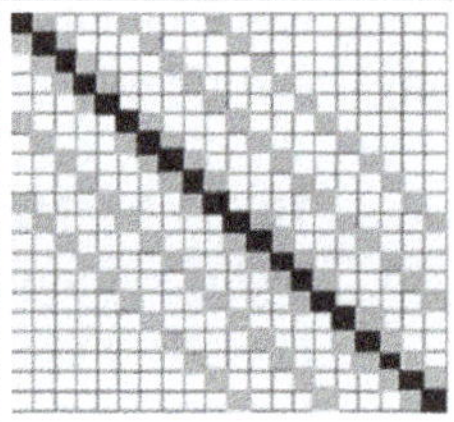	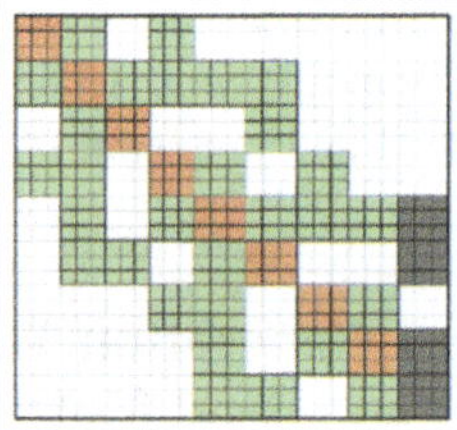

The meaning of each term in the matrix equation is shown in Table 10.2.

10.1.1.2 Mathematical Matrix Generation

Multiple media at different scales including nano/micro-pores, complex natural/artificial fractures are developed in tight sandstone reservoirs, for which complex-structured wells are applied. The coefficient matrix and the right-hand side of the matrix equation both contain the information of multiple media and complex-structured wells.

1. The generation of coefficient matrix

In the coefficient matrix of the differential equation for the nonlinear flow in tight sandstone reservoirs, complex information regarding the multiple media and complex-structured wells are considered. As aforementioned, the coefficient matrix is essentially the partial derivatives of the oil/gas/water equations for each grid with respect to three principal variables. The matrix structure is shown in Fig. 10.2.

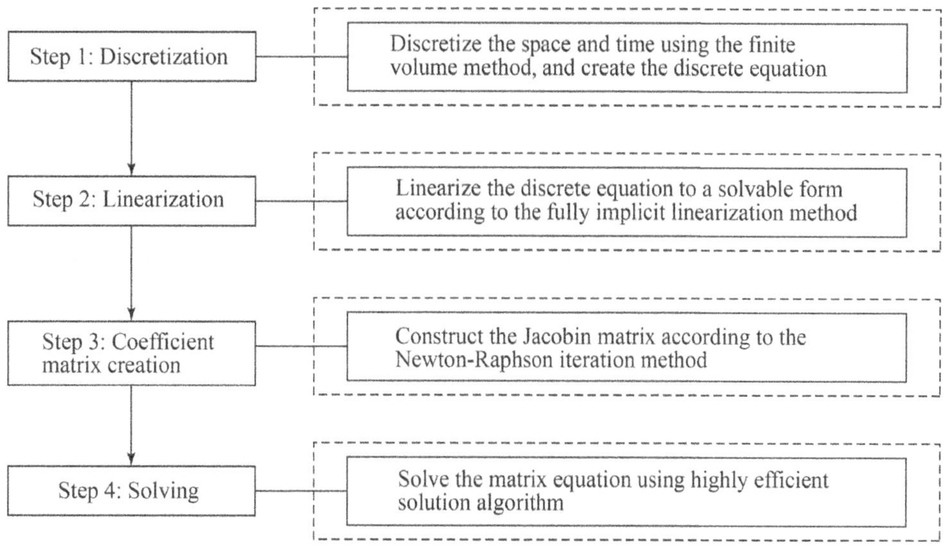

Fig. 10.1 Flowchart of equations buildup

Table 10.2 Equation composition

Term	Meaning
Coefficient matrix	The matrix can be divided into 4 parts, and each part consists of multiple submatrices, including three components: multiple media, multiple components, and variables ① Partial derivatives of mass conservation equations for discrete components with respect to each principal variable ② Each row represents the mass conservation equation for a component, and each grid corresponds to three equations for gas, oil and water ③ Each column represents the solving of a main variable that are P_o, S_g and S_w for oil, gas and water phases
Variable array	The variation of main variables of media grids and well grids in multiple media
Right hand side	Value of the equations for tight reservoir grids and wellbore grids at the n moment

where, J_{RR} is the partial derivative of a reservoir equation with respect to reservoir variables, represented as RR; J_{RW} is the partial derivative of a reservoir equation with respect to wellbore variables, represented as RW; J_{WR} is the partial derivative of a wellbore equation with respect to reservoir variables, represented as WR; J_{WW} is the partial derivative of a wellbore equation with respect to wellbore variables, represented as WW.

Tight sandstone reservoirs contain multiple media such as nano/micro-pores, micro fissures, and natural/artificial fractures, and are developed with complex-structured wells. This makes the four parts of the coefficient matrix very complicated. The generation of the coefficient matrix is illustrated with an example in Fig. 10.3. The figure is made up of eight unstructured meshes, and a multi-segment well model that is composed of four meshes, and the meshes noted with ②, ③, and ④ are the perforated.

Complete format of the mathematical matrix has been shown in Fig. 10.4.

(1) The generation of the RR matrix (reservoir equations with respect to reservoir variables)

The RR matrix is the coefficient matrix of reservoir equations with respect to the reservoir variables. Its composition includes: the derivatives of the oil/gas/water mass conservation equations with respect to variables such as reservoir pressure P_o, gas saturation S_g and water saturation S_w in a grid and the adjacent grids. The connectivity table provides relationship for the adjacent unstructured grids. Table 10.3 shows the specific information for each grid.

According to the derivation rule in the above table, and the mathematical model in Sect. 3 of Chap. 4, the general formulas of the derivatives of the variables in the RR coefficient matrix are obtained, as shown in Table 10.4.

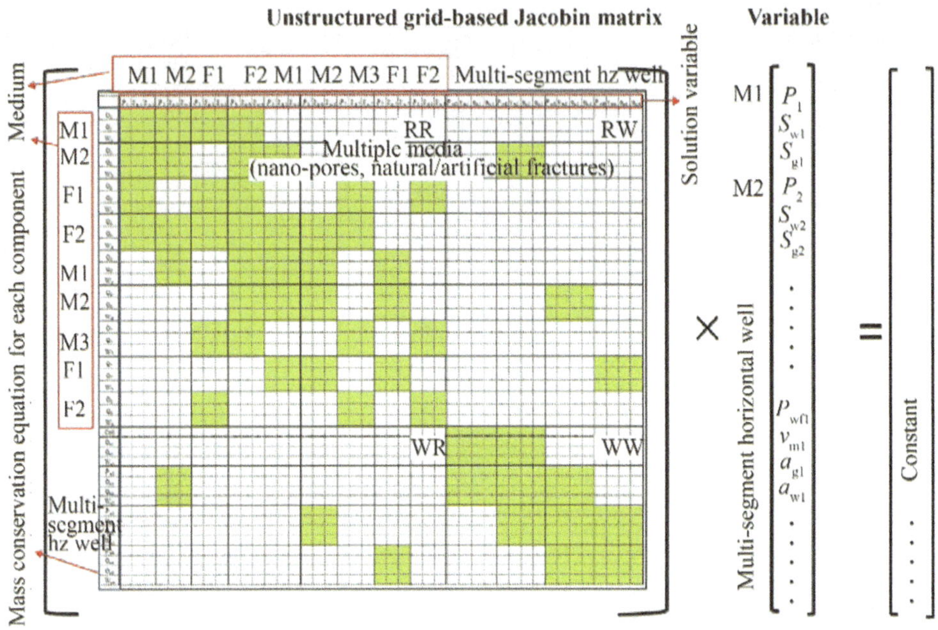

Fig. 10.2 Matrix equation based on unstructured grid

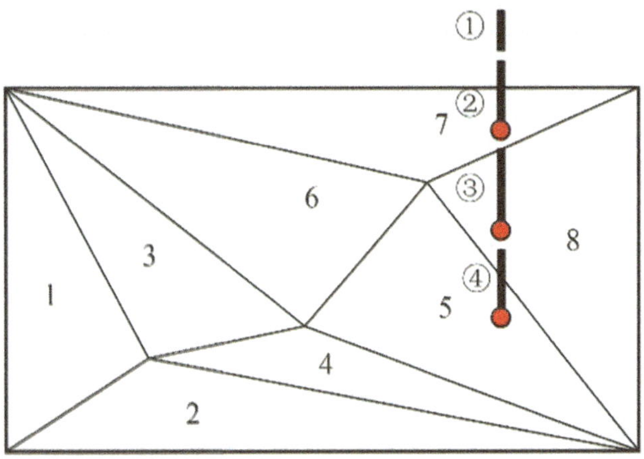

Fig. 10.3 Jacobin matrix based on unstructured grid

The RR part is shown in Fig. 10.5:

(2) The generation of the RW matrix (reservoir equations with respect to reservoir variables)

The RR matrix is the coefficient matrix of reservoir equations with respect to the reservoir variables. Its composition includes: the derivatives of the oil/gas/water mass conservation equations with respect to wellbore pressure at the perforation grids. The connectivity table provides the relationship of grids with perforation. Table 10.5 shows the information for each perforation grid.

According to the derivation rule in the above table, combing the mathematical model in Sect. 3 of Chap. 4, and taking the multi-segment wellbore treatment scheme as an example, the general formula of the derivatives of variables in the RW coefficient matrix is obtained, as shown in Table 10.6.

When the grid shape and connection are like those in Fig. 10.4, where the R7, R8 and R5 grids in the reservoir are connected to W2, W3, and W4 in multi-segment wells, respectively. The RW part is shown Fig. 10.6.

(3) The generation of the WR matrix (reservoir equations with respect to reservoir variables)

The RR matrix is the coefficient matrix of reservoir equations with respect to the reservoir variables. Its composition includes: the derivatives of the wellbore equations with respect to reservoir pressure, gas saturation, water saturation at the perforation grids. The connectivity table provides the relationship of the perforation grids. Table 10.7 shows the information for each perforation grid.

According to the derivation rule in the above table, combing the mathematical model in Sect. 3 of Chap. 4, and taking the multi-segment wellbore treatment scheme as an example, the general formula of the derivatives of the variables in the WR coefficient matrix are obtained, as shown in Table 10.8.

When the grid shape and connection relation are like those in Fig. 10.4, where the R7, R8 and R5 grids in the reservoir are connected to W2, W3, and W4 in multi-segment wells, respectively. the WR part is shown Fig. 10.7.

(4) The generation of the WW matrix (reservoir equations with respect to reservoir variables)

The WW matrix is the coefficient matrix of well equations with respect to the well variables. Its composition includes: the partial

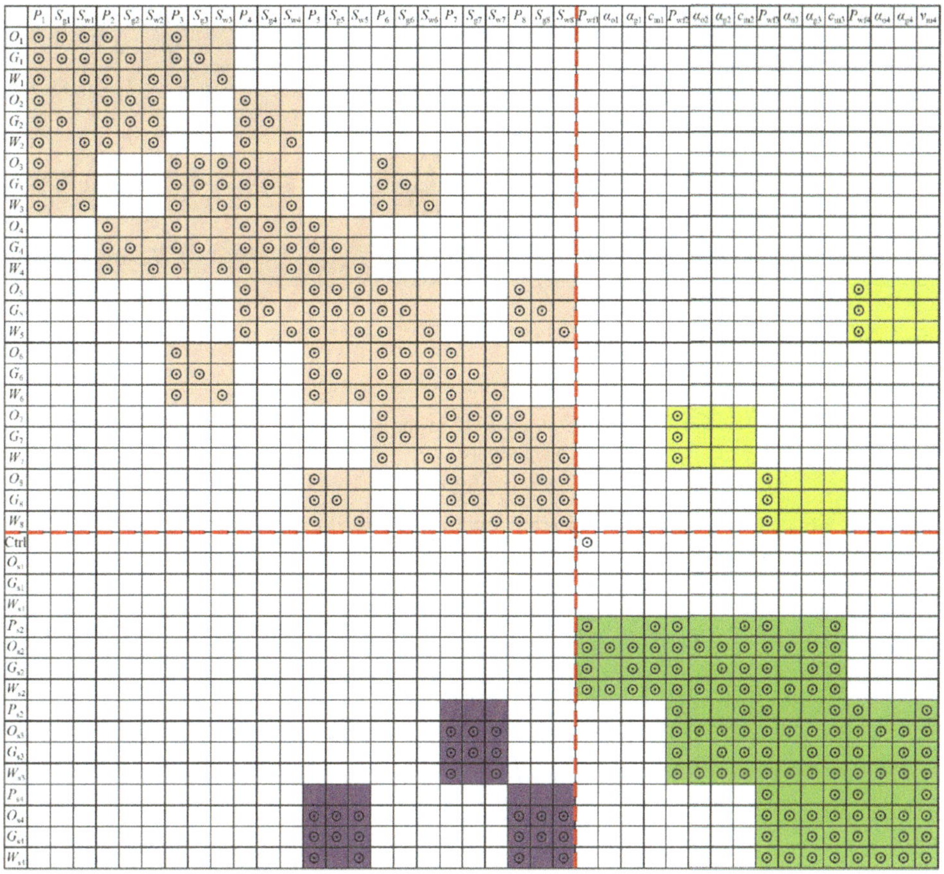

Fig. 10.4 Complete format of the mathematical matrix

Table 10.3 Variables in the RR matrix

Name	Content	Diagonal variable	Non-diagonal variable
Mass Conservation equation	Row 1: mass conservation equation for gas F_g Row 2: mass conservation equation for oil F_o Row 3: mass conservation equation for water F_w	$\begin{pmatrix} \frac{\partial F_{g,i}}{\partial P_i} & \frac{\partial F_{g,i}}{\partial S_{w,i}} & \frac{\partial F_{g,i}}{\partial S_{g,i}} \\ \frac{\partial F_{o,i}}{\partial P_i} & \frac{\partial F_{o,i}}{\partial S_{w,i}} & \frac{\partial F_{o,i}}{\partial S_{g,i}} \\ \frac{\partial F_{w,i}}{\partial P_i} & \frac{\partial F_{w,i}}{\partial S_{w,i}} & \frac{\partial F_{w,i}}{\partial S_{g,i}} \end{pmatrix}$	$\begin{pmatrix} \frac{\partial F_{g,i}}{\partial P_j} & \frac{\partial F_{g,i}}{\partial S_{w,j}} & \frac{\partial F_{g,i}}{\partial S_{g,j}} \\ \frac{\partial F_{o,i}}{\partial P_j} & \frac{\partial F_{o,i}}{\partial S_{w,j}} & \frac{\partial F_{o,i}}{\partial S_{g,j}} \\ \frac{\partial F_{w,i}}{\partial P_j} & \frac{\partial F_{w,i}}{\partial S_{w,j}} & \frac{\partial F_{w,i}}{\partial S_{g,j}} \end{pmatrix}$
Principal variable	Column 1: reservoir pressure P_o Column 2: gas saturation S_g Column 3: water saturation S_w		

derivatives of the complex-structured well, governing equations with respect to wellbore pressure in the grid and adjacent grids, the mixed flow velocity in a wellbore, the oil saturation in the wellbore, and the gas saturations in the wellbore. Table 10.9 shows the information for each perforation grid.

According to the derivation rule in the above table, combing the mathematical model in Sect. 3 of Chap. 4, and taking the multi-segment wellbore treatment scheme as an example, the general formula of the derivatives of the variables in the WW coefficient matrix are obtained, as shown in Table 10.10.

When the grid shape and connection relation are like those in Fig. 10.4, the WW part is shown in Fig. 10.8.

2. Generation of the right-hand side

According to the Newton-Raphson method, the right-hand side of the matrix equation is the residual value of the nth time step multiplied by −1; that is, $R^{(\nu)}$ at the nth step multiplied by −1, as shown in the following equation:

$$\boldsymbol{J}^{(\nu)}\delta\boldsymbol{x}^{(\nu+1)} = -\boldsymbol{R}^{(\nu)}$$

Table 10.4 General formula of the derivatives of the variables in the RR coefficient matrix

Variable	Equation for oil component	Equation for gas component	Equation for water component
$\delta P_{o,i}$	$b_2 - \sum_j T_{ij} \cdot a_3 + c_3$	$b'_2 - \sum_j T_{ij} \cdot a'_3 + c'_3$	$b''_2 - \sum_j T_{ij} \cdot a''_3 + c''_3$
$\delta S_{g,i}$	$b_3 - \sum_j T_{ij} \cdot a_4 + c_4$	$b'_3 - \sum_j T_{ij} \cdot a'_5 + c'_4$	–
$\delta S_{w,i}$	$b_4 - \sum_j T_{ij} \cdot a_5 + c_5$	$b'_4 - \sum_j T_{ij} \cdot a'_6 + c'_5$	$b''_3 - \sum_j T_{ij} \cdot a''_5 + c''_4$
$\delta P_{o,j}$	$-\sum_j T_{ij} \cdot a_2$	$-\sum_j T_{ij} \cdot a'_2$	$-\sum_j T_{ij} \cdot a''_2$
$\delta S_{g,j}$	–	$-\sum_j T_{ij} \cdot a'_4$	–
$\delta S_{w,j}$			$-\sum_j T_{ij} \cdot a''_4$

Note please refer to the model in Chap. 4 for the specific forms of the letters in the table same for the following sections

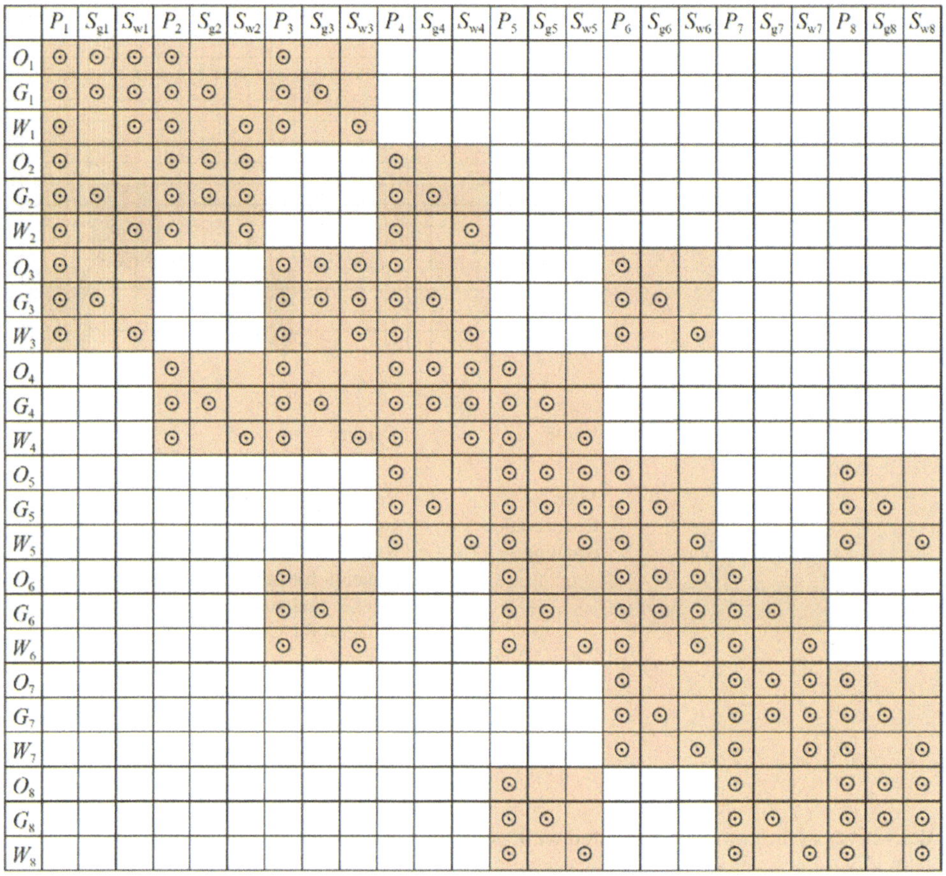

Fig. 10.5 RR coefficient matrix format. *Note* means zero value and/means non-zero at the position in the matrix

where, $J^{(v)}$ is the coefficient matrix, $J = \begin{bmatrix} \text{RR} & \text{RW} \\ \text{WR} & \text{WW} \end{bmatrix}$; $\delta x^{(v+1)} = x^{(v+1)} - x^{(v)}$ is the variation of the reservoir variable between Step $v + 1$ and Step v, x is the principal variable, $x = [P, S_g, S_w]^T$; $R^{(v)}$ is the residual of the equation at Step v, $R = [R_R, R_w]^T$.

3. Flowchart of mathematical matrix generation

Tight sandstone reservoirs are characterized by nano/micro-pores, complex natural/artificial fracture networks, complex-structured wells, and complex geological boundaries, which cause many invalid grids, complex matrix formats, and some issues that conventional grids are unable

Table 10.5 Variables in the RW matrix

Name	Content		Specific format
Mass conservation equation	Row 1: mass conservation equation for gas F_g Row 2: mass conservation equation for oil F_o Row 3: mass conservation equation for water F_w		–
Principal variable	Line-source wellbore	Column 1: wellbore pressure P^W	$\left[\frac{\partial F_{g,i}}{\partial P_k^W} \quad \frac{\partial F_{o,i}}{\partial P_k^W} \quad \frac{\partial F_{o,i}}{\partial P_k^W}\right]^T$
	Discrete wellbore grid	Column 1: wellbore grid pressure P_w Column 2: gas saturation S_g Column 3: water saturation S_w	$\begin{bmatrix} \frac{\partial F_{g,i}}{\partial P_k^W} & \frac{\partial F_{g,i}}{\partial S_{w,k}} & \frac{\partial F_{g,i}}{\partial S_{g,k}} \\ \frac{\partial F_{o,i}}{\partial P_k^W} & \frac{\partial F_{o,i}}{\partial S_{w,k}} & \frac{\partial F_{o,i}}{\partial S_{g,k}} \\ \frac{\partial F_{w,i}}{\partial P_k^W} & \frac{\partial F_{w,i}}{\partial S_{w,k}} & \frac{\partial F_{w,i}}{\partial S_{g,k}} \end{bmatrix}$
	Multi-segment wellbore	Column 1: wellbore pressure P^W	$\left[\frac{\partial F_{g,i}}{\partial P_k^W} \quad \frac{\partial F_{o,i}}{\partial P_k^W} \quad \frac{\partial F_{o,i}}{\partial P_k^W}\right]^T$

Table 10.6 General formula of the derivatives of the variables in the RW coefficient matrix

Variable	Equation for oil component	Equation for gas component	Equation for water component
δP_{wfk}	c_2	c_2'	c_2''

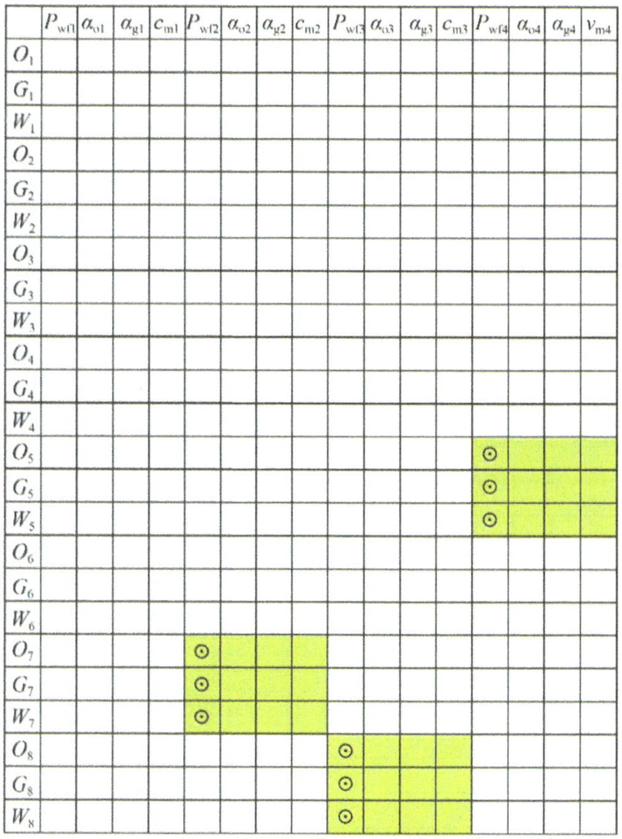

Fig. 10.6 Format of the RW coefficient matrix

to describe. On the basis of the connectivity table of unstructured grids and the sparse matrix derivation rules, a mathematical matrix generation technology for multiple media based on unstructured grids is developed. The corresponding technology flowchart is shown in Fig. 10.9.

10.1.1.3 Non-zero Structure of a Mathematical Matrix and Its Affecting Factor

The format of a coefficient matrix is influenced by various factors, such as variables, flow mechanisms, complex-structured wells, and the unstructured grid type. The computing speed and precision of linear algebraic equations corresponding to different mathematical matrix formats are rather different.

1. Effect of variables on mathematical matrix format

Variables will not affect the macroscopic form of mathematical matrix if they remain fixed in submatrices. As shown in Fig. 10.10, the RR part is composed of several submatrices. The position and arrangement of the submatrices affect the format of the mathematical matrix, but the variation of the submatrix variables has no effect on the format of the macro matrix.

Variables in submatrices: pressure and saturation.

$$\frac{\partial \boldsymbol{R}_{i,j}}{\partial \boldsymbol{x}_{l,m}} = \begin{bmatrix} \frac{\partial \boldsymbol{R}_{i,j}^{o}}{\partial P_{ol,m}} & \frac{\partial \boldsymbol{R}_{i,j}^{o}}{\partial S_{wl,m}} & \frac{\partial \boldsymbol{R}_{i,j}^{o}}{\partial S_{gl,m}} \\ \frac{\partial \boldsymbol{R}_{i,j}^{w}}{\partial P_{ol,m}} & \frac{\partial \boldsymbol{R}_{i,j}^{w}}{\partial S_{wl,m}} & \frac{\partial \boldsymbol{R}_{i,j}^{w}}{\partial S_{gl,m}} \\ \frac{\partial \boldsymbol{R}_{i,j}^{g}}{\partial P_{ol,m}} & \frac{\partial \boldsymbol{R}_{i,j}^{g}}{\partial S_{wl,m}} & \frac{\partial \boldsymbol{R}_{i,j}^{g}}{\partial S_{gl,m}} \end{bmatrix}$$

2. Effect of flow mechanisms on mathematical matrix format

When the diffusion, slipping, starting pressure gradient, high-speed nonlinear flow, stress sensitivity, imbibition and desorption are considered, the flow term and the cumulative

Table 10.7 Variables in the WR matrix

Name	Content		Element
Mass conservation equation	Line-source wellbore	Row 1: well production equation F^{W}	$\begin{bmatrix} \frac{\partial F_k^{\mathrm{W}}}{\partial P_k} & \frac{\partial F_k^{\mathrm{W}}}{\partial S_{\mathrm{w},k}} & \frac{\partial F_k^{\mathrm{W}}}{\partial S_{\mathrm{g},k}} \end{bmatrix}$
	Discrete grid wellbore	Row 1: mass conservation equation for gas $F_{\mathrm{g}}^{\mathrm{W}}$ Row 2: mass conservation equation for oil $F_{\mathrm{o}}^{\mathrm{W}}$ Row 3: mass conservation equation for water $F_{\mathrm{w}}^{\mathrm{W}}$	$\begin{bmatrix} \frac{\partial F_{\mathrm{g},k}^{\mathrm{W}}}{\partial P_k} & \frac{\partial F_{\mathrm{g},k}^{\mathrm{W}}}{\partial S_{\mathrm{w},k}} & \frac{\partial F_{\mathrm{g},k}^{\mathrm{W}}}{\partial S_{\mathrm{g},k}} \\ \frac{\partial F_{\mathrm{o},k}^{\mathrm{W}}}{\partial P_k} & \frac{\partial F_{\mathrm{o},k}^{\mathrm{W}}}{\partial S_{\mathrm{w},k}} & \frac{\partial F_{\mathrm{o},k}^{\mathrm{W}}}{\partial S_{\mathrm{g},k}} \\ \frac{\partial F_{\mathrm{w},k}^{\mathrm{W}}}{\partial P_k} & \frac{\partial F_{\mathrm{w},k}^{\mathrm{W}}}{\partial S_{\mathrm{w},k}} & \frac{\partial F_{\mathrm{w},k}^{\mathrm{W}}}{\partial S_{\mathrm{g},k}} \end{bmatrix}$
	Multi-segment wellbore	Row 1: wellbore pressure reduction F_p^{W} Row 2: mass conservation equation for gas $F_{\mathrm{g}}^{\mathrm{W}}$ Row 3: mass conservation equation for oil $F_{\mathrm{o}}^{\mathrm{W}}$ Row 4: mass conservation equation for water $F_{\mathrm{w}}^{\mathrm{W}}$	$\begin{bmatrix} \frac{\partial F_{\mathrm{P},k}^{\mathrm{W}}}{\partial P_k} & \frac{\partial F_{\mathrm{P},k}^{\mathrm{W}}}{\partial S_{\mathrm{w},k}} & \frac{\partial F_{\mathrm{P},k}^{\mathrm{W}}}{\partial S_{\mathrm{g},k}} \\ \frac{\partial F_{\mathrm{g},k}^{\mathrm{W}}}{\partial P_k} & \frac{\partial F_{\mathrm{g},k}^{\mathrm{W}}}{\partial S_{\mathrm{w},k}} & \frac{\partial F_{\mathrm{g},k}^{\mathrm{W}}}{\partial S_{\mathrm{g},k}} \\ \frac{\partial F_{\mathrm{o},k}^{\mathrm{W}}}{\partial P_k} & \frac{\partial F_{\mathrm{o},k}^{\mathrm{W}}}{\partial S_{\mathrm{w},k}} & \frac{\partial F_{\mathrm{o},k}^{\mathrm{W}}}{\partial S_{\mathrm{g},k}} \\ \frac{\partial F_{\mathrm{w},k}^{\mathrm{W}}}{\partial P_k} & \frac{\partial F_{\mathrm{w},k}^{\mathrm{W}}}{\partial S_{\mathrm{w},k}} & \frac{\partial F_{\mathrm{w},k}^{\mathrm{W}}}{\partial S_{\mathrm{g},k}} \end{bmatrix}$
Principal variable	Line-source wellbore	Column 1: reservoir pressure P_{o} Column 2: gas saturation S_{g} Column 3: water saturation S_{w}	–
	Discrete grid wellbore		
	Multi-segment wellbore		

Table 10.8 General formula of the derivatives of the variables in the WR coefficient matrix

Variable	Equation for oil component	Equation for gas component	Equation for water component
$\delta P_{\mathrm{o},i}$	c_3	c_3'	c_3''
$\delta S_{\mathrm{g},i}$	c_4	c_4'	–
$\delta S_{\mathrm{w},i}$	c_5	c_5'	c_4''

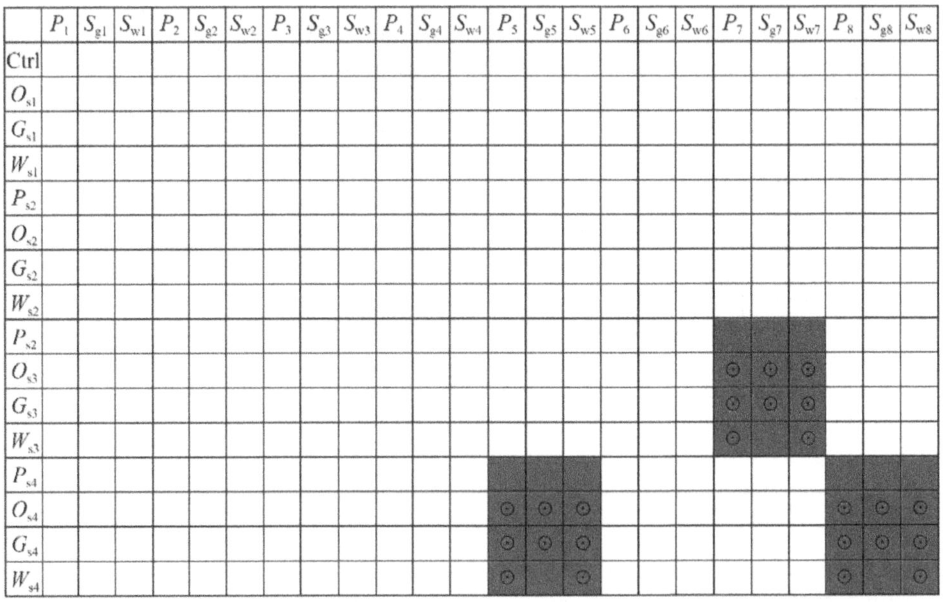

Fig. 10.7 Formation of the WR coefficient matrix

Table 10.9 Variables in the WW matrix

Name	Content		Element
Mass conservation equation	Line-source wellbore	Row 1: well production equation F^{W}	–
	Discrete grid wellbore	Row 1: mass conservation equation for gas F_{g}^{W} Row 2: mass conservation equation for oil F_{o}^{W} Row 3: mass conservation equation for water F_{w}^{W}	
	Multi-segment wellbore	Row 1: wellbore pressure reduction F_{p}^{W} Row 2: mass conservation equation for gas F_{g}^{W} Row 3: mass conservation equation for oil F_{o}^{W} Row 4: mass conservation equation for water F_{w}^{W}	
Principal variable	Line-source wellbore	Column 1: wellbore pressure P^{W}	$\frac{\partial F_{k}^{W}}{\partial P_{k}^{W}}$
	Discrete grid wellbore	Column 1: pressure in the well grid P_{w} Column 2: gas saturation in the well grid S_{g} Column 3: gas saturation in the well grid S_{w}	$\begin{bmatrix} \frac{\partial F_{g,k}^{W}}{\partial P_{k}^{W}} & \frac{\partial F_{g,k}^{W}}{\partial S_{g,k}^{W}} & \frac{\partial F_{g,k}^{W}}{\partial S_{w,k}^{W}} \\ \frac{\partial F_{o,k}^{W}}{\partial P_{k}^{W}} & \frac{\partial F_{o,k}^{W}}{\partial S_{g,k}^{W}} & \frac{\partial F_{o,k}^{W}}{\partial S_{w,k}^{W}} \\ \frac{\partial F_{w,k}^{W}}{\partial P_{k}^{W}} & \frac{\partial F_{w,k}^{W}}{\partial S_{g,k}^{W}} & \frac{\partial F_{w,k}^{W}}{\partial S_{w,k}^{W}} \end{bmatrix}$
	Multi-segment wellbore	Column 1: wellbore pressure P^{W} Column 2: mixed fluid velocity in the wellbore V_{m}^{W} Column 3: gas holdup in the wellbore α_{g}^{W} Column 3: liquid holdup in the wellbore α_{w}^{W}	$\begin{bmatrix} \frac{\partial F_{P,k}^{W}}{\partial P_{k}^{W}} & \frac{\partial F_{P,k}^{W}}{\partial V_{m,k}^{W}} & \frac{\partial F_{P,k}^{W}}{\partial \alpha_{g,k}^{W}} & \frac{\partial F_{P,k}^{W}}{\partial \alpha_{w,k}^{W}} \\ \frac{\partial F_{g,k}^{W}}{\partial P_{k}^{W}} & \frac{\partial F_{g,k}^{W}}{\partial V_{m,k}^{W}} & \frac{\partial F_{g,k}^{W}}{\partial \alpha_{g,k}^{W}} & \frac{\partial F_{g,k}^{W}}{\partial \alpha_{w,k}^{W}} \\ \frac{\partial F_{o,k}^{W}}{\partial P_{k}^{W}} & \frac{\partial F_{o,k}^{W}}{\partial V_{m,k}^{W}} & \frac{\partial F_{o,k}^{W}}{\partial \alpha_{g,k}^{W}} & \frac{\partial F_{o,k}^{W}}{\partial \alpha_{w,k}^{W}} \\ \frac{\partial F_{w,k}^{W}}{\partial P_{k}^{W}} & \frac{\partial F_{w,k}^{W}}{\partial V_{m,k}^{W}} & \frac{\partial F_{w,k}^{W}}{\partial \alpha_{g,k}^{W}} & \frac{\partial F_{w,k}^{W}}{\partial \alpha_{w,k}^{W}} \end{bmatrix}$

Table 10.10 General formula of the derivatives of the variables in the WW coefficient matrix

Unknowns	Fluid equation			Pressure falloff equation
	Oil component	Gas component	Water component	
$\delta P_{o,i}$	c_3	c_3'	c_3''	–
$\delta S_{g,i}$	c_4	c_4'	–	–
$\delta S_{w,i}$	c_5	c_5'	c_4''	–
δP_{wfk}	$b_2^{W} - a_2^{W} + c_2$	$b_2'^{W} - a_2'^{W} + c_2'$	$b_2''^{W} - a_2''^{W} + c_2''$	d_2^{W}
$\delta \alpha_{o,k}$	$b_3^{W} - a_5^{W}$	$b_3'^{W}$	$b_3''^{W} - a_5''^{W}$	–
$\delta \alpha_{g,k}$	$-a_6^{W}$	$b_4'^{W} - a_5'^{W}$	$b_4''^{W} - a_6''^{W}$	–
$\delta v_{m,k}$	$-a_4^{W}$	$-a_4'^{W}$	$-a_4''^{W}$	d_4^{W}
$\delta P_{wf,k-1}$	$-a_3^{W}$	$-a_3'^{W}$	$-a_3''^{W}$	d_3^{W}
$\delta \alpha_{o,k-1}$	$-a_8^{W}$	–	$-a_8''^{W}$	–
$\delta \alpha_{g,k-1}$	$-a_9^{W}$	$-a_7'^{W}$	$-a_9''^{W}$	–
$\delta v_{m,k-1}$	$-a_7^{W}$	$-a_6'^{W}$	$-a_7''^{W}$	–

Fig. 10.8 Format of the WW coefficient matrix

	P_{wf1}	α_{o1}	α_{g1}	v_{m1}	P_{wf2}	α_{o2}	α_{g2}	v_{m2}	P_{wf3}	α_{o3}	α_{g3}	v_{m3}	P_{wf4}	α_{o4}	α_{g4}	v_{m4}
Ctrl	⊙															
O_{s1}																
G_{s1}																
W_{s1}																
P_{s2}	⊙			⊙	⊙			⊙	⊙			⊙				
O_{s2}	⊙	⊙	⊙	⊙	⊙	⊙	⊙	⊙	⊙	⊙	⊙	⊙				
G_{s2}	⊙		⊙	⊙	⊙		⊙	⊙	⊙		⊙	⊙				
W_{s2}	⊙	⊙	⊙	⊙	⊙	⊙	⊙	⊙	⊙	⊙	⊙	⊙				
P_{s2}					⊙			⊙	⊙			⊙	⊙			⊙
O_{s3}					⊙	⊙	⊙	⊙	⊙	⊙	⊙	⊙	⊙	⊙	⊙	⊙
G_{s3}					⊙		⊙	⊙	⊙		⊙	⊙	⊙		⊙	⊙
W_{s3}					⊙	⊙	⊙	⊙	⊙	⊙	⊙	⊙	⊙	⊙	⊙	⊙
P_{s4}									⊙			⊙	⊙			⊙
O_{s4}									⊙	⊙	⊙	⊙	⊙	⊙	⊙	⊙
G_{s4}									⊙		⊙	⊙	⊙		⊙	⊙
W_{s4}									⊙	⊙	⊙	⊙	⊙	⊙	⊙	⊙

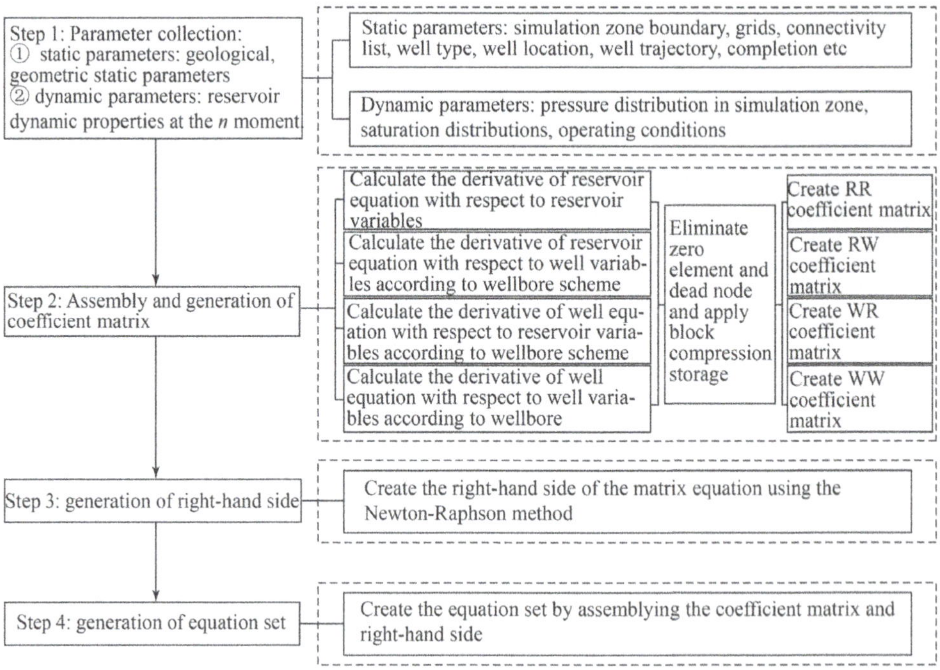

Fig. 10.9 Flowchart for the generation of the mathematical matrix based on unstructured grid

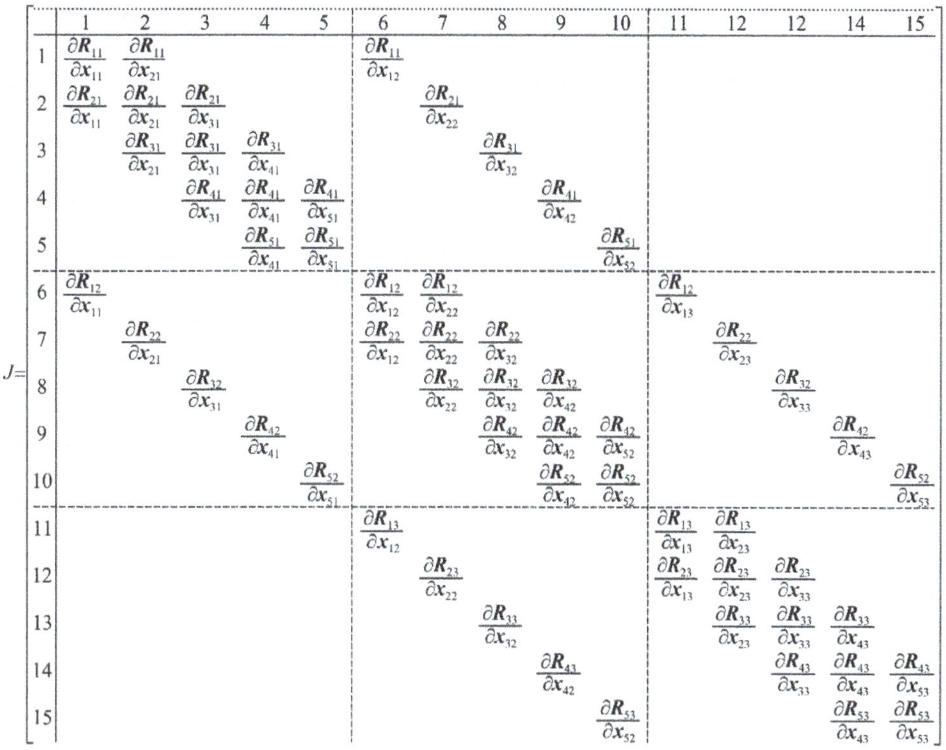

Fig. 10.10 The format of the RR matrix

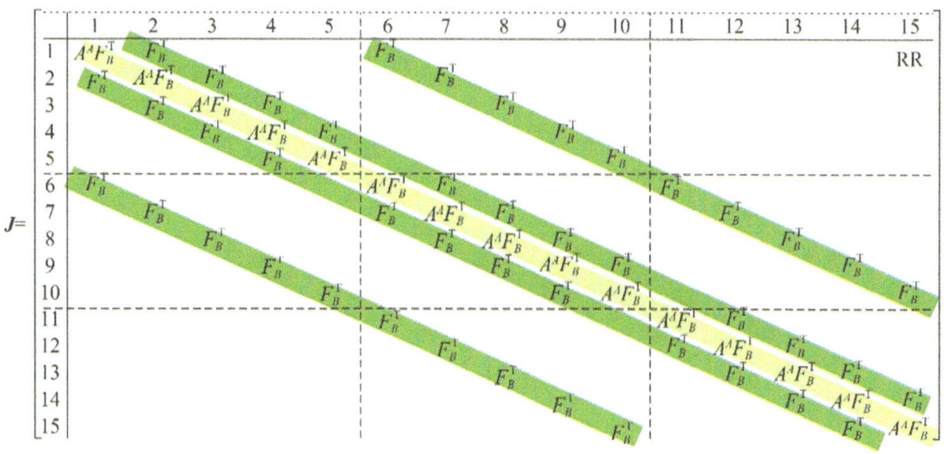

Fig. 10.11 The format of the RR matrix that considers flow mechanisms

term in the matrix change, but the matrix format stays the same (Fig. 10.11).

where, the flow term $F_{\mathrm{B}}^{\mathrm{T}}$ incorporates the diffusion, slipping, starting pressure gradient, high-speed nonlinear seepage, stress sensitivity, imbibition and desorption. It is expressed as

$$F_B^{\mathrm{T}} = -\frac{\partial \boldsymbol{F}_B^{\mathrm{D,K,G,F,M,I}}}{\partial \boldsymbol{x}} = \frac{\partial \boldsymbol{R}_{i,j}}{\partial \boldsymbol{x}_{l,m}} = \begin{bmatrix} \frac{\partial R_{i,j}^{\mathrm{o}}}{\partial P_{\mathrm{o}l,\mathrm{m}}} & \frac{\partial R_{i,j}^{\mathrm{o}}}{\partial S_{\mathrm{w}l,\mathrm{m}}} & \frac{\partial R_{i,j}^{\mathrm{o}}}{\partial S_{\mathrm{g}l,\mathrm{m}}} \\ \frac{\partial R_{i,j}^{\mathrm{w}}}{\partial P_{\mathrm{o}l,\mathrm{m}}} & \frac{\partial R_{i,j}^{\mathrm{w}}}{\partial S_{\mathrm{w}l,\mathrm{m}}} & \frac{\partial R_{i,j}^{\mathrm{w}}}{\partial S_{\mathrm{g}l,\mathrm{m}}} \\ \frac{\partial R_{i,j}^{\mathrm{g}}}{\partial P_{\mathrm{o}l,\mathrm{m}}} & \frac{\partial R_{i,j}^{\mathrm{g}}}{\partial S_{\mathrm{w}l,\mathrm{m}}} & \frac{\partial R_{i,j}^{\mathrm{g}}}{\partial S_{\mathrm{g}l,\mathrm{m}}} \end{bmatrix}$$

The coupled expression of the accumulation term and the flow term is (A^A denoted the desorption is included):

$$A^A F_B^{\mathrm{T}} = \frac{\partial \boldsymbol{A}}{\partial \boldsymbol{x}} - \frac{\partial \boldsymbol{F}_B^{\mathrm{D,K,G,F,M,I}}}{\partial \boldsymbol{x}} = \frac{\partial \boldsymbol{R}_{i,j}}{\partial \boldsymbol{x}_{l,m}} = \begin{bmatrix} \frac{\partial R_{i,j}^{\mathrm{o}}}{\partial P_{\mathrm{o}l,\mathrm{m}}} & \frac{\partial R_{i,j}^{\mathrm{o}}}{\partial S_{\mathrm{w}l,\mathrm{m}}} & \frac{\partial R_{i,j}^{\mathrm{o}}}{\partial S_{\mathrm{g}l,\mathrm{m}}} \\ \frac{\partial R_{i,j}^{\mathrm{w}}}{\partial P_{\mathrm{o}l,\mathrm{m}}} & \frac{\partial R_{i,j}^{\mathrm{w}}}{\partial S_{\mathrm{w}l,\mathrm{m}}} & \frac{\partial R_{i,j}^{\mathrm{w}}}{\partial S_{\mathrm{g}l,\mathrm{m}}} \\ \frac{\partial R_{i,j}^{\mathrm{g}}}{\partial P_{\mathrm{o}l,\mathrm{m}}} & \frac{\partial R_{i,j}^{\mathrm{g}}}{\partial S_{\mathrm{w}l,\mathrm{m}}} & \frac{\partial R_{i,j}^{\mathrm{g}}}{\partial S_{\mathrm{g}l,\mathrm{m}}} \end{bmatrix}$$

3. Effect of complex-structured well on matrix format

A complex-structured well is located at the right and bottom of the mathematical matrix "fringed symmetrical band". The format of RW and WR are only related to the perforated wellbore segments. The format of WW is related to all wellbore segments. The more the wellbore segments, the wider the band and the more complicated the mathematical matrix.

For a vertical well with four segments and three of them perforated, the mathematical matrix format is shown in Fig. 10.12, and the RR format is not affected.

For a horizontal well with eight segments and four of them perforated, the mathematical matrix's format is shown in Fig. 10.13, and the RR format is not affected.

4. Effect of unstructured grids on mathematical matrix format

The matrix format is closely related to the ordering of an unstructured grid. The matrix is symmetrical to the principal diagonal. Non-zero elements are banded on the main diagonal line, and scattered on two sides. The non-zero elements are irregularly distributed on each row.

As shown in Fig. 10.14, in a Jacobin matrix based on structured grids, the non-zero elements are regularly distributed on the principal diagonal and its two sides in banded arrangement:

As shown in Fig. 10.15, in a Jacobin matrix based on unstructured grids, the non-zero elements are banded on the principal diagonal, and irregularly distributed on two sides:

5. Effect of different types of unstructured grids on matrix format

Different grid types have different adjacent grid number, and so the of no-zero element number. In general, the more grid types, the more complex the matrix becomes. Here the triangular grid, PEBI grid and hybrid grid are compared as follows (Figs. 10.16, 10.17 and 10.18):

(1) Triangular grid
(2) PEBI grid
(3) Hybrid grid

6. Effect of grid ordering on the mathematical matrix format

The grid ordering and the number of adjacent grids determine the grid adjacency, which directly determines the matrix format. For similar grid adjacency but different grid types, the matrix formats are similar.

Comparison of Figs. 10.17 and 10.19 shows that different types but similar ordering lead to a similar matrix format. Comparison of Figs. 10.18 and 10.19 shows that same types but different ordering leads to a similar matrix format.

7. Effect of multiple media on matrix format

When the model is for multiple media, the non-zero elements of the same medium lumping together on the principal diagonal, reflects the flow exchange within the same medium. The distribution of non-zero elements on the two sides relating to the grid sorting, reflects the inter-porosity flow among different media. RW and WR reflect the coupled flow between the media and the wellbore, as shown in Fig. 10.20.

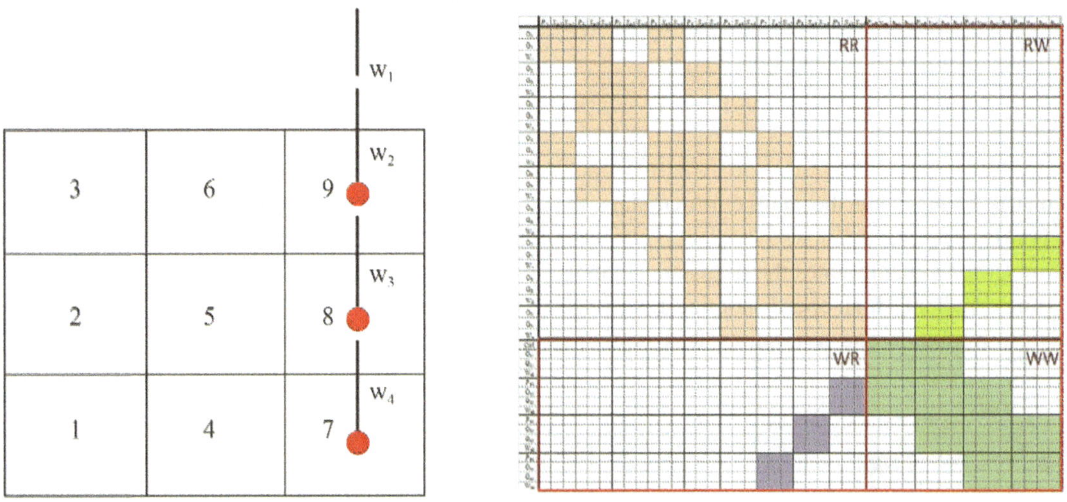

(a) Multi-segment vertical well (b) Format of coefficient matrix with multi-segment wellbore

Fig. 10.12 Multi-segment vertical well

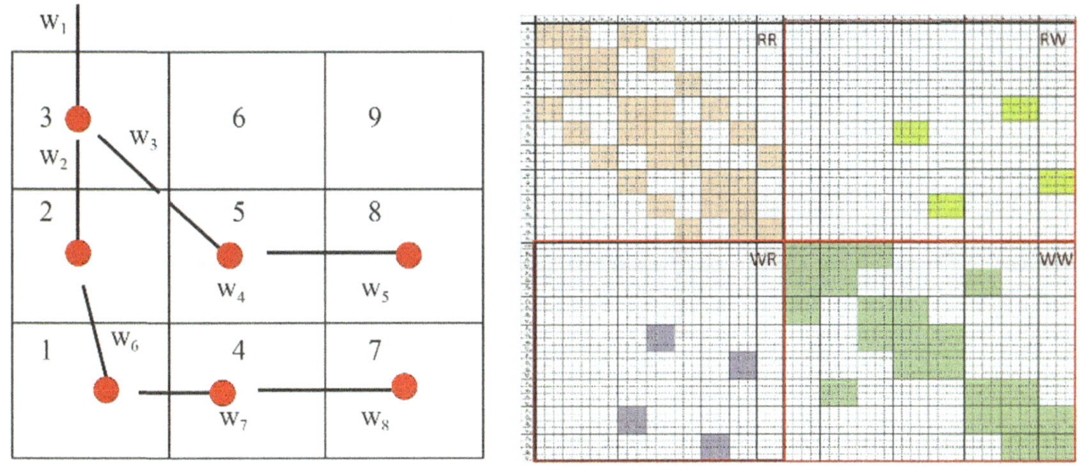

(a) Multi-segment cluster horizontal well (b) Format of the matrix with multi-segment wellbore

Fig. 10.13 Multi-segment cluster horizontal well

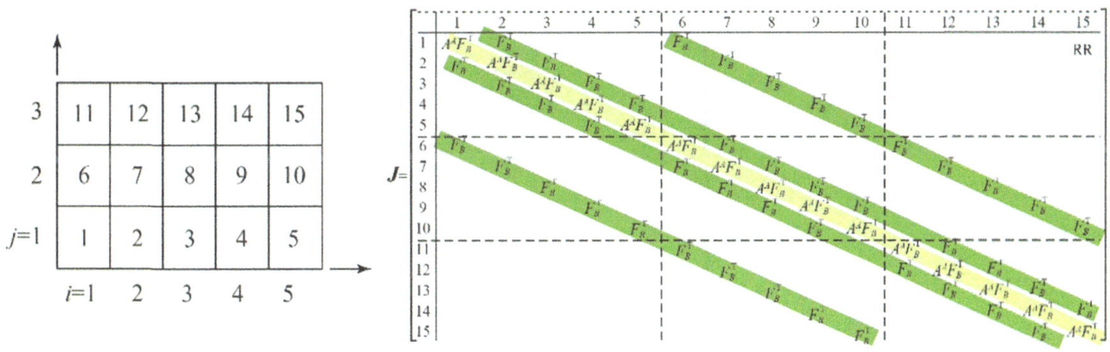

(a) Grid ordering (b) Matrix format

Fig. 10.14 Jacobin matrix based on structured grid

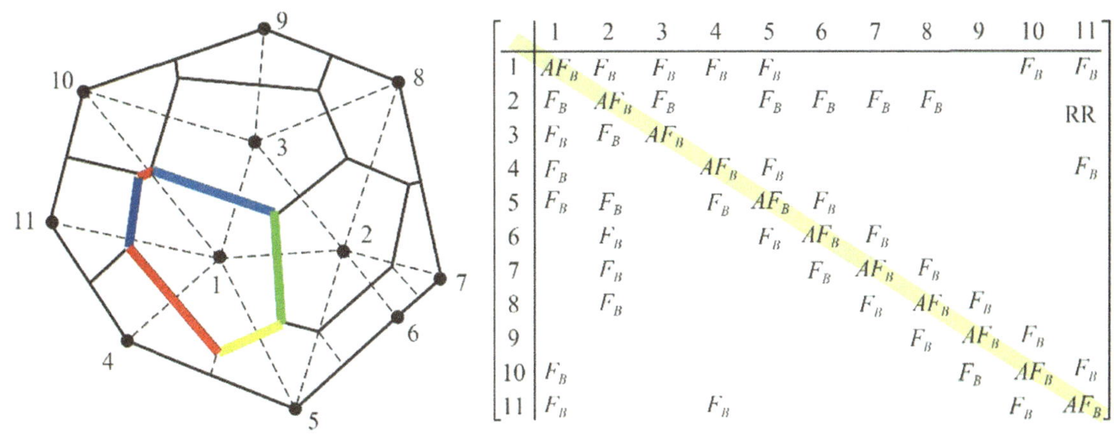

(a) Grid order (b) Matrix format

Fig. 10.15 Jacobin matrix based on unstructured grid

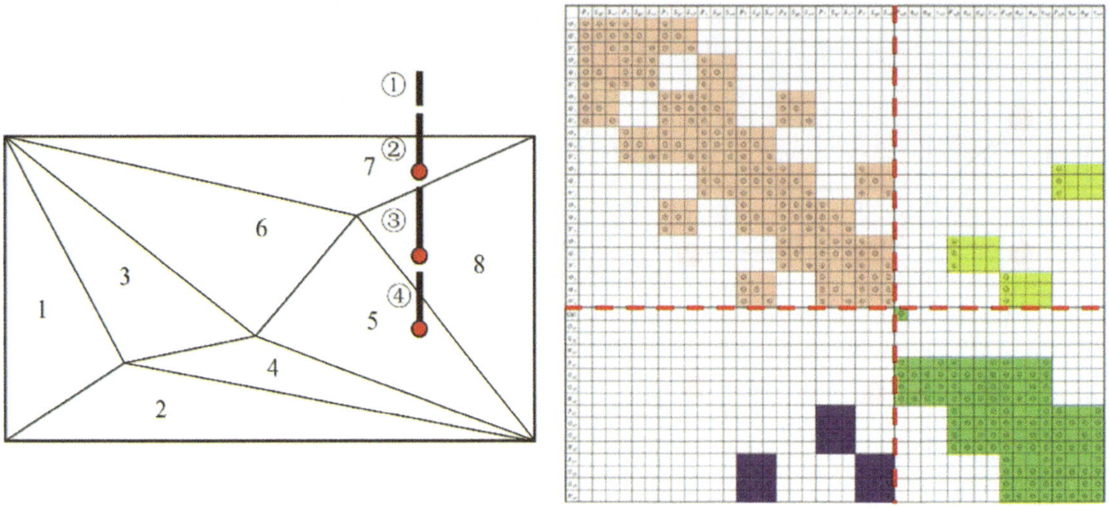

(a) Schematics of triangular grid (b) Matrix format

Fig. 10.16 Jacobin matrix based on triangular grid

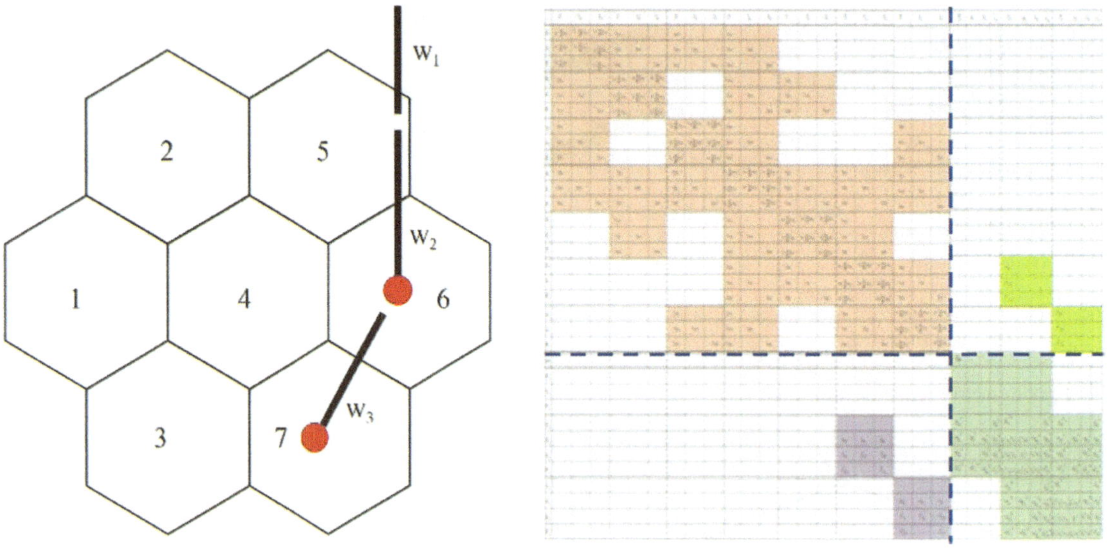

(a) Schematics of PEBI grid (b) Mathematical matrix format

Fig. 10.17 Jacobin matrix based on PEBI grid

10.1.2 Compression and Storage Technology for Complex-Structured Matrix

Tight sandstone reservoirs are characterized by nano/micro-pores, complex natural/artificial fracture networks, complex-structured wells, and complex geological boundaries, which cause increased dead nodes and zero elements (Table 10.11). This not only wastes large amount of matrix storage space, but also has longer searching time and slower computing speed (Table 10.12). Therefore, it is necessary to develop a compression storage technology applicable for a complex-structured matrix for tight sandstone oil/gas reservoirs, to save the storage space and improve the computing efficiency (Lim and Aziz, 1995; Jiang 2007; Wang et al. 2013).

The coefficient matrix of a complex flow model after discretization for a tight sandstone reservoir has the characteristic of a typical block structure (Fig. 10.21). To reduce the memory consumption and improve the calculation speed, a compression storage technology based on points and blocks with zero-element and dead-node exclusion are developed.

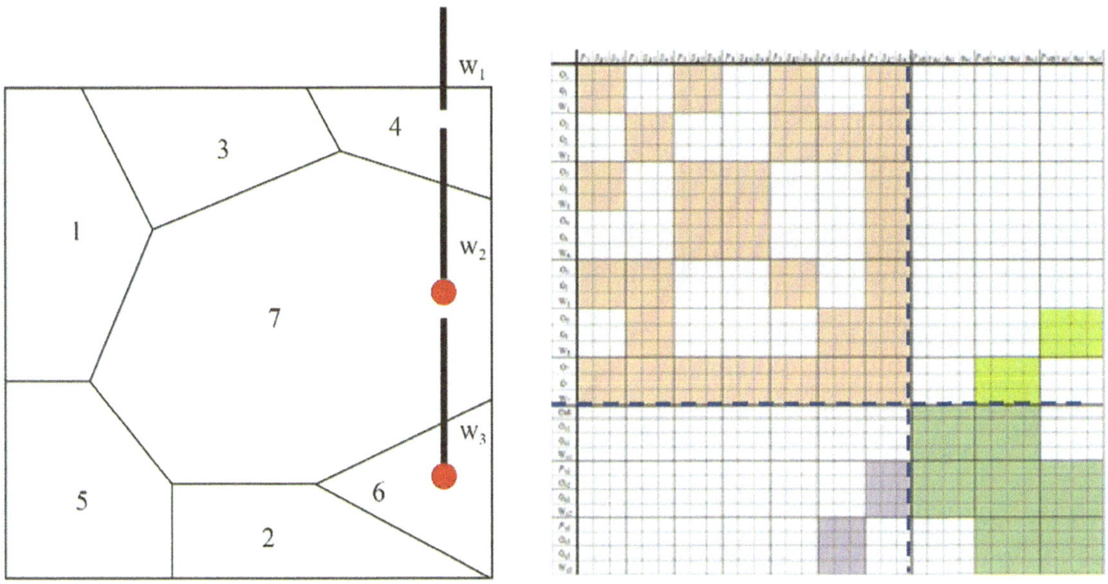

(a) Schematics of hybrid grid (b) Mathematical matrix format

Fig. 10.18 Jacobin matrix based on hybrid grids

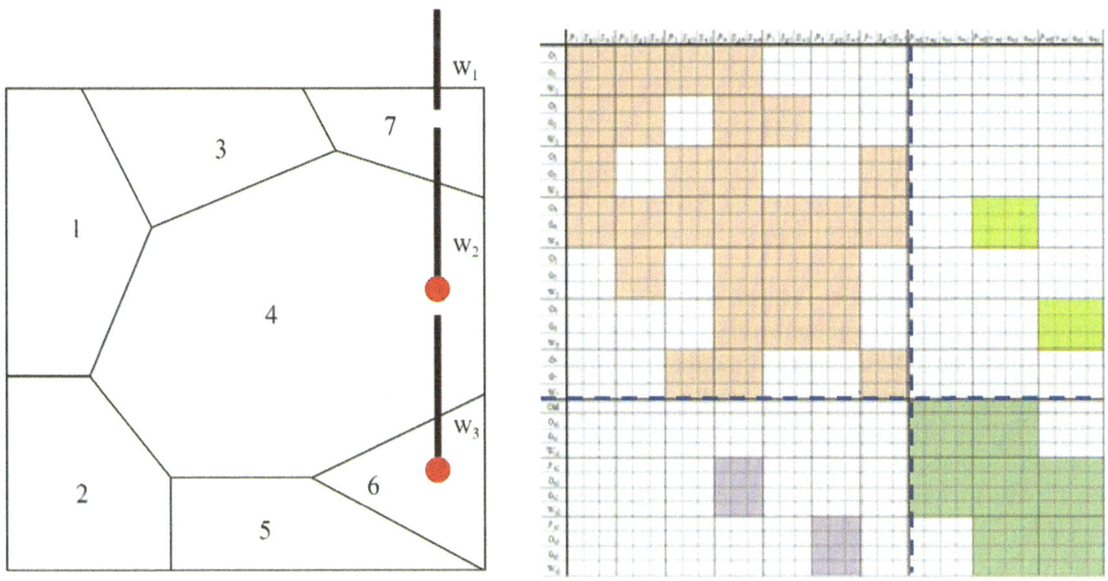

(a) Schematics of hybrid grid (b) Matrix format

Fig. 10.19 Jacobin matrix based on hybrid grids

10.1.2.1 Compression and Storage Technology Based on Point with Zero-Element and Dead-Node Exclusion

A large number of zero elements exist for dead nodes tight sandstone oil/gas reservoir and unstructured grids. Aiming at this issue, a compression storage technology based on points is developed that eliminates zero elements and dead nodes (Fig. 10.22).

1. Flowchart for compression storage technology based on point with zero-element and dead-node exclusion

For a $n_{\text{row}} \times n_{\text{col}}$ mathematical matrix with N non-zero elements, three sets of arrays are used to store the element values, line pointer addresses, and array serial numbers, respectively. For each array set, the mathematical matrix is scanned first row-by-row and then column-by-column. The

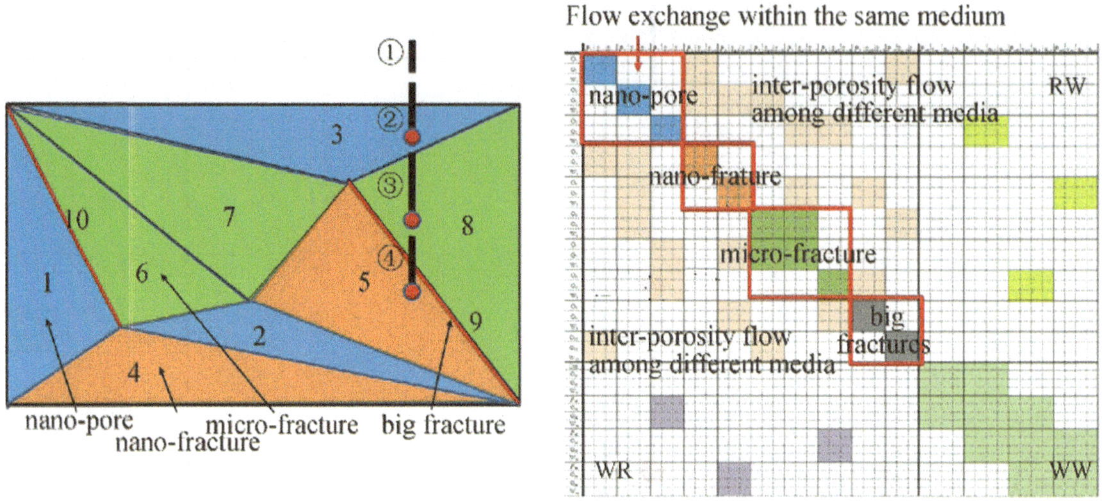

(a) Four media model (b) Matrix format

Fig. 10.20 Jacobin matrix for multiple media

non-zero elements are stored and zero elements are skipped. The purpose of this technology is to directly and quickly reinstate the coefficient matrix with the row number, the column number, and the value of each element, in preparation for iteration.

2. Example

An example in which a 5×5 mathematical matrix has 11 non-zero elements is taken to illustrate the technology (Fig. 10.23).

Table 10.13 lists the elements values and the corresponding row number and column number:

Table 10.14 lists the 1D mathematical matrix created at each step according to the method in this study:

Comparing Tables 10.14 and 10.13, it can be seen that there is no zero element in the compressed matrix, and Table 10.14 further compresses the row pointer address matrix in Table 10.13, leading the storage matrix to a minimum capacity.

10.1.2.2 Compression and Storage Technology Based on Blocks with Zero-Element and Dead-Node Exclusion

The compression storage technology based on blocks with zero-element and dead-node exclusion can reduce memory requirement, but it is still time-consuming since it needs to search each element in the compression process. For tight oil/gas reservoirs, a mathematical matrix for the complex flow model has a typical block structure due to the multiple media, multiphase flow, complex-structured wells, and multiple variables to be solved. Directing at this feature, the compression and storage technology stores the non-zero elements based on blocks, which greatly improves the compactness of the data arrangement and the computing speed. Figure 10.24 shows the compression and storage technology based on blocks with zero-element and dead-node exclusion.

1. Flowchart of compression storage technology based on points with zero-element and dead-node exclusion

During the compression and storage process of a coefficient matrix, the block is always used as the basic unit, and the mathematical matrix is stored in the order of "from diagonal to non-diagonal, form row to column, from left to right". The entire coefficient matrix is stored in a large 1D array with block as the base.

2. Example

An example is given to illustrate the technology in this section, as shown in Fig. 10.25.

Step 1: Creates a 1D mathematical matrix with blocks as the base, and stores the mathematical matrix in the order of the "from diagonal to non-diagonal, form row to column, and from left to right". The diagonal elements are stored in a large 1D array with blocks as the base (Fig. 10.26).

Table 10.11 Comparison of different compression and storage technologies for complex-structured mathematical matrix

Item	Dead node of grids		Zero element of matrix	
Definition	For the numerical simulation grid nodes with no storage capacity (porosity equals zero), no flow capacity (permeability equals zero) and no volume (thickness equals zero), the all values corresponding to the element in the sparse coefficient matrix equal zero after discretization, and these grid nodes are called dead nodes		In the sparse coefficient matrix after numerical discretization, the elements with its values equal to zero become a zero element	
Cause	If the following values of a grid equals zero: ① Porosity equals zero ② Permeability equals zero ③ Thickness equals zero The elements corresponding to the grid after discretization equal zero		① for non-adjacent grids, because there is no adjacency in space, it is necessary to fill the vacancy with zero, resulting in a large number of zero elements ② Adjacent in space but not connected (for example, due to a fault), which means that there is no material, momentum and energy exchange between the grid and its non-adjacent grids, leading to a large number of zero elements in the Jacobin matrix after the discretization	
Grid schematics	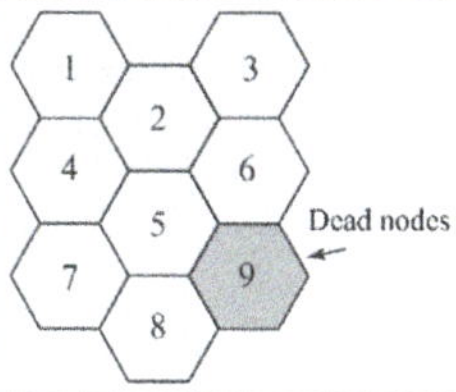		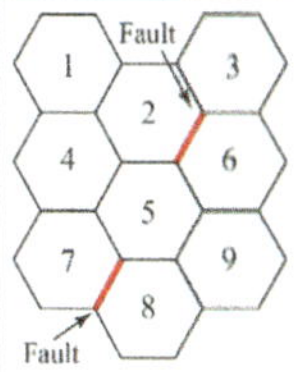	
Connectivity table (connection between grids)	Central grid	Adjacent grid	Central grid	Adjacent grid
	1	2, 4	1	2, 4
	2	1, 3, 4, 5, 6	2	1, 3, 4, 5
	3	2, 6	3	2, 6
	4	1, 2, 5, 7	4	1, 2, 5, 7
	5	2, 4, 6, 7, 8	5	2, 4, 6, 7, 8
	6	2, 3, 5	6	3, 5
	7	4, 5, 8	7	4, 5
	8	5, 7	8	5, 9
			9	5, 6, 8
Corresponding mathematical matrix format	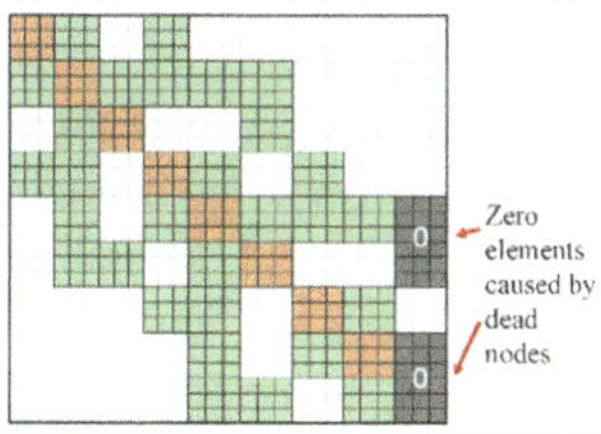		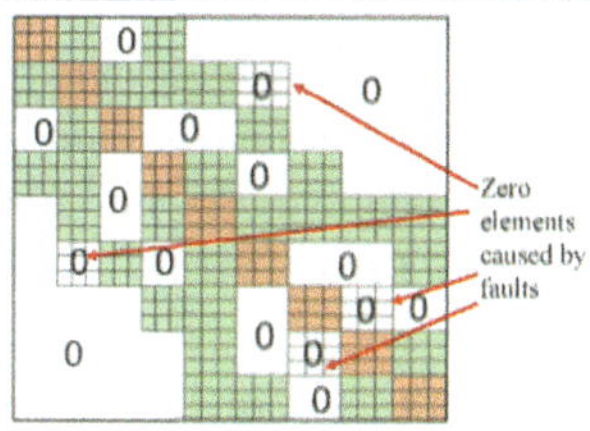	

Table 10.12 Point-versus block-compression and storage

Item	Block compression and storage	Point compression and storage	Non-compression and storage
Memory for Storage (matrix rank = 5000i)	2.3 MB	2.84 MB	190.7 MB
Non-zero element Searching times (matrix rank = 5000)	13900 times	295200 times	–
Applicability	3 phases	2 phases	All

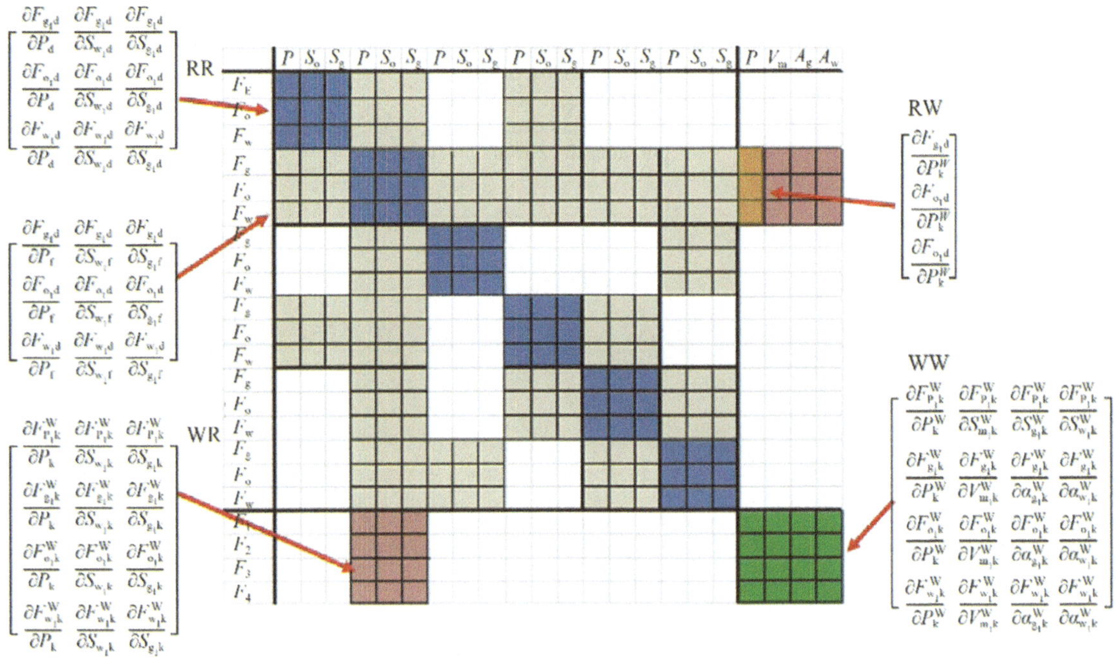

Fig. 10.21 Structure of a block coefficient matrix for tight sandstone reservoirs

It also creates a 1D mathematical matrix to store the upper triangular matrix (Fig. 10.27).

It creates a 1D mathematical matrix to store the lower triangular matrix (Fig. 10.28).

Step 2: A 1D integer array is created to store the row number and column number of the first non-zero element of the upper-triangular block matrix (Table 10.15).

Step 3: A 1D integer array is created to store the row number and column number of the first non-zero element of the triangular block matrix (Table 10.16).

Following the above steps, the elements in the blocks are arranged one after another continuously, and the grid variables are centralized to facilitate the global solution and improve the computing speed.

10.2 Efficient Solving Technology for Linear Algebraic Equations for Multiple Media Based on Unstructured Grids

The flow model for tight sandstone oil/gas reservoir couples the nonlinear partial differential equations of pressure and saturation equations and the algebraic equations of complex-structured wells (Fig. 10.29). The coefficient matrix of the matrix equations has the characteristics of strong interposition, strong nonlinearity, strong coupling, and multiple scales in space and time.

After discretization and linearization, an algebraic equation set can be obtained in the format of $Ax = B$, in which the coefficient matrix A is a large sparse matrix. The coefficient matrix changes with the change of such factors as the complex flow mathematical model, structured/unstructured

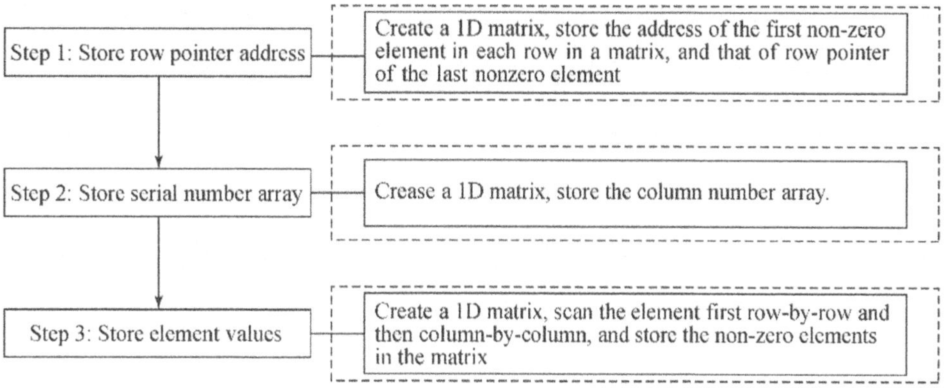

Fig. 10.22 Flowchart of the compression and storage technology based on points with zero-element and dead-node exclusion

	Column 1	Column 2	Column 3	Column 4	Column 5
Row 1	10.0①	19.0②	0.0	0.0	0.0
Row 2	8.0③	13.0④	0.0	0.0	0.0
Row 3	0.0	0.0	12.0⑤	11.0⑥	12.0⑦
Row 4	0.0	0.0	8.0⑧	20.0⑨	0.0
Row 5	10.0⑩	0.0	0.0	0.0	16.0⑪

Fig. 10.23 An example of a coefficient matrix

grids, the number of grids, complex-structured wells, and the grid ordering (Fig. 10.30) (Chen 2011; Russell 1989; Forsyth and Sammon 1986; Kazemi and Stephen 2012; Appleyard and Cheshire, 1983).

The sparse coefficient matrix is large in scale, complex in format, and numerous in conditions, which leads linear algebraic equations to have unstable solution, poor convergence and slow computation. Targeting at the above issues, this section introduces and elaborates the coefficient matrix preconditioning technology and a solving technology.

10.2.1 Brief Summary of Coefficient Matrix Preconditioning and Solving Technology

For reservoir numerical simulation, the conventional solving methods for linear algebraic equations include direct solution and iterative solution. The direct solving method is represented by the Gauss elimination method. Because of the high requirement in the matrix structure and the computer memory, the direct solution method is no longer applied in practice at present. Iterative solving methods consist of the Gauss-Seidel iteration method and the super relaxation iterative method. These methods are simple and easy to use, and demand small storage memory of computers. But for large sparse matrices, they need relatively more iterations and longer computing time, which is difficult to meet the requirement for daily use.

In the tight sandstone reservoir simulation, because of the consideration of flow mechanisms and the treatment of multiple media, the coefficient matrix of the nonlinear equations becomes complicated, and its eigenvalue distribution non-uniform, the condition number large, showing

Table 10.13 Correspondence between element value, row number and column number

Serial No.	①	②	③	④	⑤	⑥	⑦	⑧	⑨	⑩	⑪
Element value	10.0	19.0	8.0	13.0	12.0	11.0	12.0	8.0	20.0	10.0	16.0
Row number	1	1	2	2	3	3	3	4	4	5	5
Column number	1	2	1	2	3	4	5	3	4	1	5

Table 10.14 Compressed data table for three array sets

Step No.	Serial No.	①	②	③	④	⑤	⑥	⑦	⑧	⑨	⑩	⑪
Step 1	Element value matrix	10.0	19.0	8.0	13.0	12.0	11.0	12.0	8.0	20.0	10.0	16.0
Step 2	row point address matrix	1	–	2	–	3	–	–	4	–	5	6
Step 3	Column number matrix	1	2	1	2	3	4	5	3	4	1	5

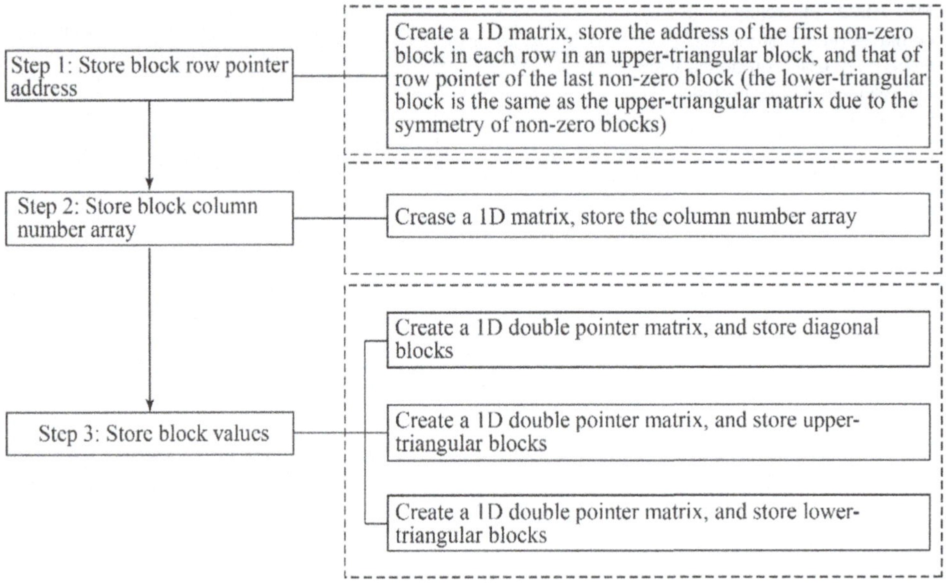

Fig. 10.24 Flowchart of the compression and storage technology based on blocks with zero-element and dead-node exclusion

Fig. 10.25 Initial mathematical matrix before block storage

	P_1	S_{g1}	S_{w1}	P_2	S_{g2}	S_{w3}	P_3	S_{g3}	S_{w3}	P_4	S_{g4}	S_{w4}	P_5	S_{g5}	S_{w5}	P_6	S_{g6}	S_{w6}	P^W	α_g	α_w	V_m
O_1	1	4	7	71	74	77				80	83	86										
G_1	2	5	8	72	75	78				81	84	87										
W_1	3	6	9	73	76	79				82	85	88										
O_2	164	167	170	10	13	16	89	92	95	98	101	104	107	110	113	116	119	122	125	128	131	134
G_2	165	168	171	11	14	17	90	93	96	99	102	105	108	111	114	117	120	123	126	129	132	135
W_2	166	169	172	12	15	18	91	94	97	100	103	106	109	112	115	118	121	124	127	130	133	136
O_3				173	176	179	19	23	25							137	140	143				
G_3				174	177	180	20	23	26							138	141	144				
W_3				175	178	181	21	24	27							139	142	145				
O_4	182	185	188	191	194	197				28	31	34	146	149	152							
G_4	183	186	189	192	195	198				29	32	35	147	150	153							
W_4	184	187	190	193	196	199				30	33	36	148	151	154							
O_5				200	203	206				209	212	215	37	40	43	155	158	161				
G_5				201	204	207				210	213	216	38	41	44	156	159	162				
W_5				202	205	208				211	214	217	39	42	45	157	160	163				
O_6				218	221	224	227	230	233				236	239	242	46	49	52				
G_6				219	222	225	228	231	234				237	240	243	47	50	53				
W_6				220	223	226	229	232	235				238	241	244	48	51	54				
Ctrl				245	249	253													55	59	63	67
O^W				246	250	254													56	60	64	68
G^W				247	251	255													57	61	65	69
W^W				248	252	256													58	62	66	70

typical pathological characteristics. The convergence of the Krylov subspace iteration method is strongly dependent on the distribution of the eigenvalues of the coefficient matrix, which results in slow convergence and even non-convergence. To solve this problem, the preconditioning method is needed to improve the property of the coefficient matrix and remove the unreasonable eigenvalues that affecting the computing. It has been proved that a good preconditioning method can greatly improve the convergence speed of an iterative method and reduce the computing time. Thereby, a preconditioning method is applied in the tight sandstone oil/gas reservoir simulation. The

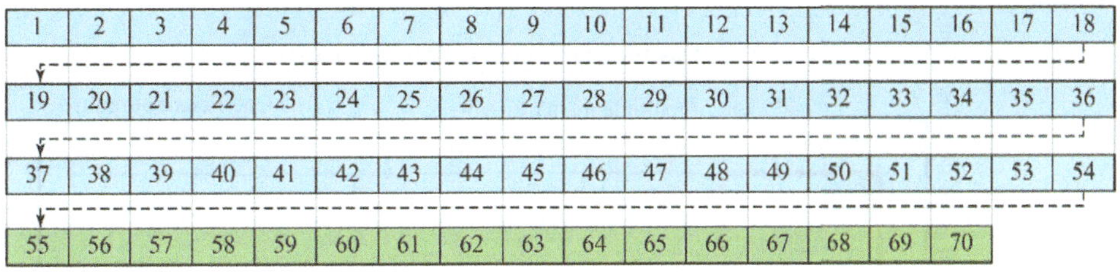

Fig. 10.26 Diagonal element array

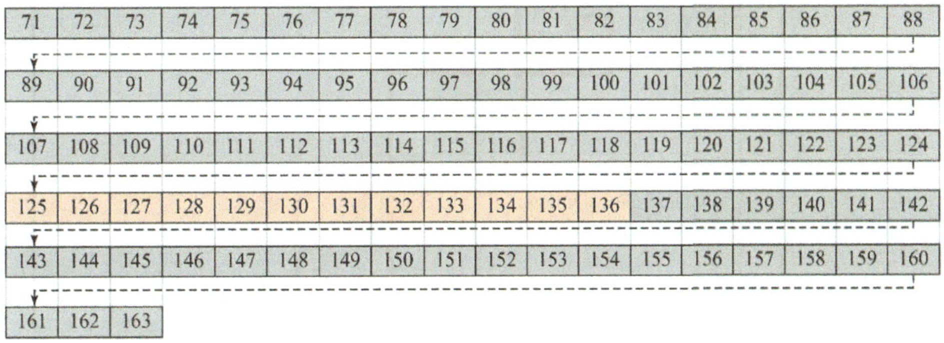

Fig. 10.27 Upper-triangular element array

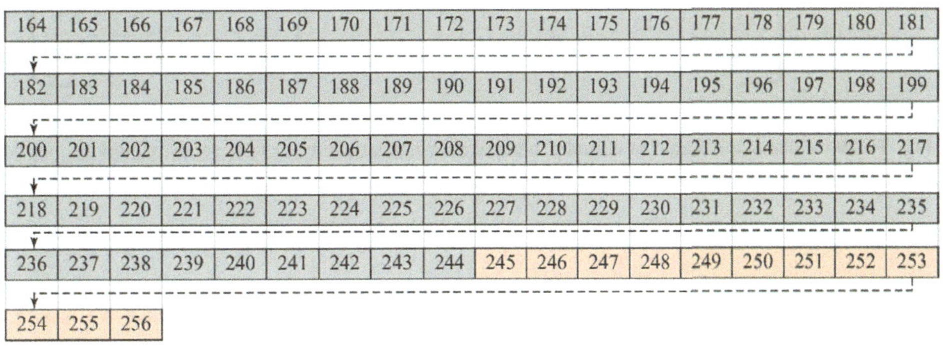

Fig. 10.28 Lower-triangular element array

Table 10.15 Row and column serial number

Serial number	71	80	137	146	155	163
Row	1	1	3	4	5	–
Column	2	4	6	5	6	–

Table 10.16 Row and column serial number

Serial number	164	173	182	200	209	218	236	245	256
Row	2	3	4	5	5	6	6	7	–
Column	1	2	1	2	4	2	5	2	–

conventional preconditioning technology includes single variable multigrid AMG preconditioning and single variable incomplete LU decomposition. For the sparse coefficient matrix for tight sandstone oil/gas reservoirs with complex physical properties and coupling properties (pressure, saturation and well coupling), the AMG method can only treat a pressure coefficient matrix and cannot solve the system directly. The BILU method is able to solve the matrix equation but usually at low convergence speeds. Because it is difficult to eliminate a low-frequency error, the performance of common single variable preconditioning methods is poor (Table 10.17). Therefore, according to the following characteristics of a sparse coefficient matrix for tight sandstone reservoirs, low frequency of an elliptical pressure equation and high frequency of a hyperbolic saturation equation, the pressure equation and the saturation equation

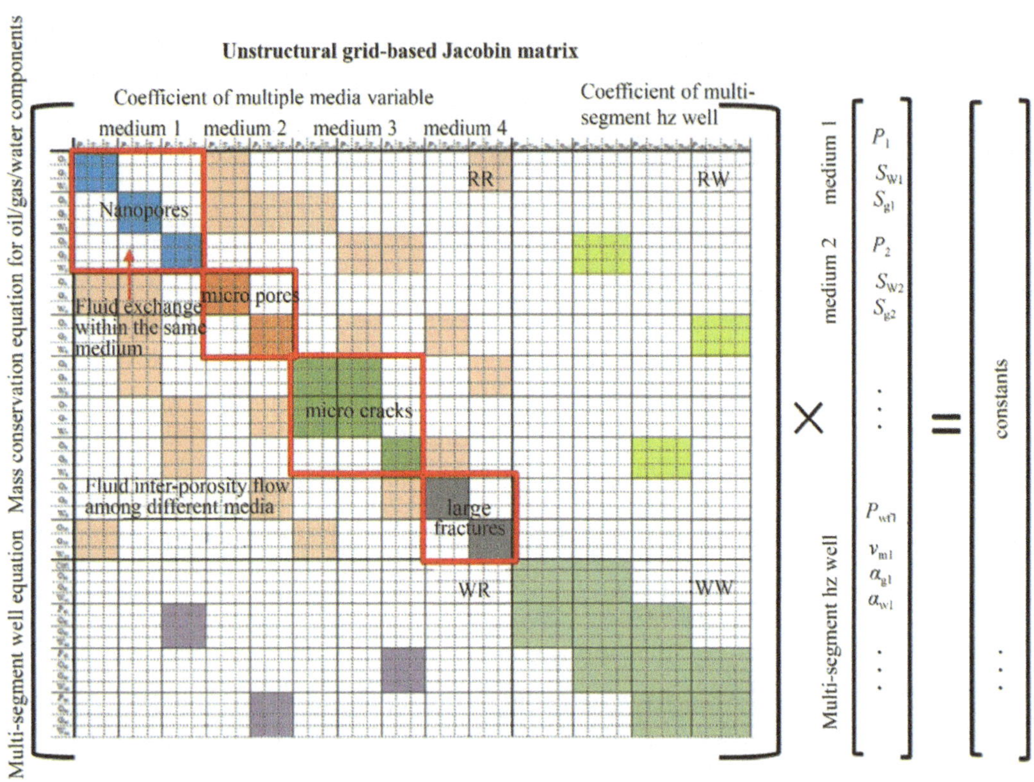

Fig. 10.29 Schematic diagram of complex numerical matrix structure for tight sandstone oil/gas reservoir

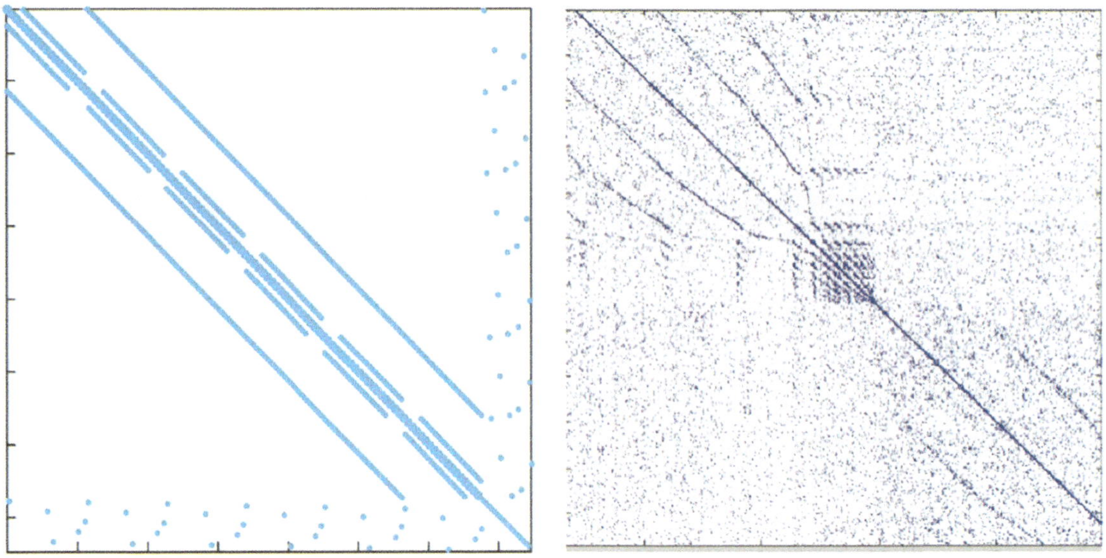

(a) Coefficient matrix format based on structured grid

(b) Coefficient matrix format based on unstructured grid

Fig. 10.30 The format of sparse matrix

are treated separately. The AMG method is used to handle pressure coefficient matrix first, and then the BILU (0) is applied to the entire coefficient matrix. Namely, a AMG + BILU multiple preconditioning technology is applied to the coefficient matrix. Meanwhile, according to the characteristics of a nonlinear flow model for tight sandstone reservoirs, the generalized minimum residual method (GMRES) is employed to solve linear algebraic equations (Mora and

Wattenbarger 2009; Guo et al. 2014; Li and Russell, 2006; Milad et al. 2013; Kalantari-Dahaghi et al. 2012; Olorode et al. 2013; Liu 2009; Shi et al. 2008; Chen et al. 2006; Li et al. 2011; Wu et al. 2012).

This section starts from the different characteristics of pressure and saturation variables in a nonlinear flow equation for a tight sandstone oil/gas reservoir, and applies the AMG + BILU method to treat the coefficient matrix. More specifically, first, the coefficient matrix is decoupled. Then, the AMG method is used to treat the pressure coefficient matrix and the BILU (0) method is used to deal with the saturation coefficient matrix. Finally, the coefficient matrices are combined and solved by the GMRES method. This method integrates multiple preconditioning technologies and can quickly solve complex matrix systems for tight sandstone reservoir simulation with high accuracy and strong stability. The detailed solution procedures are shown in Fig. 10.31.

10.2.1.1 Preconditioning Technology

The main strategy of preconditioning technology is to transform the original linear system into an equivalent linear system which is easy to solve by means of "preconditioner".

$$M^{-1}Ax = M^{-1}b \quad \text{(Left precondition method)}$$

or

$$AM^{-1}y = b,\ x = M^{-1}y \ \text{(Right precondition method)}$$

or

$$M_1^{-1}AM_2^{-1}y = M_1^{-1}b,\ x = M_2^{-1}y,$$
$$M = M_1 \cdot M_2 \ \text{(Split condition method)},$$

where, the matrices M has the same rank as that of A, named as preconditioner. The selection of a preconditioner directly determines the computing speed and solution precision of the linear equation set. A good preconditioner should have the following features:

① The preconditioner is a good approximation of the inverse mathematical matrix of the coefficient matrix in some aspect.
② The memory consumption of the preprocessor is not very large.
③ The condition number of the preconditioned mathematical matrix is far less than the condition number of the original coefficient matrix.
④ The new preconditioning linear system is easier to solve than the original linear system.

The preconditioning methods can be classified into two categories: the conventional preconditioning technology and the highly efficient preconditioning technology, which are respectively elaborated below.

1. Conventional preconditioning technology

The conventional preconditioning technology processes the initial value for an equation system by using a solution method. The common single variable preconditioning technology includes the multigrid technology and the incomplete LU decomposition technology. The advantages of this method is wide applicability and easy to use, but the disadvantage is that the computing time increases rapidly with the growth of a mathematical matrix size, which makes it unsuitable for large-size matrices with complex structures.

(1) Single variable AMG multigrid preconditioning technology

The AMG multigrid method is a fast computing method for solving a discrete system of partial differential equations. It is generally considered as the fastest method to solve a discrete system of elliptic equations. Depending on whether the geometric information is used, it is divided into two methods: geometric multigrid method and algebraic multigrid method.

The geometric multigrid method is similar to the algebraic multigrid method in the core ideas, and both eliminate the high frequency error by using a simple iterative method: the error is smoothed stepwise from fine to coarse for the coarse grids of each level, which is then restored from coarse to fine through interpolation. The geometric method requires the geometric information of the grids to construct the smooth and interpolation operators. Whereas the algebraic method only needs the matrix information to construct the operators. Therefore, the later is easier to use in practice.

The AMG method was originated in the early 1980s. With fast convergence, it is mainly used to solve elliptic equations. Specifically, it divides an error into high frequency and low frequency parts according to the Fourier components of the error. The high frequency error is a local behavior, which comes from the coupling between some of the nearby grids and is not related to the grids that are at the boundary or far away. The low frequency error is a global behavior, which comes mainly from the boundary information. Traditional iterative methods are all highly local methods, which are quick at smoothing local high-frequency errors, but slow at reducing global low-frequency errors. Considering that once converted to coarse meshes, it is easy

Table 10.17 Comparison of solving technology

Class		Technology name	Targeting problem
Preconditioning technology	Common	Single variable AMG multigrid preconditioning technology	Low frequency pressure equation for elliptical equation
		Single variable ILU decomposition preconditioning technology	High frequency saturation equation for hyperbolic equation
	Efficient	AMG + BILU multiple preconditioning	Low frequency pressure equation and high frequency saturation equation
Solving technology	Common	Gauss elimination technology	For matrix with small size and regular format
		Gauss Seidel iteration technology	For matrix with large size and regular format
	Efficient	GMRES method, stable dual conjugate gradient technology	For matrix with large size and complex format

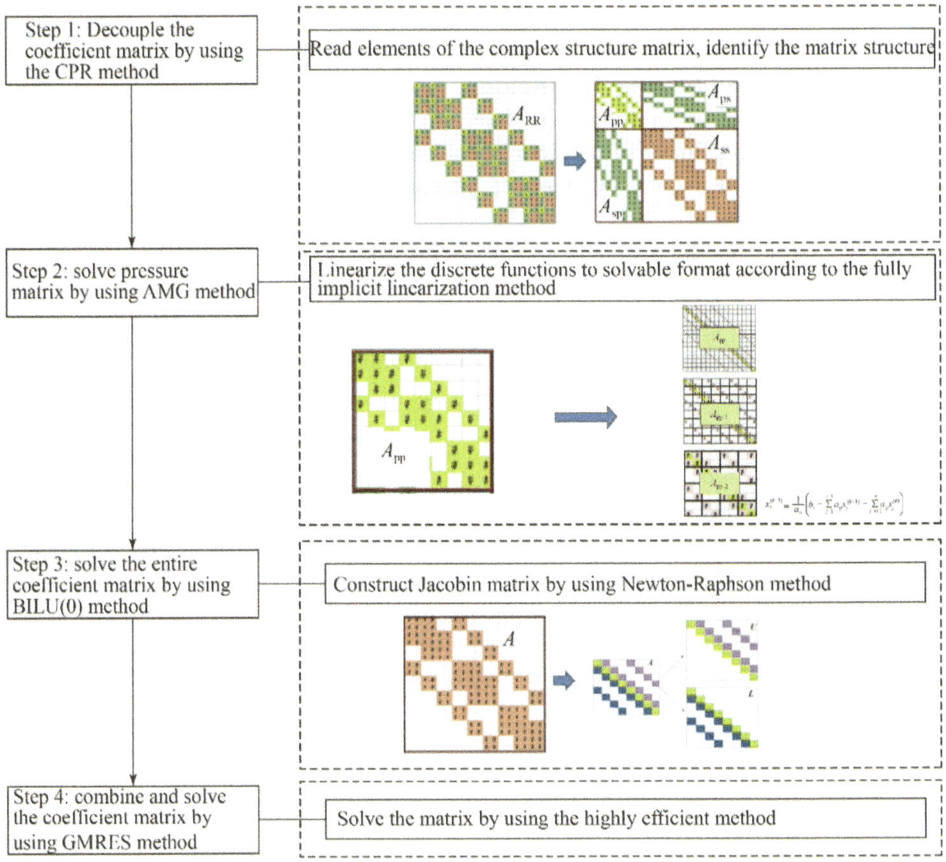

Fig. 10.31 Solution flowchart

to get and eliminate the high frequency component of a low frequency error in a fine grid, The AMG method designs a series of meshes of different sizes: it first maps an error to the coarsest meshes and then accurately solves the equation, and lastly the solution is returned to the original grid layer by layer and merged with the approximate solution to get the final solution.

(2) Single variable ILU decomposition preconditioning technology

The single variable ILU decomposition is one of the most effective preconditioning technology for solving general sparse linear equations. This type of technologies include the method proposed by Meijerink, the method proposed by

Axelsson, ILU(0), ILU(1), ILUT, MRILUT, and many other methods.

ILU(0) keeps the structure of a sparse matrix unchanged. On one hand, it is very convenient to use; on the other hand, the approximation of the coefficient matrix is too rough, which limits the validity of this method. To improve the approximation, non-zero elements are allowed to be filled, but this demands more memory. The amount of non-zero element filling depends on the elimination order. In this sense, the validity of incomplete LU decomposition depends on the order of elimination and the number of filled elements that are discarded.

Depending on the problems, it can be classified as the incomplete LU decomposition based on points and incomplete BILU decomposition based on blocks. Because the mathematical matrix in reservoir simulation is often a block matrix, if each block is regarded as a non-zero position rather than a specific non-zero element, the processing efficiency will be greatly improved. BILU replaces the single non-zero elements in ILU with small dense matrix blocks, and applies matrix operations in place of element operations. In mathematics, the two are completely equivalent, so the convergence rate will not be improved. However, due to the large reduction in dimension, the calculation efficiency of a block matrix is dramatically improved. ILU requires non-zero elements on the diagonal, corresponding to the reversibility of the diagonal blocks in BILU. The latter is more flexible, and thus able to deal with more extreme problems at better stability. The BILU series method is rather applicable in reservoir simulation.

2. Highly efficient preconditioning technology

For the tight sandstone reservoir simulation, to improve the calculation velocity, a complex global problem is decomposed into a relatively simpler pressure subsystem and a saturation subsystem depending on the mathematical and physical characteristics of the flow model. The coefficient matrix of the flow model in the fully implicit discrete format is composed of the pressure variable δ_P and the saturation variable δ_S. Therefore, the linear equation set can be rewritten as

$$\begin{pmatrix} A_{\text{pp}} & A_{\text{pS}} \\ A_{\text{Sp}} & A_{\text{SS}} \end{pmatrix} \begin{pmatrix} \frac{\delta_P}{\delta_S} \end{pmatrix} = \begin{pmatrix} \frac{f_{\text{P}}}{f_{\text{S}}} \end{pmatrix}$$

where, A_{pp} corresponds to the pressure block, which has the behavior of an elliptic equation; A_{SS} corresponds to the saturation block and has the behavior of a convective (hyperbolic) equation; A_{Sp} and A_{pS} represent the coupling relationship between pressure and saturation.

Therefore, the nonlinear flow mathematical model is a complex partial differential equation for pressure and saturation. It has both the behaviors of an elliptic equation for the pressure variable and a hyperbolic equation for the saturation variable. The pressure variable is represented by a high frequency error and the saturation variable represented by a low frequency error. The coefficient matrix after numerical discretization has complex physical properties and coupling properties (pressure, saturation and well coupling), which leads to the poor computing stability. But the AMG method alone is unable to solve the system. Although the BILU is applicable, it has slow convergence because it is difficult to eliminate the low frequency error. In short, all the traditional single variable preconditioning technology are ineffective for these tight reservoir simulations.

For the above issues, a multiple preconditioning technology is often used at present. The idea of this method is to decompose the complex nonlinear global problem into simpler linear sub-problems such as pressure and saturation depending on the mathematical and physical characteristics of the flow model. Efficient solution methods are first designed according to the mathematical and physical properties of the sub-problems, which are then reasonably combined to get a fast and stable solver.

At present, the combination of AMG and BILU is the more common multiple preconditioning technology. In which the AMG method is used to solve the elliptic equation for the pressure variable first, the BILU method is then used to solve the hyperbolic equation for the saturation variable. This way, the advantages of these two methods are both taken to accelerate the convergence, improve the computing efficiency, and handle large matrices. This method will be introduced in following chapters.

10.2.1.2 Solving Technology for Linear Algebraic Equations

There are two solving technologies for the linear algebraic equations: the conventional solving technology and the efficient solving technology.

1. Conventional solving technology

The conventional solving technologies include Gaussian elimination and iterative technology.

(1) Gaussian elimination

The Gaussian elimination technology is based on the Gauss elimination method. Specifically, first, after certain operations, some of the variables in the original equations are eliminated one by one. Then, a set of equations equivalent to the original equations are obtained. The unknown variables are solved at last. If the rounding errors during calculation are ignored, the direct solution can be seen as an exact solution, the solution of the original linear equations can be obtained all at once. It is an effective method to solve the low-order dense linear equations at present.

A Gauss elimination technology has two parts: elimination and regression. The format of the matrix equations is:

$$\begin{pmatrix} a_{11} & a_{12} & \cdots & a_{1n} \\ a_{21} & a_{22} & \cdots & a_{2n} \\ \vdots & \vdots & & \vdots \\ a_{n1} & a_{n2} & \cdots & a_{nn} \end{pmatrix} \begin{pmatrix} x_1 \\ x_2 \\ \vdots \\ x_n \end{pmatrix} = \begin{pmatrix} b_1 \\ b_2 \\ \vdots \\ b_n \end{pmatrix}$$

The principal elements are first eliminated. After n-1 steps, the above equation is transformed into an upper triangular matrix with the same solution.

$$\begin{pmatrix} a_{11}^{(1)} & a_{12}^{(1)} & \cdots & a_{1n}^{(1)} \\ & a_{22}^{(2)} & \cdots & a_{2n}^{(2)} \\ & & & \vdots \\ & & & a_{nn}^{(n)} \end{pmatrix} \begin{pmatrix} x_1 \\ x_2 \\ \vdots \\ x_n \end{pmatrix} = \begin{pmatrix} b_1^{(1)} \\ b_2^{(2)} \\ \vdots \\ b_n^{(n)} \end{pmatrix}$$

Then, substituting the knowns into the triangulated coefficient matrix yields the unknowns:

$$\begin{cases} x_n = b_n^{(n)} / a_{nn}^{(n)}, \\ x_k = \left(b_k^{(k)} - \sum\limits_{j=k+1}^{n} a_{kj}^{(k)} x_j \right) / a_{kk}^{(k)}, \ k = n-1, n-2, \cdots, 1 \end{cases}$$

The exact solution can be obtained by the direct solving method, but the elimination process is time-consuming and the computing load huge (the magnitude reaches to matrix order to the power three). When the coefficient matrix order is high, the computing takes too long, and the Gaussian elimination technology would no longer be applicable in terms of memory and computing speed.

(2) Frequently-used iterative technology

The Frequently-used iterative technology first appeared from the 1950s to the early 1970s. Derived from the angle of mathematical matrix splitting, it first transforms a problem from solving linear equations to solving its fixed points, and then constructs an iterative format, which mainly includes the Jacobi method, the Gauss-Seidel method, the SOR super relaxation method and its modified and accelerated forms.

Split the coefficient matrix A as

$$A = M - N$$

Then $Ax = b$ is equivalent to

$$Mx = Nx + b$$

Construct an iteration formula

$$Mx^{(k+1)} = Nx^{(k)} + b, \ k = 0, 1, \ldots$$

Or written as:

$$x^{(k+1)} = Bx^{(k)} + f, \ k = 0, 1, \ldots$$

where,

$$B = M^{-1}N, f = M^{-1}b$$

The above is the frequently-used iteration formula. If the iteration converges, the iteration formula becomes linear when k is large enough.

Conventional iteration methods mainly include the Jacobi method, the Gauss-Seidel method, and the SOR super relaxation method. Their respective specific formats are shown below:

The Jacobi iteration format:

$$x_i^{(k+1)} = \frac{1}{a_{ii}} \left[b_i - \sum_{j \neq i} a_{ij} x_j^{(k)} \right], \ i = 1, 2, \cdots, n, \ k = 0, 1, \cdots$$

The Gauss-Seidel iteration format:

$$x_i^{(k+1)} = \frac{1}{a_{ii}} \left[b_i - \sum_{j=i}^{i-1} a_{ij} x_j^{(k+1)} - \sum_{j=i+1}^{n} a_{ij} x_j^{(k)} \right], \ i = 1, 2, \cdots, n, \ k = 0, 1, \cdots$$

The SOR super relaxation iteration format:

$$x_i^{(k+1)} = (1 - \omega) x_i^{(k)} + \frac{\omega}{a_{ii}} \left[b_i - \sum_{j=i}^{i-1} a_{ij} x_j^{(k+1)} - \sum_{j=i+1}^{n} a_{ij} x_j^{(k)} \right], \ i = 1, 2, \cdots, n, \ k = 0, 1, \cdots$$

One of the key issues of constructing an iterative method is how to select matrix M, so a satisfactory solution can be obtained with the least number of iterations. The closer the matrix M gets to the coefficient matrix A, the fewer iterations are needed to reach the convergence criteria. But if the matrix M is selected improperly, the convergence speed can be very slow.

2. Efficient solving technology

An efficient solving technology is able to solve large sparse linear equations with fast convergence. It is mainly based on the Krylov subspace iterative method, which begins in the mid-1970s. The Krylov subspace iteration method has the advantage of smaller storage memory, small computing speed, and easier parallelism. Soit is very suitable to solve large sparse linear equations.

The Krylov subspace iteration technology need not to construct a matrix into an iterative matrix but to get the solution through minimizing the error,

$$r^{(k)} = b - Ax^{(k)}.$$

in the Krylov subspace:

$$\kappa_{\mathrm{m}} = \mathrm{span}\left\{r^{(0)}, A\, r^{(0)}, \ldots, A^{m-1}\, r^{(0)}\right\},\ m \geq 1$$

where, $r(0) = b - Ax(0)$, $x(0)$ is the initial value, and $x(k)$ is the approximate value at the kth iteration ($k \geq 0$). Given an initial vector $x(0)$, minimize the residue in a certain direction, correct the approximate value in that direction $x(k+1) = x(k) + f\,[r(k)]$, and repeat the iteration until the exact solution x^* is approached.

The Krylov subspace iteration method includes the conjugate gradient method (CG), the Bi-conjugate gradient method (BICGSTAB), the generalized minimum residual method (GMRES) and the nonlinear orthogonal minimization method (ORTHOMIN). These Krylov subspace iterative methods can effectively solve the asymmetric and non-positive definite matrix equations. They have the advantages of high precision, stability, quick convergence, and have become a successful and effective method for solving reservoir numerical simulation problems.

10.2.2 Efficient Solving Technology for Linear Algebraic Equation Set for Multiple Media Based on Unstructured Grid

In the numerical simulation of tight sandstone oil/gas reservoirs, a nonlinear flow model is a mixed partial differential equations for pressure and saturation. This equation has both the behaviors of an elliptic equation for pressure variables and the hyperbolic equation for saturation variables. The pressure variable is represented by a high frequency error, and the saturation variable represented by a low frequency error. The coefficient matrix after numerical discretization has complex physical properties and coupling properties (pressure, saturation and well coupling), which lead to the poor computing stability. The AMG method alone is unable to solve the system. Although the ILU is applicable, it is slow in convergence because it is difficult to eliminate the low frequency errors. Thus, the traditional single variable iteration methods are poor at solving these equations.

Aiming at the above problems, an integrated solving technology based on unstructured grid is adopted, i.e., a two-step preconditioning method, the Constraint Pressure Residual (CPR) method. It has been proved to be the most efficient and stable preconditioning method so far. Meanwhile, the GMRES method is used for solving the equations, which is a successful and effective method for reservoir numerical simulation. It is applicable for large asymmetric sparse linear equations and suitable for an algebraic system of multi-scale variables. It can accurately solve the algebraic equations in the numerical simulation of complex reservoirs (Cao et al. 2005).

10.2.2.1 Flowchart of CPR Technology Based on Unstructured Grids for Solving Algebraic Linear Equation for Multiple Media

There are four basic steps in the CPR method:

① Coefficient matrix decoupling. The coefficient matrix is decoupled into two parts: a submatrix for pressure variables and a submatrix for saturation variables, by using the IMPES method or the real/quasi IMPES elimination method.

② Pressure equation preconditioning by the AMG method. The decoupled pressure variable equation is pre-solved by using the AMG method (a series of mesh spaces of different sizes, mapping operators and iterative method).

③ Global equation preconditioning by the BILU method. The AMG-preconditioned global equations are pre-solved by the BILU method (an incomplete LU decomposition method based on blocks).

④ Integrated solving by the GMRES method. The pressure and saturation matrices that are block diagonally dominant after preconditioning by the AMG and BILU methods are solved as a whole by using the generalized minimum allowance (GMRES) method.

Figure 10.32 shows the detailed procedures below.

The following section elaborates the significance of each part.

10.2.2.2 Significance of Efficient Solving the Linear Algebraic Equation Set for Multiple Media Based on Unstructured Grids

1. Coefficient matrix decoupling

To prepare for preconditioning, the RR coefficient matrix needs to be decoupled into two parts: a pressure variable submatrix and a saturation variable submatrix. Taking Fig. 10.33 as an example, the yellow part is the coefficient matrix of the pressure equations and the orange is the coefficient matrix of saturation equations. The purpose is to separate the pressure and saturation equations from the fully implicit equations.

Therefore, the coefficient matrix of the original equations need to be transformed equivalently to extract the pressure equation from the fully implicit equations. At present, there are two methods. The first one obtains the linearized pressure equation through explicitly treat the saturation, and the other obtains the pressure equations by eliminating the extra elements through matrix operations. The pressure equations obtained by these two methods are similar, and both retain the important information of the fully coupled mathematical matrix. However, the former is inferior to the latter in terms of accuracy and iteration number since it adopts the explicit format of saturation, so we choose the matrix operations to get the pressure equation.

Again, there are two ways to get the pressure equation through matrix operations. One is called true IMPES elimination and the other is quasi IMPES elimination. Like IMPES, the former explicitly deals with the saturation variable in the flow term, which ignores the derivative of the flow term with respect to saturation. Then, a simple row transformation is performed to eliminate the derivative of the cumulative term with respect to the saturation in the first equation. Thus, the pressure equation is obtained. The latter obtains the pressure equation by eliminating the derivative of the cumulative term with respect to the saturation and ignoring the derivative of the flow term with respect to the saturation in the first equation through row transformation. The details are shown in Fig. 10.34.

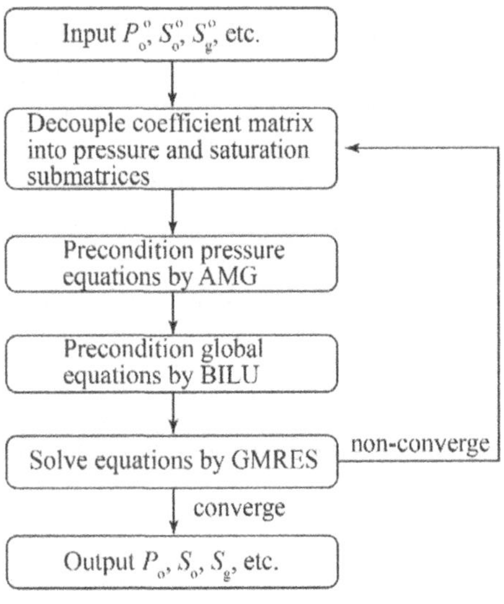

Fig. 10.32 Flowchart of CPR technology

The details in the schematic diagram are illustrated below.

First, the diagonal grids (i,i) and adjacent no-diagonal grids (i, j) are written as typical representatives, taking oil/gas/water three-phase as an example.

$$\begin{pmatrix} A_{op}+F_{op} & A_{osg}+F_{osg} & A_{osw}+F_{osw} \\ A_{gp}+F_{gp} & A_{gsg}+F_{gsg} & A_{gsw}+F_{gsw} \\ A_{wp}+F_{wp} & A_{wsg}+F_{wsg} & A_{wsw}+F_{wsw} \end{pmatrix}_{i,i} \begin{pmatrix} \hat{F}_{op} & \hat{F}_{osg} & \hat{F}_{osw} \\ \hat{F}_{gp} & \hat{F}_{gsg} & \hat{F}_{gsw} \\ \hat{F}_{wp} & \hat{F}_{wsg} & \hat{F}_{wsw} \end{pmatrix}_{i,j}$$

Coefficient matrix of diagonal grids (i,i) Coefficient matrix of non-diagonal grids (i,j)

(1) Obtain the pressure equation through explicitly processing saturation

All terms related to saturation derivatives are set to the values at the previous iteration step:

$$\begin{pmatrix} A_{op}+F_{op} & A^{n}_{osg}+F^{n}_{osg} & A^{n}_{osw}+F^{n}_{osw} \\ A_{gp}+F_{gp} & A_{gsg}+F_{gsg} & A_{gsw}+F_{gsw} \\ A_{wp}+F_{wp} & A_{wsg}+F_{wsg} & A_{wsw}+F_{wsw} \end{pmatrix}_{i,i} \begin{pmatrix} \hat{F}_{op} & \hat{F}^{n}_{osg} & \hat{F}^{n}_{osw} \\ \hat{F}_{gp} & \hat{F}_{gsg} & \hat{F}_{gsw} \\ \hat{F}_{wp} & \hat{F}_{wsg} & \hat{F}_{wsw} \end{pmatrix}_{i,j}$$

The saturation is known at the nth moment and thus can be moved to the right-hand side. The equation becomes the following format, with the pressure equation decoupled:

$$\begin{pmatrix} A_{op}+F_{op} & 0 & 0 \\ A_{gp}+F_{gp} & 0 & 0 \\ A_{wp}+F_{wp} & 0 & 0 \end{pmatrix}_{i,i} \begin{pmatrix} \hat{F}_{op} & 0 & 0 \\ \hat{F}_{gp} & 0 & 0 \\ \hat{F}_{wp} & 0 & 0 \end{pmatrix}_{i,j}$$

(2) Obtain the pressure equation through matrix operations

1) True IMPES elimination for nonlinear mass conservation equations

According to the IMPES method, the saturation variables in the flow items are explicitly processed.

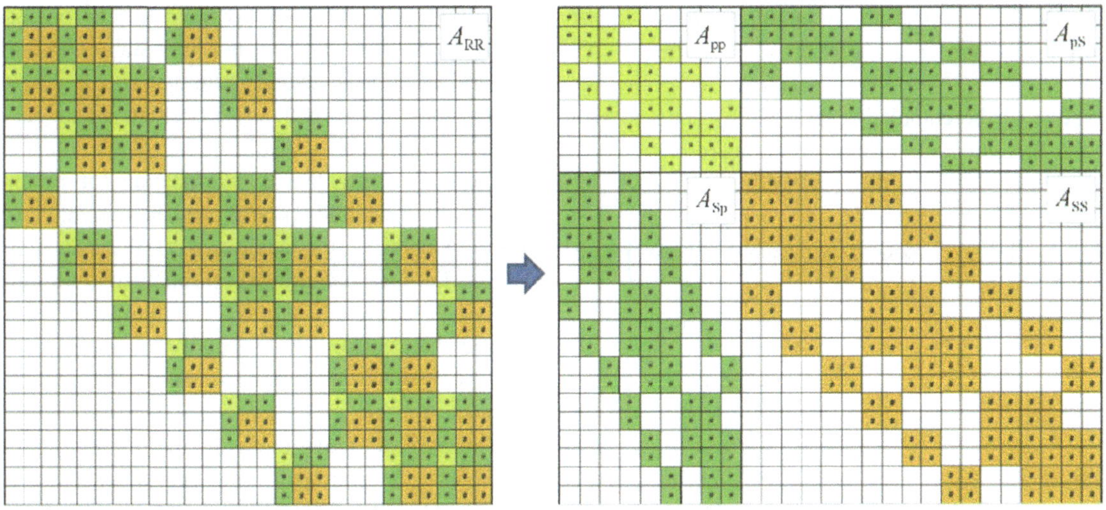

(a) Coefficient matrix before decoupling (b) Coefficient matrix after decoupling

Fig. 10.33 Schematics of the coefficient matrix before and after decoupling

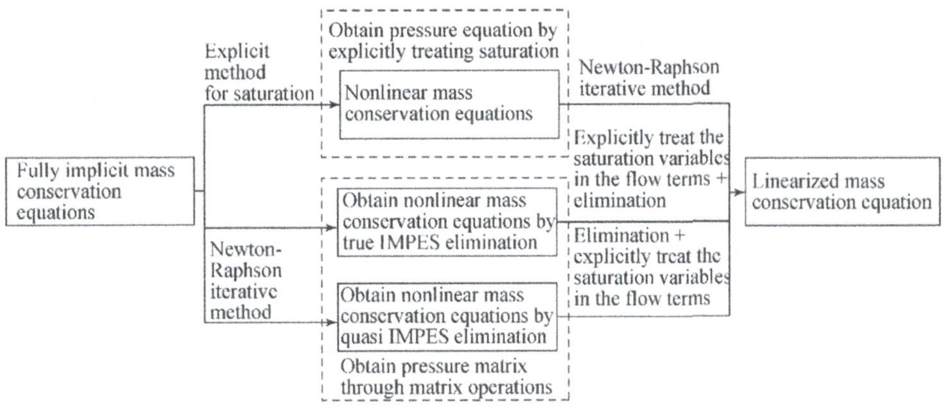

Fig. 10.34 Comparison of different treatment methods

$$\begin{pmatrix} A_{op}+F_{op} & A_{osg}+F^n_{osg} & A_{osw}+F^n_{osw} \\ A_{gp}+F_{gp} & A_{gsg}+F^n_{gsg} & A_{gsw}+F^n_{gsw} \\ A_{wp}+F_{wp} & A_{wsg}+F^n_{wsg} & A_{wsw}+F^n_{wsw} \end{pmatrix}_{i,i} \begin{pmatrix} \hat{F}_{op} & \hat{F}^n_{osg} & \hat{F}^n_{osw} \\ \hat{F}_{gp} & \hat{F}^n_{gsg} & \hat{F}^n_{gsw} \\ \hat{F}_{wp} & \hat{F}^n_{wsg} & \hat{F}^n_{wsw} \end{pmatrix}_{i,j}$$

The saturation is known at the nth moment and thus can be moved to the right-hand side, which ignores the derivative of the flow term with respect to saturation. The equation becomes the following format:

$$\begin{pmatrix} A_{op}+F_{op} & A_{osg} & A_{osw} \\ A_{gp}+F_{gp} & A_{gsg} & A_{gsw} \\ A_{wp}+F_{wp} & A_{wsg} & A_{wsw} \end{pmatrix}_{i,i} \begin{pmatrix} \hat{F}_{op} & 0 & 0 \\ \hat{F}_{gp} & 0 & 0 \\ \hat{F}_{wp} & 0 & 0 \end{pmatrix}_{i,j}$$

Simple row transformations are conducted in the equation of a grid, eliminate the derivative of the cumulative term with respect to the saturation in the first equation. Thus, the pressure equation is obtained:

$$\begin{pmatrix} J_1 - J_2 J_4^{-1} J_3 & 0 & 0 \\ A_{gp}+F_{gp} & A_{gsg} & A_{gsw} \\ A_{wp}+F_{wp} & A_{wsg} & A_{wsw} \end{pmatrix}_{i,i} \begin{pmatrix} J_5 - J_2 J_4^{-1} J_6 & 0 & 0 \\ \hat{F}_{gp} & 0 & 0 \\ \hat{F}_{wp} & 0 & 0 \end{pmatrix}_{i,j}$$

where

$$J_1 = (A_{op}+F_{op}) - A_{osw}A_{gsw}^{-1}(A_{gp}+F_{gp})$$
$$J_2 = A_{osg} - A_{osw}A_{gsw}^{-1}A_{gsg}$$
$$J_3 = (A_{gp}+F_{gp}) - A_{gsw}A_{wsw}^{-1}(A_{wp}+F_{wp})$$
$$J_4 = A_{gsg} - A_{gsw}A_{wsw}^{-1}A_{wsg}$$
$$J_5 = \hat{F}_{op} - A_{osw}A_{gsw}^{-1}\hat{F}_{gp}$$
$$J_6 = \hat{F}_{gp} - A_{gsw}A_{wsw}^{-1}\hat{F}_{wp}$$

2) Quasi IMPES elimination for nonlinear mass conservation equations

The derivative of the cumulative term with respect to saturation in the first equation through row transformations:

$$\begin{pmatrix} J_1 - J_2J_4^{-1}J_3 & 0 & 0 \\ A_{gp} + F_{gp} & A_{gsg} + F_{gsg} & A_{gsw} + F_{gsw} \\ A_{wp} + F_{wp} & A_{wsg} + F_{wsg} & A_{wsw} + F_{wsw} \end{pmatrix}_{i,i}$$

$$\begin{pmatrix} J_5 - J_2J_4^{-1}J_6 & J_7 - J_2J_4^{-1}J_9 & J_8 - J_2J_4^{-1}J_{10} \\ \hat{F}_{gp} & \hat{F}_{gsg} & \hat{F}_{gsw} \\ \hat{F}_{wp} & \hat{F}_{wsg} & \hat{F}_{wsw} \end{pmatrix}_{i,j}$$

where,

$$J_1 = (A_{op} + F_{op}) - (A_{osw} + F_{osw})(A_{gsw} + F_{gsw})^{-1}(A_{gp} + F_{gp})$$

$$J_2 = (A_{osg} + F_{osg}) - (A_{osw} + F_{osw})(A_{gsw} + F_{gsw})^{-1}(A_{gsg} + F_{gsg})$$

$$J_3 = (A_{gp} + F_{gp}) - (A_{gsw} + F_{gsw})(A_{wsw} + F_{wsw})^{-1}(A_{wp} + F_{wp})$$

$$J_4 = (A_{gsg} + F_{gsg}) - (A_{gsw} + F_{gsw})(A_{wsw} + F_{wsw})^{-1}(A_{wsg} + F_{wsg})$$

$$J_5 = \hat{F}_{op} - (A_{osw} + F_{osw})(A_{gsw} + F_{gsw})^{-1}\hat{F}_{gp}$$

$$J_6 = \hat{F}_{gp} - (A_{gsw} + F_{gsw})(A_{wsw} + F_{wsw})^{-1}\hat{F}_{wp}$$

$$J_7 = \hat{F}_{osg} - (A_{osw} + F_{osw})(A_{gsw} + F_{gsw})^{-1}\hat{F}_{gsg}$$

$$J_8 = \hat{F}_{osw} - (A_{osw} + F_{osw})(A_{gsw} + F_{gsw})^{-1}\hat{F}_{gsw}$$

$$J_9 = \hat{F}_{gsg} - (A_{gsw} + F_{gsw})(A_{wsw} + F_{wsw})^{-1}\hat{F}_{wsg}$$

$$J_{10} = \hat{F}_{gsw} - (A_{gsw} + F_{gsw})(A_{wsw} + F_{wsw})^{-1}\hat{F}_{wsw}$$

Ignoring the derivative of the flow term with respect to saturation on the non-diagonal grids in the first equation:

$$J_7 - J_2J_4^{-1}J_9 = 0$$
$$J_8 - J_2J_4^{-1}J_{10} = 0$$

The pressure equation is also obtained, as shown below:

$$\begin{pmatrix} J_1 - J_2J_4^{-1}J_3 & 0 & 0 \\ A_{gp} + F_{gp} & A_{gsg} + F_{gsg} & A_{gsw} + F_{gsw} \\ A_{wp} + F_{wp} & A_{wsg} + F_{wsg} & A_{wsw} + F_{wsw} \end{pmatrix}_{i,i} \begin{pmatrix} J_5 - J_2J_4^{-1}J_6 & 0 & 0 \\ \hat{F}_{gp} & \hat{F}_{gsg} & \hat{F}_{gsw} \\ \hat{F}_{wp} & \hat{F}_{wsg} & \hat{F}_{wsw} \end{pmatrix}_{i,j}$$

In summary, the key of the above two methods is how to assign zero to the coefficients of the derivatives of the cumulative/flow term with respect to saturation. In Method 1 (Obtain the pressure equation through explicitly processing saturation), the derivatives of the cumulative terms and flow terms with respect to saturation are directly processed at the nth moment. It is easy but not very stable in computing. In Method 2B (Quasi IMPES elimination for nonlinear mass conservation equations), some coefficients are set as zero at the last step, which also affects the calculation stability. The Method 2A (True IMPES elimination for nonlinear mass conservation equations) only uses the derivative of the flow term with respect to saturation at the nth moment, and obtains the pressure equation through rigid elimination later on, which hardly affects the computing stability. It has been proved that Method 2B is better than others to obtain the mass conservation nonlinear equations through a large number of tests.

Through the above methods, the pressure and saturation variables are separated first, and then re-combined and rearranged. That is, the pressure part of a coefficient matrix is extracted alone and arranged by points, and the saturation part is arranged by blocks (as shown in Fig. 10.35).

The extraction of the pressure equation can be expressed as:

$$A_{pp} = RA_{FIM}C$$

where, A_{pp} is the decoupled pressure coefficient matrix; A_{FIM} is the fully implicit pressure and saturation coefficient matrix; R and C are row and column operations. Meanwhile, the coefficient matrix of the derivative of the first equation with respect to saturation is $A_{ps} = 0$, and A_{sp} and A_{ss} keep the characteristics of the original matrix.

2. AMG method for preconditioning pressure equations

After decoupling the coefficient matrix, the coefficient matrix A_{pp} is obtained, which is only related to the pressure. This mathematical matrix has the characteristics of an elliptic equation and is suitable for the preconditioning by AMG method (Cao and Aziz 2002; Chen 2008; Stueben 1983; Li 2013).

Setting a grid where A_{pp} is currently located as the finest grid, a series of grids can be obtained after coarsening:

$$\Omega^1 \supset \Omega^2 \supset \cdots \Omega^l \supset \cdots \Omega^L$$

(1) Pre-smoothing

Suppose that the initial solution of linear equations $A_l x_l = f_l$ on the Ω^l grid is x_0, and calculate the residue on fine grids:

$$e_l = f_l - A_l x_l$$

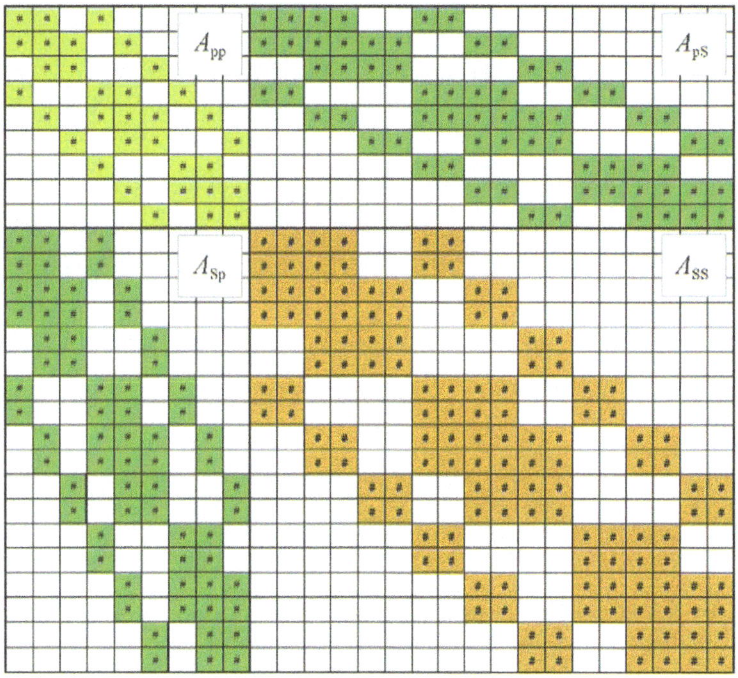

Fig. 10.35 Coefficient matrix with rearranged pressure and saturation

The solution x_l is obtained after v_1 times of iterations.

(2) Correcting coarse grids

① According to Galerkin method, a restriction operator is constructed I_l^{l+1} (l = 1,2,3,…,L−1), which can map the information on fine grids Ω^l to coarse grids Ω^{l+1}.
② In the same way, an extension operator I_l^{l+1} (l = 1,2,3, …,L-1)is constructed, which maps the information on coarse grids Ω^{l+1} to the fine grids Ω^l.
③ Based on the restriction operator and the extension operator, the coefficient matrix/on the coarse grids is generated.
④ Map the error e_l to coarse grids $e_{l+1} = \left(I_l^{l+1}\right)^{\mathrm{T}} e_l$.
⑤ Solve $A_{l+1}x_{l+1} = e_{l+1}$ on coarse grids and get the approximate solution $\tilde{x}$. Solution methods mainly include the Jacobi iteration, Gauss-Seidel iteration, and relaxation iteration. The Gauss-Seidel method is selected to solve the equations:

$$x_i^{(k+1)} = \frac{1}{a_{ii}}\left(b_i - \sum_{j=1}^{i-1} a_{ij}x_j^{(k+1)} - \sum_{j=i+1}^{n} a_{ij}x_j^{(k)}\right), (i = 1, 2, \cdots, n)$$

Figure 10.36 shows the schematics of the AMG method.

⑥ Back map the pressure from coarse grids to fine grids through an extension operator.

(3) Post-interpolation

The vector x_{p} solved in the first stage is extrapolated to the vector in the fully implicit system:

$$x_1 = Cx_{\mathrm{p}}$$

where, C is the column operation matrix, which is the same as that in the above equation. For three-phase problems, x_1 and x_{p} are in the following formats:

$$x_1 = \left[\delta p_1, 0, 0, \delta p_2, 0, 0, \delta p_3, 0, 0, \cdots, \delta p_n, 0, 0, \delta p^{\mathrm{W}}\right]^{\mathrm{T}}$$
$$x_p = \left[\delta p_1, \delta p_2, \delta p_3, \cdots, \delta p_n, \delta p^{\mathrm{W}}\right]^{\mathrm{T}}$$

The above methods can be used to develop the AMG algorithms in multiple layers and multiple types. x_1 obtained in this stage is used to update the right-hand side:

$$M_2x_2 = b - Ax_1$$

To sum up, the flowchart of the AMG method that is composed of the 1st and (l + 1)th layers' calculations is shown in Fig. 10.37.

3. BILU preconditioning the global equation

After the coefficient matrix decoupling and the AMG preconditioning, the coefficient matrix has the characteristics of a hyperbolic equation, which is suitable for the LU decomposition. It is observed when the coefficient matrix

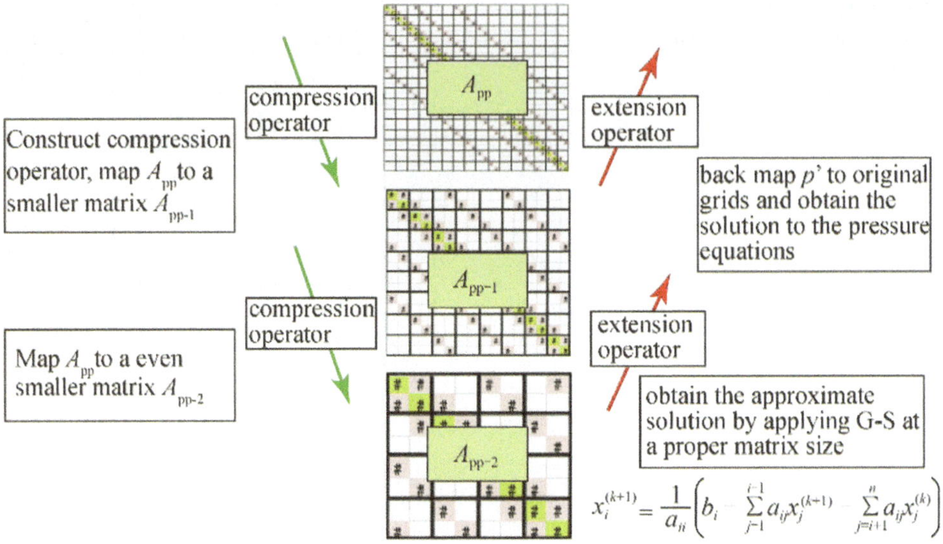

Fig. 10.36 Schematics of the AMG method

has fewer non-zero elements and distributed in certain format, the lower triangular unit matrix L and the upper triangular unit matrix U produced by the LU decomposition generally cannot maintain the same sparse mode as the original matrix. That is, if the original coefficient matrix is a banded matrix, the corresponding L and U are added by a number of non-zero elements and are no longer in the banded format, which increases the computing load and complexity (Saad 1994; Lin 1987).

Aiming at the above situation, the incomplete LU (ILU) decomposition method is employed. The method approximately decomposes a sparse matrix into a product of a lower triangular spare matrix L and a upper triangular sparse matrix U that are in a specific structure.

$$A \approx L(G)U(G),$$

where, G is a non-zero index set, and all elements except those located at G position are set as zero in L and U.

When the non-zero index set is exactly the same as the non-zero structure of A, it is ILU (0) decomposition, which is a simple and fast preconditioning method. Adding the non-zero index in G can get different preconditioning methods such as ILU(k) and ILU(τ). With the increase in the non-zero index, the structure of LU gets closer to A and the iterated result approaches closer to the exact solution. But the calculation load and complexity are also increased.

Because a nonlinear flow mathematical model for tight sandstone oil/gas reservoirs is characterized by multiple scales, multiple media, and multiple flow regimes, the coefficient matrix usually is block diagonally dominant instead of point diagonally dominant, which leads to a lot of elements with small absolute values on the diagonal line. The computing time can increase drastically if using Gauss elimination to conduct the ILU decomposition for the coefficient matrix. Therefore, the incomplete LU decomposition technology based on block is used to preprocess the coefficient matrix.

Block ILU(0) decomposition (Fig. 10.38) treats the governing equations and unknown variables on each grid as a whole and applies ILU(0) decomposition to the coefficient matrix. This method can quickly eliminate the high singularity of the coefficient matrix and speed up the iteration. BILU(0) decomposition is similar to ILU(0) decomposition, but replaces the point operations by block operations in the latter (as shown in Fig. 10.38). Compared with ILU(0), the diagonals of L and U are still blocks. For example, the blocks on the main diagonal are two-order submatrices for a two-phase black oil model. The blocks on the diagonal are three-order submatrices for a three-phase black oil model. The block ILU decomposition only requires submatrices reversible on the diagonal, which is weaker than point ILU decomposition. Thus, the block ILU decomposition is more robust and faster than the point ILU decomposition. The block ILU preconditioning reduces the coupling between grids and decomposes a complex coefficient matrix into two simple matrices, which reduces the computation load, saves the calculating time, and reaches quick convergence.

In addition, in the block ILU decomposition process, p^2 elements (p is the number of phases) are operated in each elimination, and the searching number is fewer than that of point ILU by $1/p^2$. This makes full use of the computer cache and also improves the computing speed.

It is block operation in the block ILU decomposition. That is, the corresponding elements in blocks are eliminated. Specifically, block elimination actually operates an element in one block and another one in the same position in other blocks.

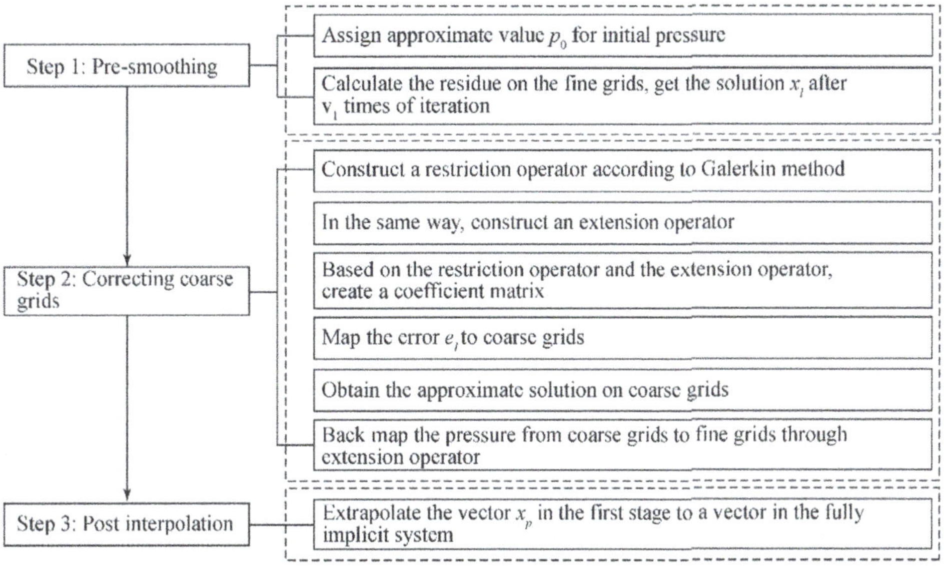

Fig. 10.37 flowchart of the AMG algorithm

Figure 10.39 shows the technical flowchart of BILU(0) decomposition:

where, $L_S(i)$ is the total number of non-zero elements of the ith block and p is the number of the black oil phases.

4. GMRES method for integrated solving

The preconditioned coefficient matrix is block diagonally dominant and is solved by using the generalized minimum allowance (GMRES) method for better stability and convergence and less computing time (Saad and Gmres 1986; Li et al. 2005a, b).

Suppose that A is a symmetric positive definite matrix, and consider how to solve a minimum value problem:

$$\begin{aligned} f(x) &= \min_{x \in x_0 + \kappa_m(A, r_0)} E(x) \\ &= \min_{x \in x_0 + \kappa_m(A, r_0)} [A(x - x^*), (x - x^*)]^{1/2} \end{aligned}$$

where, $r_0 = b - Ax_0$; x_0 is an arbitrary vector in r_n; x^* is the true solution to $Ax = b$; $\kappa_m(A, r_0)$is the m-order Krylov submatrix for A and r_0:

$$\kappa_m(A, r_0) = span\{r_0, Ar_0, \cdots, A^{m-1} r_0\}$$

The method is the generalized minimum residual solution (GMRES). Based on its definition, the true solution is approximated by obtaining the vector $x_m \in \kappa_m$ that minimizes residues. In other words, the solution obtained from the minimum value problem is closer to the true solution than x_0.

Suppose $V_m = (v_1, v_1, \ldots, v_m)$ is a set of standard or thonormal basis in the Krylov subdomain, and

$$\begin{aligned} AV_m = V_{m+1}\overline{H}_m = V_m H_m + h_{m+1,m} v_{m+1} e_m^{\mathrm{T}}, \\ V_m^{\mathrm{T}} A V_m = H_m \end{aligned}$$

where $v_1 = r_0 / \|r_0\|_2$ and $\overline{H}_m$ is the m-order Heisenberg matrix.

From the above formula, it is known that the vector x_m on κ_m can be written as $x_m = v_m y$, where $y \in r_m$, then $x \in x_0 + \kappa_m$ can be expressed as $x = x_0 + V_m y$, and

$$\begin{aligned} E(x) = \|b - Ax\|_2 = \|b - A(x_0 + V_m y)\|_2 = \|r_0 - AV_m y\|_2 \\ r_0 = \beta v_1 (\beta = \|r_0\|_2) \end{aligned}$$

which is

$$\begin{aligned} E(x) = \|\beta v_1 - V_{m+1}\overline{H}_m y\|_2 = \|V_{m+1}(\beta e_1 - \overline{H}_m y)\|_2 = \|\beta e_1 - \overline{H}_m y\|_2 \\ f(x) = \min_{y \in R^m} \|\beta e_1 - \overline{H}_m y\|_2 \end{aligned}$$

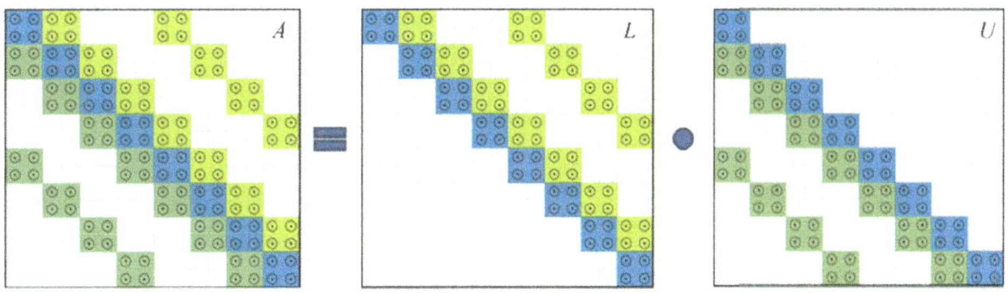

Fig. 10.38 Schematics of block ILU(0) decomposition

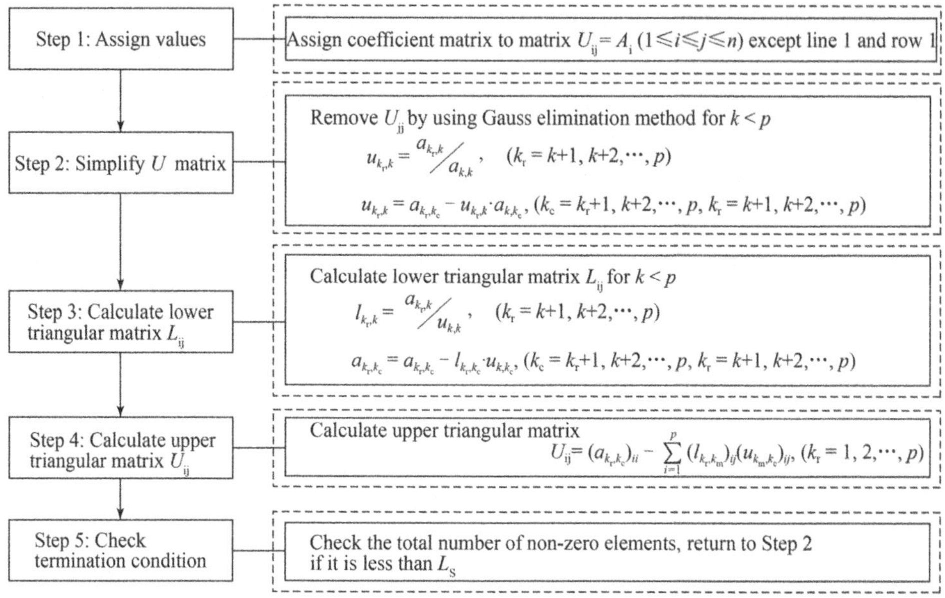

Fig. 10.39 Flowchart of BILU (0) preconditioning

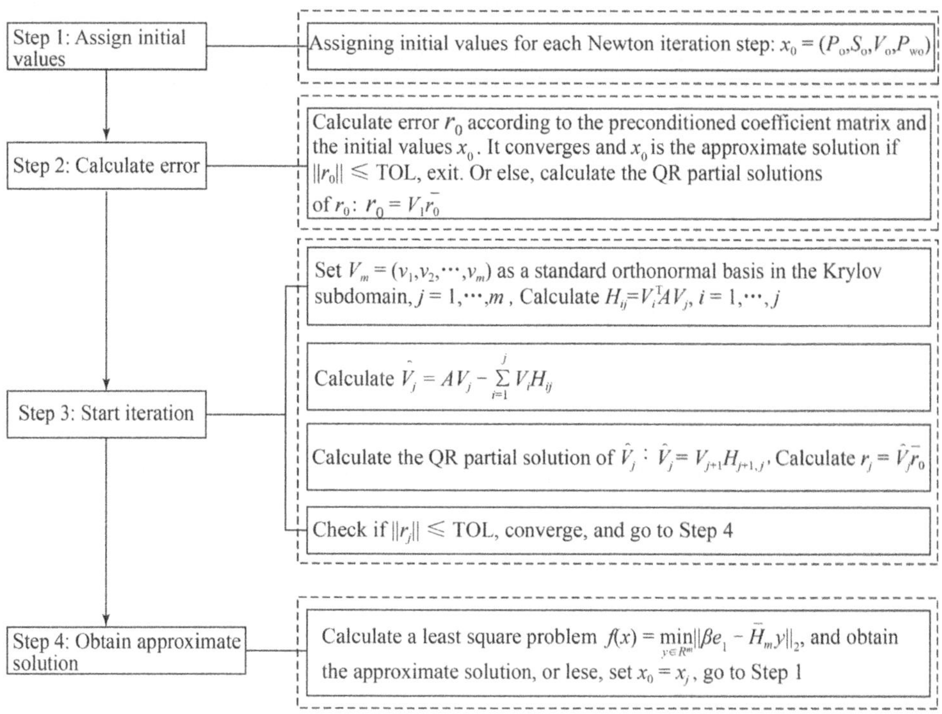

Fig. 10.40 Flowchart of the GMRES iterative method

Figure 10.40 shows the flowchart of the GMRES method. where, the matrix-vector multiplication is modified from the point-to-point operation to the block-to-block operation. For example, in the three-phase black oil model (n grids), the product of a coefficient matrix and a vector Ax is $\sum_{j=1}^{3n} a_{ij} x_j$ based on point operations, but is $\sum_{j=1}^{n} A_{ij} x_j$ based on block operations.

$$A_{ij} x_j = \begin{pmatrix} a_{11} & a_{12} & a_{13} \\ a_{21} & a_{22} & a_{23} \\ a_{31} & a_{32} & a_{33} \end{pmatrix}_{ij} \begin{pmatrix} x_1 \\ x_2 \\ x_3 \end{pmatrix}_j = \begin{pmatrix} \sum_{k=1}^{3} a_{1k} x_k \\ \sum_{k=1}^{3} a_{2k} x_k \\ \sum_{k=1}^{3} a_{3k} x_k \end{pmatrix}_j$$

The vector-vector operation is also modified from the point-to-point operation to the block-to-block operation. For example, an inner product is revised from/to $\sum_{i=1}^{n} X_i Y_i$, where

$$X_i Y_i = (x_1, x_2, x_3) \begin{pmatrix} y_1 \\ y_2 \\ y_3 \end{pmatrix} = x_1 y_1 + x_2 y_2 + x_3 y_3$$

From the calculation steps of the GMRES iteration method, it is found that calculations in this method are mostly matrix-to-vector and vector-to-vector calculations, which are benefited from a small computing load, less storage memory, high precision, good stability, and quick convergence.

References

Appleyard JR, Cheshire IM (1983) Nested factorization, SPE 12264

Cao H (2002) Development of technology for general purpose simulation, PhD thesis, Stanford University

Cao H, Aziz K (2002) Performance of impsat and impsat-aim models in compositional simulation, SPE 77720, presented at the SPE Annual Technical Conference and Exhibition, San Antonio, Texas

Cao H, Tchelepi HA, Wallis JR, Yardumian H (2005) Parallel scalable cpr type linear solver for reservoir simulation, SPE 96809, presented at the SPE Annual Technical Conference and Exhibition, Dallas, Texas

Chen S (2008) Algebraic multiple grid method and its application in the Krylov preconditioner, master thesis from Zhejiang University, 11–28

Chen Z (2011) The finite element method, World Scientific

Chen Z, Huan G, Ma Y (2006) Computational methods for multiphase flows in porous media. Soc Ind Appl Math, Philadelphia, pp 1–2

Forsyth PA, Sammon PH (1986) Practical considerations for adaptive implicit methods in reservoir simulation. J Comp Phys 62:265–281

Guo C, Wei M, Chen H, He X, Bai B (2014) Improved numerical simulation for shale gas reservoirs, SPE OTC 24913-MS

Han D, Chen Q, Yan C (1999) Reservoir simulation basics. Beijing: Petroleum Industry Press: 34–44, 241–242

Jiang Y (2007) Technology for modeling complex reservoirs and advanced wells, PhD thesis, 27–78

Kalantari-Dahaghi S, Esmaili S, Mohaghegh D (2012) Fast track analysis of shale numerical models, SPE 162699

Karimi-Fard M, Durlofsky LJ (2012) Accurate resolution of near-well effects in upscaled models using flow-based unstructured local grid refinement. SPE J

Kazemi A, Stephen KD (2012) Schemes for automatic history matching of reservoir modeling: A case of nelson oilfield in the UK. Petrol Exp Dev 39(3):326–337

Li X (2013) Fast solving of the pressure equation in reservoir simulation, master thesis from Tsinghua University, 10–23

Li Y, Russell TJ (2006) Rapid flash calculations for compositional simulation. SPE Reserv Eval Eng, 521–529

Li W et al (2005a) Comparision of the gmres and orthomin for the black oil model in porous media. Int J Numer Meth Fluids 48 (5):501–519

Li W, Chen Z, Ewing RE, Huan G, Li B (2005b) Comparison of the GMRES and ORTHOMIN for the black oil model in porous media. Int J Numer Meth Fluids 48(5):501–519

Li X et al (2011) Numerical simulation of pore-scale flow in chemical flooding process. Theor Appl Mech Lett 1(2):1022–1028

Lim KT, Aziz K (1995) A new approach for residual and Jacobian array construction in reservoir simulators, SPE 84228, SPE Computer Applications

Lin X (1987) Bilucg algorithm for 2-d multiphase reservoir simulation. Petrol Exp Dev 14(1):76–84

Liu Y (2009) Solving method to Flow model of low permeability oil reservoirs and its application. Master thesis from China University of Petroleum, 30–38

Milad B, Civan F, Devegowda D, Sigal RF (2013) Modeling and simulation of production from commingled shale gas reservoirs, SPE 168853/URTeC 1582508

Mora CA, Wattenbarger RA (2009) Comparison of computation methods for cbm performance. J Can Pet Technol 48(4):42–48

Odeh AS (1981) Comparison of solutions to a three dimensional black oil reservoir simulation problem, SPE

Olorode CM, Freeman GJ, Moridis T, Blasingame A (2013) High-resolution numerical modeling of complex and irregular fracture patterns in shale-gas reservoirs and tight gas reservoirs, SPE

Russell TF (1989) Stability analysis and switching criteria for adaptive implicit methods based on the cfl condition, SPE 18416, Proceedings Of The 10th SPE Symposium On Reservoir Simulation, Houston, TX

Saad Y (1994) ILUT: A dual threshold incomplete lu factorization. Numer Linear Algebra Appl 1(4):387–402

Saad Y, Gmres MS (1986) A generalized minimal residual algorithm for solving nonsymmetric linear systems. SIAM J Sci Stat Comput 7(3):856–869

Shi Y, Li Y, Yao J (2008) Solution and application to the dual-permeability model of the naturally fractured reservoirs. Inner Mongolia Petroch Ind 69–72:77

Stueben K (1983) Algebraic multigrid (AMG): Experiences and comparisons proceedings of the international multigrid conference

Wang B et al (2013) Block compressed storage and computation in large-scale reservoir simulation. Petrol Exp Dev 40(4):462–467

Wu S et al (2012) Steam flooding to enhance recovery of a waterflooded light-oil reservoir. JPT 1:64–66

11 Application of Numerical Simulation in the Development of Tight Oil/Gas Reservoirs

This chapter introduces the special functions of the numerical simulation software for unconventional tight oil/gas reservoirs, UnTOG v1.0. The software has been applied for production performance simulation of different-size pore and fracture media, multiple flow regime identification and self-adaptive simulation of complex flow mechanisms, horizontal well and hydraulic fracturing optimization, coupled multiphase flow-geomechanics simulation during fracturing-injection-production processes, and simulation of unstructured grids of different types. The software has advanced functions and strong practicability in the aspects of tight reservoir characterization and geological modeling, horizontal well and hydraulic fracturing optimization, dynamic simulation of production mechanisms, technological policy optimization, and prediction of development index (index).

11.1 Numerical Simulation Software of Unconventional Tight Oil/Gas Reservoirs

Based on the mathematical models of discontinuous and multiple media and key technologies of numerical simulation, simulation software for unconventional tight oil/gas reservoirs, UnTOG v1.0, is developed independently. It is composed of three modules: pre-processing, simulators, and post-processing, and is specially featured with unstructured grid generation for multiple media under complex reservoir conditions, modeling and simulation of discontinuous multi-scale discrete multiple media, coupled multi-phase flow-geomechanics simulation during fracturing-injection-production processes, multiple flow behavior identification and self-adaptive simulation of complex flow mechanisms, dynamic coupled simulation of hydraulic fractured horizontal wells. It is able to handle tight reservoir characterization and geological modeling, horizontal well and hydraulic fracturing optimization, and dynamic simulation of production mechanisms, technological policy optimization, and prediction of development index.

11.1.1 Basic Functions of the Software

The simulation software for unconventional tight oil/gas reservoirs, UnTOG v1.0, is composed of three modules: pre-processing, simulators, and post-processing.

The pre-processing section mainly consists of the generation of unstructured grids for multiple media under complex reservoir conditions, the modeling of discontinuous and discrete multiple media at different scales, and the input and management of the static and dynamic parameters of reservoirs and wells.

The simulators mainly include modules of the numerical simulation of discontinuous and discrete multiple media at different scales, the coupled flow-geomechanics simulation of multiple media during fracturing-injection-production processes, the self-adaptive simulation of multiple media phase identification the complex flow mechanisms, dynamic coupled simulation of hydraulic fractured horizontal wells, and the generation of the numerical discrete efficient matrix for discontinuous multiple media and the corresponding solution module.

The post-processing part is mainly composed of statistical analysis of dynamic simulation laws for multiple media at different scales, and 2D/3D graphic display modules. The function of each module is described in Table 11.1.

The UnTOG v1.0 software achieved an integrated workflow of geological modeling and numerical simulation for tight oil/gas reservoirs, including the macroscopic and microscopic characterization of multiple media heterogeneity, the unstructured grid generation under the complex reservoir conditions, the modeling and simulation of discontinuous and discrete multiple media at different scales, optimum design of horizontal well and hydraulic fracturing,

Q. Ran, *Unconventional Tight Reservoir Simulation: Theory, Technology and Practice*,
https://doi.org/10.1007/978-981-32-9848-4_11

Table 11.1 Main module and function description

Type	Module	Function description
Pre-processing	Unstructured grid generation for multiple media under complex reservoir conditions	① Variable-scale single grid generation, including triangular mesh, quadrilateral mesh and Pebi mesh ② Hybrid grid generation for different grids characterizing different objects: different types of unstructured meshes can be used to describe different geological boundaries in tight formations, horizontal wells, natural/artificial fractures of different scales ③ Multiple media processing by hybrid grids: nested grids and interactive grids can be used to handle multiple media at micro scales
	Modeling module of discontinuous and discrete multiple media at different scales	① Natural/artificial fracture discrete modeling: discrete modeling of natural fractures that considers roughness and filling characteristics, and discrete modeling of artificial fracture that considers proppant concentration and supporting mode ② Modeling of discrete multiple media at multiple scales ③ Upscaling and equivalent modeling of multiple media at different scales
	Input and management module for static and dynamic parameters of reservoirs and wells	Input and processing functions for rock and fluid PVT parameters, phase permeability/capillary force parameters, well and production dynamic parameters
Simulator	Numerical simulation module of discontinuous and discrete multiple media at different scales	① Dynamic simulation of discrete natural/artificial fractures at different scales ② Dynamic simulation of interactive/relay supply &drainage discrete multiple media ③ Dynamic simulation of discontinuous and discrete mixed multiple media
	Coupled flow-geomechanics modeling of multiple media during fracturing-injection-production processes	① Simulation of the dynamic changes of physical parameters of multiple media at different scales during fracturing-injection-production processes ② Simulation of the effects of dynamic changes of physical parameters of multiple media at different scales during fracturing-injection-production processes on the conductivity and well index
	Self-adaptive simulation module of multiple media phase identification and complex flow mechanisms	① Automatic identification of flow pattern through medium geometry, fluid property, and pressure gradient ② Selection of dynamic equations corresponding to the recognition results of flow pattern, and self-adaptive simulation of complex flow mechanisms.
	Dynamic coupling for hydraulic fractured horizontal wells	① Simulation of horizontal well production performance after hydraulic fracturing with line-source, discrete grids, multi-section wells treatment ② Simulation of horizontal well production performance after hydraulic fracturing that considers the effect of coupled flow-geomechanicss effects on the well index near wellbore ③ Simulation of horizontal well production performance after hydraulic fracturing that considers the effect of different media in flow behavior change on the well index
	Generation and solution of numerical discrete efficient matrix for discontinuous multiple media	① Partitioning and storing of reservoirs and multiple wells by zero elements and dead nodes elimination and block compression and storage technology ② Efficient matrix solving with preconditioning, which transforms the complex-structured matrix into an equivalent system that is easy to solve
Post-processing	Statistical analysis module of simulation dynamic laws for multiple media at different scales	① Statistics of the output data (productions, reserves and recoveries) of multiple media at different scales ② Statistics for the contributions of different flow behavior to the production in different media
	2D/3D graphic display module	Display of the calculation results through 2D curve, 2D figures, and 3D figures

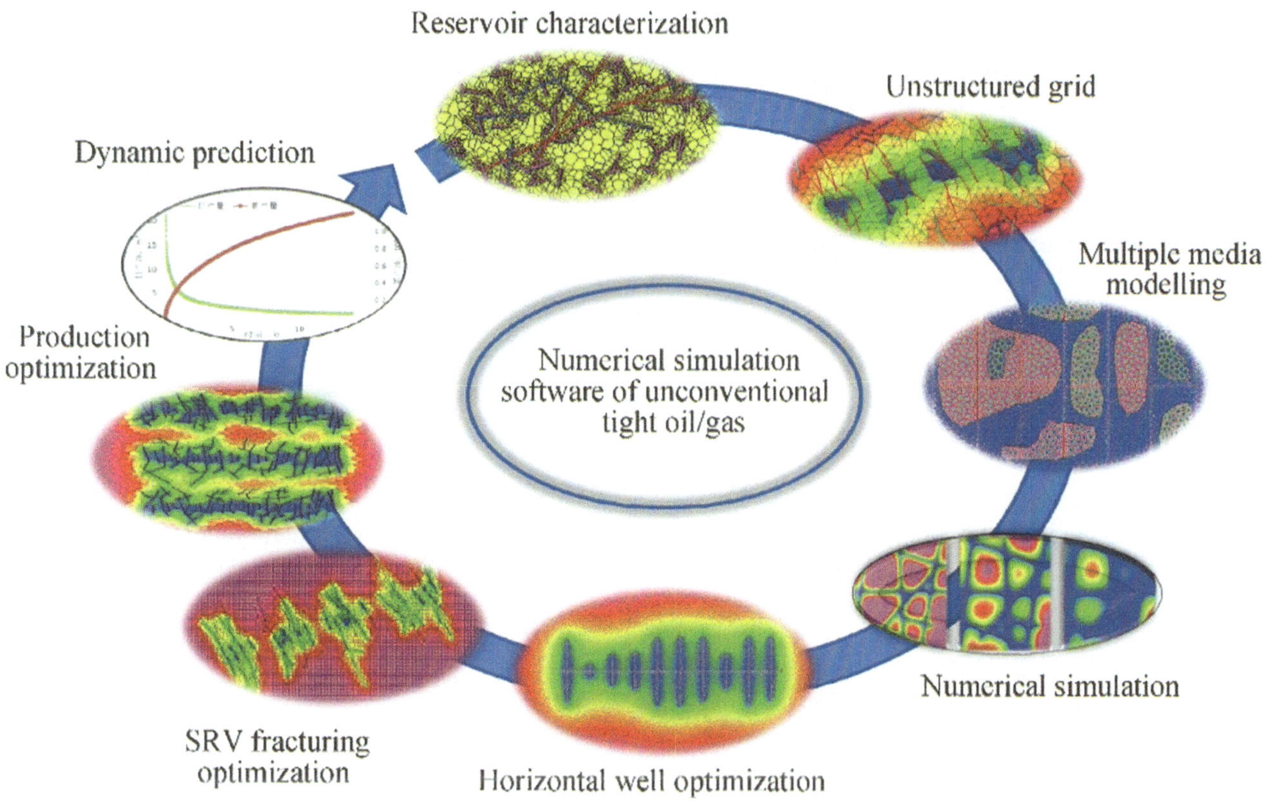

Fig. 11.1 Integrated workflow of geological modeling and numerical simulation for multiple media (UnTOG v1.0)

optimum design of the tight oil/gas production, and dynamic prediction of development index (Fig. 11.1).

11.1.2 Special Functions of the Software

The UnTOG v1.0 is featured with the simulation for discontinuous, multiple flow behaviors in multiple media at multiple scales (Fig. 11.2). It includes: unstructured grid generation under complex reservoir conditions, numerical simulation of discontinuous and discrete multiple media at multiple scales, coupled flow-mechanics modeling of multiple media during fracturing-injection-production processes, flow pattern identification of multiple media and self-adaptive simulation of nonlinear complex flow mechanisms, and dynamic coupled simulation of horizontal wells after hydraulic fracturing.

11.1.3 Optimization Simulation Function for Tight Oil/Gas Reservoir Development

UnTOG v1.0 is developed specifically for the geological characteristics and development mode of tight oil/gas reservoirs. It is able to handle practical development problems including tight formation characterization and multiple media modeling, optimization of horizontal well and hydraulic fracturing parameters, dynamic simulation of production mechanisms, development optimization design and dynamic prediction (Table 11.2 and Fig. 11.3).

11.2 Dynamic Simulation of Multiple Media at Different Scales in Tight Oil/Gas Reservoirs

Multiple-scale multiple media are developed in tight oil/gas reservoirs (Du et al 2014; Du 2016). Different geometric characteristics, physical properties, oil bearing, fluid properties and existence state at different scales lead to a large difference in flow behaviors and mechanisms, which affects the dynamic production characteristics of multiple media with different-scale pores and fractures. This section presents the dynamic simulation of multiple-scale pore media, dynamic simulation of fluid properties and flow characteristics in multiple-scale pore media, dynamic simulation of multiple-scale natural fractures, and self-adaptive dynamic simulation of fluid behaviors and complex flow mechanisms, which reveals the production and recovery rules in multiple media with multiple-scale pores and fractures.

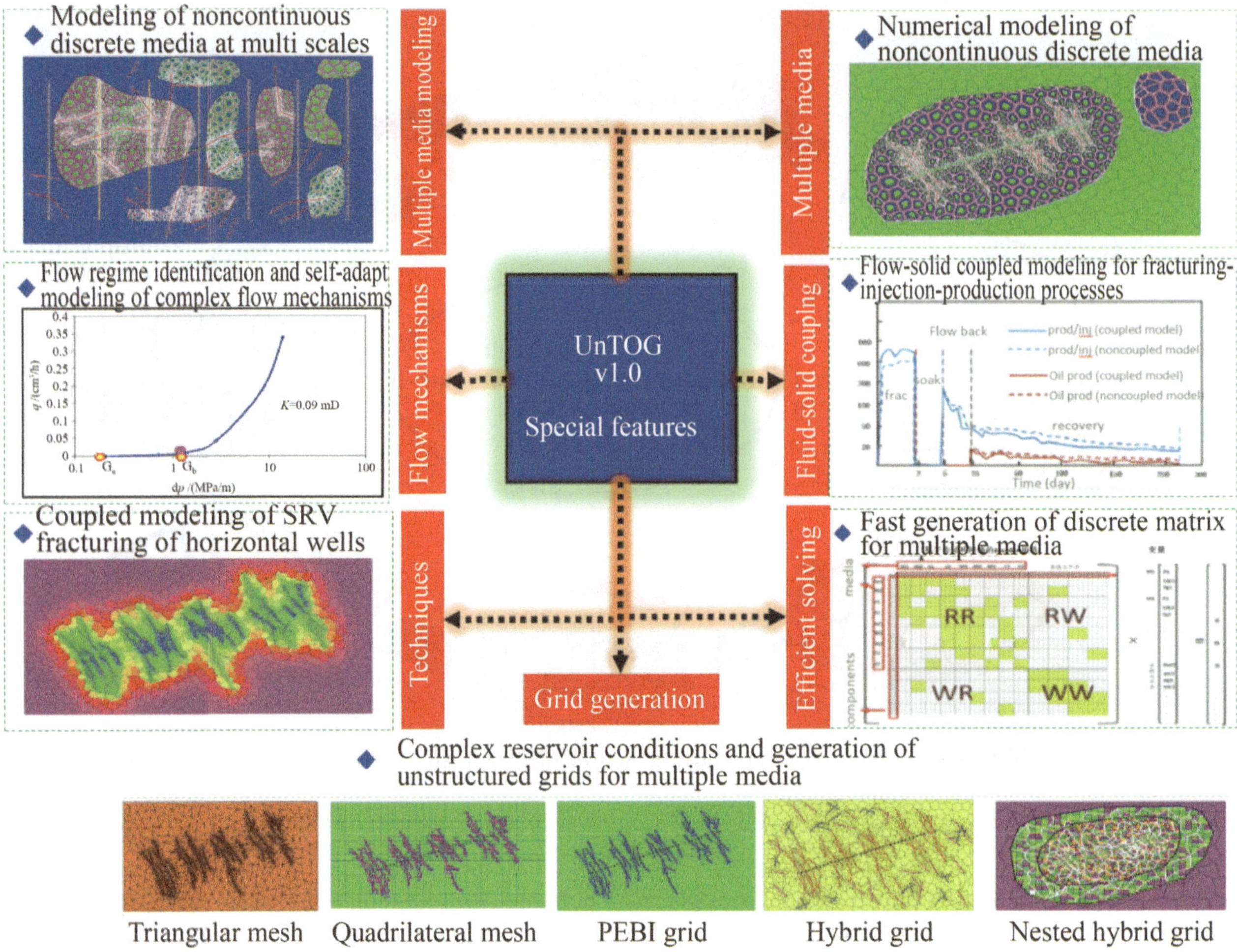

Fig. 11.2 Special features of UnTOG v1.0

Table 11.2 Main applications and development optimization simulation functions of UnTOG v1.0

Application	Development optimization simulation functions
Formation characterization and geological modeling	Characterization of natural fractures, artificial fractures, and pore media at different scales Discrete crack modeling, discrete multiple media modeling, and automatic unstructured grid generation
Horizontal well and hydraulic fracturing	Well location/layer position/well trajectory optimization, horizontal well length and drilling-encounter ratio optimization, well completion mode optimization, classification and perforation optimization Hydraulic fracturing mode optimization, fracturing parameter optimization, artificial fracture distribution and shape simulation
Modeling of production mechanisms	Simulation of pore media and natural fractures at different scales, adaptive simulation of flow behaviors and flow mechanisms, coupled flow-mechanics simulation of multiple media at different scales during fracturing-injection-production processes, dynamic simulation of horizontal well and hydraulic fracturing, dynamic simulation of unstructured grids of different types
Development optimization and dynamic prediction	Well pattern and spacing optimization, dynamic reserve and producing area calculation, optimization of development technique strategy Productivity evaluation, decline curve analysis, and development index prediction

11.2.1 Production Simulation of Pore Media at Different Scales

The differences in geometric scale, physical property and oil-bearing property of pore media at different scales lead to big differences in the availability and recoverability. Meanwhile, the numeric composition (quantity proportion) and spatial distribution of pore media at different scales, and reservoir matrix block sizes and physical properties also have a big influence on the production performance.

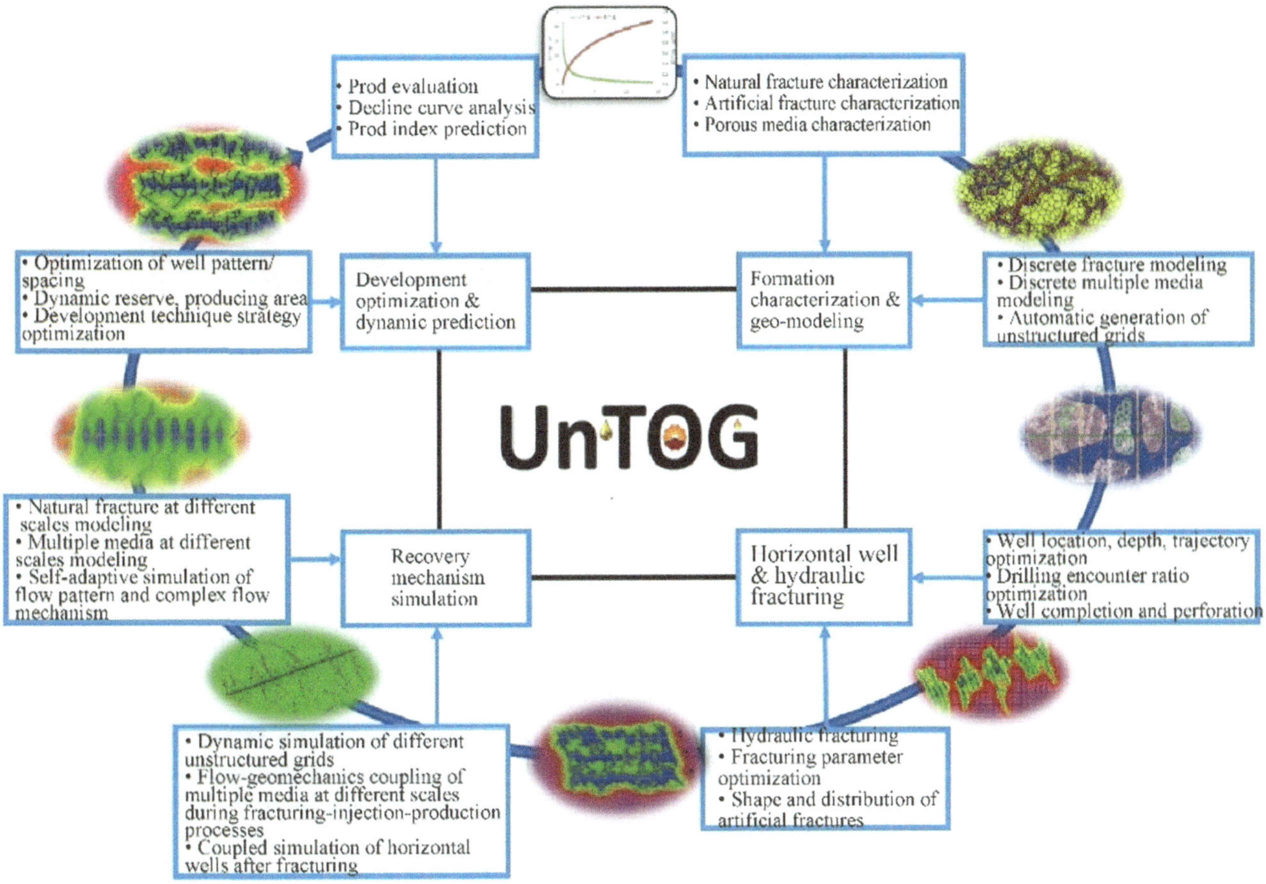

Fig. 11.3 UnTOG application

Table 11.3 Physical properties of pore media at different scales

Type	Porosity/%	Permeability/mD	Oil saturation/%	Residual oil saturation/%	Recoverable oil saturation/%
Micro-pores	15	0.2	90	20	70
Nano/micro-pores	10	0.05	70	30	40
Nano-pores	5	0.01	50	40	10

11.2.1.1 Simulation of the Effect of Numeric Composition of Multiple-Scale Pore Media on Production Performance

Discrete pore media are developed in tight formations. Influenced by its macro and micro heterogeneity, The distributions of pore media at different scales are distinct, and its numeric composition largely has a large effect on the production performance.

To compare the effects of the types and quantities of multiple-scale pore media on the production performance, the simulation of different numeric compositions is carried out. The physical properties of multiple-scale pore media are shown in Table 11.3, and the simulation results are shown in Figs. 11.4, 11.5, 11.6, 11.7 and Table 11.4.

1. Pores at different scales having different oil recoveries

Micro-pores are relatively larger, in which the crude oil is easy to flow and the recovery can be higher than 12%. Nano/micro-pores are relatively smaller, in which the crude oil is more difficult to flow, and the recovery is 2%–5% (Fig. 11.7).

2. Numeric composition of pore media having large effect on recovery

If the pore media are mainly micro-pores (>70%), the overall development is good and the recovery can be higher than 12%. If there are mainly nano/micro-pores (70%), the

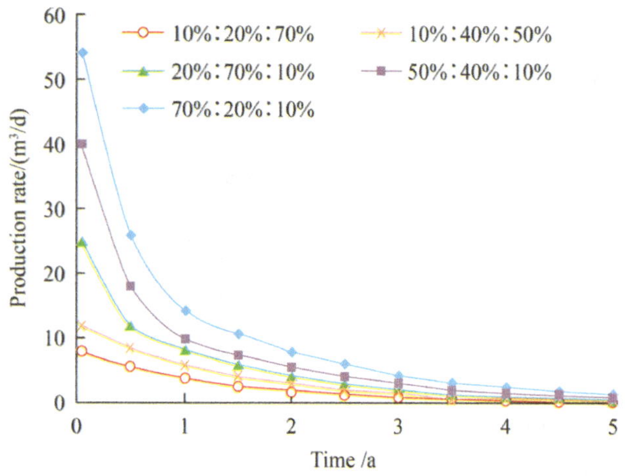

Fig. 11.4 Daily oil production versus numeric composition

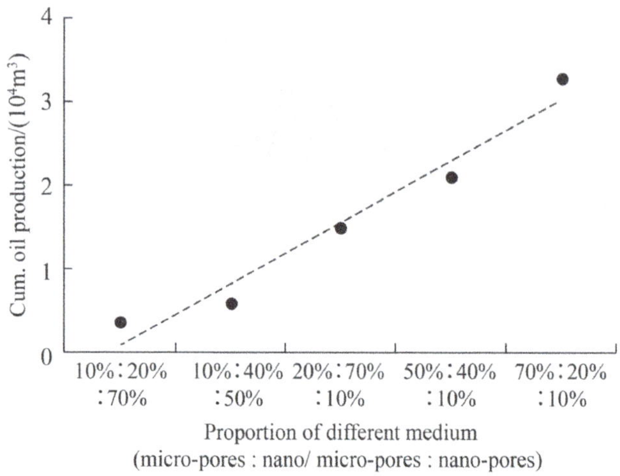

Fig. 11.5 Cumulative oil production versus numeric composition

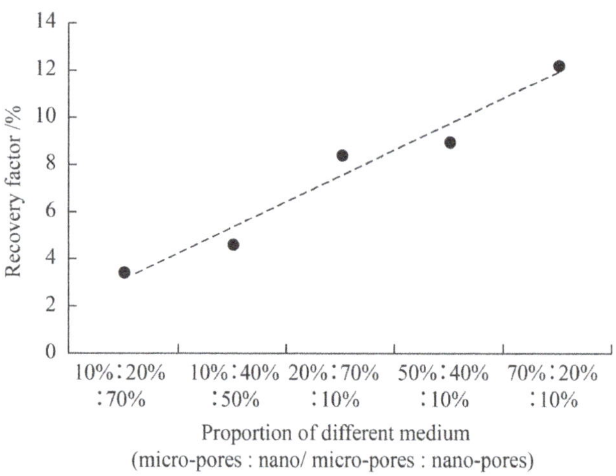

Fig. 11.6 Oil recovery versus numeric composition

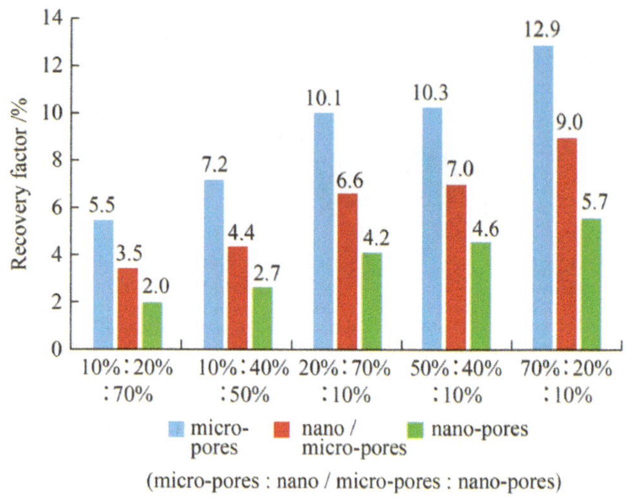

Fig. 11.7 Oil recovery versus pore media types

recovery is about 8%. If there are mainly micro-pores (<70%), the development is the worst, and the recovery can be less than 4%.

11.2.1.2 Simulation of the Effect of Spatial Distribution of Multiple-Scale Pore Media on Production Performance

Under different geological conditions, there are great difference in the spatial distribution characteristics of pore media at different scales, which greatly influences the production performance and recovery.

1. Effect of micro scale distribution of multiple-scale pore media on production performance

Influenced by the lithofacies, lithology, and physical properties of tight reservoirs, the spatial distribution of pore media at different scales varies largely, and it is commonly classified into two categories: relay supply & drainage distribution and interactive distribution. The two distribution patterns can be further subdivided into several types according to different distribution of pore media at different scales.

(1) Simulation of different relay supply & drainage patterns

There lay supply & drainage is one pattern in which pore media of different scales have a zonal distribution in certain proportions within the unit. Depending on the distribution of micro-pores, micro-nanometer pores and nanometer media, it can be subdivided into three categories: i. media size increasing from inside to outside; ii. media size randomly changing; and iii. media size decreasing from inside to outside (Fig. 11.8).

Table 11.4 Reserve and recovery for different pore media

Characteristics	Pore media	OOIP	Remaining reserve	Cumulative production	Recovery (%)
Mainly nano-pores	Micro-pores (10%)	3.23	3.053	0.177	5.48
	Nano/micro-pores (20%)	3.29	3.175	0.115	3.50
	Nano-pores (70%)	4.19	4.107	0.083	1.98
	Sum	10.71	10.335	0.375	3.50
Mainly nano/micro-pores and nano-pores	Micro-pores (10%)	3.19	2.961	0.229	7.18
	Nano/micro-pores (40%)	6.66	6.367	0.293	4.40
	Nano-pores (50%)	2.99	2.9098	0.0802	2.68
	Sum	12.84	12.2378	0.6022	4.69
Mainly nano/micro-pores	Micro-pores (20%)	10.23	9.2	1.03	10.07
	Nano/micro-pores (70%)	6.34	5.92	0.42	6.62
	Nano-pores (10%)	1.13	1.083	0.047	4.16
	Sum	17.7	16.203	1.497	8.46
Mainly nano/micro-pores and nano-pores	Micro-pores (50%)	16.01	14.36	1.65	10.31
	Nano/micro-pores (40%)	6.72	6.25	0.47	6.99
	Nano-pores (10%)	0.59	0.567	0.023	3.90
	Sum	23.32	21.177	2.143	9.19
Mainly micro-pores	Micro-pores (70%)	22.7	19.77	2.93	12.91
	Nano/micro-pores (20%)	3.32	3.03	0.29	8.73
	Nano-pores (10%)	0.57	0.538	0.032	5.61
	Sum	26.59	23.338	3.252	12.23

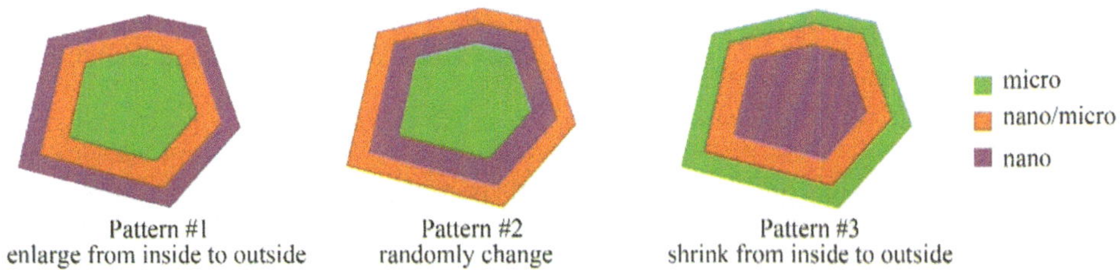

Fig. 11.8 Different distribution pattern within a unit under the relay supply and drainage mode

The simulation results of different relay patterns are shown in Figs. 11.9 and 11.10.

Under the relay supply & drainage mode, fluid flows in a single serial relationship in one unit. Given that the media scale increasing from inside to outside, the medium with the largest scale and the best physical properties is located at outside and recovered first, leading the fluid in the inner media of smaller scale and poor physical properties to flow out in turn; that is, big pores drive small pores, leading to good production and best recovery. Given that the media scale decreasing from inside to outside, the medium with the smallest scale and the worst physical properties is located inside and recovered first, leading the fluid in the outer media to flow in turn; that is, small pores drive the large pores, resulting in poor production and worst recovery. In the case of a random distribution pattern, the production and recovery are the intermediate.

(2) Interactive distribution patterns

Influenced by the geological rules, pore media at different scales are usually distributed in an interactive pattern, which can be classified into four categories (Fig. 11.11) and their effects on the production are investigated given fixed numerical composition.

Given fixed numeric composition of pore media at different scales, under different interactive distribution modes, the contact and flow relation between media are different,

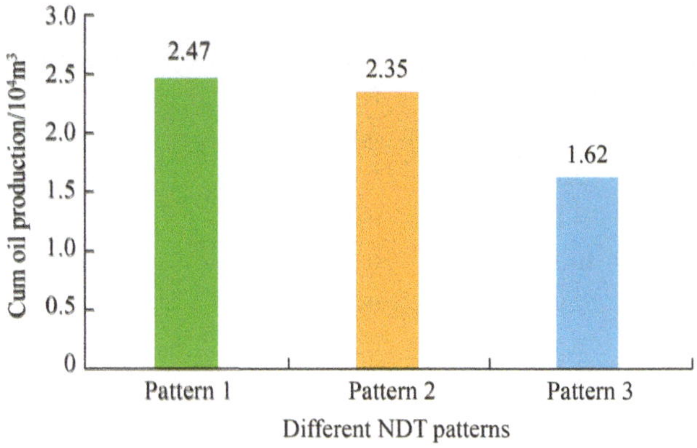

Fig. 11.9 Cumulative production for different NDT patterns

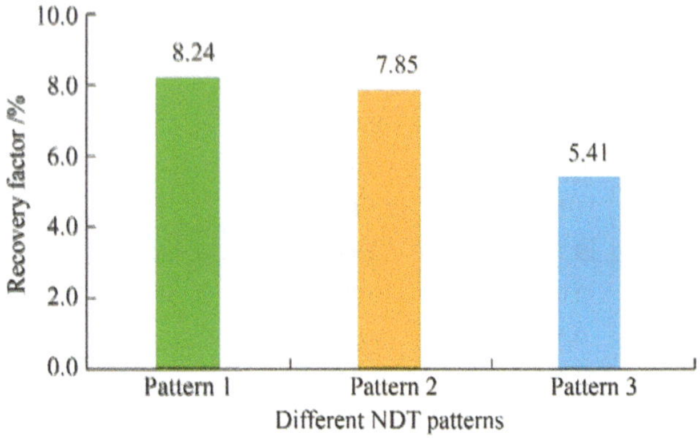

Fig. 11.10 Recovery for different NDT patterns

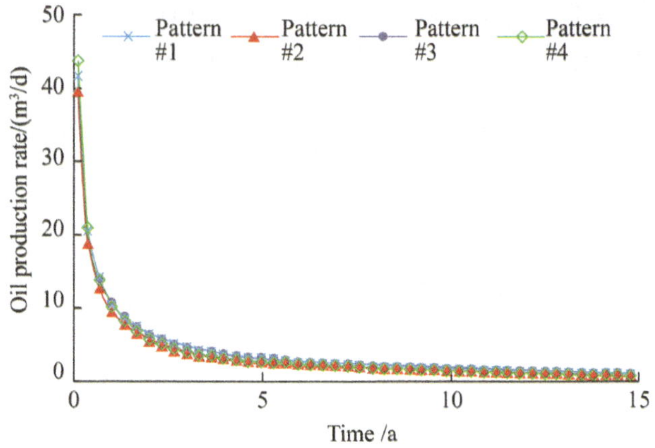

Fig. 11.12 Oil production versus different interactive distribution patterns

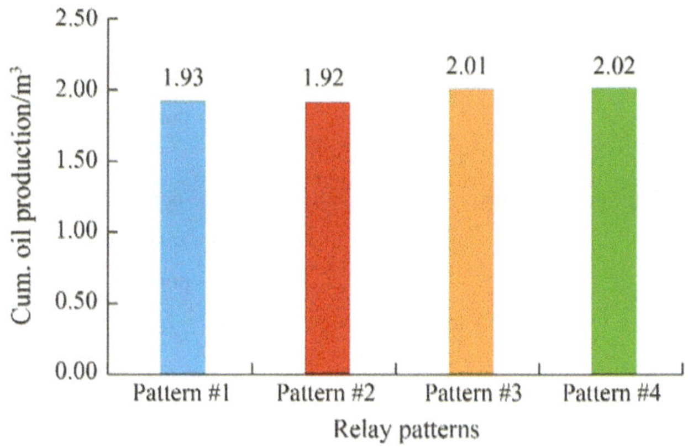

Fig. 11.13 Cumulative production for different interactive patterns

which lead to different orders of production, as well as the production rate and recovery (Figs. 11.12, 11.13 and 11.14).

2. Coupled simulation of macroscopic heterogeneity and microscopic multiple media of tight oil/gas reservoirs

Tight reservoirs have strong macroscopic and microscopic heterogeneity, multiple scales, and multiple media in the distributions of lithology, lithofacies, and formation types. The integration of macroscopic heterogeneity, discrete fracture media, and micro multiple media can achieve the differentiation of macro heterogeneity, coupled modeling of natural/artificial fractures and micro-scale multiple media, and reveal the effects of macro heterogeneity and micro-scale multiple media on the production performance.

According to the distribution of reservoir lithology, a reservoir can be divided into five lithological regions: coarse medium sandstone, medium fine sandstone, fine sandstone, fine powder sandstone and siltstone (Fig. 11.15). The media types corresponding to different lithology and their respective volume percentage are shown in Table 11.5.

According to the macroscopic distribution of reservoir lithology and the corresponding multiplicity and volume fraction of the multiple media, the reservoir is meshed by

Fig. 11.11 Different interactive distribution patterns

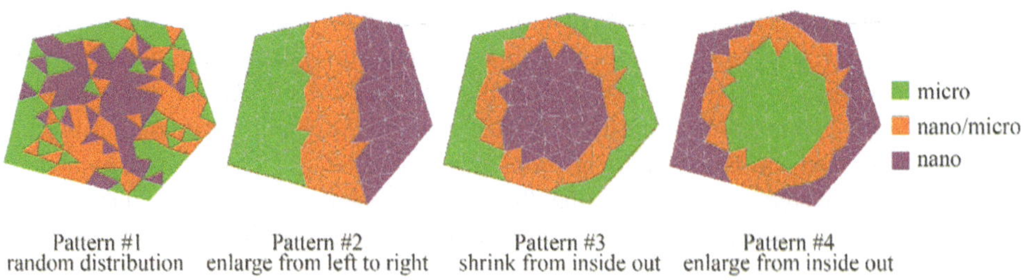

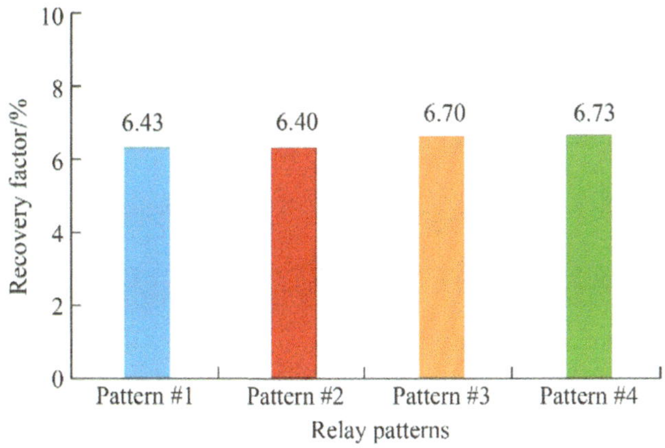

Fig. 11.14 Recovery for different interactive patterns

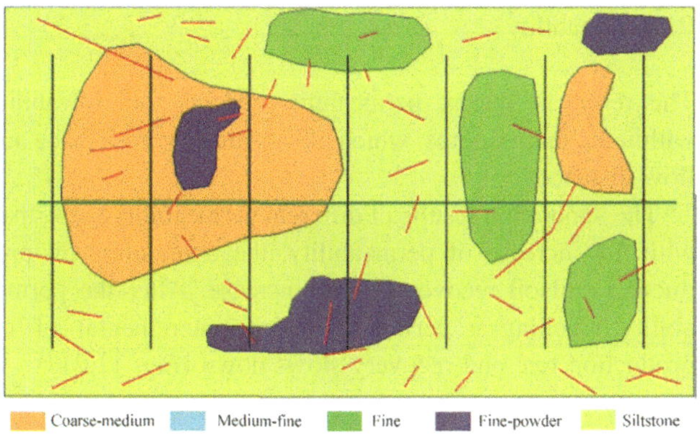

Fig. 11.15 Macroscopic heterogeneity

Table 11.5 Volume fraction of pores at different scales corresponding to different lithology

Lithology	Micro-pores/%	Nano/micro-pores/%	Nano-pores/%
Coarse-medium sandstone	70	30	–
Medium fine sandstone	50	30	20
Fine sandstone	30	50	20
Fine powder sandstone	–	45	55
Siltstone	–	–	100

nested grids based on the relay supply & drainage distribution mode of the multiple media, as shown in Fig. 11.16.

① Controlled by the macroscopic zoning and natural/artificial fractures, the pressure distribution varies greatly in different zones: it has the largest drop near natural/artificial big fractures, where it achieves good recovery. The pressure drop in different lithological zones varies, so is the recovery (Fig. 11.17).

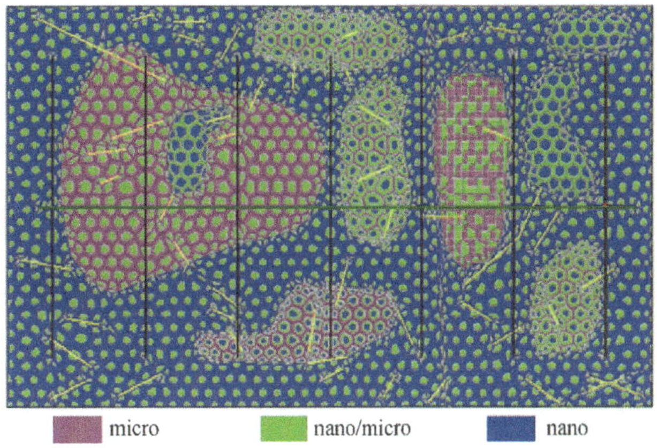

Fig. 11.16 Coupled simulation of macro zoning, natural/hydraulic fractures at different scales, and micro multiple media at different scales

② Influenced by small-scale heterogeneity and multiple media, the pressure change and recovery vary in different characteristic units within the same zone (Fig. 11.18).

③ Influenced by microscopic multiple media, the pressure drop and recovery vary in the same characteristic unit (Fig. 11.19).

In summary, the multiple-scale heterogeneity of tight reservoirs, i.e., the variations in macroscopic lithology, lithofacies, and formation types, and the media types, numeric composition, and distribution of microscopic multiple media, all would make a big difference in the well production performance.

Fig. 11.17 Zonal pressure distribution for macro-micro coupled simulation

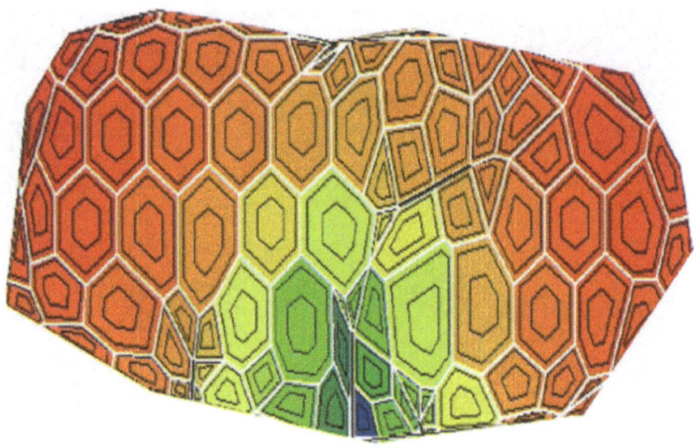

Fig. 11.18 Sub-unit pressure distribution for macro-micro coupled simulation

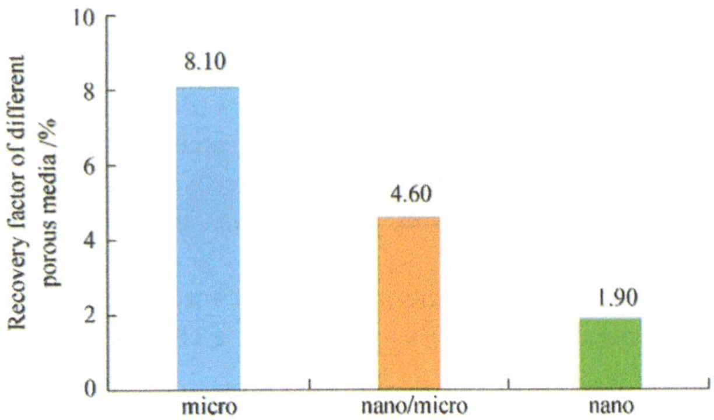

Fig. 11.19 Recovery for pore mediaat different scales

11.2.1.3 Simulation of the Effect of Physical Properties of Reservoir Matrix on Production Performance

The storage capacity, permeability, and oil-bearing capacity of the tight reservoir matrix greatly influence the well performance. This section analyzes the effect of such factors as porosity, permeability, and oil saturation on the production performance.

1. Porosity

Porosity reflects the storage capacity of tight reservoirs, and its value affects the recharge capacity of reservoirs. The effect of different porosities on production performance is simulated.

With the increase in porosity, oil producing increases. The higher the porosity, the stronger the supply capacity of the reservoir and the higher the oil production. When the porosity of matrix is 2%, the cumulative oil production is only 0.64×10^4 m^3. When the porosity of the matrix increases to 10%, the cumulative oil production is 1.43×10^4 m^3, which is enhanced by more than two times (Fig. 11.20).

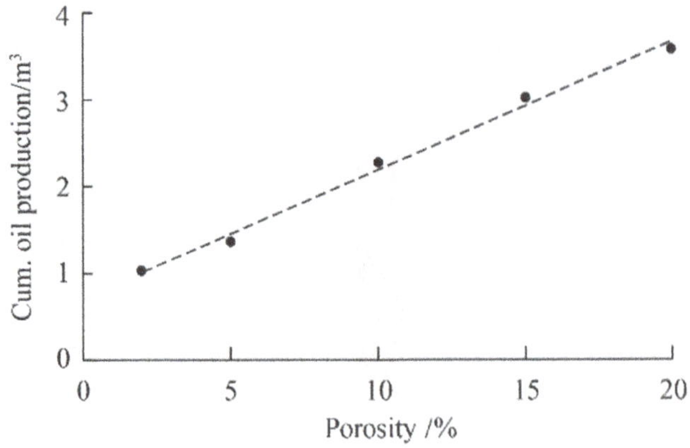

Fig. 11.20 Effect of reservoir matrix porosity on cumulative oil production

2. Permeability

The reservoir matrix permeability reflects the flowability within the matrix pores, which affects the flow resistance and flow distance.

The simulation results of different permeability show that, with the increase of permeability, the cumulative oil production and oil recovery factor increase. After the permeability reaching a certain level, the incremental of oil production rate and recovery slows down (Fig. 11.21).

3. Oil saturation

Oil saturation reflects the reservoir matrix capacity to store oil. The oil saturation in a conventional reservoir has little change above the water-oil contact. But the oil saturation in tight reservoirs depends on the lithology and physical properties, and varies greatly in the spatial distribution, showing very strong heterogeneity.

The simulation results show that with the increase of oil saturation, the oil production increases. As the oil saturation

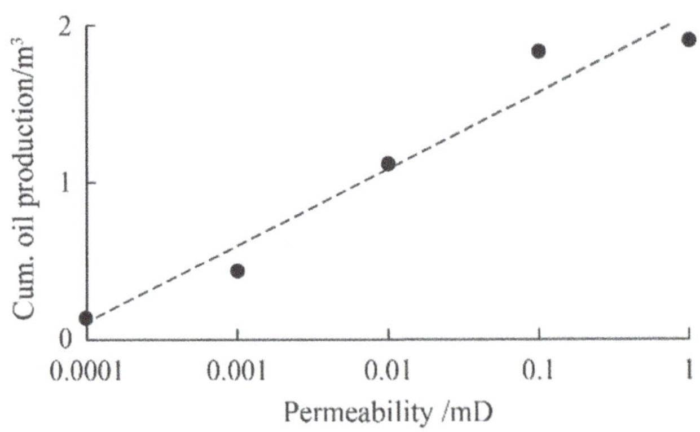

Fig. 11.21 Effect of reservoir matrix permeability on cumulative oil production

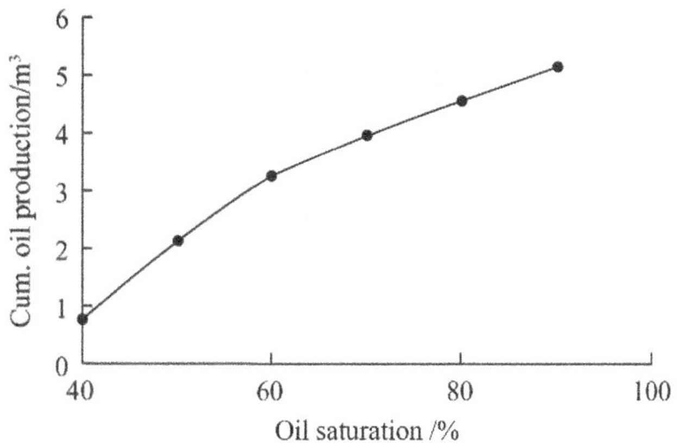

Fig. 11.22 Effect of matrix permeability on cumulative oil production

increases, the recoverable oil saturation increases and the oil-phase mobility is enhanced, so the oil production rate increases. Meanwhile, the increase of oil saturation enlarges the oil reserve, which enhances the supply and the cumulative oil production (Fig. 11.22).

Therefore, the porosity, permeability, and oil saturation of the reservoir matrix all have significant effects on the well production performance.

11.2.1.4 Simulation of the Effect of Matrix Block Size on Production Performance

Natural fractures communicate with artificial fractures in tight reservoirs to form complex fracture networks, cutting a reservoir into fragments of various sizes. The size of matrix blocks determines the contact area between the fractures and the reservoir and the distance from the matrix to the fractures, thus affects the inner recovery of oil of the reservoir matrix. Different sizes of matrix blocks are simulated and the corresponding results are shown in Figs. 11.23, 11.24 and 11.25.

Simulation results show that the size of a matrix block has a significant effect on the recovery: smaller matrix blocks have larger fracture area and larger contact area between

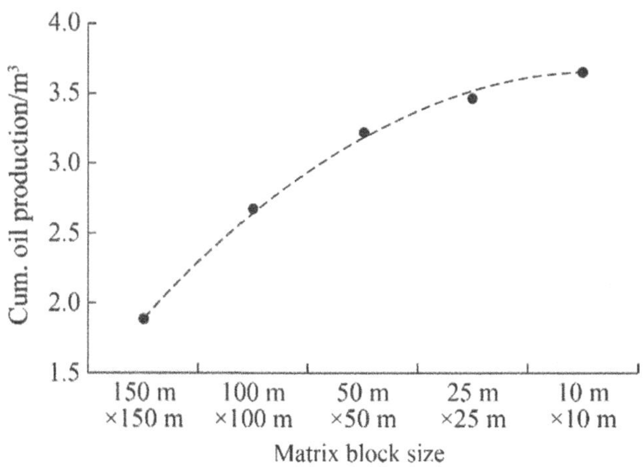

Fig. 11.23 Cumulative oil production for different matrix blocks

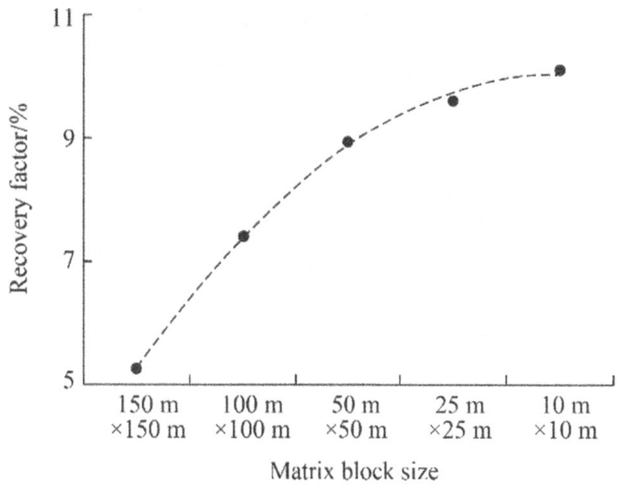

Fig. 11.24 Recovery for different matrix blocks

fractures and the matrix. It indicates that better developed fracture networks result in higher inner recovery and larger production rate and recovery. However, when the block size is equivalent to the effective flow distance of the matrix, the matrix supply capacity has reached its maximum level, and the production rate would not increase greatly with the decrease of the matrix block size, the production enhancement would be inconspicuous.

Thus, the smaller the rock size within a certain range, the higher the oil production but there should exist an optimal value.

11.2.2 Simulation of Fluid Properties and Flow Characteristics of Pore Media at Multiple Scales

Under reservoir temperatures and pressures, the fluid composition varies greatly in pore media at different scales in tight reservoirs, resulting in distinct fluid properties such as viscosity, the solution gas-oil ratio, and high-pressure high-temperature PVT properties. Besides, different sizes of pore throats lead to different capillary forces, as well as the flow characteristics of oil/gas/water, which further affects the flowability, recovery and production performance of these fluids in multiple-scale pore media.

11.2.2.1 Simulation of the Effect of Different Fluid Properties on Production Performance

1. Viscosity

Viscosity influences the fluid flowability. Since tight reservoirs have low permeability, the variation in viscosity can

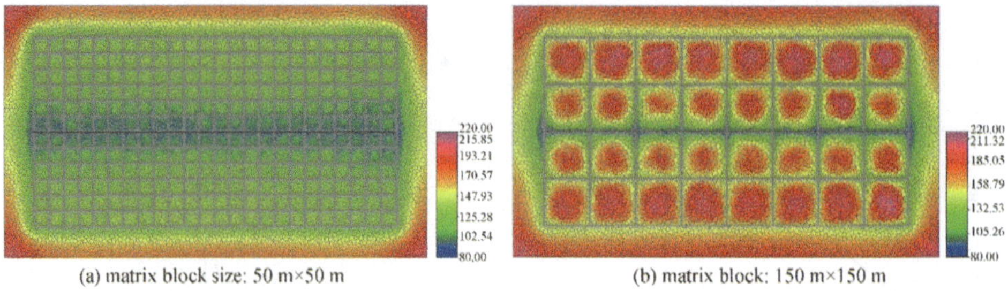

Fig. 11.25 Pressure distribution after 5-year production for different matrix block sizes

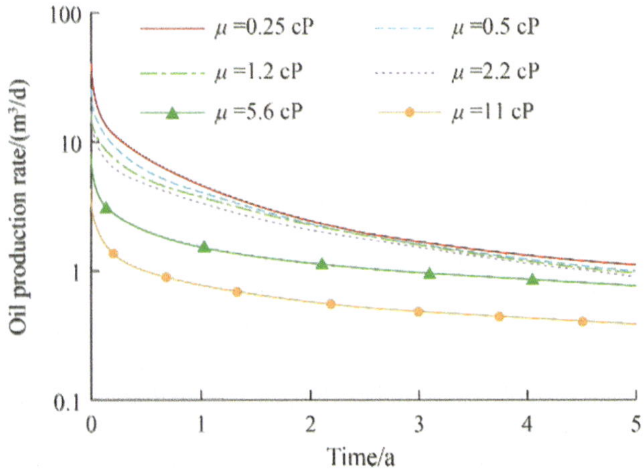

Fig. 11.26 Daily production versus viscosity

cause a great change in mobility. Thus, viscosity of fluid has a great impact on its production rate and recovery.

Different viscosities of tight oil are simulated and the results show that with the increase of viscosity, the initial production declines, the decline rate accelerates with time, and the cumulative production and recovery drops (Figs. 11.26, 11.27 and 11.28). When the oil viscosity is reduced from 11 cP to 1.2 cP, single well's cumulative production and recovery can be improved by more than 3

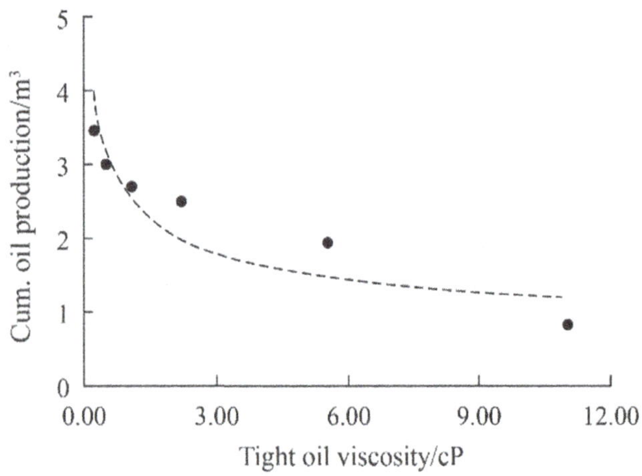

Fig. 11.27 Cumulative production versus viscosity

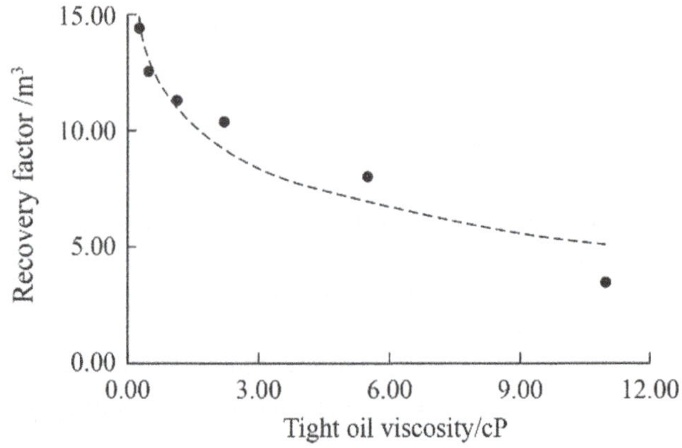

Fig. 11.28 Recovery versus viscosity

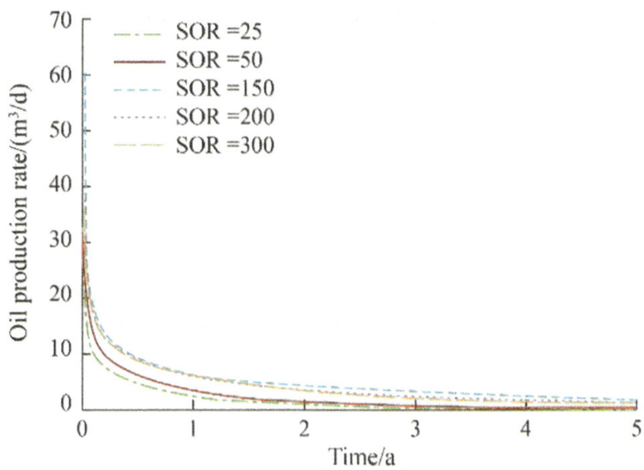

Fig. 11.29 Daily production for different solvent gas-oil ratio

times, and the recovery can be improved from 2.16% to 7.06%, which is nearly 5% higher.

Therefore, viscosity is an important factor in tight oil production and recovery.

2. Solution gas-oil ratio

The solution gas-oil ratio affects the energy of the solution gas drive and the degassed oil viscosity. The solution gas-oil ratio varies greatly with different types of tight oil. In Xinjiang, Changqing, Songliao, and Sichuan oilfields, the solution-gas oil ratio ranges in 20–150 m^3/m^3.

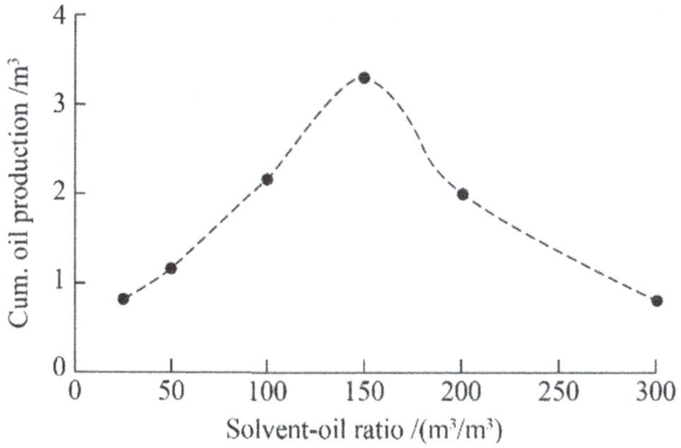

Fig. 11.30 Cumulative oil production for different solvent gas-oil ratio

Simulation results show that both oil production and recovery increase when the solution gas-oil ratio increases within a certain range. when the solution gas-oil ratio is too large, oil production drops with the increase of solution gas-oil ratio. When the solution gas-oil ratio increases within a certain range, the increased solution gas can supplement the formation energy, leading to greater production rate and recovery. When the solution gas-oil ratio is too large, the amount of degassing increases significantly, the oil viscosity rises dramatically and its mobility decline, leading to reduced production rate and recovery (Figs. 11.29, 11.30 and 11.31).

It can be seen that the solution gas-oil ratio has a great influence on production performance. It is necessary to consider its comprehensive effects on the formation energy and fluid mobility comprehensively.

3. Pressure coefficient

A pressure coefficient reflects the magnitude of the formation energy. It affects the recovery of primary depletion production and the subsequent strategy of energy supplementation. The pressure coefficient of tight reservoirs varies with different regions due to generation and trapping conditions. They are typically 1.3–1.5 in such representative ultra-high pressure tight reservoirs as Bakken and Eagle Ford in North America, and 0.75–0.85 in such low-pressure tight reservoirs as the Chang 7 reservoir in Ordos Basin.

Simulation results of different pressure coefficients show that the higher the pressure coefficient, the more sufficient the formation energy, the higher the initial production rate, and the higher the cumulative oil production and recovery (Figs. 11.32, 11.33, 11.34 and 11.35). When the pressure coefficient increases from 0.7 to 1.8, the cumulative oil production increases from 1.57×10^4 m^3 to 2.54×10^4 m^3, and the recovery increases from 3.8% to 6.7%, nearly doubled.

It can be seen that the value of formation pressure coefficient directly affects the oil recovery. Effective supplementation of formation energy is the key to improve the production performance of low-pressure tight oil reservoirs.

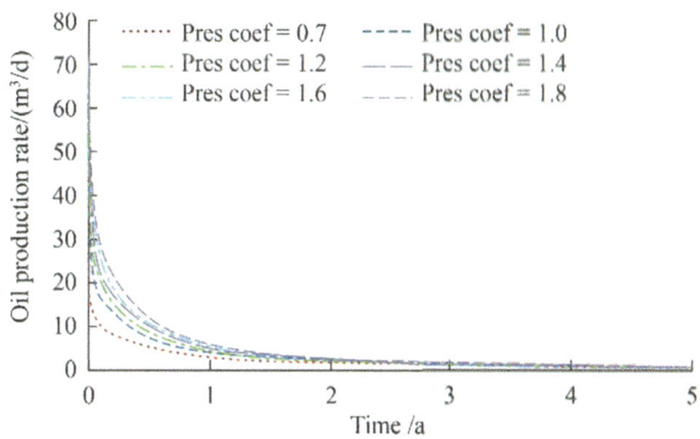

Fig. 11.32 Daily production versus pressure coefficient

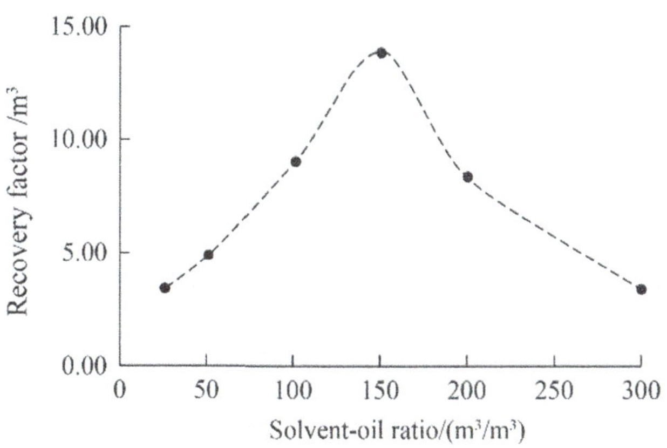

Fig. 11.31 Recovery factor for different solvent gas-oil ratio

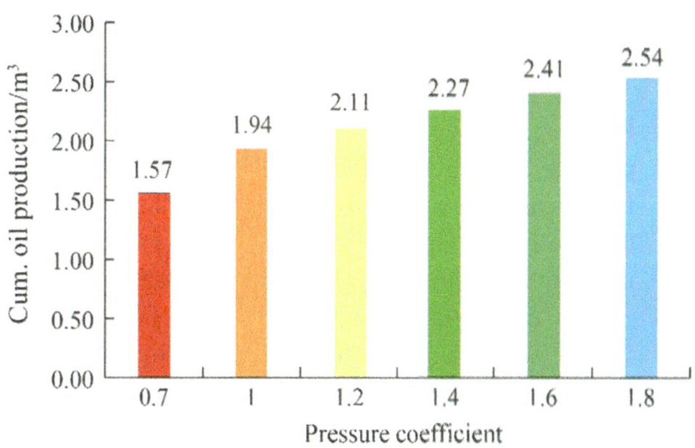

Fig. 11.33 Cumulative production versus pressure coefficient

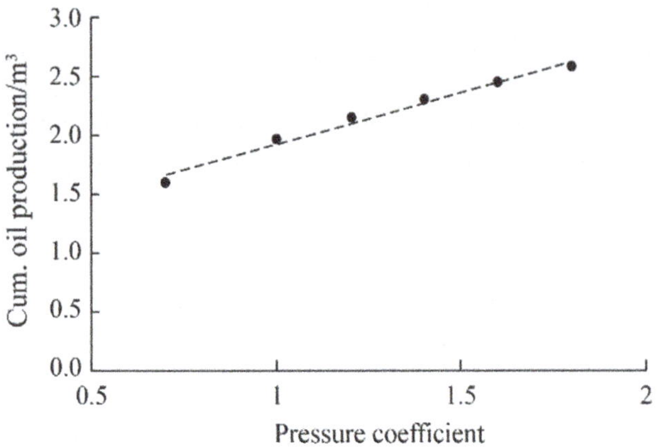

Fig. 11.34 Cumulative production versus pressure coefficient

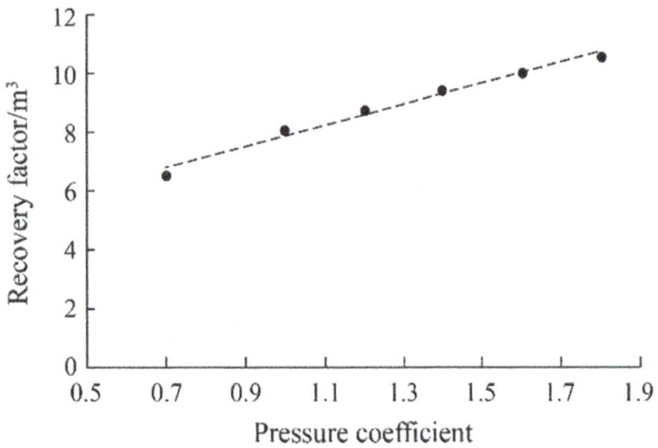

Fig. 11.35 Recovery versus pressure coefficient

11.2.2.2 Simulation of Phase Change in Multiple-Scale Pore Media

Under reservoir temperatures and pressures, fluid compositions in pores at different scales vary greatly, leading to various solution gas-oil ratios, saturation pressure, and PVT properties in multiple-scale media.

Three pore media (micro-pores, nano/micro-pores, and nano-pores) are simulated in two cases (with the same and different PVT properties). Results show that:

① In both cases, the cumulative production and formation pressure are slightly different (Figs. 11.36 and 11.37).

② Degassing time and degassing rate are rather different in the two cases (Fig. 11.38); the recovery varies for pore media at different scales (Fig. 11.39). With the decrease of pressure, crude oil in the large-scale micro-pores is first degassed, and the fluid is transformed from two phases to three phases, which is the first and most easily to be produced. Therefore, the recovery increases.

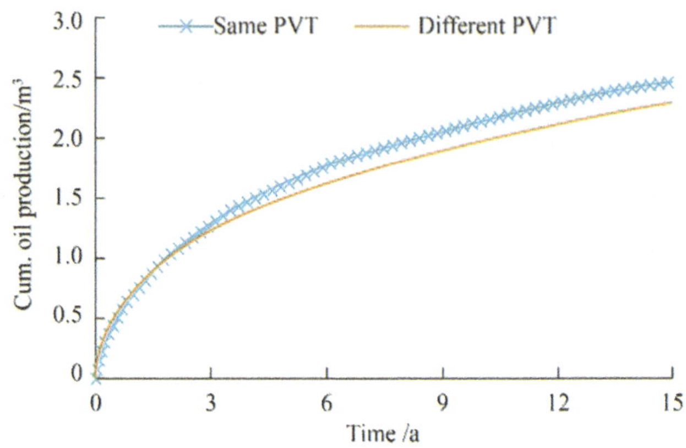

Fig. 11.36 Cumulative production for same and different PVT properties

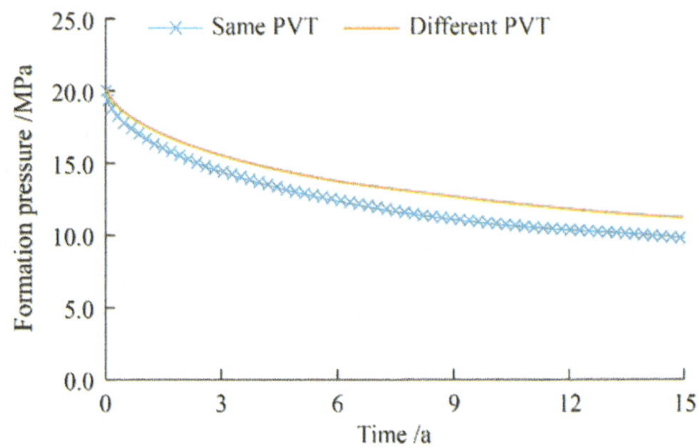

Fig. 11.37 Reservoir pressure for same and different PVT properties

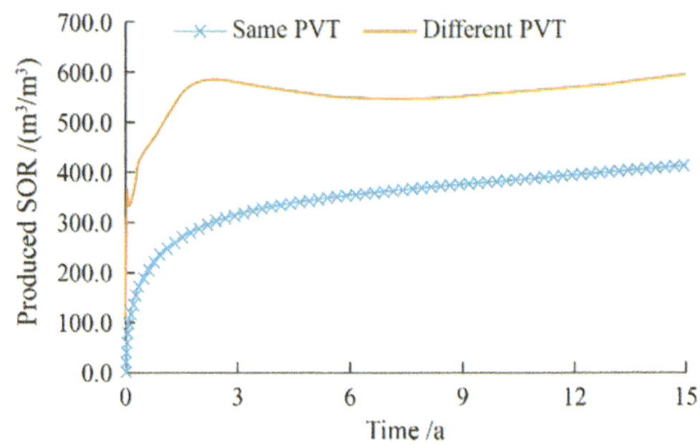

Fig. 11.38 Produced SOR for same and different PVT properties

It can be seen that the PVT properties in different-scale pore media are different, which affect the mobility and recovery in pore media at different scales. The differences in PVT properties of different-scale pore media should be fully considered to more accurately predict the production and clarify their respective contributions.

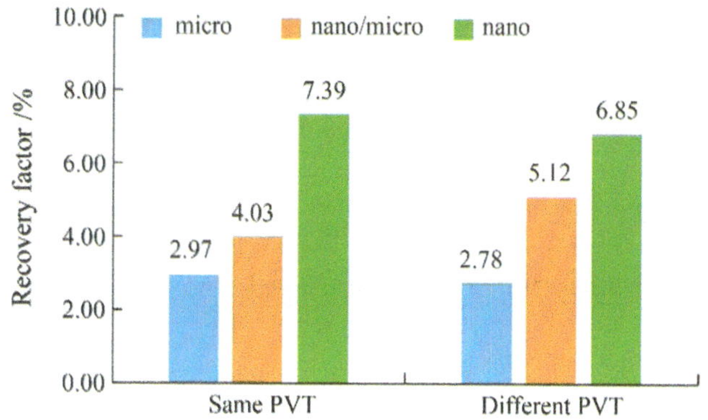

Fig. 11.39 Recovery factors for the two cases from different pore media

11.2.2.3 Simulation of Flow Characteristics in Multiple-Scale Pore Media

The oil bearing and occurrence in the pore media at different scale due to lithology and physical properties. The large-medium pores are mainly filled with flowable oil, and the micro- and nano-pores mainly contain capillary oil that exist mostly in membrane and absorbed oil. Pore media vary dramatically in tight reservoirs, and the flow characteristics largely depend on the geometric scale of the pores. Compared with conventional reservoirs, pore throats in tight reservoirs have larger variation at different scales, and so is the capillary forces (Wu and press 1988).

Two scenarios are simulated: i. different flow parameters (phase and capillary forces) and saturation parameters are used for pore media at different scales; ii. averaged flow and saturation parameters are used for all media. The relative permeability and capillary force curves for different media are shown in Figs. 11.40, 11.41, 11.42 and 11.43.

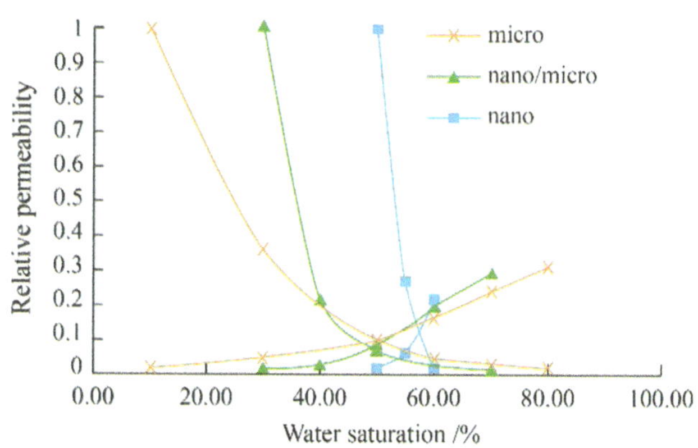

Fig. 11.40 Relative permeability curves

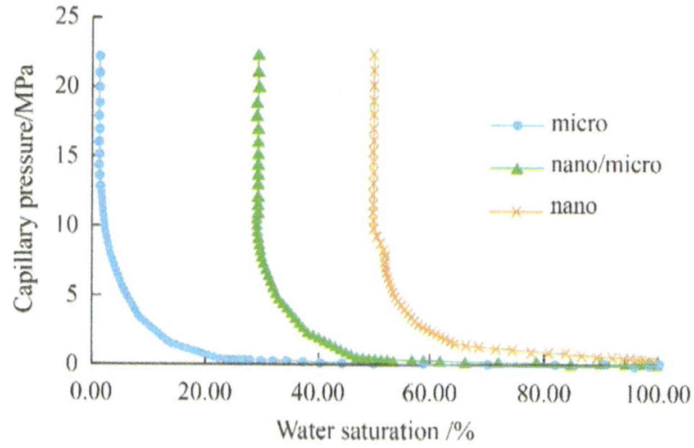

Fig. 11.41 Capillary force curves

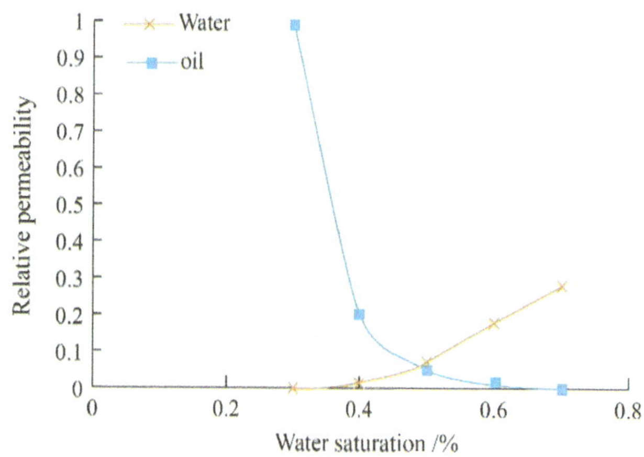

Fig. 11.42 Averaged relative permeability

Results show that the difference in the recoveries increases for using different relative permeability and capillary force curves (Fig. 11.44 and Table 11.6).

Compared with those with averaged relative permeability, the first scenario can more accurately describe the differences in oil-bearing properties and recovery at different scales: micro-pores bear more oil with higher recoverable oil saturation, resulting in higher recovery. Nano/micro-pores bear less oil with lower oil recoverable oil saturation, resulting in smaller recovery. Therefore, different relative permeability curves for different pore media can highlight the differences in the oil recovery. In contrast, using the same relative permeability and capillary force for all media weakens the differences between different pore media.

So, there are large differences in the fluid occurrence and flow parameters (relative permeability and capillary force) in pore media at different scales, resulting in large differences in their respective availabilities and recoveries.

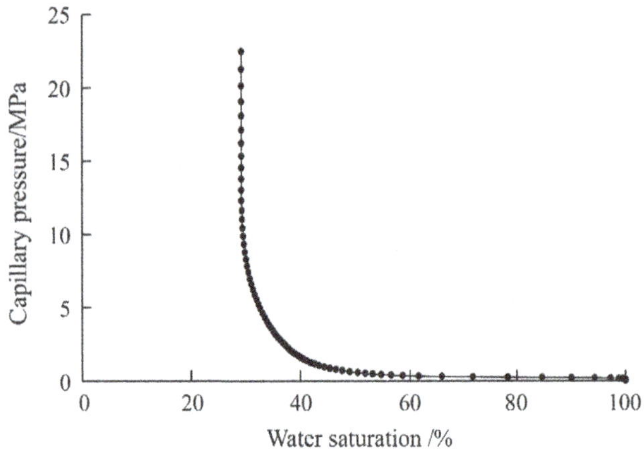

Fig. 11.43 Averaged capillary force curves

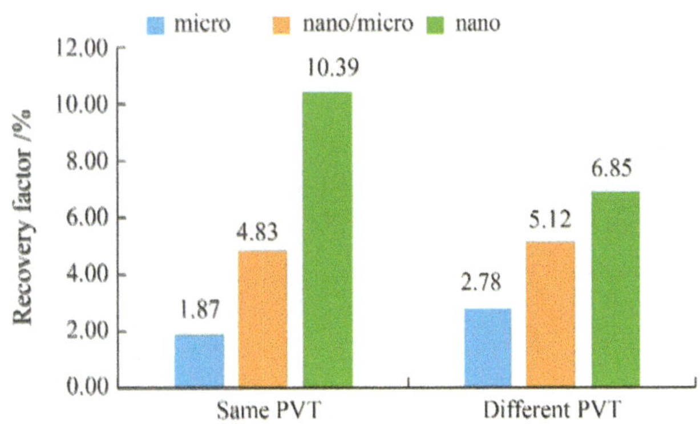

Fig. 11.44 Recovery for different pore media

11.2.3 Simulation of Multiple-Scale Natural Fracture Media

The size, number, spatial distribution, connectivity, and natural fracture complexity in tight reservoirs affect the fluid flow in the matrix block, and exerts great effects on the recovery of matrix block, reservoir dynamics and productivity.

11.2.3.1 Effect of Multiple-Scale Natural Fracture Media on Production Performance

The geometric characteristics and properties of natural fracture media at different scales are different, exerts great effects on the recovery and production performance of matrix block. Natural fracture media of different scales are combined in various ways, and their respective effects and contributions to the fluid production are compared and analyzed.

Simulation results show that:

① The combination of natural fractures of different scales has large effects on daily production, cumulative production, and recovery factor.

The initial productions are close since they are mainly controlled by artificial fractures. Under different natural fracture combinations, their respective cumulative productions and recoveries are rather different. Their cumulative productions and recoveries are relatively lower given that only large-media or small-micro fissures exist. Their cumulative productions and recoveries reach the highest level when large-media and small-micro fissures co-exist (Figs. 11.45, 11.46, 11.47 and 11.48).

② Fractures of different scales play different roles in the production.

Large and medium fractures have far extensions, strong conductivity and a large control area for single fracture, which affect the initial production. Small and micro fractures have short extensions and a limited control area, and communicates large-medium fracture and reservoir matrix, which affects the recovery of reservoir matrix. The fractures of different scales are effectively coupled, the better the communication of fracture network, the higher the production performance (Fig. 11.49).

It can be seen that when the artificial/natural fractures at different scales are coupled with the reservoir matrix, more

Table 11.6 Results for the same and different relative permeability curves for three pore media

	Pore media	OOIP/10^4 m^3	Remaining reserve/10^4 m^3	Cumulative production/10^4 m^3	Recovery (%)
Same relative permeability for all media	Nano-pores	5.10	4.96	0.14	2.78
	Nano/micro-pores	13.80	13.09	0.71	5.12
	Micro-pores	26.90	25.06	1.84	6.85
different relative permeability for different media	Nano-pores	5.10	5.00	0.10	1.87
	Nano/micro-pores	13.80	13.13	0.67	4.83
	Micro-pores	26.90	24.11	2.79	10.39

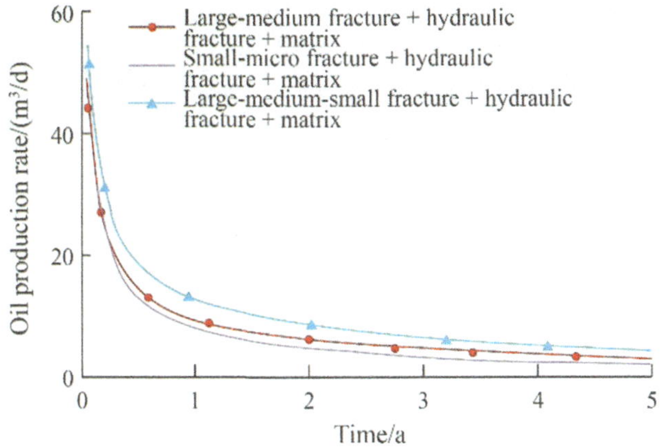

Fig. 11.45 Daily production versus multiple-scale natural fracture combination

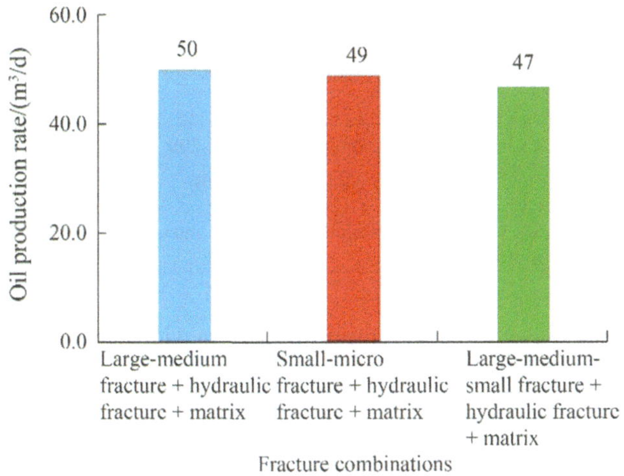

Fig. 11.46 Primary production versus multiple-scale natural fracture combination

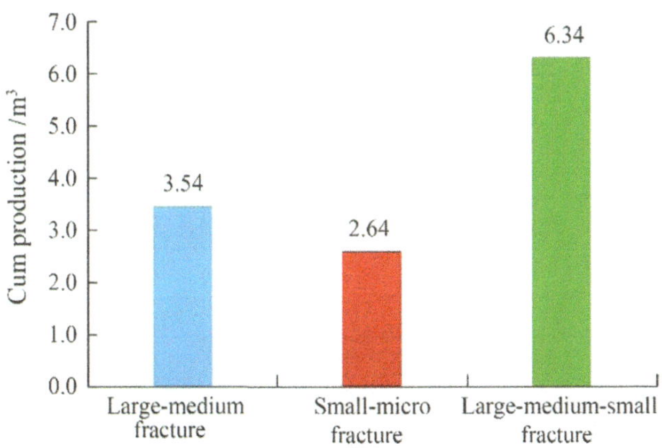

Fig. 11.47 Cumulative production versus multiple-scale natural fracture combination

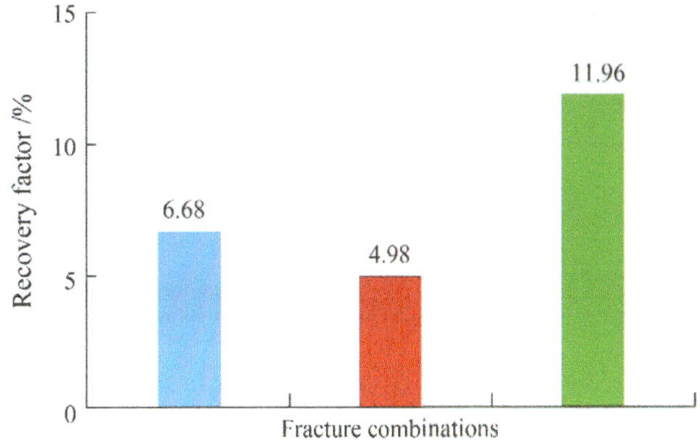

Fig. 11.48 Recovery factor versus multiple-scale natural fracture combination

oil can be really touched and the production can be evidently enhanced, and the single-well production rate can be improved.

11.2.3.2 Effect of Spatial Distribution Pattern of Natural Fractures on Production Performance

The spatial distribution of natural fractures mainly refers to the density and communication of fractures. The density (quantity) of fractures and the communication among them (intersection points) reflect the development and complexity of a fracture network (Karimi-Fard et al. 2003), and determine the size of matrix blocks, and affects the recovery and production performance there of.

① Natural fractures have the same intersection points and different fracture numbers.

Several cases with the same number of intersections but different numbers of fractures are simulated. Results show that the more fractures, the better development of fracture network, the stronger the communication, and the higher the production and recovery. But when the fracture number grows over a certain level, the matrix block has been intersected sufficiently, and the production enhancement weakens for the increase of fracture number (Figs. 11.50, 11.51 and 11.52).

② Natural fractures have different intersection points and same fracture numbers.

Cases with the same number of fractures but different number of intersections are simulated. Results show that given a fixed number of fractures, the more intersection points, the more complex a fracture network, the higher the production and recovery. But with the increase of fractures,

Fig. 11.49 Pressure distribution for different natural fracture combinations after 15-year production

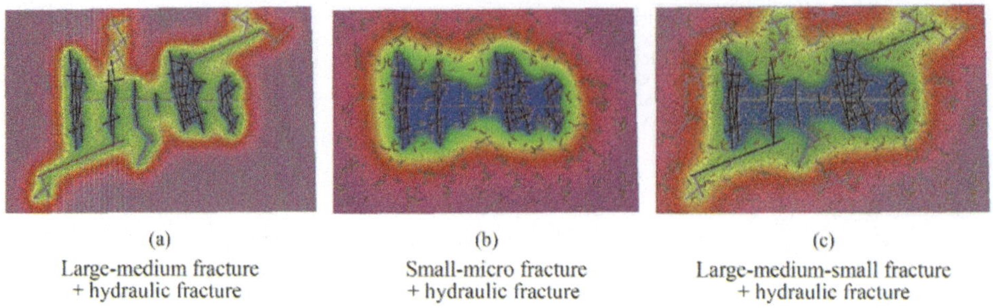

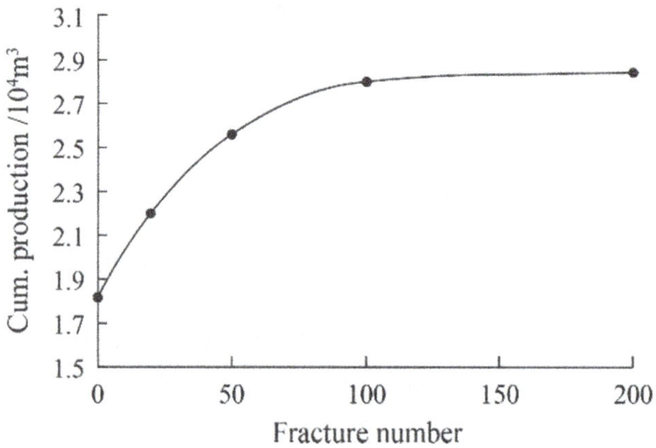

Fig. 11.50 Cumulative production for different fracture numbers

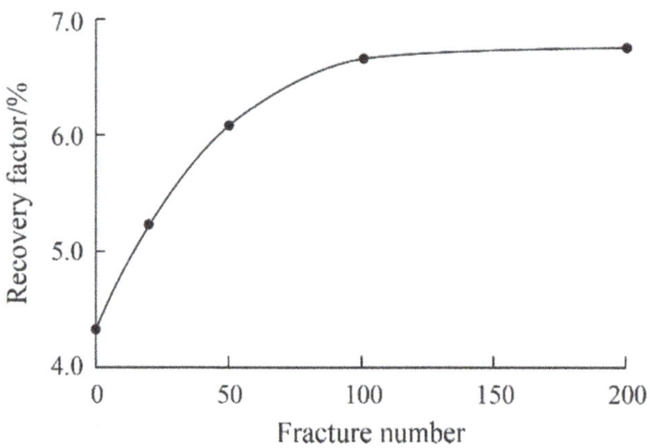

Fig. 11.51 Recovery for different fracture numbers

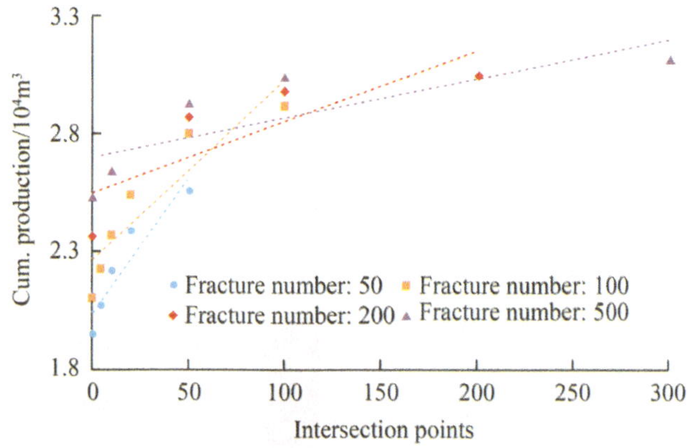

Fig. 11.53 Cumulative production versus intersection points for different fracture numbers

the production enhancement slows down with the increase of intersection points (Figs. 11.53 and 11.54).

③ Natural fractures have different intersection points and different fracture numbers.

When the number of fractures and the number of intersections both increase, the complexity of a fracture network doubles, and its ability to communicate the reservoir matrix becomes stronger. So the production rate and recovery incremental become more evident. When the number of fractures and the number of intersections both increase over

Fig. 11.52 Pressure distribution for different fracture numbers after 15-year production

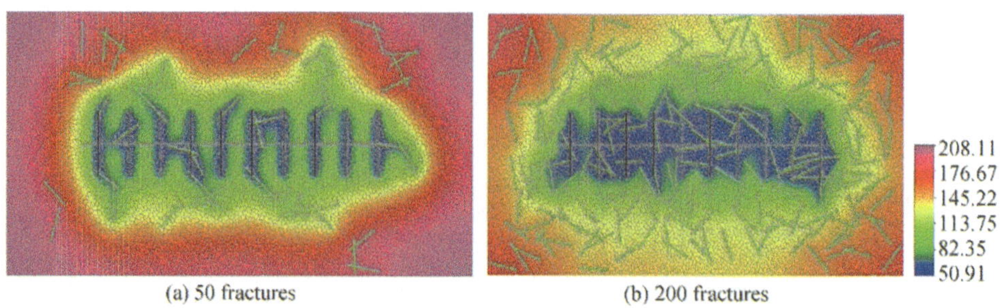

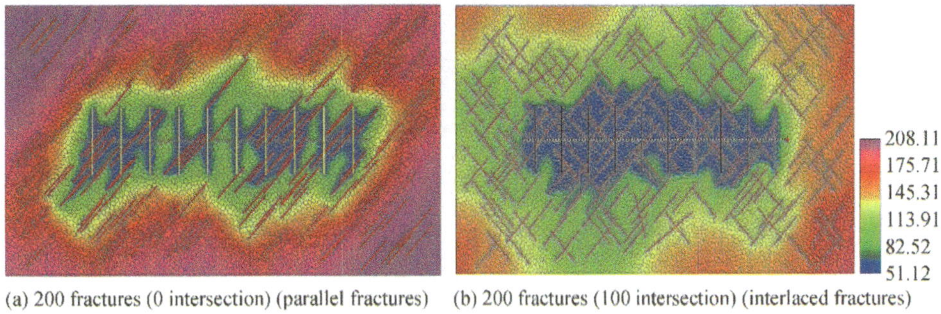

Fig. 11.54 Pressure distribution for same fracture number and different intersection points after 15-year production

certain numbers, the production enhancement weakens (Figs. 11.55, 11.56 and 11.57).

In summary, for denser fractures and better communication, the reservoir matrix can be better touched, resulting in higher production and recovery. The more complex the fracture network, the smaller the matrix block can be intersected, and the shorter the flow distance, the better the production.

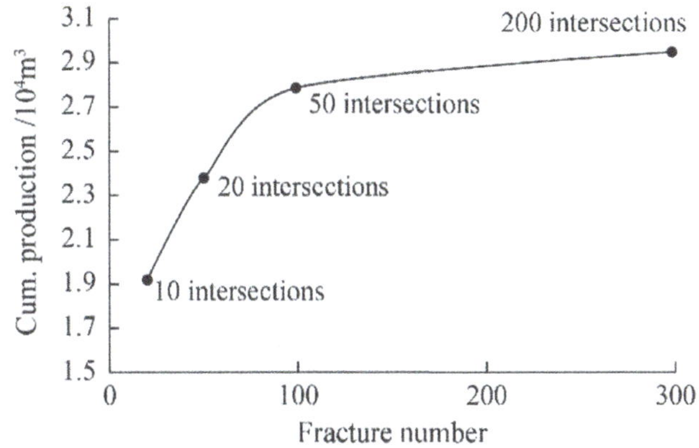

Fig. 11.55 Cumulative production for different fracture numbers and different intersections

11.2.4 Self-adaptive Simulation of Flow Behaviors and Complex Flow Mechanisms in Multiple Media

The geometric characteristics, physical properties, and flow behaviors of multiple-scale multiple media in tight reservoir are rather distinct, which lead to different fluid behaviors and flow mechanisms. Fluid behaviors and flow mechanisms vary greatly at different production stages in the same medium, which influence flow of multiple media and the reservoir dynamics, as well as the productivity.

11.2.4.1 Simulation of Tight Oil Flow Behaviors and Nonlinear Flow Mechanisms

1. Identification of flow behaviors and self-adaptive simulation of complex flow mechanisms

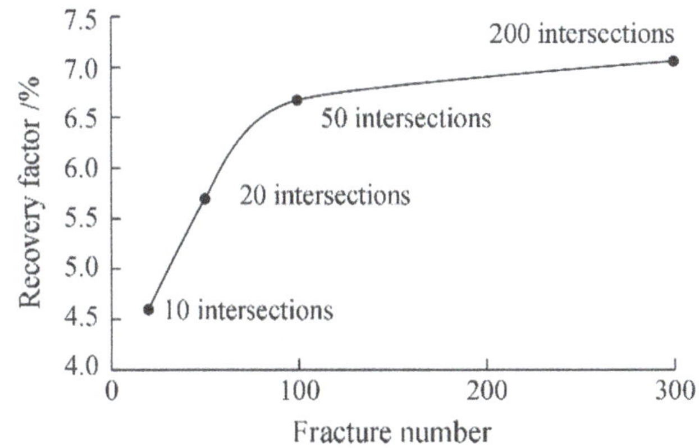

Fig. 11.56 Recovery for different fracture numbers and different intersections

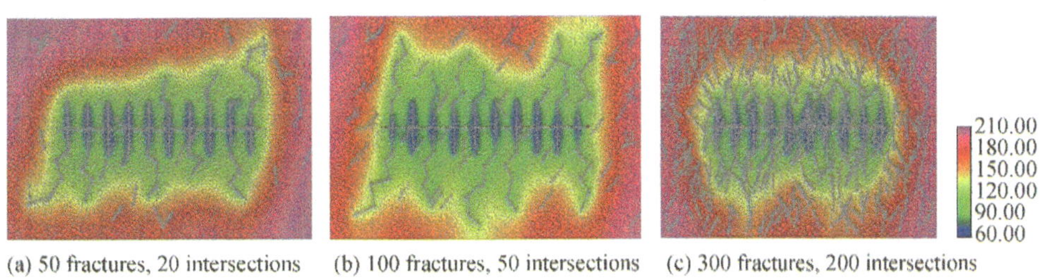

Fig. 11.57 Pressure distribution for different fracture numbers and different intersection points after 15-year production

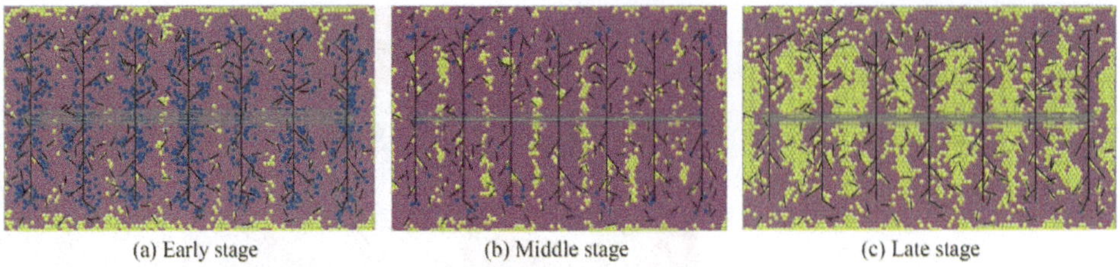

Fig. 11.58 Fluid flow throughout different production stages

Throughout the real production process, due to the change of production conditions at different stages, the flow behaviors and mechanisms dynamically change in different media. Normally, the flow behaviors can be classified into four categories: high-speed nonlinear flow, linear flow, pseudo-linear flow, and low-speed nonlinear flow. Different flow behaviors correspond to different reservoir dynamics and productivity. Therefore, the flow behaviors can be automatically identified through critical flow parameters and identification criteria, based on which the dynamic equations that are suitable for the flow behaviors and flow mechanisms can be selected to perform the dynamic simulation, and thus reveals the transition of flow behaviors and flow mechanisms of multiple media and their effects on the production performance.

(1) Simulation of flow behavior transition of multiple-scale multiple media through different production stages

1) Spatial flow behaviors of multiple-scale multiple media

The simulation of self-adaptive identification of flow behaviors and flow behavior transition show that in the same production stage, flow behaviors are significantly different in different locations and different multiple-scale multiple media. In different production stages, flow behaviors are significantly different in different multiple-scale multiple media. In the early stage (Fig. 11.58a), near the large-medium artificial/natural fractures, there is a large pressure gradient and fast flow mainly in the high-speed non-Darcy flow behavior. The flow is mainly pseudo-linear in pore media and low-speed nonlinear flow far from wells and fractures. In the middle stage (Fig. 11.58b), the differential pressure becomes smaller and the flow slows down. Most of the high-speed non-Darcy flow changes to a pseudo-linear flow, and the pseudo-linear flow transitions to a low-speed nonlinear flow. In the late stage (Fig. 11.58c), it is mainly quasi-linear flow and low-speed nonlinear flow.

2) Temporal flow behaviors of multiple-scale multiple media

The transitions of temporal flow behaviors are analyzed through self-adaptive identification and dynamic simulation.

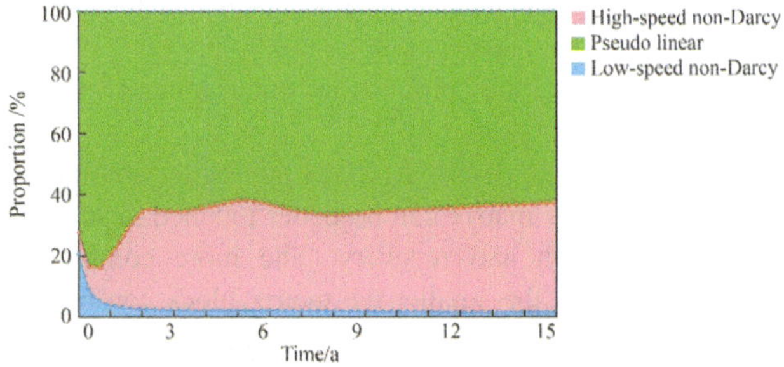

Fig. 11.59 Flow Change of flow behaviors with time

It is found that during the production process, the pore pressure and pressure gradient of multiple-scale multiple media vary extremely, which leads the dramatic change in the flow behaviors. Different flow behaviors change drastically at different production stages, and their respective contributions to the production vary with time (Fig. 11.59). As shown in the figure, in the early stage, the high-speed non-Darcy flow behavior has a larger contribution. As time goes, the pressure difference diminishes and the flow slows down. Part of the high-speed non-Darcy flow shifts to other behaviors. As an important part in the early stage, pseudo-linear flow changes to low-speed non-Darcy flow later on. As a small part in the early stage, due to the transformation from other flow regimes, low-speed non-Darcy flow becomes important gradually in the late stage. The pseudo-linear and low-speed non-linear flow dominant in the middle and late stages.

The flow behaviors in pore media and fracture media are similar to the overall flow behavior of the multiple media. But in the fracture media, the high-speed nonlinear flow and quasi-linear flow account for relatively larger proportions, and the low-speed nonlinear flow account for relatively smaller proportions (Fig. 11.60).

(2) Simulation with different flow-behavior identification parameters and identification criteria

For the macroscopic flow characteristics of tight oil, the pressure gradient or Reynolds number can be used to identify the flow behaviors. Therefore, two different

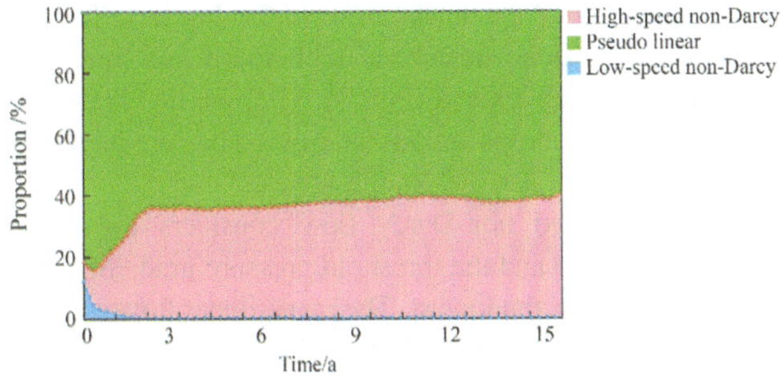

(a) Flow behaviors in porous medium

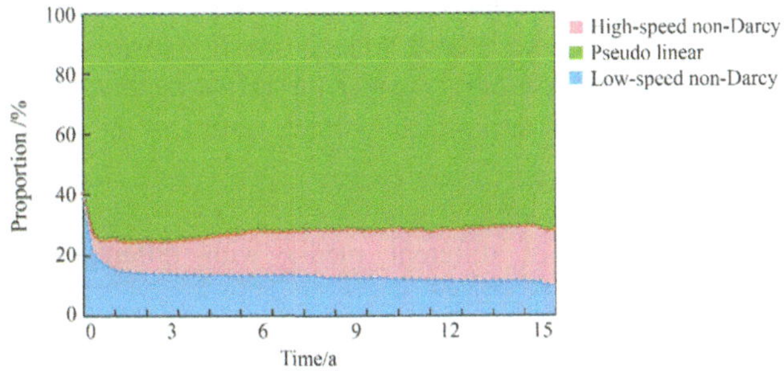

(b) Flow behaviors in fractured medium

Fig. 11.60 Change of flow behaviors with time

identification parameters and identification criteria are used in the simulation.

1) Using pressure gradient as identification parameter

The pressure gradient is used as the parameter to identify, and the effect of flow behavior change on the production is simulated and compared in three scenarios: i.Different pore media having their corresponding pressure gradient limits and nonlinear flow parameters; ii.All media having the same pressure gradient limits and nonlinear flow parameters corresponding to small-scale pore media; iii.All media having the same pressure gradient limits and nonlinear flow parameters corresponding to large-scale pore media.

It shows that when different pressure gradient limits and non-linear flow parameters are applied for different media, the differences in flow behavior changes are obvious, which has a relatively larger on the production performance. When the pressure gradient limits and non-linear flow parameters of small-scale media are applied for all media, the pressure gradient limits and threshold pressure gradient become larger, and the effects of low speed non-linear flow behavior strengthens, and the production decreases. When the pressure gradient limits and non-linear flow parameters of the large-scale media are applied to all media, the pressure gradient limits and threshold pressure gradient become smaller,

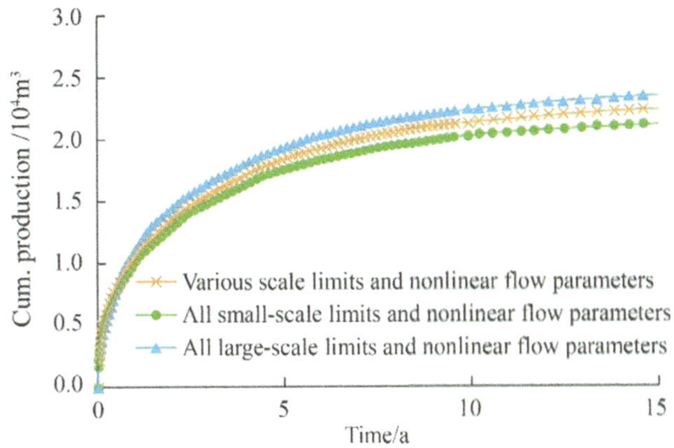

Fig. 11.61 Effects of pressure gradient limits and non-linear flow parameters on oil production

and the effects of low speed non-linear flow behavior weakens, and the production increases (Fig. 11.61).

2) Using Reynolds number as identification parameter

The Reynolds number is used as the parameter to identify, and the effect of flow behavior change on the production is simulated and compared in three scenarios: i. Different pore media having their corresponding Reynolds number limits and nonlinear flow parameters; ii. All media having the same Reynolds number limits and nonlinear flow parameters corresponding to small-scale pore media; iii. All media having the same Reynolds number limits and nonlinear flow parameters corresponding to large-scale pore media.

It shows that when different Reynolds number limits and non-linear flow parameters are applied for different media, the differences in flow behavior changes are obvious, which has a relatively larger influence on the production performance. When the Reynolds number limits and non-linear flow parameters of small-scale media are applied for all media, the Reynolds number limits and threshold pressure gradient become larger, and the effects of low speed non-linear flow behavior strengthens, and the production decreases. When the Reynolds number limits and non-linear flow parameters of the large-scale media are applied to all media, the Reynolds number limits and threshold pressure gradient become smaller, and the effects of low speed non-linear flow behavior weakens, and the production increases (Fig. 11.62).

It can be seen that different identification parameters and identification criteria lead to a big difference in the identification results. The corresponding kinetic equations and

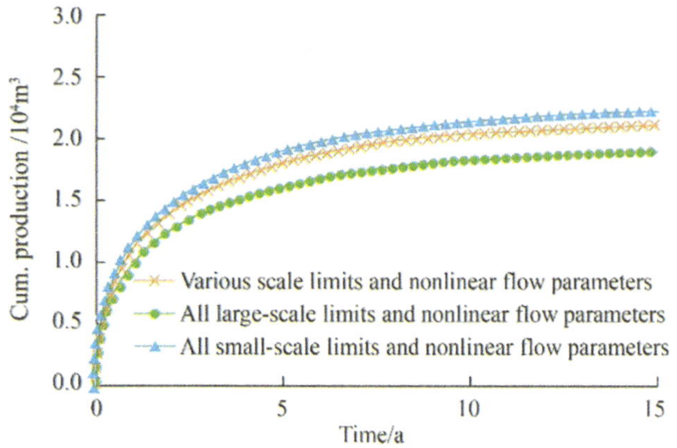

Fig. 11.62 Effects of pressure gradient limits and non-linear flow parameters on oil production

nonlinear flow parameters are also different, which affect the production performance and productivity.

2. Effect of different flow behaviors of multiple-scale multiple media on production performance

In multiple-scale media with pores and natural/hydraulic fractures, the dynamic change of flow behaviors have an extreme influence on the well productivity.

(1) Simulation of different media with same flow behavior

Suppose that different multiple media (big fractures, small fractures, micro-pores, nano/micro-pores, and nano-pores) have the same flow behavior. Different flow behaviors, including high-speed non-linear flow, linear flow, pseudo-linear flow, and low-speed nonlinear flow, are simulated and their effects on the production performance are compared.

Simulation results show that under different flow behaviors, there is a significant difference in the production rate: the production of the linear flow is the highest; The production of the quasi-linear and high-speed non-linear flow decrease to some extent (Fig. 11.63), but the high-speed nonlinear flow drops more sharply. The production of low-speed non-linear flow is the lowest.

The low-speed non-Darcy flow considers the surface effects (*n* values) and the threshold pressure gradient, and its production rate is the lowest. The quasi-linear flow considers the influence of the quasi-linear critical threshold pressure gradient, and its production fall somewhat; the high-speed non-Darcy flow considers the disturbance of turbulence that increases the flow resistance, so its production lowers. While the linear flow does not consider the above factors, and its flow resistance is insignificant, which results in the highest production.

(2) Simulation of different media with different flow regimes

Different multiple media with pores and fractures are assigned to different flow behaviors: large fractures (including artificial fractures) are assigned to high-velocity nonlinear flow, micro-cracks are designated as linear flow, micro-pores are pseudo-linear flow, and nano/micro-pores and nano-pores as low-speed nonlinear flow. The simulations of different media with different flow regimes are carried out and the results are compared with those for different media with the same flow behavior.

From the comparison results, the production for different media with different flow behaviors is higher than that for different media with low-speed nonlinear flow, but is lower than those of the linear, quasi-linear, and high-speed nonlinear flows. Obviously, due to the change of flow behaviors, the corresponding flow mechanisms differ. Thus, the flow behavior has a significant impact on the production of tight oil production performance (Fig. 11.64).

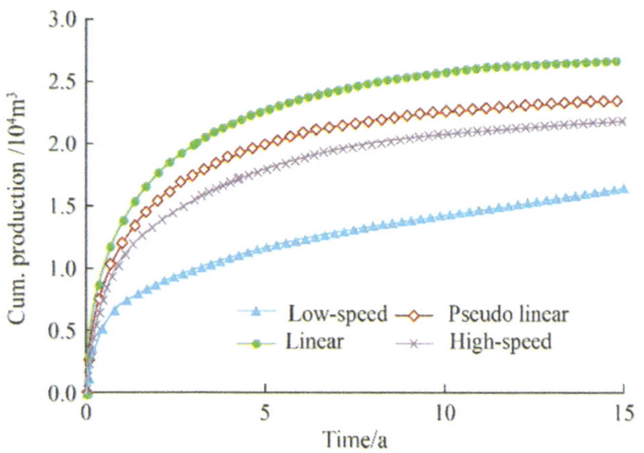

Fig. 11.63 Effect of flow behaviors on the production

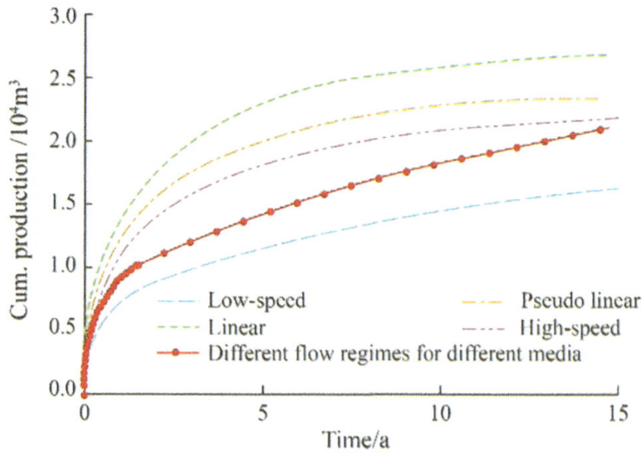

Fig. 11.64 Effect of different flow behaviors on the tight oil production

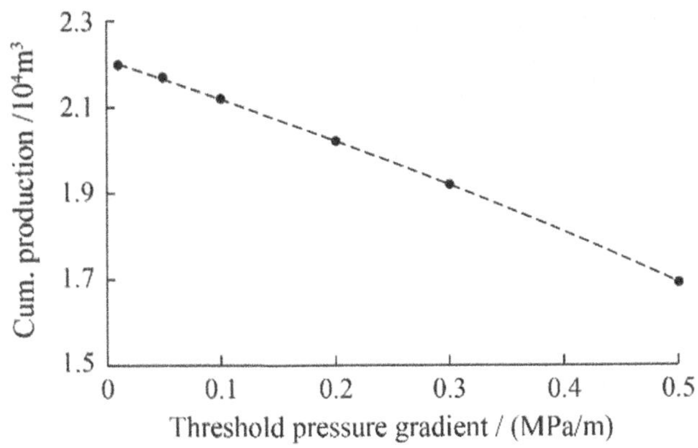

Fig. 11.65 Effect of threshold pressure gradient on the production

3. Effect of nonlinear flow mechanisms on tight oil production performance

(1) Effect of threshold pressure gradient on tight oil production performance

The threshold pressure gradient reflects the flow resistance the fluid needs to overcome. To reveal its effect on the production, different threshold pressure gradients are simulated (Fig. 11.65).

As shown in the figure, as the threshold pressure gradient increases, the more resistance the fluid needs to overcome, and the lower the production. When the threshold pressure gradient increases from 0.01 MPa/m to 0.5 MPa/m, the corresponding cumulative production decreases from 2.2×10^4 to 1.69×10^4 m^3, and the production rate declines by 24%.

(2) Effect of n value on the production performance

The n value reflects the strength of surface effect and difficulty to flow. Different n values are simulated to understand its effect on the production performance (Fig. 11.66).

As can be seen from the figure, as the n value increases, the weaker the surface effect, the easier the fluid is to flow, and the higher the production. The reasonable range of the n value is

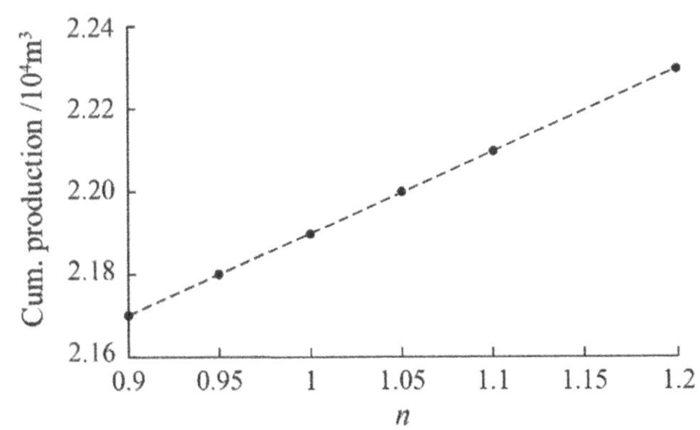

Fig. 11.66 Effect of n value on the production

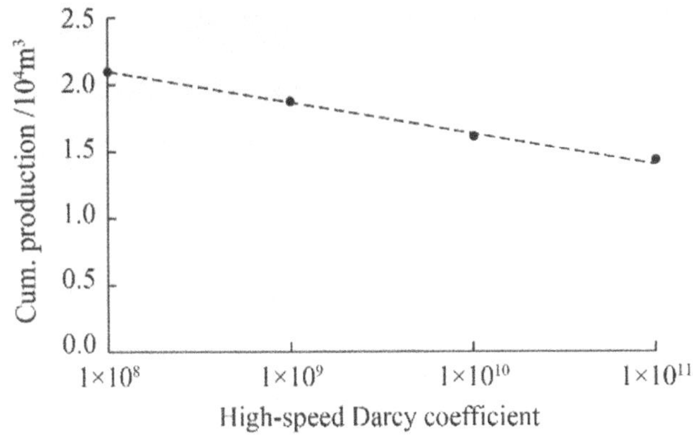

Fig. 11.67 Effect of high-speed non-Darcy turbulence coefficient on the production

usually 0.9–1.2. When it increases from 0.9 to 1.2, the corresponding cumulative production increases from 2.17 to 2.23×10^4 m^3, and the production rate increases by 2.6%.

(3) Effect of high-speed non-Darcy turbulence coefficient on tight oil production performance

The high-speed nonlinear coefficient β reflects the effect of high-speed non-Darcy flow on fluid flow. Simulations are conducted to analyze the effect to different β values on the production performance (Fig. 11.67).

As it can be seen from the simulation results, the larger the high-speed non-Darcy coefficient, the lower the production rate, which is because larger high-speed non-Darcy coefficient results in greater disturbance from high-speed turbulence flow to fluid flow. The range of a high-speed non-Darcy coefficient is usually 1×10^8–1×10^{11} m^{-1}. When it increases from 1×10^8 to 1×10^{11} m^{-1}, the cumulative production decreases from 2.1 to 1.5×10^4 m^3, and the production rate increases by 28%.

11.2.4.2 Simulation of Tight Gas Flow Behavior and Nonlinear Flow Mechanisms

1. Self-adaptive simulation of tight gas flow behavior and complex flow mechanisms

Throughout the gas production process, three flow behaviors exist such as high-speed nonlinear flow, pseudo-linear flow, and low-speed nonlinear flow. Meanwhile, in the case of small porosity, low pressure, and small permeability, slippery and diffusion occur. In addition, desorption is also an important flow mechanism in the tight gas production. Different flow behaviors exert great effects on the tight gas production performance and productivity. Therefore, the flow behaviors can be automatically identified through

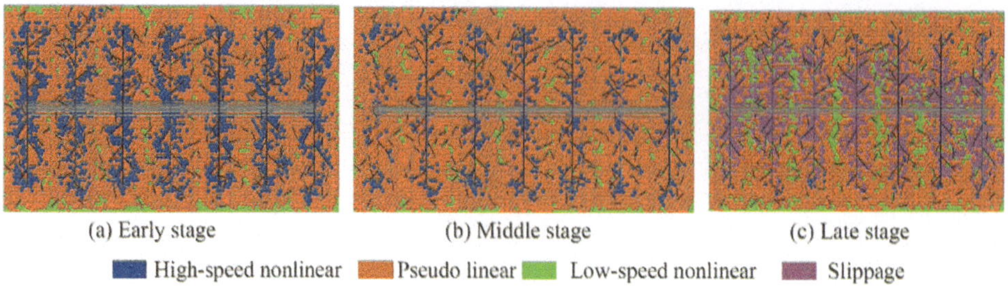

Fig. 11.68 Fluid flow throughout different production stages

critical flow parameters and identification criteria, based on which the dynamic equations that are suitable for the flow behaviors and flow mechanisms can be selected to perform the dynamic simulation, and thus reveals the transition of flow behaviors and flow mechanisms of multiple media and their effects on the production performance.

(1) Spatial flow behaviors of multiple-scale multiple media

The simulation of self-adaptive identification of flow behaviors and flow behavior transition show that in the same production stage, flow behaviors are significantly different in different locations and different multiple-scale multiple media. In different production stages, flow behaviors are significantly different in different multiple-scale multiple media. In the early stage (Fig. 11.68a), the pressure difference is large and fluids flow fast near fractures, and fluid flows mainly in the high-speed non-Darcy behavior. The flow is mainly pseudo-linear in pore media and low-speed nonlinear flow far from wells and fractures. In the middle stage (Fig. 11.68b), the differential pressure becomes smaller and the flow slows down. Most of the high-speed non-Darcy flow changes to a pseudo-linear flow, and the pseudo-linear flow transitions to a low-speed nonlinear flow. In the late stage (Fig. 11.68c), formation pressure decreases, and slippery occur under low pressure conditions, fluid flows mainly in quasi-linear flow, low-speed nonlinear flow, and slippage.

(2) Temporal flow behaviors of multiple-scale multiple media

The transitions of temporal flow behaviors are analyzed through self-adaptive identification and dynamic simulation. It is found that during the production process, the pore pressure and pressure gradient of multiple-scale multiple media vary extremely, which leads the dramatic change in the flow behaviors. Different flow behaviors change drastically at different production stages, and their respective

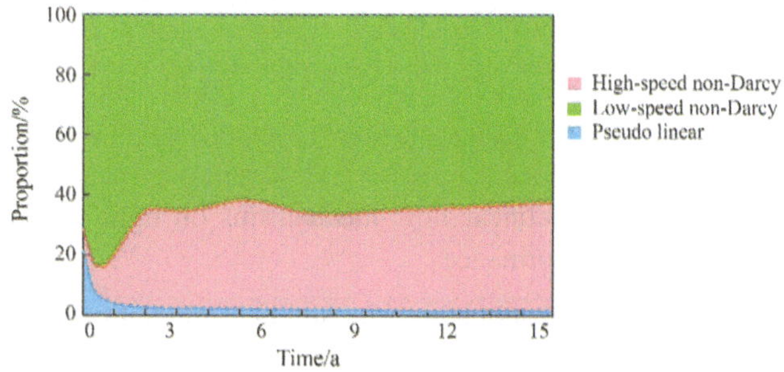

Fig. 11.69 Flow change of flow behaviors with time

contributions to the production vary with time (Fig. 11.69). As shown in the figure, overall, the high-speed non-Darcy flow behavior has a smaller proportion, the low-speed non-Darcy flow behavior has a relatively larger proportion, but the pseudo-linear flow has the largest proportion. In the early stage, the high-speed non-Darcy flow behavior has a smaller proportion. As time goes, the pressure difference diminishes and the flow slows down. Part of the high-speed non-Darcy flow shifts to other behaviors. As an important part in the early stage, pseudo-linear flow changes to low-speed non-Darcy flow later. As a small part in the early stage, due to the transformation from other flow regimes, low-speed non-Darcy flow becomes important gradually in the late stage. The pseudo-linear and low-speed non-linear flow dominant in the middle and late stages.

For pore media (Fig. 11.70a), the initial pressure difference is large and fluid flows relatively faster and mainly in the high-speed nonlinear and the pseudo-linear flow regimes. In the middle stage, it falls in the pseudo-linear, low-speed nonlinear, and slippage behaviors. In the late stage, flow rate decreases due to pressure falloff, slippage occurs and the flow is mainly pseudo linear (low speed) and slippage.

For fracture media (Fig. 11.70b), the initial pressure difference is large and fluid flows fast, mainly in the high-speed non-Darcy flow; in the late stages, flow velocity decreases due to pressure falloff, and the low-speed non-Darcy and pseudo-linear flows gradually strengthen.

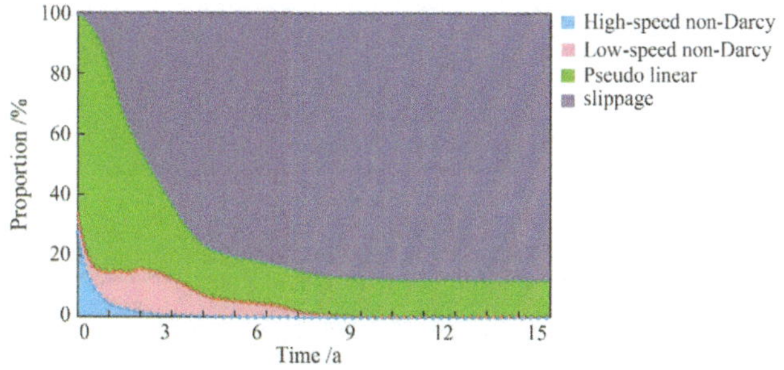

(a) Flow regimes in porous medium

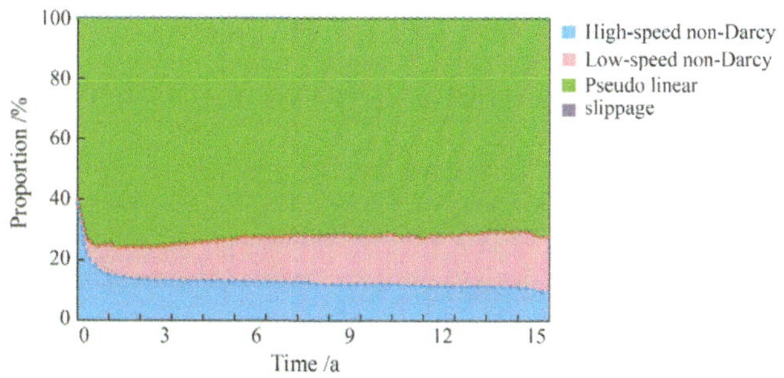

(b) Flow regimes in fractured medium

Fig. 11.70 Change of flow behaviors with time

2. Effect of different flow behaviors of multiple-scale multiple media on production performance

In multiple-scale multiple media of tight gas reservoirs, the dynamic change of flow behaviors have an extreme influence on the well productivity.

(1) Simulation of different media with same flow behavior

Suppose that different multiple media (big fractures, small fractures, micro-pores, nano/micro-pores, and nano-pores) have the same flow behavior. Different flow behaviors, including high-speed non-linear flow, linear flow, pseudo-linear flow, and low-speed nonlinear flow, slippage and diffusion are simulated and their effects on the production performance are compared.

From the simulation results (Fig. 11.71), it can be seen that under different flow behaviors, there is a significant difference in the cumulative gas production. Compared with the linear flow, the production of low-speed nonlinear, pseudo-linear, and high-speed nonlinear flow decreases to some extent. Whereas under slippage and diffusion, the cumulative production slightly increases.

(2) Simulation of different media with different flow regimes

Different multiple media with pores and fractures are assigned to different flow behaviors: large fractures (including artificial fractures) are assigned to high-velocity

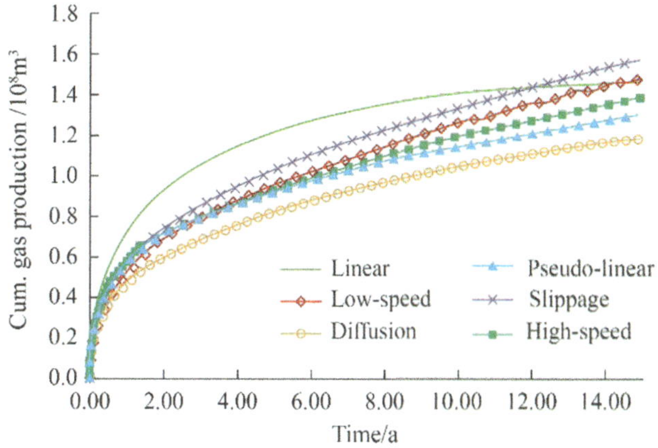

Fig. 11.71 Effect of flow behaviors on the production

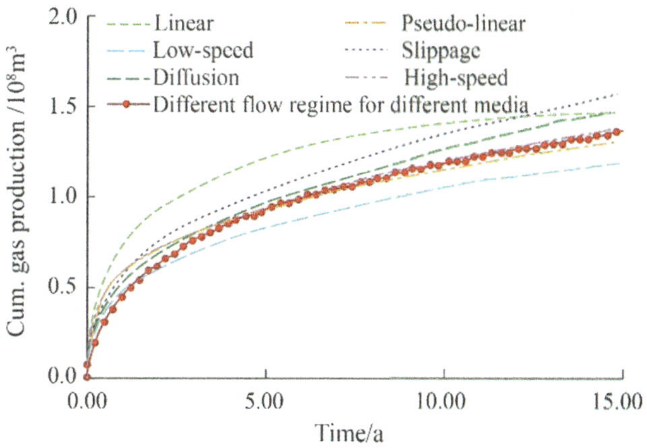

Fig. 11.72 Effect of different flow behaviors on tight gas production

nonlinear flow, micro-cracks are designated as pseudo linear flow, micro-pores are low-speed linear flow, and micro-pores as slippage, and nano-pores as diffusion. The simulations of different media with different flow regimes are carried out and the results are compared with those for different media with the same flow behavior.

From the comparison results, the production for different media in different flow behaviors are higher than that for different media in the low-speed nonlinear, pseudo-linear, and high-speed nonlinear flows, but is lower than that for slippage, diffusion, and linear flows. Thus, the flow behavior has a significant impact on the production of tight oil production performance (Fig. 11.72).

3. Effect of nonlinear flow mechanisms on tight gas production performance

(1) Effect of threshold pressure gradient on tight gas production performance

Compared with tight oil, the threshold pressure gradient of tight gas is much smaller, usually in the range of 0.001–

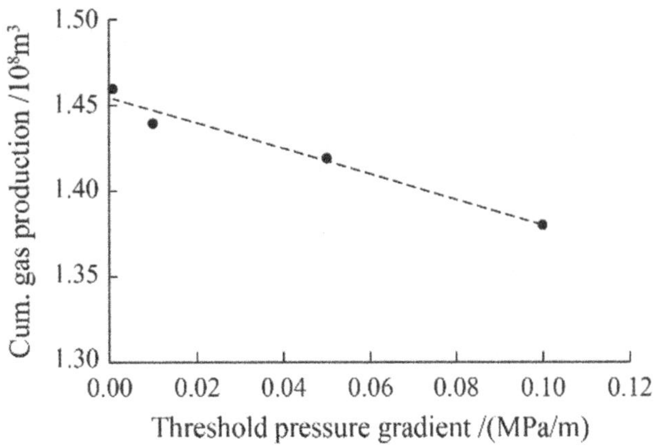

Fig. 11.73 Effect of threshold pressure gradient on the production

0.1 MPa/m. Different threshold pressure gradients are simulated to reveal its effect on the production performance (Fig. 11.73).

As shown in the figure, as the threshold pressure gradient increases, cumulative gas production decreases gradually. When it increases from 0.001 MPa/m to 0.1 MPa/m, the corresponding cumulative production declines from 0.73×10^8 m^3 to 0.69×10^8 m^3, and the production rate drops by 5%.

The threshold pressure gradient has some influence on the tight gas production. However, since the threshold pressure gradient of tight gas is relatively smaller, its effect on the production is relatively insignificant.

(2) Effect of *n* value on the production performance

For tight gas, the *n* value also reflects the strength of the surface effect and the difficulty to flow. Different *n* values are simulated to understand its effect on the production performance (Fig. 11.74).

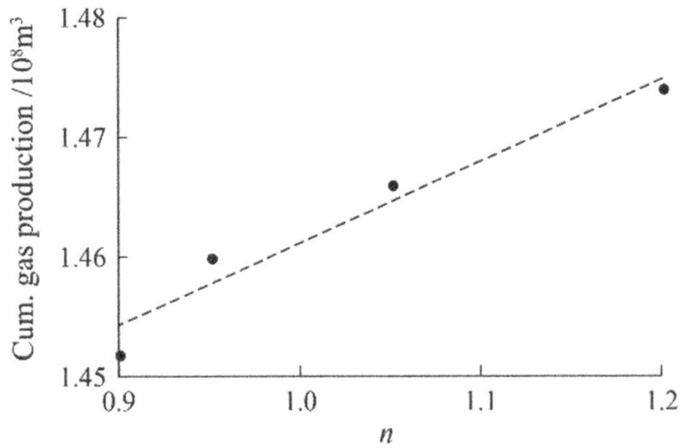

Fig. 11.74 Effect of *n* value on the production

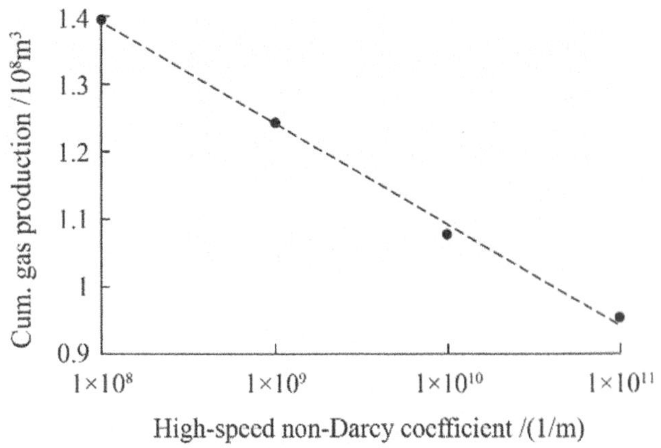

Fig. 11.75 Effect of high-speed non-Darcy turbulence coefficient on the production

As can be seen from the figure, as the *n* value increases, the weaker the surface effect, the easier the fluid flows, and the higher the production. The reasonable range of the *n* value is usually 0.9–1.2. When it increases from 0.9 to 1.2, the corresponding cumulative production increases from 0.726 to 0.737×10^4 m^3, and the production rate increases by 1.5%.

Therefore, the *n* value affects the production of tight gas to some extent, but not very large.

(3) Effect of high-speed non-Darcy turbulence coefficient on tight gas production performance

The high-speed nonlinear coefficient β reflects the effect of high-speed non-Darcy flow on fluid flow. Simulations are conducted to analyze the effect of different β values on the production performance (Fig. 11.75).

Simulation results show that as the high-speed non-Darcy coefficient increases, the production decreases. When the coefficient decreases from 1×10^8 to 1×10^{11} m^{-1}, the cumulative production decreases from 0.7 to 0.48×10^8 m^3, and the production rate increases by 31%.

Therefore, the high-speed non-Darcy flow has a large influence on the tight gas production.

(4) Effect of slippage on tight gas production performance

The slippage effect reflects the effect of pore media on gas flow under low pressure and low permeability conditions. The stronger the slippage effect, the higher the gas production rate. Different pressure and permeability (throat and pore radii) conditions are simulated to reveal the effect of slippage on tight gas production performance.

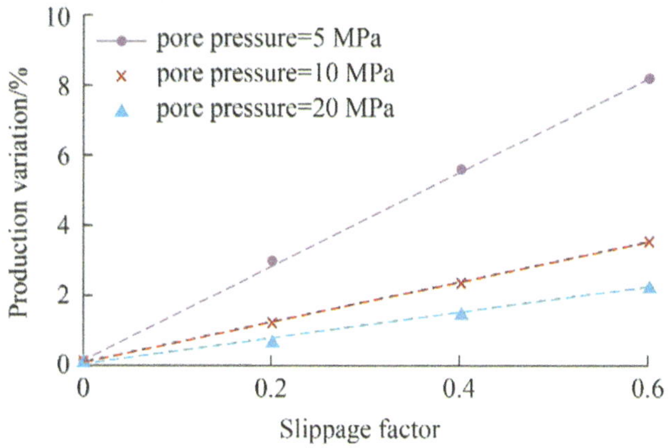

Fig. 11.76 Effect of slippage on tight gas production under different pore pressure

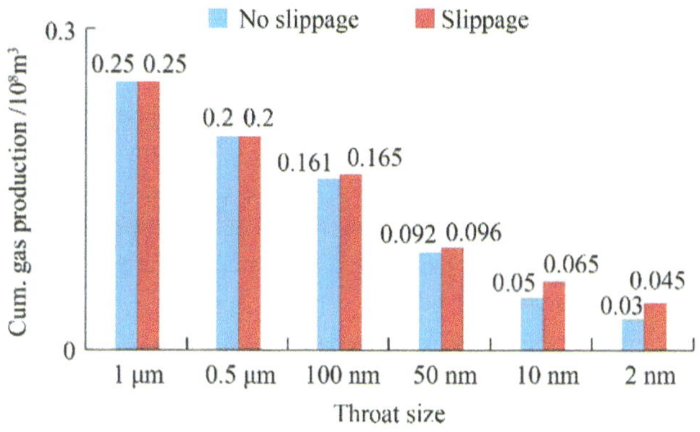

Fig. 11.77 Productions with and without slippage effect

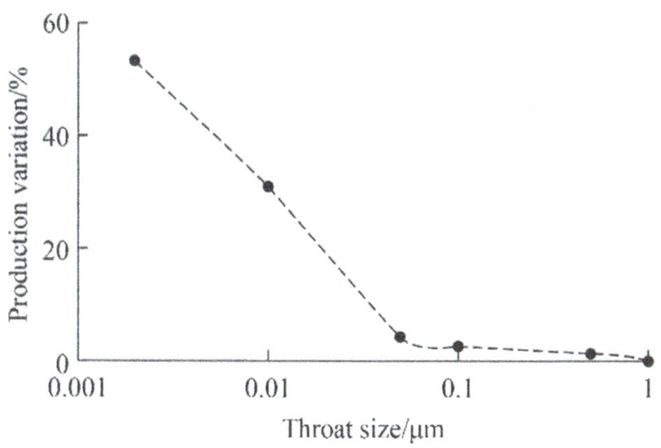

Fig. 11.78 Relative production enhancement with and without slippage effect

1) Effect of slippage effect on production performance under different pore pressures

It can be seen that under the same pore pressures, the production increases with the enhancement of the slippage effect. For a fixed slippage effect, the lower the pore pressure is, the greater the effect of slippage on the production is. At a pore pressure of 5 MPa, the slippage factor increasing from 0 to 0.6 MPa, the cumulative gas production increases by 8%. At a pore pressure of 20 MPa, the slippage factor increasing from 0 to 0.6 MPa, and the cumulative gas production increases by 2% (Fig. 11.76).

2) Effect of slippage effect on production performance under different multiple-scale pore and throat radii

The smaller the radii of the pores and throats is, the greater the slippage effect is. Different multiple-scale pore and throat radii are simulated to analyze the effect of slippage effect on the gas production.

Simulation results (Figs. 11.77 and 11.78) show that the slippage effect is negligible at a pore pressure of 5 MPa and pore-throat radii of greater than 100 nm. The slippage effect is significant at the pore-throat radii of less than 100 nm. The cumulative gas production increases by 50% with slippage at pore-throat radii of less than 2 nm.

(5) Effect of diffusion on tight gas production performance

Under the conditions of small porosity, low permeability, and low pressure, diffusion exists in the gas flow. A stronger diffusion leads to a larger production. Different pressure and permeability (throat and pore radii) conditions are simulated to reveal the effect of diffusion on tight gas production performance.

Simulation results (Figs. 11.79 and 11.80) show that diffusion is negligible at a pore pressure of 5 MPa and pore-throat radii of greater than 100 nm. The diffusion has somewhat influence at the pore-throat radii of less than 100 nm. Diffusion contributes about 9% of the production at the pore-throat radii of less than 2 nm.

(6) Effect of desorption on tight gas production performance

During the tight gas production process, with the decrease of formation pressure, the adsorbed gas gradually becomes free

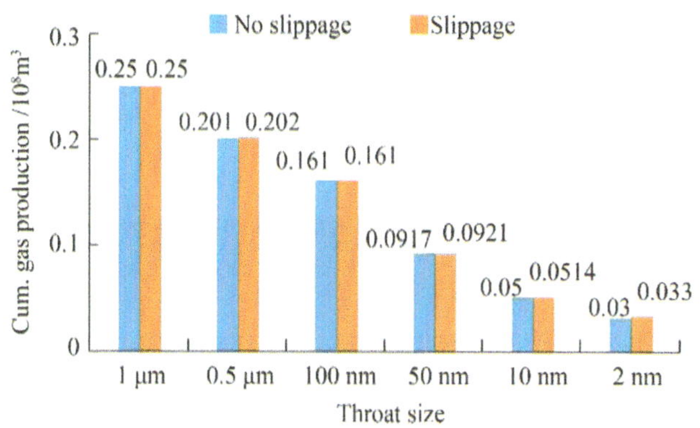

Fig. 11.79 Productions with and without diffusion

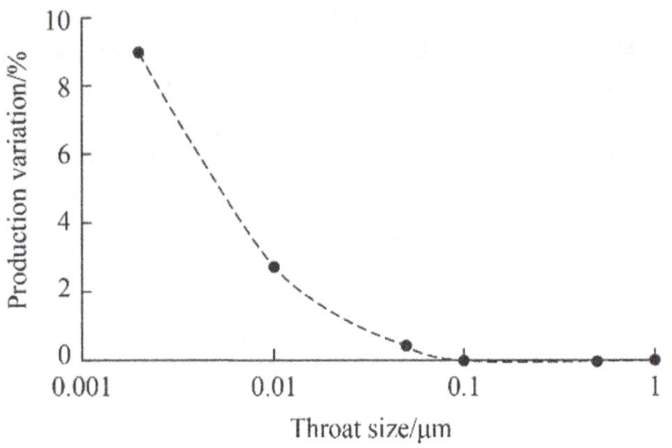

Fig. 11.80 Relative production enhancement with and without diffusion

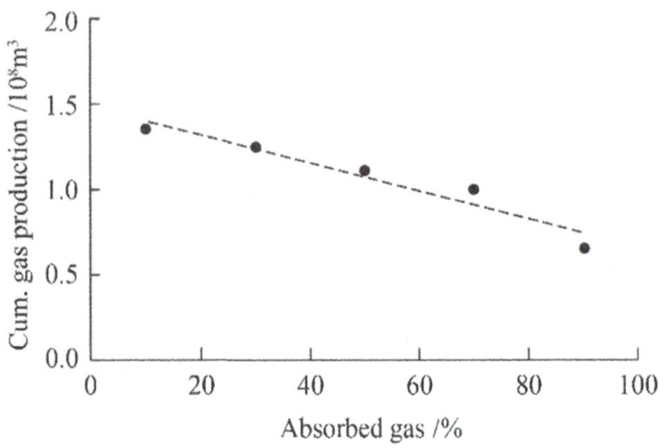

Fig. 11.81 Effect of adsorbed gas content on the cumulative gas production

gas under the effect of desorption, affecting the production performance.

1) Effect of adsorbed gas content on tight gas production performance

The proportions of adsorbed gas and free gas affect the productivity of tight gas. Different adsorbed gas contents (the ratio of adsorbed gas to total gas) are simulated to analyze its effect on the tight gas production performance.

Simulation results show that the higher the adsorption gas content and the lower the free gas content, the lower the cumulative production (Fig. 11.81).

2) Effect of diffusion coefficient on tight gas production performance

The gas in the reservoir matrix block (including free gas and desorption gas) is transported to fractures through diffusion, and the diffusion coefficient directly affects the gas supply

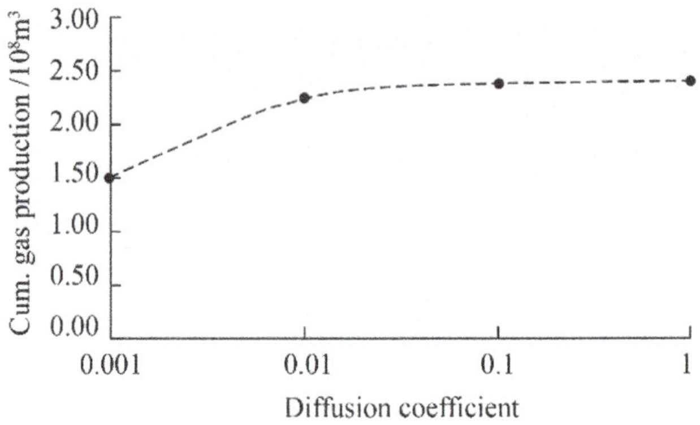

Fig. 11.82 Effect of diffusion rate on production

rate to the fracture system (Yu et al. 2012). Under the constraint of desorption rules, different diffusion coefficients are simulated to analyze its effect on the production performance.

It can be seen that the greater the diffusion coefficient, the greater the production. When the diffusion coefficient increased from 0.001 to 1 m^2/d, the cumulative gas production rises from 1.5 to 2.41 $\times$ 10^8 m^3. Therefore, the gas diffusion coefficient has a large influence on the productivity of tight gas (Fig. 11.82).

3) Effect of desorption on tight gas production performance

The desorption process of adsorbed gas is mainly controlled by the desorption rules and pore pressure (Yu et al. 2012). Under the same adsorption gas content, different desorption rules (Zhang et al. 2014) are simulated to understand the influence of desorption rules on the tight gas productivity.

Three different adsorption isothermal curves (Fig. 11.83) are selected with the same Langmuir volumes (the maximum volume of adsorbed gas) and different

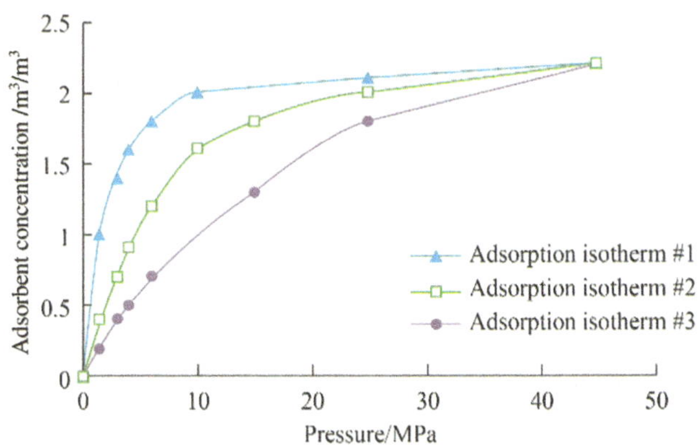

Fig. 11.83 Different isothermal curves

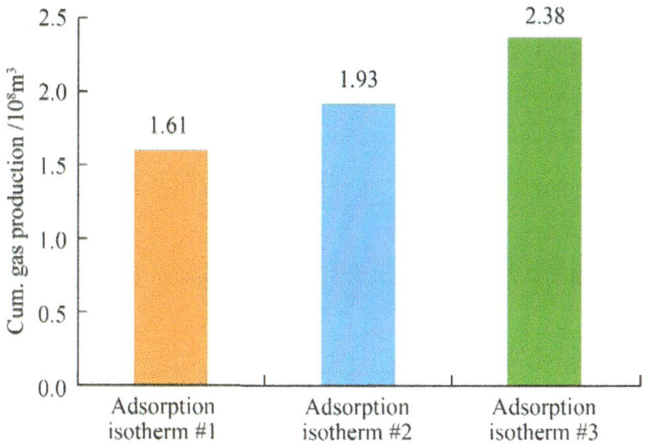

Fig. 11.84 Effect of different isothermal curves on cumulative gas production

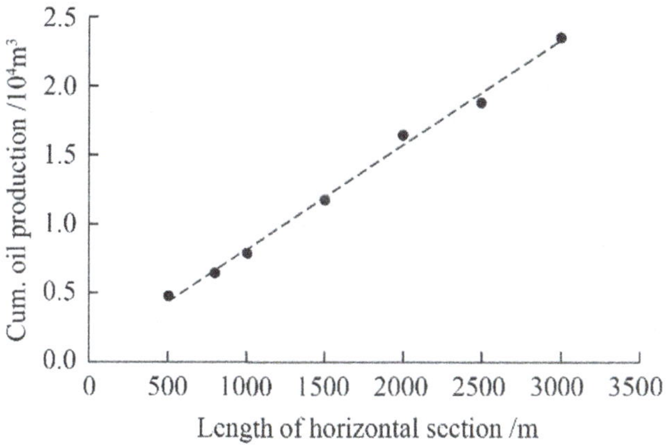

Fig. 11.85 Effect of the length of horizontal section on the production performance

Langmuir pressures (the gas adsorption amount reached the maximum value of 50%). Curve 1 has the largest curvature and lowest Langmuir pressure; Curve 2 has an intermediate curvature and intermediate Langmuir pressure; Curve 3 has the smallest curvature and the largest Langmuir pressure.

It can be seen that a higher Langmuir pressure leads to a smaller curvature of the adsorption isotherm curve and a higher gas production, vice versa (Fig. 11.84).

Because the higher the Langmuir pressure, the smaller the curvature of adsorption isothermal curve, which indicates that more gas is adsorbed in the high-pressure area, and it can be more easily desorbed in the production process, resulting in a higher production.

11.3 Simulation of Horizontal Wells and Hydraulic Fracturing in Tight Oil/Gas Reservoirs

Developed with horizontal wells and hydraulic fracturing in tight oil/gas reservoirs, contact and flow behaviors between reservoir matrix and well bore are complicated. The horizontal well production performance and productivity are largely influenced by the well properties and stimulation conditions. In this section, horizontal wells and hydraulic fracturing are simulated to investigate the effects of horizontal well lengths, drilling encounter ratios, well completion mode, fracture geometry, hydraulic fracturing parameters the productivity and performance.

11.3.1 Simulation of Horizontal Wells in Tight Oil/Gas Reservoirs

11.3.1.1 Simulation of Horizontal Well Length and Drilling Encounter Ratio

To develop tight reservoirs, horizontal wells are commonly applied, and the horizontal section is extended as far as possible to increase the well's control reserve. However, the drilling encounter ratio in high-quality formations should be considered in the optimization of horizontal well length.

1. Effect of the length of horizontal section on production performance

Given weak rock heterogeneity in lithology, physical properties, oil saturation, and recoverability, the longer the horizontal segment, the higher the production is (Fig. 11.85) at technically feasible and economic viable conditions. Because a longer horizontal well controls more reserve. Through effective fracturing technology, its control area and communication can both be improved, leading to good production.

2. Simulation of effect of drilling encounter ratio on production performance

Given strong rock heterogeneity and big variations of lithology, physical properties, oil saturation, and recoverability in different horizontal well segments, there is tremendous differences in different horizontal well segments' contribution to the production.

With a fixed length of the horizontal section, different drilling encounter ratios are simulated. The results show that a drilling encounter ratio has a great influence on the production performance.

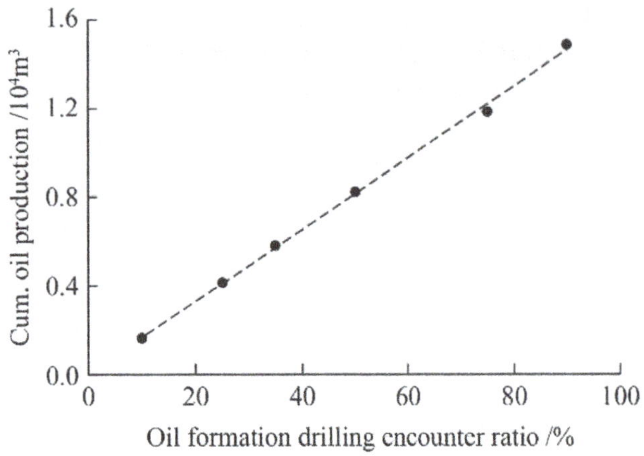

Fig. 11.86 Effect of drilling encounter ratio on the cumulative oil production

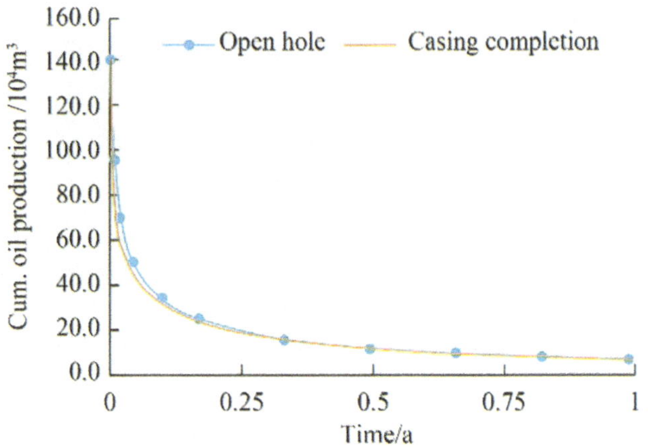

Fig. 11.88 Effect of different completions on the daily production in reservoirs with undeveloped natural fractures

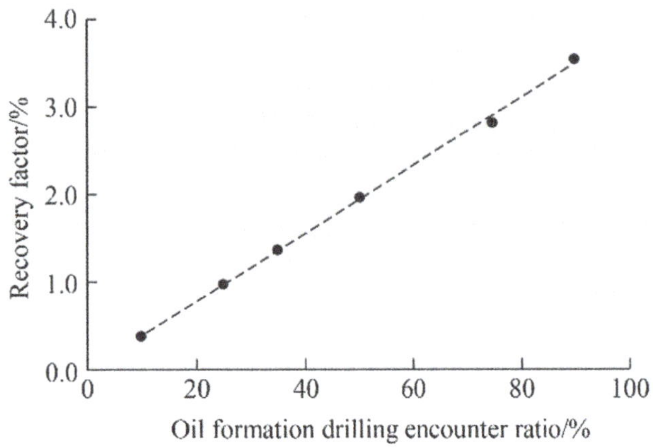

Fig. 11.87 Effect of drilling encounter ratio on the oil recovery factor

For the same length of the horizontal section, the greater the drilling encounter ratio, the higher the single well production and recovery (Figs. 11.86 and 11.87).

It can be seen that the drilling encounter ratio of high-quality reservoirs plays an important role in the tight oil production performance. It is necessary to have a deeper understanding of the rock heterogeneity, to improve the drilling encounter ratio and increase the tight oil production through sweet spot evaluation and well location and trajectory optimization.

Not only the length of horizontal section, but also the length through high-quality reservoirs is what a horizontal well design should be optimized.

11.3.1.2 Simulation of Different Horizontal Well Completion Modes

Developed with horizontal wells and hydraulic fracturing, different well completion modes (perforation and open hole) have the contact and flow behaviors different, influencing the production performance and productivity. For perforation completion, reservoir fluids flow first into the artificial fractures and then to the wellbore. For open-hole completion, artificial fractures, reservoir matrix pores of different scales, and natural cracks all directly contact a horizontal wellbore. Fluids in the reservoir can flow to the wellbore with or without passing through the artificial fractures.

1. Effect of different well completions on production with undeveloped naturally fractures

Two different completion methods are simulated in reservoirs with undeveloped natural fractures. Results show that the two completion methods (open-hole and perforation) make little difference in the production. Because the flowability in tight matrix is extremely low, without natural fractures, very little fluids can directly flow from pore media into a wellbore. Artificial fractures are the main flow channels between the pore media and the wellbore. So, the simulation results showed little difference for the two completion modes (Figs. 11.88 and 11.89).

2. Effect of different well completions on production with developed naturally fractures

Two different completion methods are simulated in reservoirs with developed natural fractures. Results show that the two completion methods (open-hole and perforation) made a marked difference in the production. Because a considerable amount of fluids flow from matrix pores to a wellbore through artificial fractures that is an effective communication. The more developed the natural fractures, the larger the difference between the two completion methods (Figs. 11.90 and 11.91).

In conclusion, contact and flow behaviors between the reservoir and a horizontal wellbore under different completion methods are different, which influences the production

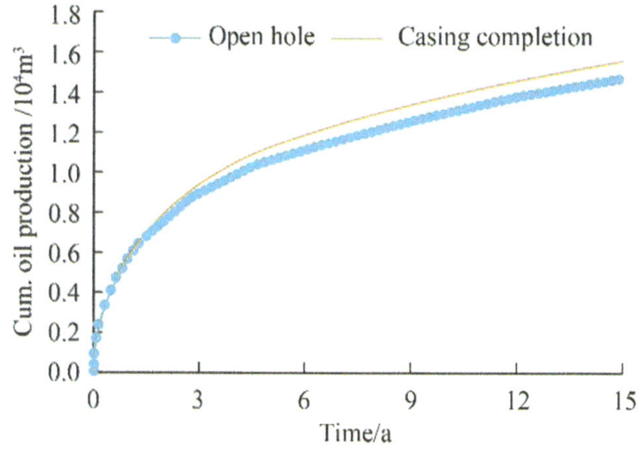

Fig. 11.89 Effect of different completions on the cumulative production in reservoirs with undeveloped natural fractures

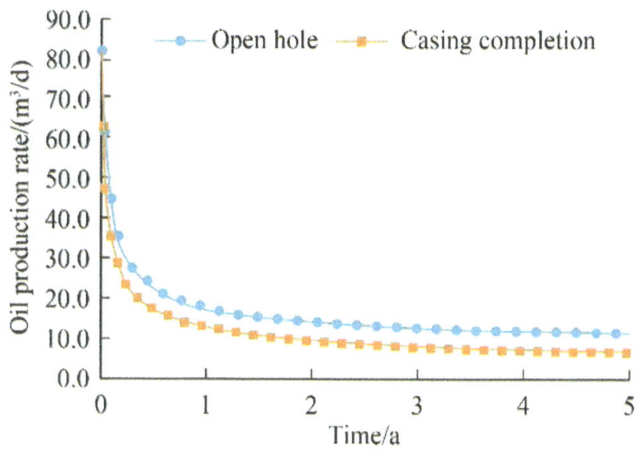

Fig. 11.90 Effect of different completions on the daily production in reservoirs with developed natural fractures

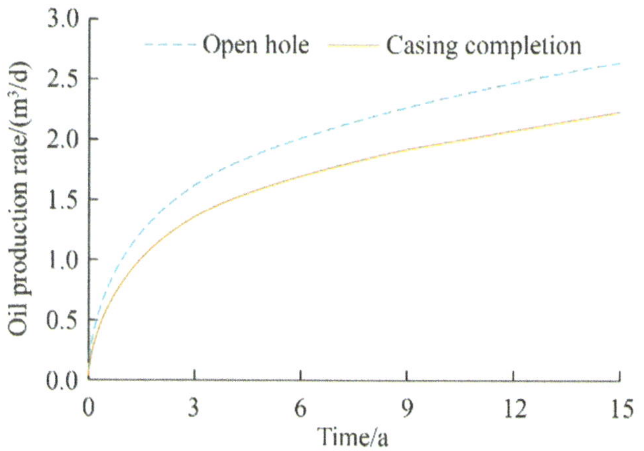

Fig. 11.91 Effect of different completions on the cumulative production in reservoirs with developed natural fractures

performance and productivity. It cannot be ignored, especially in the reservoirs with developed natural fractures.

11.3.1.3 Simulation of Different Wellbore Treatment Modes

Depending on the treatment schemes and factors need to be considered, horizontal wells can be treated in three ways: a line source, a discrete wellbore and a multi-segment well. The line-source treatment is the most common method and it regards a wellbore as a system consisting of several point sources, in which the well index is used to express the flow between a wellbore and the formation. A discrete wellbore treats a horizontal wellbore as a system composed of several discrete grid blocks, in which the conductivity between well grids and reservoir grids is used to describe the fluids flow between wellbore and reservoir. This method can treat the contact relationship between wellbore and reservoir under open-hole completion mode. The multi-segment well considers the friction, gravity, multiphase pipe flow inside the wellbore and the fluid exchange and pressure loss between wellbore segments. The fluid flow between wellbore segments is treated as a pipe flow.

Given all other conditions are the same, different wellbore treatment methods are simulated and the results are shown in Figs. 11.92 and 11.93.

1. Close productions for discrete wellbores and line source wellbores

For a discrete wellbore treatment, when the wellbore grid size is equivalent to the real wellbore diameter, the transmissibility between well grids and reservoir grids nears the well index for the line-source wellbore treatment. Therefore, the simulation results are close to each other under the two treatments.

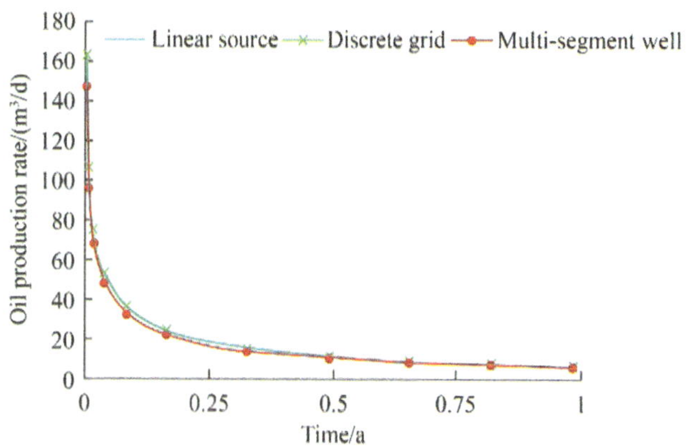

Fig. 11.92 Effect of wellbore treatments on the daily oil production

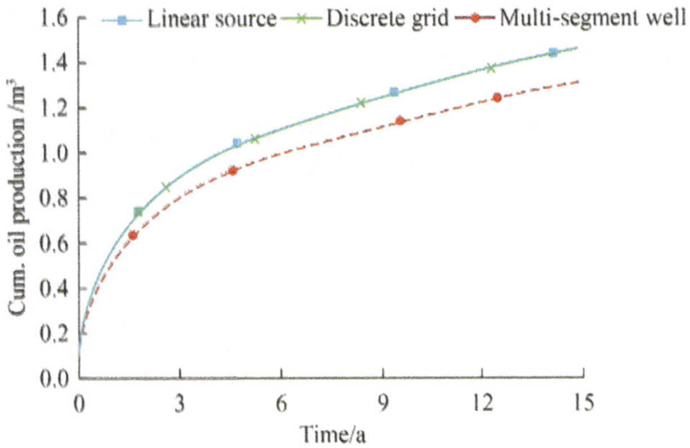

Fig. 11.93 Effect of wellbore treatments on the cumulative oil production

2. Lower production for multi-section than for other two treatment modes

For a multi-section wellbore treatment, the production is lower than those for the other two treatments because it considers the friction inside the wellbore, the multiphase pipe flow, and the fluid exchange and pressure loss between the wellbore segments.

11.3.2 Simulation of Hydraulic Fracturing in Tight Oil/Gas Reservoir

11.3.2.1 Simulation of Different Fracture Geometries in Horizontal Wells

During the hydraulic fracturing of horizontal wells, the shape of artificial fractures vary with the geological, geo-mechanical characteristics and fracturing procedures. Artificial fracture scan be classified into four categories based on their geometry and complexity: i. symmetrical pattern with double wings; ii. asymmetric pattern with different half length; iii. simple strip pattern; and iv. complex fracture networks. Fractures geometries have great influences on the reservoir dynamics and productivity. Different fracture shapes are simulated and the results are shown in Figs. 11.94, 11.95, 11.96 and 11.97.

1. Significantly different daily and cumulative production for different fracture geometries

The symmetrical and asymmetric patterns have simple single fracture distributions with small fracture density and large spacing, the production are relatively low, and their respective daily/cumulative production are close. Meanwhile, the simple strip pattern and complex fracture networks have complex distributions with big fracture density, small spacing, and a large fracture area, and their daily and cumulative productions are remarkably higher than the previous two patterns.

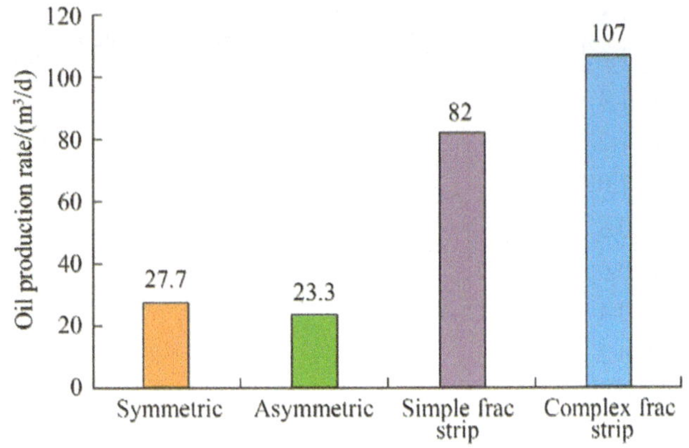

Fig. 11.94 Effect of fracture geometry on the initial daily oil production

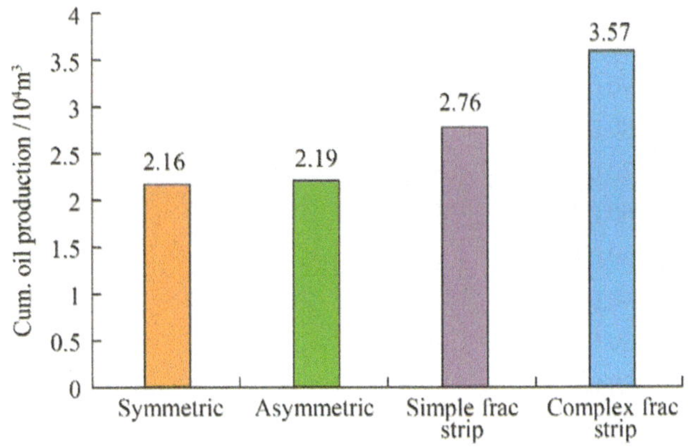

Fig. 11.95 Effect of fracture geometry on the initial cumulative oil production

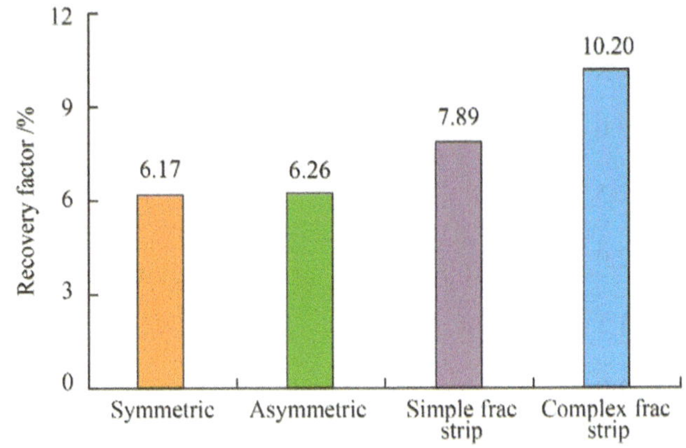

Fig. 11.96 Recovery factors for different fracture geometries

Fig. 11.97 Pressure distribution under different hydraulic fracture situations

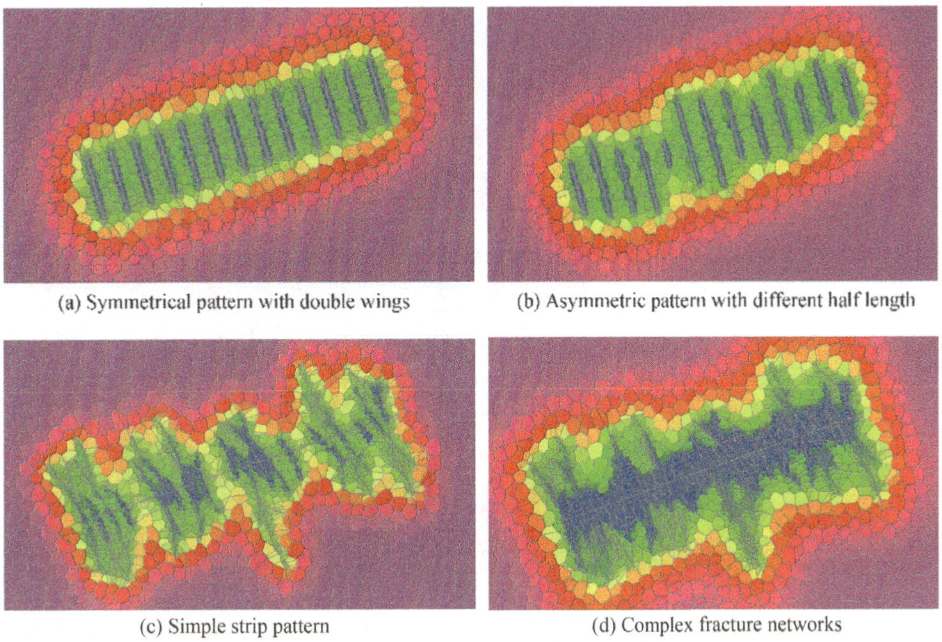

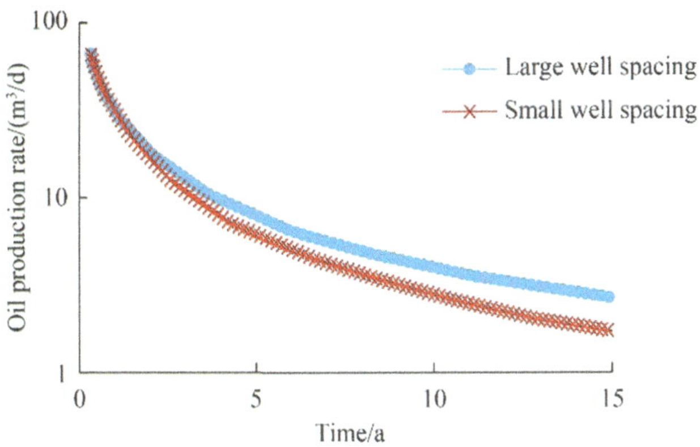

Fig. 11.98 Effect of well spacing on the well-pad daily production

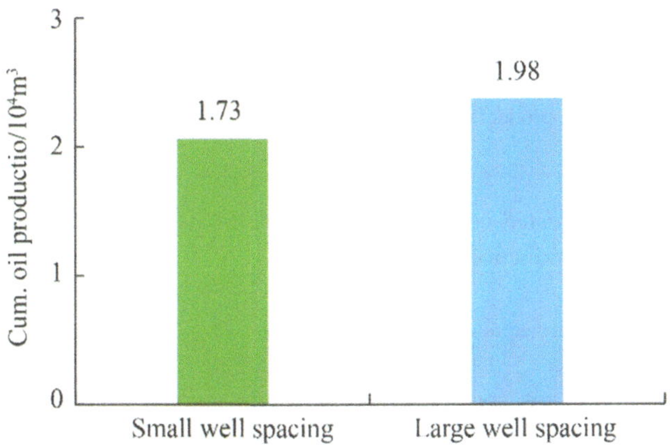

Fig. 11.99 Effect of well spacing on the well-pad cumulative production

2. Smaller reservoir matrix blocks and higher production for more complex fractures

More complex fracture lead to smaller fracture spacing, denser matrix blocks, smaller matrix blocks, and a shorter flow distance from the matrix to fissures. More oil in reservoir are touched, which improves the production performance.

11.3.2.2 Simulation of Well Pad Performance with Different Well Spacing

The optimum well pads of horizontal wells depends on the physical properties of matrix blocks, fluid viscosity, and magnitude of artificial fracture networks, which should not only avoid well interference but ensure the availability between wells. Well pad performances with different well spacing are simulation to investigate the effect of well spacing and fracture magnitude on the well performance (Figs. 11.98, 11.99 and 11.100).

1. Certain difference in daily and cumulative production for different well spacings

Given all other conditions are the same, different well spacings lead to the same initial production because different well spacings have different controlling reserves and thus the distinct production decline rates. In the middle and late stages, small well spacings have well interference and weaker supply capacity and thus quick production decline, whereas large well spacings have relatively large controlling reserves and thus slow production decline. Therefore, small well spacing has less cumulative production than large well spacing.

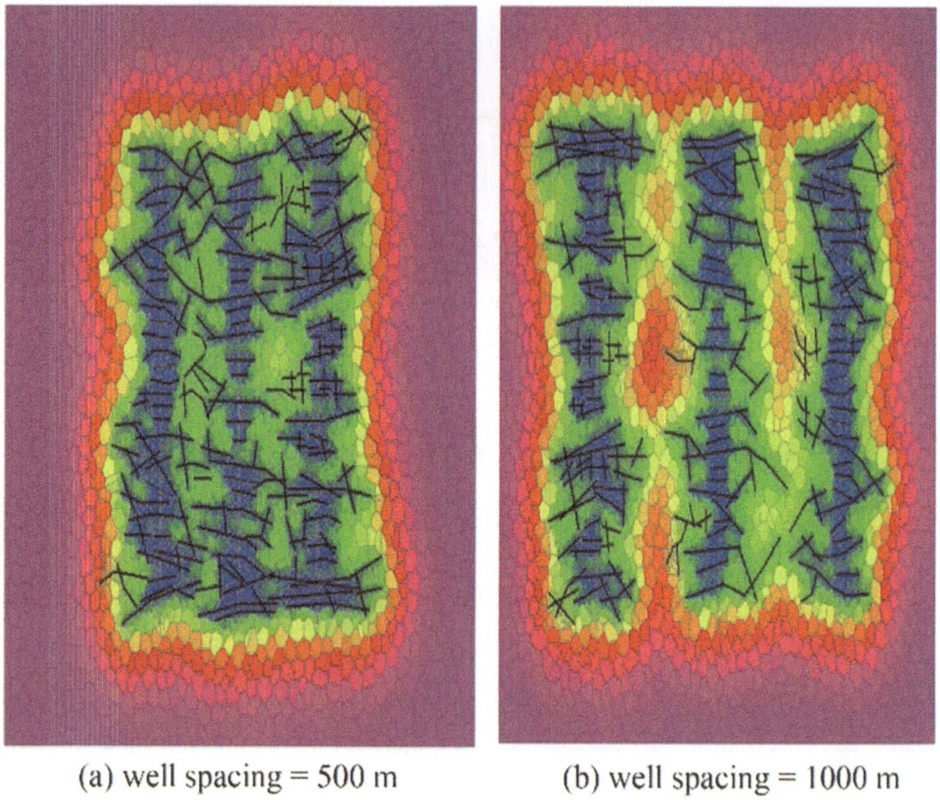

Fig. 11.100 Effect of well spacing on the well-pad pressure distribution

2. Certain difference in recovery for well spacings

For small well spacing, fractures easily join and cause well interference in middle and late stages, which lead to high availability. For large well spacings, an untapped zone is easy to occur, which causes low availability.

11.3.2.3 Optimization of Hydraulic Fracturing Parameters

The magnitude of hydraulic fracturing has an important impact on the production performance and productivity of tight oil/gas reservoirs. Fracture length, well spacing, conductivity, and fracturing efficiency are simulated to investigate their respective effects of the tight oil/gas production performance and productivity.

1. Fracture length

The length of hydraulic fracture determines the stimulation volume of artificial fractures, thus have a significant effect on the production. Different fracture lengths are simulated to investigate their effects on the production performance (Figs. 11.101, and 11.102).

It can be seen from the simulation results that longer fractures lead to higher production rates and recovery factors. Because longer fractures have larger fracture area and communicated volume, and thus a larger producing area and higher production rate.

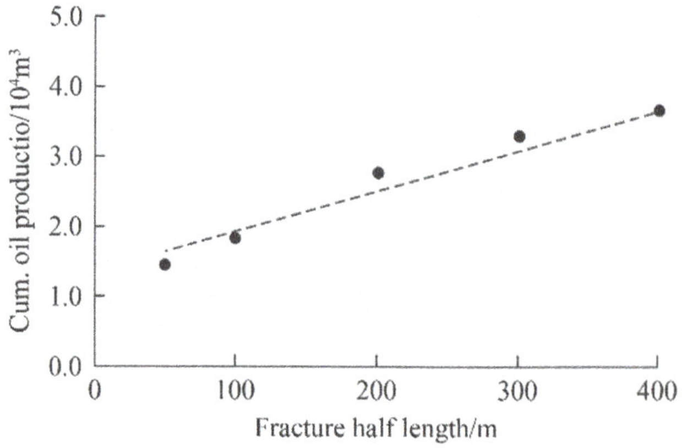

Fig. 11.101 Effect of different hydraulic fracture lengths on the cumulative production

2. Fracture spacing

The fracture spacing determines the number and area of cracks and the flow distance in matrix blocks. Different fracture spacings are simulated to reveal their effect on the production performance (Figs. 11.103 and 11.104).

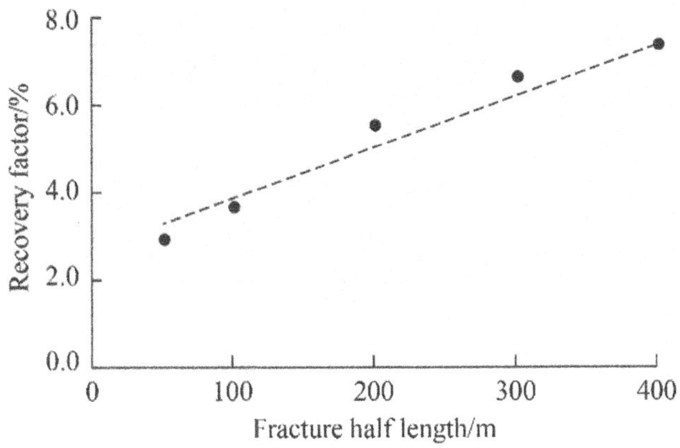

Fig. 11.102 Effect of different hydraulic fracture lengths on the recovery factor

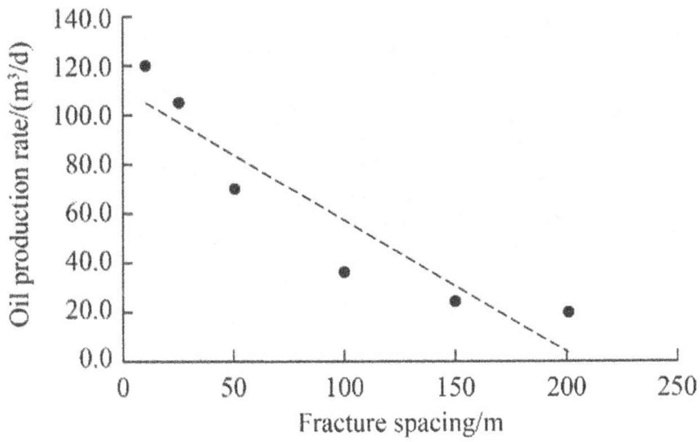

Fig. 11.103 Effect of fracture spacings on the daily production

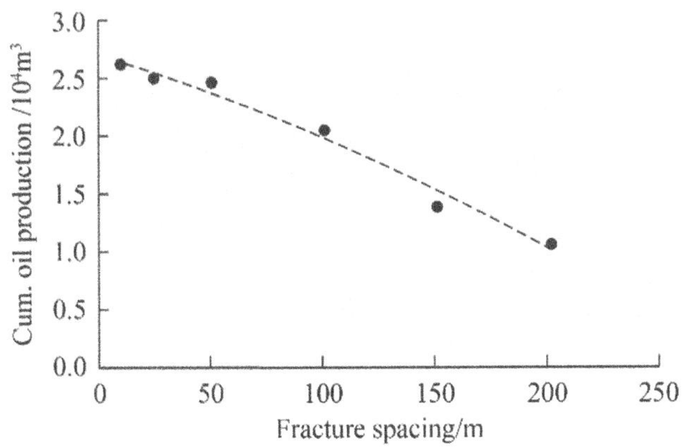

Fig. 11.104 Effect of fracture spacing on the cumulative production

The simulation results show that the fracture spacing has great influence on the production (Fig. 11.105).

① The shorter the fracture spacing, the bigger the fracture number, the larger the fracture area, the smaller the flow distance in the matrix blocks, and the higher the initial daily and cumulative productions.

② If the well spacing is too large, there may easily be an untapped zone, which leads to low recovery in-between fractures. When the fracture spacing is small enough, the control reserves of single fractures are limited and fracture interference exists, which decreases the production incremental. Therefore, the fracture spacing needs to be optimized.

3. Fracture conductivity

The conductivity of artificial fractures is greatly influenced by the proppant type, particle sizes, concentration, and supporting mode. Different conductivities lead to different communication abilities of fractures and the productivity of horizontal wells. Different fracture conductivity is simulated to reveal its effects on the production performance (Figs. 11.106 and 11.107).

Simulation results show that with the increase of fracture conductivity, communication is improved, so are the daily and cumulative production. When the fracture conductivity is enhanced to a certain level, the production incremental diminishes. Therefore, fracture conductivity needs to be optimized.

4. Efficiency of hydraulic fractures

Fracture efficiency varies greatly depending on the reservoir heterogeneity, in-situ stress, rock brittleness, and fracturing techniques. Fracture efficiency affects the production performance and productivity of horizontal wells. Different fracture efficiencies are simulated to reveal their effects on the production performance (Figs. 11.108 and 11.109).

According to the differences between the actual fracture length, opening, height and conductivity and their design values, fractures can be classified into three categories: (I) hydraulic fracture efficiency >70%; (II) hydraulic fracture efficiency between 40% and 70%; (III) hydraulic fracture efficiency between 10% and 40%.

(1) Big difference in daily and cumulative productions for distinct hydraulic fracture efficiencies

The daily and cumulative production drops with the decrease of hydraulic fracture efficiency. For a fracture efficiency of 100%, the production reaches up to 56 m^3/d and cumulative production to 2.31 × 10^4 m^3. If the Type III fracture dominates, the production rate decreases to 36.4 m^3/d and the cumulative production to 0.98 × 10^4 m^3, the production rate declines by 35% and the cumulative production drops by 57% (Fig. 11.110).

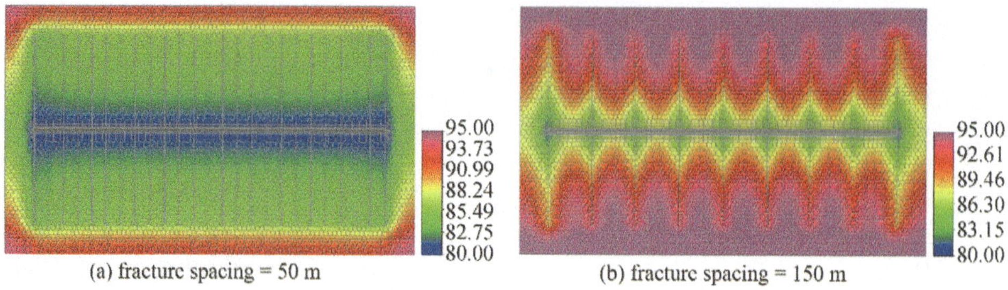

Fig. 11.105 Pressure distribution for fracture spacing

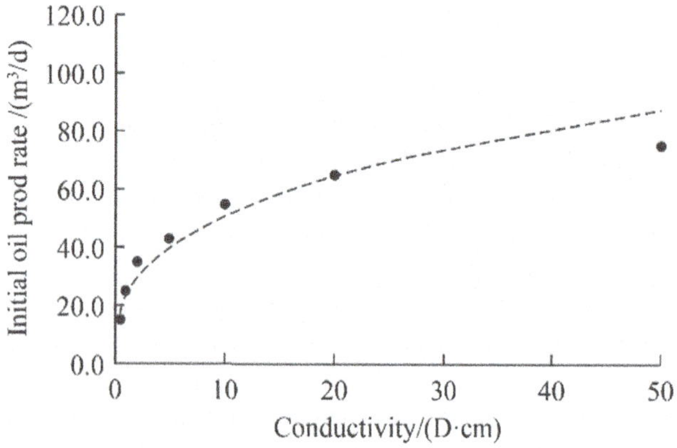

Fig. 11.106 Effect of conductivity on the daily production

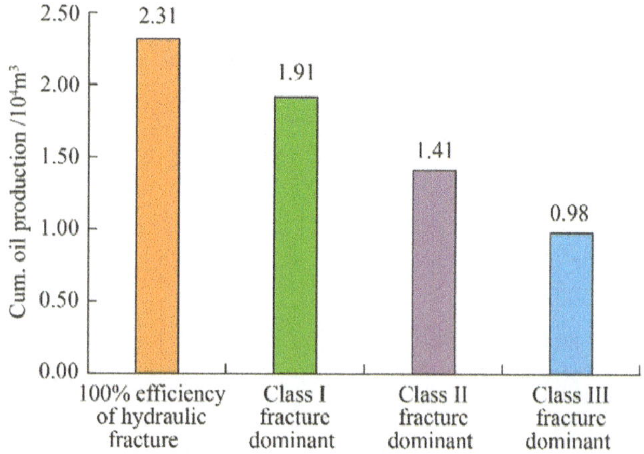

Fig. 11.108 Effect of fracture efficiency on the daily production

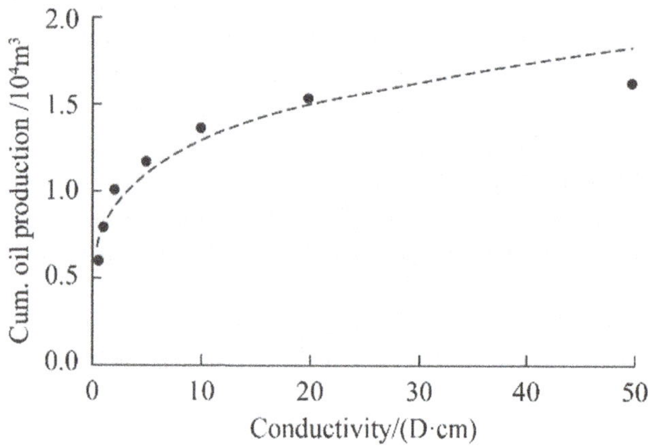

Fig. 11.107 Effect of conductivity on the cumulative production

(2) Some difference in availability and recovery for distinct hydraulic fracture efficiencies

With the decrease of the hydraulic fracture efficiency, the producing area and recoverable reserves gradually decrease, so are the availability and recovery (Fig. 11.111).

11.4 Coupled Simulation of Flow-Geomechanicss Coupling for Multiple-Scale Multiple Media During Fracturing-Injection-Production Processes

During the processes of fracturing, injection and production, pore pressures and in-situ stresses of artificial/natural fractures and pore media at different scales vary with time, which causes significant deformation of pores, natural and artificial fractures, and leads to the changes of geometric scales, properties, conductivities in-between media, and well index, which drastically impacts the reservoir dynamics and productivity characteristics. It is very important to simulate the effects of the dynamic change of multiple media on the reservoir and the production performance.

In this section, the deformation, geometric characteristics, and physical properties of multiple media at different scales during the fracturing-injection-production processes are simulated. The corresponding formation conductivity and well index are calculated to analyze the effects of flow-geomechanicss coupling on the productivity characteristics.

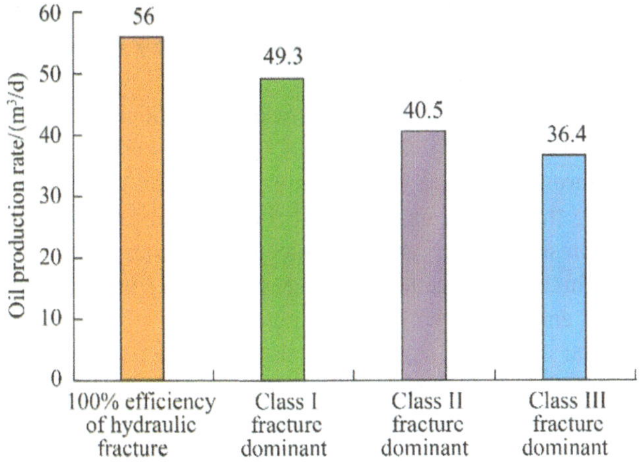

Fig. 11.109 Effect of fracture efficiency on the cumulative production

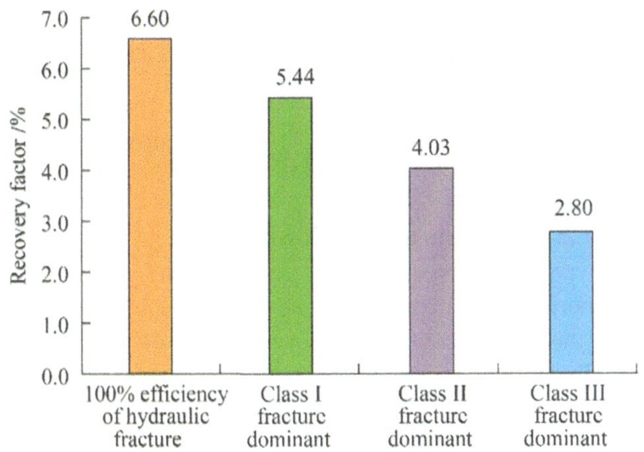

Fig. 11.110 Recovery factor for different fracture efficiency

11.4.1 Integrated Simulation of Fracturing-Injection-Production Processes

The fracture dynamics during the fracturing-injection-production processes are simulated through flow-geomechanicss coupling module, including multistage fracturing, artificial fracture generation, natural fracture opening, and fracturing-injection-integration integrated simulation.

11.4.1.1 Simulation of Stepwise Generation of Artificial Fractures During Multistage Fracturing

The number, geometric shape, and spatial distribution of hydraulic fractures are determined by the horizontal well hydraulic fracturing design, micro-seismic data, fracturing monitoring data and production data. At the same time, the stepwise generation of artificial fractures during a multistage fracturing process is simulated based on the fluids and sands amounts in each stage (Fig. 11.112).

11.4.1.2 Opening, Propagation and Extension of Artificial Fractures at Each Stage

According to the amounts of fracturing fluids and sands at each stage, the fracture evolution during rapid fracturing is simulated: when the pore pressure surpasses the fracturing pressure due to fracturing fluids injection, an artificial fracture opens. It continues opening and expending with the continuous injection of the fracturing fluids and sands. When the pressure exceeds the extension pressure, the fracture extends forward (Fig. 11.113).

Fig. 11.111 Pressure distribution for different fracture efficiencies

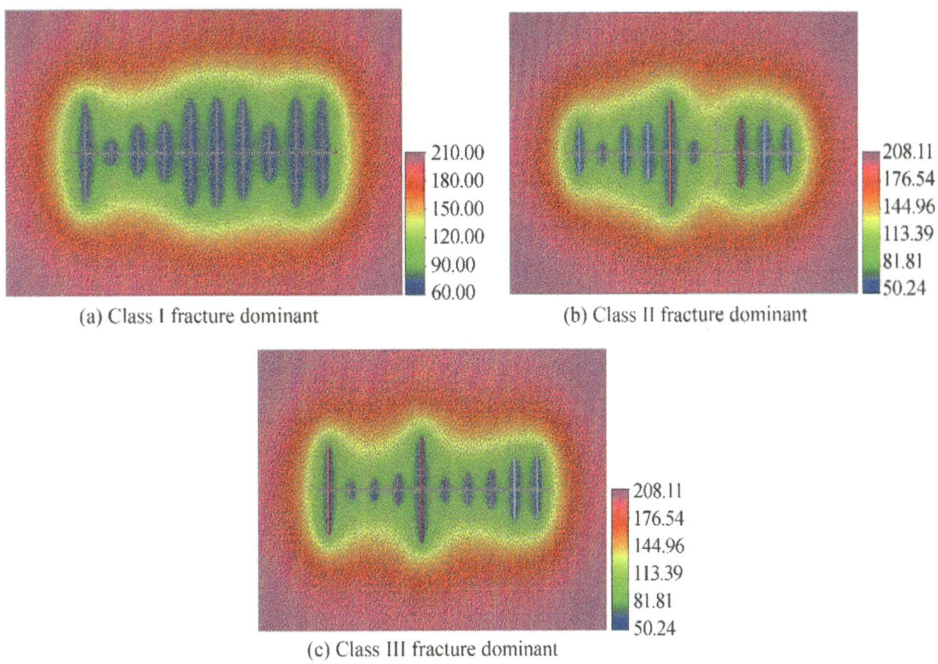

11.4.1.3 Simulation of Opening, Propagation and Extension of Natural Fractures

When an artificial fracture extends forward and meet natural fractures, fracturing fluids enter the natural fractures. The natural fractures open when the pressure surpasses their opening pressure and expand and extend as the pressure continues to rise (Fig. 11.114).

11.4.1.4 Integrated Simulation of Fracturing, Soak, Flowback, and Production

An integrated simulation of fracturing, soak, flowback, and production is conducted. First, the multistage fracturing is simulated according to the fluid volume, sand quantity and fracturing time at each stage. Meanwhile, the pressure diffusion process during the soak period is simulated. Then, fluid flow and pressure variation during the flowback process are simulated according to the flowback operating conditions and the corresponding fluid volumes. On top of this, the production process is simulated according to the operating conditions and production data (Fig. 11.115).

11.4.2 Simulation of Pore Pressure Variation During the Fracturing-Injection-Production Processes

The pore pressure in multiple-scale multiple media during the fracturing-injection-production processes is simulated, which shows that the pore pressure changes significantly in multiple-scale multiple media during the fracturing-injection-production processes.

11.4.2.1 Pressure Variation in Artificial Fractures, Natural Cracks, and Reservoir Matrix Pores

In the fracturing stage, the pressure in artificial fractures, natural cracks and reservoir matrix pores increases rapidly as fracturing fluids are injected into a formation. As the fracturing fluid enters natural cracks through artificial fractures, the pressure in the artificial fractures becomes higher than that in natural fractures. At the same time, the fracturing fluid enters reservoir matrix pores through artificial and natural fractures, as it flows slowly, the pressure in the matrix pore is the lowest. In the soak stage, the high-pressure fluid in the cracks diffuses into the surrounding reservoir matrix pores, so the pressure in fractures decreases somewhat and the pressure in the reservoir matrix pores increases continuously. During the flowback and production stages, the pressure in the artificial/natural fractures and reservoir matrix pores decreases significantly with the extraction of fracturing fluid and oil/gas. Because of the natural cracks connect directly to the wellbore, there is good conductivity and the pressure declines most significantly. Meanwhile, due to the low permeability in reservoir matrix pores, the pressure there declines the slowest (Fig. 11.116).

11.4.2.2 Pressure Variation in SRV

The variation of pore pressure in SRV is consistent with that the pore media of different scales. In the fracturing stage, with the rapid injection of fracturing fluids, the pressure rises rapidly (when the cumulative injection reaches 11730.2 m^3, the pressure in SRV increases to 4.89 MPa). In the soak stage, the pressure in the hydraulic basically stabilizes since

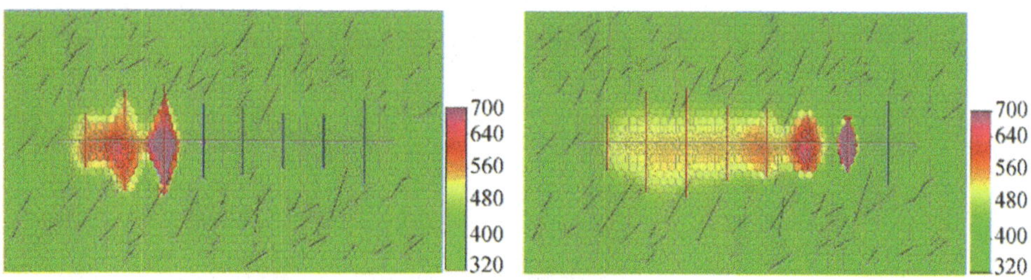

Fig. 11.112 Multistage fracturing simulation

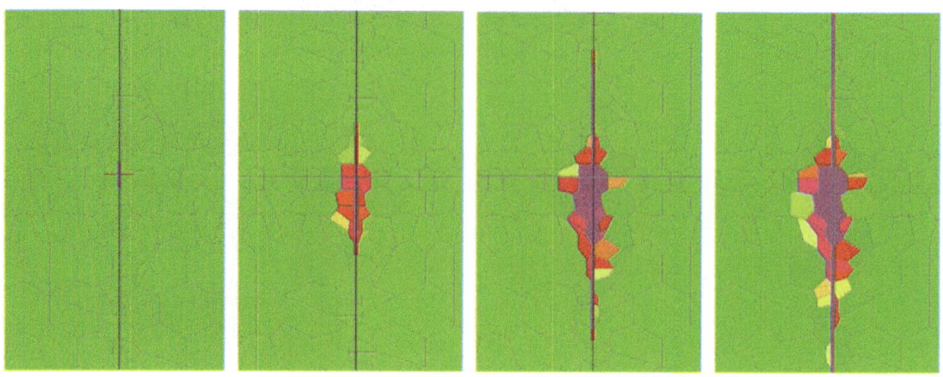

Fig. 11.113 Opening, propagation and extension of hydraulic fracture

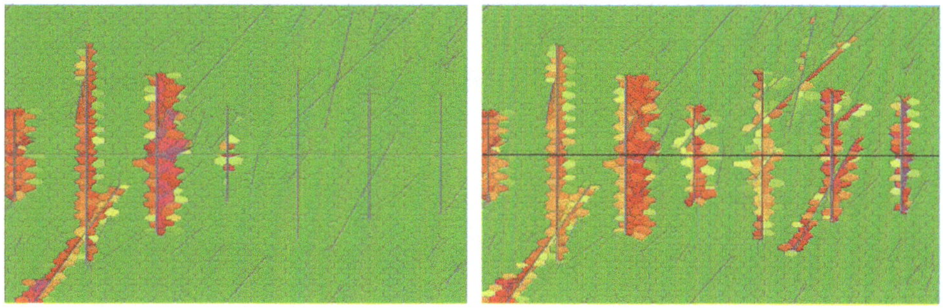

Fig. 11.114 Opening and extension of natural fractures

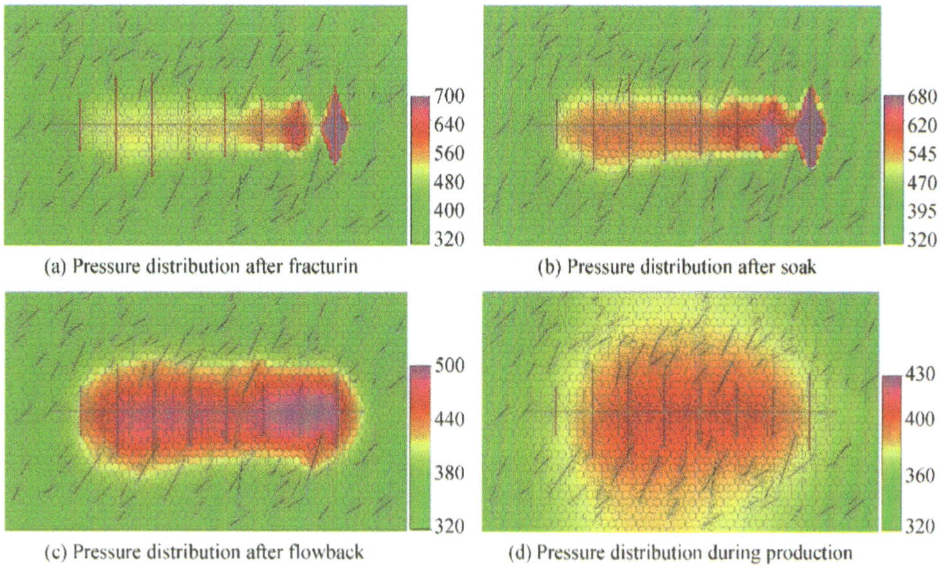

Fig. 11.115 Integrated simulation of fracturing-injection-production

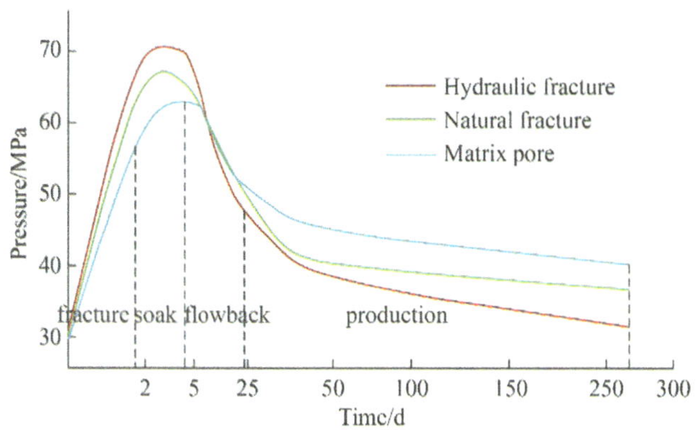

Fig. 11.116 Pressure variation in multiple media during the fracturing-injection-production process

its diffusion occurs among different media. In the flowback stage, the pressure in SRV decreases rapidly (the return volume of the fracturing fluid is 2533.7 m^3, the return rate is 21.6%, and the pressure reduction in SRV is 2.07 MPa); In the production stage, the pressure continues to decline with the production of oil/gas (the cumulative liquid production is 1790 m^3 and the pressure reduction in SRV is 1.75 MPa) (Fig. 11.117).

11.4.3 Simulation of Physical Property Variation of Multiple Media During the Fracturing-Injection-Production Processes

The dynamic variation of physical properties (porosity and permeability) of multiple media during the fracturing-injection-production processes is simulated, and the variation is found to be significant in each of the processes.

11.4.3.1 Variation of Porosity in Artificial Fractures, Natural Cracks, and Reservoir Matrix Pores

The porosity of artificial fractures is close to that of the matrix before breakdown. During the fracturing stage, after the opening of artificial fractures, the proppant particles are

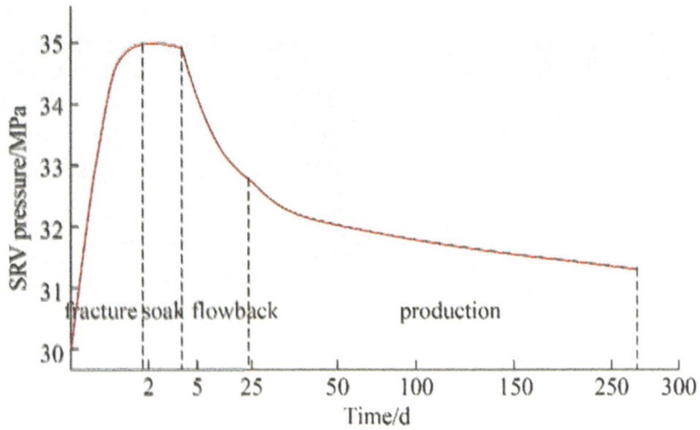

Fig. 11.117 Pressure variation in SRV during the fracturing-injection-production process

suspended in the fracture, and the porosity of fracture increases significantly. During the soak stage, the porosity of fractures declines due to a pore pressure drop. During the flowback and production stages, fractures deform and narrow down, and fracture porosity decreases gradually (Fig. 11.118).

The initial porosity of natural fractures is higher than that of the matrix due to the roughness. during the fracturing stage, the porosity of fractures increases sharply as the fractures open and widen. During the soak period, the porosity increases slightly. During the flowback and production stages, the supporting strength of natural fractures is the lowest, which leads to a dramatic decline in fracture width due to deformation. Thus, the decrease of porosity is most significant.

In the fracturing stage, the injection of a fracturing fluid makes the pore pressure rise rapidly, and the matrix porosity grows fast as the reservoir matrix pores expand. In the soak period, the matrix porosity increases slightly. In the flowback and production stages, the porosity gradually decreases due to the compression and deformation.

11.4.3.2 Variation of Permeability in Artificial Fractures, Natural Cracks, and Reservoir Matrix Pores

The initial permeability of artificial fractures is similar to that of matrix. With the injection of fracturing fluids, artificial fractures open and gradually widen, leading their permeability to increase rapidly. During the soak stage, the permeability of fractures declines somewhat due to the slight drop of pore pressure. During the flowback and production stages, artificial fractures deform and narrow, and fracture permeability gradually decreases but remains at a relatively high level due to the supporting of proppants (Fig. 11.119).

During the fracturing stage, natural fractures open and gradually widen as the pressure increases, leading their permeability to increase rapidly. During the soak period, the permeability of natural fractures increases slightly due to the pressure increase. During the flowback and production stages, as the quick reduction of pressure, natural fractures have the most serious deformation and closure, thus they narrow down drastically, leading the permeability declines most significantly.

During the fracturing stage, the injection of the fracturing fluid makes the pore pressure rise rapidly, and the matrix permeability grows fast as the reservoir matrix pores expand. In the soak period, the matrix permeability increases slightly due to a pressure increase. During the flowback and production stages, the permeability gradually decreases due to a pressure reduction and the resultant compression and deformation.

11.4.4 Simulation of Conductivity and Well Index Variation During the Fracturing-Injection-Production Processes

Throughout the fracturing-injection-production process, the deformation and variation of geometry and properties of multiple-scale multiple media cause the conductivity among

Fig. 11.118 Porosity variation during the fracturing-injection-production process

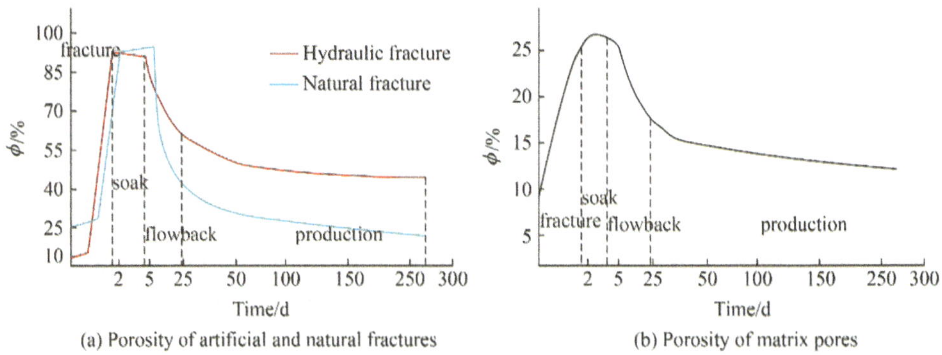

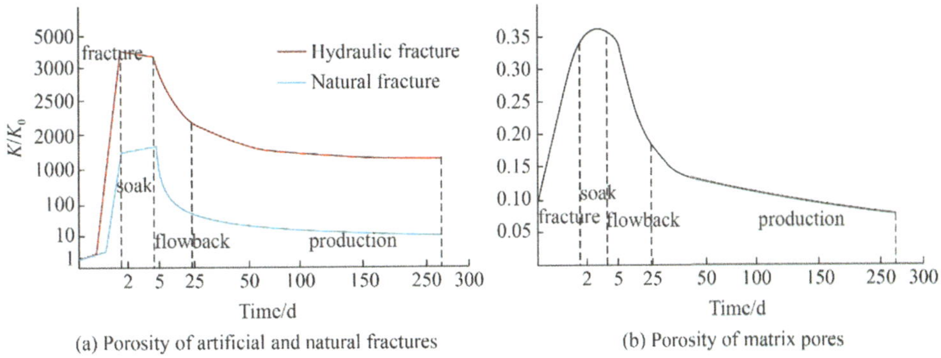

Fig. 11.119 Permeability variation during the fracturing-injection-production process

different media and well index between formation and wellbore to change tremendously. Therefore the conductivity and well index during the fracturing-injection-production processes are simulated in this section.

11.4.4.1 Variation of Conductivity Among Different Multiple Media

In the fracturing stage, the conductivity among different media increases rapidly with the permeability of the multiple media. In the soak stage, it decreases slightly. During the flowback and production stages, it drops largely due to significant decline in permeability of the multiple media (Fig. 11.120).

11.4.4.2 Variation of Well Index Between Formation and Wellbore

During the fracturing stage, the formation permeability rapidly increases due to the increase of pressure in the wellbore grids, which causes the well index between formation and wellbore to grow. During the soak stage, high-pressure fluids in the wellbore grids diffuse into the surrounding formation media, causing the pressure in the wellbore grids to drop. So, both the permeability and well index drop. During the flowback and production stages, permeability decreases due to dramatic falloff of the pressure in the wellbore grids, permeability declines, so is the well index (Fig. 11.121).

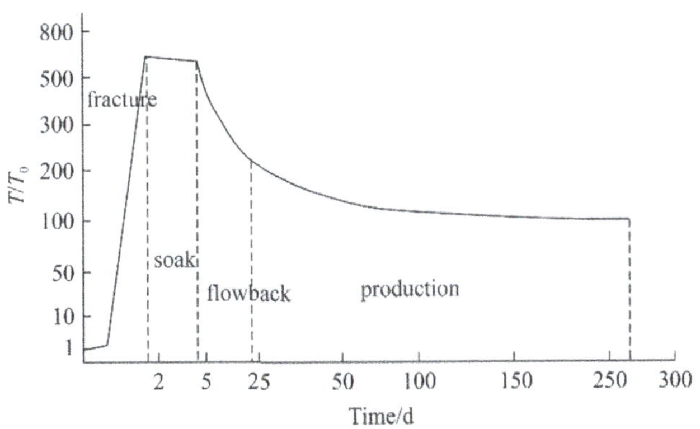

Fig. 11.120 Variation of conductivity among different multiple media

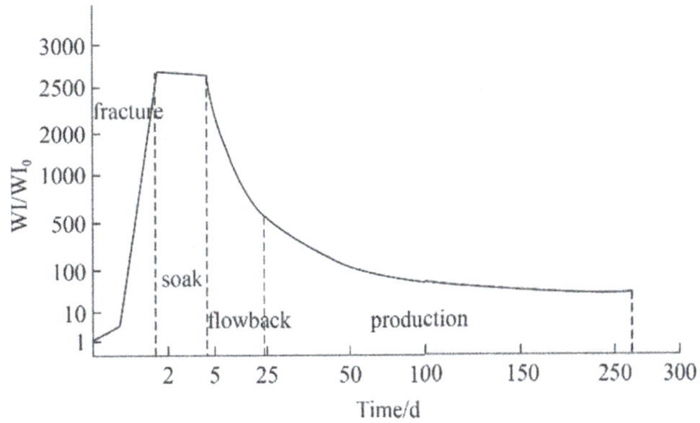

Fig. 11.121 Variation of well index between formation and wellbore

11.4.5 Simulation of the Impact of Coupled Flow-Geomechanicss Mechanism on Production

Throughout the fracturing-injection-production process, the deformation and variation of geometry and properties of multiple-scale multiple media cause the conductivity among different media and well index between formation and wellbore to change accordingly, which largely affect the production performance and the productivity. Therefore, the effects of flow-geomechanics coupling on the production under different development modes are simulated in this section.

11.4.5.1 Depletion Production Mode

Under the depletion mode, the formation pressure gradually decreases with the production of oil/gas. This causes the shrinkage of reservoir matrix pores and the closure of natural and artificial fractures, leading the conductivity and well index to decrease, and eventually the daily and cumulative production to drop. Comparative analysis shows that when

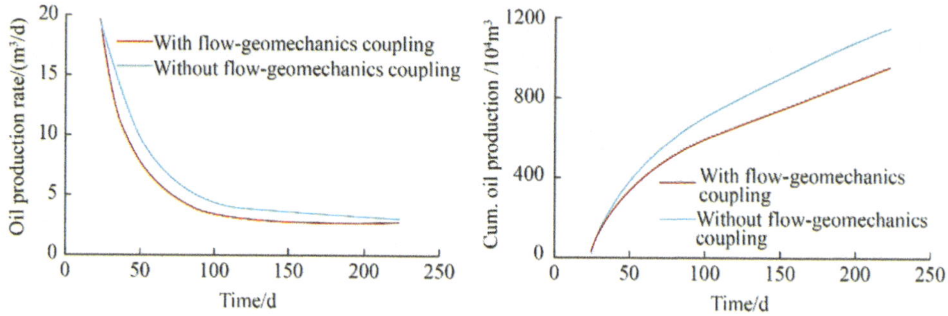

Fig. 11.122 Effect of coupled flow-geomechanics simulation on the production under depletion mode

the flow-geomechanics coupling is ignored, the daily and the cumulative productions are larger than the actual data, the deviation can be up to 17.8% (Fig. 11.122).

11.4.5.2 Injection-Production Mode

After a certain period of depletion production (about 200 days), water is injected to supplement the formation energy (injection for 30 days at an injection rate of 36 t/d). Then, an injection-production period of about 180 days is applied. During the water injection, the pore pressure of multiple media increases rapidly, so are the physical properties. During the production stage, the physical properties decrease due to the falling of pore pressure in the multiple media (Figs. 11.123, 11.124, 11.125 and 11.126).

The production decreases rapidly in the depletion-production stage. It is improved to some extent through supplementing the formation energy, and the decline rate is slowed down and cumulative production increased. This shows that the energy supplement through water injection can improve the recovery. Comparative analysis shows that when the effects of flow-geomechanics coupling are considered, the daily production decreases, and the cumulative production declines by 18.6% (Fig. 11.127).

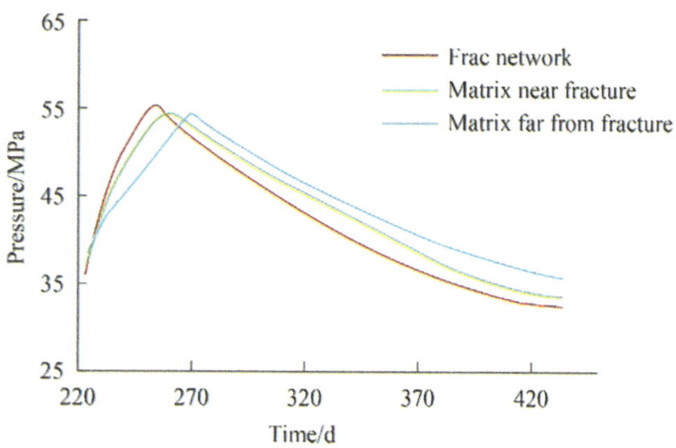

Fig. 11.123 Pressure of mesh pore media with injection-production mode

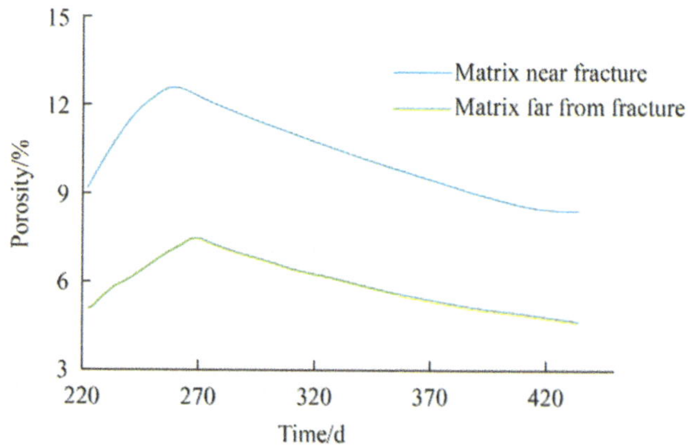

Fig. 11.124 Porosity of mesh pore media with injection-production mode

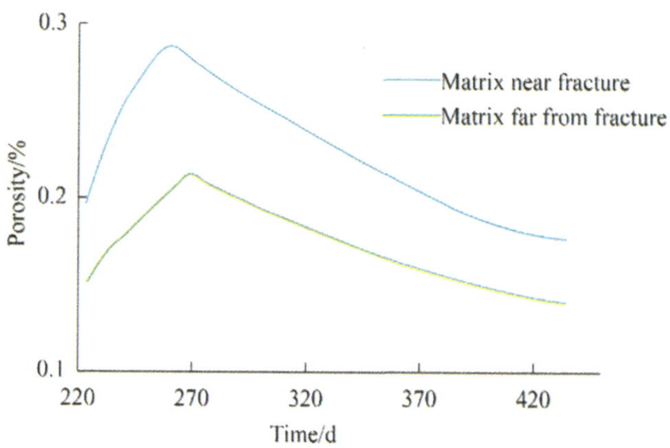

Fig. 11.125 Permeability of mesh pore media with injection-production mode

11.4.5.3 Huff-N-Puff Production Mode

After a certain period of depletion production mode, huff-n-puff production mode is applied for multiple cycles. In each cycle, the pore pressure and physical parameters increase with an increase in the injected volume during the injection stage, and decrease with the increase of the produced volume in the production stage (Figs. 11.128, 11.129, 11.130 and 11.131). The average pressure and physical

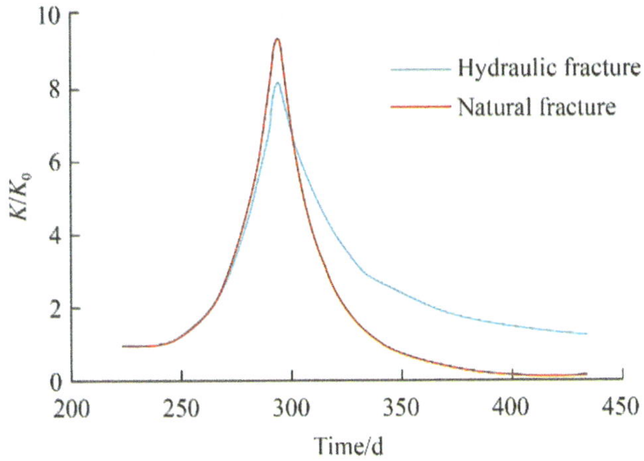

Fig. 11.126 Permeability of different fractures with injection-production mode

properties of each cycle gradually decrease since the produced fluids is greater than the injected fluids in each cycle.

After the initial depletion production, production can be improved through multiple cycles of the huff-n-puff mode. Comparative analysis is done for the effects of flow-geomechanics coupling on production. It is found when the flow-geomechanics coupling is considered, the daily output decreases and the cumulative oil production declines by 19.3% (Fig. 11.132).

11.4.5.4 Comparison of Different Development Modes

Through the comparison of three modes (depletion, injection-production, and huff-n-puff), it is found that the daily production decreases fastest and the cumulative production is the lowest with the depletion mode. The oil production rate and recovery are improved through formation energy supplement by the injection-production mode. Compared with the depletion mode, the cumulative production increases by 7.9%. Under the huff-n-puff mode, the formation energy and imbibition effect are further strengthened, and cumulative production and recovery are further improved. Compared with the depletion mode, the cumulative production increases by 18.9% (Figs. 11.133 and 11.134).

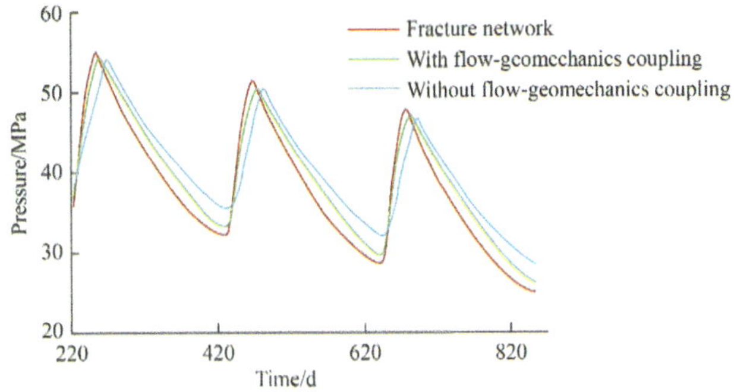

Fig. 11.128 Pressure of mesh pore media with huff-n-puff mode

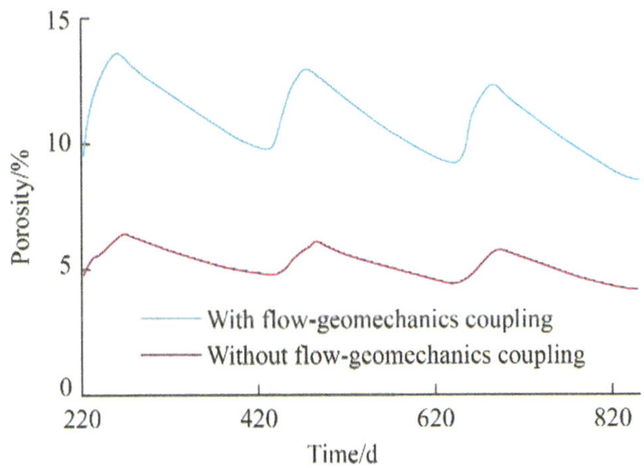

Fig. 11.129 Porosity of mesh pore media with huff-n-puff mode

11.5 Simulation of Production Performance by Different Types of Unstructured Grids

Aiming at the complex geological conditions, diversified boundary conditions, and developed multiple media at multiple scales, different types of unstructured grids and the combinations thereof are applied. In this way, an integrated simulation of macro zones, small-scale units, and micro sub-multiple media can be realized, and the calculation

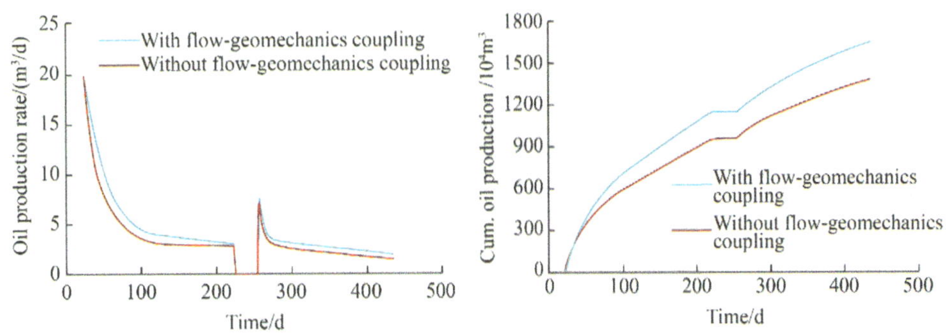

Fig. 11.127 Effect of flow-geomechanics coupling on the production performance

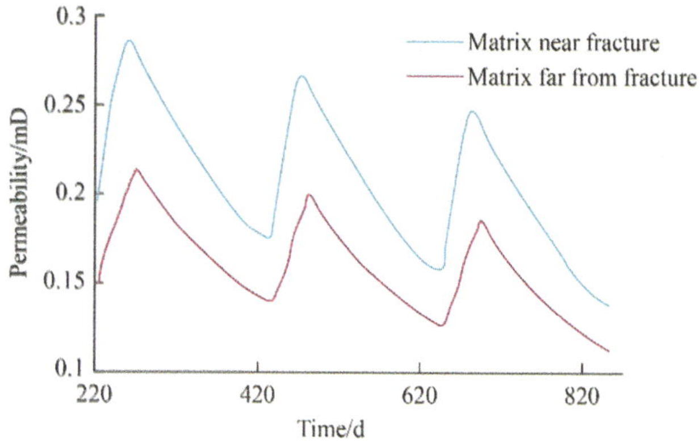

Fig. 11.130 Permeability of mesh pore media with huff-n-puff mode

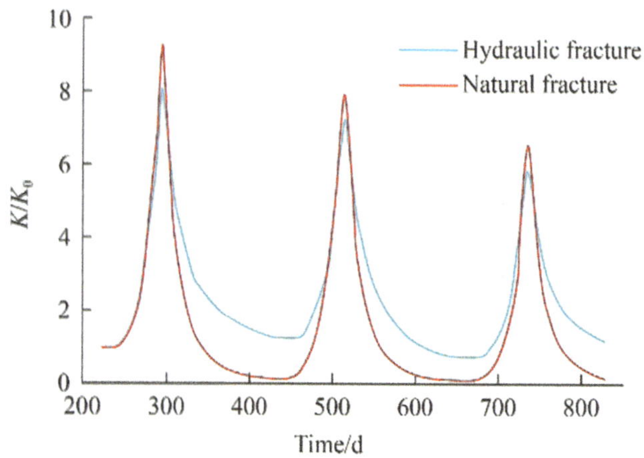

Fig. 11.131 Permeability of different fractures with huff-n-puff mode

accuracy and speed are greatly improved. different types of single scale-varying grids, mixed treatment of different types of grids, and the hybrid grids for multiple media are simulated to analyze the effect of grid types on the fineness of characterization, simulation precision and computing speed.

11.5.1 Simulation of Production Performance by Single-Type Scale-Varying Grids

Triangular meshes, quadrilateral meshes, and PEBI meshes are used to characterize the complex inner and outer boundaries, such as the geological conditions of tight reservoirs, the multiple media of different scales, and hydraulic fractured horizontal wells. The grid scale can be automatically optimized according to the geometric characteristics of the object, which not only improves the fineness of characterization but reduces the number of grids.

Given the same reservoir geological conditions, fractures, and inner and outer boundary conditions of a horizontal well, the triangular meshes, quadrilateral meshes and PEBI meshes are respectively employed for self-adaptive and scale-varying grid simulation. The corresponding fineness of characterization, calculation accuracy, and computing speed for different scale-varying grids are compared and analyzed.

11.5.1.1 Effect of Grid Types on Characterization Fineness for Complex Fracture

Through scale-varying, the distribution and shape of complex fractures can be flexibly handled by triangular meshes and PEBI meshes with high precision. The characterizations are closer to actual situations. However, it is difficult to precisely describe the distribution and shape of the fractures by quadrangle meshes, and the characterizations deviates far from the actual situations (Fig. 11.135).

11.5.1.2 Effect of Grid Types on Calculation Accuracy

Simulations with different types of scale-varying meshes show that the results by triangular meshes and PEBI meshes are close to the actual situation. While by quadrilateral meshes, due to the poor characterization, the accuracy of results are unsatisfactory and hardly match the real data (Figs. 11.136 and 11.137).

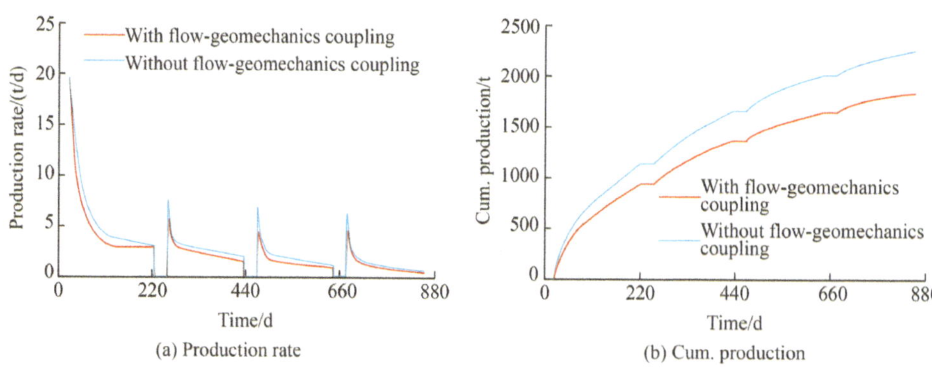

Fig. 11.132 Effect of coupled flow-geomechanics simulation on the production

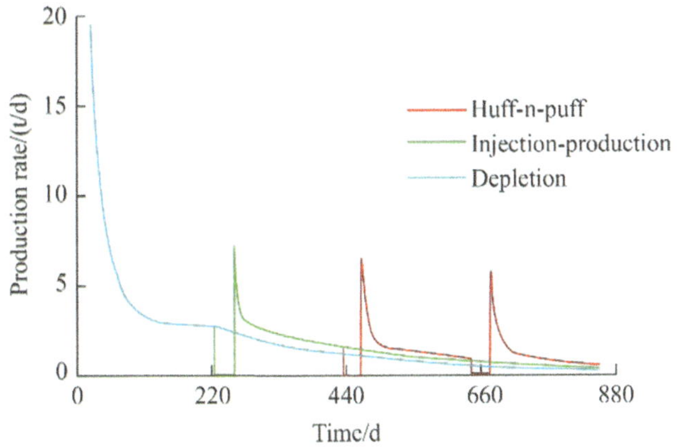

Fig. 11.133 Daily oil production for different modes

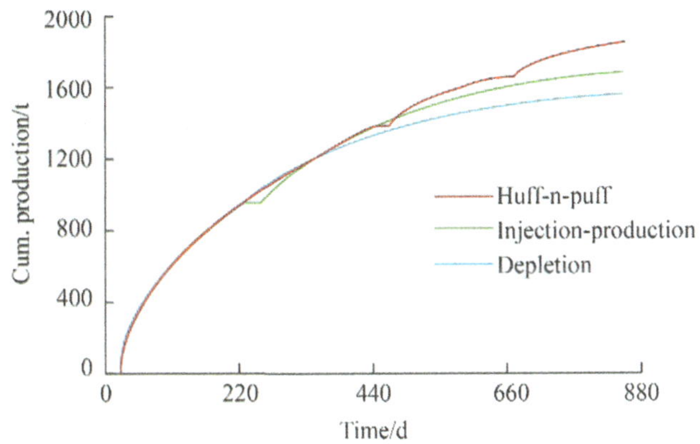

Fig. 11.134 Cumulative oil production for different modes

11.5.1.3 Effect of Grid Types on Grid Number and Computation Speed

With PEBI meshes, the number of grids is the largest and the precision is the highest, but its computing time is the shortest. Although triangular meshes result in high characterization precision, but need longest calculation time. The number of quadrilateral cells is the least and its computing is fast yet its characterization precision is the worst. Therefore, the PEBI mesh is the best choice to deal with complex geological conditions and complex fractures in terms of characterization precision and computing time (Figs. 11.138 and 11.139).

11.5.2 Simulation of Production Performance by Different Types of Hybrid Grids

Natural fractures, artificial fractures and horizontal wells in tight reservoirs have distinct geometric characteristics and complex spatial distributions. Aiming at the geometric characteristics and spatial distributions of such media, different types of meshes are mixed with scale-varying processing, which can not only precisely describe the geometric features and spatial distribution of different media, but also improve the simulation precision and calculation accuracy.

Different types of scale-varying meshes are used to represent and simulate different media (pores, natural fractures, artificial fractures, and horizontal wells). Artificial/natural fractures and horizontal wells have distinctive geometric features and distribution patterns, which significantly affects the fluid flow. Quadrangular long strip meshes can be

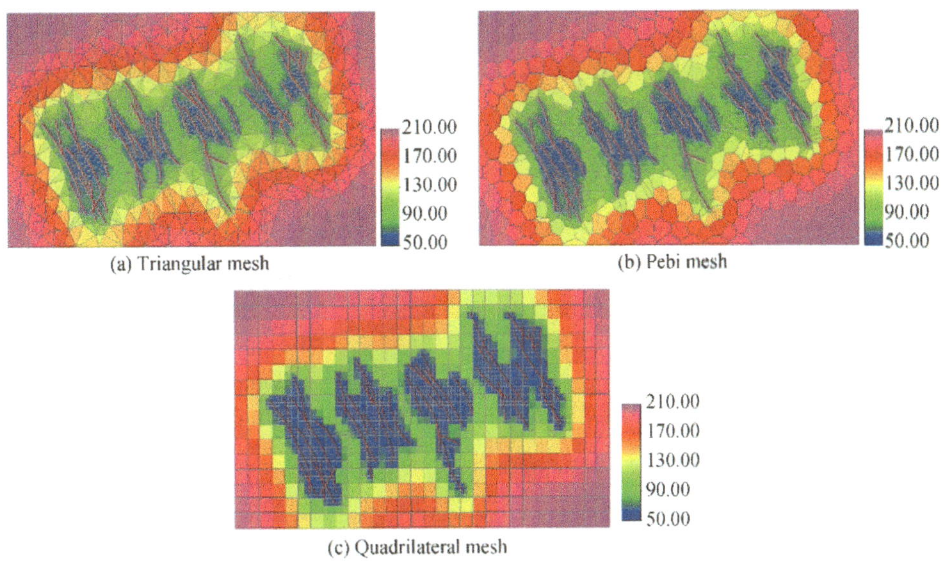

Fig. 11.135 Effect of grid types on the characterization of complex fractures

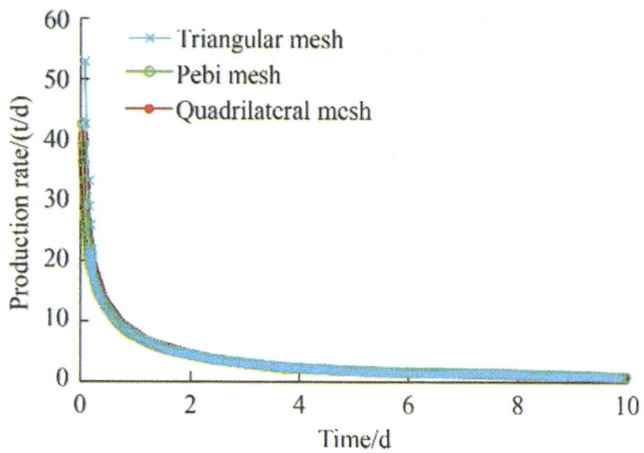

Fig. 11.136 Daily production with different types of mesh

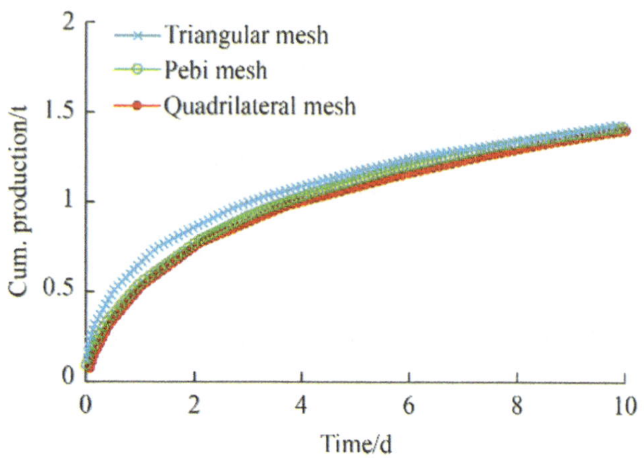

Fig. 11.137 Cumulative production with different types of mesh

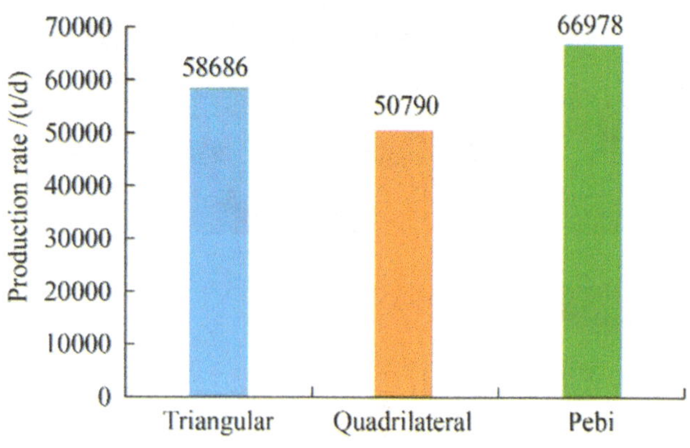

Fig. 11.138 Number for different types of mesh

applied to artificial/natural fractures to describe their geometric characteristics and flow behavior. At the same time, runway meshes are employed to represent the geometric

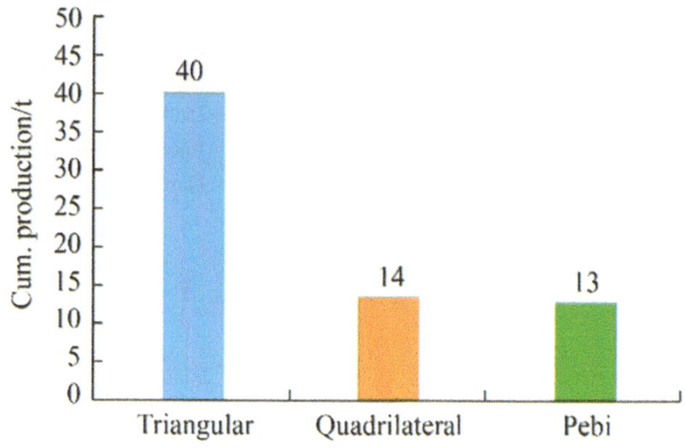

Fig. 11.139 Computation speed for different types of mesh

characteristics of horizontal wells and the corresponding flow patterns. For matrix blocks, the flow has no clear direction due to its small scale of pores, for which the PEBI mesh, triangular mesh, quadrilateral mesh, and different combinations thereof can be used to describe the geometric properties, spatial distribution, and fluid flow.

To analyze the influence of hybrid grids on the production performance, based on long-strip quadrangular meshes for artificial/natural fractures and runway meshes for a horizontal well, four cases are simulated: PEBI mesh, triangular mesh, triangular mesh near fractures and quadrilateral mesh for other parts, and PEBI mesh near the fractures (Fig. 11.140).

1. Effect of mesh combinations on characterization fineness and simulation accuracy

As the long-strip meshes and runway meshes can fairly describe the geometric features and flow behaviors in natural/artificial fractures and horizontal wells, respectively. PEBI and triangular meshes can flexibly handle various complicated boundary conditions. But for the matrix block, with unclear flow direction due to the small scale of pores, PEBI, triangular, and quadrilateral meshes, and their combinations can be applied to describe the geometric and distribution characteristics and the fluid flow inside.

Therefore, for such complex objects as pore media, natural fractures, artificial fractures, and horizontal wells, different types of scale-varying meshes are employed for mixed representation and simulation, which can improve the characterization fineness of natural/artificial fractures, horizontal wells, and matrix blocks. Simulation results (Figs. 11.141 and 11.142) show that after optimization, the simulation results of different combinations are close to each other, indicating that the simulation results are reliable and meet the requirements of simulation precision.

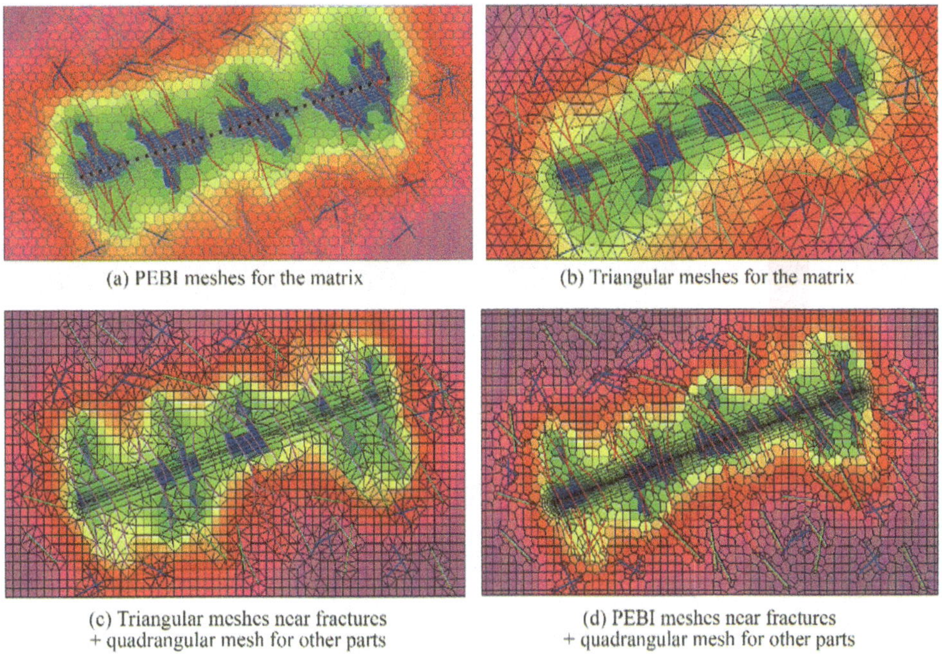

Fig. 11.140 Simulation of different hybrid meshes

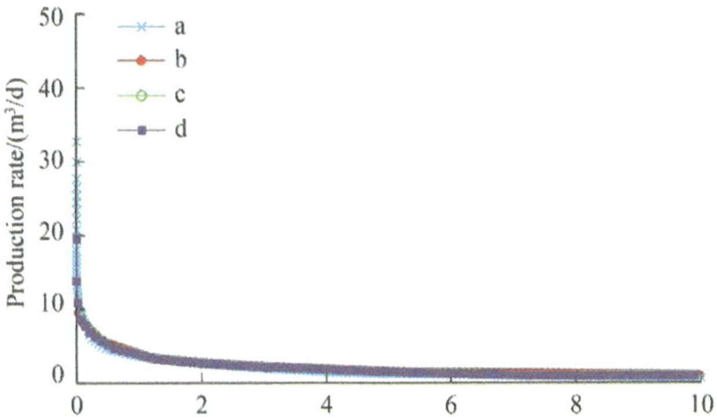

Fig. 11.141 Daily production for different hybrid meshes

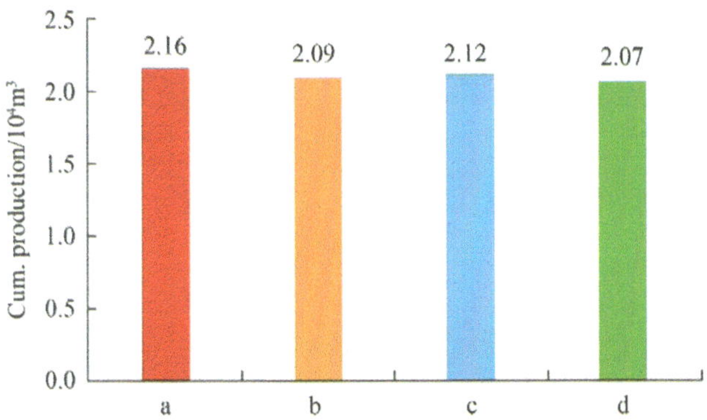

Fig. 11.142 Cumulative production for different hybrid meshes

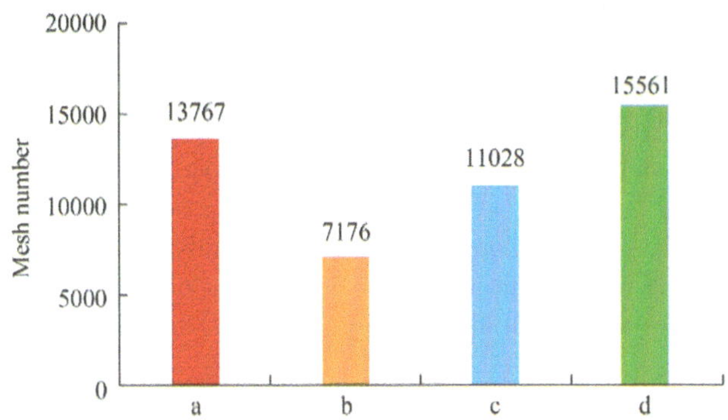

Fig. 11.143 Grid number for different hybrid grids

2. Effect of mesh combination for mesh quantity and computing speed

Simulation results show that the number of meshes is consistent with the computing speed, the fewer the number of meshes, the faster the computation (Figs. 11.143 and 11.144).

11.5.3 Simulation of Macroscopic Heterogeneity and Multiple Media by Hybrid Grids

Tight reservoirs have strong macroscopic heterogeneity, and the reservoir properties are rather different among different regions. At the same time, because of the differences in lithology, lithofacies, and reservoir types, there are strong

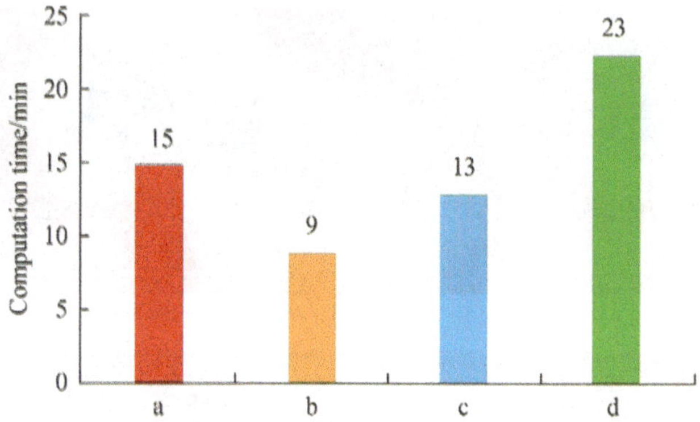

Fig. 11.144 Computation time for different hybrid grids

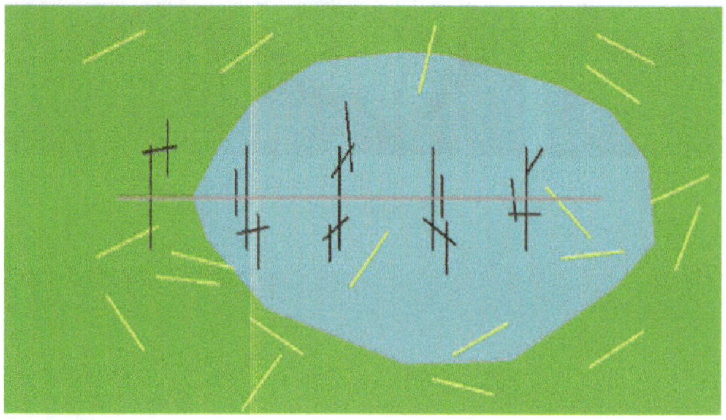

Fig. 11.145 Reservoir macroscopic heterogeneity zoning

heterogeneity among the small-scale units, and multiple media in different unit has different types, quantity compositions and spatial distributions. Therefore, it is necessary to conduct the integrated numerical simulation of large-scale zoning, small-scale sub-uniting, and micro-scale sub-multiple media. At the same time, since tight formations has large- and meso-scale natural/artificial fractures, small-scale fractures and micro- and nano-pores, their geometric characteristics and physical properties are significantly different, and flow mechanisms are even more complicated. Therefore, the discrete macro and micro fractures need to be integrated as a whole, to carry out non-continuous, discrete, and hybrid simulation for multiple media (Fig. 11.145).

Therefore, the hybrid meshes are employed in the zoning simulation at different scale, and discrete, and hybrid simulation for multiple media at different scales, and the simulation precision and computing speed are significantly improved.

To reveal the effects of different hybrid grids on the simulation of zoning, sub-units, and sub-media, based on the long-strip meshes for large fractures and PEBI meshes for small units, three cases are applied to simulate the micro-scale multiple media: (a) nested meshes for the relay supply & drainage, (b) interactive meshes, and (c) nested + interactive meshes (Fig. 11.146).

1. Effect of different types of hybrid grids on flow behavior and production performance between multiple media

It can be seen from the simulation results that given the same formation heterogeneity and grid types, and the same multiplicity and quantity of multiple media, simulation results are

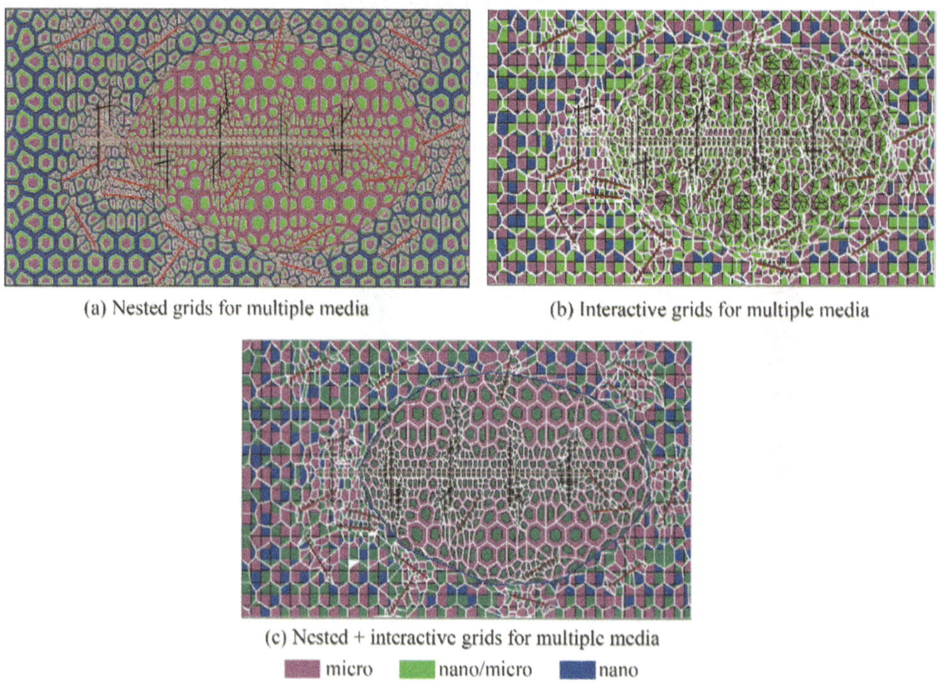

Fig. 11.146 Dynamic simulation of partitioned, sub-unit and sub-media with hybrid meshes

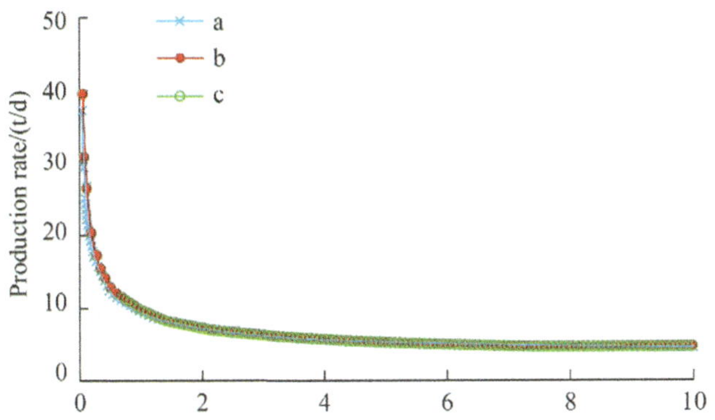

Fig. 11.147 Daily production for different types of hybrid meshes

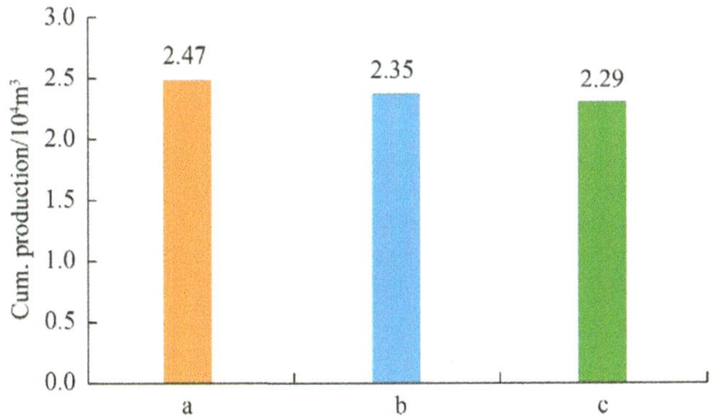

Fig. 11.148 Cumulative production for different types of hybrid meshes

somewhat different with different types of hybrid grids for flow behavior and production performance between multiple media. This shows that the spatial distribution patterns and flow behaviors of multiple media have something to do with its production performance (Figs. 11.147 and 11.148).

2. Effect of different types of hybrid grids on computing speed

Different types of hybrid grids are used to simulate the flow behaviors in the multiple media, and the resultant computing

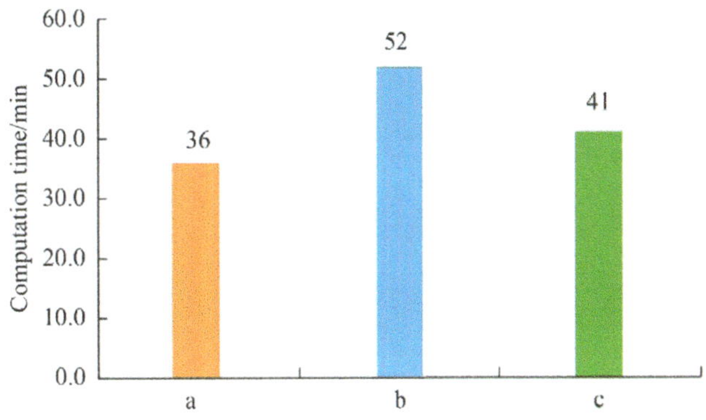

Fig. 11.149 Computation time for different types of hybrid grids

speeds are rather different (Fig. 11.149). When nested meshes being used for the relay supply & drainage, the contact and flow between different media inside one unit are relatively simple, and the computing is fast. When interactive meshes being used, different media inside one unit contact each other randomly, and the flow behaviors are complicated, which results in a heavy computing load and a long computing time.

References

Du J (2016) Continental tight oil in China. Petroleum Industry Press, Beijing

Du J, He H, Yang T, Li J, Huang F, Guo B, Yan W (2014) Progress in china's tight oil exploration and challenges. China Petroleum Expl 19(1):1–8

Wang H et al (2016) Experimental study on slippage effects of gas flow in compact rock. Chinese J Geotechn Eng 38(5):185–777

Yu R, Zhang X, Bian Y, Li Y, He M (2012) Flow mechanisms of shale gas and productivity analysis. Nat Gas Ind 32(9):10–15

Zhang L, Tang H, Chen G, Li Q, He J (2014) Adsorption capacity and controlling factors of the lower Silurian Longmaxi shale play in southern Sichuan basin. Nat Gas Ind 34(12):63–69

Karimi-Fard M, Durlofsky LJ, Aziz K (2003) An efficient discrete fracture model applicable for general purpose reservoir simulators. Texas, USA, SPE Reservoir Simulation Symposium

Wu Y, Press K (1988) A multiple-porosity method for simulation of naturally fractured petroleum reservoirs. SPE Reservoir Engineering

12 Trend and Prospects of Numerical Simulation Technology for Unconventional Tight Oil/Gas Reservoirs

At present, the numerical simulation for unconventional tight oil/gas reservoirs has made breakthroughs in theory and technology and been applied in practice. In future, the following aspects should receive more attention: i. integration of unconventional geological modeling and numerical simulation, ii. real-time dynamic4D geological modeling of fluid distribution, geostress, and geomechanical properties for multiple media of different scales, based on the modeling of multiple-scale multiple media with pores and fractures, iii. integrated numerical simulation technologies for fracturing-injection-production, coupled simulation of flow-geomechanics-temperature, numerical simulation for multiple components, complicated phase behaviors, and dynamic flow parameters; iv. efficient solving technologies for numerical simulation of unconventional resources.

12.1 Geological Modeling for Unconventional Tight Oil/Gas Reservoirs

Based on conventional geological models, unconventional tight oil/gas reservoir simulation emphasizes the establishment of models, including geological models for multiple media at multiple scales, fluid models for multiple media with pores and fractures, geostress and geomechanical models. For the multiple-scale nano/micro-pores and natural/artificial fractures, the modeling technologies for discrete fractures, and discrete multiple media, mixed multiple media at multiple scales have been developed. For the discontinuous distribution of reservoir fluids, oil bearing characteristics and occurrence in multiple-scale multiple media with pores and fractures, and the discontinuous variation of fluid properties, distributed modeling technologies need to be developed for multiple-scale multiple media in the future. The geostress distribution varies greatly in space and value, and it changes dynamically and redistributes during the production process, so the modeling technologies for the geostress and rock mechanics would be the future targets. In addition, the parameters of geological models, fluid distributions, geostress and rock mechanics dynamically change in the production process, so the establishment of the real-time 4D geological model is also a trend in the future.

12.1.1 Modeling Technology of Geostress and Geomechanical Parameters

The conventional oil/gas reservoirs have natural productivity and can be effectively developed without fracturing. Therefore, in the geological modeling of conventional oil/gas reservoirs, it is generally unnecessary to consider the modeling of geostress and geomechanical parameters. Unconventional tight oil/gas reservoirs need fracturing for effective production. The geomechanical parameters and the distribution and strength of geostress have a critical influence on the post-fracturing/refracturing performance. Depending on the reservoir structure, formation lithology, and buried depth, the distribution, direction, and strength of geostress and the horizontal geostress and the difference of horizontal geostress varies greatly in different regions, zones, and well segments (Ge et al. 1998; Zeng and Wang 2005). Moreover, such activities as hydraulic fracturing/refracturing, water injection, gas injection and production will cause the change of pore pressure and temperature, and generate the induced geostress, which redistributes the geostress field, and enlarges the heterogeneity of geostress distribution, and impacts the shape, direction, and complexity of artificial fracture network during the hydraulic fracturing/refracturing process. In addition, the dynamic changes of geostress field and mechanical parameters of reservoirs greatly affect the deformation of pore and fracture media at different scales, and thus affect the fluid flow and its production. Therefore, in the future, it is necessary to develop the geostress field and geomechanical parameter modeling technology for the

Q. Ran, *Unconventional Tight Reservoir Simulation: Theory, Technology and Practice*,
https://doi.org/10.1007/978-981-32-9848-4_12

unconventional tight oil/gas reservoirs, establish unconventional geological models for the distribution, direction, and strength of geostress and geomechanical parameters, and assist the optimization design of the horizontal well hydraulic fracturing. Meanwhile, it is imperative to take account of the redistribution of geostress and geomechanical parameters in the production process in the geological modeling, to realize the coupled flow-geomechanics numerical simulation of the geostress field and the flow field.

12.1.2 Modeling Technology of Fluid Distribution in Multiple-Scale Multiple Media

Without evident oil-water and gas-water contacts in unconventional tight oil/gas reservoirs, but there are large oil-water and gas-water transition zones. Given large oil-bearing areas in tight formations, due to the differences in reservoir lithology, sand bodies, physical properties, pore structures as well as petroleum accumulation mechanism and the collocation between source rocks and reservoirs, great distinctions exist in oil bearing and original oil/water/gas saturations in different blocks, zones, and well segments. Nano/micro-pores and natural/artificial fractures are discontinuously distributed in space, which leads to large differences in the occurrence state of the fluids and their composition and properties. This eventually causes the discontinuous distribution of fluid properties and parameters in the multiple media with pores and fractures at multiple scales.

The conventional fluid modeling is based on the oil-water and gas-water contacts, and static and continuous flow models are established by using the equilibrium method. The fluid distribution and parameters are constant (Wang and Yao 2009; Zhang 2010), which is unable to describe the discontinuous distribution of reservoir fluids, oil bearing and occurrence state, and flow parameters. Thus, it is necessary to develop distributed modeling technology for the fluid distribution at different scales in the future.

12.1.3 Real-Time Dynamic 4D Geological Modeling Technologies

During the fracturing-injection-production processes of unconventional tight oil/gas reservoirs, the dynamic changes of fluid flow, geostress, and temperature fields lead to the redistribution of pressure, saturation, geostress and temperature in reservoirs. The flow-geomechanics coupling causes the deformation of multiple media with pores and fractures at multiple scales, and the change in geometric and physical parameters. The variation of fluid flow and temperature causes the shifting of the geostress and pressure fields, leading to the redistribution of rock mechanics parameters. The variation of the temperature field causes the change of occurrence, composition, and flow parameters of the reservoir fluids (Zhou et al. 2008).It can be seen that during the production process of tight oil/gas reservoirs, the geological model (physical properties of different media), the flow model (pressure field, saturation field, fluid property and flow parameter), the geostress model (geostress distribution, rock mechanics parameter), and the temperature model all dynamically change. So it is necessary to develop a real-time dynamic 4D geological model for an unconventional tight oil/gas reservoir, which would be a main direction in the future.

12.2 Integratednumerical Simulation of Fracturing-Injection-Production

The physical properties of unconventional tight reservoirs are poor, which need hydraulic fracturing for commercial production; also the refracturing may be necessary to improve the performance and recovery. The fracturing and production of unconventional tight oil/gas reservoirs is a continuous process. The integrated numerical simulation of fracturing, injection and production processes is the main trend in the future for unconventional oil/gas reservoirs (Zou et al. 2015; Lv 2012; Gao et al. 2015; Yao et al. 2016).

At present, a fracturing simulator and a reservoir simulator are independent simulators for unconventional reservoir development. The two simulators are based their models on the rock mechanics and flow mechanics, respectively, which develops two numerical simulation technologies and simulators. The hydraulic fracturing simulator and the reservoir simulation simulator use independent geological models to simulate the fracturing process and the production process. In reservoir simulation, the artificial fracture information obtained by the fracturing simulator is incorporated into the geological model, but the change of pressure and saturation fields from the fracturing simulator are neglected.

The fracturing, soak, flowback, injection, and production of unconventional reservoirs is a continuous dynamic process. Integrated simulation of these processes is going to be a trend in the future for the numerical simulation of unconventional tight oil/gas reservoirs. First, an initial model is developed that includes a geological model for multiple media with pores and fractures, a fluid distribution model, and a geomechanics and geostress model, as shown in Fig. 12.1. Second, the multi-stage fracturing, soak, and flowback processes are simulated to describe the generation and extension of artificial fractures, the opening and extension of natural fractures, and expansion of matrix pores. This way, it is able to obtain the shape and spatial distribution and

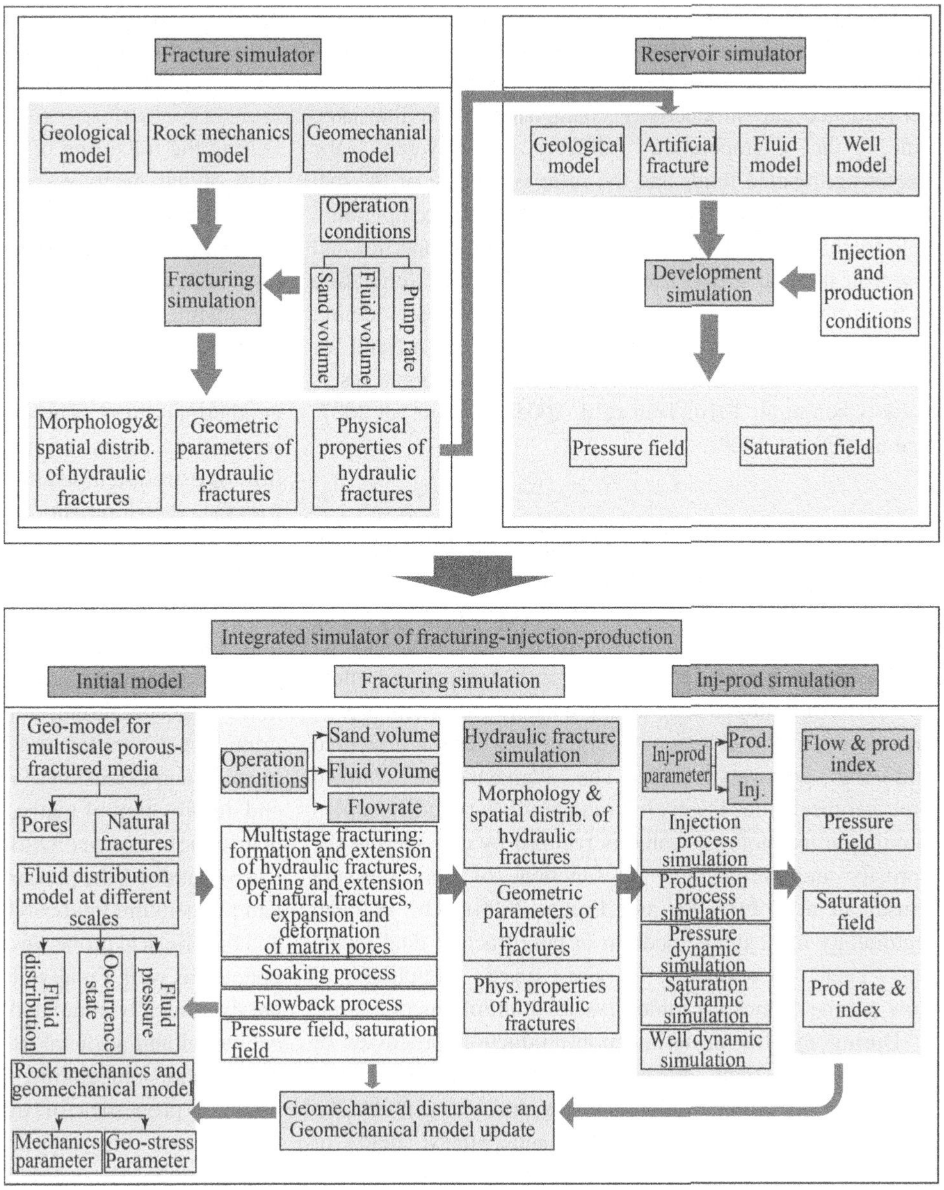

Fig. 12.1 Simulation procedures for conventional and unconventional reservoirs

geometric parameters, physical properties of artificial fractures, and update the fluid distribution model and the initial geostress model on the basis of the variation of pressure field, saturation field, and stress field during the fracturing process. Finally, the injection and production processes are simulated according to the initial geological model and artificial fracture model, to realize the integrated simulation of the multistage fracturing, soak, flowback, injection, and production of unconventional tight oil/gas reservoirs.

12.3 Coupled Flow-Geomechanics Simulation of Fracturing-Injection-Production Processes

During the fracturing-injection-production process of unconventional tight reservoirs, there is strong coupling relationship between the flow field and the geostress field. The natural/artificial fractures and pore media of different scales change dynamically, causing the geometric scale, property parameters, inter-media conductivity, and well indices to vary accordingly, which greatly affects the well performance and productivity characteristics. At the same time, the temperature difference between the fracturing/injected fluids and the reservoir fluids, or thermal recovery techniques, can induce the reservoir temperature field to change, leading to the change of fluid properties and reservoir thermal stress. Therefore, the integrated simulation of fracturing-injection-production and the coupled modeling of flow geostress field-temperature field will be a trend for numerical simulation of unconventional tight oil/gas reservoirs (Chen et al. 2010; Tran et al. 2005; Abass et al. 2007; Tae and David 2009).

12.3.1 Coupled Flow-Geomechanics Simulation of Multiple Media During Fracture-Injection-Production Process

During the production process, the pore pressure and effective stress of multiple-scale multiple media varies drastically, and there is extremely strong coupling between the flow field and the geostress field. The conventional coupled flow-geomechanics reservoir simulation is a pseudo-coupled simulation technology, which is realized by considering the porosity and permeability as functions of pore pressure (Susan et al. 1993; Wu and Pruess 2000). The coupled technology for a single medium or pore/fracture dual media has been developed to the quasi coupling of flow-geomechanics for multiple-scale multiple media.

During the fracturing-injection-production processes of unconventional tight oil/gas reservoirs, the dynamic coupling of fluid-geomechanics changes drastically under the coupled effects of flow flied and stress field of artificial/natural fracture and pore media of different scales. Based on the deformation mechanisms and the dynamic physical properties of multiple-scale multiple media during different production processes, an coupled flow-geomechanics numerical simulation for fracturing-injection-production process is developed for the multiple media of different scales.

The conventional pseudo-coupled flow-geomechanics simulation technologies only account for the effect of flow fields on the physical properties, are unable to obtain dynamic change of the stress field during the fracturing-injection-production process to meet practical requirement of oil/gas field development. Integrated fracturing-injection-production and fully coupled flow-geostress simulation is the future of numerical simulation technologies for unconventional tight oil/gas reservoirs.

12.3.2 Coupled Flow-Geostress-Temperature Simulation of Multiple Media During Fracturing-Injection-Production Process

During the fracturing or injection process of unconventional tight oil/gas reservoirs, the temperature difference between the fracturing/injected fluids and the reservoir fluids cannot only shrink the rock and decrease its physical properties, but also increase the fluid viscosity and reduce the mobility. Meanwhile, for tight reservoirs with a viscous oil, the application of thermal recovery processes can elevate the reservoir temperature and change the fluid properties and geothermal stress. Therefore, it is necessary to develop an integrated fracturing-injection-production and coupled flow-geostress-temperature numerical simulation technology for the development of such unconventional tight oil reservoirs.

First, an initial model is developed that includes a geological model for multiple media, a fluid model for pore-fracture media, a rock mechanics and geostress model, and a temperature model, as shown in Fig. 12.2. Second, the multi-stage fracturing, soak, and flowback are simulated to mimic the generation and extension of artificial fractures, the opening and extension of natural fractures, and the expansion and deformation of matrix pores. Thus, the shape and spatial distribution, and geometric parameters of artificial fractures can be obtained, the pressure field, saturation field, stress field, and temperature field can be updated based on their changes, and the physical properties of multiple media and fluids can be updated as well. Finally, the injection and production process are dynamically simulated according to the updated geological model and artificial fracture model. The pressure, saturation, stress, temperature fields are updated, as well as the physical properties of multiple media and fluid property parameters. This way, the integrated fracturing-injection-production and coupled flow-geomechanics-temperature numerical simulation can be carried out.

12.4 Simulation of Multi-Component and Complex-Phase for Multiple-Scale Multiple Media

For unconventional tight oil/gas reservoirs, gas injection and property modification can be applied to modify the occurrence state, the rock-fluid interface properties, and fluid properties, to improve the recovery of adsorbed and filmed oil in nano/micro-pores. The existing mathematical models of numerical simulation are unable to describe the multi-component and complex-phase change in multiple media of different scales, as well as the variation of oil/gas occurrence state, rock-fluid interface properties, and fluid

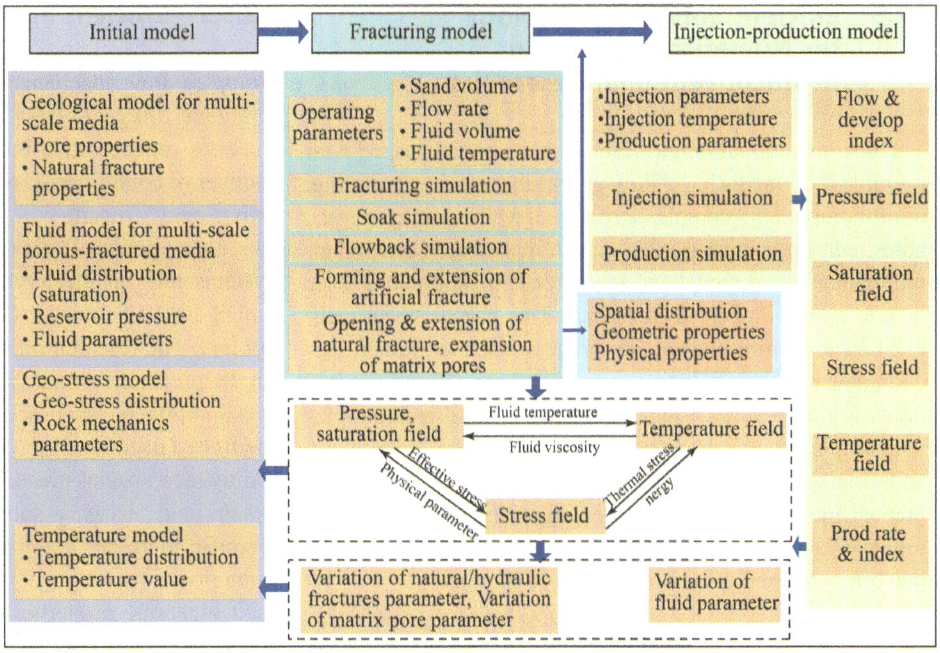

Fig. 12.2 Integrated fracturing-injection-production and coupled flow-geomechanics-temperature simulation forunconventional oil/gas reservoirs

properties in nano/micro-scale multiple media. Therefore, it is desirable to develop a numerical simulation technology to describe the multiple media, multi-component, complex-phase, fluid properties, and dynamic parameters for unconventional tight oil/gas reservoirs under different development methods.

12.4.1 Simulation of Multi-Component Complex-Phase for Multiple-Scale Multiple Media

A conventional multi-component numerical simulator can describe gas injection mechanisms and phase behaviors, but it is only applicable to single or dual media oil/gas reservoirs (Zhou 2012; Li et al. 2006). For unconventional tight oil/gas reservoirs, the numerical simulation technologies have been developed for discontinuous multiple media including nano/micro-pores and natural/artificial fractures. Because the fluid component, occurrence state, phase change, and flow mechanisms vary greatly under different recovery processes, especially, gas injection techniques, the numerical simulation technology for discontinuous multiple media can only handle black oil production but unable to characterize the difference of component and its dynamic change in multiple-scale multiple media, neither to describe the flow mechanisms of oil-gas mixed phase and unmixed phases during gas-injection processes. Therefore, it is necessary to establish a mathematical model to accurately describe the multi-component and complex-phase behaviors in multiple media at different scales, and develop numerical simulation technologies for multiple-components and complex-phases of discontinuous multiple media.

12.4.2 Simulation of Fluid Properties and Flow Parameters for Multiple-Scale Multiple Media

For unconventional oil/gas reservoirs, recovery technologies can be applied to modify the occurrence state, rock-fluid interface properties, and fluid properties to improve the recovery of adsorbed and filmed oil in nano/micro-pores and fractures. The numerical simulation technology of discontinuous multiple media is capable of the dynamic simulation of multiple media with different parameters, but unable to describe the variation of oil/gas occurrence state, rock-fluid interface properties (wettability, interfacial tension, and capillary pressure), fluid properties and flow parameters (density and viscosity) under property-modification production processes. The conventional EOR simulation technology can only simulate the EOR processes in single or dual media reservoirs. Therefore, it is necessary to build a mathematical model that can describe the occurrence state of oil/gas, rock-fluid interface properties, dynamic fluid properties and flow parameters, and develop the numerical simulation technology of property-modification recovery processes for unconventional tight oil/gas reservoirs.

12.5 Efficient Solving Technology for Numerical Simulation of Unconventional Reservoirs

The unconventional oil/gas reservoirs are characterized with multiple media, flow-geomechanics coupling, multi-component, complex-structured wells, multiple variables, and unstructured grids, which makes the corresponding mathematical model and Jacobin coefficient matrix large in size and complicated in format. This causes some technical issues during the solving process, such as tremendous computer memory and long computing time. Therefore, it is necessary to develop compression and storage technologies for complex matrices, efficient preconditioning technology, and heterogeneous parallel technology to make the mathematical models easy to solve, enlarge the computing magnitude of unconventional numerical simulation, and improve the computing efficiency of numerical simulation.

12.5.1 Compression and Storage Technology for Complex Matrix

Multiple-scale multiple media are developed in unconventional oil/gas reservoirs, which are featured with multi-component, multi-flow regimes, flow-geomechanics coupling, and complex structured wells. This makes the corresponding mathematical models and Jacobin coefficient matrices very complicated. Moreover, unstructured grids are required to treat the complex fracture networks, horizontal wells, and boundary conditions, which further increases the number of invalid meshes and aggravate the complexity of matrices. To effectively reduce the computer memory and total storage, and improve the efficiency of iterative calculation, it is necessary to develop an efficient compression and storage method for large matrices (Sun et al. 2007). There exist a number of compression methods: CSR, CSC, COO, HYB, CSR5, and CSRL. The optimum storage structure depends on the types of the problem, machine conditions, and parallel computing. It is anticipated that more storage formats for sparse matrix will be developed with the development of parallel framework. Therefore, it is necessary to develop matrix compression technologies for the increasingly heterogeneous and fragmentized parallel computing settings, to meet the requirements for unconventional reservoirs that are characterized by multiple media, multi-component, multi-flow regimes, unstructured grids, and develop compression and storage technology for complex matrix under increasingly heterogeneous and fragmented parallel computing settings.

12.5.2 Efficient Preconditioning Technology

Complex flow mechanisms and multiple media of unconventional oil/gas reservoirs lead to complex coefficient matrices, uneven distribution of eigen values, and a large number of conditions. Using the traditional iterative methods only may result in slow convergence or even non-convergence. The existing large sparse linear algebraic systems rely on the Krylov subspace solution methods, in which GMRES and BiCGstab methods are most commonly applied. The research on solving methods is mainly focused on the preconditioning, from the early single-stage universal preconditioners (such as the block ILU preconditioner and the nested decomposition preconditioner) to the two-stage or multi-stage special preconditioners (such as CPR preconditioners and auxiliary subspace correction preconditioners). Because there are many different types of problems in the numerical simulation of unconventional oil/gas reservoirs, their algebraic properties of discrete systems are quite different in complexity, so are the difficulties of the solving methods. Therefore, aiming at the complex coefficient matrices and uneven distribution of eigen values, it is imperative to develop an efficient preconditioning technology to make mathematical models easy to solve, and improve the efficiency and robustness of numerical simulation.

12.5.3 Heterogeneous Parallel Technology

Technical issues of numerical simulation for unconventional oil/gas reservoirs, such as complex mathematical models and coefficient matrices, integrated modeling of fracturing-injection-production, and coupled flow-geomechanics simulation, cause a large magnitude of computing, multiple media, and high-definition of inner/outer boundary fine grids, which further leads to a large computing load and a long computing time. Thus, it is desirable to develop efficient parallel solvers for the numerical simulation of unconventional oil/gas reservoirs. At present, parallel hardware architecture technology is being developed rapidly, from the early distributed parallel CPU architecture to the nowadays heterogeneous parallel architectures that are assisted by the CUDA/MIC acceleration computing cards (Zhang et al. 2006). The future parallel hardware architecture will still be fragmented with tremendous uncertainty for a long time, but is almost certain to be the heterogeneous parallel hardware architecture. The existing single MPI or CUDA parallel technology cannot meet the needs in efficient solving for the numerical simulation of unconventional oil/gas reservoirs. Therefore, it longs to develop parallel architecture technologies with algorithms, data structures, and solving methods increasingly compatible to heterogeneity, to enhance the size, speed, and solution precision of the numerical simulation problems for unconventional reservoirs.

References

Abass HH, Ortiz I, Khan MR, Beresky JK, Aramco S, Sierra L (2007) Understanding geostress dependent permeability of matrix, natural fractures. Hydraulic Fractures in Carbonate Formations. SPE110973

Chen HY, Teufe IL, Wand L (2010) Coupled fluid flow and geomechanics in reservoir study-l. Theory and governing equations. SPE30752

Gao D et al (2015) Research progress and development trend of numerical simulation technology for unconventional wells. Appl Math Mech 36(12):1239–1254

Ge H, Lin Y, Wang S (1998) Geostresses determination technology and its applications in petroleum exploration and develop. J Univ Pet China 22(1):94–99

Li Q, Kang Y, Luo P (2006) Multi-scale effect in tight sandstone gas reservoir and production mechanism. Nat Gas Ind 26(2):111–113

Lv X (2012) A study on unconventional hydrocarbon reservoir modeling and a/d integration technology. Coal Geol China 24 (8):85–91

Sun Z, Lu H, Sun Z (2007) Upscale method for the fine 3d geological model of hydrocarbon reservoirs and its application effectiveness. J Geomechan 13(4):368–375

Susan EM, et al (1993) Staggered in time coupling of reservoir flow simulation and geomechanical deformation: step I-one-way coupling. SPE 51920

Tae HK, David SS (2009) Estimation of fracture porosity of naturally fractured reservoirs with no matrix porosity using fractal discrete fracture networks. SPE Reserv Eval Eng 110720:232–242

Tran D, Nghiem L, Buchanan L (2005) An overview of iterative coupling between geomechanical deformation and reservoir flow. SPE 97879

Wang Z, Yao J (2009) Analysis of characteristics for transient pressure of multi-permeability reservoirs. Well Test 18(1):10–12 2010, 32 (6):112–120

Wu YS, Pruess K (2000) Integral solutions for transient fluid flow through a porous medium with pressure-dependent permeability. Int J Rock Mechan Product Sci 37(2):51–61

Yao J, et al (2016) Scientific engineering problems and development trends in unconventional oil/gas reservoirs. Pet Chin Sci Bull 01 (1):128–142

Zeng L, Wang G (2005) Distribution of earth stress in Kuche Thrust belt, Tarim Basin. Pet Explor Dev 32(3):59–60

Zhang L, Chi X, Mo Z (2006) Introduction to parallel computing. Tsinghua University Press, Beijing

Zhang D, Li J, Wu Y (2010) Influencing factors of the numerical well test model of the triple-continuum in fractured vuggy reservoir. J Southwest Pet Univ (Science & Technology Edition) 32 (6):112-120

Zhou N (2012) The research and application of flash black oil model and simulator considering the character of nonlinear seepage in low permeability reservoirs. Southwest Petroleum University

Zhou X, Tang D, Zhang C (2008) Present situation and growing tendency of detailed reservoir simulation technology. Spec Oil Gas Reserv 15(4):1–6

Zou C, et al (2015) Progress in China's unconventional oil & gas exploration and development and theoretical technologies. Acta Geologica Sinica 89(6):979–1007

The manufacturer's authorised representative in the EU is Springer
Nature Customer Service Centre GmbH, Europaplatz 3, 69115 Heidelberg,
Germany. If you have any concerns regarding our products, please
contact ProductSafety@springernature.com

Printed and bound by CPI Group (UK) Ltd, Croydon, CR0 4YY
30/09/2026
02225271-0004